Power Plant Synthesis

MECHANICAL AND AEROSPACE ENGINEERING
A Series of Textbooks and Reference Books

Founding Editor
Frank Kreith

Power Plant Synthesis
Dimitris Al. Katsaprakakis

Air Distribution in Buildings
Essam E. Khalil

Nanotechnology: Understanding Small Systems, Third Edition
Ben Rogers, Jesse Adams, Sumita Pennathur

Introduction to Biofuels
David M. Mousdale

Energy Conversion
D. Yogi Goswami, Frank Kreith

CRC Handbook of Thermal Engineering
Raj P. Chhabra

Energy Efficiency and Renewable Energy Handbook, Second Edition
D. Yogi Goswami, Frank Kreith

Energy, the Environment, and Sustainability
Efstathios E. Michaelides

Principles of Sustainable Energy Systems, Third Edition
Charles F. Kutscher, Jana B. Milford, Frank Kreith

Fluid Power Circuits and Controls
Fundamentals and Applications, Second Edition
John S. Cundiff, Michael F. Kocher

For more information about this series, please visit: https://www.crcpress.com/Mechanical-and-Aerospace-Engineering-Series/book-series/CRCMECAERENG

Power Plant Synthesis

By Dimitris Al. Katsaprakakis

CRC Press
Taylor & Francis Group
Boca Raton London New York

CRC Press is an imprint of the
Taylor & Francis Group, an **informa** business

First edition published 2020
by CRC Press
6000 Broken Sound Parkway NW, Suite 300, Boca Raton, FL 33487-2742
and by CRC Press
2 Park Square, Milton Park, Abingdon, Oxon, OX14 4RN

ISBN: 978-1-138-05384-7 (hbk)

Typeset in Times by Cenveo® Publisher Services

Visit the CRC Press Web site: http://crcpress.com/9781138053847

Dedication

To my beloved son, the divine gift of my life and eternal source of light, hope, and strength in my path.

Contents

Preface

Since the Industrial Revolution, energy has been, apart from a physical magnitude, a commercial good, essential for the prosperity and the harmonic co-existence of human societies. With the technical and technological evolution, the dependency on energy sources for the implementation of a large amount of activities increased continuously and with gradually higher intensity, starting with the simple and regular daily tasks of a household and expanding to the execution of missions in space, leading rapidly to the exhaustion of nonrenewable energy sources and the degradation in many cases of natural landscapes and the environment in general. The last two negative consequences, combined with the emerging inability of conventional energy sources to fulfill the coverage of the increasing energy needs in the approximate future, constituted the required prerequisites for the investigation and the development of alternative power production technologies for electricity, thermal, and mechanical energy production. These new technologies, almost entirely, fall into the cluster of renewable energy source (RES) technologies, with two of the most fundamental of them, namely wind power and solar radiation, being characterized with a major and considerable drawback: their stochastic availability. Trying to face this unfavorable feature, several alternative power plant concepts have been and are still being designed and developed, aiming to ensure guaranteed and secure power production from RES plants, or to maximize the efficiency of conventional power plants, approaching the so-called rational use of energy (RUE).

In this book, a highly ambitious effort was implemented: to develop a scientific and technical resource that would cover as many different implementations of power production plants as possible, which are expected to play a critical role in energy's transition from the existing fossil fuel–based energy environment, to an alternative future of extensive RES penetration and RUE. To this end, a variety of power production plants is covered and thoroughly presented, most of them accompanied by detailed technical analysis, operation algorithms for computational simulation, and presentation of the results from executed case studies. More specifically, the book covers conventional thermal power plants for electricity production, hybrid power plants for electricity production with alternative energy storage technologies (pumped storage, compressed-air energy storage, and electrochemical batteries), combined heat and power cogeneration systems including the concept of trigeneration, hybrid power plants for thermal power production, and, finally, the book delivers a short reference on the new, emerging technologies for decentralized energy production, storage, and management, integrated in the frame of smart grids. Appendices to the book are also available as e-resources on the CRC Press website: http://www.crcpress.com/9781138053847

This book was initially written in Greek, in a much shorter and simplified version, aiming to cover the syllabus of the synonymous undergraduate course in the Mechanical Engineering Department of the Hellenic Mediterranean University. Despite its initial target, the book, already from its initial version, expanded beyond the syllabus of an undergraduate course, to cover subjects and concepts at the current state-of-the-art of the corresponding topics. While finishing the Greek version, the approach with Taylor & Francis Group came up absolutely incidentally. At this point, I should deeply thank Jonathan Plant, for his initial interest and his warm willingness to proceed with the publishing of that material in English. The initial 400 pages of the Greek book became more than 600 pages in this new English book. It is so conceivable that practically a brand new book was produced through this process with discrete new elements, additional case studies, new sections (e.g., on solar thermal power plants), and an entire new chapter on smart grids. The book is almost totally based on personal experience gathered through the involvement on real case projects. This is how this large number of case studies has been developed and integrated in the material of the book. Most of them are based, as they are presented, on the results of real projects or were adapted or inspired from real accomplished projects.

Given the approach clarifications, the book hopes to serve as a resource for all levels of engineers: undergraduate or postgraduate students, researchers and PhD candidates, academics and professional engineers, and designers of energy production plants. Any comments, hints for further improvement, or enhancements in a potential future version of the book are warmly welcome at my personal e-mail address. My wish and hope for this book is to satisfy the expectations and the anticipation of its readers and to become a useful reference for their works and projects.

Dimitris Al. Katsaprakakis
Heraklion, Crete, Greece
dkatsap@hmu.gr, dkatsap@gmail.com

Acknowledgments

My warmest thanks and my deep appreciation for their contribution to this book are owed to the following exceptional people:

- Jonathan Plant, executive editor in Engineering at Taylor & Francis Group, for his belief and warm interest in my work and his willingness to proceed with publication of this book.
- Ms. Bhavna Saxena, editorial assistant at Taylor & Francis Group, for her endless patience and constant support throughout the publishing process.
- Ms. Kyra Lindholm, editor at Taylor & Francis Group on Thermal Fluids, Aerospace/Aviation, Energy, Nuclear Engineering, for her support throughout the publishing process.
- Ms. Madhulika Jain and Ms. Julie Searls, for peer-reviewing and suggesting typographical and linguistic corrections during the composition process of the book.
- Dr. Dimitris Christakis, full professor in Mechanical Engineering Department at Hellenic Mediterranean University, for the supply of the examples for the energy transformers' efficiency calculation connected either in-series or in-parallel (Chapter 1), for the name of the undergraduate course in our department that also gave the title of this book, and, most importantly, because he has been and still is the pylon in my academic career and a continuous source of inspiration. I remain always deeply grateful.
- Mr. Bjarti Thomsen, renewable energy advisor at Environment Agency, Faroe Islands, for the provided data for the Faroe Islands case study in Chapter 3 and his photo offered for Chapter 7.
- Dr. Laurent Sam, energy engineer in Public Utilities Corporation of Seychelles, for the provided data for the corresponding case study presented in Chapter 2.
- Prof. Giacomo Falcucci, associate professor in Department of Enterprise Engineering in Tor Vergata University of Rome, for his assistance on the presentation of the theoretical background of compressed-air energy storage systems.
- Prof. Ioannis Templalexis, associate professor in Hellenic Air Force Academy, for his contribution on the simulation of the operation of compressed-air energy storage systems.
- Prof. Ioannis Nikolos, full professor in school of production engineering and management in Technical University of Crete, for the revision of the CAES section.
- Ms. Vasiliki Drossou, Dr. mechanical engineer and Head of Solar Thermal Systems Department in the Centre of Renewable Energy Sources of Greece, for her contribution on trigeneration, solar cooling plants, and absorption chillers.
- All individuals, research and academic institutes, public organizations, editors, and others for their permissions provided to use figures, photos, and tables already published in their works and publications. All these are in detail cited and mentioned in the reference lists of each chapter. My sincere appreciation and gratitude to all of them.

About the Author

Dimitris Al. Katsaprakakis was born in Athens in 1973 and was raised in Crete. He got his first degree in Mechanical Engineering from the National Technical University of Athens (1991–1997). He accomplished his PhD thesis at the same University on "Wind parks penetration in insular autonomous systems" (2002–2007). Currently he serves as a full professor in Hellenic Mediterranean University, Department of Mechanical Engineering.

His scientific work and interests, in general, fall in the field of energy transition and rational use of energy, with emphasis on insular systems. He started in the early years of his carrier with the study of advanced power plants based on the combined operation of electricity production technologies from renewable energy sources and energy storage technologies. The same concept was then expanded for thermal energy systems. Extensive work has been also accomplished on the energy performance upgrade of buildings and facilities, with studies covering a range from large public buildings to national stadiums. Advanced combined systems have been also executed on combined heat and power plants. His field of expertise is integrated with the study of impacts of renewable energy sources projects on natural and human environment and the approach of social and economic development for human communities through energy transition.

As an academic he has participated in several research programs, representing his University. His scientific work has been published in the most internationally acclaimed scientific journals, with tens of articles. He has also done considerable work in the consulting field, with a pioneering role supporting Greek and European islands to approach energy transition, expressed through a number of hybrid power plants studies, on behalf of public organisations, private firms, energy cooperatives and the European Commission. He also actively supports local communities to claim energy transition for their regions. He is a member of the Sifnos Energy Cooperative and a founding member of the Minoan Energy Community, the first and largest energy community in Crete.

He lives permanently in Heraklion, Crete, Greece.

1 Introductory Concepts

1.1 ENERGY AND POWER

1.1.1 Energy

Energy is one of the most well-known physical magnitudes, while at the same time it remains one of the most difficult concepts to understand. It necessarily accompanies every change in the physical world, from the simplest ones, such as throwing a stone, to the most complicated ones, like various biological procedures. Energy is not perceptible from our senses, unlike other common physical attributes, such as temperature, volume, or length of an object. Unlike these magnitudes, energy cannot be seen, cannot be caught, cannot be heard, and cannot be tasted. The only way energy can be perceptible is through its results.

Yet, what is energy? How can energy be defined? Often mentioned in scientific or educational texts, the following definition covers to a large extent the concept of energy:

Energy is the ability of a body or system to produce work or to cause a change.

How exactly is this definition understood? Before trying to answer this question, let's recall some fundamental concepts from physics. When a force $\vec{F}$ applied to an object makes it move for a distance $\vec{S}$, with an angle φ between the directions of the vectors $\vec{F}$ and $\vec{S}$, then we say that this force $\vec{F}$ produces work W which is equal to the internal product of $\vec{F}$ and $\vec{S}$ (Figure 1.1):

$$W = |\vec{F}| \cdot |\vec{S}| \cdot \cos\varphi \tag{1.1}$$

The units of work arise from the product of the units of the involved magnitudes in the above definition relationship:

$$[W] = [F] \cdot [S] \Leftrightarrow [W] = Nt \cdot m = Joule \tag{1.2}$$

Consequently, work and energy have the same units; hence, work and energy are equivalent magnitudes. Specifically, the production of work requires the consumption of at least an equal amount of energy. Having recalled the concepts of energy and work and having understood that the work production implies the consumption of at least equal energy, we can now try to present some physical interpretations of the energy definition given above.

A moving object produces work; hence, it must have or it must be provided with at least equal energy. Therefore, a human who walks and covers a specific distance produces work, which obviously is given by the internal product of the covered distance with a total resultant force applied in the human body. In this case, the force is applied by the human's muscular system, mainly by the muscles of the legs, and should be as high as required in order to overcome the ground and the air resistances and the body's weight component opposite to the direction of movement (Figure 1.2). Where does this energy come from? In this case, it is provided by the energy that the human body has either stored in the form of fat, or has recently received from foods. Is there any case where a human body is not able to produce work due to not having enough energy availability? Certainly there is, although these conditions that can make a human body unable to move because of its remarkably degraded energy level, in current times and in most parts of the modern world, are

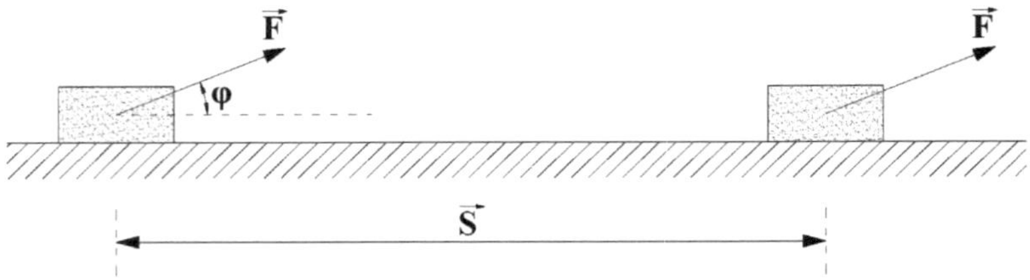

FIGURE 1.1 Work production.

uncommon. Yet, it is clear that for a human body that is not provided with the required energy for a long time period (starvation), the body will reach a point at which it will not be able to perform any kind of movement. Hence, it will not have the ability to produce work.

Similar to a human body, a moving car that covers a distance produces work, which equals the internal product of the covered distance with the resultant applied force to this, in order to overcome the ground and the air resistances and the car's weight component opposite to the direction of movement. In this case, the applied force is provided by the car's engine. Work is produced as a result of the car's available energy content, practically the chemical energy contained in the stored fuel in the car's fuel tank. Obviously a car with an empty tank cannot move; hence, there is no possibility for work production precisely because there is no energy available.

An object dropped (without any initial push) from the roof of a building to perform a free fall to the level of the ground will cover a distance equal to the height h of the building. During this free fall, the only force applied to the object is its weight $\vec{B}$. The vectors of the applied force and the covered distance have the same direction, namely the angle between these vectors will be 0°. Because the object moves, there will be work produced, which is given by the relationship:

$$W = \left|\vec{B}\right| \cdot \left|\vec{h}\right| \cdot \cos\varphi \Rightarrow W = m \cdot g \cdot h \cdot \cos 0° \Leftrightarrow W = m \cdot g \cdot h \tag{1.3}$$

where m is the mass of the object and g the acceleration of gravity. For this movement there was neither an initial boost nor any fuel consumed to provide the required energy. What was the source of energy consumed for the production of the above calculated work? The answer is provided by

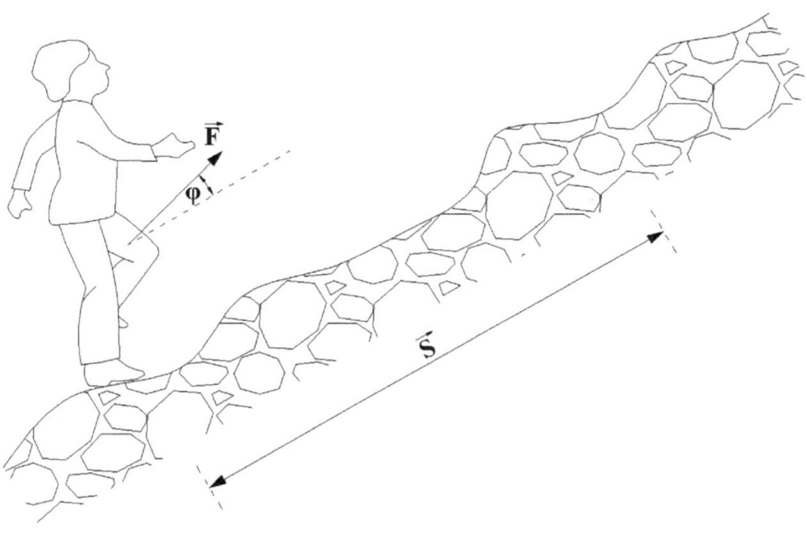

FIGURE 1.2 Work production from a human body.

Equation 1.3: The dynamic energy due to the gravitational field of earth that the object has due to its position at the roof of the building versus the ground level. Indeed, the produced work equals the dynamic energy of an object with a mass m located at height h versus the reference level. Consequently, the work production in this case is possible exactly due to the dynamic energy of the object. During the object's free fall, the initial dynamic energy is totally consumed for the production of equal work.

Energy is a physical magnitude that can be perceptible only from its results. We may say that it is well-hidden in bodies and systems, such as the fuel in the car's reservoir, and is revealed only when bodies and systems take part in physical or chemical processes. During these processes, energy is usually converted from one form to another, causing, thus, conceivable results to the surroundings. For example, nobody can realize the dynamic energy contained in the flowerpot at the sill of a balcony. If, however, a clumsy hand forces this flowerpot to change position and land on the roof of the parked car below the balcony, making a dent, the initial dynamic energy will be conceivable from this result.

1.1.2 Power

Power equals the rate that energy is produced or consumed by a body or a system, or, more accurately, the rate of energy transferred or converted:

$$P = \frac{dE}{dT} \tag{1.4}$$

Having grasped the concept of energy, power seems to be more easily understood and can be explained with a number of examples. Let's assume two runners with identical physical constitutions and the same mass, competing in a 100 m race. At the end of the race, both runners will have covered the same distance and, through their muscular systems, they will have applied exactly the same force to their bodies, given the assumption that both of them have the same mass and physical shape, so both will be pushing against the same resistance by the air. Hence, the two runners will produce exactly the same work and, consequently, they will have consumed the same chemical energy E, which will have been converted to mechanical energy E_m. If one of the two runners reaches the finish faster than the other, he produced the required mechanical energy in a shorter time interval $t_1 < t_2$. The average power with which the two runners cover the race will be given by the equation:

$$P_1 = \frac{E_m}{t_1} \ \& \ P_2 = \frac{E_m}{t_2} \tag{1.5}$$

Given that $t_1 < t_2$ it will be $P_1 > P_2$. Namely, the chemical to mechanical energy conversion rate of the first runner was higher, a feature owed to the ability of his muscular system. This is the reason the first runner covered the distance faster. From this example, we may conclude that speed races in athletics are essentially a competition of the runners' muscular systems as mechanical power production machines, because the winner will be the athlete who possesses the muscular system with the ability to convert chemical to mechanical power in shorter time intervals, in other words with the higher energy conversion rate or, equivalently, with higher mechanical power.

In another relevant example, let's assume that we want to heat a particular mass of water m from an initial temperature T_1 (e.g., the temperature of the water supply network) to a final temperature T_2 (e.g., 80°C). The required thermal energy Q for this process will be:

$$Q = m \cdot c_p \cdot (T_2 - T_1) \tag{1.6}$$

where c = 4,184 J/(kg·K) the specific heat of water.

Let's also assume that this water mass m is heated twice alternatively with the same utensil and with the same heat source (e.g., a hotplate), starting from and ending at the same initial T_1 and final temperature T_2 both times. However, at the first time, the source heat is set at half capacity, while at the second time the source operates at full capacity. Neglecting, in favor of the example's simplicity, the heat losses to the surrounding air and the utensil itself, the energy required to heat the water both times will be exactly the same, because, according to Equation 1.6, it depends on the heated medium's specific heat, its mass, and the initial and final temperatures, namely on parameters that remain the same between the two alternative experiments.

Nevertheless, it is obvious that the first attempt will take longer to reach the final required temperature T_2, hence it will be $t_1 > t_2$. What changes between the two alternative executions of the same experiment, given that the required final heat remains the same? Obviously, the rate with which heat is transferred from the heat source to the water, or, in other words, the thermal power, which in the first case is lower than the second one.

The power of a body, a machine, or a system expresses its ability to convert energy from one form to another with a low or a high rate. In the example with the runners, we concluded that speed races are practically a competition of the mechanical power of their muscular systems. Consequently, the magnitude that characterizes the athletes is the power of their muscular systems and not the energy that they consume while running in a race. Similarly, all the energy conversion machines and systems are characterized and classified based on their maximum possible rate that they are able to convert energy from one form to other, namely according to their maximum power, named as *nominal power*. Hence, it is power that defines an energy system and not the energy converted during a particular process and during a specific time period.

This is sensible if we compare a car and a human. Both of them can cover a distance of 1,000 m; however, the car, although heavier, will most certainly cover this distance in a shorter time, namely it will consume more mechanical energy in a shorter time interval. Both the car and the human can produce the mechanical energy required to cover the distance of 1,000 m. However, it is not this mechanical energy that characterizes their ability and classifies them as energy machines. On the contrary, this magnitude will be their power, which defines how fast they can convert the initial form of chemical energy to the final mechanical energy.

Thus, the energy capacity of an energy convertor is always expressed through its nominal power and never by the converted amount of energy during an energy process. Hence, we say that a car has a nominal power of 100 HP, the nominal power of a diesel generator is 50 MW, and a heater has a nominal capacity of 50,000 kcal/h. These features characterize the energy machines and accompany them as nominal specifications.

It would not make sense to try to evaluate the energy capacity of an energy convertor, e.g., a diesel generator, saying, for example, that it produces electricity equal to 100 MWh. Such information would not provide any indication for the generator's capacity, because the crucial feature is not only the total electricity production but also the time required from the machine to produce this energy amount. If, for example, the nominal power of the diesel generator is 50 MW, then the 100 MWh will be produced in:

$$P_{el} = \frac{E_{el}}{t_1} \Leftrightarrow t_1 = \frac{E_{el}}{P_{el}} = \frac{100\ \text{MWh}}{50\ \text{MW}} \Leftrightarrow t_1 = 2\ \text{h}$$

Yet, if the nominal power of this generator is 100 kW, then for the production of 100 MWh, the required time will be:

$$P_{el} = \frac{E_{el}}{t_2} \Leftrightarrow t_2 = \frac{E_{el}}{P_{el}} = \frac{100\ \text{MWh}}{0.1\ \text{MW}} \Leftrightarrow t_2 = 1{,}000\ \text{h}$$

TABLE 1.1
Typical Power Values of Biological and Technical Energy Systems

Insect flying	0.001 W
Human heart	1 W
Human working	75 W
Incandescent lamp	100 W
Electrical refrigerator	150 W
Horse galloping	1,000 W
Electrical water heater	3 kW
Medium-size car engine	75 kW
Boeing 707 engine	21 MW
Conventional thermal or nuclear power plant	1,000 MW

Every energy system, technical or physical, is characterized by its power, namely its ability versus time to convert energy from one form to another. Indicative power values for some characteristic biological, mechanical, and electrical energy systems are presented in Table 1.1.

1.1.3 Energy and Power Evolution versus Time

During an energy conversion process, both energy and power change. Power may sometimes be higher and sometimes lower. This means that energy can be converted at a slower or faster rate, namely with lower or higher power. Yet, no matter what the energy conversion rate may be, the transformed energy will always increase, as the conversion process evolves.

A characteristic example of the above procedure is an exercise machine in a gym (e.g., stationary bike, cross trainer, stepper). Everyone who has used such a machine has observed the control panel with its different screens, providing detailed information on the performed exercise. Among these screens, most probably there is one with the indication Wand another one with the indication kcal. The symbol W designates the unit Watt for power in SI (Le Système International d'unités), while the symbol kcal designates the unit kilocalories for energy. An extensive reference on power and energy units is presented in the next section. Based on information on the exercise machine's screens, the user can gain information on the current power with which the exercise is performed—this is actually the provided power by his/her muscular system—and the total energy consumed from the beginning of the training until the current moment. During the training, the depicted values in the screen with the indication W, namely the athlete's current mechanical power, may increase or decrease, versus the performed exercise intensity. An increase of the athlete's mechanical power imposes a relevant increase of the heart rate, just like an increase of a car's power imposes a corresponding increase on the motor's rotational speed. On the contrary, the depicted value in the screen with the indication kcal, namely the total energy consumed during the whole training program, continuously increases. When the athlete's mechanical power increases, the total consumed energy will increase at a higher rate, and vice versa. When the user ends the training program, the value in the power screen will be zero, while the energy screen displays a constant value presenting the total energy consumed during the entire training session.

Figure 1.3 presents the electrical power consumption for a typical household during a 24-hour period. The depicted power values correspond to hourly time intervals and designate the hourly average power consumption by the involved electrical devices of the household. As shown in Figure 1.3, the power consumption increases and decreases during the examined time period, exhibiting a minimum value (0 W) at the fourth hour and a maximum value (1,600 W) at the twelfth and the

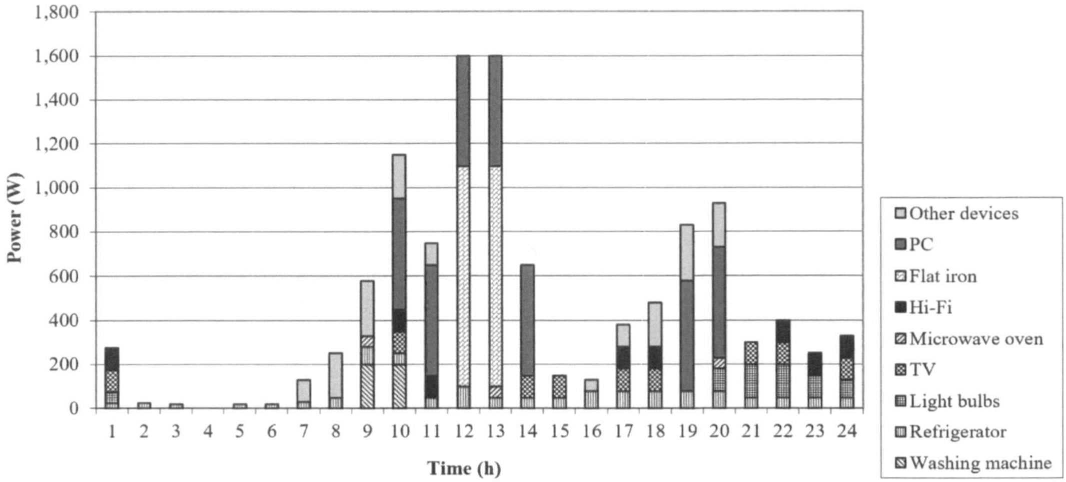

FIGURE 1.3 Electrical power consumption fluctuation for a typical household during a 24-hour period.

thirteenth hour of the 24-hour period. At the first hour, the consumed power equals 275 W. The corresponding consumed electricity during this hour will be:

$$P_{el} = \frac{E_{el}}{t} \Leftrightarrow E_{el} = P_{el} \cdot t \Rightarrow E_{el} = 275\ \text{W} \cdot 1\ \text{h} \Leftrightarrow E_{el} = 275\ \text{Wh}$$

This energy is graphically depicted by the area of the first vertical rectangle in the diagram, with a height of 275 W and a width of 1 h. Similarly, the electricity consumption can be calculated during the subsequent hours, depicted graphically by the area of the second, third, and so on, vertical rectangles in the diagram. The sum of the rectangles' areas in the diagram, until a specific hour, represents the total electricity consumption in the household until this hour, which will increase continuously versus time, regardless of whether or not the current electrical power consumption increases.

The total electricity consumption during the whole 24-hour period will be given by the sum of the products of the hourly average power values with the hourly time intervals for all the 24 hours of the examined time period, and it will be graphically depicted by the total area of the diagram's vertical rectangles:

$$E_{24h} = \sum_{i=1}^{24} P_i \cdot t \Leftrightarrow E_{24h} = t \cdot \sum_{i=1}^{24} P_i \tag{1.7}$$

where in the specific example t = 1 h (constant).

Equation 1.7 constitutes the arithmetic expression of the following analytical one:

$$E_{24h} = \int_{t=0}^{t=24h} P(t) \cdot dt \tag{1.8}$$

according to which the total electricity consumption over the whole examined time period is given by the integral of the consumed power function P(t) versus time t. This energy will be graphically

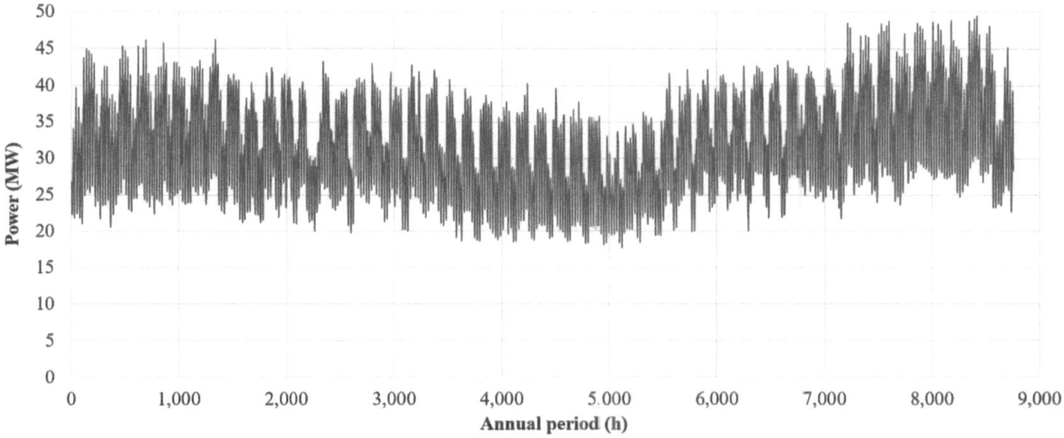

FIGURE 1.4 Annual power consumption curve for the autonomous electrical system of the Faroe Islands.

depicted in a P(t) – t diagram by the total area defined by the curve P(t), the time axis (horizontal axis), and the vertical lines starting from the start and end time points at the time axis.

Figure 1.4 presents the annual power consumption curve for 2015 for the autonomous electrical system of the Faroe Islands, located in the North Atlantic Ocean, between Scotland and Iceland. From this figure, we may get some indicative conclusions on the variation and the evolution versus time of the human commercial and residential activities in the country, defined, to a large extent, by the prevailing climate conditions, as well as on how these activities affect the electricity demand. The power increases during the winter period, due to adverse weather conditions characterized by storm winds and temperatures close to 0°C. During this period, and specifically on December 17, the power demand exhibits its maximum annual value, equal to 49.5 MW. On the other hand, the mild weather conditions recorded for a 2-month period during summer lead to a corresponding reduction in the electricity demand. Indeed, the minimum annual power demand equals 17.8 MW and happens on August 2. We may also conclude that during summer there is no considerable tourist activities, despite the exceptional landscape and the available activity options of the Faroe Islands, which could have kept the electricity consumption high. As in the earlier example, the total area defined in the graph below the power demand curve depicts graphically the total annual electricity consumption in the Faroe Islands in 2015, equal to 278,555.90 MWh.

1.1.4 Energy and Power Units

The unit of energy in SI is Joule. Joule comes from the definition of work, given the equivalence between energy and work. One Joule equals the work produced, or the energy required to be consumed for this work production, when a force of 1 Nt acts on an object in the direction of its motion through a distance of 1 m:

$$1 \text{ Joule} = 1 \text{ Nt} \cdot 1 \text{ m} \Leftrightarrow 1 \text{ Joule} = 1 \text{ kg} \cdot 1\frac{\text{m}}{\text{s}^2} \cdot 1 \text{ m} \Leftrightarrow 1 \text{ Joule} = 1 \text{ kg} \cdot 1\frac{\text{m}^2}{\text{s}^2} \tag{1.9}$$

The unit of power in SI is Watt, defined by the relationship between energy and power. Hence, 1 Watt equals the power that corresponds to an energy conversion rate of 1 Joule/s:

$$1 \text{ Watt} = \frac{1 \text{ Joule}}{1 \text{ s}} \tag{1.10}$$

The variety, the large number, and the different sizes of energy systems observed in both the technical and physical world impose the use of different energy and power units, so the corresponding energy and power amounts can be expressed with figures easily intuitive for human perception.

Perhaps the most common form of energy is electricity. The unit employed for the measuring of electricity production or consumption is the Watt-hour (Wh) and its multiples. The equivalence between Wh and Joule arises from the fundamental relationship between energy and power and is calculated as shown below:

$$1 \text{ Joule} = 1 \text{ Watt} \cdot 1 \text{ s} \Leftrightarrow 1 \text{ Joule} = 1 \text{ Watt} \cdot \frac{1}{3{,}600} \text{ h} \Leftrightarrow 3{,}600 \text{ Joule} = 1 \text{ Watt} \cdot 1 \text{ h} \Leftrightarrow 1 \text{ Wh} = 3{,}600 \text{ J} \tag{1.11}$$

From Equation 1.11, we can also define the multiples of Wh, which are more commonly used for the measuring of electricity:

1 kWh (kilowatt-hour): 10^3 Wh = $3.6 \cdot 10^6$ J
1 MWh (megawatt-hour): 10^6 Wh = $3.6 \cdot 10^9$ J
1 GWh (gigawatt-hour): 10^9 Wh = $3.6 \cdot 10^{12}$ J

The use of kWh for the measuring of electricity is sensible, because it arises from the power of the electrical devices usually used by final consumers and from a time unit, such as hour, much closer to the human's time scale activities than seconds. For example, a common household in the western world can consume 300 kWh of electricity during a month (approximately 10 kWh per day). If Joule was used for electricity measuring, instead of kWh, the above electricity consumption would be expressed as $1{,}080 \cdot 10^6$ Joule, namely approximately 1 billion Joule, a figure much more difficult to comprehend. The unit of kWh is closer to the human standard of measurement because it is defined as the product of 1 kW, namely a power unit that approaches the installed electrical power in a household, with 1 h, a time unit that corresponds to human living time standards (we say for example that a day consists of 24 hours or an academic course lasts 2 hours instead of 7,200 s). On the contrary, 1 Joule is a very small energy unit to describe the electricity consumption from commonly used electrical devices, because it comes from the product of 1 Watt, namely a relatively small power unit (as shown in Table 1.1, 1 Watt is the power of the human heart) with 1 second, a time unit too short to express the use of the electrical apparatus of a household (e.g., we never turn on the lights or the washing machine for only some seconds).

It must be noted at this point that, by observing Equation 1.11, we may conclude that Joule is given by the product of Watt with seconds, while kWh is given by the product of kW with hours. Conclusively, in both cases, the energy unit is given by the product of a power unit (W or kW) and a time unit (second or hour). This general rule must be understood and memorized:

The product of a power unit with a time unit gives always an energy unit.

Another important energy unit, often used in thermal applications, is the calorie (cal) or *small calorie*, and its multiple, the kilocalorie (kcal) or *large calorie*. Calorie is a French term originating from the Latin word *calor,* which means "heat." One calorie is the approximate amount of energy needed to increase the temperature of 1 gr of pure, distilled water by 1°C at a pressure of 1 atmosphere, with the note that this temperature rise should be between 15°C and 16°C. This energy is approximately equal to 4.184 Joule, namely:

$$1 \text{ cal} = 4.184 \text{ Joule} \tag{1.12}$$

For human activities and applications, usually the kilocalorie is used, which equals the heat required to increase the temperature of 1 kg of distilled water by 1°C. Equation 1.13 is a direct consequence of Equation 1.12:

$$1 \text{ cal} = 4.184 \text{ Joule} \ \& \ 1 \text{ kcal} = 4.184 \text{ kJoule} \quad (1.13)$$

The kcal is commonly used for the measurement of the heat required for indoor space heating or hot water production. Also, this unit is used for the description of the energy consumption from a human body, as well as the energy received with foods. A middle-aged human, with a weight around 80 kg, a height of 1.80 m, and a common daily urban life (sedentary pursuit, lack of any particular physical exercise) consumes around 2,000 kcal per day. On the contrary, a human running with a medium speed of 5–10 km/h can consume 500–700 kcal after an hour of training, while a football player needs 1,500–2,000 kcal during a football match. From the above figures, we see how important physical exercise is for the attaining and the maintaining a good physical shape and fitness.

Beyond the above two energy units, employed for the measurement of two fundamental energy forms met in daily human activities (electricity and heat), the variety of energy conversion processes on the planet implies the introduction of additional energy units.

Hence, in the microcosm of nuclear fission, the energy unit introduced to describe the heat released from the fission of one nucleus of nuclear fuel is the electronvolt (eV), which is generally used in atomic and subatomic applications. By definition, 1 eV is the amount of energy gained (or lost) by the charge of a single electron moving across an electric potential difference of 1 volt. The equivalence between eV and Joule is given by the following:

$$1 \text{ eV} = 1.6021765314 \cdot 10^{-19} \text{Joule} \quad (1.14)$$

The thermal energy released with the fission of one nucleus of U235 is equal to 180 MeV.

Two additional energy units often used are:

- Btu (British thermal units): A traditional energy unit of the British Imperial System, equal to 1,054.35 Joule. It is defined as the amount of thermal energy required to increase the temperature of 1 pound of water by 1°F.
- Toe (Tonne of Oil Equivalent): The amount of energy released by burning 1 ton of crude oil. 1 toe equals to $41.86 \cdot 10^9$ Joule.

Table 1.2 presents the basic energy units described above, along with their conversion factors to Joule.

TABLE 1.2
Energy Units and Conversion Factors to Joule

Energy Unit	Conversion Factor to Joule
Joule	—
cal/kcal	1 cal = 4.184 Joules / 1 kcal = 4.184 kJoules
kWh	1 kWh = 3,600,000 Joules
Btu	1 Btu = 1,055.05 Joules
toe	1 toe = $41.86 \cdot 10^9$ Joules
eV	1 eV = $1.6021765314 \cdot 10^{-19}$ Joules

The fundamental power units arise from the above energy units. Regarding electricity production, Watt and its multiples, depending on the size of the electrical device or system, are used:

- 1 kW = 10^3 Watt
- 1 MW = 10^6 Watt
- 1 GW = 10^9 Watt
- 1 TW = 10^{12} Watt

Additionally, from Btu and kcal we also have the following corresponding power units:

- Btu/h = 0.293071 Watt
- kcal/h = 1.16222 Watt

We may easily conclude the above equivalences from the corresponding relationships between the involved energy units. For example:

$$1\,\frac{\text{kcal}}{\text{h}} = \frac{4.184\text{ kJ}}{3{,}600\text{ s}} = 1.162222\cdot10^{-3}\,\frac{\text{kJ}}{\text{s}} \Leftrightarrow 1\,\frac{\text{kcal}}{\text{h}} = 1.162222\cdot10^{-3}\text{ kW} = 1.162222\text{ W} \tag{1.15}$$

From the above relationships, it is observed that = several power units come as the ratio of an energy unit (Joule or Btu or kcal) over a time unit (hour or second). Similarly with the energy units, this is also a general rule that must be understood and memorized:

The ratio of an energy unit over a time unit gives always a power unit.

There is another power unit often used, so-called horsepower, designated with the symbol HP. This unit does not arise as the ratio of some of the above presented energy and time units, but it is a legacy of older power measurement approaches, adopted and used in the past. One HP equals 746 Watts. The term *horsepower* was introduced in the late eighteenth century by the famous Scottish engineer James Watt (the fundamental power unit was named after him) to compare the power output of steam engines with the power of horses. Later, the term was also expanded to include the output power of internal combustion engines, gas or hydro turbines, electrical motors, etc. The definition of this unit can be different depending on the geographical area. Hence, the German definition of this unit introduces the symbol PS, from the corresponding German word *pferdestärke* (horsepower). The units HP and PS are not exactly equal to each other. Specifically, 1 PS = 736 Watts, hence 1 HP = 1.015 PS.

Table 1.3 presents the basic power units described above, along with their conversion factors to Watt.

TABLE 1.3
Power Units and Conversion Factors to Watt

Power Unit	Conversion Factor to Watt
Watt	—
kcal/h	1 kcal/h = 1.16222 Watts
Btu/h	1 Btu/h = 0.293071 Watts
HP	1 HP = 746 Watts
PS	1 PS = 736 Watts

1.2 ENERGY FORM CLASSIFICATION AND TRANSFORMATION

1.2.1 Energy Form Classification

In the physical and technical world, energy appears in a number of different forms. In this section, we will try to distinguish and classify them.

The energy of the free motion of an object or a particle is called *mechanical energy*. The energy of the chaotic movement and interaction of elements in material macrosystems will be called *heat*. The amount of heat that can be released and transformed to other forms of energy, giving temperature differences, will be called *thermal energy*.

The energy that holds the nucleus of an atom will be called *nuclear energy*. It can be obtained through reactions of fission and fusion of an atomic nucleus. The energy released from chemical reactions as the result of the rearrangement of the valence electrons of the molecules of elements will be called *chemical energy*.

If we consider the various stress fields created by the interaction of masses between them, we take the corresponding potential forms of energy. The electromagnetic interaction determines the potential energy of objects inside electrical or magnetic fields, which will be respectively called *electrostatic* or *magnetic energy*. A special case of magnetic energy is *gravitational energy*, which is the energy of the earth's magnetic field. Another special case of electromagnetic energy is *light energy*, which refers to the energy carried by waves in the visible region of the electromagnetic spectrum. Finally, the potential energy of electrically charged objects inside an electrostatic field will be called *electrostatic energy*.

The energy that comes from any kind of tactical movement of electrons inside conductors will be called *electrical energy*, or *electricity*, or *electrodynamic energy*, a term rather unusual, although more precise.

If a metallic string is stressed or a gas is compressed in constant temperature, then energy will be stored in the form of *elastic energy*. Finally, the energy contained and transferred in electromagnetic radiation will be called *electromagnetic energy*.

Below is a summary of the main energy forms, based on the above analysis:

1. Mechanical energy: energy derived from the free motion of a particle or an object.
2. Heat: energy of the chaotic movement and the interaction of particles in macrosystems.
3. Thermal energy: amount of heat that can be released and transformed to other forms of energy, leading to temperature differences.
4. Electrical or electrodynamic energy: energy of the tactical movement of electrons inside conductors.
5. Gravitational energy: dynamic energy derived from the interaction forces between material objects, due to the magnetic field of earth.
6. Nuclear energy: captured energy in the heavy nuclei, released with their fusion and the creation of new lighter nuclei.
7. Chemical energy: energy released with a chemical reaction, mainly from the combustion of hydrocarbons.
8. Electrostatic-magnetic energy: potential energy of electrically charged or magnetized objects inside an electrostatic or magnetic field, respectively.
9. Elastic energy: energy of a stressed spring or a compressed gas in constant temperature, corresponding to energy storage. This form of energy can be the result of mechanical, electromagnetic, thermal, or gravitational interaction.
10. Electromagnetic energy: energy transferred with electromagnetic radiation.

After outlining the above classification of energy forms, a typical question can be: Which is the role of each of them? The criterion to answer is to examine their practical value and their usability for human activities and natural processes. The practical value of the above forms of energy varies, depending on the following parameters:

- their availability and their natural resources on earth
- whether or not they can be considered as renewable
- their concentration—or their energy density—on earth
- their possibility to be exploited directly, namely without having to be converted to another form
- their possibility to be stored and conserved
- their possibility to be efficiently transported over long distances
- their possibility to be converted in practically used forms of energy
- any potential consequences or impacts from their exploitation on human activities and the natural environment.

The one and only producer of primary energy sources is nature, which distinguishes them in renewable and nonrenewable energy sources. As a renewable energy source (RES), solar radiation is directly exploited. Being converted in heat and chemical energy in plants and trees creates the fundamental prerequisites for the development and the existence of any kind of life on earth. Of the remaining primary RES (energy of the rivers, thermal energy of the earth, wind energy), only a small percentage is exploited in its initial form, namely as it is available in nature. The largest amount of these energy forms is converted to other forms of energy, which are most commonly used in several applications met in human activities. These forms of energy are presented below:

- thermal energy, corresponding to 75% of the total annual energy consumption globally
- mechanical energy, corresponding to 24% of the total annual energy consumption globally
- electrical and light energy, corresponding to the last 1% of the total annual energy consumption globally.

Thermal energy demand appears both in industry (melting, drying, etc.) and the residential sector (indoor space heating, hot water production). Thermal energy is also met as a primary energy source in nature (e.g., in geothermal fields).

Electricity is used, for example, in the aluminum production process. Nevertheless, electricity is mainly used as the most appropriate form of energy for transportation over long or short distances.

Finally, nonrenewable energy sources are mainly met in nature with only two forms: chemical energy, met in solid, liquid, and gas fossil fuels, and nuclear energy.

From the above analysis, we make the following conclusions regarding the role and the usage of various forms of energy:

- Nuclear, chemical, mechanical, thermal, and electromagnetic forms of energy constitute the primary forms of energy, met in the natural environment.
- Thermal, mechanical, electromagnetic, and electrical forms of energy are practically used in daily applications.
- Finally, the remaining forms of energy, namely gravitational, electrostatic, magnetic, and elastic forms of energy, constitute the energy storage forms. These are potential forms of energy with which energy can be stored. Chemical and nuclear energy also constitute energy storage forms, because, actually, by storing fossil or nuclear fuels, energy is eventually stored.

Figure 1.5 summarizes the above conclusions.

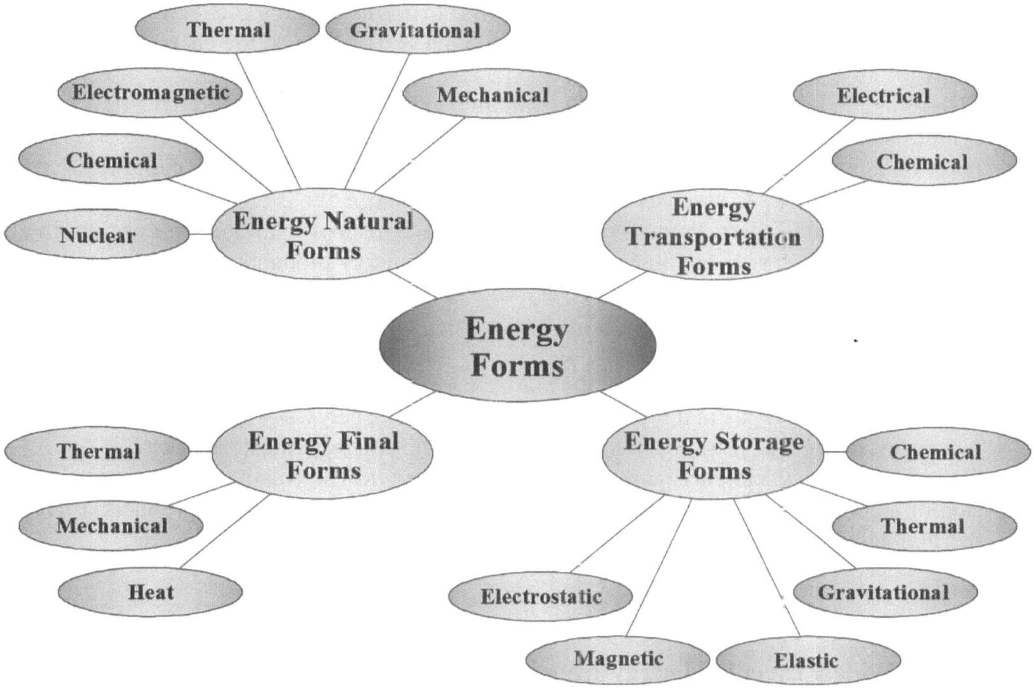

FIGURE 1.5 Energy form classification chart.

1.2.2 Energy Form Transformations

The study of the various energy transformation processes leads to the conclusion that for any energy transformation, two principal prerequisites should be satisfied:

1. an adequate power concentration level (to be transformed) should be available, and
2. a working medium, characterized by fundamental properties, should also be available.

For example, compressed cold air, namely of low specific enthalpy, hence with low thermal energy density, cannot be exploited in a gas turbine for power production. Furthermore, regardless of any possible fluctuations of the magnetic flow inside a magnetic field, an electricity conductor should be introduced inside the field to convert magnetic energy into electricity.

Still, in many cases, even if these principal prerequisites are fulfilled, the direct transformation of one form of energy to another is not feasible. For example, nuclear fission will not be performed for a uranium nucleus without any neutron bombardment, even though it may exhibit higher mass than the critical one. Furthermore, there is no way to directly produce electricity from diesel oil, namely through a direct transformation of the initial (primary) chemical energy to electricity. The only way to do this is through a series of consecutive transformations, from chemical to thermal energy, then to mechanical energy, and finally to electricity.

Based on the principal forms of energy classification analyzed in the previous section, the most popular direct transformations between them, summarized in Table 1.4, are:

- Nuclear to thermal energy: the energy form transforms during nuclear fission. Nuclear energy is released and is transformed in thermal energy.
- Chemical to thermal energy: the energy form transforms in a combustion chamber of a piston engine or a gas turbine or a central heat burner, widely performed either for heat or electricity production.

- Chemical to electrical energy: the energy form transforms inside a fuel cell, constituting a most promising approach towards the introduction of hydrogen to the energy production sector.
- Electromagnetic to chemical energy: the fundamental energy transforms during the photosynthesis process, being the source of life in the natural kingdom.
- Electromagnetic to electricity: the energy transforms in a photovoltaic (PV) cell. Electricity production from PVs is based exactly on this mechanism.
- Electromagnetic to thermal energy: the energy transforms in a solar collector or even in a microwave oven. A very popular process for thermal energy production.
- Gravitational to mechanical energy: a very widely used energy transformation in both the technical and natural world. This is the first energy transformation applied for electricity production, with hydro power plants.
- Elastic to mechanical energy: one of the main energy transformations during the expansion of a compressed hot gas in a turbine or a piston engine. Energy storage with compressed-air energy storage (CAES) systems exploits exactly this energy transformation.
- Thermal to mechanical energy: the second main energy transformation during the expansion of a compressed hot gas in a turbine or a piston engine. Electricity production from thermal generators is mainly based on this transformation.

TABLE 1.4
Possible and Widely Executed Energy Transformations

No.	Energy Form	Role	1	2	3	4	5	6	7	8	9	10
1	Nuclear	NS								a		
2	Chemical	NS, SF, TF							b	c		
3	Electromagnetic	NS, FF		d					e	f		
4	Gravitational	SF, NS									g	
5	Elastic	SF									h	
6	Electrostatic—magnetic	SF										
7	Electrical	TF		i	j			k			l	m
8	Thermal	NS, SF, FF									n	o
9	Mechanical	NS, FF					p		q			
10	Heat	FF										

Notation on energy form roles:	Energy transformation processes or technologies:	
• NS: Natural source • SF: Storage form • TF: Transportation form • FF: Final form	a. Nuclear fission and fusion b. Electrochemical batteries c. Combustion d. Photosynthesis e. Photovoltaics f. Solar collectors, microwave ovens g. Hydro power plants h. Compressors i. Electrochemical storage	j. Microwave ovens, medical applications k. Capacitors, electrical magnets l. Electrical motors m. Electrical devices for heat production n. Piston engines and turbines o. Cogeneration p. Compressors q. Induction generators

- Mechanical to electrical energy: the last energy transformation in the whole sequence for electricity production with inductive generators.
- Electricity to mechanical, thermal, light, electromagnetic energy: the very popular and widely used energy transformations, performed by the corresponding invented electrical devices to provide motion (e.g., elevators, electrical tools, air conditioning equipment), heat (e.g., electrical resistances, thermal convectors, electrical ovens), light, and other applications.

1.3 ENERGY SOURCES

From a scientific point of view, the commonly used term *energy sources* is not a valid one, because, due to the energy conservation law, energy is neither created nor destroyed. It simply changes forms. Hence, the term *energy sources* describes the initially available forms of energy that can be transformed to other forms.

Begrudgingly keeping this term, we may claim that some of the above presented energy forms are naturally found in the physical environment, and as such, they can be considered primary energy sources. With the development of relevant technologies, it is possible to collect, to concentrate, to store, to transport, and to transform the primary energy sources to energy forms that can be used by humans (e.g., mechanical, electricity, thermal). For example, the sails of a sailing boat convert the wind's kinetic energy (wind energy) to kinetic energy directly exploited by the boat. A solar collector converts solar radiation to thermal energy exploited for indoor space heating or hot water production. Finally, mining of so-called fossil fuels (e.g., lignite, coal, oil, natural gas, and nuclear fuels) enables their transportation and storage, in order to be used according to the electricity or thermal energy needs of a specific community.

Energy sources are classified in two main categories:

- nonrenewable energy sources
- renewable energy sources (RES).

1.3.1 Nonrenewable Energy Sources

Primary energy sources that are not regenerated from physical processes, or they are regenerated very slowly compared to their consumption rates, are called nonrenewable energy sources. Coal, oil, and natural gas are the most used nonrenewable energy sources, commonly known as fossil fuels. Of course, nature does not stop creating either coal or oil. However, each year humans consume a total amount of fossil fuels that will take 1,000 years to be regenerated with physical processes. This extraordinary rate of consumption puts into perspective the concept of renewability. A brief description of the main fossil fuels is given in the next paragraphs.

1.3.1.1 Coal

The term *coal* characterizes the organic sediments created from vegetative residues through a series of carbonation processes. These processes result in the natural residues' enrichment with coal. A plant's conversion to turf and the transition from turf (first stage of carbonation) to anthracite (last stage of carbonation) is a function of time, temperature, and pressure. The conversion of vegetative material to coal started 400 million years ago and, of course, is still executed. It is estimated that a layer of 2.5 m thickness of vegetative material is required for the creation of a 30 cm thickness layer of coal.

The grading of coals is based on their heat capacity, along with the chemical analysis of their organic compound. Coals with high carbon and low hydrogen or oxygen content are considered of high quality, while, as the carbon content decreases, the coal quality is degraded. Based on the carbonation extent, coals are classified in turf, lignite, subtarred/tarred coals, and anthracite.

With the transition from turf to anthracite, coal's heat capacity increases, along with its quality as a primary energy source.

1.3.1.2 Oil

Oil is found in liquid form in underground cavities. It was created from animal or vegetative microorganisms, mainly marine life, after their concentration with sea streams at the bottoms of physical basins, where they were pressed down from soil or other processes. There, in the absence of air, they were converted to oil during a period of thousands of years. The contained energy in oil comes from the energy that the initial microorganisms, from which it was created, had gathered from sun and food. Nowadays oil is pumped from its underground deposits. The main contents of oil are alkanes (paraffin), cyclohexanes, aromatic hydrocarbons, and, in lower quantities, oxygen, nitrogen, and sulfur compounds. Oil represents the most important mineral for the global economy, because it constitutes the main primary source of energy and the raw material for the production of a huge range of products (e.g., plastics and rubbers, medicines, cosmetics, detergents, explosives, synthetic materials for clothing and constructions).

1.3.1.3 Natural Gas

Natural gas is a mixture of hydrocarbons, particularly of methane and, in considerably lower amounts, of ethane, propane, butane, and pentane. The chemical compound of natural gas is mainly affected by its geographical origination, specifically, whether it originates from pure natural gas deposits or it comes from oil deposits. The commercial exploitation of natural gas began approximately in 1810, when it was used as the required fuel for lighting lamps. After World War II, the first networks for natural gas transportation and distribution were constructed. Among its main advantages as a source of energy are its transportation ability over long distances through pipelines and, of course, being less environmentally damaging than other fossil fuels.

1.3.1.4 Nuclear Fuels

Practically, the term *nuclear energy* is used to designate the energy released in huge amounts from:

a. nuclear fission, namely the fission of heavy atoms' nuclei to lighter ones, and
b. nuclear fusion, namely the fusion of light atoms' nuclei to form heavier ones.

Uncontrolled nuclear reactions take place during the explosion of an atomic bomb or a hydrogen bomb. Controlled nuclear reactions are used for electricity production, as well as for the production of mechanical energy from nuclear motors. Until 1995, the mechanical energy production from nuclear fuels was restricted mainly to marine applications (e.g., army ships, submarines, icebreaker ships), while there were also efforts to develop nuclear rocket motors. However, of higher importance for the global economy is considered the use of nuclear energy as a primary source of energy for electricity production, by employing special equipment and infrastructure that formulate nuclear reactors.

1.3.2 Available Reserves of Nonrenewable Energy Sources

According to the data provided by BP Statistical Review of World Energy 2018 [1] and World Energy Council (WEC) [2], the globally available reserves of nonrenewable energy sources at the end of 2015 (or of 2007 with regard to nuclear energy) were configured as presented in Table 1.5. In addition, the table shows the annual consumption of 2015 (or 2007) and the reserve over production

TABLE 1.5
Available Reserves of Nonrenewable Energy Sources

	Reserves in 2015 or 2007 for Uranium	Annual Production in 2015 or 2007 for Uranium	Reserve to Production Ratio
Oil (billions barrels)	1,697.6	33.5	50.7
Natural gas (trillions Nm^3)	186.9	3.5	53.8
Coal (millions tonnes)	891,531	7,820	114.0
Uranium (thousand tonnes recoverable at up to 130 US$/kgU)	3,338.3	41.3	80.8

ratio, an indicator that provides a rough estimation of the expected exhaustion time period in years for each nonrenewable energy resource.

The life period of nonrenewable energy sources is obviously determined by their availability and their demand. If the demand cannot be satisfied by the availability, then the prices of the nonrenewable energy sources increase, so the demand decreases, and, as a result, the total life period of the remaining resources is prolonged. The peak availability of oil, for example, is estimated to coincide with the consumption of 50% or 70% of the total initial reserves, according to Hubbert's peak oil theory [3, 4]. The maximization of the corresponding fossil fuel production implies that, from that time point and on, the available production will not be able to fulfill the current demand, with all the reasonably expected social and economic consequences.

In a more optimistic scenario, the nonrenewable energy resource consumption rates can be stabilized or even reduced in the next years, as a result of the globally running campaign during the last decades for energy saving and increases of the RES share in production processes. Given this approach, the exhaustion time point of nonrenewable energy resources may be prolonged until the beginning of the twenty-second century. This considerable extension is due to power demand reduction from the side of the large-size consumers (mainly industries, transportation sector, and commercial buildings), which will decelerate the oil consumption rates of South America, the Middle East, and Western Africa. With similar efforts, the consumption rates of coal and natural gas can be also reduced. Given these assumptions and the documented oil, coal, and natural gas resources, we may achieve a postponement of their exhaustion time point for some decades, or even until the end of the twenty-first century.

Finally, we must not overlook that, due to the continuous improvement of oil and natural gas detection and pumping technologies, new deposits, the exploitation of which was considered not economically feasible until recently, become now approachable. These technologies include highly accurate detection of oil deposits with the support of satellites as well as horizontal perforation of the continental shelf, which renders unnecessary new, expensive investigative drillings.

Nevertheless, the prediction tends to lean more towards the more pessimistic scenario regarding the exhaustion time period of nonrenewable energy reserves, mainly because of energy consumption increases of currently underdeveloped or developing countries. The annual energy consumption share of these countries in 2015 was lower than 30% versus the global annual energy consumption, despite the fact that the vast majority of the earth's population lives in them. This energy global consumption share is expected to be reversed in the next 20 years, as presented in Figures 1.6–1.9 [1].

Given the increase of the energy consumption in the developing countries depicted in these figures, the exhaustion time point of the nonrenewable energy reserves should be expected much earlier than it may be estimated today. This perspective makes urgent the necessity for working toward an alternative energy future.

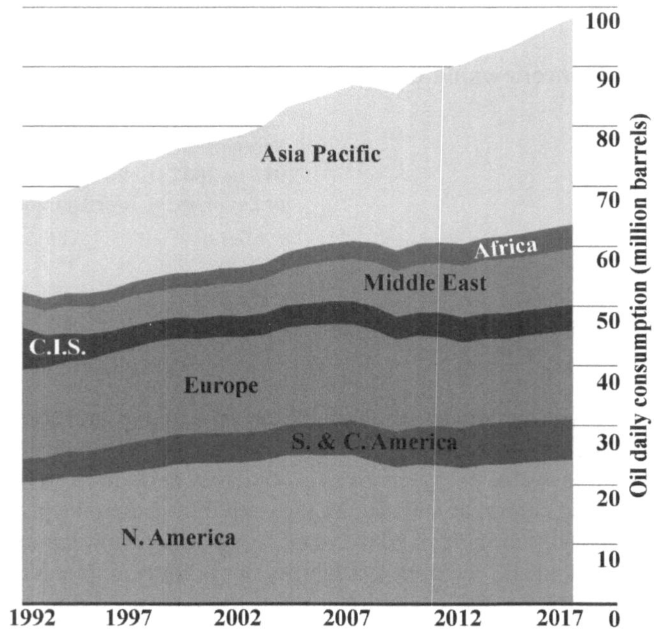

FIGURE 1.6 Oil daily consumption evolution versus time by geographical region.

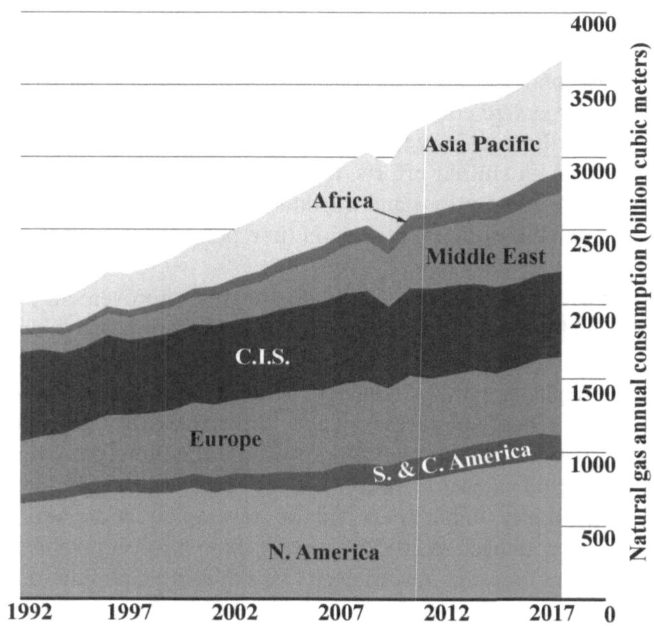

FIGURE 1.7 Natural gas annual consumption evolution versus time by geographical region.

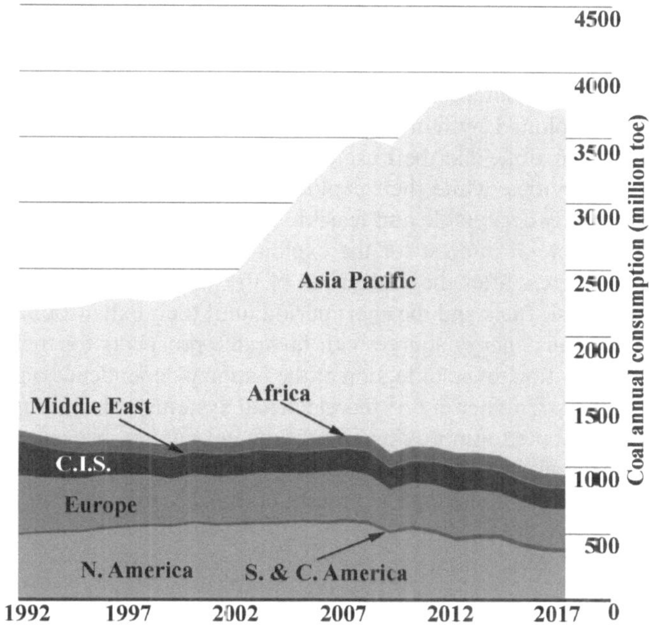

FIGURE 1.8 Coal annual consumption evolution versus time by geographical region.

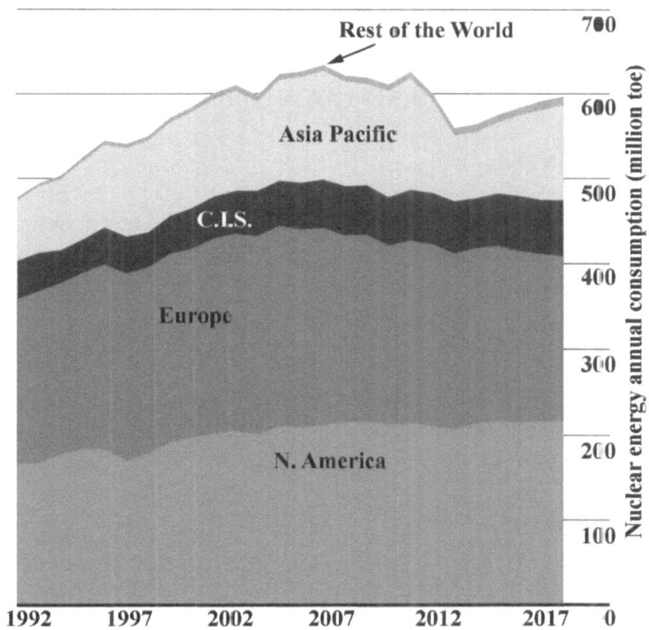

FIGURE 1.9 Nuclear energy annual consumption evolution versus time by geographical region.

1.3.3 Renewable Energy Sources

Energy sources that are abundantly available in the natural environment and regenerated annually with higher rates than their consumption are known as renewable energy sources (RES). RES are the energy sources first exploited by humans, before the turn to the intensive use of fossil fuels. RES are practically inexhaustible, and their use does not impose serious impacts on either natural environment or human activities, while their exploitation is restricted only by the development of economically and technically acceptable and feasible technologies aimed at the harvesting of their available potential. Commercial interest for the exploitation of RES first appeared after the first oil crisis in 1974. Interest grew after the realization of the global and considerable environmental impacts from the use of fossil fuels and the short period until their exhaustion. For many countries, RES constitute an indigenous energy source with favorable prospects for their energy production share, including the contribution to the reduction of the national dependence on expensive, imported energy sources and to the strengthening of the electrical systems' energy supply security. At the same time, RES also contribute to improving environmental quality, because it is widely accepted that the energy sector is the main culprit responsible for environmental pollution. In Figure 1.10 the contribution of the main energy consumption sectors in the U.S. greenhouse gas annual emissions is presented [5].

The main forms of RES are:

- sun: solar radiation
- wind: wind energy
- hydrometeors: hydraulic or hydrodynamic energy
- the earth's ground: geothermal energy
- biomass: the thermal or chemical energy coming from so-called biofuels, such as the residues from carpentry or agricultural activities (woods and branches), or the exploitation of industrial, urban, and stock farming residues
- the sea: the energy of waves, the tidal energy, and the oceans' energy coming from the temperature differences of water between the sea surface and high underwater depths.

The main advantages of RES are:

- They are practically inexhaustible and they contribute to the restriction of the dependence on the gradually exhausted fossil fuel reserves. Table 1.6 presents the annual available amounts of RES on earth [6].

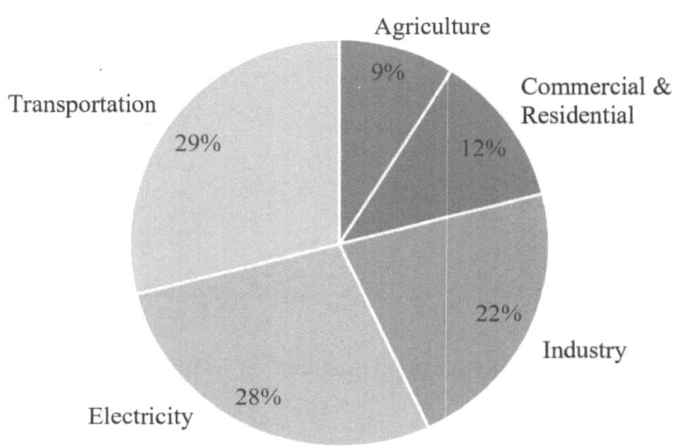

FIGURE 1.10 Total US greenhouse gas emissions by economic sector in 2017.

TABLE 1.6
Annually Available Energy Quantities of RES on Earth

Energy Form	Annual Available Amount (kWh)
Solar radiation	$580{,}000 \cdot 10^{12}$
Waves and tidal energy	$70{,}000 \cdot 10^{12}$
Wind energy	$1{,}700 \cdot 10^{12}$
Hydraulic energy	$18 \cdot 10^{12}$

If we take into account the annual primary energy consumption of 13,699 Mtoe = $159{,}288.93 \cdot 10^9$ kWh on earth in 2014 [7], it is conceivable that the planet receives every year from RES a total energy amount a thousand times higher than the total global annual energy needs.

- RES, as a consequence of their nature, are always indigenous energy sources and contribute to the support of national energy independence and energy supply security.
- Again, as a consequence of their nature, they are geographically dispersed and they lead to the development of decentralized energy systems.
- They usually exhibit low operation cost, which, furthermore, is not affected by the fluctuations of the international economy and, particularly, by fossil fuel prices.
- They can constitute in many cases the core for the invigoration of economically and socially deprived communities and the lever to claim a sustainable development, through the implementation of electricity production projects from RES and energy saving works with active involvement of local communities.
- They are friendly for the environment and human communities.

Nevertheless, most RES unfortunately also exhibit some negative features, which restrict their high contribution of the corresponding technologies in energy production balance. Such disadvantages are:

- Their dispersed potential is difficult to be concentrated to give large power availability.
- RES exhibit low energy and power density and, consequently, for large power plants the capturing of extensive land areas is required.
- Their availability often exhibits considerable fluctuations, which may be of extensive duration, implying the requirements of alternative, supporting production technologies or expensive energy storage plants and causing, in cases of high RES contribution to the energy production balance, security and stability problems, especially on weak electrical systems of medium or small size.
- Their fluctuating availability often leads to a relatively low capacity factor for the corresponding power production plants.

1.3.4 Power Density of Energy Sources

The term *power density* expresses the density of the energy content of a source of energy. In solid, liquid, and gas fossil fuels, their energy content density is expressed per unit of mass or volume and is defined as heat capacity. More specifically, the heat capacity of fossil fuels is defined as the thermal energy released with the combustion of a unit of mass or volume of the fossil fuel under normal conditions (absolute pressure of 1 atm, absolute temperature of 273 K). It is classified in upper and lower heat capacity.

TABLE 1.7
Heat Capacity of Fossil Fuels and Biofuels

Fuel	Lower Heat Capacity			
Lignite	19,700	kJ/kg	5.47	kWh/kg
Heavy fuel	40,600	kJ/kg	11.28	kWh/kg
Diesel for electricity production	35,200	kJ/lt	9.78	kWh/lt
Central heating diesel	36,250	kJ/lt	10.07	kWh/lt
Gasoline	29,900	kJ/lt	8.31	kWh/lt
Liquefied petroleum gas (LPG) (60% C_3H_8–40% C_4H_{10})	26,800	kJ/lt	7.44	kWh/lt
Natural gas (με 93% CH_4)	40,200	kJ/m^3	11.17	kWh/m^3
Wood (0% humidity)	18,800	kJ/kg	5.22	kWh/kg
Biomass pellets	20,900	kJ/kg	5.79	kWh/kg

1.3.4.1 Upper Heat Capacity

When the water produced from the combustion of a fossil fuel is disposed in liquid phase, namely it has not absorbed any part of the thermal energy released with the combustion as latent heat in order to be vaporized, the fossil fuel's heat capacity is considered the upper heat capacity.

1.3.4.2 Lower Heat Capacity

When the water produced with the combustion of a fossil fuel is disposed in gas phase (vapors), then it has absorbed a part of the thermal energy released from the combustion as latent heat. In this case, the fossil fuel's heat capacity, which will obviously have a lower value than the upper heat capacity, will be called its lower heat capacity.

For most energy calculations, the fossil fuels' lower heat capacity is used. Table 1.7 presents the lower heat capacity of some characteristic fossil fuels and biofuels.

The power density, yet, is not only defined for the solid, liquid, or gas fossil fuels. It is also certainly defined for the RES. For RES, power density may be defined with two alternative ways:

i. Based on the power density of the primary energy source versus unit of interaction area. For example, the power density of the total solar radiation (direct, diffused, and reflected) for a specific geographical area is defined per unit of area on the horizontal plane. Indicatively, the incident solar radiation power density in a 60° inclined surface versus the horizontal plane with south orientation in the center of Paris (48.5°N, 2.2°E) varies from 17/5 to 25/5 as graphically depicted in Figure 1.11. Respectively, the annual average power density of wind energy is defined per unit of perpendicular area on the wind's blowing direction and it may be, for a specific site, for example equal to 800 W/m^2.
ii. Based on the required land area per unit of nominal power for the installation of a specific RES power plant. For example, for the installation of 1 MW of a PV power plant, 2.0–2.5 hectares of land are required. Respectively, for the installation of a 3 MW wind turbine, 2,000 m^2 of land are required. From the above it is obvious that this second approach is not defined on the basis of the available energy content density of the primary energy source, but on the power density of a particular technology employed for the exploitation of the primary energy source. Hence, it might be more precise to define this magnitude, for example, as PVs' power density or wind parks' power density. Although this approach does not refer to the primary energy source itself, it is often used especially in cases of development of electricity production projects from RES, when a main design or investment parameter is the required land for the installation of particular nominal power of the RES power plant.

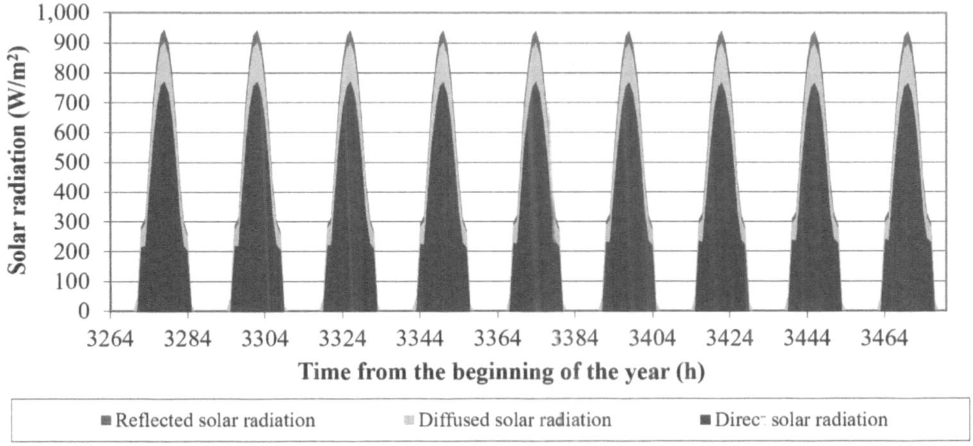

FIGURE 1.11 Fluctuation of the total incident solar radiation on a 60° inclined surface versus the horizontal plane in Paris, from 17/5 to 25/5 on a horizontal surface.

Closing this section, we must underline the significantly higher power density of fossil fuels versus RES. This feature, combined with the ease of their storage and transportation, has led to their dominance in the energy sector. Yet, on the other hand, their short exhaustion time period and their global impacts on the natural environment, have imposed the reduction of their consumption and their replacement with the exploitation of RES technologies.

1.4 ENERGY EFFICIENCY AND TRANSFORMATION

1.4.1 Energy Efficiency

As mentioned in the previous section, primary energy sources are rarely used as they are available in nature. This is because humans usually need either a different form of energy to consume, or they need it in a different location from the place it is naturally provided. As a result, the primary energy sources must be transformed either to the required form to be consumed or to a proper form to be transported to the consumption location, where they will be eventually transformed to the final desired form.

Every energy transformation from one form to another implies energy losses. This means that during the transformation process, it is not possible to achieve 100% conversion of the initial energy form to the final one. An amount of the initially available energy is not transformed; hence, it either remains in its initial form or it is transformed to another form, different from the required one, which may or may not be possible to be exploited.

The amount of energy that is successfully transformed from the initial to the final form is expressed as the transformation efficiency. Generally, the energy transformation efficiency η from an initial energy amount E_{in} to a final energy amount E_{out} is defined with the following relationship:

$$\eta = \frac{E_{out}}{E_{in}} \tag{1.16}$$

From the above relationship, we see that the efficiency of an energy transformation is defined as the ratio of the final energy E_{out} transformed to the required form over the initially available energy E_{in}.

1.4.2 Energy Transformation

As explained in the first sections of this chapter, during the evolution of a transformation process the total energy transformed from one form to the other gradually increases. From the initially available energy E_{in} an amount is transformed to the required final energy output E_{out} and, probably, to a number ν of other energy forms E_i, which are either disposed of in the environment or exploited in parallel applications. Because, according to the first thermodynamic law, the total energy of the system is conserved, it must be:

$$E_{in} = E_{out} + \sum_{i=1}^{\nu} E_i \quad (1.17)$$

Hence, during an energy transformation, energy, while changing forms, is always conserved. This is why the correct term to describe this process is *energy transformation,* which designates the transition from the one form to the other, and not *energy production,* since energy pre-exists and is not produced.

The various energy transformations are accomplished either physically, namely in nature without the intervention of any technical device, or technically. For example, we saw in a previous section that a fundamental energy transformation on which the existence of life on earth depends is the transformation of solar radiation to chemical energy in plants and trees, known as photosynthesis. This is a characteristic example of a physical energy transformation. On the other hand, as the wind strikes the blades of a wind turbine, it carries its kinetic energy to the turbine's rotor, which through the gearbox and the electrical generator, is finally transformed to electricity. Hence, in this case we have a technical energy transformation. Physical and technical energy transformations are realized in physical or technical mechanisms, which should be correctly referred to as *energy transformers*. Plants, trees, animals, and humans are physical energy transformers, while wind turbines and diesel generators are technical energy transformers.

Usually in the physical and technical world we do not have single energy transformation but, rather, sequels of energy transformations in series, during which the one energy transformation is followed by the other, starting from the initial form of energy and ending at the last one. In cases of technical energy transformation, the last energy form constitutes the required final form of energy for usage (consumption, transportation, or storage).

A characteristic example is given by a car's conventional piston engine. The initial form of energy provided for the energy transformer, the car's motor, is the fuel's chemical energy. The final form of energy exploited by the user is the mechanical energy in the car's wheels. The overall energy transformation from the initial to the final form of energy is not a direct process, but it is integrated in a number of interim stages. Specifically, the initial chemical energy, through the fuel's combustion in the combustion chamber, is first transformed to thermal energy, disposed in the combustion's produced gases. The energy content of the produced gases, expressed through their specific enthalpy, is then transformed to mechanical energy, through the gases' exhaustion in the internal combustion engine's piston. Finally, the piston's mechanical energy, through the crankshaft connecting rod system, is transferred to the car's main shaft and, eventually, to its wheels.

For an electricity production diesel generator, the required final form is not mechanical energy in the diesel engine's shaft, but electricity. Consequently, in order to reach the final product, there is one more required energy transformation, namely the transformation of the diesel engine's mechanical energy to electricity through an inductive electrical generator.

Summarizing the above, in a car, in order to take the final required mechanical energy from the initial chemical energy of the stored fuel in the car's reservoir, two energy transformations in series intervene. In a diesel generator, there are three energy transformations from the initial chemical energy to the final required electricity. Both the above energy transformation sequels are presented graphically in Figure 1.12.

Every intermediary energy transformation, which constitutes a part of the overall transformation, is characterized by a separate efficiency, which is certainly lower than unity. In other words, in every intermediary energy transformation, an amount from the initial energy is not transformed to

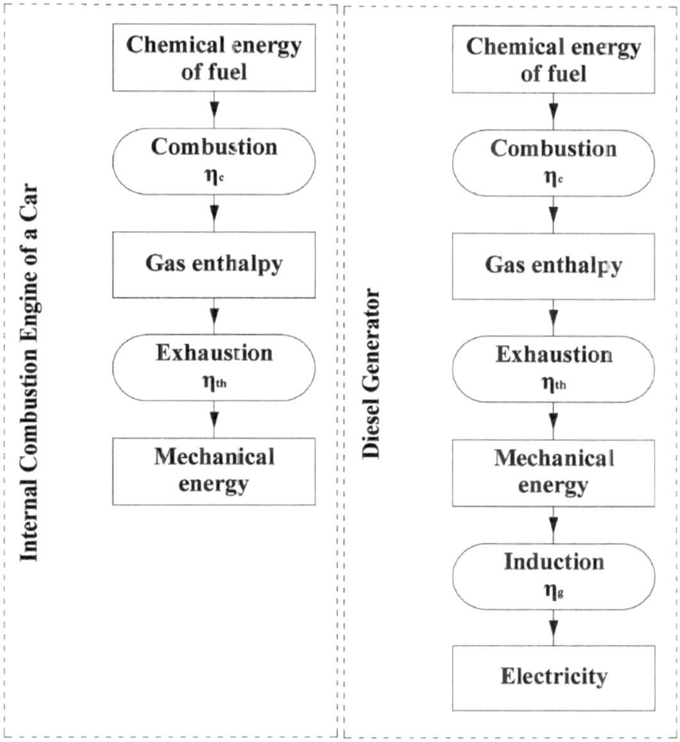

FIGURE 1.12 Energy transformation sequels in a car's piston engine and in a diesel generator.

the required form, but is lost. The energy losses, appearing in every intermediary transformation, are aggregated to configure the total losses, and consequently, the total efficiency of the overall energy transformation.

In the above examples, the efficiencies characterizing the involved intermediary energy transformations are the combustion efficiency η_c, the thermodynamic efficiency η_{th} of the internal combustion engine (describes the thermal to mechanical energy transformation), and, additionally only for the diesel generator case, the electrical generator's efficiency η_g. Because the involved energy transformations in a car's piston engine or in a diesel generator are in series, namely one is executed right after the other by accepting as energy input the energy output from the previous transformation, as it will be shown in the next section, the total efficiency of the overall transformation is given by the product of the efficiencies of the intermediary transformations:

$$\eta_{ce} = \eta_c \cdot \eta_{th} \tag{1.18}$$

$$\eta_{dg} = \eta_c \cdot \eta_{th} \cdot \eta_g \tag{1.19}$$

where η_{ce} and η_{dg} are the total efficiencies of the overall transformation in a car's engine and in a diesel generator respectively.

For the above example, in the theoretical assumption that we could use exactly the same diesel engine for both the car and the diesel generator, it would be sensible to expect higher total efficiency for the case of the car, given that for electricity production with the diesel generator, one more energy transformation would be involved in the overall sequel from the initial to the final energy form, namely one more efficiency would be introduced in the calculation product of the total efficiency.

Based on the above analysis, in order to maximize the exploitation of a primary energy source, the number of intermediary energy transformations involved in the overall sequel from the primary to the final energy form should be minimized. It is then sensible, if, for example, we have to use oil for heating production (e.g., for indoor space heating), this should be used directly in a central heating burner. On the contrary, from the rational use of energy point of view, it is completely unreasonable, illogical, and uneconomical to consume the available oil in a diesel generator to produce electricity and then to consume the produced electricity in electrical resistances or heat pumps to produce the required heating. In the first case (heating production directly with an oil burner) there is only one energy transformation (combustion) from the initial chemical energy contained in oil to the final heating production, with an overall efficiency that may be higher than 85%. On the contrary, in the second case, according to Figure 1.12, there are three energy transformations involved from the initial chemical energy to the electricity production, with a total efficiency being calculated from Equation 1.18. In this efficiency, we should also take into account the electricity transportation losses from the thermal power plant to the electricity consumption destination, as well as the heating production efficiency of the electrical resistances or the heat pump. The overall efficiency in this second case for the final heating production would be around 35%.

Energy transformers employed for electricity "production" with the fewest intermediary energy transformations involved from the primary available energy source to electricity exhibit the highest total efficiencies. These transformers are hydro turbines and wind turbines with only one energy transformation involved (mechanical energy from the machine's rotor to electricity) and another essential efficiency referring to the transition of mechanical energy from the water or the wind to the machine's rotor. For these machines, the total efficiency may exceed 90% for hydro turbines, while for wind turbines it ranges between 40%–45%.

On the contrary, regarding thermal generators (steam turbines, diesel generators, gas turbines), at least three energy transformations are involved from the initial chemical to the final electrical energy production. The total efficiencies for these transformations can be 30%–35% for steam turbines and gas turbines and 40%–45% for diesel generators. The relatively low total efficiencies of thermal generators are mainly configured by the thermodynamic efficiency, referring to the thermal to mechanical energy transformation, commonly lower than 50%. Indeed, the theoretical Otto and diesel thermodynamic cycles that describe the operation of an internal combustion piston engine exhibit efficiencies not higher than 50%–55%.

1.4.3 Average and Instant Efficiency

In general energy transformation processes, the input and the output rate of the initial and the final energy form respectively in and from the energy transformer, namely the input and the output power, are not constant. On the contrary, both of them change following the power demand, as configured by the current energy needs.

We will now define the instant efficiency of an energy transformation, with regard to a specific time point, as the ratio of the power output P_{out} over the power input P_{in} of the transformer, at this specific time point:

$$\eta = \frac{P_{out}}{P_{in}} \tag{1.20}$$

In general, the instant efficiency of an energy transformation is not constant, but it changes as the operation point of the employed transformer alters too, of course on the condition that the transformer's operation in different operation points implies different efficiencies. This last prerequisite is characteristic for almost all the available energy transformers (e.g., thermal generators, pumps, PV panels). Yet, there are cases of energy transformers, such as some particular hydro turbine models (Pelton, Kaplan), with the capability of maintaining almost constant efficiency for a wide operation range.

Apart from the instant efficiency of an energy transformation, we can also define the average efficiency over a specific time period:

$$\bar{\eta} = \frac{E_{out}}{E_{in}} \tag{1.21}$$

The average efficiency is defined for a time period and not for a time point. During this time period, the transformer's power input and output may change, so the instant efficiency may change too. Obviously the average efficiency cannot be calculated as the ratio of instant power inputs and outputs. Rather, it is calculated as the ratio of the total energy output over the total energy input from and to the transformer respectively, during the under consideration time period.

In Figure 1.13, in a common power versus time diagram, the output and input power fluctuations versus time of an energy transformation are presented for a total time period t_o. At a random time point t_i the instant efficiency of the energy transformation is given by the relationship:

$$\eta(t_i) = \frac{P_{out}(t_i)}{P_{in}(t_i)} \tag{1.22}$$

The total energy output during the whole time period t_o is depicted graphically by the area below the power output curve, the horizontal axis and the vertical lines starting from the beginning and the end of the time period (dark area). Respectively, the total energy input during the whole time period t_o is depicted graphically by the area below the power input curve, the horizontal axis and the vertical lines starting from the beginning and the end of the time period (dark and light areas together). The ratio of the dark area over the total dark and light area together gives the average efficiency of the energy transformation for the time period t_o, according to Equation 1.21.

Knowing the instant efficiency is helpful for the optimization effort of the transformation procedure, which aims at minimizing the initial energy consumption per unit of outgoing final energy. Hence, the instant efficiency knowledge provides an indication of the distance between the transformer's current operation point from the optimum one, namely the one with the maximum

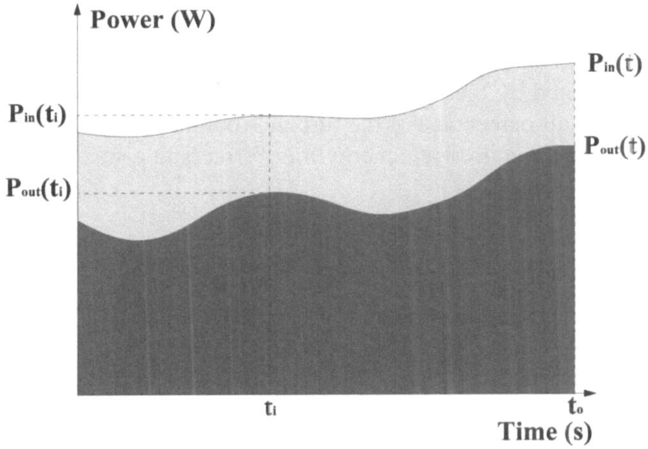

FIGURE 1.13 Instant and average efficiency for an energy transformation.

efficiency. Additionally, it may be used as an indicator to approach a higher efficiency by changing the transformer's operating conditions, if, of course, such a task is not restricted by other parameters.

The average efficiency of an energy transformation integrated over a time period constitutes a typical indicator of the whole process, which characteristically expresses the efficiency of the energy transformation, allowing also the calculation of additional features of the process, such as the total consumed primary energy and the corresponding energy production cost.

1.5 EFFICIENCY OF TRANSFORMERS CONNECTED IN SERIES AND IN PARALLEL

Energy transformers in energy systems can use two basic connection types: in series and in parallel. We can also have combinations of these two basic connection types. The selection of the transformer connection for a specific energy transformation is imposed by the type of the transformation, the size of each employed transformer with regard to the total size of the energy system, and by a number of involved parameters (e.g., transformation cost, efficiency maximization, special requirements of power demand).

For example, in a thermal power plant we usually have several thermal generators operating in parallel. At a specific time, all the dispatched units are synchronized at the electrical grid's fundamental frequency while operating independently. In other words, the operation of each thermal generator neither affects, nor is affected by, the operation of any other generator, apart from the requirement that the total power output from all of them must be equal to the power demand.

Yet, in a thermal generator there may coexist more than one energy transformer, which, while connected in series, aim to transform the input energy form to electricity. Hence, for example, in a steam turbine, the involved energy transformers connected in series are:

- the steam boiler, aiming at the input chemical energy transformation to thermal energy of the working medium (steam)
- the compressor, aiming at the transformation and the transition of the mechanical energy of the compressor's rotor to the working medium, expressed by its specific enthalpy
- the turbine, where the steam's contained enthalpy is transformed to mechanical energy through its exhaustion in the turbine's rotor
- the electrical generator, where the turbine's mechanical energy is transformed to electricity.

These four transformers involved in a steam turbine are connected in series. The power output of each transformer constitutes the power input for the next transformer and, consequently, each transformer's operation is affected by the operation of the precedent one, apart from the compressor, whose input power is a small percentage of the turbine's output power. The energy transformation sequel and the connection of the involved energy transformers in a steam turbine are presented in Figure 1.14.

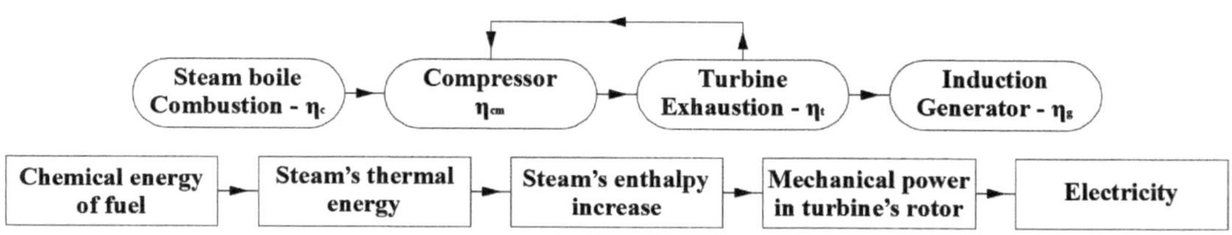

FIGURE 1.14 Transformers' connection and energy transformation sequel in a steam turbine.

From the above examples, we may conclude that a thermal power plant consists of a number of fundamental energy transformers which, at the first level, are connected in series between them to form integrated energy transformers, the so-called thermal generators. These generators are connected in parallel and cooperate to fulfill the current power demand of the electrical grid.

The above connection example of energy transformers in series and in parallel is not only met in electrical systems, but is a common case for most energy systems.

The calculation of the total instant efficiency is often required for a particular energy transformer's connection layout. As already mentioned in the previous section, the knowledge of the instant efficiency of a power system, along with the specifications of the employed energy transformers, enables the regulation of the system's operation to approach the maximization of the overall efficiency, implying, in turn, as direct impacts, the minimization of the primary energy consumption and the transformation process cost. In the following subsections, the analytical procedure is presented for the calculation of the total instant efficiency of the two main energy transformer connection types in series and in parallel.

1.5.1 Energy Transformers Connected in Series

For a number v of energy transformers connected in series, we will develop below the analytical relationship for the calculation of the total efficiency of the overall system. Let's assume $P_{in,i}$, and $P_{out,i}$, the power input and output respectively in the energy transformer T_i and η_i the instant efficiency of the same transformer at this particular operation point. The instant efficiency of the transformer T_i is given by the relationship:

$$\eta_i = \frac{P_{out,i}}{P_{in,i}}$$

Also, let's designate with P_{in} and P_{out} the total power input and output of the whole system (see Figure 1.15).

The total instant efficiency η_{tot} of the whole system will be:

$$\eta_{tot} = \frac{P_{out}}{P_{in}}$$

Yet, it is:

$$P_{out} = P_{out,v} = \eta_v \cdot P_{in,v} \Leftrightarrow P_{out} = \eta_v \cdot P_{out,v-1}$$

because it is

$$P_{out,v-1} = P_{in,v}$$

namely the power output from the transformer T_{v-1} coincides with the power input in the transformer T_v.

Yet it is:

$$P_{out,v-1} = \eta_{v-1} \cdot P_{in,v-1}$$

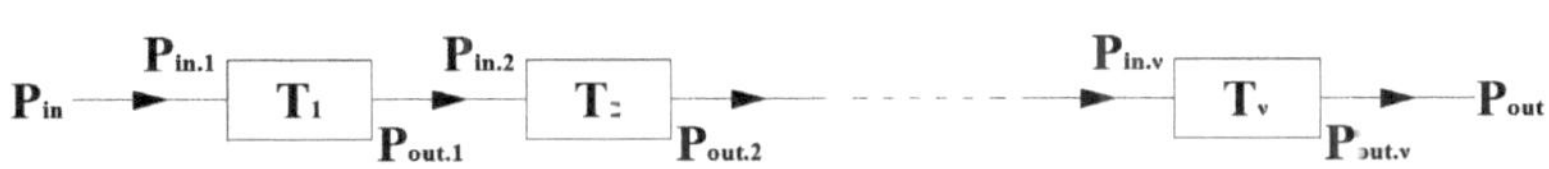

FIGURE 1.15 Energy transformers connection in series.

hence:

$$P_{out} = \eta_v \cdot \eta_{v-1} \cdot P_{in,v-1} \Leftrightarrow \cdots \Leftrightarrow P_{out} = \eta_v \cdot \eta_{v-1} \cdot \ldots \cdot \eta_2 \cdot P_{in,2} \Leftrightarrow P_{out} = \eta_v \cdot \eta_{v-1} \cdot \ldots \cdot \eta_2 \cdot P_{out,1}$$

$$P_{out} = \eta_v \cdot \eta_{v-1} \cdot \ldots \cdot \eta_2 \cdot \eta_1 \cdot P_{in,1} \Leftrightarrow P_{out} = \eta_v \cdot \eta_{v-1} \cdot \ldots \cdot \eta_2 \cdot \eta_1 \cdot P_{in}$$

because it is obviously $P_{in,1} = P_{in}$.

Eventually, the last relationship gives:

$$P_{out} = \eta_v \cdot \eta_{v-1} \cdot \ldots \cdot \eta_2 \cdot \eta_1 \cdot P_{in} \Leftrightarrow \frac{P_{out}}{P_{in}} = \eta_v \cdot \eta_{v-1} \cdot \ldots \cdot \eta_2 \cdot \eta_1 \Leftrightarrow \eta_{tot} = \eta_v \cdot \eta_{v-1} \cdot \ldots \cdot \eta_2 \cdot \eta_1 \quad (1.23)$$

Namely, we conclude that the total instant efficiency of the overall connected in series transformers equals the product of the instant efficiencies of the involved transformers.

1.5.2 Energy Transformers Connected in Parallel

For the following system of transformers connected in parallel (Figure 1.16), we will, similarly to the previous section, try to develop an analytical relationship for the calculation of the total instant efficiency of the whole system. For this purpose, let's assume $P_{in,i}$ and $P_{out,i}$ the input and output power respectively of the T_i transformer and η_i its instant efficiency for this particular operation point.

Let's also designate again as P_{in} and P_{out} the total input and output power respectively of the whole system.

We will now define the ratio x_i of the power output $P_{out,i}$ from the transformer T_i over the total power output P_{out} of the whole system, namely:

$$x_i = \frac{P_{out,i}}{P_{out}} = \frac{P_{out,i}}{\sum_{i=1}^{v} P_{out,i}}$$

because the total power out P_{out} of the whole system obviously equals to the sum of the power outputs $P_{out,i}$ from each transformer T_i separately, namely:

$$P_{out} = \sum_{i=1}^{v} P_{out,i}$$

The instant efficiency η_i of the transformer T_i is given by the relationship:

$$\eta_i = \frac{P_{out,i}}{P_{in,i}}$$

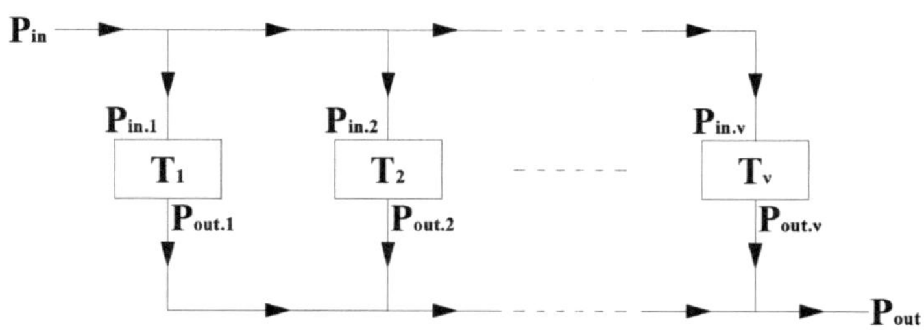

FIGURE 1.16 Energy transformers connection in parallel.

hence:

$$x_i = \frac{P_{out,i}}{\sum_{i=1}^{\nu} P_{out,i}} = \frac{\eta_i \cdot P_{in,i}}{\sum_{i=1}^{\nu} P_{out,i}} \Leftrightarrow \frac{x_i}{\eta_i} = \frac{P_{in,i}}{\sum_{i=1}^{\nu} P_{out,i}}$$

The total instant efficiency η_{tot} of the whole system will sensibly be equal to:

$$\eta_{tot} = \frac{P_{out}}{P_{in}} = \frac{\sum_{i=1}^{\nu} P_{out,i}}{\sum_{i=1}^{\nu} P_{in,i}} \Leftrightarrow \sum_{i=1}^{\nu} P_{out,i} = \eta_{tot} \cdot \sum_{i=1}^{\nu} P_{in,i}$$

By combining the last two relationships we take:

$$\frac{x_i}{\eta_i} = \frac{P_{in,i}}{\sum_{i=1}^{\nu} P_{out,i}} \Leftrightarrow \frac{x_i}{\eta_i} = \frac{P_{in,i}}{\eta_{tot} \cdot \sum_{i=1}^{\nu} P_{in,i}}$$

If we now write the last relationship ν times, once for each energy transformer, and then we aggregate both parts of the arisen ν relationships, we take:

$$\sum_{i=1}^{\nu} \frac{x_i}{\eta_i} = \sum_{i=1}^{\nu} \frac{P_{in,i}}{\eta_{tot} \cdot \sum_{i=1}^{\nu} P_{in,i}} \Leftrightarrow \sum_{i=1}^{\nu} \frac{x_i}{\eta_i} = \frac{\sum_{i=1}^{\nu} P_{in,i}}{\eta_{tot} \cdot \sum_{i=1}^{\nu} P_{in,i}} \Leftrightarrow \eta_{tot} = \left[\sum_{i=1}^{\nu} \frac{x_i}{\eta_i}\right]^{-1} \quad (1.24)$$

The last relationship gives the total instant efficiency of ν energy transformers connected in parallel. Equation 1.24 reveals that the total instant efficiency is a function of the instant efficiencies of the involved transformers and the contribution of each transformer's power output versus the total power production of the overall system.

1.6 CONVENTIONAL AND HYBRID POWER PLANTS

In recent years, particularly after the vast introduction of RES exploitation technologies, the terms *hybrid power plants* or *hybrid energy systems* often appear in the relevant scientific and technical literature. In this section we will try to contrast a conventional, common energy system with a hybrid power plant. For the rest of the book, we will conventionally use the terms *energy production* or *production units*, which are common for the description of energy transformation and energy transformers.

1.6.1 Conventional Energy Systems

In a conventional energy system, the energy production is based on guaranteed power production units. We define guaranteed power production units as the energy transformers for which the availability and, in turn, the supply of the initial energy form can be predicted and controlled by the user. The ability to adjust and control the supply of the initial energy form practically implies the possibility to control the energy transform rate to the final energy form at will and, essentially, according to the power demand's requirements. For example, the operation of a thermal generator, given the availability of the consumed fuel, can be certainly adjusted and controlled at will, based on its technical specifications and in combination with any other probably available thermal generators, in order to totally fulfill the electricity needs of the electrical system. Similarly, the operation schedule of a central heating burner, again given the availability of the consumed fuel (e.g., oil, biomass pellets) in the system's storage tank, can be absolutely regulated and controlled at will, aiming to ensure thermal comfort conditions for the conditioned indoor space.

Usually, the guaranteed power production units are the energy transformers based on the consumption of nonrenewable energy sources, mainly fossil fuels. Apart from fossil fuels, energy transformers exploiting biomass fuels, hydrogen, geothermal potential, etc. can also be considered as guaranteed power production units, on the condition that the availability of the initial energy form is known and can be predicted, independently, whether it is enough or inadequate with regard to a specific power demand.

At this point it must be clarified that the classification of an energy transformer as a guaranteed power production unit is not determined by its adequacy to cover the power needs at a specific time point. Rather, an energy transformer can be considered as a guaranteed power production unit when the energy transformation can be accomplished following a schedule predefined and controlled by the end user. For example, a diesel generator will in any case be a guaranteed power production unit, because, given the availability of the consumed fuel (heavy fuel, diesel oil, or natural gas), its production schedule can be controlled at will by the electrical system's operator. This classification is not affected by whether this diesel generator is installed in a large, mainland thermal power plant, where the fuel's supply can be considered secure and guaranteed, or in a non-interconnected, remote, and not easily accessible island, where a possibly extended period of adverse weather conditions can potentially put under risk the adequacy of the available fuel reserves for the electricity production in the autonomous, insular system.

We will additionally define nonguaranteed power production units as the energy transformers in which the availability and, in turn, the supply of the initial energy form cannot be predicted or controlled by the user. In these cases, the energy transformation rate is most commonly determined by the availability of the primary energy source, which, unlike in guaranteed generators, is stochastic and cannot be predicted with the required accuracy for the secure operation of a power system. The fact that the availability of the primary energy source is not guaranteed and is unpredictable, makes the corresponding energy transformation also stochastic and uncontrolled by the user. Logically, then, it follows that the nonguaranteed power production units are mainly the various RES exploitation technologies. The stochastic nature of the primary energy source, such as the wind energy, the wave potential, or the solar radiation, classifies the corresponding energy transformers (wind turbines, solar collectors, PV panels) as nonguaranteed power production units.

From our practical experience, or perhaps from any specialized awareness we may have obtained, we can realize that energy transformation to final, consumed energy forms is globally based, until today, mainly on conventional energy systems and on guaranteed power production units. We know, for example, that electricity production is based particularly on guaranteed power production units, with the consumed primary energy sources being provided by solid, liquid, or gas fossil fuels (coal, oil, natural gas, and nuclear fuels). Similarly, thermal or mechanical energy production is also based on guaranteed power production units, such as internal combustion engines as car motors, gas turbines as airplane motors, and central heating burners with the primary energy source again being provided by fossil fuels (gasoline, diesel oil, kerosene, LPG). All these energy systems, which have dominated in the energy production sector, based on guaranteed power production units, will be classified as conventional energy systems.

Obviously in conventional energy systems, together with the guaranteed power production units, nonguaranteed power production units can also be available. For example, the overall electricity system of Denmark has a total installed power of 12.7 GW, including 9.5 GW of thermal generators, consuming natural gas, coal, and oil, and 3.1 GW of wind parks, with a contribution to the annual electricity production in 2015 of 42%, the highest RES national annual penetration percentage globally. Similarly, in an autonomous central heating system of a building, apart from the central burner (guaranteed production unit), a number of solar collectors may also be installed, which can contribute up to a maximum percentage, based on the available solar radiation, to the annual coverage of the building's heating loads. In both examples, the final energy production is based on guaranteed power production units (thermal generators or central burner) and, consequently, both systems are classified as conventional energy systems. The role of the RES technologies, namely the nonguaranteed power production units, in these systems is auxiliary or supplementary. More specifically, they aim to penetrate as much as possible at the coverage of the energy demand, given the availability of the RES potential (wind energy or solar radiation), contributing, at the same time, respectively to the reduction of the guaranteed units production and the consumption of nonrenewable energy reserves. Yet, in both systems, the conventional units still remain the main power production units with the highest contribution share in the annual energy production balance.

1.6.2 "Hybrid" Energy Systems

What happens if we decide to change the roles of guaranteed and nonguaranteed units in an energy system? Specifically, what happens if we try to operate the nonguaranteed units as the fundamental production units and restrict the guaranteed units to a supplementary role? Is it possible to achieve a secure and stable operation for an energy system under such a scenario, given the unpredictable and stochastic nature of the power production from nonguaranteed units?

For the successful integration of such energy systems, in order to achieve the ultimate target of continuous and secure power production from nonguaranteed units, the stochastic, nonguaranteed power production should be adapted according to the current power demand. The only way to achieve this goal is by combining operation of the nonguaranteed units with energy storage technologies. A simplified approach of the nonguaranteed plus storage units combined operation algorithm can be described as:

- when the produced power by the nonguaranteed units exceeds the current power demand, the available energy surplus is stored in the storage units
- when the power demand is higher than the available power production from the nonguaranteed units, the energy production shortage is compensated by the storage units.

The above described operation algorithm is graphically presented in Figure 1.17, which depicts the electricity power demand and the power production from wind parks in an imaginary electrical system, during a period of 10 days.

The light areas in this diagram designate electricity production surplus from the wind parks (nonguaranteed units), provided for storage in the system's storage units. The dark areas in the diagram designate energy provided for the power demand by the storage units, every time the power demand is higher than the available power production from the wind parks. In this way, the storage units contribute to the adaptation of the wind parks' stochastic power production to the nonnegotiable power demand.

To ensure uninterrupted and secure operation, the system should be able to maintain the power production even in cases of power production shortage from the nonguaranteed units, with regard to the power demand, and concurrent inadequate charge level of the storage units. In such cases, the only way to avoid the interruption of the power production is to equip the system with a guaranteed unit, e.g., a diesel generator in case of an electrical system.

An energy system, operating with the above philosophy, aiming at the maximization of RES penetration in the system and at the restriction of guaranteed units as backup units, is widely known

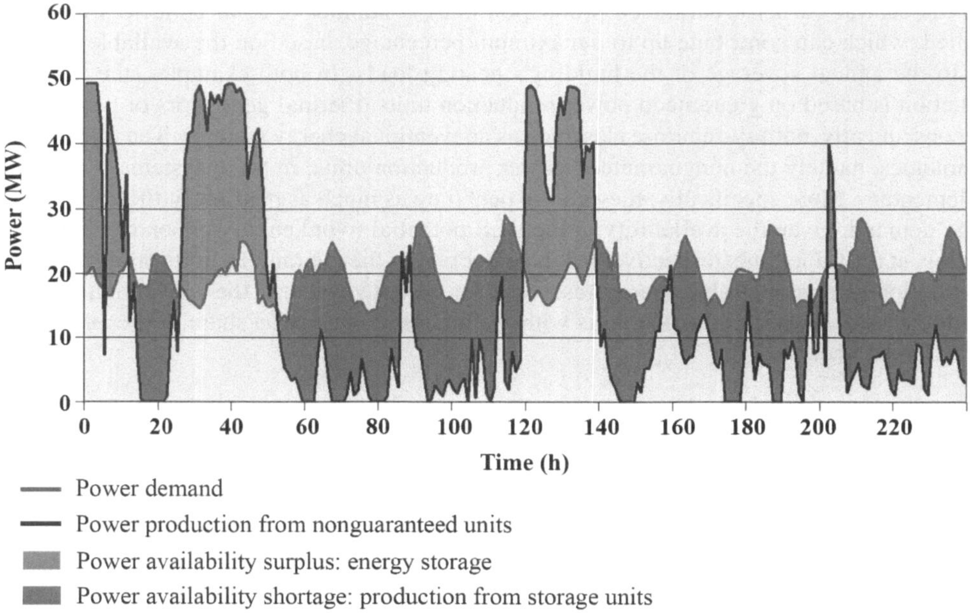

FIGURE 1.17 Graphical depiction of the basic operation concept of a hybrid power plant.

as a hybrid power plant. From the above description, we can conclude that a hybrid energy system consists of three discrete components:

- Nonguaranteed power production units characterized as "base units" because the power production is mainly based on them, namely they are the basic production units. The main objective of the hybrid energy system is the maximization of the exploitation of the nonguaranteed energy produced by the base units. In an electrical system, the base units can be wind parks or small wind turbines and PV stations. In thermal energy systems, the base units can be solar collectors (flat, of selective coating or concentration).
- Storage units, which aim, through the bidirectional power flow from and to them, to adapt the nonguaranteed power production from the base units, determined by the availability of the RES potential, to the power demand. In electrical systems, the storage units can be electrochemical batteries of several types, fuel cells cooperating with hydrogen production electrolysis units, pumped hydro storage systems, and CAES systems. In thermal energy systems, the storage unit can be storage water tanks, rock beds, pressurized water thermal tanks, etc.
- Backup units, aiming at the coverage of the power demand with guaranteed power production in instances where the base or the storage units are not able to provide enough power (low availability of RES potential and simultaneous low charge level of the storage units). The backup units exhibit ultimate priority, namely they are dispatched last in the power production, only when there is no possible production from the base or storage units. In electrical systems, the backup units can be diesel generators of large or small size and fuel cells, on the condition that the availability of the consumed fuel can be considered as guaranteed. In thermal energy systems, the backup units can be conventional oil, LPG, or biomass burners, heat pumps, etc.

The term *hybrid* energy system is improperly used, because the word *hybrid* originates from the ancient Greek word *yvris*, which means insult, hubris. The term *hybrid* was initially introduced to describe energy systems in the same sense as it was used for the description of genetically modified agricultural products (e.g., corn or soy hybrids). These agricultural hybrid products were developed violating the physical processes; namely, they essentially constitute a hubris for nature. However, a hybrid power plant cannot in any case be considered as a hubris for any physical or even technical process. The term is improperly used to describe an energy system that differentiates from the operation concept of conventional systems. This attitude, yet, does not imply that this alternative energy production approach can be characterized as hybrid, namely as a hubris for any energy transformation. On the contrary, we may claim precisely the opposite, namely that conventional energy systems, which are based on the waste of nonrenewable energy natural resources, constitute a hubris for nature and a bane to environmental conservation. Not to mention that these natural resources are also employed in various parallel uses, apart from the energy sector, for the production of a large number of goods used in several applications in our modern daily lives. Perhaps it is enough to consider that the oil amount required for the production of 40 lt of gasoline, namely the quantity to fill up the reservoir of a small-size car, is enough to provide the required raw material for the production of all the necessary plastics to fully equip a modern household.

Maybe a more precise term for the description of hybrid power plants could be the term *RES–storage combined power plants* or something similar. Nevertheless, for the description of these systems in this book, the term *hybrid* will be used, because it has been widely accepted, although incorrectly.

1.7 THE BOOK'S LAYOUT

This book aims to cover a focused syllabus on the technical and economic features, the operation, the design, the development, and the study of power plants for electricity and thermal energy production. It anticipates to constitute a scientific and technical handbook on the design and the development of power plants for students of mechanical engineering and electrical engineering, professional engineers, and post-graduate students. The contents are organized in the following chapters.

Chapter 1: Introductory Concepts

In this introductory chapter, integrated with this final section, the basic definitions were given for fundamental magnitudes and concepts involved in the study of power plants and energy transformations. Specifically:

- The magnitudes of energy and power were defined, and their concepts were explained with a number of examples. The most commonly used units for the measurements of energy and power were presented and the conversion factors between them were also provided.
- The various energy forms were classified and the role of each of them was described. The most important energy transformations were also presented.
- The basic primary energy sources were listed and classified as renewable and nonrenewable energy sources. There was also a reference on the available reserves of nonrenewable energy forms and the annually available renewable energy resources on the planet. The power density for an energy source was defined and characteristic values were indicatively given for the most popular energy sources.

- The definitions of energy transformation and energy transformers were given. The average and the instant efficiency of an energy transformation were also defined.
- The basic connections of energy transformers, namely in series and in parallel, were presented and the analytical relationships for the calculations of the total instant efficiency for both connection layouts were developed.
- Power plants were classified as conventional and hybrid power plants. Indicative, explanatory examples were presented.
- The first chapter is integrated with an overview of the book's content (in this very section!).

Chapter 2: Conventional Power Plants for Electricity Production

The conventional power plants for electricity production are studied in Chapter 2, consisting of thermal and RES power plants. The chapter is restricted to the production sector, without being extended to electricity transportation and distribution topics. The chapter's topics include:

- A comprehensive description of electrical systems (power stations, transportation and distribution grids, power substations). Classification of electrical systems as interconnected and autonomous. Peculiarities met in non-interconnected systems regarding their security, stability, and the electricity production cost.
- The definitions for dynamic security and the stability of electrical grids are given and supplemented with a number of examples for secure and insecure systems' operation. The process of spinning reserve is also described, aiming at the support of the electrical systems' dynamic security.
- The rules and the restrictions for nonguaranteed power production unit penetration in conventional electrical systems are given. Special reference is given on the nonguaranteed unit penetration in autonomous electrical systems and their potential impacts on the systems' dynamic security.
- The available thermal generator technologies employed for electricity production in thermal power plants are presented (steam turbines, diesel generators, gas turbines, combined cycles). Their basic technical specifications are given, such as their response rate versus the power demand fluctuations, their efficiency variation versus the power output, the possibly consumed fossil fuels, and their final production specific cost.
- The available main electricity production technologies from RES (wind turbines, PV panels, hydro turbines) are also presented along with their main operation features.
- The dispatch criteria of thermal generators in electrical systems are given. The thermal generators are classified in base and peak production units.
- Finally, for a better understanding of the above concepts and procedures, computational simulations of the annual operation of two characteristic power plants of small and large size are thoroughly presented. By taking into account all the restrictions and operation directives of conventional electricity systems presented in this chapter, the systems' simulation aims to estimate the power production synthesis for every time calculation step (hour) and for a whole annual operation period. These tasks are integrated with the calculation of overall characteristic annual results describing the systems' operation, such as the annual average overall efficiency of the involved thermal generators along with their production-specific cost, the fuels' consumption, and the overall systems' production cost.

Chapter 3: Electricity Production Hybrid Power Plants

In Chapter 3, the concept of hybrid power plants for electricity production is presented. The topics examined in this chapter are:

- The components of hybrid power plants for electricity production are distinguished. The fundamental difference of hybrid from conventional power plants for electricity production is explained.
- Centralized and decentralized hybrid power plants are presented, namely the hybrid power plants of small and large size. The available RES and energy storage technologies that can be employed as the hybrid plant's base and storage components are presented. The possible syntheses of hybrid power plants are also given, as imposed mainly by the hybrid plant's size and the available parameters in the installation site.
- The fundamental arithmetic methodologies for the dimensioning optimization of hybrid power plants are then thoroughly presented. All the examined methods are based on the annual computational simulation of the hybrid plants operation, with hourly, daily, or monthly calculation steps. Different optimization approaches are investigated, based on RES penetration maximization or economic criteria. The examined procedures are applied in a number of characteristic cases of hybrid power plants of small and large size, employing different RES and energy storage technologies.
- All the above will be further explained with four characteristic case studies on the design, dimensioning, and operation simulation of hybrid power plants with different involved RES and energy storage technologies and different sizes.

Chapter 4: Hybrid Plants for Thermal Energy Production

Chapter 4 presents hybrid power plants for thermal energy production (thermal hybrid plants). Given the difficulties with thermal energy transportation, thermal hybrid plants are mainly decentralized systems. Consequently, the examined plants in this chapter refer to decentralized consumptions, such as buildings, swimming pools, and industries. The chapter's syllabus is:

- Definition of the hybrid plant for thermal energy production.
- Available technologies for thermal energy production by exploiting solar radiation (various types of solar collectors) and thermal energy storage.
- Methodologies and procedures for the calculation of the thermal power production from different solar collectors' types. Efficiency of solar collectors.
- Thermal energy storage. Available technologies and calculation processes.
- Heat exchangers involved in thermal energy transition between different hydraulic networks. Efficiency calculation methodology.
- Dimensioning optimization methods for thermal hybrid plants, based on RES production maximization or economic criteria. Computational simulation procedure of the annual operation of thermal hybrid plants.
- Solar thermal power plants for electricity production. Given that the power production calculation from concentrated solar collectors has been presented previously in this chapter, a description of the technical specifications of the involved components and the way they are integrated to form a solar thermal power plant is provided.

The chapter is again integrated with two characteristic case studies of thermal hybrid plants dimensioning and operation simulation for different applications, specifically the heating of swimming pools and the heating of buildings' indoor conditioned space.

Chapter 5: Cogeneration Power Plants

Chapter 5 presents the subject of electricity and heat cogeneration for centralized and decentralized plants. Practically the chapter presents the theoretical approach and the practical procedures for the exploitation of the rejected thermal energy from thermal power plants. The chapter covers the following subjects:

- The available different cogeneration technologies are presented (steam turbines, gas turbines, reciprocating engines, combined cycles, etc.).
- The main indicators for the evaluation of cogeneration plants are defined, such as the overall cogeneration efficiency, the power to heat ratio, and the fuel saving ratio.
- The mathematical background for the energetic and the exergetic analysis of a cogeneration process is presented. The concept of exergy is introduced.
- The essential design layout and concept of district heating and cooling systems is analyzed. Characteristic success stories from implemented district heating and cooling systems are presented.
- The concept of trigeneration systems is also given, with special focus on solar cooling systems. Technical and mathematical analysis is provided, along with a proposed algorithm for the computational simulation of a solar cooling system.

The chapter is also integrated with two characteristic case studies, one for a combined heat and power system for the heating of algae cultivation ponds and another one referring to the design and dimensioning of a solar cooling system.

Chapter 6: Smart Grids

Chapter 6 aims to provide a general reference on the essential smart grid topics. First, the concept of smart grids is introduced with their main functionalities of smart grids and a reference on their evolution versus time. The main benefits expected from the realization of smart grids, as well as the main barriers towards fast and adequate integration in electrical systems, are listed.

An extensive presentation of the demand side management concept is given, along with the main demand side management strategies. The main programs for the realization of these strategies are discussed. A reference on the expected benefits from demand side management realization is also made.

The chapter is then expanded from the theoretical to the technological section of smart grids. The required technologies for the realization of smart grids are classified in control systems and devices, in monitoring systems, and in communication systems.

The chapter is integrated with a short presentation of some very first attempts on the development of smart grids in different geographical regions.

Chapter 7: Energy as a Consumptive Product

Apart from the presentation of specialized scientific and technical issues, a major target of this book is the configuration of critical thought and opinion on the available technologies for electricity and thermal energy production. For this reason, in this last chapter an evaluation of the different production technologies presented in this book is attempted, based on economic, social, and environmental criteria. The reader will have the chance to form an integrated opinion on the investigated systems, not only as technical solutions towards an alternative energy future, but also as potential means for the social development of human communities and environmental conservation. The combined provision of technical training and documented critical opinion contribute to the balanced and integrated composition of professional engineers and researchers.

REFERENCES

1. BP Statistical Review of World Energy 2018. https://www.bp.com/content/dam/bp/business-sites/en/global/corporate/pdfs/energy-economics/statistical-review/bp-stats-review-2018-full-report.pdf (Accessed on August 2019).
2. World Energy Council, "Survey on Energy Resources, Interim update 2009." https://www.worldenergy.org/wp-content/uploads/2012/10/PUB_Survey-of-Energy-Resources_Interim_update_2009_WEC.pdf (Accessed on August 2019).
3. Hubbert, M. K. (1979). Hubbert estimates from 1956 to 1974 of US oil and gas. *Methods and Models for Assessing Energy Resources*, 370–383.
4. Hubbert, M. K. (1976). Survey of world energy resources. *Energy and the Environment Cost-Benefit Analysis*, 3–38.
5. United States Environmental Protection Agency. https://www.epa.gov/ghgemissions/sources-greenhouse-gas-emissions (Accessed on August 2019).
6. Alekseev, G. A. (1986). *Energy and Entropy*. Moscow: Mir Publishers.
7. Wikipedia: World Energy Consumption. https://en.wikipedia.org/wiki/World_energy_consumption (Accessed on August 2019).

2 Conventional Power Plants for Electricity Production

2.1 ELECTRICAL SYSTEMS

2.1.1 LAYOUT OF AN ELECTRICAL SYSTEM

An electrical system consists of the following discrete components:

- power production plants
- transmission lines of extra-high, high, and medium voltage
- distribution lines of medium and low voltage
- voltage transformation sub-stations
- consumption.

The combined operation of the power production plants, the sub-stations, and the transmission and distribution lines aims at the uninterrupted cover of the electricity demand, with no power production curtailments and without deviations from the system's predefined operation standards, regarding the nominal values for the fundamental frequency and the RMS voltage. Each one of the above components of an electrical system plays an important role, which we discuss below.

2.1.1.1 Electricity Production Power Plants

The initially available energy resource is transformed into the final desirable form, namely electricity, in power production plants. This transformation is traditionally described as "electricity production," even though this is a slight misnomer. Recall from Chapter 1, energy pre-exists in nature in different forms, consequently it is not produced; on the contrary, it is transformed from one form to another. Electricity production power plants can be:

- Thermal power plants, where electricity constitutes the final product of an overall conversion sequence, with thermal energy appearing as an intermediate, involved form of energy. The primary source of energy, from which the whole conversion process begins, can be the chemical energy of fossil fuels (lignite, coal, oil, natural gas) or the nuclear energy of nuclear fuels (uranium, plutonium). In the latter case, the thermal power plant is called a nuclear power plant. Thermal generators, which will be thoroughly presented in this chapter, are all synchronous generators, namely their rotational speed is analogous to the fundamental frequency of the produced alternative current (60 Hz in the US and Canada, 50 Hz in Europe).
- Hydro power plants, where electricity constitutes the final product of a conversion sequence starting from the gravitational energy of water, due to its storage or natural availability in higher absolute altitudes compared to the location of the hydro power plant. The generators employed for this conversion are called hydro turbines, and they are coupled with synchronous generators too.

- Several technologies for the production of electricity from renewable energy sources (RES), such as wind turbines, PV (PV) panels, geothermal power plants, biomass plants, solar thermal power plants, have been gaining a continuously increasing percentage of the annual electricity production share globally, since the 1980s and thereafter.

More on electricity production power plants will be presented below in this chapter.

2.1.1.2 Electricity Grids

The basic layout of an electrical system is presented in Figure 2.1.

For electricity transportation from power production plants to final consumers, two types of grid are involved, depending on the grid's nominal voltage: the transmission grid and the distribution grid.

The transmission grid is responsible for the electricity transfer from power production plants to sub-stations. The transmission grid operates under high voltage (150 kV or 66 kV for small insular systems) and extra-high voltage (400 kV), in order to keep the transportation losses under an upper acceptable limit, in cases of long transferred distances. Electricity with the appropriate voltage value must be provided for the final medium-voltage (15 kV or 20 kV) or low-voltage (220 V or 380 V) consumers. Consequently, the transmission grid ends at the high/medium voltage substations, where the inlet high voltage of the transmission grid is transformed to the medium voltage of the distribution grid. The voltage sub-stations constitute nodes of the electrical system.

The main components of the transmission grid are:

- pylons or towers for the bearing of the transmission lines,
- insulators, for the mounting of the transmission lines on the pylons, and
- conductors, mainly from copper or aluminum.

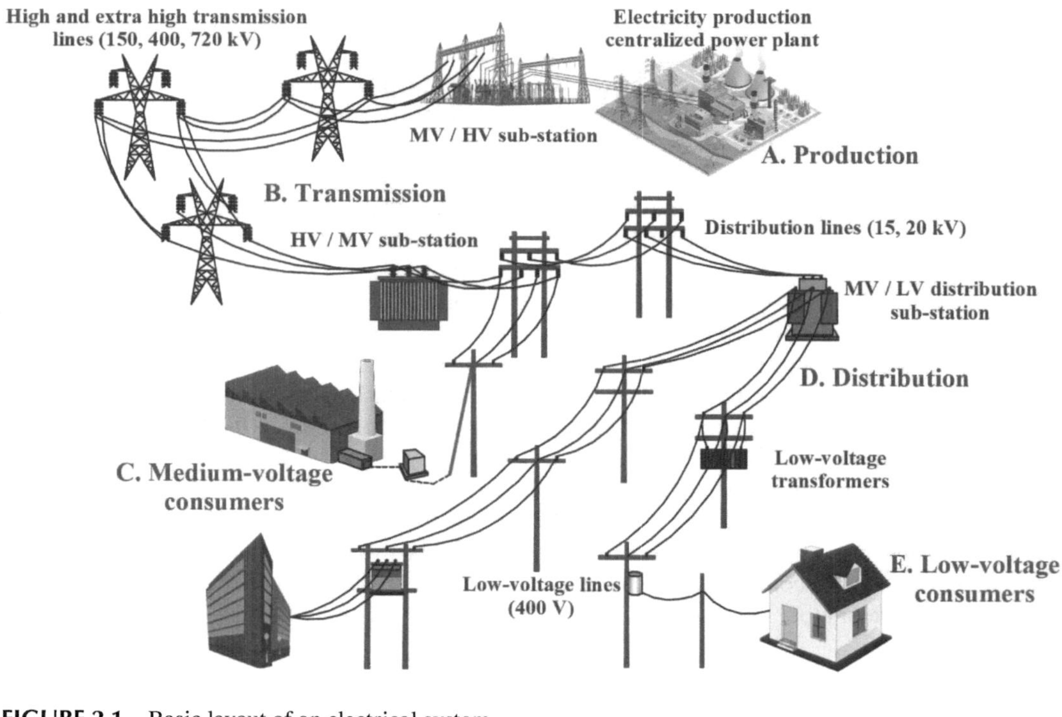

FIGURE 2.1 Basic layout of an electrical system.

The distribution grid starts from the high/medium voltage sub-stations and ends at the final low-voltage consumers. It consists of:

- the medium-voltage grid (15 kV or 20 kV), employed for the electricity distribution from the high/medium voltage sub-stations to the medium/low voltage transformers
- the distribution sub-stations, employed to degrade the grid's voltage from medium to low
- the low-voltage grid (220 V or 380 V), which distributes the electricity to the final consumers.

2.1.1.3 Voltage Sub-Stations

Voltage sub-stations are introduced in the transmission and distribution grids. These sub-stations constitute nodes in the electrical system, at which the electricity transmission and distribution lines may end or begin. The basic role of the voltage sub-stations in an electrical system is the voltage transformation to the appropriate nominal voltage of either the distribution or the transmission grids. A voltage upgrade in an electrical system is required for electricity transmission over long distances, in order to reduce the transmission losses. A voltage degrade is performed for the electricity distribution to the final consumers.

The electrical losses P_L for the transmission of active power P under voltage V are given by the relationship:

$$P_L = I^2 \cdot R \tag{2.1}$$

where I is the transmitted current and R the total ohmic resistance of the transmission line.

The total resistance of an electricity conductor is given by the relationship:

$$R = L \cdot \rho \tag{2.2}$$

where L is the conductor's length and ρ its specific resistance (in Ω/km).

The transmitted current for a three-phase circuit is:

$$I = \frac{P}{\sqrt{3} \cdot V} \tag{2.3}$$

From Equations 2.2 and 2.3, Equation 2 1 becomes:

$$P_L = \frac{P^2}{3 \cdot V^2} \cdot L \cdot \rho \tag{2.4}$$

from which it is revealed that the electricity transmission losses are inversely proportional to the second power of the grid's voltage [1].

From the above, it is concluded that the principal component in the voltage sub-station is the voltage transformers. In addition to them, a voltage sub-station includes buses, switches, fuses, etc. A sub-station in an electrical system can be:

- voltage upgrade sub-station, introduced for the voltage upgrade from the power production plants to the transmission grid (66, 150, 400 kV)
- voltage degrade sub-station, introduced for the voltage degrade from the transmission grid to the medium-voltage distribution grid (15, 20 kV).

Finally, the distribution stations in an electrical system are introduced to degrade the distribution grid's medium voltage (15, 20 kV) to the low voltage (220, 380 V) for the final electricity supply for consumers.

2.1.2 Interconnected and Non-Interconnected Electrical Systems

An electrical system is considered as interconnected when it is electrically connected with one or more neighboring electrical systems, e.g., of other neighboring states or geographical territories. Today, the continental national electricity grids are interconnected with one another. Additionally, some insular electricity grids are also interconnected to mainland national grids, such as the case of Sicily and Sardinia in Italy or the case of the national electrical grid of Denmark, which is interconnected with all Danish insular systems and the neighboring national grids of Norway, Sweden, and Germany, as presented in Figure 2.2.

An electrical system that is not electrically interconnected with other electrical systems is defined as an isolated or autonomous or non-interconnected electrical system. Practically, an autonomous electrical system covers all the electricity demand exclusively with power plants installed in its geographical territory, without any support from other electrical systems. Usually, non-interconnected electrical systems are used in remote geographical areas, with low population density, such as Greenland or Siberia, and, of course, islands, such as the ones in the Caribbean Sea, the Mediterranean Sea, the Atlantic, the Pacific, and the Indian Ocean. Croatia constitutes a characteristic case with several autonomous electrical systems. Due to the large number of islands in Croatia, their relatively low electricity demand, and their long distances from the mainland country, some of them remain still non-interconnected, as presented in Figure 2.3.

Interconnected systems can support each other, by way of the electricity transmission between them. Electricity export and import between national transmission grids is a common and widely applied process, within the frames of commercial agreements, aiming either at the reduction of

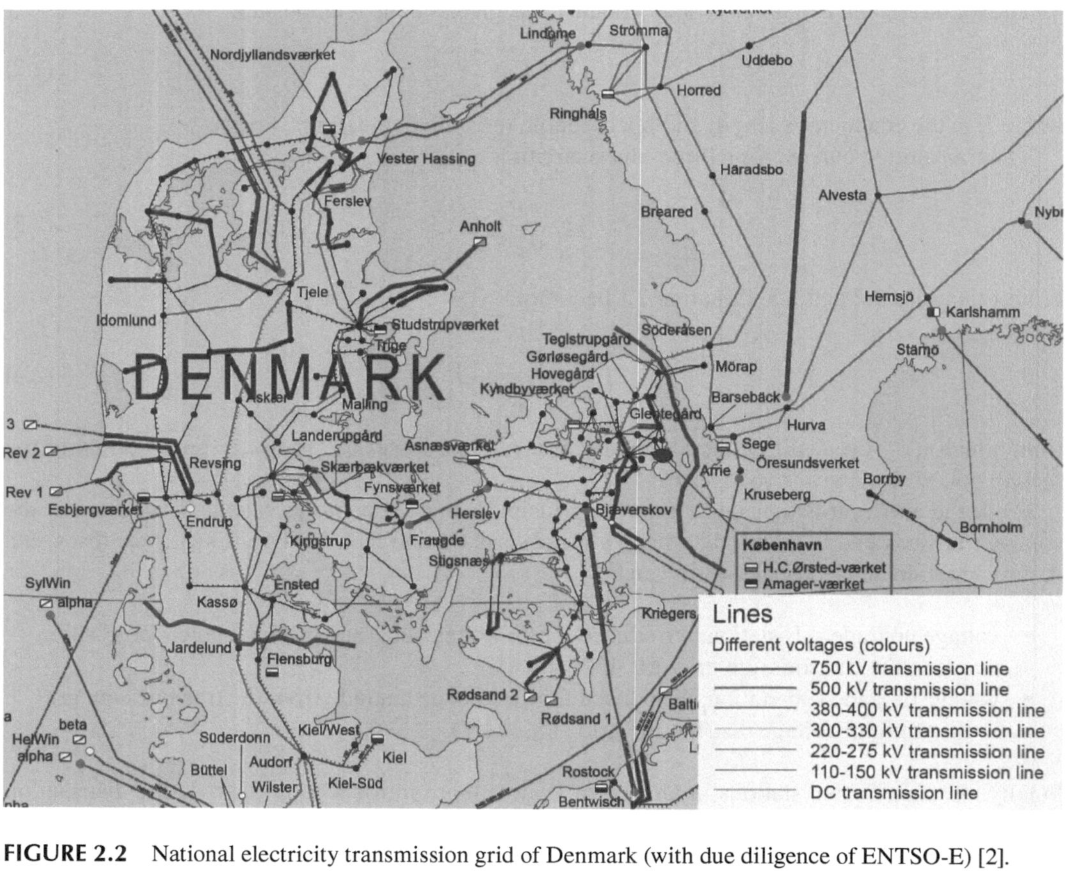

FIGURE 2.2 National electricity transmission grid of Denmark (with due diligence of ENTSO-E) [2].

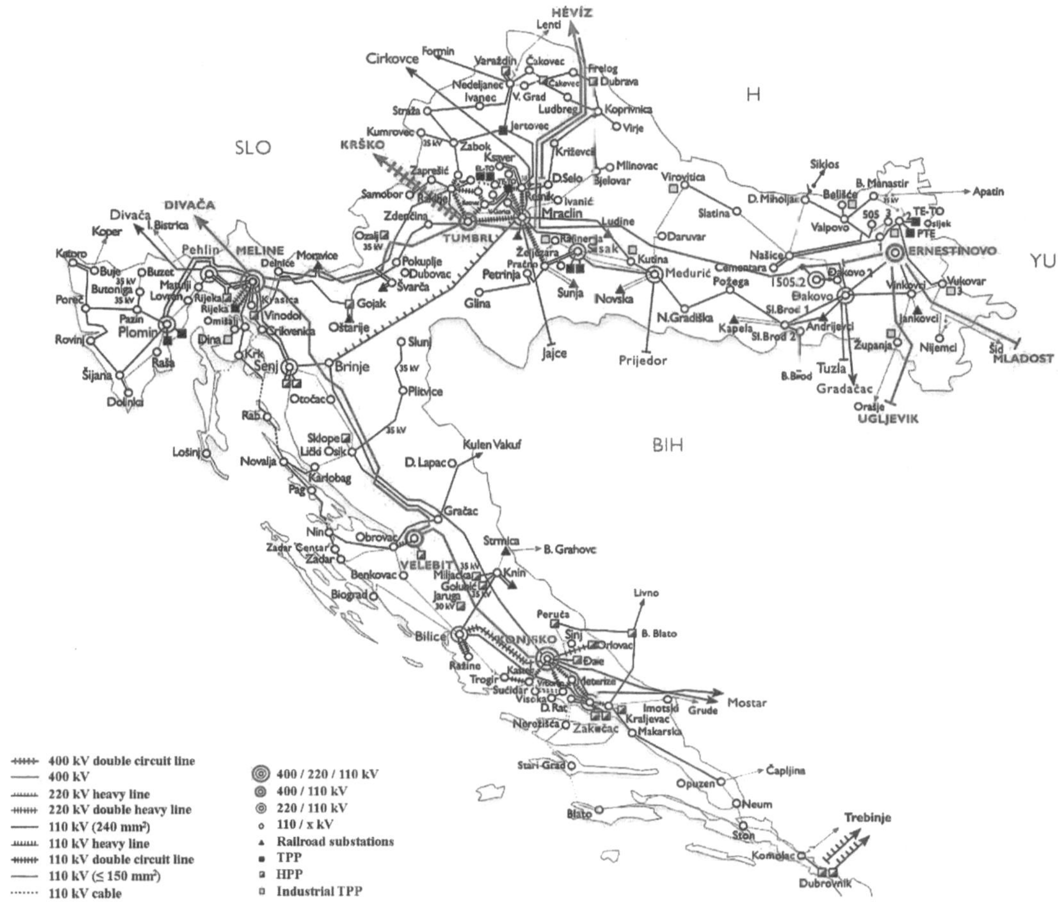

FIGURE 2.3 National electricity transmission grid of Croatia [3].

electricity production costs or at the improvement of the electrical system stability. The latter feature constitutes the most important advantage of interconnected systems: They exhibit the ability to support each other in emergency cases, such as a sudden loss of power production that may jeopardize the delicate, dynamic balance between power demand and production. In such contingencies, the ability of the interconnected systems to support each other may protect them from a probable partial or total black-out.

On the other hand, the above favorable features are not available in non-interconnected systems. A first direct consequence is that, generally, the autonomous grids are "weak" electrical systems, namely they are not able to react adequately to retain the dynamic balance between power demand and production after a considerable contingency, such as a significant power production loss, abrupt fluctuations of power production (e.g., from RES plants), or a sudden significant increase in power demand. Such events often can lead to serious dynamic security issues in small autonomous systems. Generally, the smaller the autonomous system, the more vulnerable in such kinds of disturbances it is. The stability in autonomous systems is improved with the availability of spinning reserve from thermal generators with fast response, a technique that leads to higher electricity production-specific cost. Indeed, the electricity production-specific cost in small autonomous systems is often several times higher than in interconnected systems. More details on spinning reserves and increased electricity production cost are presented later in this chapter.

2.1.3 Electrical System Security

The term *electrical system security* describes the ability of the system to maintain the dynamic balance between power demand and production, namely the ability of the power production system to adequately respond to power demand fluctuations, keeping the nominal RMS voltage and the fundamental frequency of the system within the predefined limits.

The fundamental frequency of an electrical system is directly related to the active power demand, while the nominal amplitude of the alternative voltage waveform is directly related to the reactive power demand, according to the equations presented below [1, 4]:

$$\Delta f = -\frac{\Delta P_L}{P_n} \cdot f_n \cdot R \tag{2.5}$$

$$\Delta Q = E_i \frac{\Delta V}{X_d} \cdot \cos(\delta) - \frac{(\delta V)^2}{X_d} \tag{2.6}$$

where

Δf: frequency fluctuation due to the active power demand change
ΔP_L: active power demand change
P_n: initial total active power demand
f_n: grid's fundamental frequency under nominal operation
R: speed droop of the system
ΔQ: reactive power demand change
E_i: excitation voltage of the generator
ΔV: voltage fluctuation due to the reactive power demand change
X_d: inductive resistance of the generator
δ: power angle of the generator.

The dynamic security in an electrical system practically depends on the availability of the active and reactive power production as required by the consumption, with adequate response rates, so the arisen frequency and voltage changes due to the active and reactive power demand fluctuations, according to Equations 2.5 and 2.6, always remain within the acceptable limits, imposed by the requirement to ensure secure operation of the involved power generators.

In interconnected electrical systems, where large power production plants, transmission, and distribution grids coexist and cooperate, the successful conservation of the system's dynamic security constitutes a much easier task than in autonomous electrical systems. This is due to the following facts:

- The power demand fluctuations in large interconnected electrical systems are as intensive as those in small autonomous systems, and, consequently, they are not easily experienced by the power generators, because a probable power demand drop at a specific grid point may be compensated by a concurrent power demand rise in a different location; hence, in large interconnected systems, the power production generators rarely have to face intensive load fluctuations.
- In case of extreme contingencies (e.g., abrupt power demand changes, generator or grid malfunctions), there is always the possibility for the large interconnected systems to be supported by their neighboring ones through power transition from one to the other.

The above favorable operational features of interconnected electrical systems are not available in autonomous ones. In small systems, power production may be supplied by a single synchronous generator. In this case, the grid's fundamental frequency is based only on the angular velocity of the generator's rotor. The dynamic security of the whole electrical system depends on the technical features of this very specific generator, such as its droop[1] and its response rate to load fluctuations. In most autonomous systems, the only way to improve the dynamic security is the availability of spinning reserve, a technique which will be described in the next section. One drawback to this approach is the electricity production-specific cost increase.

As the size of an autonomous electrical system increases, its dynamic security depends on the combined operation of a number of synchronous generators. Usually, in large autonomous electrical systems, the fundamental frequency is regulated by some specific thermal generators—the so-called base generators—while the load's fluctuations are undertaken by some other generators, the so-called peak generators.

Practically, the dynamic security of an electrical system can be seriously affected by significant voltage and frequency drops in the grid. A voltage or a frequency drop can be caused by a sudden increase of the power demand or a power production loss after a generator's malfunction. The reaction of the electrical system depends on:

- the current penetration percentage of nonguaranteed power production units versus the total power production
- the technical features of the dispatched thermal generators, especially their response rates to the power demand fluctuations
- the percentage versus the power demand and the type of the spinning reserve (explained in the next section)
- the tolerances of the dispatched generators to the grid's voltage and frequency fluctuations.

Theoretically, in case of a voltage or frequency drop in an electrical system, the following events may occur:

- The system recovers successfully without any power production interruption. This can be the case after a mild contingency or if there is adequate spinning reserve maintained by guaranteed power generators with fast response rates.
- The system recovers with a partial interruption of the electricity supply for a specific part of the grid. In this case, the available spinning reserve is not enough to support the adequate system's reaction after the occurred contingency. The safety mechanisms of the grid cut off power to specific portions of the electricity consumers, either randomly or according to a predefined reaction scenario, until the dynamic balance between power production and demand is restored for the remaining part of the grid. The power is provided back for the detached part once there is additional power production available from newly dispatched thermal generators.
- The system collapses (black-out). This is usually the case after serious errors in power plants, followed by generator production losses (generators' tripping) with high power production share. The system's black-out comes after a chain of generators' tripping, due to the activation of their protection mechanisms. The generators' tripping begins with the more sensitive generators on the grid's voltage and frequency fluctuations, which, in turn, affect the operation of the others, and, eventually, of the whole production system.

At this point it must be noted that all the electricity generators exhibit maximum tolerances on the grid's voltage and frequency fluctuations, as far as both the size and the duration of these fluctuations are concerned. If these limits are violated, the generators are disconnected from the grid, interrupting their operation, for self-protection reasons.

---

[1] A synchronous generator's droop is a speed control mode of a prime mover driving a synchronous generator connected to an electrical grid. It is equal to the ratio of the grid's frequency change versus the corresponding power change produced by the generator as a result of the prime mover action [2, 5].

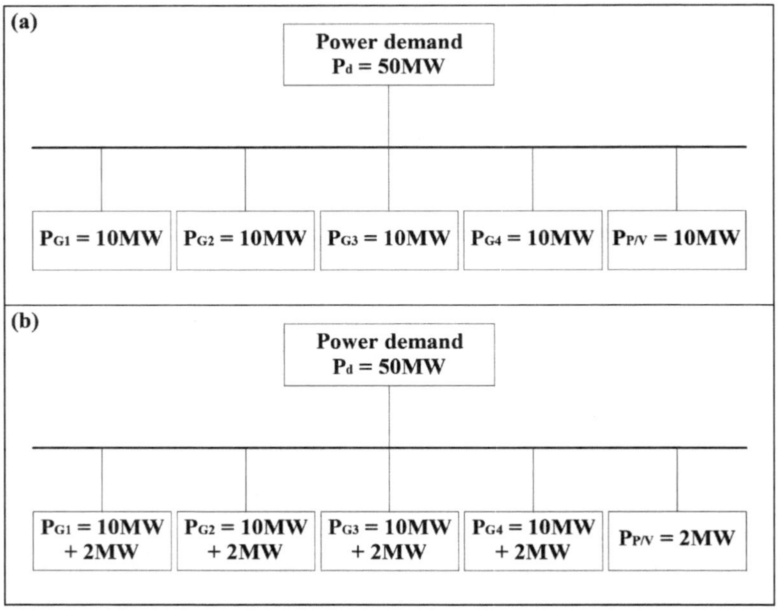

FIGURE 2.4 Graphical depiction of (a) the power production synthesis of an electrical system under constant state operation (b) its possible reaction after 8 MW pv power production loss.

In the example presented in Figure 2.4, the total power demand of 50 MW is covered initially by a group of four thermal generators and one PV station with 10 MW power production from each of them.

Abruptly, the power production from the PV station reduces to 2 MW, due to a considerable reduction of the available solar radiation. Consequently, there is an instant power production loss of 8 MW, which should be compensated from the other dispatched generators. Practically, the total power production loss should be now allocated in the rest of the four dispatched thermal generators, in order to restore the dynamic balance between power production and demand. Perhaps the most sensible approach would be the allocation of the 8 MW evenly in the available four generators, which implies a 2 MW power production increase for each thermal generator.

The power production loss and the subsequent request for immediate power production increase will be experienced by the thermal generators' rotors as a sudden electromagnetic drag, which, in turn, will lead to a direct rotational speed deceleration and a corresponding drop of the electrical system's fundamental frequency. Depending on the size of the frequency drop, the thermal generators will be able to remain connected to the grid for a maximum time period, determined by their type and technical specifications. In case the grid's nominal frequency has not been restored by the end of this maximum time interval, the thermal generators will then serially start disconnecting from the grid, causing a total system collapse.

To avoid the above possibility, the following two prerequisites should be in effect:

- a total power production increase margin of 8 MW, namely equal to the power production loss, should be in total available from all the four dispatched thermal generators
- the dispatched thermal generators should be characterized with fast response rate to power demand fluctuations, adequate to ensure a quick power production increase, in order to compensate for the power production loss within the available time margin before the beginning of the generators' tripping.

The above two prerequisites generally define the concept of the spinning reserve, which will be thoroughly analyzed in the following section.

2.1.4 Spinning Reserve

Every thermal generator is characterized by a lower and an upper power production limit, which determine its power production range. The lower limit is known as *technical minimum* and represents the lowest possible power production from the thermal generator. The upper limit is known as *maximum capacity*, which either coincides with the generator's nominal power or is a little higher than this (usually up to 10%).

Let's assume an electrical system with four available thermal generators, with the following technical minimum and maximum capacities:

- Generator 1: technical minimum 4 MW, maximum capacity 9 MW
- Generator 2: technical minimum 4 MW, maximum capacity 8 MW
- Generator 3: technical minimum 3 MW, maximum capacity 6 MW
- Generator 4: technical minimum 2 MW, maximum capacity 4 MW.

Four different operation scenarios, regarding the power demand and the power production synthesis for this particular electrical system are presented in Figure 2.5. All these scenarios will be examined below, focusing particularly on the achieved system's dynamic security.

- Scenario 1

  The power demand equals 16 MW and is covered with the production of 9 MW from Generator 1 and 7 MW from Generator 2. The most severe contingency that may occur in the electrical system under this operation condition is the loss of the dispatched generator with the highest maximum capacity, namely Generator 1. The ability of the other dispatched generator to increase its contribution from the current power production to its maximum capacity equals 1 MW. Consequently, a potential tripping of Generator 1 cannot be covered by the possible power production increase of the other dispatched generator. Such a contingency should be faced with the dispatching of more generators, practically of

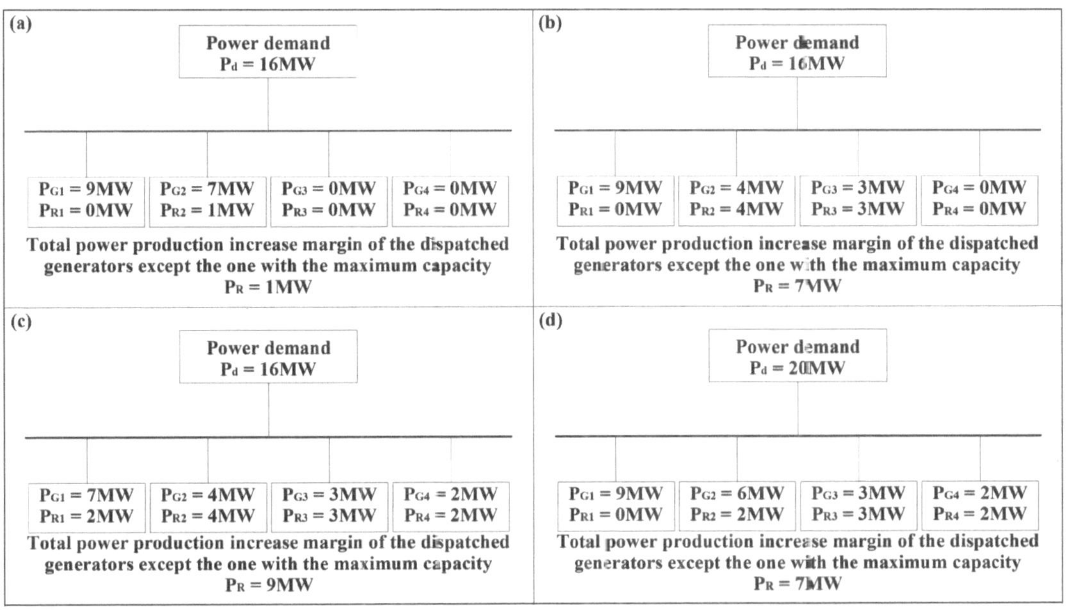

FIGURE 2.5 Graphical depiction of four alternative examined scenarios regarding the power demand, the power production synthesis, and the maintained spinning reserve of a theoretical electrical system.

Generators 3 and 4, a process most likely not fast enough to avoid a system black-out, due to the frequency drop, which will cause Generator 2's tripping too.

- Scenario 2
  The power demand equals 16 MW and is covered by the production of 9 MW from Generator 1, 4 MW from Generator 2, and 3 MW from Generator 3. The ability for power production increase of all the dispatched generators, apart from the generator with the highest maximum capacity (Generator 1), namely of Generators 2 and 3, is 7 MW. Consequently, a potential tripping of the generator with the highest maximum capacity, namely Generator 1, will be able to be covered by the remaining two dispatched generators, with a simultaneous 2 MW reduction of the power production, until the fourth available generator is prepared to contribute to the power production synthesis. It is conceivable that this power production synthesis can be considered as more secure than the one examined in Scenario 1.
- Scenario 3
  The power demand still equals 16 MW. This time it is covered by the production of 7 MW from Generator 1, 4 MW from Generator 2, 3 MW from Generator 3, and 2 MW from Generator 4. The ability for power production increase of all the dispatched generators, apart from the generator with the highest maximum capacity (Generator 1), namely of Generators 2, 3 and 4, is 9 MW. Consequently, a potential tripping of Generator 1 will be totally covered by the other currently dispatched generators. This operation scenario is the most secure compared to the previous ones.
- Scenario 4
  In this last scenario, the power demand is assumed at 20 MW. Although the power production has been allocated to all the available generators (9 MW from Generator 1, 6 MW from Generator 2, 3 MW from Generator 3, and 2 MW from Generator 4), the ability for power production increase of all the dispatched generators, apart from the generator with the highest maximum capacity (Generator 1), namely of Generators 2, 3, and 4, is 7 MW. Consequently, a potential tripping of Generator 1 cannot be covered by the remaining generators. It is understood that, although the power production has been allocated to all the available thermal generators, it is not possible to maintain a fully secure operation of the electrical system because of the size of the power demand, given also the total maximum capacity of the available generators.

The total ability of the dispatched thermal generators, apart from the one with the highest maximum capacity, to increase their power production up to their maximum capacity, is known as *spinning reserve*. The availability of spinning reserve in an electrical system aims to provide a power backup from the dispatched thermal generators, which can be exploited to compensate a potential power production loss. As revealed from the above scenarios, the spinning reserve is achieved with the allocation of the total power demand to an adequate number of thermal generators, in order to maintain the power production from each one of them lower than its maximum capacity.

So, how much spinning reserve should be maintained during a specific operation of an electrical system, so the system's operation status can be considered secure? The spinning reserve policy usually adopted in electrical systems is that the total power production increase capacity of the dispatched thermal generators, apart from the one with the highest maximum capacity (spinning reserve), should be equal at least to the maximum capacity of the dispatched generators. However, it should be understood that even if this policy is kept, the system's security cannot be ensured against all potential contingencies. As already mentioned, the dynamic security of electrical systems depends on the total maintained spinning reserve, the response rate of the dispatched generators, weather conditions, the nonguaranteed power plant penetration, and the size of the power demand versus the available thermal generators.

In the examples presented previously, the above mentioned spinning reserve policy is achieved only in Scenario 3. In Scenarios 1 and 2, there is not enough spinning reserve due to inappropriate power production synthesis. In Scenario 4, the above spinning reserve policy cannot be achieved, due to the increased power demand, which implies increased power production from the thermal generators and reduced margins until their maximum capacities.

As we will see in a following section, the efficiency of the thermal generators decreases for part-load operation, namely for power production lower than their nominal power. Hence, the spinning reserve conservation, because it is approached by allocating the total power demand in more thermal generators and by forcing them to operate in partial loads, imposes operation with lower efficiency and, hence, increasing production cost.

2.1.5 RES Power Plants and Dynamic Security of Electrical Systems

The most widely introduced electricity production technologies from RES are wind parks and PV stations. The power production from both technologies is not constant, because it is affected by the fluctuations of the primary energy resource, namely the wind and the sun. So, both technologies are considered as nonguaranteed power production technologies, as defined in Chapter 1. Wind turbines and PV stations, as nonguaranteed power production units, exhibit particular impacts on the electrical systems' dynamic security. The power production of these particular RES technologies cannot be controlled at will by the system's operator. On the contrary, it depends on the availability of the primary energy resource, which usually varies. Fluctuations of the wind potential and the solar radiation versus time cause corresponding variations on the produced power, which must be compensated by the other dispatched guaranteed power production units (e.g., thermal generators), so that the total power production from all the generators will always equal the power demand. An extreme case of power production fluctuations from a wind park or a PV station is a sudden significant power production drop or complete loss, which may be the result of several causes, such as the abrupt reduction of the available RES potential (e.g., abrupt reduction of solar radiation due to cloud cover) or a malfunction on the RES plants themselves (e.g., a lightning strike on a wind turbine's blade). In such events, the higher the initial RES units' penetration is, the more difficult it will be for the remaining dispatched generators to face a corresponding power production loss.

Beyond the varying, nonguaranteed power production, wind turbines, specifically, are characterized as vulnerable generators with regard to their relatively limited tolerance on the grid's nominal voltage and frequency fluctuations. This fact implies that the dispatched wind turbines in an electrical system are the first in range generators expected to be disconnected from the grid when a serious grid voltage or frequency disturbance is detected. Characteristic tolerances of modern wind turbines on grid voltage and frequency variations are presented in Table 2.1. The values presented in this table are indicative and may differ for different wind turbine manufacturers. Certainly, as the technology of wind turbines develops, their performance is also improved. However, the majority of the wind turbines currently installed worldwide are characterized, more or less, by tolerances similar to the ones presented in Table 2.1 [5, 6].

TABLE 2.1
Wind Turbine Tolerances on Grid Nominal Voltage and Frequency Fluctuations

Disturbance	Fluctuation (%)	Time Delay (s)
Voltage drop	90.00	60.0
Overvoltage	110.00	60.0
Frequency drop	−6.00	0.2
Over-frequency	+2.00	0.2

The nonguaranteed power production units' penetration in electrical systems, especially non-interconnected ones, introduces difficulties regarding dynamic security conservation. For this reason, the spinning reserve maintenance from the dispatched thermal generators, adequate to compensate for a potential RES power production loss, is regularly adopted as a remedy towards the achievement of acceptable security operation status, with a direct impact, of course, on the increase of the electricity production cost. The alternative approach, obviously, can be the significant limitation of the RES technology penetration in the power production synthesis, in order to avoid any potential negative impacts on the system security caused by their fluctuating power production.

Generally, the effect of an RES unit on the dynamic security of an electrical system depends on the following:

- the RES penetration percentage in the power production synthesis
- the technical specifications of the dispatched thermal generators, particularly, their response rates versus the power demand fluctuations
- the available spinning reserve
- the wind turbines' tolerances on the grid's voltage and frequency fluctuations
- the grid's topology (general layout and siting of the transmission and distribution lines and the power plants)
- the weather conditions.

Regarding the RES penetration percentage, relevant simulation studies of autonomous electrical systems have indicated that for penetration percentages up to 10%, no serious effects on the system's dynamic security should be expected, even after a total loss of the RES power production. On the other hand, for RES penetration percentages higher than 35%, the electrical system will hardly recover after a corresponding power loss [7].

Nevertheless, the above maximum percentages do not guarantee the secure operation of electrical systems. Generally, the maximum secure RES penetration percentage is also affected by the parameters mentioned previously, such as the availability of spinning reserve, the grid's topology, and the weather conditions. These parameters impose different upper limits of secure RES penetration percentage. Practically, the maximum secure RES penetration percentage may be between 10% and 15% for small weak systems under unfavorable operation conditions (low power demand, adverse weather conditions, slow spinning reserve), while, for larger systems it may exceed 50%.

In order to evaluate the dynamic security of electrical systems under different RES penetration scenarios, dynamic security assessments are executed, based on the computational simulation of the electrical systems, realized with specialized software applications. The results of such an assessment, executed for an imaginary electrical system, are presented in Figure 2.6.

In Figures 2.6, the system's fundamental frequency variation is presented for two different initial operation states and two alternative introduced contingencies in the under examination electrical system. Specifically, in Figure 2.6a the total power demand equals 258 MW, covered by 178 MW power production from thermal generators and 80 MW power production from wind parks (31% penetration percentage). Additionally, 30 MW spinning reserve from steam turbines is available, namely from thermal generators with slow response rate. In Figure 2.6b the power demand and the wind parks' total production remain the same; however, in this scenario there are 159 MW of spinning reserve offered by diesel generators and gas turbines, namely by thermal generators with fast response rates. In both operation simulations, the following two hypothetical contingencies are introduced:

- contingency A: 80 MW of wind park power production loss
- contingency B: 25 MW of steam turbine power production loss.

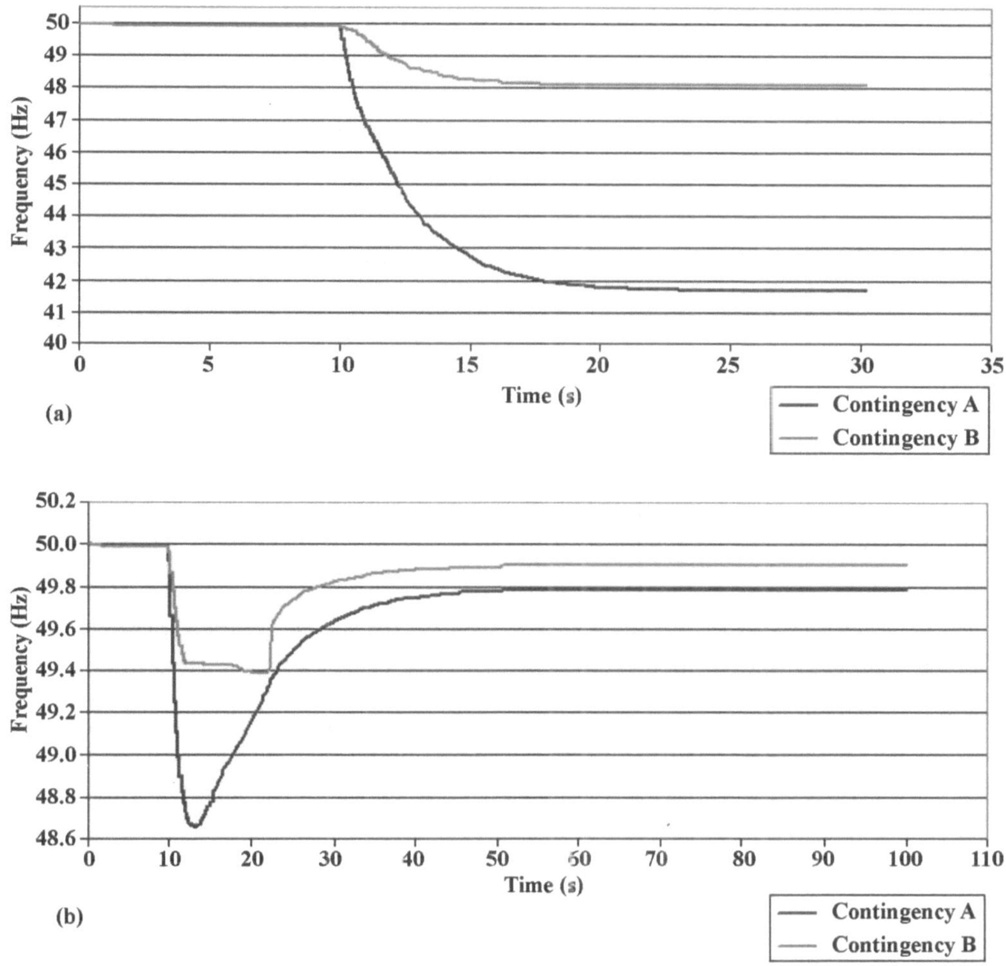

FIGURE 2.6 Simulation results of an electrical system's reaction after the power production loss for alternative power production synthesis and spinning reserve scenarios.

In the first examined operation state, the system collapses with both contingencies. Even with the loss of 25 MW of steam turbine production, although lower than the maintained spinning reserve of 30 MW, the system is not able to recover, because the maintained spinning reserve is based on thermal generators with slow response rates. In the second scenario, the high and fast maintained spinning reserve enables the system to react successfully after both contingencies. Finally, as obviously expected, contingency B is handled more easily than contingency A, due to the lower power production loss.

This characteristic example reveals that the secure operation of an electrical system is a multi-parametric task, affected by a series of different parameters, such as the size and the type of the spinning reserve, the severity of the contingencies, and the RES unit penetration percentage.

2.1.6 Electrical System Power Demand

The power demand in modern electrical systems is automatically recorded with supervisory control and data acquisition (SCADA) systems for regular time intervals (e.g., average values for every 10 minutes) and for annual time periods. The responsible power utilities or the grid operators usually publish annual technical bulletins, with all the relevant figures regarding system operation, such as the annual power demand variation, the annual electricity consumption, the electricity production from the available thermal generators and RES power plants, the fossil fuel consumption, the annually average efficiencies, any special events recorded during the system's annual operation (e.g., power production faults, load cut-off), and, eventually, annual electricity production cost and specific cost per production unit and in total. From this overall inspection procedure it is generally observed that the main power demand features usually remain almost the same from year to year, exhibiting perhaps slight changes, such as small annual electricity consumption increase or decrease rate, of course on the condition that there are no significant changes on the overall production and consumption system, like the addition of new or the removal of old generators, the substitution of a fuel type (e.g., heavy fuel) with a new one (e.g., natural gas), or a considerable change in the employed fossil fuel prices.

In Figure 2.7 the annual power demand time series is presented for the island of Samos, a Greek island of medium size, with a permanent population of 33,000 inhabitants, located in the eastern Aegean Sea. The annual maximum power demand is met during the summer period, of course due to tourist activities, and is configured close to 35 MW, while the minimum power demand is recorded during autumn, lower than 10 MW. The maximum annual power demand in an electrical system is generally known as annual peak demand. During the winter period, the power demand also increases compared to the spring or autumn seasons, yet without reaching the values exhibited during summer.

The power demand annual profile presented in Figure 2.7 is indicative for a large number of insular and mainland electrical systems from different parts of the world, where there is a significant increase in human activities during summer, most commonly due to tourism. The maximization of human activities in the summer season imposes a corresponding maximization of the power demand. On the other hand, the power demand is minimized during the seasons when, first, there are no considerable professional or any other type of activities and, secondly, the outdoor climate conditions approach the standard thermal comfort indoor space conditions, minimizing, thus, the electricity consumption for indoor space conditioning. These seasons, for most climates in the world, are certainly spring and autumn. Finally, during winter, electricity consumption increases, mainly due to the increasing energy needs for heating, as well as, in some cases, due to the execution of specific seasonal activities, such as the operation of energy-intensive olive oil mills, a common winter process for the whole Mediterranean basin.

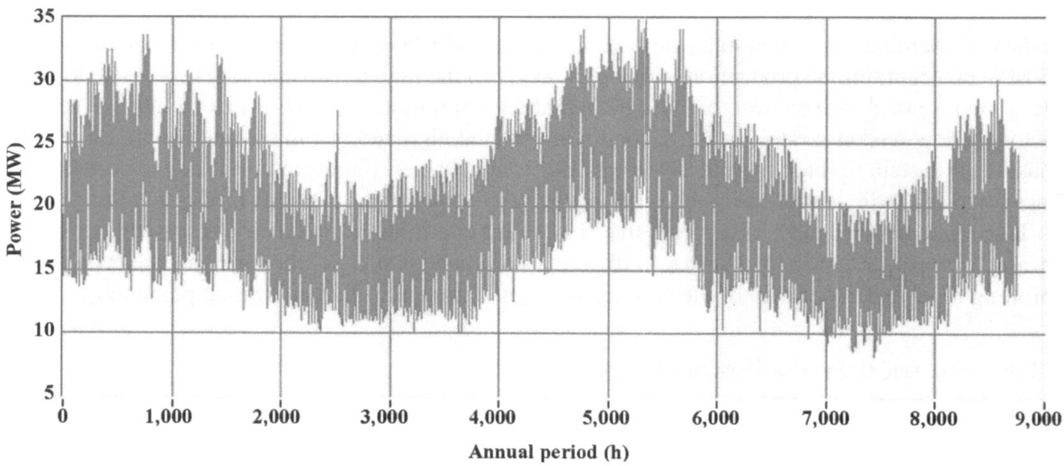

FIGURE 2.7 Annual power demand variation for the Greek island of Samos.

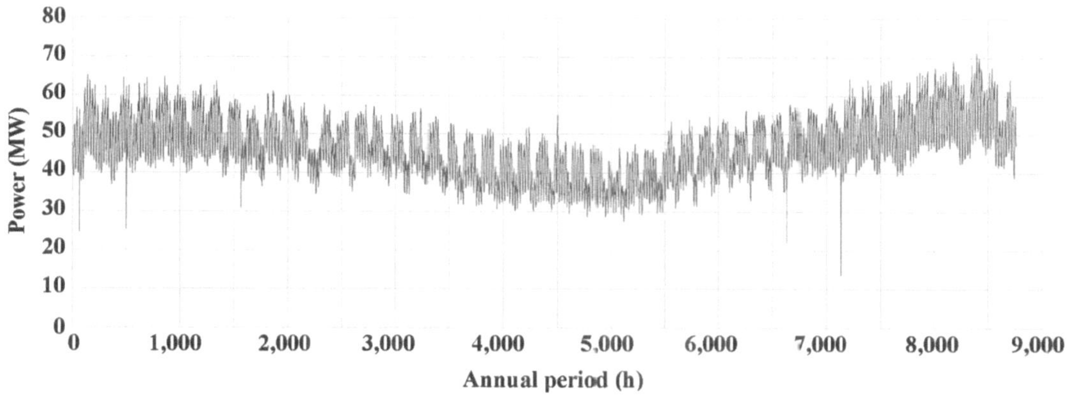

FIGURE 2.8 Annual power demand variation of the Faroe Islands.

Another interesting case is the power demand annual variation of the Faroe Islands, presented in Figure 2.8. The Faroe Islands is an independent state, consisting of 18 islands (17 inhabited) and located in the Atlantic Ocean, between Scotland and Iceland (62°N). The permanent population in the state reaches 50,000 people [8]. The power demand annual variation is mainly configured by the adverse weather conditions that dominate the area for 9 months annually, namely from September to May (temperatures lower than 10°C and stormy ocean winds). During this period the power demand is maximized and remains almost constant, due to the increasing need for indoor space heating. Furthermore, because human activities do not exhibit any significant seasonal variation, the mild weather during summer imposes a corresponding power demand drop.

After the above analysis, it is revealed that by examining the annual power demand profile of a specific geographical region, we may gain information on:

- the size of the electrical system and the standards of living of the corresponding geographical region,
- the seasonal variation of the professional activities of the local population,
- the structure of the local economy,
- the peak period of professional and social activities, and
- the weather conditions during the year.

Beyond the characteristic features depicted in annual power demand curves, similar information can also be observed in daily power demand profiles. Figure 2.9 presents the daily power demand profiles for the island of Samos for four 24-hour periods during the last 10 days of July.

By observing the daily power demand profiles in Figure 2.9 we may conclude the following:

- The daily power demand profile exhibits a clear repeatability from day to day.
- From the 24-hour period shift, namely at 24:00, until the sunrise and the beginning of daily human activities, the power demand follows a decreasing monotony. During this period, the power demand exhibits the daily minimum values.
- With the sunrise the power demand begins to increase, following the human activities, reaching the first daily power peak demand around noon.
- After noon and towards the early afternoon hours, the power demand decreases, exhibiting the minimum value during daytime.
- In the early evening hours, the power demand increases again, reaching the second daily peak demand, which, usually, constitutes the maximum power demand over the whole 24-hour period.

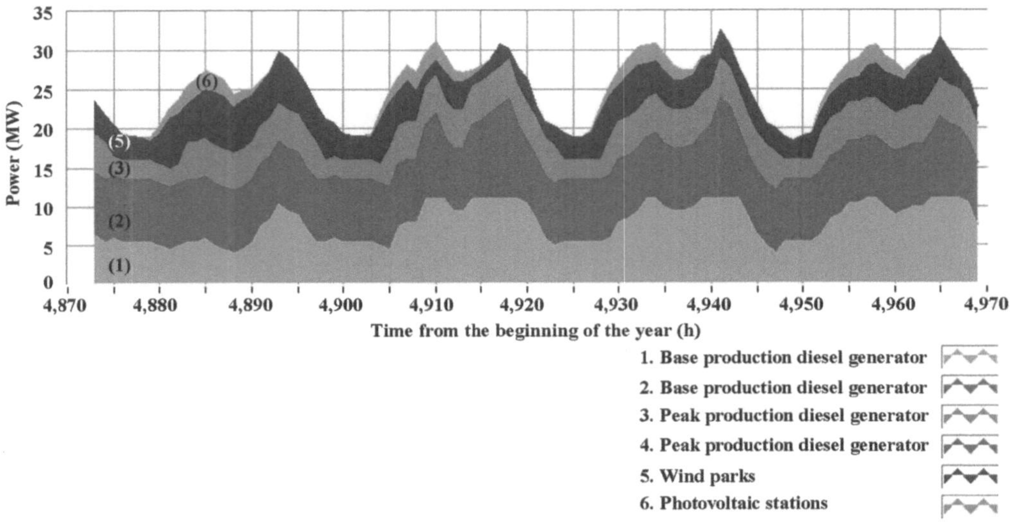

FIGURE 2.9 Daily power demand variation curve for the island of Samos for four 24-hour periods during the last 10 days of July.

- The maximum power demand recorded at noon and in the evening constitute the *daily power demand peaks* and the corresponding time periods are the *power demand peak periods.*
- The periods during the early morning or the early afternoon hours, during which the power demand profile exhibits its minimum values, are known as *power demand valleys* or *power demand off-peak periods.*
- In winter, the overall daily power demand profile remains the same, yet the power demand peak periods are usually recorded earlier.

The different colors in Figure 2.9 designate production from different types of involved generators. Given that this graph is plotted on a power versus time Cartesian coordinate system, the area covered by each different color represents graphically the electricity from each different type of the involved generators for the specific time period. These graphs, presenting the power production from the dispatched generators during a specific time period, are known as power production synthesis graphs. The plotting order of the generators' production curves in the graph designates their dispatching order in the power production process, which is based on certain, predefined parameters, aiming to ensure the secure operation of the electrical system with the minimum possible production cost. More on the dispatching order of electricity generators in conventional electrical systems is presented in the following section.

2.2 POWER GENERATORS

Electricity is produced by transforming an initial, primary source of energy. For this transformation, different generators may be employed, depending on the form of the initial available energy source. Briefly, the basic electricity generators most widely employed for electricity production are thermal generators, hydro turbines, wind turbines, and PV panels.

1. Thermal generators.
   The name of these generators declares that for the conversion of the initial available source of energy to electricity, thermal energy appears in an intermediate stage within the

conversion sequence, transformed through a thermodynamic process to mechanical power which, eventually, is converted to electricity in an inductive generator. The initial, primary source of energy introduced in thermal generators is, usually, the chemical energy of solid, liquid, or gas fossil fuels (i.e., coal, oil, natural gas), as well as the nuclear energy contained in the nucleus of nuclear fuels. During recent decades, the first solar thermal power plants have been developed in which thermal energy is produced by the concentration of solar radiation in parabolic solar collectors. Regardless of the available primary energy form (chemical energy or solar radiation), thermal energy is converted to electricity with one of the following thermal generator types:

- steam turbines
- diesel generators
- gas turbines
- combined cycles.

2. Hydro turbines.
   These are employed in hydroelectric power plants for the conversion of hydrodynamic energy (dynamic energy of water due to the earth's gravitational field) to electricity.
3. Wind turbines.
   These are used for the conversion of wind energy (kinetic energy of wind) to electricity.
4. PV panels.
   These are used for the direct conversion of solar radiation to electricity, without the appearance of an interim form of energy.

Apart from the above technologies, within the efforts for the exploitation of further available primary energy sources, there are several inventions developed for the production of electricity from sea waves or tidal energy, as well as fuel cells for electricity production from stored hydrogen. These efforts remain in the research stage, exhibiting high production costs and low technical maturity levels. Only a few applications of such systems are currently operating globally, with relatively low total installed power, funded by focused research or innovative actions and operating under a demonstrative (pilot) mode rather than as commercial projects. For these reasons, these technologies will not be investigated in this chapter. For the fuel cells particularly, there is an extended reference in the next chapter, approached as a special technology of electrochemical storage.

In the following sections the basic technologies for electricity production are presented, with a thorough analysis of the technical and economic features that determine their role in an electrical system and the way they are dispatched and cooperate in them.

2.2.1 Steam Turbines

The production of mechanical power in a steam turbine is performed through the expansion of super-heated, pressurized, high-enthalpy steam in a turbine. The modern form of steam turbines was introduced in 1884 by the British Sir Charles Parsons (1854–1931). Steam turbines are widely used for electricity production. More than 85% of the electricity consumed worldwide is produced by steam turbines.

The size of steam turbines may vary from small compact units of 0.75 kW nominal power, used to power pumps, compressors, and small electricity generators, to large units with nominal power of 1.5 GW, employed for electricity production in large thermal or nuclear power plants.

Steam turbines consist of a number of turbines embedded in a common shaft with the inductive generator. At the beginning of the shaft, the turbines' sequel starts with the high-pressure stage, followed by an intermediate-pressure stage to end, eventually, at one or two low-pressure turbines (low-pressure stage) before the inductive generator. As steam moves through the turbines, delivering continuously mechanical power, its pressure and thermal energy gradually decrease, leading to a corresponding volume increase. The increasing volume requires rotors with higher diameter and blade lengths to capture

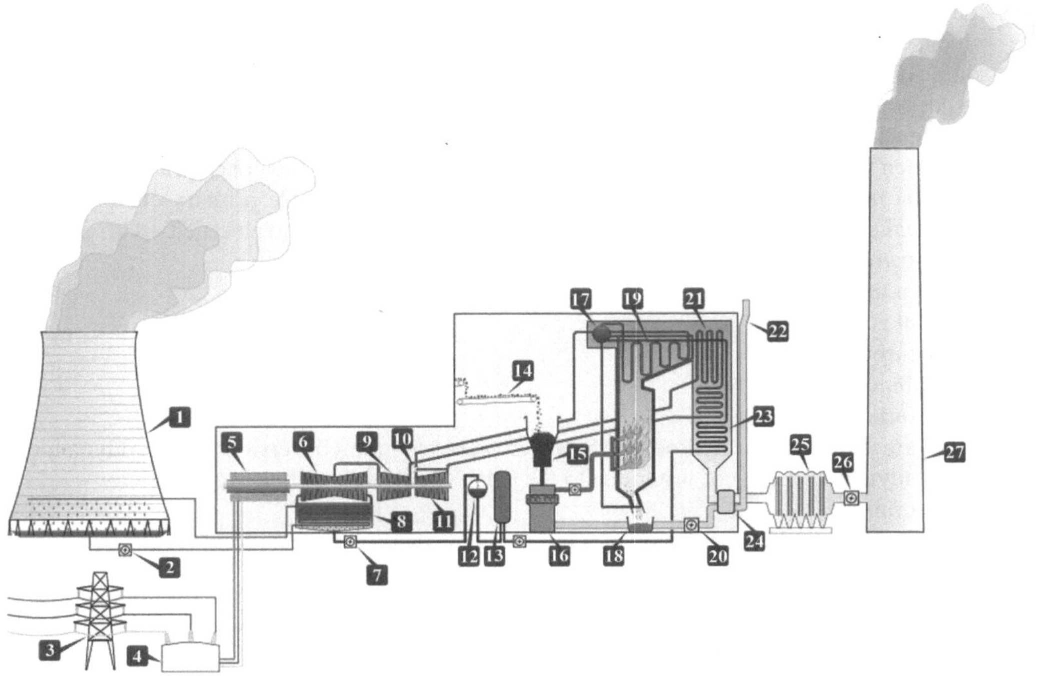

1. Cooling tower
2. Cooling water pump
3. Transmission grid
4. Unit transformer
5. Inductive generator
6. Low pressure turbine
7. Boiler feed pump
8. Condenser
9. Intermediate pressure turbine
10. Steam governor
11. High pressure turbine
12. Deaerator
13. Feed heater
14. Coal conveyor
15. Coal hopper
16. Pulverized fuel mill
17. Boiler drum
18. Ash hopper
19. Superheater
20. Forced draught fan
21. Reheater
22. Air intake
23. Economizer
24. Air preheater
25. Precipitator
26. Induced draught fan
27. Chimney stack

FIGURE 2.10 General layout of a steam-electric power station [9].

the remaining energy of steam. The total rotating mass of a steam turbine may exceed 200 tn with a total length of 30 m. To avoid any chance of axis bending and loss of symmetry, due to the huge mass and weight of the turbines, the rotation of a steam turbine is continuously maintained, even with very low rotational speeds (as low as 3 rpm) during periods that it remains out of duty.

It is easily to see that the size of the rotational masses, together with the combustion chamber and the steam boiler overall construction, make the whole installation particularly voluminous and complicated. Such an installation is presented in Figure 2.10 [9].

Electricity production in a steam turbine can be distinguished in three stages:

- steam preparation stage
- steam expansion and mechanical power production stage
- steam restoration in liquid stage after its expansion.

Following Figure 2.10, the operation of a steam turbine can be analyzed as follows [10].

2.2.1.1 Steam Preparation Stage

- The fuel (usually solid) is transferred with conveyors (14) from its storage space to the pulverized fuel mill (16), where it is grinded and converted to small grains. If the employed fuel is not solid (e.g., heavy fuel), there is no conveyor or pulverized mill; in these cases, the fuel is pumped from its storage tank and led directly to the combustion chamber.

- From the pulverized mill, the grinded fuel is led to the combustion chamber, where it is blended together with ambient air and preheated in the air preheater (24). The preheated mixture is led to the combustion chamber with the support of an air fan (20).
- The already heated air-fuel mixture is ignited inside the combustion chamber.
- The released thermal energy from the combustion is transferred to distilled water, which flows perpendicularly to the boiler's walls.
- The water is vaporized and transferred to the boiler's drum (17), where any possible liquid remains are removed.
- Pure steam is then led to the superheater (19), where its pressure and temperature are raised up to 200 bar and 570°C respectively. At this point, the steam preparation stage is integrated.

2.2.1.2 Steam Expansion Stage

- First, steam is led to the high-pressure turbine (11), where the first transformation of the steam's enthalpy to mechanical power is executed. The turbine's manual and automatic control is performed by the steam governor (10).
- The outgoing steam from the high-pressure turbine, with reduced pressure and temperature compared to its conditions while debouching the superheater (19), is now led to the reheater (21), where it is reheated.
- The reheated steam passes through the intermediate-pressure turbine (9) and, then, directly through the low-pressure turbine (6). At this point, the mechanical power transmission to the steam turbine's shaft is integrated.

2.2.1.3 Steam Restoration Stage

- After the low-pressure stage, the expanded steam, with temperature slightly higher than the water's boiling point, is led to the condenser (8), where it is cooled down and liquefied, by coming in contact, through a separate closed loop, with cold water pumped from the cooling tower (1). The cooling water returns back to the cooling tower for the disposal of the heat received through the steam's cooling process.
- The condensed water is led back to the boiler's feed pump (7) and the deaerator (12) for the disposal of any contained gases. Afterwards, once passed through the feed heater (13), it is finally led back to the boiler, through the economizer (23).

Finally, the electricity production is performed with the inductive generator (5), whose axis is coupled with the common axis of all three turbine stages.

The fundamental technical and economic features that mainly characterize steam turbines as power generators and determine their role in an electrical system are:

- Steam turbines are the only type of thermal generators that can consume any type of solid, liquid, or gas fossil fuel. Usually steam turbines operate with solid fuels, such as lignite and coal, due to their low cost. This advantage is owed to the fact that steam turbines are external combustion engines, namely the fuel's combustion is performed in a chamber entirely irrelevant to the machine's components where the mechanical power is produced (turbines).
- In nonconventional thermal power plants, such as nuclear power plants, geothermal stations of high-enthalpy fields, and solar thermal power plants, steam turbines constitute the only possible technology that can be used for the transformation of the produced, or naturally available, thermal energy to mechanical power. This ability is also a direct result of the fact that steam turbines are external combustion engines.

- Generally steam turbines are characterized with low overall efficiency, which usually varies from 30% to 35%. The highest energy losses within the involved energy forms conversions are detected in the transformation of the steam's enthalpy to the shaft's mechanical power. This transmission is expressed as a thermodynamic efficiency in the range of 40%–45%. The other energy losses refer to the internal electricity consumption of the steam-electric power plant (5%–10%), the combustion losses (5%–10%), the mechanical losses (<2%) on the several rotating parts (shaft, turbines, bearings), and the inductive generator's losses (<2%).
- The electricity production-specific cost of a steam turbine (production cost per produced unit of electricity) depends, obviously, on the type of the consumed fuel. For heavy fuel, the production-specific cost of a steam turbine is configured around 0.10–0.12 $/kWh, while in case of locally available solid fuel (lignite or coal), the production-specific cost drops close to 0.05 $/kWh or even lower. For nuclear power plants, the electricity production-specific cost is around 0.08–0.10 $/kWh [11, 12].
- The procurement and installation cost of a steam turbine ranges versus the size and the type of the power plant. In case of conventional thermal power plants, the total specific setup cost (setup cost per unit of installed power) can be around 1,000–1,500 $/kW. The setup costs for nuclear power plants can exceed 4,000 $/kW [11, 12].
- Due to their large rotating masses, steam turbines exhibit high inertia torque. For the same reason, steam turbines are not flexible generators, with regard to their slow response rates versus the power demand fluctuations.
- Due to the required steam preparation process, which precedes the mechanical power production stage, it usually takes considerable time for a steam turbine to be fully prepared to be dispatched in an electrical system. This preparation time period can be of some hours for large-size units. For this reason, steam turbines are considered to be units with a slow startup process.

The role of steam turbines in an electrical system, configured by their above presented technical and economic specifications, will be analyzed in a later section, where the corresponding features of other generators will be summarized as well.

2.2.2 Diesel Generators

A diesel generator consists of a diesel internal combustion engine and an inductive generator. Piston engines are generally characterized as extensively wide use and long-term service life, covering a wide range of output power from some kWs to tens of MWs. As a consequence, the procurement and installation of a diesel generator is considered cost-effective and fast. Additionally, the advanced technology and, particularly, the fact that internal combustion engines operate by exhausting a working medium (combustion gas) under considerably high pressure and temperature, enable performance with high efficiency and, as a result, low operating cost.

In piston engines, ambient air or a mixture of fuel and air is suctioned inside the combustion chamber and compressed up to a maximum pressure, increasing simultaneously its temperature. At the end of the compression stage, the fuel-air mixture, either suctioned in the combustion chamber (in an Otto engine), or created by injecting fuel into the compressed air (diesel engine), is ignited. With the combustion of the fuel-air mixture, the chemical energy contained in the injected fuel is added to its elastic energy, already gained with the compression, thus, further increasing its temperature and pressure. The high-enthalpy working medium is then exhausted, pushing the engine's piston and delivering, in this way, the produced mechanical power, a small portion of which is self-consumed in the preceding compression, while the rest of it constitutes the delivered mechanical

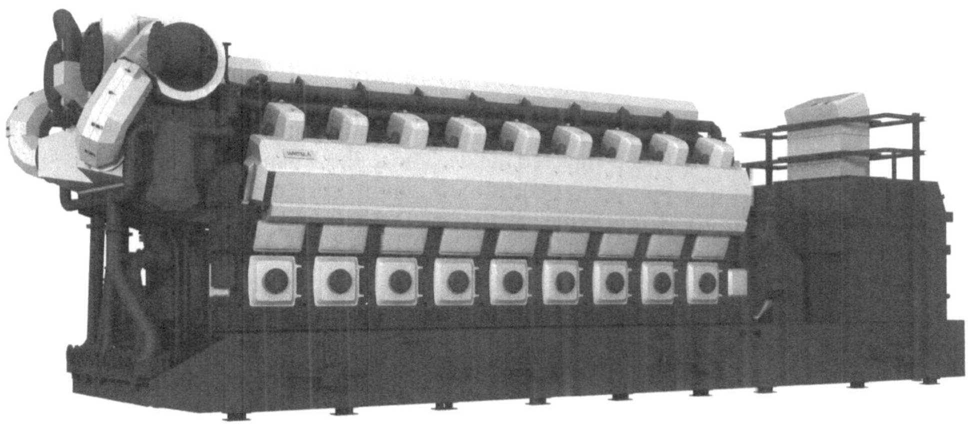

FIGURE 2.11 Diesel generating set with Wärtsilä 50 SG gas engine (Image courtesy of Wartsila [13]).

power output. From the above description, it reasons that in a piston engine all the sequential strokes, the suction (or intake), the compression, the ignition/combustion, and the exhaust, are executed in the same place, namely inside the combustion chamber. The combustion chamber is located at the same part of the reciprocating engine where the mechanical power is produced. Due to this fact, piston engines are classified as internal combustion engines.

In electrical systems usually two-stroke diesel generators are used, to achieve higher power production density for a specific cubic capacity of the internal combustion engine. The output power of diesel generators can vary from some kWs, in cases of small diesel generator sets (diesel gensets), used as backup units in hospitals, commercial centers, etc., to some tens of MWs, employed for electricity production in large thermal power plants.

In Figure 2.11 an image of a specific diesel generator commercial model is presented for electricity production, operating with natural gas.

The technical and economic features that configure and determine the role of diesel generators in an electrical system are:

- They exhibit the highest overall efficiency among the other thermal generators, which, generally, exceeds 40%.
- They can operate with all the different forms of liquid or gas fossil fuels (diesel oil, natural gas), as well as with heavy oil after a preheating process.
- The ability of diesel generators to operate with heavy fuel, combined with their overall relatively high efficiency, implies a low production-specific cost, which ranges around 0.10 \$/kWh. This cost is doubled in case of diesel generators of small size operating with diesel oil.
- Diesel generators of small size exhibit high response rates versus power demand fluctuations. This ability is relatively restricted for diesel generators of large size.
- They can be dispatched in the electrical system following a direct, fast startup process.
- Due to the reciprocating motion of several components (pistons, connecting rods, valves), diesel generators usually exhibit more wear and require more frequent and demanding maintenance, imposing, thus, a higher maintenance cost.

The above features determine the role of diesel generators in an electrical system, which will be presented in a following section.

2.2.3 Gas Turbines

Gas turbines are also classified as internal combustion engines. They consist of a main shaft, with one or, usually, two compressor stages at the one end, and one or, usually, two turbine stages at the other end. As the shaft rotates, ambient air is suctioned by the compressor, compressed, and led to the combustion chamber, where fuel is injected inside. The high-pressure air-fuel mixture is ignited. The produced high-enthalpy gas is firstly led to the high-pressure turbine stage and, then, to the low-pressure turbine stage, where it is exhausted, producing, thus, mechanical power. A small percentage of the produced power is employed for the delivery of the required power to the compressor, while the rest of it constitutes the delivered mechanical power. A graphical representation of a gas turbine is given in Figure 2.12.

Unlike piston engines, the required stages for mechanical power production in gas turbines are executed in three separate components: the compressor, the combustion chamber, and the turbine. A comparison graph between a four-stroke piston engine and a gas turbine is provided in Figure 2.13.

The basic technical and economic specifications of gas turbines, which determine their role in an electrical system are:

- They exhibit low efficiency, which may reach up to 35%, though it is typically closer to 30% or even lower. The gas turbines' efficiency decreases during summer, when the ambient temperature is high. It is reminded that the maximum efficiency η_C of a Carnot thermodynamic cycle is given by the relationship:

$$\eta_C = 1 - \frac{T_2}{T_1} \tag{2.7}$$

  where T_1 and T_2 are the temperatures of the working medium right after its compression and exhaustion respectively. Because the exhausted gas is disposed of in the ambient environment, the latter temperature is strongly affected by the ambient temperature. As the ambient temperature increases, T_2 increases too, leading to lower efficiencies η_C.
- They can operate either with diesel oil or natural gas.
- In cases of operation with diesel oil, given the low efficiency and the fuel's high procurement price, the electricity production-specific cost considerably increases, exceeding 0.30 \$/kWh. In cases of operation with natural gas, the electricity production-specific cost may drop to 0.20 \$/kWh.

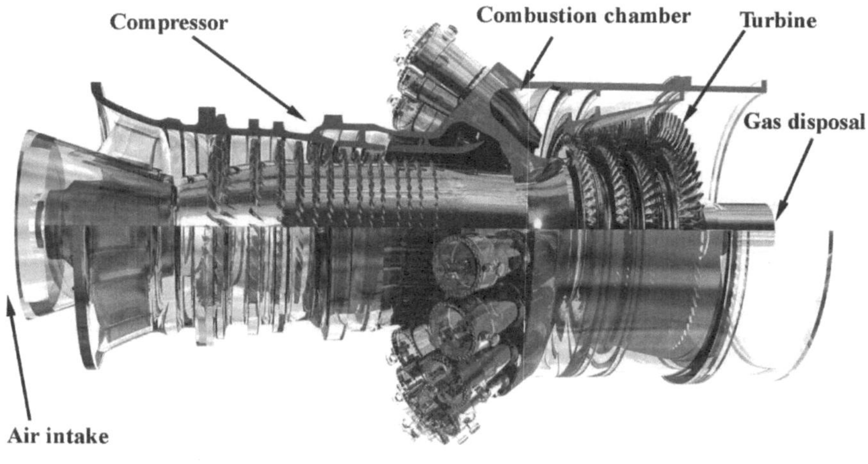

FIGURE 2.12 Graphical representation of a gas turbine's fundamental parts [14].

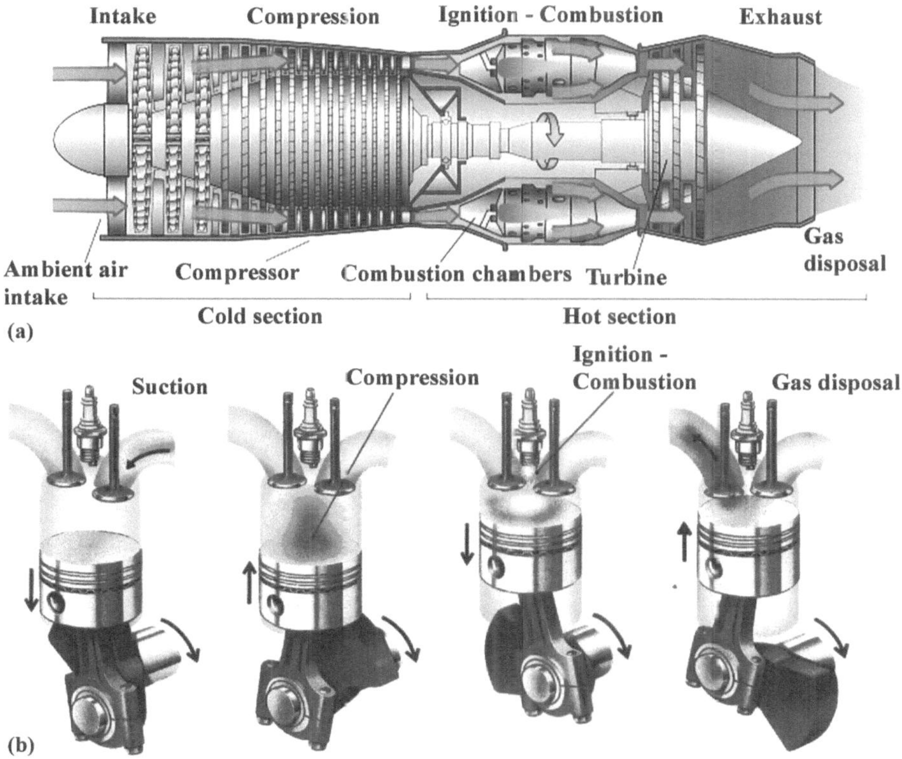

FIGURE 2.13 Comparison graph between (a) a gas turbine and (b) a four-stroke piston engine [15, 16].

- They are fast machines, exhibiting both quick startup procedures, which may last only a few minutes, and high response rates to power demand fluctuations.
- They have relatively low procurement costs, they are compact, and, as such, they are easily installed.

The above features determine the role of gas turbines in an electrical system, which will be presented in a following section.

2.2.4 Combined Cycles

Apart from the three fundamental thermal generator types, there is also the possibility for the combined operation of some of them, aiming, ultimately, to increase the overall efficiency and to reduce the corresponding production cost. Implementations formulated by the combined operation of the thermal generator basic types are known as *combined cycles*.

A combined cycle consists of two parts:

- the base production units, which in most cases are gas turbines, while they can also be diesel generators, and
- the secondary unit, which must be a steam turbine.

In a combined cycle there are usually two base units, while there is always the possibility of only one base unit as well. On the other hand, there is, most commonly, only one secondary unit.

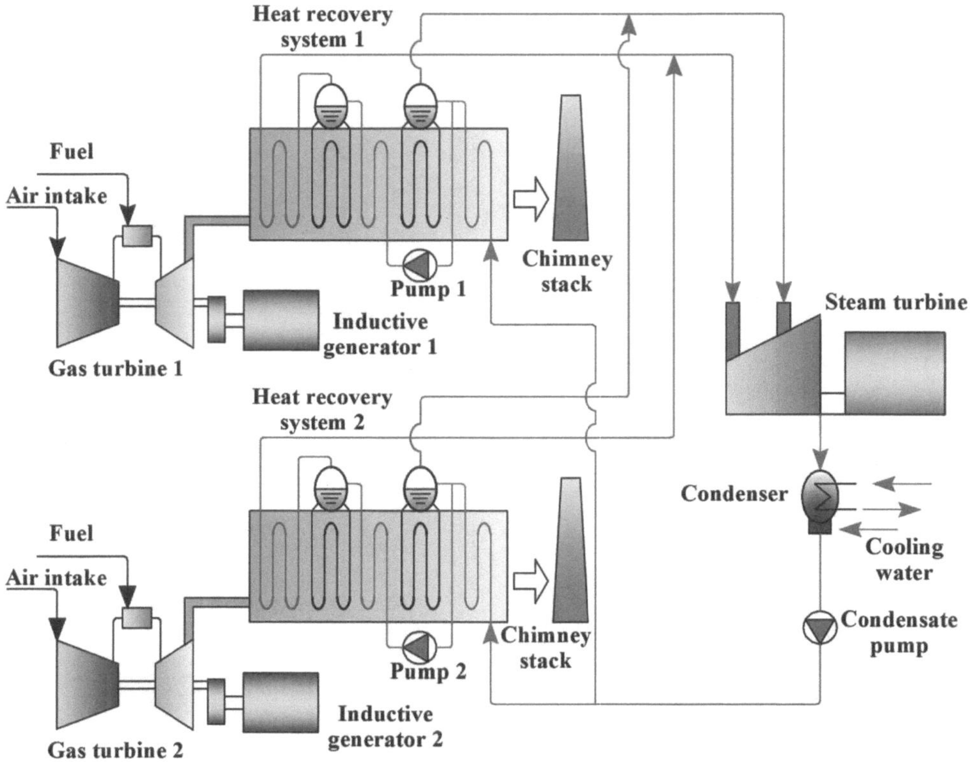

FIGURE 2.14 General layout and operation concept of a combined cycle, consisting of two gas turbines and one steam turbine [17].

The general layout and the operation concept of a combined cycle are presented in Figure 2.14. The hot exhausted gas from the two gas turbines, which constitute the combined cycle's base units, before being disposed in the ambient, passes through the heat recovery exchangers, where a part of the contained thermal energy is transmitted to condensed water cross-flown in a separate closed hydraulic circuit, forcing it to vaporize. The low-enthalpy gas, after the heat recovery exchangers, is disposed in the ambient, while the produced steam is led to the steam turbine (combined cycle's secondary unit) for the production of additional electricity, which is aggregated to the electricity independently produced by the gas turbines.

The additional electricity production from the steam turbine is achieved without the consumption of extra fuel. If $\dot{m}_{f1}$ and $\dot{m}_{f2}$ are the fuel's mass flow rates in gas turbine 1 and 2 respectively, H_u the consumed fuel's heat capacity, and P_{GT1}, P_{GT2}, P_{ST} the electrical power production from the two gas turbines and the steam turbine respectively, then:

- the total average efficiency of the two gas turbines exclusively, in the case of sole operation, namely without the combined operation with the steam turbine, would be:

$$\eta_{GT} = \frac{P_{GT1} + P_{GT2}}{(\dot{m}_{f1} + \dot{m}_{f2}) \cdot H_u} \tag{2.8}$$

- the total average efficiency of the combined cycle is:

$$\eta_{GT} = \frac{P_{GT1} + P_{GT2} + P_{ST}}{(\dot{m}_{f1} + \dot{m}_{f2}) \cdot H_u} \tag{2.9}$$

The increase of the combined cycle's overall efficiency is characteristically expressed with the above two equations.

The basic technical and economic features that determine the role of combined cycles in electrical systems are:

- Combined cycles exhibit high efficiencies, up to 50%–55%, precisely due to the extra power production from the steam turbine without any additional fuel consumption.
- Obviously, the electricity production-specific cost is considerably reduced. In case of diesel oil consumption by the gas turbines, the production-specific cost of combined cycles is around 0.20 $/kWh, while for natural gas consumption, it can be as low as 0.10 $/kWh.
- As a category, given the existence of the steam turbine, combined cycles are characterized by slow startup procedures. Yet, due to the partial power production from the gas turbines, combined cycles also exhibit fast response rates versus the power demand fluctuations.

The role of combined cycles in electrical systems will be analyzed in a following section.

2.2.5 Technical Minimum, Nominal Power, and Efficiency of Thermal Generators

The operation of thermal generators is determined by a minimum and maximum potential final power production. As already mentioned, the minimum possible power production from a thermal generator is known as *technical minimum*. The thermal generator cannot operate with power output lower than its technical minimum. Consequently, this is the lowest power production with which the thermal generator can remain dispatched in an electrical system.

The maximum power production from a thermal generator represents the maximum possible power that can be produced, known as *maximum capacity*. Most commonly, the maximum capacity of a thermal generator coincides with its nominal power. Nevertheless, quite often there are thermal generators with maximum capacity slightly higher than their nominal power, usually up to 110%.

Thermal generators are characterized by their overall instant efficiency, defined as the ratio of the final power output over the product of the fuel mass (or volume) consumption rate with its heat capacity, or, in other words, over the initial chemical energy consumption rate, according to the relationship:

$$\eta_{el} = \frac{P_{el}}{\dot{m}_f \cdot H_u} = \frac{P_{el}}{P_{ch}} \quad (2.10)$$

where:

P_{el}: final electrical power production
$\dot{m}_f$: fuel mass (or volume) consumption rate
P_{ch}: initial chemical power input
H_u: heat capacity of the consumed fuel
η_{el}: overall efficiency of the thermal generator.

The overall efficiency of a thermal generator does not remain constant versus the final power output. Generally, for all types of thermal generators, the overall efficiency is minimized for operation close to technical minimum, while as the power production approaches the generator's nominal power, the efficiency increases. The overall efficiency is maximized for operation at nominal power.

The overall efficiency variation curve versus the power output constitutes a characteristic feature of the thermal generator, provided by the generator's manufacturer. Usually, this feature is provided in the form of the consumed fuel mass or volume per unit of electricity produced, e.g., in gr of

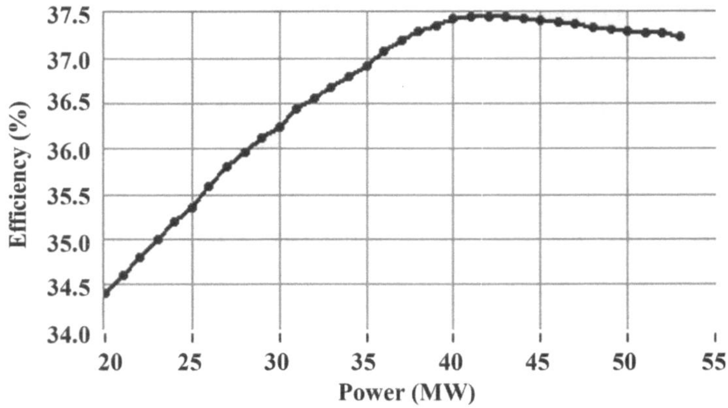

FIGURE 2.15 Characteristic efficiency curve of a steam turbine with nominal power 52 MW and technical minimum at 20 MW.

consumed fuel mass (m_f) per kWh of produced electricity (E_{el}). In this case, the overall efficiency of the thermal generator is calculated according to the following:

$$\eta_{el} = \frac{E_{el}}{m_f \cdot H_u} = \frac{E_{el}}{E_{ch}} \tag{2.11}$$

Indicative efficiency diagrams are presented in Figures 2.15, 2.16, 2.17, and 2.18 for a steam turbine, a diesel generator, a gas turbine and a combined cycle. All these diagrams come from real thermal generators from thermal power plants in Crete.

The above presented efficiency curves show that the overall efficiency of a thermal generator increases as the power output approaches the nominal power. This implies that, given the fact that the minimization of electricity production costs constitutes a fundamental target of the operation optimization of an electrical system, the thermal generators should not operate close to their technical minimums. On the contrary, in order to approach cost-effective operation, the thermal generators should operate as close to their nominal power as possible.

Another crucial parameter, with regard to the electrical system flexible operation, is the technical minimums of the dispatched thermal generators. Generally, it is desirable for a thermal generator to have as low technical minimum as possible. This enables a corresponding reduction of the power output, in case this is required by the system's operator, in order, for example, to allow higher

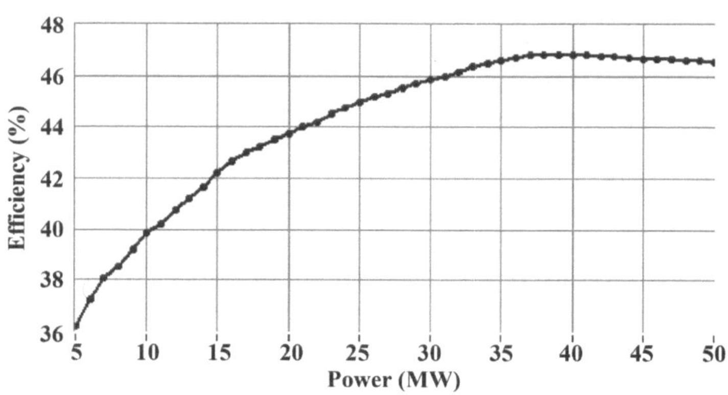

FIGURE 2.16 Characteristic efficiency curve of a diesel generator with nominal power 50 MW and technical minimum at 5 MW.

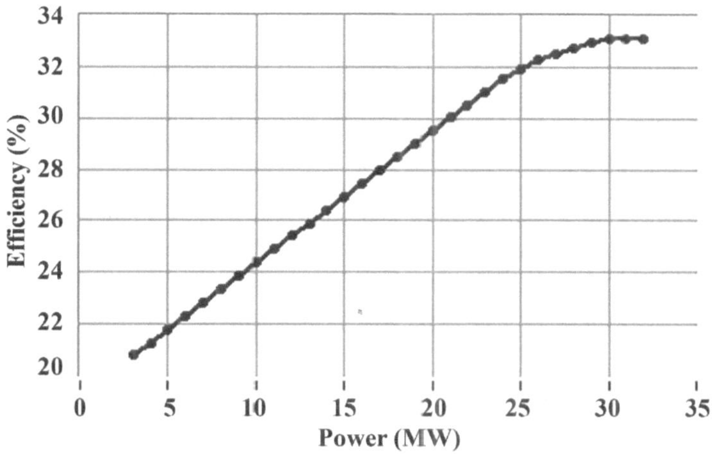

FIGURE 2.17 Characteristic efficiency curve of a gas turbine with nominal power 32 MW and technical minimum at 3 MW.

RES plant penetration or during time periods of respectively low power demand. High technical minimums restrict the operation flexibility of the thermal generator, particularly if this is also characterized by slow startup procedures (steam turbines of combined cycles), and, consequently, the shutdown and the startup of the unit is not feasible for short time periods. In such cases, the system's operator is obliged to maintain the operation of these units close to their technical minimums, restricting a potential penetration of RES power plants or alternative power production from other thermal generators with lower production costs.

On the other hand, it was shown in Section 2.1.4, that in order to approach the secure operation of an electrical system, the spinning reserve maintenance implies the operation of dispatched thermal generators not close to their nominal power, so a power production increase margin is always available to undertake any potential power production loss. However, as documented with the above presented efficiency curves, as the thermal generator's power output diverges from the nominal power, the overall efficiency drops and, consequently, the electricity production cost increases. It is so revealed that the secure and cost-effective operation of an electrical system can be two conflicting tasks, especially when the electricity production is based on conventional thermal generators.

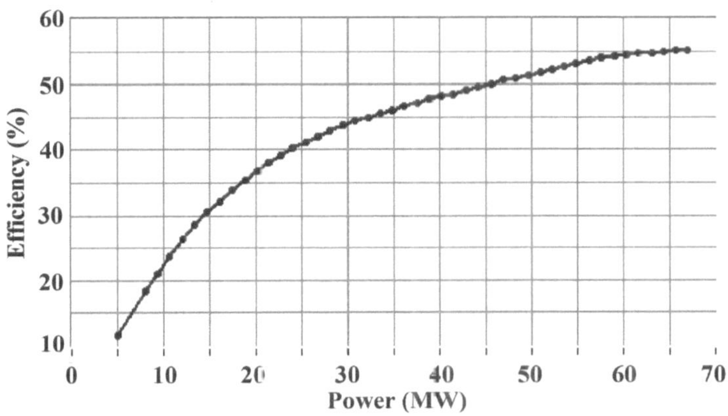

FIGURE 2.18 Characteristic efficiency curve of a combined cycle operating with natural gas, with nominal power 64 MW, consisting of two gas turbines with nominal power 25 MW each and a steam turbine with nominal power at 14 MW.

2.2.6 Hydro Turbines

Hydro turbines are among the first machines employed for mechanical power production. The first references on the exploitation of hydraulic energy for the operation of mills are found right after the dawn of the first Christian era. Yet, in the work "Da architectura" of the Roman writer, architect, and engineer Marcus Vitruvius Pollio (80 BC–15 BC), written around 25 BC, there are enough details that reveal the engine's design and operating concept. It consists, as Vitruvius says, of a vertical shaft forced to rotate by a water stream running below the shaft and striking continuously on a series of axial blades. This description given by Vitruvius is clear, without ambiguities, and able to deliver a clear picture for the machine that 2,000 years later is still used for the same purpose.

The truth is that, according to the available evidence, hydraulic energy was not used for three whole centuries. The first archaeological, literary, and drawing sources begin to present its usage in Rome, in Athens, in France, in Syria, and in Britain from the fourth century and later. From the sixteenth century, the machine described by Vitruvius, with water striking the rotating wheel from below, is gradually replaced with a similar one, in which the water is driven to the upper side of the machine's rotor. The latter version exhibits twice higher efficiency, as already proven from the eighteenth century.

Beginning in the nineteenth century, the steam engine was capable of producing more power for the needs of Britain during the Industrial Revolution. Since then, hydraulic energy will never again be the fundamental primary energy source for power production. Several traditional types of watermills used until the nineteenth century were not able to exploit a hydraulic head considerably higher than the rotor's diameter. The first hydroelectric power plant was constructed in 1895 in Niagara Falls, in the US, following the development of modern and innovative (for the specific era) hydro turbine types. It consisted of two hydro turbines with nominal power of 4.1 MW each.

With hydro turbines, the mechanical power of water is transferred to the machine's rotor and, eventually, to the rotating shaft. The working medium is natural water available from the physical flow of a stream or a river and the exploited primary energy source is the dynamic energy of water due to the gravitational field of earth, which is analogous to the height difference of the absolute altitudes between an initial and a final position of the water's route (e.g., the positions of a reservoir and the hydro power station). The dynamic energy of water, called *hydrodynamic energy*, is renewable, given that the water's flow in rivers and streams comes as a result of the hydrological cycle.

The flow of a water stream, due to gravity, follows a route to continuously decreasing absolute altitudes, ending either in the sea or, in case of a closed hydrological basin, in a natural lake. During this natural water's flow, its dynamic energy is continuously degraded and converted to thermal energy and mechanical energy through mechanisms of hydraulic losses, turbulence, and debris transportation.

The layout of a hydro power plant with rerouting of the natural river is given in Figure 2.19. In this power plant, the pressure side consists of an open tunnel and a closed pipeline. In Figure 2.20 a typical layout of a small hydro power plant is presented, with low available geostatic head and the power station embodied inside the dam [18, 19].

The possible configurations of small hydro power plants are not restricted to the basic layouts presented in these figures. Taking into account the variety of the natural terrain morphology, it stands to reason that for the optimum design of a hydro power plant, the personal experience and creativity of the designer/engineer plays a fundamental role, in order to achieve the simplest, cheapest, and safest solution, given the available earth morphology and the existing geological features.

Exactly due to this variety of the available geomorphology and geological conditions, a series of different hydro turbine models have been developed. Before any further differentiation regarding their design or characteristic features, the hydro turbines are classified into two distinct types:

- reaction turbines
- impulse or constant-pressure turbines.

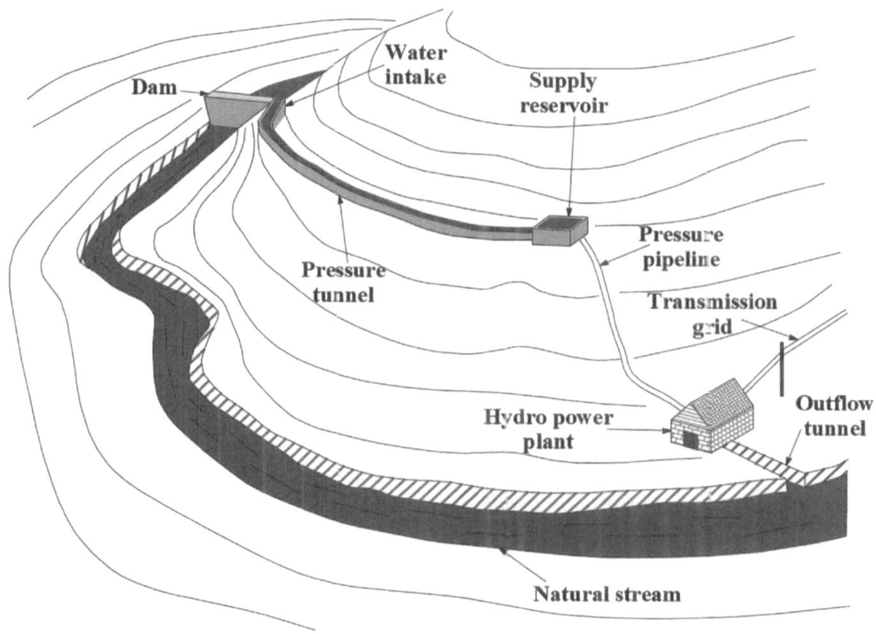

FIGURE 2.19 Typical layout of a hydraulic plant with an open pressure tunnel and a supply reservoir (with due diligence of Tsotras Editions) [18].

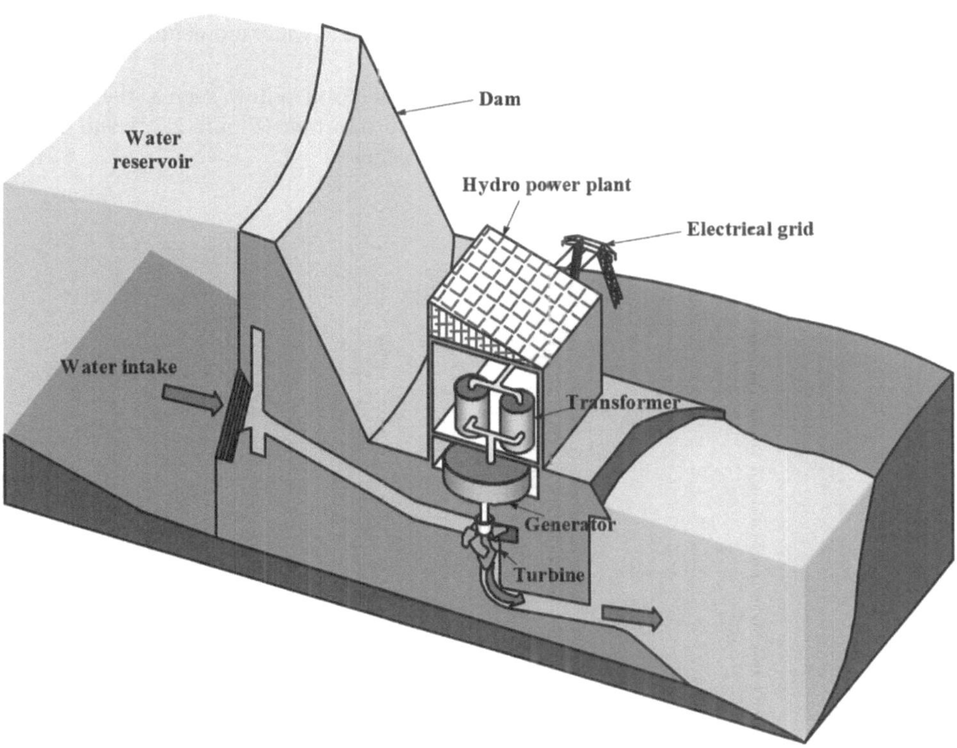

FIGURE 2.20 Typical layout of a small hydraulic plant, with low geostatic head and the power station embodied inside the dam.

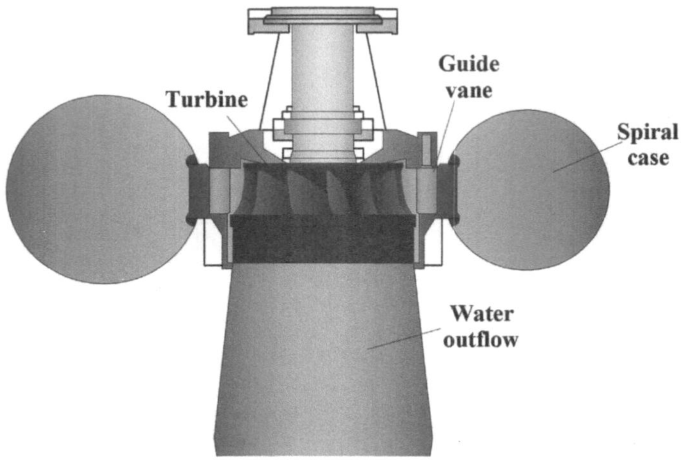

FIGURE 2.21 Typical design of a vertical axis Francis hydro turbine [20].

In reaction turbines the runner operation is axisymmetric. For this reason the casing of the machine must enclose the runner, ensuring, thus, axisymmetric flow conditions in the pressure side of the turbine. On the contrary, in impulse hydro turbines, flow is directed through a nozzle to impact only a portion of buckets attached to the periphery of the runner. Consequently, every moment only a portion of the runner's buckets contributes to the energy transfer from the water to the turbine.

In reaction hydro turbines, energy is transferred from the water to the runner due to the momentum and the hydrostatic pressure change of the fluid, as it passes through the runner. On the other hand, in impulse hydro turbines, the energy transfer is only due to water momentum change, because it is accomplished under constant atmospheric pressure.

Reaction hydro turbines are further divided into radial and mixed flow, such as the Francis hydro turbine (Figure 2.21); in axial flow, such as the Kaplan hydro turbine (Figure 2.22); and other variations of axial flow turbines (e.g., Deriaz, bulb, tube, crossflow).

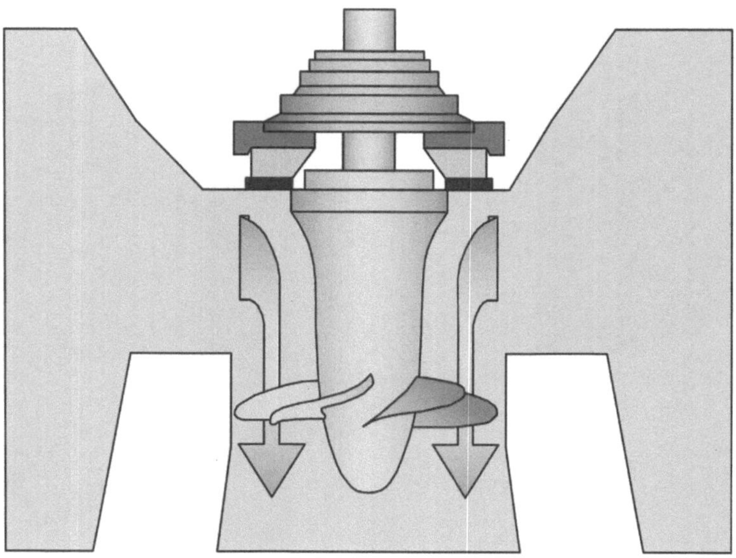

FIGURE 2.22 Typical design of a vertical axis Kaplan hydro turbine.

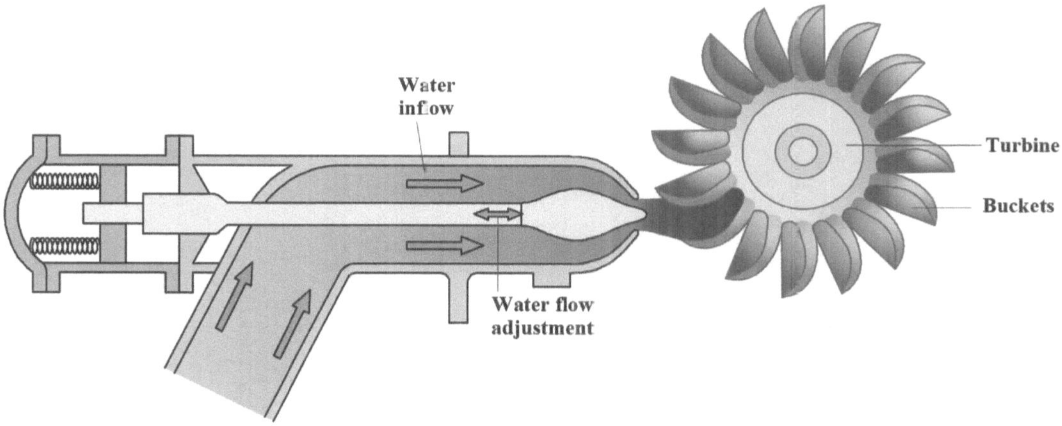

FIGURE 2.23 Typical design of a Pelton hydro turbine.

The differences between the Francis and the Kaplan hydro turbines are depicted in Figures 2.21 and 2.22.

The design of the Kaplan runner approaches the shape of a vessel's propeller. The Francis runner consists of two parallel hoops with the blades embedded between them. In both designs, the water flows through a peripheral arrangement of regulating blades, the so-called wicket gate or guide vane, aiming at the regulation of the water flow rate through the turbine's runner. This is achieved by rotating the regulating blades around their axis in order to adjust the gap between them and, in this way, the water flow rate through the turbine's runner.

The Kaplan hydro turbine is appropriate for high water flows and low heads, while the Francis hydro turbine is used for medium water flows and heads. More on the operation field of the most popular hydro turbine types is provided below.

The most characteristic type of impulse hydro turbine is the Pelton turbine (Figure 2.23). The Turgo and cross-flow turbines are variations of the Pelton turbine. The differences in the Pelton turbine from the Francis and Kaplan are obvious by comparing Figures 2.21, 2.22, and 2.23. Instead of hydrodynamically designed blades, as in Francis and Kaplan runners, the Pelton runner consists of a hoop of buckets peripherally embedded in a disc. The water flow strikes on these buckets, directed through one or more nozzles. The nozzles' opening, regulated by a needle, defines the water flow rate and, in turn, the produced power by the turbine.

More than all the other designs, the simple and robust construction of the Pelton turbine approaches the early inventions developed for the exploitation of hydrodynamic energy. The solidity of the Pelton turbine against high hydrostatic pressure makes it appropriate for operation under high geostatic heads.

The selection of the appropriate hydro turbine model for installation in a specific hydro power plant is fundamentally based on the available geostatic head and water flow. Secondary selection parameters may also be considered, such as the procurement and setup cost, the turbine's efficiency under partial-load operation, and the requirement for reverse operation (namely as a pump). Generally, the hydro turbine type selection for a specific hydro power plant, given the available geostatic head and water flow, is based on graphs similar to the one presented in Figure 2.24. According to this figure:

- The Pelton hydro turbine is used for high geostatic heads. Practically, it is the only hydro turbine type appropriate for operation under geostatic heads higher than 400 m. On the other hand, its operation field is restricted in relatively low water flow rates.

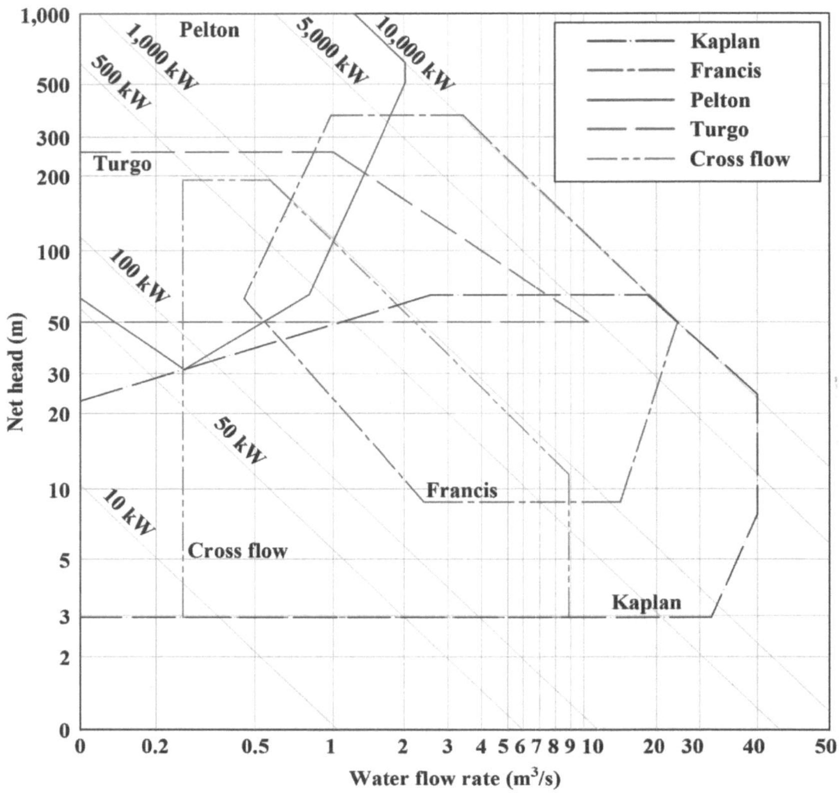

FIGURE 2.24 Hydro turbine type selection versus the available water flow rate and the geostatic head [21].

- The Turgo hydro turbine constitutes a variation of Pelton turbine, adapted in order to be able to operate with higher water flow rates.
- The Francis hydro turbine covers the operation field of medium geostatic heads (10–350 m) and water flows (0.5–25 m^3/s). It is the most commonly installed hydro turbine type globally.
- The Kaplan hydro turbine is used for high water flow rates (up to 50 m^3/s) and low geostatic heads (lower than 60 m).
- The Deriaz hydro turbine constitutes a variation of Kaplan turbine, adapted in order to approach operation with higher available geostatic heads, keeping, at the same time, the ability to operate with as high water flow rates as the Kaplan turbine. Hence, it covers the operation field of the Kaplan turbine, while being expanded also upwards, inside the Francis operation field.
- Another variation of Kaplan turbine is the bulb hydro turbine, developed to operate under even higher water flow rates (up to 100 m^3/s).

Another particularly important diagram is presented in Figure 2.25. In this diagram, the fluctuation of the hydro turbine's overall efficiency for partial-load operation is presented for different turbine types. According to Figure 2.25, the following may be concluded:

- The maximum overall efficiency for all hydro turbine types exceeds 80%. This feature is due to the fact that the electricity production from a hydro turbine is executed only in two distinct phases: the mechanical energy transfer from the water to the turbine's runner and

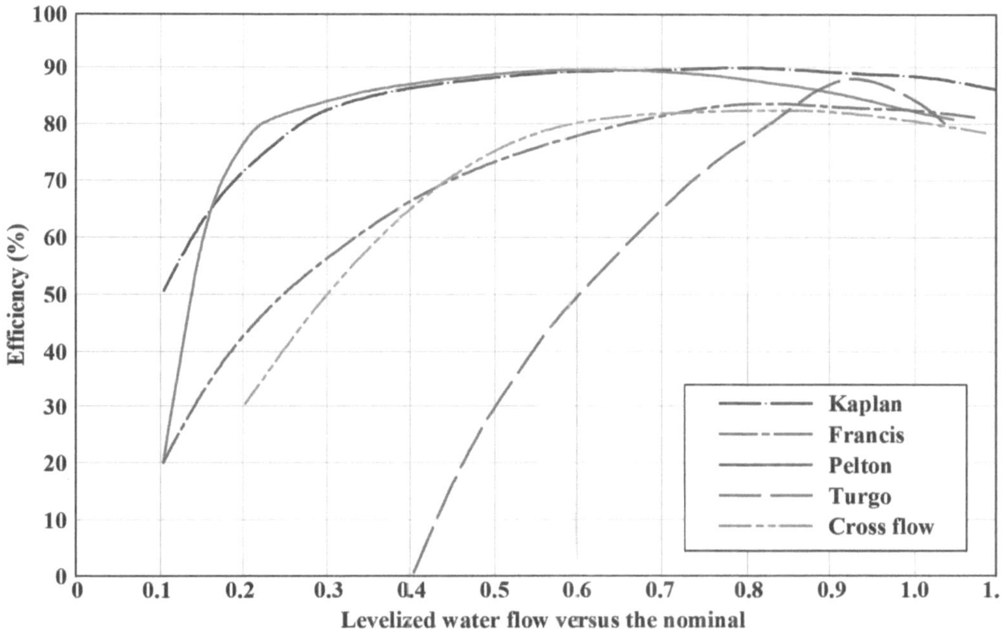

FIGURE 2.25 Overall efficiency fluctuation for partial-load operation of different hydro turbine types [22].

the mechanical energy conversion to electricity. It is so understood that the energy conversion process starts directly from the mechanical power, without any thermal to mechanical energy conversion stage intervened, as in thermal generators, which, actually, exhibits the highest energy conversion losses among the energy transformation sequels involved in thermal generators. With this energy conversion stage being totally absent from the electricity production process in hydro turbines, the overall turbines' efficiency is remarkably increased.

- The overall efficiency for some particular hydro turbine types, such as the Pelton, the Kaplan, and the Deriaz, increases and remains higher than 80% for a wide turbine operation range, starting from 20% and ending at 110% of the turbine's nominal flow.

The importance of the above presented features with regard to the hydro turbines' effective operation and flexibility is conceivable. Generally, the hydro turbines' operation in electrical systems is determined by the following parameters:

- As already mentioned, hydro turbines exhibit high efficiency, which is maintained for a wide part of their operation range.
- The electricity production-specific cost of a hydro power plant, given the absence of any fuel consumption, which constitutes the fundamental production cost component of thermal generators, is mainly configured by the amortization of the power plant's setup cost over its life period. As documented previously, the construction of a hydro power plant is highly dependent on the locally met geomorphological and geological conditions, as well as the available water flow rate, and may be approached with alternative configurations. For example, a hydro power plant may be supported by a water reservoir or not, the power station may be above-ground, underground, or embedded in the reservoir's dam, there may be a supply reservoir or not, the dam itself may be a gravity or an arch dam, etc. All these alternative possible implementations of a hydro power plant impose

different setup specific costs (per installed unit of power). Hence, for small hydro power plants, with nominal power lower than 10 MW, the setup specific cost is configured between 1,200 and 4,000 $/kW. In large hydro power plants the fluctuation of the setup specific cost is not so intensive, ranging indicatively between 1,800 and 2,800 $/kW. The setup specific cost fluctuation implies a corresponding fluctuation on the electricity production-specific cost. For small hydro power plants, the production-specific cost may vary from 0.03 to 0.10 $/kWh, while for large hydro power plants it may be from 0.02 to 0.05 $/kWh.

- Hydro turbines exhibit the highest response rate versus the power demand fluctuations compared to any other electricity generator. They are also characterized with fast startup processes (it may be in the range of seconds) and high flexibility regarding their operation mode and the power output.
- The operation of a hydro power plant is determined by the availability of the working medium, namely of the employed water flow. In cases of abundant water flow (e.g., in central and west Europe, in Scandinavia, in specific hydro power plants in China and the US), hydro turbines operate continuously, practically as base units. On the contrary, in cases of intensive seasonal fluctuations of the water flow during the year, the most commonly selected practice is to store water in a reservoir in order to use it during high power demand periods, avoiding, thus, the operation of thermal generators with high production-specific cost (e.g., gas turbines).
- Hydraulic energy is renewable and environmentally friendly.

The role and the dispatch concept of hydro turbines in electrical systems will be further developed in a later section.

2.2.7 Nonguaranteed Power Production Units

The fundamental nonguaranteed power production units, adequately evolved with regard to their technical maturity and economic competitiveness, with already significant contribution to the global electricity production share, are wind parks and PV stations.

The technological state of modern wind turbines, regarding particularly their interaction with the electrical grid, has been considerably improved. Additionally, modern wind turbines are lighter than the older ones and exhibit higher efficiency, as high as 45%–50% for sites with remarkable wind potential (average annual wind velocity in the range of 10 m/s), leading to overall, annual capacity factors higher than 40% or even 45%.

With regard to PV stations, the technology of monocrystalline modules has dominated as the one with the maximum efficiency, which for fixed installations may be up to 15%–17%. For areas with annual solar irradiation higher than 1,700 kWh/m^2, the capacity factor of a monocrystalline PV panel installed in a sun orbit's tracker, aiming to continuously maintain a 90° angle of solar radiation incidence on the panel's surface, may exceed 20%.

The main features that determine the role of nonguaranteed power production units in electrical systems are:

- The primary energy sources (wind potential or solar radiation) are renewable and the corresponding technologies are environmentally friendly, because their operation does not impose any gas emissions or any other type of waste.
- The primary energy sources are always locally available in the installation area of the power plant. Consequently, the electricity production from these technologies is performed by exploiting indigenous energy sources, leading to a corresponding reduction of the capital outflows for energy source imports and strengthening the local and the national economies.

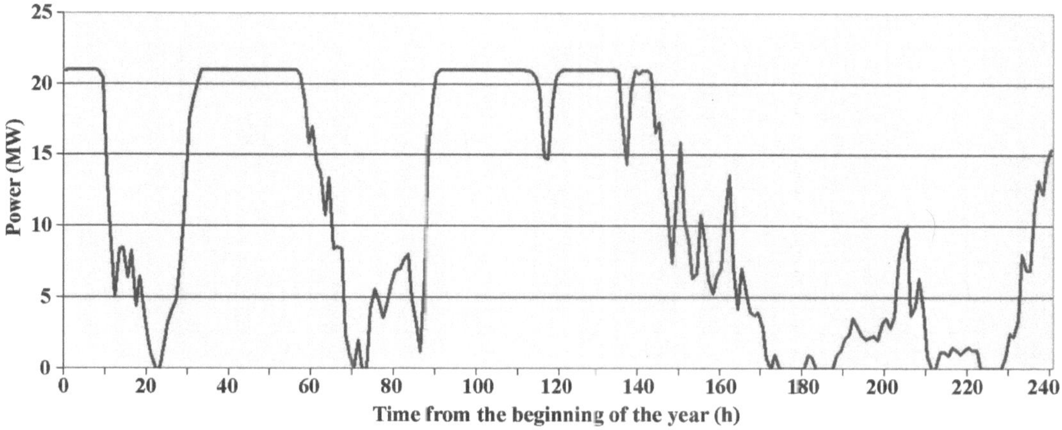

FIGURE 2.26 Net power production fluctuation (after losses) from a 24 MW wind park during the first 10 days of the year.

- The availability of the primary energy source is uncertain and the final power production cannot be controlled according to the system's operator will (nonguaranteed power production units). Indicatively, the net power production fluctuation (after losses) from a 24 MW wind park and a 4.4 MW PV station during the first 10 days of the year are respectively presented in Figures 2.26 and 2.27 (simulation results based on wind potential and solar radiation measurements from the Mediterranean basin).

Intensive power production fluctuation can be observed for both technologies in these figures. Practically, during the examined time period, any prediction on the anticipated power production from either the wind park or the PV station is not possible. On the other hand, the power production from the same PV station is presented in Figure 2.28 for the second 10-day period of June. During this period, a certain repeatability is observed on the PV power production, enabling the secure and accurate prediction of the expected production availability. A corresponding improved performance may be also observed for wind parks, in cases of wind blowing with limited fluctuations. Such cases

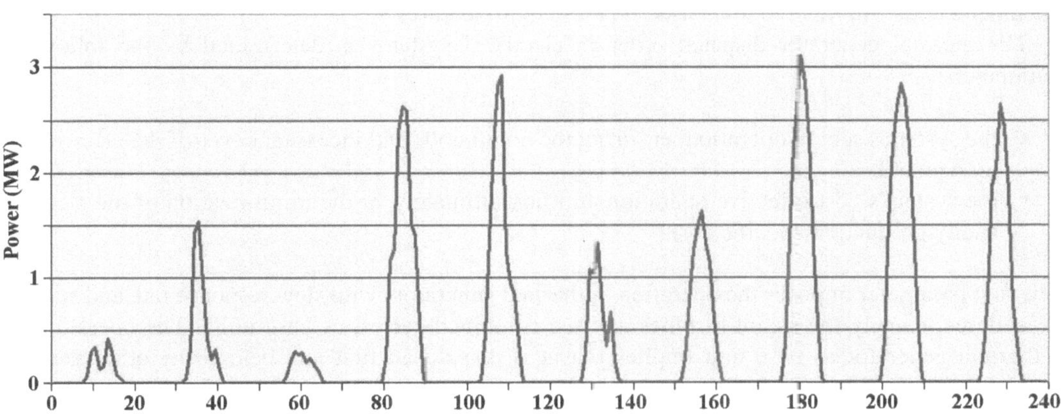

FIGURE 2.27 Net power production fluctuation (after losses) from a 4.4 MW PV station during the first 10 days of the year.

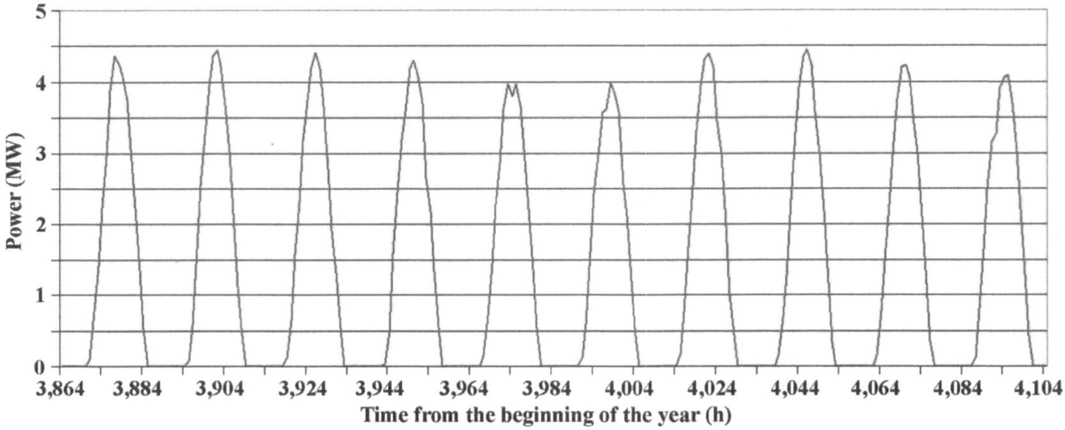

FIGURE 2.28 Net power production fluctuation (after losses) from a 4.4 MW PV station during the second 10-day period of June.

may appear due to local phenomena mainly during the summer period, in the absence of extreme weather conditions, such as storms.

- In most geographical areas, the total turnkey setup cost for both technologies is configured from 1,000 to 1,300 \$/kW.
- The electricity production-specific cost, given the absence of fuel consumption, is configured by the amortization of the setup cost over the power plant's life period and the annual electricity production (the higher the electricity production, the lower the production-specific cost). For wind park and PV station capacity factors at 40% and 17% respectively, and for a setup specific cost of 1,000 \$/kW for both technologies, the production-specific cost is configured around 0.04 \$/kWh for the wind park and 0.08 \$/kWh for the PV station.

2.2.8 Power Production Generators Dispatch Order

The main features of the electricity generators presented in the earlier sections are summarized in Table 2.2. The dispatch order of each generator technology and its role in electrical systems will be documented later in this section, based on these main features.

The thermal generator dispatch order in electrical systems is determined by the following parameters:

- the system's secure operation, ensuring the continuous and incessant cover of the electricity demand
- the system's cost-effective operation, aiming, ultimately, at the minimization of the electricity production-specific cost.

The first parameter imposes the operation of thermal generators with slow response rate and startup procedures, namely the steam turbines and the combined cycles, as base units. The operation of a thermal generator as base unit implies that it is dispatched first and before any other generator. Additionally, due to their slow startup procedure, steam turbines and combined cycles operate continuously; they are never shut down, except for malfunctions or scheduled maintenance. Even in cases of low power demand, these units are kept on-duty, producing power at least equal to their technical minimums, perhaps forcing the power production drop of other generators with possibly lower production-specific costs. Such cases are mainly met during low power demand periods.

TABLE 2.2
Summary of Fundamental Features of Electricity Production Technologies

Technology	Efficiency (%)	Consumed Fuel	Response Rate/ Startup Process	Production-Specific Cost (S/kWh)
Steam turbine	30–35	Solid fuels	Slow/slow	0.05
		Heavy oil		0.10–0.12
		Nuclear fuel		0.13–0.15
Diesel generator	40–45	Heavy oil	Fast/fast(2)	0.08
		Diesel oil		0.25
		Natural gas		0.05
Gas turbine	30–35	Diesel oil	Fast/fast	0.35
		Natural gas		0.20
Combined cycle	45–55	Diesel oil	Fast/slow(3)	0.20
		Natural gas		0.10
Hydro turbine	80–90	—	Fast/fast	0.02–0.05(4)
				0.03–0.10(5)
Wind turbine	40–50(1)	—	—	0.04(1)
PV panel	15–17(1)	—	—	0.08(1)

Comments:

1. The efficiencies and the production-specific costs for the wind turbines and the PV panels refer to operation with available potential that corresponds to annual final capacity factors in the range of 40% for the wind park and 15% for the PV station and for set-up prices of 2020.
2. Fast response of diesel generators mainly refers to units of small size (nominal power lower than 20 MW).
3. Fast response of combined cycles refers exclusively to the response rates of the involved gas turbines.
4. For hydro power plants of large size.
5. For hydro power plants of small size (nominal capacity up to 10 MW).

The second parameter imposes the following thermal generator dispatch order:

1. combined cycles operating with natural gas
2. diesel generators operating with natural gas
3. steam turbines operating with solid fuels
4. diesel generators operating with heavy oil
5. steam turbines operating with heavy oil
6. combined cycles operating with diesel oil
7. diesel generators operating with diesel oil
8. gas turbines operating with natural gas or diesel oil.

The above dispatch order is kept when there are not any restrictions or special measures regarding the electrical system's security and stability, with the steam turbines and the combined cycles being always operative with power production at least equal to their technical minimums, regardless of their dispatch order.

While dispatching the available generator technologies in the above order, attention should also be paid to the maintenance of adequate spinning reserve, in order to approach as secure as possible operation of the electrical system.

Hydro turbines can be dispatched either as base units, substituting power production from the base thermal generators, or as peak units, substituting power production from the peak thermal generators, depending on the availability of the hydrodynamic potential.

Nonguaranteed units can penetrate up to a maximum penetration percentage versus the current power demand, imposed by the necessity to maintain and ensure the system's stability and security (see Section 2.1.5).

A more detailed description on the dispatch order and the role of each electricity production technology is provided in the next paragraphs.

2.2.8.1 Steam Turbines

Steam turbines are introduced in electrical systems for basically two reasons:

- in case there is a primary energy source that cannot be exploited by any other thermal generator type (e.g., solid fuels, nuclear fuel, geothermal potential, solar thermal power plants)
- to undertake the task of regulating and keeping constant the system's fundamental frequency, given their high available moment of inertia, due to their huge-mass rotating components (compressor and turbine stages, shaft).

On the other hand, due to the size and the complexity of the whole installation, as well as the subsequent cost, the introduction of steam turbines is usually feasible only for medium and large electrical systems, practically with average annual power demand higher than 100 MW.

The regulation of the system's fundamental frequency by the steam turbines, combined with their time-consuming startup process and their slow response rate versus power demand fluctuations, impose their usage as base units. The steam turbines' operation as base units practically means that they operate continuously. Their power output during low power demand periods cannot be reduced below their technical minimums. Apart from unpredictable malfunctions, steam turbines are shut down only for scheduled maintenance, executed regularly, usually once per year.

2.2.8.2 Diesel Generators

Diesel generators constitute the most flexible thermal generators. They can operate with natural gas, heavy fuel, or diesel oil. They exhibit the highest efficiency of all the other basic thermal generators (apart from combined cycles). In cases of operation with heavy fuel, they also exhibit the minimum production-specific cost among the basic thermal generator technologies. Their installation is easy, quick, and cost-effective. They exhibit fast startup processes and fast response rates, especially the medium- or small-size engines (nominal power lower than 20 MW). Their basic drawbacks, compared to the other thermal generator types, are their extremely noisy operation (a feature also common for gas turbines) and their needs for frequent maintenance, due to the involved reciprocating components.

Given the above features, diesel generators can be employed:

- as base units, for large-size generators (nominal power in the range of 50 MW or higher), operating with natural gas or heavy fuel, aiming to minimize the electricity production cost
- as peak units, for small-size generators, regardless of the type of consumed fuel.

Diesel generators are the most commonly used thermal generators for small- or medium-size insular non-interconnected systems. They can be found with nominal power from 100–200 kW to 5–10 MW, operating either with heavy fuel or diesel oil. In cases of small insular systems, diesel generators usually undertake the role of both base and peak units. The electricity production-specific cost characteristically ranges from 0.15 $/kWh, in cases of heavy fuel consumption, to 0.30 $/kWh, in cases of diesel oil consumption.

2.2.8.3 Gas Turbines

Gas turbines, regardless of the type of the consumed fuel (natural gas or diesel oil), exhibit high production-specific costs, configured mainly by their low efficiency and the high procurement price of the consumed fuel, especially diesel oil. Their most favorable features are their fast startup processes and their high response rate against the power demand fluctuations. All the above characteristics lead to the operation of gas turbines exclusively as peak units, namely only during peak power demand periods and once the power production capacity of the other technologies available in the electrical system, with lower production-specific costs, has been exhausted.

2.2.8.4 Combined Cycles

A combined cycle is characterized, as an entity, with a slow startup process, due to the involvement of the steam turbine. Consequently, combined cycles are considered and employed as base units. However, combined cycles have the peculiarity, due to the involvement of the gas turbines, to exhibit high response rates against power demand fluctuations, at least as far as the gas turbines' power output is concerned. Conclusively, it could be stated that, as a whole, combined cycles have all the features of base units, operating continuously at least at their technical minimums, just like steam turbines, and never being shut down apart from scheduled maintenance or malfunctions. Yet, at the same time, they have the ability to adequately follow the power demand fluctuations, exhibiting adequate response rates, due to the gas turbine operation.

Given their high overall efficiency, the dispatch order of combined cycles is practically determined by the goal of minimizing the production-specific cost of the electrical system, eventually affected by the type of the consumed fuel. Particularly, in cases of natural gas, combined cycles will be dispatched with absolute priority against any other thermal generator, given the fact that they exhibit the operation features of the base units with low electricity production cost, configured by their high efficiency and low fuel price. In cases of diesel oil consumption, the combined cycle production will gradually be increased above its technical minimum once the production capacity from the available thermal generators with lower production-specific costs has been exhausted (e.g., steam turbines with solid fuels or heavy fuel, diesel engines with heavy fuel).

2.2.8.5 Nonguaranteed Power Production Units

The role of nonguaranteed power production units in electrical systems is mainly determined by their limited direct penetration, imposed for security and stability requirements, as documented in Section 2.1.5. Nonguaranteed power production units can directly penetrate to cover the power demand up to a maximum percentage versus the current power demand, configured by a cluster of parameters, such as the size of the system, the weather conditions, the type and the number of the dispatched thermal generators, and the type and the size of the maintained spinning reserve. By interpreting the above, nonguaranteed power production units, given:

- their renewable and environmentally friendly features,
- the selling price of the electricity produced by them, which is generally lower than the production-specific cost of the conventional thermal generators, and
- their contribution to the local and national economies,

are dispatched in the power production synthesis of an electrical system, whenever and as much as it is feasible, aiming, ultimately, at reducing the consumption of exhaustible primary energy sources, subsequent gas emissions, and the system's production-specific cost.

2.2.8.6 Hydro Turbines

Hydro turbines exhibit excellent performance regarding their fast startup process and their fast response rate according to power demand fluctuations. These operating features, combined with their low production-specific cost, make hydro turbines ideal units for electricity production, contributing to the system's secure and cost-effective operation, namely to both principal parameters involved in the determination of the power production synthesis. Given the above facts, the dispatch order of hydro turbines is configured by the availability of the working medium, namely of the available hydrodynamic potential over the year. In case of abundant water flow during the whole annual period, hydro turbines can be employed as base units, taking over the electricity demand cover for long time periods and maximizing, in this way, their annual penetration in the electricity production share. On the other hand, when the availability of water flow is practically restricted during the hydrological reach season (from autumn to spring), water is most commonly stored in reservoirs to be utilized during summer, especially when the power demand is maximized during the summer season, a very frequently met feature in southern countries with extensive summer tourist activities. In this way, hydro turbines are employed to cover the annual power peaks, thus avoiding the use of thermal generators with high production costs.

For a better understanding of the above guidelines, characteristic power production synthesis graphs are presented in Figures 2.29 and 2.30 for the non-interconnected electricity system of Crete, Greece, derived from the computational simulation of its annual operation. These graphs refer to two different time periods during the year, one of high power demand (August 10, 2016 to August 20, 2016) and one of low power demand (March 1, 2016 to March 10, 2016). The installed thermal generators are steam turbines and diesel generators, consuming heavy fuel; gas turbines; and a combined cycle with two gas turbines and a steam turbine, consuming diesel oil.

In the summer season (Figure 2.29), during the first hours of the new 24-hour period, power demand gradually decreases, imposing a corresponding reduction of the diesel generator production. The power production of both steam turbines and the combined cycle is almost constantly maintained close to their technical minimums. Diesel generators raise their power production before the other generators, because they exhibit lower production-specific cost. After them, steam turbines will come next, followed by the combined cycle (due to the consumption of diesel oil) and, lastly, the gas turbines. Steam turbines, diesel generators, and combined cycle reach their nominal capacity only once in the whole examined 10-day period, during an evening peak power demand. Gas turbines are employed only for a few hours during the same evening peak power demand, once the maximum production capacity from the other thermal generators has been exhausted, given the lack of PV production (night time) and the low contribution of the wind parks. The power production from the PV stations is obviously restricted to daytime periods only.

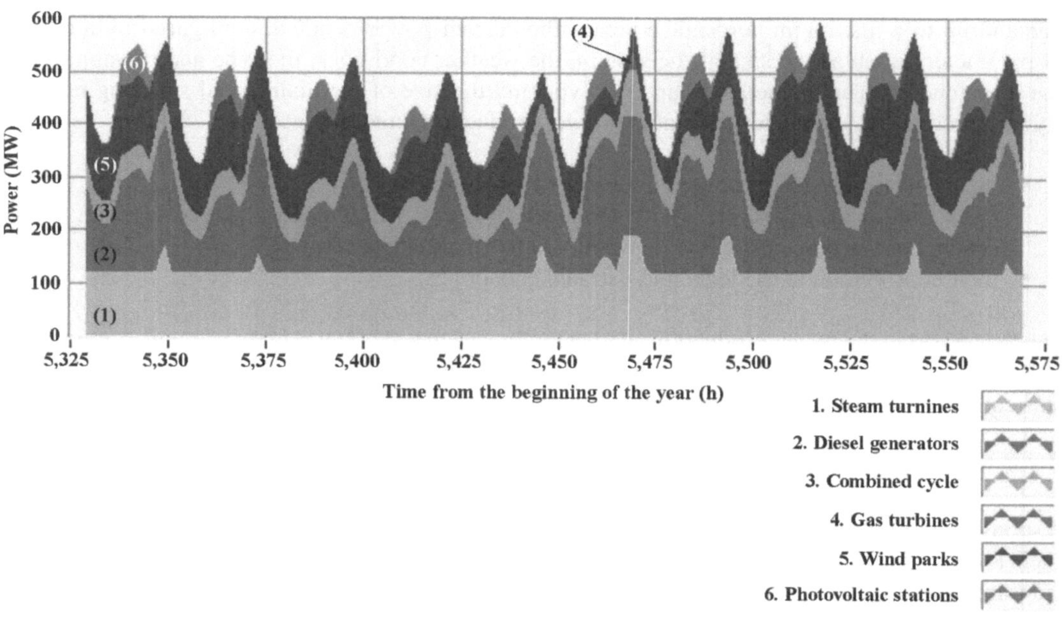

FIGURE 2.29 Power production synthesis graph for the electrical system of Crete, Greece, from August 10, 2016 to August 20, 2016.

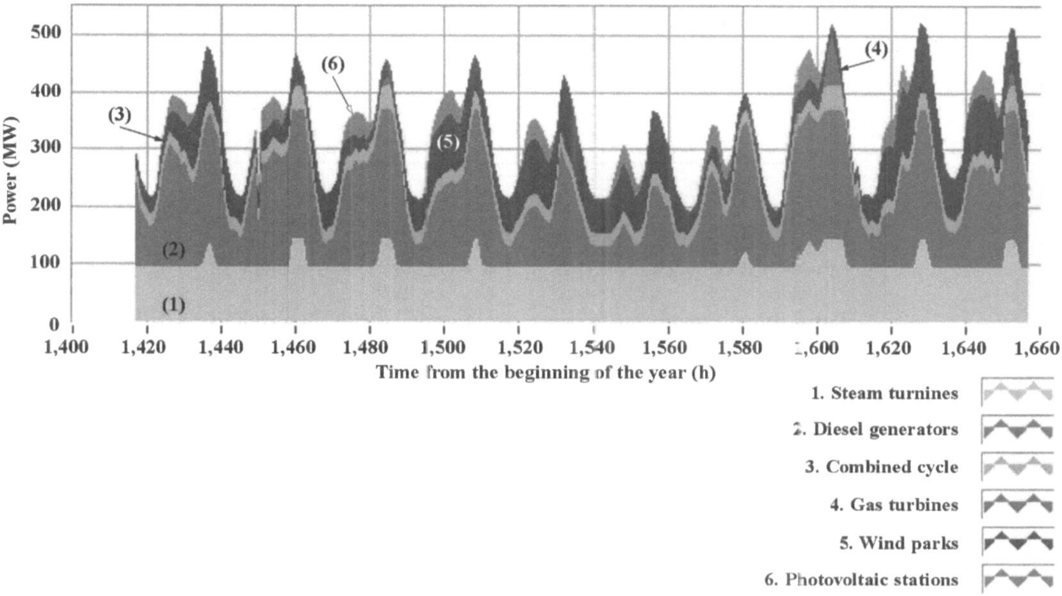

FIGURE 2.30 Power production synthesis graph for the electrical system of Crete, Greece, from March 1, 2016 to March 10, 2016.

During the autumn period (Figure 2.30), the steam turbines and the combined cycle operations are almost always restricted to their technical minimums, while the diesel generators rarely reach their maximum capacity, following the power demand fluctuations. The maximum capacity of diesel generators during the autumn period is approximately 50 MW lower than that of the summer period. This is due to the fact that one diesel generator of 52 MW nominal power is out of duty for scheduled maintenance. Gas turbines are still employed exclusively during the peak power demand periods.

In both time periods, the maximum wind power penetration is restricted below 30% versus the current power demand, in order to facilitate the system's dynamic security.

2.2.9 Spinning Reserve Policy

Spinning reserve is maintained in electrical systems, according to the general approach presented in Section 2.1.4, as well as following any special rules introduced particularly in each system. A common strategy is that the overall spinning reserve should at least equal the nominal power of the dispatched generator with the maximum capacity. The spinning reserve is allocated among a sensible number of involved thermal generators, taking also into account the system's cost-effective operation.

The spinning reserve requirements can possibly affect the generators' dispatch order. For example, when the maintained spinning reserve is less than the required, the power output of some generators should be reduced, even until their technical minimums, in order to create the margin to involve additional thermal generators in the power production synthesis and increase the total provided spinning reserve. Depending on the available thermal generators in an electrical system and the current power demand, the theoretically optimum dispatch order, as presented in the previous section, may be affected, in order to satisfy the spinning reserve requirements.

An indicative example is provided in Figure 2.31. Let's assume an electrical system with four thermal generators installed in total, two similar diesel generators with nominal power 50 MW and

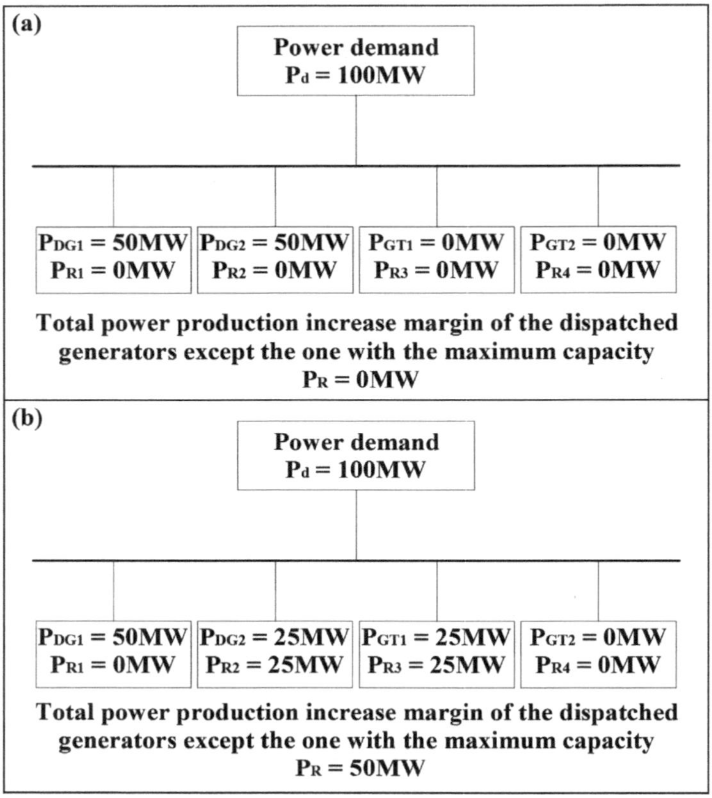

FIGURE 2.31 Power production synthesis scenarios in an electrical system with and without the maintenance of spinning reserve.

technical minimum 10 MW each, and two similar gas turbines with nominal power 50 MW and technical minimum 5 MW each. At a specific time moment the power demand is 100 MW. By taking into account exclusively the system's cost-effective operation, the gas turbines should remain off and the power demand should be covered by dispatching only the two available diesel generators, each one of them operating at its nominal power of 50 MW (Scenario A). With this power production synthesis scenario, the operation of the gas turbines is avoided and the involved diesel generators exhibit maximum efficiency, because they operate at their nominal power. Consequently, this is the optimum scenario regarding the cost-effective operation of the system, because the resulting power production-specific cost will be the minimum possible for the particular power demand and the available thermal generators. However, the total maintained spinning reserve is null; the power production total increase margin of the dispatched generators is zero. Under these operation conditions, a potential malfunction of one of the dispatched diesel generators or a sudden power demand increase could not be accommodated by the system, which would be mathematically led to a partial or total collapse.

An alternative power production synthesis, again for the power demand of 100 MW, could be the following (Scenario B):

- Diesel Generator 1: power production $P_{DG1} = 50$ MW, spinning reserve $P_{R1} = 0$ (dispatched unit with the maximum capacity),
- Diesel Generator 2: power production $P_{DG2} = 25$ MW, spinning reserve $P_{R2} = 25$ MW,
- Gas Turbine 1: power production $P_{GT1} = 25$ MW, spinning reserve $P_{R3} = 25$ MW,
- Gas Turbine 2: power production $P_{GT2} = 0$, spinning reserve $P_{R4} = 0$.

In this power production synthesis scenario, a total spinning reserve of 50 MW is achieved, provided by Diesel Generator 2 (25 MW) and Gas Turbine 1 (25 MW). This total spinning reserve equals the nominal power of the dispatched unit with the maximum capacity (Diesel Generator 1), so the corresponding fundamental spinning reserve strategy is satisfied. Gas Turbine 2 remains off, so it does not contribute either to the power production or to the total spinning reserve. The system's operation state is considered secure. Yet, at the same time, it is obvious that this alternative and absolutely safer power production synthesis results in increasing production-specific cost, due to both the involvement of a gas turbine in the power production synthesis, and the fact that two out of the three dispatched generators, namely Gas Turbine 1 and Diesel Generator 2, operate at partial load (half of their nominal power), exhibiting, thus, reduced efficiency.

From the above, we can see that spinning reserve maintenance implies an increase in the power production-specific cost for two reasons:

- The transitory operation of thermal generators with increased production-specific cost is often required, while, at the same time, the possibility of power production from more cost-effective units has not been exhausted.
- Spinning reserve is basically configured with a number of thermal generators under partial-load operation, some of them perhaps close to their technical minimums. In such cases, the efficiency of these units decreases, leading to increasing fuel-specific consumption (in gr or l per kWh produced) and production-specific cost.

A reasonable question arises: Which criterion should be more crucial in the configuration of a power production synthesis, the system's cost-effective or secure operation? The theoretical answer is obvious:

The secure system's operation and the incessant power production versus the power demand constitutes the ultimate target of electrical systems, consequently it exhibits highest priority regarding the configuration of the power production synthesis, compared to the requirement for the minimization of the electricity production cost.

Generally, the electrical system operation is based on the above basic principle. Nevertheless, there are frequently cases when, under favorable operational conditions, such as mild weather conditions, moderate power demand fluctuations, high response rate of the dispatched thermal generators, the grid's operators can take the initiative, based on their practical experience, to deviate from the above rule, for the benefit of reducing production costs. These cases cannot be written down or further analyzed, because they are based on the invaluable long-term experiences of the system operators, which enable them occasionally to break the spinning reserve rules and reduce the production-specific cost without affecting the system's security and stability.

2.3 POWER PRODUCTION SYNTHESIS EXAMPLES

For a better understanding of the precedent sections, indicative power production synthesis examples will be presented in this section. In all examined cases, the power production syntheses will be configured aiming at, first, the uninterrupted and secure power production and, second, the minimization of the power production-specific cost.

Let us assume a non-interconnected electrical system, with the thermal generators installed as presented in Table 2.3, along with their nominal power and technical minimums.

It is assumed, for simplicity, that all the installed steam turbines and gas turbines are similar to each other. The same stands with the two types of diesel generators (with nominal power at 50 and 10 MW) and the two gas turbines of the combined cycle. The overall efficiencies of the available thermal generators in this hypothetical system are presented for three specific operating points in Table 2.4.

TABLE 2.3
Installed Thermal Generators in the Under Examination Hypothetical Non-Interconnected Electrical System

No.	Thermal Generator Type	Number of Units	Technical Minimum (MW)	Nominal Power (MW)
1	Steam turbine	5	6.00	20.00
2	Diesel generator	2	10.00	50.00
3	Diesel generator	4	3.00	10.00
4	Combined cycle	2 gas turbines	7.00	40.00
		1 steam turbine	6.00	40.00
5	Gas turbine	5	5.00	40.00

TABLE 2.4
Thermal Generator Efficiencies for Three Specific Operating Points

		Operating Point (% over the nominal power)		
No.	Thermal Generator Type	30	70	100
1	Steam turbine	32.44	34.43	35.17
2	Diesel generator	42.22	46.62	46.54
3	Combined cycle	30.91	40.74	44.18
4	Gas turbine	21.16	31.11	35.59

Finally, the fuel consumed by each thermal generator along with its heat capacity and an indicative price in Europe in 2015 are given in Table 2.5. In addition to the above presented thermal generators, it is also assumed that in this specific autonomous electrical system there are 170 MW of wind parks installed and connected to the grid.

The scope of this example is to approach the optimum power production syntheses of the electrical system, for the three operating cases presented in Table 2.6, regarding the power demand and the power production availability from the wind parks. It is also assumed that the values presented in Table 2.6 represent average hourly operation states.

The optimum power production syntheses will be configured by taking into account the system's both secure and cost-effective operation. To ensure the secure operation, the nonguaranteed power production unit (namely the wind parks) penetration is restricted maximum at p_{max} = 30% versus the current power demand. Additionally, there will always be the target to maintain a total spinning reserve equal to the nominal power of the dispatched thermal generator with the maximum capacity, at a first stage, or, if possible, to the total RES penetrating power.

TABLE 2.5
Consumed Fuel Type, Heat Capacity, and Indicative Price in Europe in 2015

No.	Thermal Generator Type	Fuel Type	Heat Capacity	Price
1	Steam turbine	Heavy fuel	11.36 kWh/kg	0.50 €/kg
2	Diesel generator	Heavy fuel	11.36 kWh/kg	0.50 €/kg
3	Combined cycle	Diesel oil	10.07 kWh/l	0.95 €/l
4	Gas turbine	Diesel oil	10.07 kWh/l	0.95 €/l

TABLE 2.6
Investigated Hourly Average Operation Cases of the Electrical System

No.	Power Demand (MW)	Power Production Availability from the Wind Parks (MW)
1	250.00	110.00
2	410.00	100.00
3	520.00	125.00

Case 1:
According to the given data:

power demand P_d	: 250 MW
available power production from the wind parks P_w	: 110 MW
maximum possible power penetration of nonguaranteed units P_{wp}	: 0.30×250 MW = 75 MW

The maximum possible power penetration of nonguaranteed units $P_{wp} = 75$ MW is lower than the available total power production from the wind parks. Consequently, the wind power penetration will be equal to the maximum possible one, namely:

$$P_{wp} = 75 \text{ MW}.$$

The wind power rejection will be:

$$P_{rej} = 110 \text{ MW} - 75 \text{ MW} = 35 \text{ MW},$$

which corresponds to a rejection percentage versus the available wind power:

$$r = \frac{P_w - P_{wp}}{P_w} = \frac{P_{rej}}{P_w} = 31.82\%$$

The instant RES power penetration percentage will obviously be equal to:

$$p = \frac{P_{wp}}{P_d} = 30.00\%$$

After the calculation of the wind parks' penetration (nonguaranteed power production units), a power demand of 250 – 75 = 175 MW still remains to be covered by the thermal generators. To configure the optimum power production synthesis, the steps below are followed:

- First, the total technical minimums are calculated for the thermal generators, which should be continuously kept dispatched, due to their slow startup process. For this specific example, these generators are the steam turbines and the combined cycle, exhibiting total technical minimums equal to:

  $$5 \times 6 \text{ MW} + 2 \times 7 \text{ MW} + 6 \text{ MW} = 50 \text{ MW}.$$

  By subtracting the above calculated total technical minimums from the remaining power demand after the wind parks' penetration, we take: 175 – 50 = 125 MW.

- The first thermal generators that should be put on duty are the ones with the lowest production-specific cost. These are the diesel generators, given their efficiency (higher than 40%) and the cheapest consumed fuel (heavy fuel). The total nominal power of the diesel generators is 140 MW, which is higher than the remaining to be covered power demand of 125 MW. Among the available diesel generators, there are two with high nominal power (50 MW each), which should be first dispatched as base units, because they are expected to exhibit relatively lower response rates than the remaining diesel generators of smaller size. The remaining 25 MW will be produced by dispatching three out of the four diesel generators of small size. Two will operate at their nominal power (10 MW) and the third at 5 MW.
- Having covered the total power demand, the next step is to calculate the maintained spinning reserve with the formulated power production synthesis, as presented below:

steam turbines (slow spinning reserve)	: 5 units × (20 – 6) MW = 70 MW
diesel generators (fast spinning reserve)	: 1 unit × (10 – 5) MW = 5 MW
combined cycle's gas turbines (fast spinning reserve)	: 2 units × (40 – 7) MW = 66 MW
combined cycle's steam turbines (slow spinning reserve)	: 1 unit × (40 – 6) MW = 34 MW
total fast spinning reserve	: 71 MW
total slow spinning reserve	: 104 MW.

The total fast spinning reserve exceeds the nominal power of the dispatched thermal generator with the maximum capacity (50 MW) and approaches the penetrating wind power. The total maintained spinning reserve (provided by both slow and fast units) should be considered capable to ensure the system's reaction in case of a potential contingency. The operation state can be considered safe.

The results from the above analysis are summarized in Table 2.7.

Case 2:

According to the given data:

power demand P_d	: 410 MW
available power production from the wind parks P_w	: 100 MW
maximum possible power penetration of nonguaranteed units P_{wp}	: 0.30 × 410 MW = 123 MW.

The maximum possible power penetration of nonguaranteed units P_{wp} (123 MW) is higher than the available total power production from the wind parks (100 MW). Consequently, the wind power penetration will equal the total wind power available production, namely:

$$P_{wp} = 100 \text{ MW}.$$

The wind power rejection will be null.

The instant RES power penetration percentage will be equal to:

$$p = \frac{P_{wp}}{P_d} = 24.39\%$$

After the calculation of the wind parks' penetration (nonguaranteed power production units), the remaining power demand of 410 – 100 = 310 MW should be covered by the thermal generators. To configure the optimum power production synthesis, the steps below are followed:

- As previously, first the total technical minimums are calculated of the thermal generators, which should be continuously kept dispatched, due to their slow startup process, namely the steam turbines and the combined cycle:

$$5 \times 6 \text{ MW} + 2 \times 7 \text{ MW} + 6 \text{ MW} = 50 \text{ MW}.$$

TABLE 2.7
Optimum Power Production Synthesis of the Case 1 Operation State

No.	Thermal Generator	Technical Minimum (MW)	Nominal Power (MW)	Power Production (MW)	Spinning Reserve (MW)
1	Steam turbine 1	6.00	20.00	6.00	14.00
2	Steam turbine 2	6.00	20.00	6.00	14.00
3	Steam turbine 3	6.00	20.00	6.00	14.00
4	Steam turbine 4	6.00	20.00	6.00	14.00
5	Steam turbine 5	6.00	20.00	6.00	14.00
6	Diesel generator 1	10.00	50.00	50.00	0.00
7	Diesel generator 2	10.00	50.00	50.00	0.00
8	Diesel generator 3	3.00	10.00	10.00	0.00
9	Diesel generator 4	3.00	10.00	10.00	0.00
10	Diesel generator 5	3.00	10.00	5.00	5.00
11	Diesel generator 6	3.00	10.00	0.00	0.00
12	Combined cycle – gas turbine 1	7.00	40.00	7.00	33.00
13	Combined cycle – gas turbine 2	7.00	40.00	7.00	33.00
14	Combined cycle – steam turbine	6.00	40.00	6.00	34.00
15	Gas turbine 1	5.00	40.00	0.00	0.00
16	Gas turbine 2	5.00	40.00	0.00	0.00
17	Gas turbine 3	5.00	40.00	0.00	0.00
18	Gas turbine 4	5.00	40.00	0.00	0.00
19	Gas turbine 5	5.00	40.00	0.00	0.00
20	Wind parks			75.00	0.00
	Total:			**250.00**	**175.00**

By subtracting the above calculated total technical minimums from the remaining power demand after the wind parks' penetration, we take: 310 – 50 = 260 MW.

- The thermal generators with the lowest production-specific cost are the diesel generators, with a total installed capacity of 140 MW, which is not enough to cover all the remaining power demand, after the wind parks' penetration. Consequently, the diesel generators will operate at their nominal power. The remaining power demand after dispatching the available diesel generators will be 260 – 140 = 120 MW.
- The next more cost-effective thermal generators are the steam turbines, with total additional available capacity of 5 × 14 MW = 70 MW, calculated by subtracting from their nominal power their total technical minimums, because they actually have already been taken into account. After dispatching the steam turbines, the remaining power demand is 120 – 70 = 50 MW.
- The only available choice to reach the total power demand, before dispatching the peak power production units (gas turbines), is to increase the combined cycle's production, which already operates at its technical minimum, equal to 20 MW. Hence, to cover the remaining 50 MW of power demand, the total power production of the combined cycle should become in total equal to 70 MW. This power production comprises the production of 24 MW from each gas turbine (48 MW in total from both of them) and an additional power production of 22 MW from the steam turbine, given that the steam turbine's power production from a combined cycle is usually equal to half of the gas turbines' total production.

- Having covered the total power demand, the next step is to calculate the maintained spinning reserve with the formulated power production synthesis, as presented below:

combined cycle's gas turbines (fast spinning reserve)	: 2 units × (40 – 24) MW = 32 MW
combined cycle's steam turbines (slow spinning reserve)	: 1 unit × (40 – 22) MW = 18 MW
total fast spinning reserve	: 32 MW
total slow spinning reserve	: 18 MW.

The total maintained spinning reserve (provided by both the slow and fast thermal generators) is equal to the nominal power of the dispatched unit with the maximum capacity. The system's security level is considered satisfactory. The requirement for additional spinning reserve implies the dispatch of gas turbines, which will lead to a significant increases in the production cost. For this reason, this choice is rejected.

The results from Case 2 are summarized in Table 2.8.

Case 3:

According to the given data:

power demand P_d	: 520 MW
available power production from the wind parks P_w	: 125 MW
maximum possible power penetration of nonguaranteed units P_{wp}	: 0.30 × 520 MW = 156 MW.

TABLE 2.8
Optimum Power Production Synthesis of the Case 2 Operation State

No.	Thermal Generator	Technical Minimums (MW)	Nominal Power (MW)	Power Production (MW)	Spinning Reserve (MW)
1	Steam turbine 1	6.00	20.00	20.00	0.00
2	Steam turbine 2	6.00	20.00	20.00	0.00
3	Steam turbine 3	6.00	20.00	20.00	0.00
4	Steam turbine 4	6.00	20.00	20.00	0.00
5	Steam turbine 5	6.00	20.00	20.00	0.00
6	Diesel generator 1	10.00	50.00	50.00	0.00
7	Diesel generator 2	10.00	50.00	50.00	0.00
8	Diesel generator 3	3.00	10.00	10.00	0.00
9	Diesel generator 4	3.00	10.00	10.00	0.00
10	Diesel generator 5	3.00	10.00	10.00	0.00
11	Diesel generator 6	3.00	10.00	10.00	0.00
12	Combined cycle – gas turbine 1	7.00	40.00	24.00	16.00
13	Combined cycle – gas turbine 2	7.00	40.00	24.00	16.00
14	Combined cycle – steam turbine	6.00	40.00	22.00	18.00
15	Gas turbine 1	5.00	40.00	0.00	0.00
16	Gas turbine 2	5.00	40.00	0.00	0.00
17	Gas turbine 3	5.00	40.00	0.00	0.00
18	Gas turbine 4	5.00	40.00	0.00	0.00
19	Gas turbine 5	5.00	40.00	0.00	0.00
20	Wind parks			100.00	0.00
	Total:			**410.00**	**50.00**

The maximum possible power penetration of nonguaranteed units P_{wp} (156 MW) is higher than the available total power production from the wind parks (125 MW). Consequently, the available wind power production will totally penetrate the electrical grid, namely:

$$P_{wp} = 125 \text{ MW}.$$

No wind power will be rejected.

The instant RES power penetration percentage will be equal to:

$$p = \frac{P_{wp}}{P_d} = 24.04\%$$

After the calculation of the wind parks' penetration (nonguaranteed power production units), the remaining power demand of 520 – 125 = 395 MW should be covered by the thermal generators. To configure the optimum power production synthesis, the steps below are followed:

- The total technical minimums of the steam turbines and the combined cycle are calculated:

$$5 \times 6 \text{ MW} + 2 \times 7 \text{ MW} + 6 \text{ MW} = 50 \text{ MW}.$$

  By subtracting the above calculated total technical minimums from the remaining power demand after the wind parks' penetration, we take: 395 – 50 = 345 MW.
- After dispatching the total nominal power of the available diesel generators (140 MW), namely the thermal generators with the lowest production-specific cost, the remaining power demand is 345 – 140 = 205 MW.
- The next most cost-effective thermal generators are the steam turbines, with total additional available production capacity of 5 × 14 MW = 70 MW, calculated by subtracting from their nominal power their total technical minimums, which actually have been already taken into account. The remaining power demand after dispatching the steam turbines is 205 – 70 = 135 MW.
- The next most cost-effective thermal generator is the combined cycle, with a nominal capacity of 120 MW. The combined cycle already operates at its technical minimum, equal to 20 MW. Hence, there is 100 MW of available power capacity still to be produced by the combined cycle, which is not enough to cover the remaining power demand of 135 MW. After dispatching the combined cycle, there will be 135 – 100 = 35 MW of power demand still to be covered.
- The last available choice for the coverage of the remaining power demand is the use of gas turbines. For the 35 MW, two gas turbines will be dispatched, at 15 MW and 20 MW each.
- Having covered the total power demand, the next step is to calculate the maintained spinning reserve provided by the formulated power production synthesis, as presented below:

gas turbine 1 (fast spinning reserve)	: 1 unit × (40 – 15) MW = 25 MW
gas turbine 2 (fast spinning reserve)	: 1 unit × (40 – 20) MW = 20 MW
total fast spinning reserve	: 45 MW
total slow spinning reserve	: 0 MW

The total spinning reserve (45 MW of fast generators) approaches the nominal power of the dispatched unit with the maximum capacity (50 MW), yet, without exceeding it. To achieve this, another gas turbine should be dispatched, imposing a subsequent reduction of the power production of the other two dispatched gas turbines, which, in turn, will lead to a drop of their efficiency and

TABLE 2.9
Optimum Power Production Synthesis of the Case 3 Operation State

No.	Thermal Generator	Technical Minimums (MW)	Nominal Power (MW)	Power Production (MW)	Spinning Reserve (MW)
1	Steam turbine 1	6.00	20.00	20.00	0.00
2	Steam turbine 2	6.00	20.00	20.00	0.00
3	Steam turbine 3	6.00	20.00	20.00	0.00
4	Steam turbine 4	6.00	20.00	20.00	0.00
5	Steam turbine 5	6.00	20.00	20.00	0.00
6	Diesel generator 1	10.00	50.00	50.00	0.00
7	Diesel generator 2	10.00	50.00	50.00	0.00
8	Diesel generator 3	3.00	10.00	10.00	0.00
9	Diesel generator 4	3.00	10.00	10.00	0.00
10	Diesel generator 5	3.00	10.00	10.00	0.00
11	Diesel generator 6	3.00	10.00	10.00	0.00
12	Combined cycle – gas turbine 1	7.00	40.00	40.00	0.00
13	Combined cycle – gas turbine 2	7.00	40.00	40.00	0.00
14	Combined cycle – steam turbine	6.00	40.00	40.00	0.00
15	Gas turbine 1	5.00	40.00	15.00	25.00
16	Gas turbine 2	5.00	40.00	20.00	20.00
17	Gas turbine 3	5.00	40.00	0.00	0.00
18	Gas turbine 4	5.00	40.00	0.00	0.00
19	Gas turbine 5	5.00	40.00	0.00	0.00
20	Wind parks			125.00	0.00
	Total:			**520.00**	**45.00**

a corresponding increase of the fuel's specific consumption and the production-specific cost. For this reason, this scenario is rejected, avoiding, in this way, the production-specific cost increase and achieving a spinning reserve equal to 90% of the nominal power of the dispatched unit with the maximum capacity. The security level can be considered satisfactory.

The results from Case 3 are summarized in Table 2.9.

2.4 HOURLY CALCULATION OF AN ELECTRICAL SYSTEM

2.4.1 Operation with Wind Park Penetration

As a follow up of the above examples, in this section we will calculate the fundamental features that define and characterize the electrical system's operation. The calculations will be executed for Case 2 above, namely for an hourly average power demand of 410 MW and wind power availability of 100 MW. For this operation case, the optimum power production synthesis is presented in Table 2.8.

The characteristic features calculated in this section are:

a. the total heavy fuel consumption (in kg) and the total diesel oil consumption (in l)
b. the consumed fuel total procurement cost
c. the consumed fuel and purchased wind energy total cost, assuming an electricity selling price from the wind parks' owners to the utility company at 0.095 €/kWh

d. the hourly average electricity production-specific cost in €/kWh for:

- each different type of thermal generator
- all the involved thermal generators as a whole
- the total power production system (thermal generators and wind parks).

Only the consumed fuel and wind energy procurement costs will be involved in the calculation of the above costs. This implies that, particularly for the thermal power plant, any other involved costs will be neglected, such as equipment amortizations, thermal generator maintenance costs, staff salaries, and gas emission costs. Besides, it should be mentioned that the electricity production cost from a thermal power plant is mainly configured by the consumed fuel procurement cost. The contribution of the other involved cost components is rather low. For the execution of the above calculation, all the data provided in the previous section will be adopted regarding the thermal generator efficiencies, the consumed fuels heat capacities, and procurement prices as presented in Tables 2.4 and 2.5.

a. Heavy fuel and diesel oil total consumption

If a thermal generator produces final electrical power P_{el} for a time internal t with efficiency η_{el} and heat capacity H_u, the fuel mass m_f or volume V_f consumption (depending on whether the fuel's heat capacity is given per unit of fuel's mass or volume respectively) is given by the following relationship, which actually is derived from Equation 2.10:

$$m_f \ (\text{or } V_f) = \frac{P_{el} \cdot t}{\eta_{el} \cdot H_u} \tag{2.12}$$

Given the concluded power production synthesis presented in Table 2.8:

– Steam turbines

All the steam turbines operate at their nominal power, hence, according to Table 2.4, their efficiency will be 35.17%.

Their total power production equals 100 MW and the heavy fuel heat capacity is given as 11.36 kWh/kg in Table 2.5. The hourly heavy fuel consumption is given by Equation 2.12:

$$m_{f\text{-ST}} = \frac{P_{el} \cdot t}{\eta_{el} \cdot H_u} \Rightarrow m_{f\text{-ST}} = \frac{100 \cdot 10^3 \text{ kW} \cdot 1\text{h}}{0.3517 \cdot 11.36 \text{ kWh/kg}} \Leftrightarrow m_{f\text{-ST}} = 25{,}029.33 \text{ kg}$$

– Diesel generators

The diesel generators also operate at their nominal power. Their efficiency, according to Table 2.4, is equal to 46.54%.

The total power production from the diesel generators is 140 MW and the consumed heavy fuel heat capacity equals 11.36 kWh/kg, as presented in Table 2.5. The corresponding hourly heavy fuel consumption is also given by Equation 2.12:

$$m_{f\text{-DG}} = \frac{P_{el} \cdot t}{\eta_{el} \cdot H_u} \Rightarrow m_{f\text{-DG}} = \frac{140 \cdot 10^3 \text{ kW} \cdot 1\text{h}}{0.4654 \cdot 11.36 \text{ kWh/kg}} \Leftrightarrow m_{f\text{-DG}} = 26{,}480.33 \text{ kg}$$

– Combined cycle

The total power production from the combined cycle is 70 MW. This value corresponds to 58.33% of its nominal power. In Table 2.4 the combined cycle's efficiencies are given for partial loads at 30% (30.91%) and at 70% (40.74%). With a linear interpolation between these two values, the efficiency for partial-load operation at 58.33% is calculated at 37.87%.

The heat capacity of diesel oil is given as 10.07 kWh/l in Table 2.5. The hourly diesel oil consumption is given by Equation 2.12:

$$V_{f\text{-}CC} = \frac{P_{el} \cdot t}{\eta_{el} \cdot H_u} \Rightarrow V_{f\text{-}CC} = \frac{70 \cdot 10^3 \text{ kW} \cdot 1\text{h}}{0.3787 \cdot 10.07 \text{ kWh/l}} \Leftrightarrow V_{f\text{-}CC} = 18{,}355.80 \text{ l}$$

Summarizing the above calculations, we get:

- heavy fuel total consumption:

$$m_{f\text{-}DG} = m_{f\text{-}ST} + m_{f\text{-}DG} \Rightarrow m_f = (25{,}029.33 + 26{,}480.33) \text{ kg} \Leftrightarrow m_f = 51{,}509.66 \text{ kg}$$

- diesel oil total consumption:

$$V_f = V_{f\text{-}CC} \Rightarrow V_f = 18{,}355.80 \text{ l}.$$

b. Consumed fuels total cost

The procurement prices of the consumed fuels are given in Table 2.5 at 0.50 €/kg for heavy fuel and 0.95 €/l for diesel oil. Based on these prices, the consumed fuel total procurement costs are calculated as follows:

– Steam turbines

$$C_{f\text{-}ST} = p_{f\text{-}m} \cdot m_{f\text{-}ST} \Rightarrow C_{f\text{-}ST} = 0{,}50 \text{ €/kg} \cdot 25{,}029.33 \text{ kg} \Leftrightarrow C_{f\text{-}ST} = 12{,}514.67 \text{ €}.$$

– Diesel generators

$$C_{f\text{-}DG} = p_{f\text{-}m} \cdot m_{f\text{-}DG} \Rightarrow C_{f\text{-}DG} = 0{,}50 \text{ €/kg} \cdot 26{,}480.33 \text{ kg} \Leftrightarrow C_{f\text{-}DG} = 13{,}240.16 \text{ €}.$$

– Combined cycle

$$C_{f\text{-}CC} = p_{f\text{-}d} \cdot V_{f\text{-}CC} \Rightarrow C_{f\text{-}CC} = 0{,}95 \text{ €/l} \cdot 18{,}355.80 \text{ l} \Leftrightarrow C_{f\text{-}CC} = 17{,}438.01 \text{ €}.$$

Summarizing the above results, the consumed fuel total procurement cost is calculated equal to:

$$C_f = C_{f\text{-}ST} + C_{f\text{-}DG} + C_{f\text{-}CC} \Rightarrow C_f = (12{,}514.67 + 13{,}240.16 + 17{,}438.01) \text{ €} \Leftrightarrow C_f = 43{,}192.84 \text{ €}.$$

c. Consumed fuels and wind energy total procurement cost

The total electricity production from the wind parks is 100 MWh (100 MW power penetration for a time interval of 1 hour). Given the selling price of 0.095 €/kWh for the electricity produced from the wind parks, the total procurement cost is calculated as:

$$C_{WE} = p_{WE} \cdot E_{WP} \Rightarrow C_{WE} = 0{,}095 \text{ €/kWh} \cdot 100{,}000 \text{ kWh} \Leftrightarrow C_{WE} = 9{,}500.00 \text{ €}.$$

The consumed fuels and wind energy total procurement cost is eventually calculated at:

$$C_{TOT} = C_f + C_{WE} \Rightarrow C_{TOT} = (43{,}192.84 + 9{,}500.00) \text{ €} \Leftrightarrow C_{TOT} = 52{,}692.84 \text{ €}.$$

d. Hourly average production-specific costs

The hourly average production-specific cost c of a single generator or of a whole electrical system is given by the ratio of the production cost C of the final electricity produced during the examined hourly time period from the generator or the system over this particular electricity production E_{el}:

$$c = \frac{C}{E_{el}} = \frac{C}{P_{el} \cdot t} \quad (2.13)$$

The requested hourly average production-specific costs for each thermal generator separately, for the thermal power plant, and the whole electrical system are calculated with Equation 2.13:

– Steam turbines

$$c_{ST} = \frac{C_{f\text{-}ST}}{E_{el\text{-}ST}} = \frac{C_{f\text{-}ST}}{P_{el\text{-}ST} \cdot t} \Rightarrow c_{ST} = \frac{12{,}514.67\ €}{100 \cdot 10^3\ \text{kWh} \cdot 1\text{h}} \Leftrightarrow c_{ST} = 0.1251\ €/\text{kWh}$$

– Diesel generators

$$c_{DG} = \frac{C_{f\text{-}DG}}{E_{el\text{-}DG}} = \frac{C_{f\text{-}DG}}{P_{el\text{-}DG} \cdot t} \Rightarrow c_{DG} = \frac{13{,}240.16\ €}{140 \cdot 10^3\ \text{kWh} \cdot 1\text{h}} \Leftrightarrow c_{DG} = 0.0946\ €/\text{kWh}$$

– Combined cycle

$$c_{CC} = \frac{C_{f\text{-}CC}}{E_{el\text{-}CC}} = \frac{C_{f\text{-}CC}}{P_{el\text{-}CC} \cdot t} \Rightarrow c_{CC} = \frac{17{,}438.01\ €}{70 \cdot 10^3\ \text{kWh} \cdot 1\text{h}} \Leftrightarrow c_{CC} = 0.2491\ €/\text{kWh}$$

– Total thermal power plant

The total power production P_{TH} from all the dispatched thermal generators has been calculated at 310 MW, according to Table 2.8. The production-specific cost is calculated at:

$$c_{TH} = \frac{C_f}{E_{el\text{-}TH}} = \frac{C_f}{P_{el\text{-}TH} \cdot t} \Rightarrow c_{TH} = \frac{43{,}192.84\ €}{310 \cdot 10^3\ \text{kWh} \cdot 1\text{h}} \Leftrightarrow c_{TH} = 0.1393\ €/\text{kWh}$$

– Total electrical system

The total production-specific cost is calculated by taking into account the overall electricity production and the corresponding cost:

$$c_{TOT} = \frac{C_{TOT}}{E_{el\text{-}TOT}} = \frac{C_{TOT}}{P_{el\text{-}TOT} \cdot t} \Rightarrow c_{TOT} = \frac{52{,}692\ 84\ €}{410 \cdot 10^3\ \text{kWh} \cdot 1\text{h}} \Leftrightarrow c_{TOT} = 0.1285\ €/\text{kWh}$$

Having integrated the above calculations, we can make the following observations:

- Diesel generators exhibit the lowest production-specific cost.
- On the other hand, the highest production-specific cost is calculated for the combined cycle, due to diesel oil consumption and partial-load operation.
- Apart from the diesel generators, all the other thermal generators exhibit production-specific cost higher than the wind parks' electricity selling price. This fact is depicted in the calculated production-specific cost for the thermal power plant in total and for the whole electrical system (including wind parks). The latter is calculated lower than the first one, a fact that expresses the wind parks' contribution to the reduction of the electrical system's production-specific cost.
- The above calculated figures can be considered as typical regarding the operation of conventional electrical systems operating with liquid fossil fuels.

2.4.2 Operation without Wind Parks

It is worthwhile to investigate the operation of the same electrical system and for the same power demand (410 MW), however this time without the availability of any power production from wind parks. In this case, the power production synthesis should be built again:

- Accounting for the requested power demand of 410 MW and with the wind power penetration of 100 MW, both the steam turbines and the diesel generators operate at their nominal power, it comes that for this new investigated case, namely without the power availability from the wind parks, the steam turbines and the diesel generators will be required to operate at their nominal power again. The total power production from these two types of thermal generators will be:

$$100 \text{ MW} + 140 \text{ MW} = 240 \text{ MW}.$$

For the total power demand of 410 MW, the remaining power to be covered will be:

$$410 \text{ MW} - 240 \text{ MW} = 170 \text{ MW}.$$

- The combined cycle's nominal power (120 MW) is lower than the remaining power demand, hence the combined cycle must operate at its nominal power too. After the combined cycle's dispatch, the remaining power demand will be:

$$170 \text{ MW} - 120 \text{ MW} = 50 \text{ MW}.$$

- The only available choice for the production of additional power of 50 MW is the gas turbines. Taking also into account, for spinning reserve maintenance reasons, that the nominal power of the dispatched thermal generator with the maximum capacity is 50 MW (diesel generator), three gas turbines will have to be dispatched, operating at 15 MW, 15 MW, and 20 MW and providing a total spinning reserve calculated at:

$$2 \times (40 - 15) \text{ MW} + (40 - 20) \text{ MW} = 70 \text{ MW}.$$

a. Heavy fuel consumption and cost, steam turbines and diesel generators production-specific costs

Because the operation state of both steam turbines and diesel generators has not changed, compared to the previously investigated electrical system's operation with the wind parks' support, the heavy fuel consumptions and costs, as well as the steam turbines' and the diesel generators' production-specific costs remain the same.

b. Diesel oil consumption

– Combined cycle

The combined cycle operates at its nominal power, consequently, according to Table 2.4, its overall efficiency will be 44.18%.

The total power production equals 120 MW and the diesel oil heat capacity is given as 10.07 kWh/l in Table 2.5. The diesel oil hourly consumption is calculated using Equation 2.12:

$$V_{f\text{-}CC} = \frac{P_{el} \cdot t}{\eta_{el} \cdot H_u} \Rightarrow V_{f\text{-}CC} = \frac{120 \cdot 10^3 \text{ kW} \cdot 1\text{h}}{0.4418 \cdot 10.07 \text{ kWh/l}} \Leftrightarrow V_{f\text{-}CC} = 26{,}972.80 \text{ l}$$

– Gas turbines

Two of the three dispatched gas turbines operate at 15 MW, while the third one operates at 20 MW. Given the nominal capacity of the gas turbines at 40 MW, the above power productions correspond to partial-load operation at 37.5% and 50% respectively. In Table 2.4

the gas turbines' efficiency is given for partial load at 30% (21.16%) and at 70% (31.11%) over the turbines' nominal power. By applying two linear interpolations, between these two operation points, the gas turbines' efficiency for partial-load operation at 37.50% is calculated at 23.03% and at 26.14% for partial-load operation at 50%.

Given the above efficiencies, 30 MW (2 units × 15 MW each) out of the 50 MW power production by the gas turbines are produced with 23.03% efficiency. The remaining 20 MW are produced with 26.14% efficiency. The diesel oil heat capacity is given as 10.07 kWh/l in Table 2.5. The diesel oil hourly consumption is calculated using Equation 2.12:

$$V_{f\text{-}GT1} = \frac{P_{el1} \cdot t}{\eta_{el} \cdot H_u} \Rightarrow V_{f\text{-}GT1} = \frac{30 \cdot 10^3 \text{ kW} \cdot 1\text{h}}{0.2303 \cdot 10.07 \text{ kWh/l}} \Leftrightarrow V_{f\text{-}GT1} = 12{,}935.94 \text{ l}$$

$$V_{f\text{-}GT2} = \frac{P_{el2} \cdot t}{\eta_{el} \cdot H_u} \Rightarrow V_{f\text{-}GT2} = \frac{20 \cdot 10^3 \text{ kW} \cdot 1\text{h}}{0.2614 \cdot 10.07 \text{ kWh/l}} \Leftrightarrow V_{f\text{-}GT2} = 7{,}597.92 \text{ l}$$

The gas turbine total diesel oil consumption is:

$$V_{f\text{-}GT} = V_{f\text{-}GT1} + V_{f\text{-}GT2} \Rightarrow V_{f\text{-}GT} = (12{,}935.94 + 7{,}597.92) \text{ l} \Leftrightarrow V_{f\text{-}GT} = 20{,}533.86 \text{ l}.$$

The total diesel oil consumption from the combined cycle and the gas turbines is:

$$V_f = V_{f\text{-}CC} + V_{f\text{-}GT} \Rightarrow V_f = (26{,}972.80 + 20{,}533.86) \text{ l} \Leftrightarrow V_f = 47{,}506.66 \text{ l}.$$

c. Consumed fuels total costs

Recall that the heavy fuel and the diesel oil procurement prices are given as 0.50 €/kg and 0.95 €/l respectively in Table 2.5. Additionally, the consumed heavy fuel cost has been calculated in the previous investigated operation scenario, with the wind power penetration, and, as explained above, it remains the same in this examined operation case.

– Steam turbines

$$C_{f\text{-}ST} = 12{,}514.67 \text{ €}.$$

– Diesel generators

$$C_{f\text{-}DG} = 13{,}240.16 \text{ €}.$$

– Combined cycle

$$C_{f\text{-}CC} = p_{f\text{-}d} \cdot V_{f\text{-}CC} \Rightarrow C_{f\text{-}CC} = 0{,}95 \text{ €/l} \cdot 26{,}972.80 \text{ l} \Leftrightarrow C_{f\text{-}CC} = 25{,}624.16 \text{ €}.$$

– Gas turbines

$$C_{f\text{-}GT} = p_{f\text{-}d} \cdot V_{f\text{-}GT} \Rightarrow C_{f\text{-}GT} = 0{,}95 \text{ €/l} \cdot 20{,}533.86 \text{ l} \Leftrightarrow C_{f\text{-}GT} = 19{,}507.17 \text{ €}.$$

The consumed fuels total cost, which in this case, in the absence of any electricity production from the wind parks, actually coincides with the total production cost of the electrical system, is given by the total sum of the above calculated costs:

$$C_f = C_{f\text{-}ST} + C_{f\text{-}DG} + C_{f\text{-}CC} + C_{f\text{-}GT} \Rightarrow$$

$$C_f = (12{,}514.67 + 13{,}240.16 + 25{,}624.16 + 19{,}507.17) \text{ €} \Leftrightarrow C_f = 70{,}886.16 \text{ €}.$$

By comparing the above result with the corresponding figure in the first investigated scenario, the total production cost of the electrical system without the wind parks' support increases to:

$$\Delta C_f = (70{,}886.16 - 52{,}692.84)\ € \Leftrightarrow \Delta C_f = 18{,}193.32\ €$$

which corresponds to a percentage increase of 34.53%.

d. Hourly average production-specific costs

The production-specific costs for the steam turbines and the diesel generators have been calculated in the previous investigated operation scenario and remain the same:

– Steam turbines

$$c_{ST} = 0.1251\ €/\text{kWh}.$$

– Diesel generators

$$c_{DG} = 0.0946\ €/\text{kWh}.$$

The production-specific cost for the combined cycle and the gas turbines are calculated below:

– Combined cycle

$$c_{CC} = \frac{C_{f\text{-}CC}}{E_{el\text{-}CC}} = \frac{C_{f\text{-}CC}}{P_{el\text{-}CC}\cdot t} \rightarrow c_{CC} = \frac{25{,}624.16\ €}{120\cdot 10^3\ \text{kWh}\cdot 1\text{h}} \Leftrightarrow c_{CC} = 0.2135\ €/\text{kWh}$$

– Gas turbines

$$c_{GT} = \frac{C_{f\text{-}GT}}{E_{el\text{-}GT}} = \frac{C_{f\text{-}GT}}{P_{el\text{-}GT}\cdot t} \rightarrow c_{GT} = \frac{19{,}507.17\ €}{50\cdot 10^3\ \text{kWh}\cdot 1\text{h}} \Leftrightarrow c_{GT} = 0.3901\ €/\text{kWh}$$

– Total thermal power plant

The power production P_{TH} from the thermal generators is equal to the power demand of 410 MW. The production-specific cost for the whole electrical system is:

$$c_{TH} = \frac{C_f}{E_{el\text{-}TH}} = \frac{C_f}{P_{el\text{-}TH}\cdot t} \rightarrow c_{TH} = \frac{70{,}886.16\ €}{410\cdot 10^3\ \text{kWh}\cdot 1\text{h}} \Leftrightarrow c_{TH} = 0.1729\ €/\text{kWh}$$

– Gas emission reduction due to wind power penetration

Assuming the following CO_2 specific emissions per unit of consumed fuel:

- $m_{gas\text{-}m} = 3.175$ kg CO_2 per kg of consumed heavy fuel
- $m_{gas\text{-}d} = 3.142$ kg CO_2 per l of consumed diesel oil

and because the heavy fuel consumption does not change between the two investigated operation scenarios, the gas emission reduction due to the wind power penetration during the hourly operation of the electrical system is calculated as:

$$\Delta m_{gas} = m_{gas\text{-}d}\cdot\Delta V_f \Rightarrow \Delta m_{gas} = 3.142\ \frac{\text{kg } CO_2}{\text{l}}\cdot(47{,}506.66 - 18{,}355.80)\ \text{l} \Leftrightarrow$$

$$\Delta m_{gas} = 91{,}592.01\ \text{kg } CO_2$$

Evaluating the above results, we may come to the following conclusions:

- With the wind power contribution in the electricity production share, given the lower wind energy price compared to the production-specific cost of the available thermal generators, a percentage reduction on the total production cost of 34.5% is achieved.

- A total reduction on the CO_2 emissions of 91.6 tn is also achieved due to the wind power penetration.
- The production-specific cost of the combined cycle is reduced in the second investigated scenario from 0.2491 €/kWh to 0.2135 €/kWh, due to its operation at full load and the corresponding increase of its overall efficiency, instead of its 58% partial-load operation in the first scenario.
- The significantly high production-specific cost of the gas turbines is also notable, a fact that implies the restriction of the operation of these thermal generators to the maximum possible extent.
- The production-specific cost of a thermal generator practically depends on its operation point, which, actually, determines the generator's overall efficiency and the consumed fuel's mass or volume rate.

2.5 COMPUTATIONAL SIMULATION OF THE ANNUAL OPERATION OF AN ELECTRICAL SYSTEM

In the above examples, the optimum power production syntheses were configured for different operation scenarios of the electrical system. Additionally, for two average hourly operation cases, the fundamental features that define and describe the system's operation were calculated, such as the fuels' consumptions and costs, the generators' efficiencies, and the production-specific costs.

By applying the same procedure, analyzed in the previous section only for specific operation cases, to the annual operation of an electrical system and with a predefined time calculation step, typically hourly in such simulations, the precise calculation of the above characteristic features may be accomplished step-by-step for the whole annual period. This task can be called the *computational or arithmetic simulation of the electrical system's annual operation* and aims to predict, when executed for future years, or to evaluate, when executed for past years, how the system is going to operate or how efficiently the system has operated, respectively, given the thermal generators dispatch order basic rules, the spinning reserve maintenance policies, etc.

For the execution of the computational simulation of an electrical system's annual operation, the following data are required:

1. annual power demand time series with at least hourly average values
2. power plants synthesis (number and type of available thermal generators)
3. for each available thermal generator the following features
   - technical minimum
   - nominal power
   - consumed fuel type
   - overall efficiency variation graph versus the power output
   - dispatch order in the power production synthesis
   - fuel type and consumption for the startup procedure
4. heat capacity for each type of consumed fuel
5. spinning reserve maintenance policies
6. average annual procurement price for each type of consumed fuel
7. installed power of nonguaranteed power production plants (usually PV stations and wind parks)
8. power production availability annual time series of the nonguaranteed power production plants
9. procurement prices for the electricity produced from the RES power plants.

By individually investigating the above required data, we see that all of them were used for the calculations and the optimum power production syntheses analysis presented in the previous sections.

The basic results arising from the implementation of computational simulation of electrical systems' operation are:

- total annual electricity production per type of involved generator and for the whole electrical system
- average annual production-specific costs and efficiencies for each different thermal generator type and for all the involved thermal generators and the overall electrical system
- fuels total consumptions and costs
- annual penetration and rejection of the nonguaranteed power production units
- annual gas emissions
- conclusions on the achieved dynamic security and stability of the electrical system.

The above described procedure cannot be accomplished manually, like the previously presented examples. However, this procedure is possible and relatively simple with the development of relevant computational applications, realizing the electrical system's operation algorithm. Examples of the computational simulation of the annual operation of the autonomous electrical systems of the island of Crete, Greece, and the island of Praslin, Seychelles, are presented in the following sections of this chapter.

2.6 COMPUTATIONAL SIMULATION OF THE ANNUAL OPERATION OF CRETE'S AUTONOMOUS ELECTRICAL SYSTEM

2.6.1 Crete Power Demand

Crete, Greece, constitutes a large island, with a permanent population above 600,000 thousand people. It is the third largest non-interconnected island in the Mediterranean, with regard to its electricity consumption, after Cyprus and Corsica. The annual operation of the autonomous electrical system of Crete will be simulated for the year 2016. The simulation will follow the above described procedure and it will be executed following an hourly time calculation step. The first task is to introduce the power demand annual time series, as it was provided by the grid's operator. The annual power demand time series for 2016 is plotted in Figure 2.32. In Table 2.10, some fundamental features of the introduced annual power demand time series are presented.

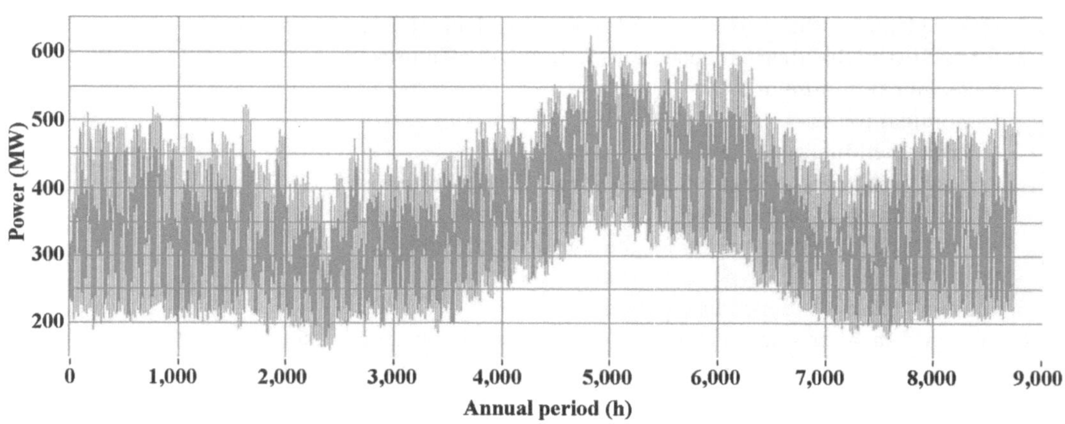

FIGURE 2.32 Annual power demand time series for the autonomous electrical system of Crete, Greece in 2016.

TABLE 2.10
Characteristic Features of the Annual Power Demand for the Autonomous Electrical System of Crete in 2016

Annual Peak Power Demand (MW)	Annual Minimum Power Demand (MW)	Annual Electricity Consumption (MWh)	Daily Average Electricity Consumption (MWh)
623.57	160.32	3,074,690.00	8,423.81

In Figure 2.32 we can observe the characteristic features already stated in previous sections, namely the power demand increase during summer, due to tourist activities, the power demand minimization during spring and autumn, and its slight increase during winter. The size of Crete's autonomous electrical system is also revealed by the data presented in both Figure 2.32 and Table 2.10. With the annual peak power demand exceeding 600 MW, the particular system is classified as a large-size electrical system.

2.6.2 Crete Thermal Power Plants

The power production system in Crete consists of three thermal power plants, located in three different prefectures of the island. Their syntheses are presented in Tables 2.11, 2.12, and 2.13. The combined cycle consists of two gas turbines and a steam turbine, following the operation concept presented in Section 2.2.4.

TABLE 2.11
The Thermal Power Plant Synthesis Located in Heraklion Prefecture

Steam Turbines	Technical Minimum (MW)	Maximum Capacity (MW)	Diesel Generators	Technical Minimum (MW)	Maximum Capacity (MW)	Gas Turbines	Technical Minimum (MW)	Maximum Capacity (MW)
1	1.8	6.2	1	3.0	11.0	1	3.0	14.0
2	8.0	14.0	2	3.0	11.0	2	3.0	14.0
3	8.0	14.0	3	3.0	11.0	3	3.0	43.0
4	18.0	24.0	4	3.0	11.0	4	3.0	13.0
5	18.0	24.0				5	3.0	32.0
6	18.0	24.0						
Totals:	71.8	106.2		12.0	44.0		15.0	116.0

TABLE 2.12
The Thermal Power Plant Synthesis located in Chania Prefecture

Gas Turbines	Technical Minimum (MW)	Maximum Capacity (MW)	Combined Cycle	Technical Minimum (MW)	Maximum Capacity (MW)
1	3.0	12.0	Gas turbine 6	12.0	37.0
4	3.0	13.0	Gas turbine 7	12.0	37.0
5	5.0	29.0	Steam turbine	21.0	36.0
11	8.0	55.0			
12	8.0	55.0			
13	3.0	32.0			
Totals:	30.0	195.0		45.0	110.0

TABLE 2.13
The Thermal Power Plant Synthesis Located in Lasithi Prefecture

Steam Turbines	Technical Minimum (MW)	Maximum Capacity (MW)	Diesel Generators	Technical Minimum (MW)	Maximum Capacity (MW)
1	25.0	45.0	1	12.0	45.0
2	25.0	45.0	2	12.0	45.0
Totals:	50.0	90.0		24.0	90.0

From the available thermal generators presented in Tables 2.11, 2.12, and 2.13, the steam turbines and diesel generators consume heavy fuel, while gas turbines and, consequently, the combined cycle consume diesel oil.

The dispatch order of the thermal generators is most commonly defined by the following parameters:

- the dynamic security and the stability of the electrical system
- the system's cost-effective operation.

The dispatch criterion imposes the continuous operation of the thermal generators with slow response rates and slow startup procedures. These thermal generators are the steam turbines and the combined cycle.

The second parameter imposes the following dispatch order of the available thermal generators:

1. diesel generators
2. steam turbines
3. combined cycle
4. gas turbines.

The above dispatch order is adopted whenever no other restrictions are introduced by the system's secure operation. Steam turbines and the combined cycle remain always dispatched, at least at their technical minimums, in cases of low power demand, regardless of the above dispatch order.

For the calculation of the fuels annual consumption, the thermal generators' efficiency curves are required. Such curves are presented in Figures 2.33 and 2.34 for the steam turbines installed in the Heraklion prefecture thermal power plant and for the gas turbines installed in the Chania prefecture thermal power plant.

To ensure the secure operation of the electrical system of Crete, spinning reserve is always maintained equal to the nominal power of the dispatched thermal generator with the maximum capacity.

2.6.3 Crete Nonguaranteed Power Production Plants

The total installed wind parks' power in Crete in 2016 was 221 MW.

The wind power penetration in the autonomous electrical system is limited by:

- the maximum instant wind power penetration percentage versus the current power demand
- the total technical minimums of the dispatched thermal generators, which define the minimum thermal power production.

Additionally, the total PV station power installed on the island reaches 90 MW.

For the calculation of the annual power production time series from the wind parks and the PV stations, annual wind velocity and solar radiation time series are employed, measured in different and characteristic locations of the island. Additionally, the power curves of the installed wind

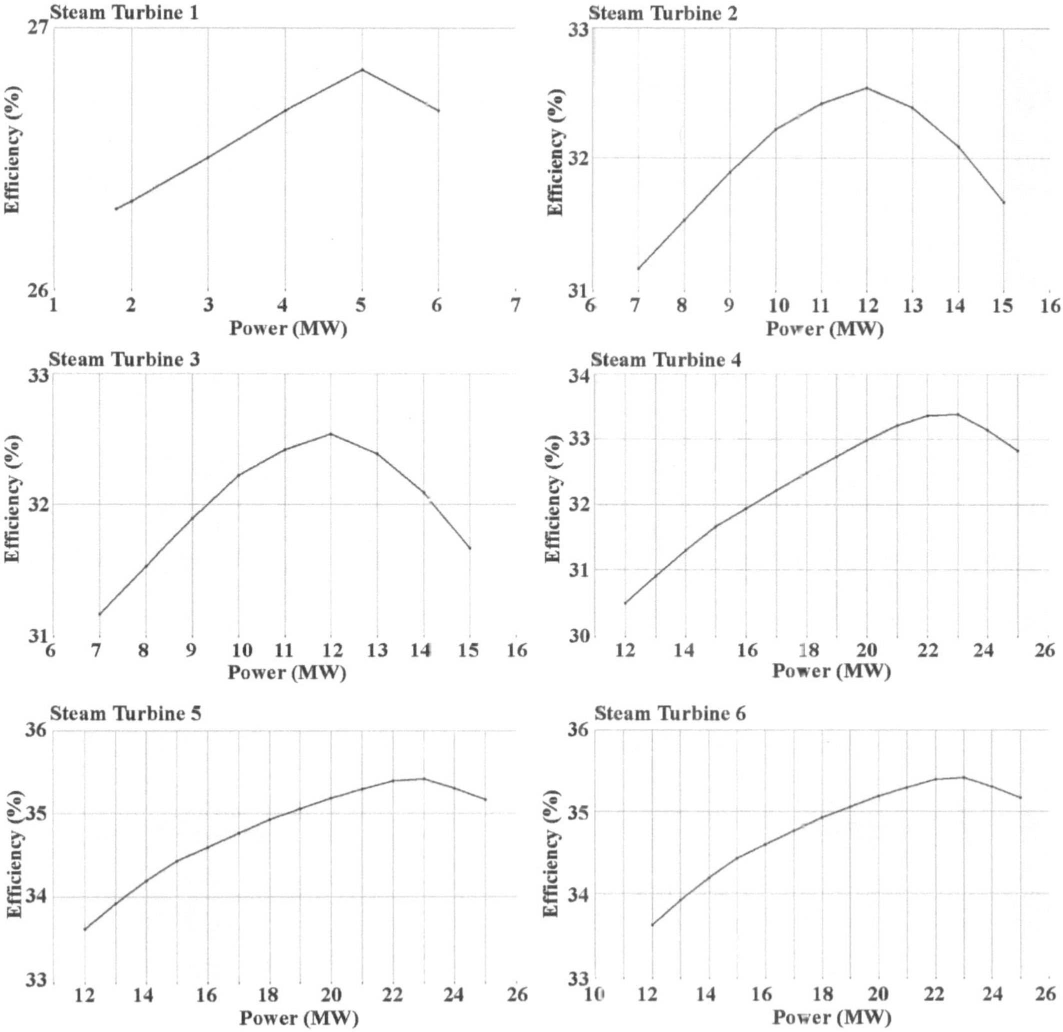

FIGURE 2.33 Efficiency curves versus the power output for the Heraklion thermal power plant steam turbines.

turbine models are introduced. The fundamental procedures for the calculation of the power production from the wind parks and the PV stations are given in the relevant bibliography [23, 24]. This whole calculation process is performed for hourly time steps over an annual time period.

Following the above process, the annual power production time series for the wind parks and the PV stations are calculated for 2016. The first one is presented in Figure 2.35 and the second one in Figure 2.36.

2.6.4 Crete Fuels

As mentioned previously, the consumed fuels are heavy fuel and diesel oil. The following minimum heat capacities are adopted:

- minimum heat capacity of heavy fuel with low sulfur content: 11.45 kWh/kg
- minimum heat capacity of diesel oil: 11.92 kWh/kg = 9.77 kWh/l.

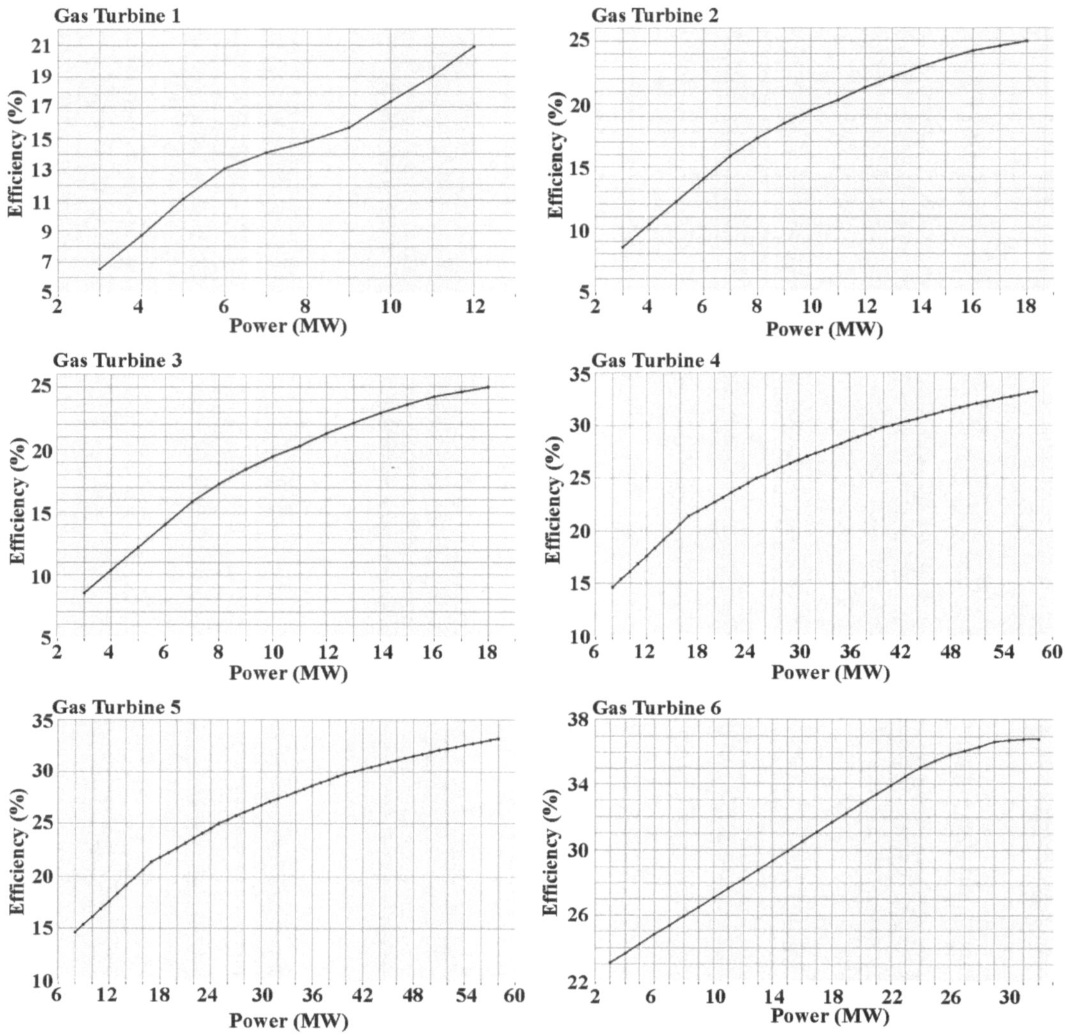

FIGURE 2.34 Efficiency curves versus the power output for the Chania thermal power plant gas turbines.

2.6.5 Crete Simulation Results

Given the above data and assumptions, the simulation of Crete's electrical system is executed with the development of a relevant software tool. Some of the most characteristic results are presented below.

2.6.5.1 Power Production Synthesis Graphs

Characteristic power production synthesis graphs are presented in the following figures. Specifically:

- Figure 2.37: annual power production synthesis graph
- Figure 2.38: power production synthesis graph from August 10, 2016 to August 20, 2016 (high power demand season)
- Figure 2.39: power production synthesis graph from March 1, 2016 to March 10, 2016 (low power demand season).

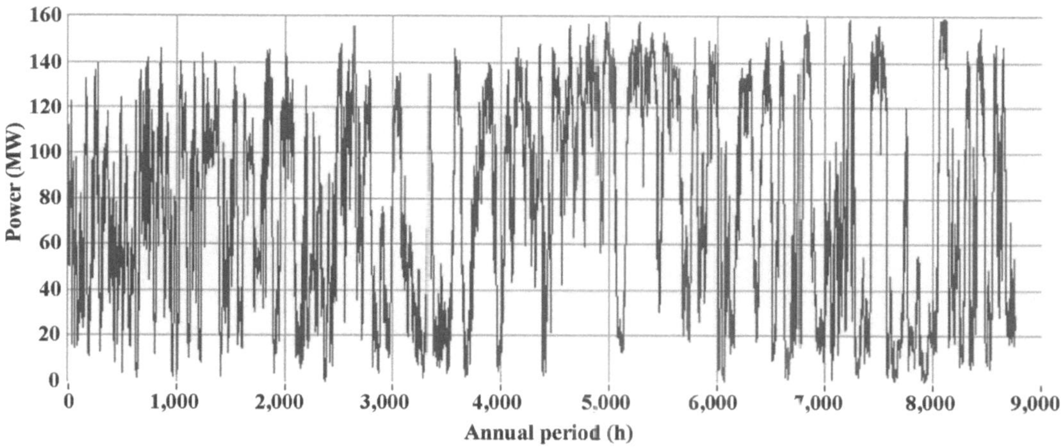

FIGURE 2.35 Annual power production time series from the installed wind parks in Crete in 2016.

From these figures, the overall concept for the power production synthesis formulation in the examined electrical system can be conceivable. Steam turbines and the combined cycle are always on-duty. In cases of low power demand, they operate at their technical minimums. They are switched off only for regular maintenance. For this reason, the total technical minimums of the dispatched steam turbines are observed reduced in the annual power production synthesis graph (Figure 2.37) during spring and autumn. Diesel generators are the first thermal generators that reach their nominal power and the combined cycle is the last one before the gas turbines. The gas turbines operation is considerably restricted mainly in summer, namely during the high power demand season, and during winter, although the power demand is not as high as in summer, because some base thermal generators are not available (switched off for regular maintenance). The total power penetration from the wind parks and the PV stations is restricted to 30% of the power demand, to ensure the system's dynamic security and stability

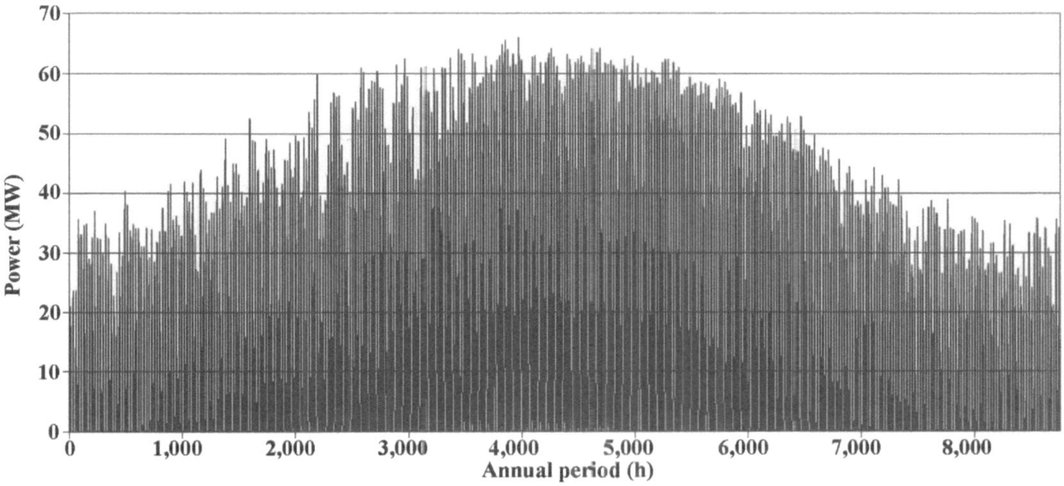

FIGURE 2.36 Annual power production time series from the installed PV stations in Crete in 2016.

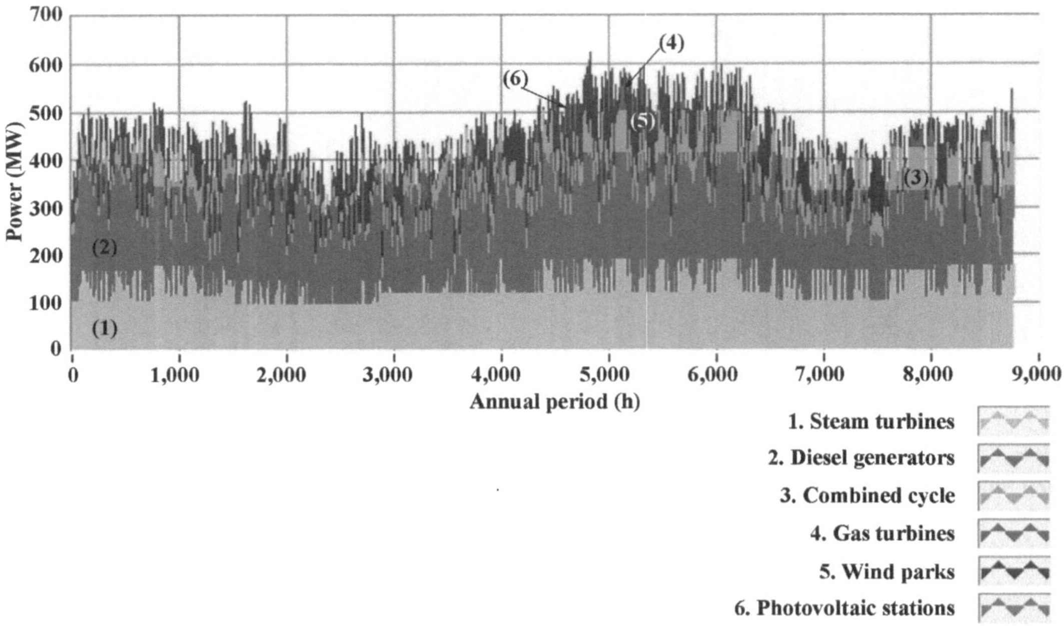

FIGURE 2.37 Annual power production synthesis graph for 2016.

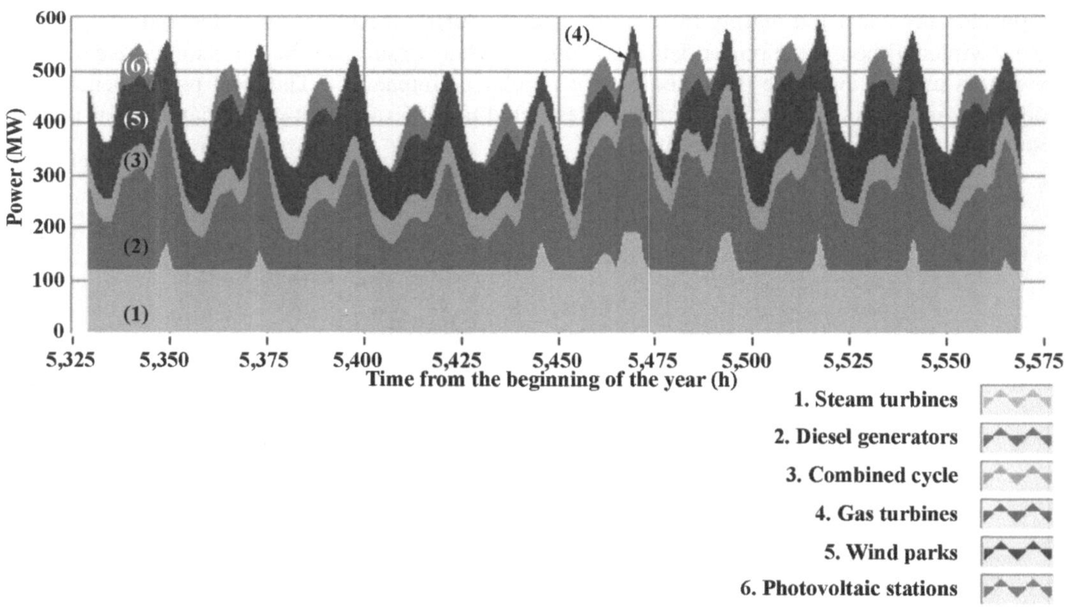

FIGURE 2.38 Power production synthesis graph from August 10, 2016 to August 20, 2016 (high power demand period).

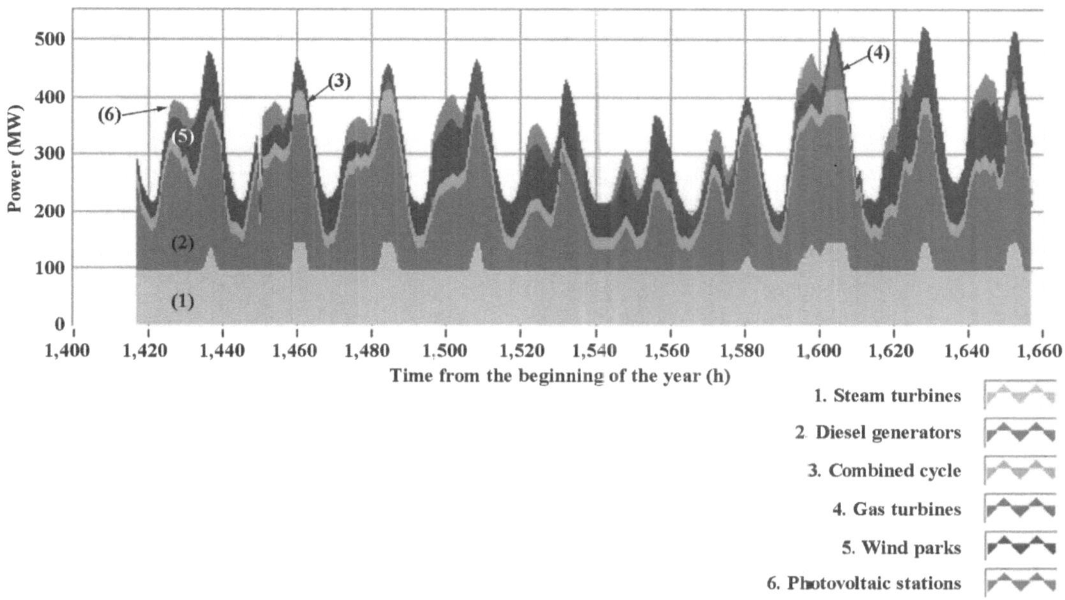

FIGURE 2.39 Power production synthesis graph from March 1, 2016 to March 10, 2016 (low power demand period).

2.6.5.2 Wind and PV Power Penetration

Figure 2.40 presents the annual power demand together with the dispatched thermal generators total technical minimums time series. The difference between these two curves defines the wind and PV power penetration margin during the year.

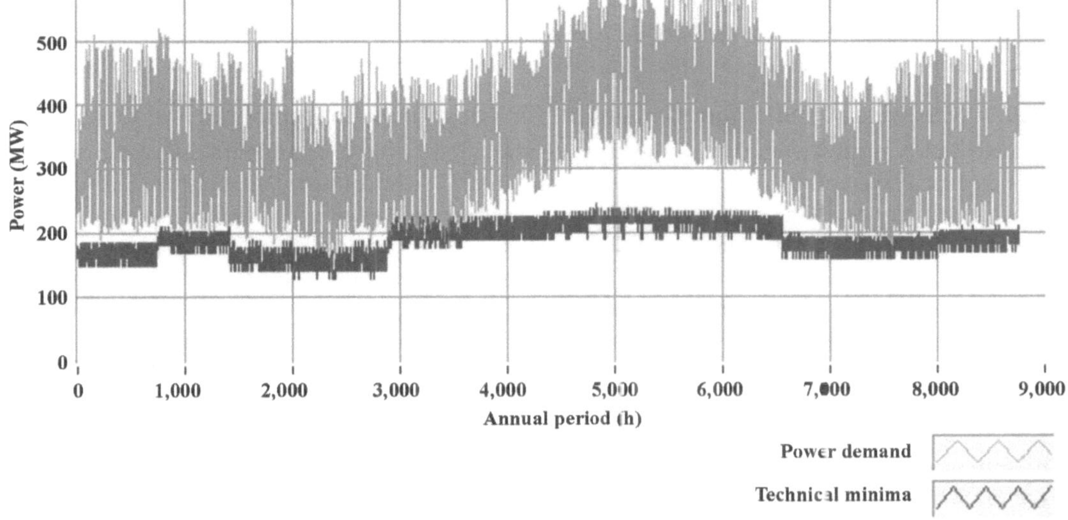

FIGURE 2.40 Annual power demand and total technical minima of the dispatched thermal generators time series.

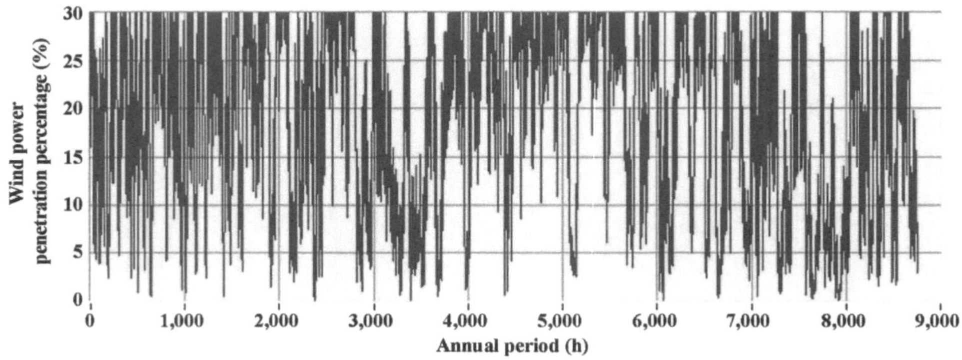

FIGURE 2.41 Annual variation of the wind power penetration percentage.

TABLE 2.14
Wind Power Penetration Annual Results

Total annual electricity production from the wind parks (MWh)	664,765.44
Total annual wind park's electricity rejection (MWh)	103,088.33
Annual rejection percentage (%)	15.51
Total annual electricity penetration (MWh)	561,677.08
Contribution percentage to the annual electricity consumption (%)	18.27
Rejection due to maximum instant penetration percentage (%)	13.05
Rejection due to total technical minimums (%)	2.46

In Figure 2.41, the annual variation of the wind power penetration percentage is presented. The maximum value is set at 30%. When the available wind power exceeds this limit, the corresponding power surplus is rejected.

Table 2.14 gives the results regarding the calculation of the wind power penetration.

Due to the relatively low power production from the PV stations with regard to the power demand (as seen in Figure 2.36, the total PV power production range is 25–30 MW in winter with an average power demand at 250 MW and at 60–65 MW in summer with an average power demand at 550 MW), no penetration restrictions are introduced for PV power production in the executed simulation.

2.6.5.3 Electricity Production, Generator Efficiency, Fuels Consumption, and Costs

The calculation results from the computational simulation of the annual operation of the autonomous electrical system of Crete are presented in Table 2.15. Specifically, among these results the annual electricity production, the overall annual average efficiencies, the fuels consumption and costs, and the production-specific costs are presented for each production unit separately and for the total electrical system. Additionally, the total fuels consumption and costs required for the thermal generators' startup procedures are also presented.

For the consumed fuels costs calculation the following procurement prices were adopted:

- annual average heavy fuel price: 487.083 €/tn,
- annual average diesel oil price: 990.607 €/klt.

The selling price for the electricity produced from the wind parks and the PV stations are set at 0.095€/kWh and at 0.195€/kWh, respectively.

The annual electricity production share percentage graph for the involved power production units is presented in Figure 2.42.

Finally, Table 2.16 summarizes results regarding the electricity production total and specific cost.

TABLE 2.15
Electricity Production, Generator Efficiencies, Fuels Consumptions and Costs for the Autonomous Electrical System of Crete in 2016

Unit/Process	Electricity Production (MWh)	Efficiency (%)	Production-Specific Cost (€/kWh)	Heavy Fuel Consumption (tn)	Diesel Oil Consumption (klt)	Heavy Fuel Cost (M€)	Diesel Oil Cost (M€)
Steam turbines	1,018,969	34.63	0.123	256,950	—	125.156	—
Diesel generators	1,009,943	46.46	0.092	189,852	—	92.474	—
Combined cycle	356,539	38.31	0.265	—	95,271	—	94.376
Gas turbines	4,222	25.04	0.405	—	1,726	—	1.710
Wind parks	561,677	—	0.095	—	—	—	—
PV stations	123,425	—	0.195				
Startup process	—	—	—	522	30	0.254	0.030
Totals:	3,074,775	40.16*	0.1273	447,324	97,027	217.884	96.116

* This is the overall average efficiency for all the thermal generators.

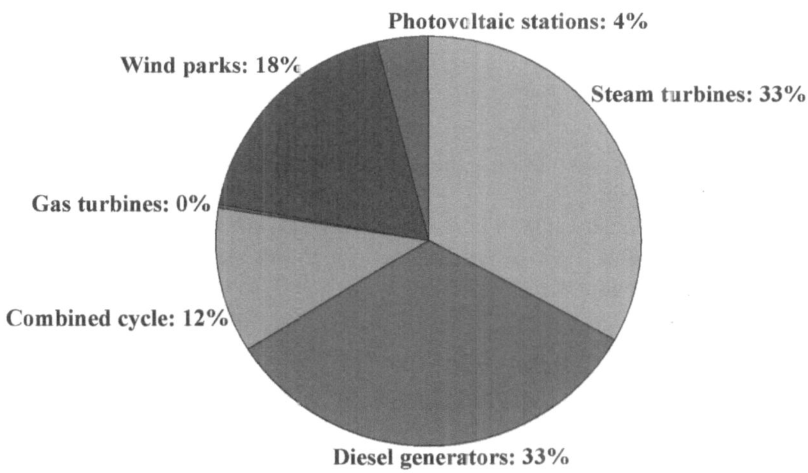

FIGURE 2.42 Annual electricity production share percentage graph for the involved power production units.

TABLE 2.16
Electricity Production Costs for the Electrical System of Crete in 2016

Total consumed fossil fuels (heavy fuel and diesel oil) cost (M€)	314.000
Total wind parks electricity cost (M€)	53.359
Total PV stations electricity cost (M€)	24.068
Total electricity production cost (M€)	391.427*
Total electricity production-specific cost (€/kWh)	0.1273*

* Only the fossil fuels and the wind and PV electricity procurement costs are included. Any other cost component is neglected (e.g., thermal generator maintenance, staff salaries, equipment amortizations).

TABLE 2.17
Gas-Specific Emissions from Fossil Fuel Combustion

	Gas-Specific Emissions (gr/kg of consumed fuel)					
Fossil Fuel Type	**CO_2**	**SO**	**CO**	**NO**	**HC**	**Particles**
Heavy fuel No. 1 (1,500) with low sulfur content	3,175	14	0,585	5.363	0.188	1.832
Heavy fuel No. 1 (1,500) with high sulfur content	3,109	70	0,553	6.251	0.184	1.832
Heavy fuel No. 3 (3,500) low sulfur content	3,175	14	0,585	5.363	0.188	1.832
Heavy fuel No. 3 (3,500) high sulfur content	3,091	80	0,550	5.221	0.183	1.832
Diesel oil	3,142	6	0,572	2.384	0.191	0.286
Liquefied petroleum gas (LPG)	3,030	0	0,332	2.102	0.080	0.100
Natural gas	2,715	0	0,332	2.102	0.080	0.100

TABLE 2.18
Annual CO_2 Emissions

Annual CO_2 emissions due to heavy fuel consumption (tn)	1,420,254
Annual CO_2 emissions due to diesel oil consumption (tn)	304,859
Total annual CO_2 emissions (tn)	1,725,113

2.6.5.4 CO_2 Emissions

For the annual CO_2 emission calculations, the specific emissions per unit of consumed fuel presented in Table 2.17 are adopted.

Assuming the diesel oil density at 0.85 kg/l, the annual CO_2 emissions from the electrical system of Crete are presented in Table 2.18, given the heavy fuel and diesel oil annual consumptions as calculated and presented previously.

2.7 COMPUTATIONAL SIMULATION OF THE ANNUAL OPERATION OF PRASLIN'S AUTONOMOUS ELECTRICAL SYSTEM

2.7.1 Praslin–La Digue Power Demand

The Seychelles archipelago consists of three main islands, the island of Mahé, which is the largest, with the capital of the country Victoria and an overall population of 77,000 inhabitants, and the islands of Praslin and La Digue, with permanent populations of 7,500 and 2,800 inhabitants respectively. The Seychelles archipelago is divided into two discrete, autonomous electrical systems, the larger one installed and covering the electricity needs in Mahé and the smaller one installed in Praslin, covering the electricity needs in Praslin and La Digue. In this last section of this chapter, the computational simulation of the annual operation of the smallest insular electrical system in the islands of Praslin and La Digue will be presented.

In Figure 2.43 the annual power demand fluctuation for the under consideration electrical system is presented, as provided by the Public Utilities Corporation, the local state grid operator in the country.

As seen in Figure 2.43, the annual power demand fluctuation in Praslin–La Digue exhibits significant differences with regard to the corresponding curves presented throughout this chapter. First of all, there are no intensive seasonal fluctuations, as seen, for example, in the power demand annual profile for the case of Crete. Secondly, the peak power demand is recorded during spring and autumn, instead of summer. A short peak power demand period is also observed at the very

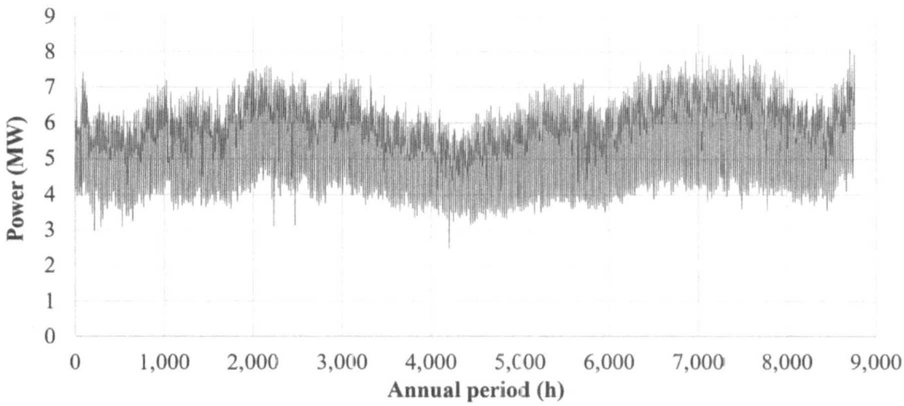

FIGURE 2.43 Annual power demand time series for the autonomous electrical system of Praslin–La Digue, Seychelles, in 2016.

TABLE 2.19
Characteristic Features of the Annual Power Demand for the Autonomous Electrical System of Praslin–La Digue in 2016

Annual Peak Power Demand (MW)	Annual Minimum Power Demand (MW)	Annual Electricity Consumption (MWh)	Daily Average Electricity Consumption (MWh)
8.09	2.50	47,228.80	129.39

last and the very first days of the year. This is because tourist activities are maximized particularly during spring and autumn, as well as during the Christmas holiday season. The mild climate conditions met in winter do not impose any power demand increase, for indoor space heating. The reduced tourist activity during summer compensates for the increasing needs for indoor space cooling, leading to the flattening of the annual power demand profile. To conclude, the annual power demand fluctuation in the investigated insular system is again configured by the tourist sector, which is the main economic activity in the islands, as well as the climate conditions.

The fundamental features of the annual power demand in the islands of Praslin and La Digue are presented in Table 2.19.

2.7.2 Praslin–La Digue Thermal Power Plant

Electricity production in the islands of Praslin and La Digue is based on an autonomous, thermal power plant, installed on the island of Praslin. The synthesis of this thermal power plant is presented in Table 2.20. As seen in this table, diesel generators are the only technology involved in Praslin's thermal power plant.

All the available thermal generators consume diesel oil. Because there are no different technologies, diesel generators will undertake both the roles of base and peak power production units. As defined in earlier sections of this chapter, the aim of the dispatch order is:

- to guarantee the dynamic security and the stability of the electrical system, and
- to minimize the electricity production cost.

In this case, the dispatch order is practically defined by the size of the involved generators, with the largest generators being dispatched first, undertaking the duty of base units. Deviations from this

TABLE 2.20
The Thermal Power Plant Synthesis in Praslin

Diesel Generator	Number of Units	Technical Minimum (MW)	Maximum Capacity (MW)
1	1	1.8	3.0
2	1	1.8	3.0
3	1	1.5	2.5
4	1	0.9	1.5
5	1	0.8	1.4
6	1	0.7	1.2
7	1	0.6	1.0
8	4	0.4	0.7
Totals:	11	8.5	14.3

order can be induced in cases of low power demand, combined with RES penetration, which practically may impose the reduction of the total power production from the thermal generators. This fact, combined with the requirement for spinning reserve maintenance, may lead to the distribution of the total power production in more thermal generators of smaller size, reversing the usual general dispatch order mentioned above.

Typical efficiency fluctuation curves are introduced for the involved diesel generators, employed for the calculation of the fuel annual consumption. Such curves are presented in Figure 2.44.

To ensure the secure operation of the electrical system in Praslin and La Digue, spinning reserve is always maintained equal to the nominal power of the dispatched thermal generator with the maximum capacity.

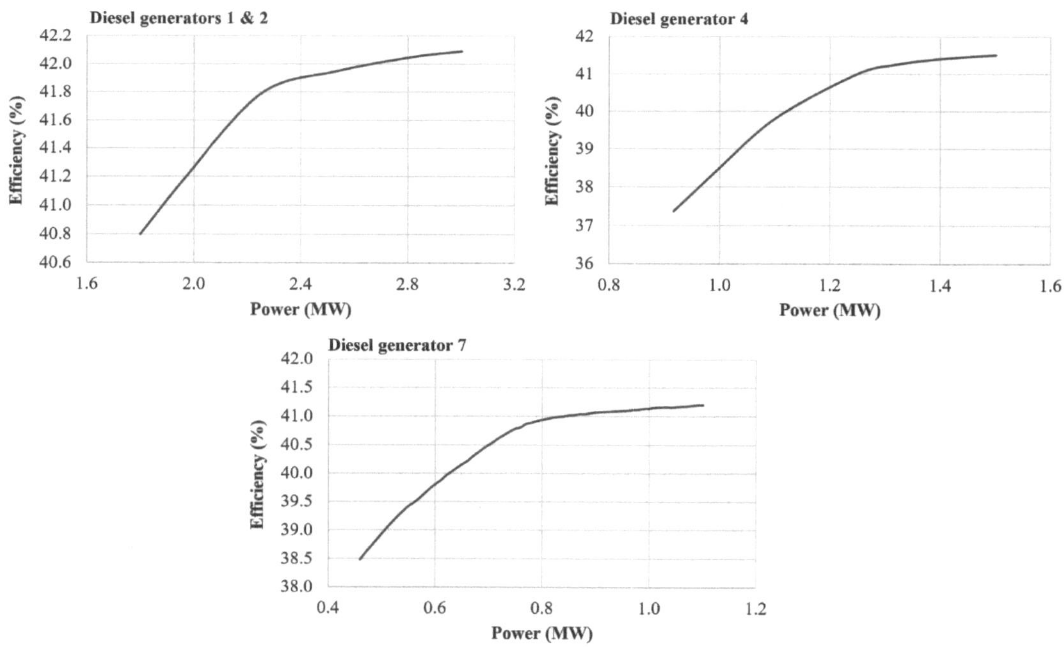

FIGURE 2.44 Indicative efficiency curves versus the power output for some of the diesel generators in Praslin–La Digue thermal power plant.

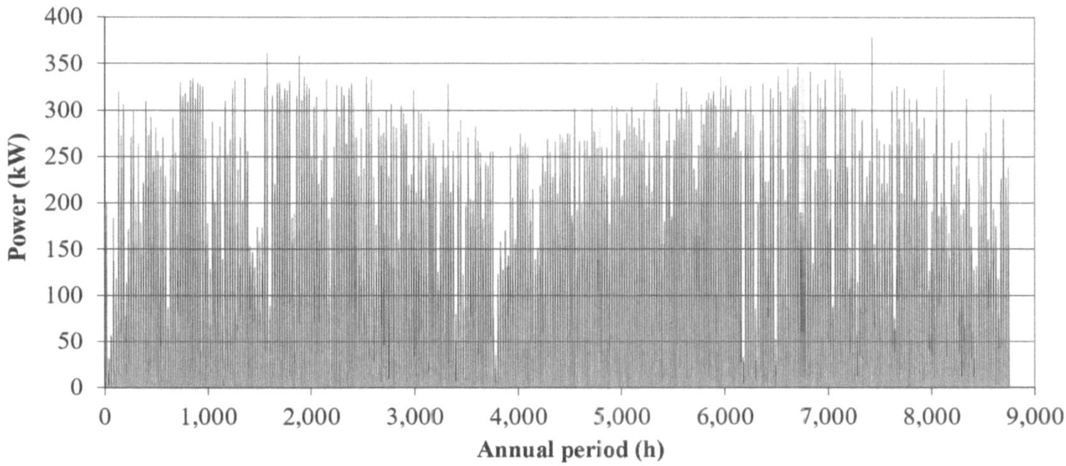

FIGURE 2.45 Annual power production time series from the installed PV stations in Praslin–La Digue in 2016.

2.7.3 Praslin–La Digue Nonguaranteed Power Production Plants

No wind parks or centralized PV stations are currently installed in the islands of Praslin and La Digue. The only RES electricity production technology involved in the electrical grid are PV stations installed on building roofs, with a total nominal power of 419 kWp.

As with the previous case study, annual solar radiation time series is employed for the calculation of the annual power production time series from the PV stations. Additionally, the annual wind velocity and ambient temperature time series are introduced for the calculation of the heat transfer from the PV panels to the ambient environment and their overall operation efficiency. The calculation process is performed for hourly time steps over the annual time period.

The annual power production time series for the PV stations is calculated for 2016, presented in Figure 2.45. The annual electricity production is calculated at 557.6 MWh, imposing a total, annual capacity factor of 15.2%.

2.7.4 Praslin–La Digue Fuels

As mentioned previously, the only consumed fuel type in the thermal power plant in Praslin is diesel oil, with a minimum heat capacity adopted at 11.92 kWh/kg = 9.77 kWh/l.

2.7.5 Praslin–La Digue Simulation Results

Given the above data and assumptions, the simulation of the electrical system in Praslin–La Digue is executed with the development of a relevant software tool. Some of the most characteristic results are presented below.

2.7.5.1 Power Production Synthesis Graphs

Characteristic power production synthesis graphs are presented in the following figures. Specifically:

- Figure 2.46: annual power production synthesis graph
- Figure 2.47: power production synthesis graph from October 1, 2016 to October 10, 2016 (high power demand season)
- Figure 2.48: power production synthesis graph from June 21, 2016 to June 30, 2016 (low power demand season).

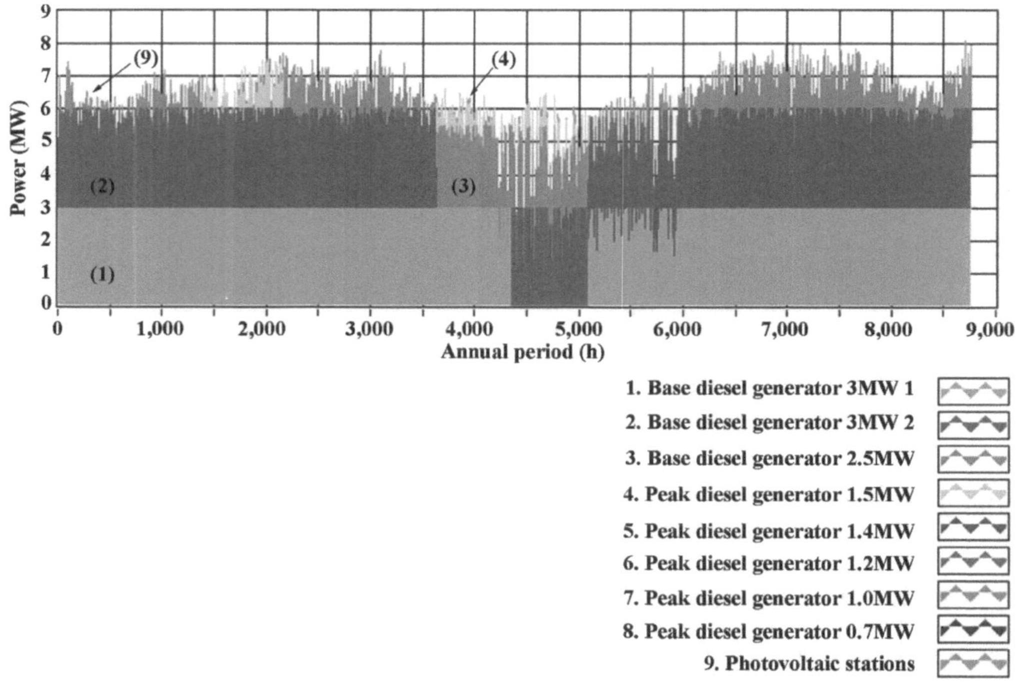

FIGURE 2.46 Annual power production synthesis graph for 2016.

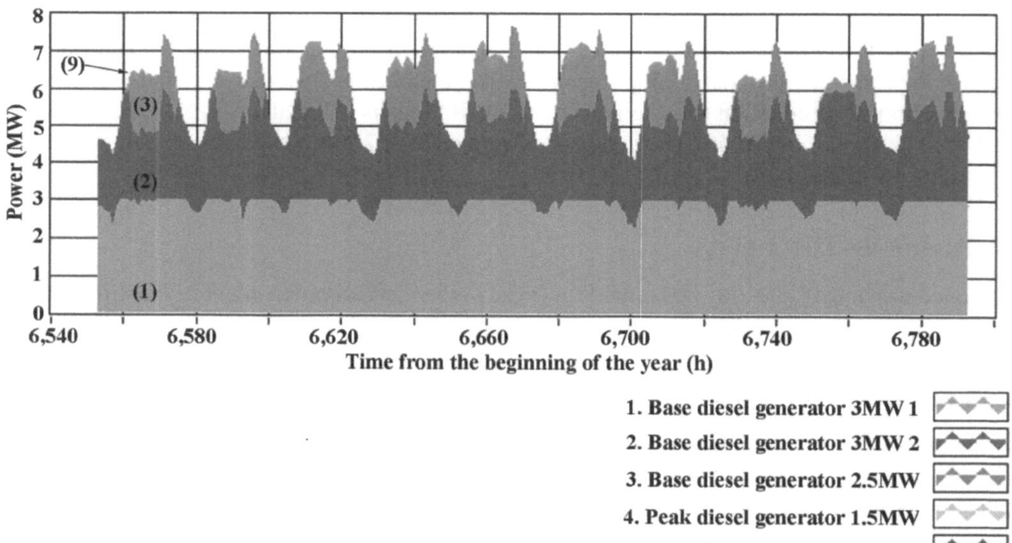

FIGURE 2.47 Power production synthesis graph from October 10, 2016 to October 10, 2016 (high power demand period).

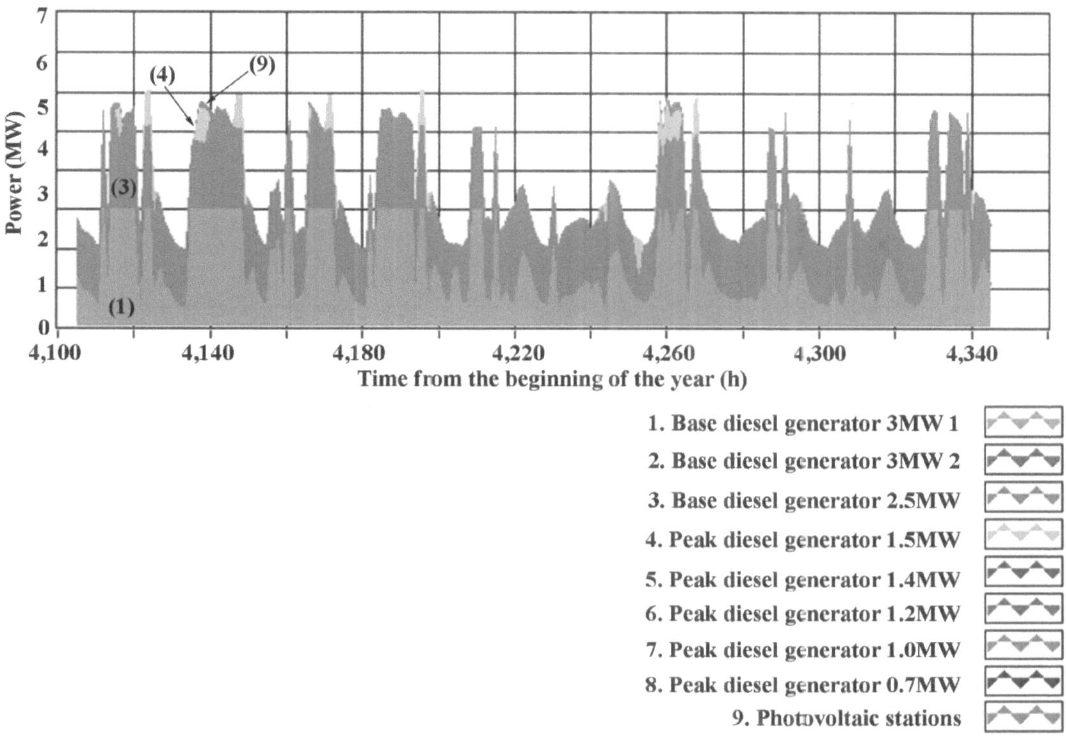

FIGURE 2.48 Power production synthesis graph from June 21, 2016 to June 30, 2016 (low power demand period).

The power production synthesis concept is characteristically depicted in these figures. First, we see that the two largest base diesel generators exhibit the highest annual electricity production contribution. The thermal generators of smaller size are dispatched mainly during the high power demand period and particularly when the largest production units are under maintenance, which is sensibly scheduled during the low power demand period (June and July, Figure 2.48). The significant contribution of the PV stations is also clear in these figures.

In Figure 2.49 the annual power demand together with the dispatched thermal generators total technical minimums time series are presented.

2.7.5.2 Electricity Production, Generator Efficiency, Fuels Consumption, and Costs

The calculation results from the computational simulation of the annual operation of the autonomous electrical system of Praslin–La Digue are presented in Table 2.21. Specifically, among these results the annual electricity production, the overall annual average efficiencies, the fuels consumption and costs, and the production-specific costs are presented for each production unit separately and for the total electrical system. Additionally, the total fuels consumption and cost required for the thermal generators' startup procedures are also presented.

For the consumed fuel cost calculation, the diesel oil procurement price of 0.57851 \$/l was adopted, particularly for the geographical territory under consideration. The PV stations are assumed to be integrated in the power system under net-metering mode, namely the annual electricity production and consumption are compensated on an annual basis.

The annual electricity production share percentage graph for the involved power production units is presented in Figure 2.50.

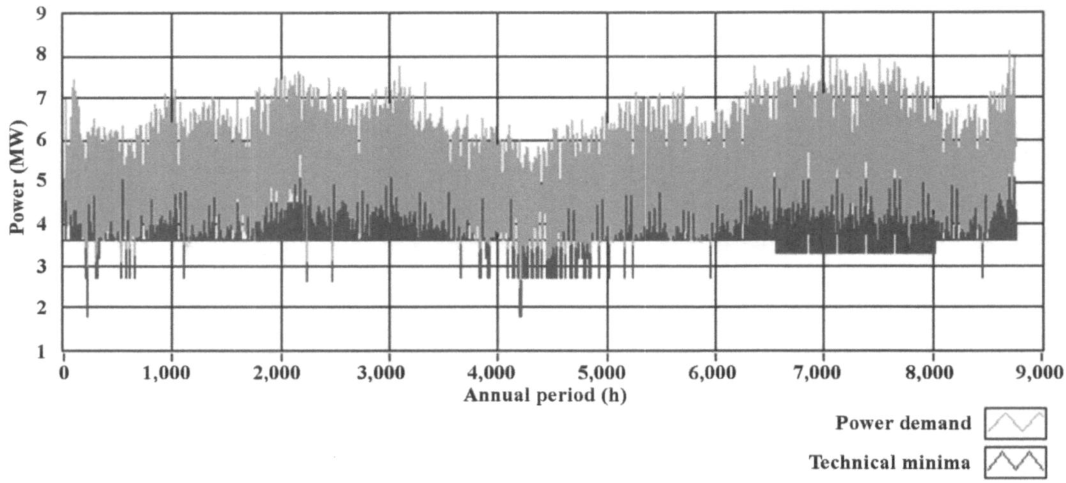

FIGURE 2.49 Annual power demand and total technical minima of the dispatched thermal generators time series.

TABLE 2.21
Electricity Production, Generator Efficiencies, Fuels Consumption, and Costs for the Autonomous Electrical System of Praslin–La Digue in 2016

Unit/Process	Electricity Production (MWh)	Efficiency (%)	Production-Specific Cost (€/kWh)	Diesel Oil Consumption (klt)	Diesel Oil Cost (M€)
Base DG 1	21,866	41.19	0.1438	5,146	2.977
Base DG 2	17,833	41.19	0.1437	4,196	2.427
Base DG 3	6,341	41.18	0.1438	1,492	0.863
Peak DG 1	624	41.03	0.1443	147	0.085
Peak DG 2	5	41.04	0.1443	1	0.001
Peak DG 3	0	—	—	0	0
Peak DG 4	0	—	—	0	0
Peak DG 5	0	—	—	0	0
PV stations	557	—	—	0	0
Startup process	—	—	—	4	0.002
Totals:	47,226	41.19 (*)	0.1438	10,986	6.355

* This is the overall average efficiency for all the thermal generators.

2.7.5.3 CO_2 Emissions

Finally, assuming the diesel oil density at 0.85 kg/l and the CO_2 emission factor at 3,142 gr CO_2 per kg of consumed diesel oil [24], the annual CO_2 emissions due to the electricity production in Praslin–La Digue is calculated at 29.340 tn.

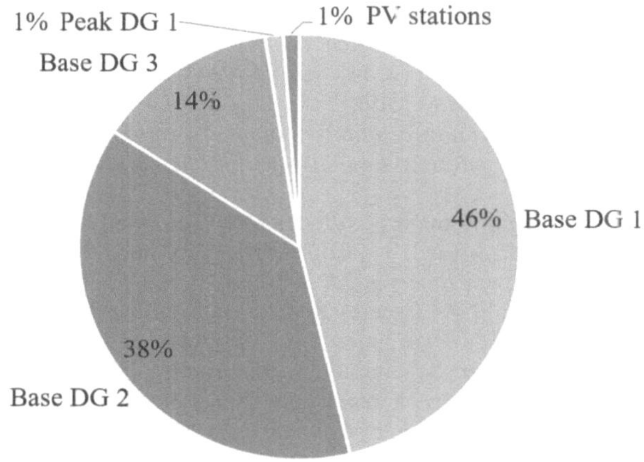

FIGURE 2.50 Annual electricity production share percentage graph for the involved power production units.

REFERENCES

1. Bobrow, L. S. (1996). *Fundamentals of Electrical Engineering (The Oxford Series in Electrical and Computer Engineering) (2nd ed.)*. Oxford University Press.
2. European Network of Transmission System Operators. https://docstore.entsoe.eu/Documents/Publications/maps/2019/Map_Northern-Europe-3.000.000.pdf (Accessed on May 2019).
3. Global Energy Network Institution—National Energy Grid, Croatia. http://www.geni.org/globalenergy/library/national_energy_grid/croatia/ (Accessed on May 2019).
4. Hambley, A. R. (2014). *Electrical Engineering: Principles and Applications* (6th ed.). Upper Saddle River, New Jersey: Pearson.
5. Dialynas, E. N., Hatziargyriou, N. D., Koskolos, N., & Karapidakis, E. (1998, August 30–September 5). Effect of high wind power penetration on the reliability and security of isolated power systems. Paper 38-302, 37th session, CIGRÉ.
6. Hatziargyriou, N., & Papadopoulos, M. (1998, September 18–19). Consequences of high wind power penetration in large autonomous power systems. Neptum, Romania: CIGRÉ Symposium.
7. Margaris, I. D., Hansen, A.D., Sørensen, P., & Hatziargyriou, N.D. (2011). Dynamic security issues in autonomous power systems with increasing wind power penetration. *Electric Power Systems Research*, 880–887.
8. Katsaprakakis, D. A., Thomsen, B., Dakanali, I., & Tzirakis, K. (2019). Faroe Islands: towards 100% RES penetration. *Renewable Energy*, *135*, 473–484.
9. Wikiversity, Steam Power, Part 2. https://en.wikiversity.org/wiki/Power_Generation/Steam_Power/Part2 (Accessed on March 2019).
10. Economou, L., Karvouniari, D., & Malamou, A. (2014) *Electrical Systems: Comprehensive Theory and Laboratory Exercises* (2nd ed.). Athens: Tziolas Edition.
11. Thomas, S. (2005, December). The economics of nuclear power. Nuclear issues paper No. 5. No 2, Heinrich Böll Stiftung. http://www.nirs.org/c20/atommythen_thomas.pdf (Accessed on July 2017).
12. Nuclear Energy Agency. The Economics of Long-term Operation of Nuclear Power Plants. Organisation For Economic Co-Operation And Development 2012. ISBN 978-92-64-99205-4. http://www.oecd-nea.org/ndd/reports/2012/7054-long-term-operation-npps.pdf (Accessed on July 2017).
13. Wartsila diesel engines. www.wartsila.com (Accessed on May 2019).
14. GE Power/9HA.01/02 Gas Turbine (50 Hz). https://www.ge.com/power/gas/gas-turbines/9ha (Accessed on May 2019).
15. Wikimedia Commons: File:Jet engine.svg. https://commons.wikimedia.org/wiki/File:Jet_engine.svg (Accessed on October 2019).
16. Mechanics Tips / Four Stroke Engine. http://mechanicstips.blogspot.com/2015/06/four-stroke-engine.html (Accessed on May 2019).

17. ABB T&D L.L.C. Power Systems Division. (2014). Simulator Training on Power Plant. State of the art system & training courses to develop a competent operations & maintenance team. https://library.e.abb.com/public/b2ab2687515e324ac1257cd90044fe8e/2VAA004177_A_en_S349_-_Simulator_training_on_Power_Plant.pdf (Accessed on May 2019).
18. Papantonis, D. (2016). *Small Hydro Power Plants.* Athens: Tsotras Editions.
19. Leyland, B. (2014). *Small Hydroelectric Engineering Practice* (1st ed.). Boca Raton, Florida: CRC Press.
20. Voith Hydro Turbines. http://voith.com/corp-en/index.html (Accessed on May 2019).
21. Hatata, A.Y., El-Saadawi, M. M., & Saad, S. (2019). A feasibility study of small hydro power for selected locations in Egypt. *Energy Strategy Reviews, 24*, 300–313.
22. Kumar, A., Schei, T., Ahenkorah, A., Caceres Rodriguez, R., Devernay, J.-M., Freitas, M., Hall, D., Killingtveit, Å., & Liu, Z. (2011). Hydropower. In Edenhofer, O., Pichs-Madruga, R., Sokona, Y., Seyboth, K., Matschoss, P., Kadner, S., Zwickel, T., Eickemeier, P., Hansen, G., Schlömer, S., & von Stechow C. (eds.), *IPCC Special Report on Renewable Energy Sources and Climate Change Mitigation.* Cambridge, UK and New York, USA: Cambridge University Press. Figure No. 5.9.
23. Corke, T., & Nelson, R. (2018). *Wind Energy Design* (1st ed.). Boca Raton, Florida: CRC Press.
24. Herold, A. (2003, July). Comparison of CO_2 Emission Factors for Fuels Used in Greenhouse Gas Inventories and Consequences for Monitoring and Reporting Under the EC Emissions Trading Scheme. European Topic Centre on Air and Climate Change Technical Paper 2003/10. https://acm.eionet.europa.eu/docs/ETCACC_TechnPaper_2003_10_CO2_EF_fuels.pdf (Accessed on March 2019).

3 Electricity Production Hybrid Power Plants

3.1 THE CONCEPT OF THE HYBRID POWER PLANT

The hybrid power plant concept for electricity or thermal energy production is presented in Section 1.6.2 of Chapter 1. This section reviews this concept, for integrity reasons. The ultimate scope of a hybrid power plant is to maximize the renewable energy source (RES) penetration in an energy system. To this end, a hybrid power plant aims to guarantee power production from primary energy sources with stochastic availability, such as wind power or solar radiation, namely from nonguaranteed power plants. This target is not approached with the operation philosophy of a conventional electrical system. More specifically, as explained in Chapter 2, nonguaranteed power plants in conventional electrical systems (e.g., a wind park or a PV station) penetrate up to a maximum percentage versus the total power demand, implied by the system's dynamic security and stability requirements and configured by specific parameters, such as the system size, its interconnection with other grids, the size and the type of the maintained spinning reserve, and current weather conditions. Given this restriction, regardless of the instantly achieved RES penetration percentage, the nonguaranteed power plants always have a complementary role in conventional electrical systems, aiming at:

- the maximization of the exploitation of the locally available RES potential, leading to a corresponding reduction of the consumption of usually imported exhaustible energy resources,
- the reduction of greenhouse gas emissions due to fossil fuel combustion, and
- the reduction of electricity production costs, given that, in most cases, the selling price of the electricity produced by an RES power plant is lower than the production-specific cost of thermal generators.

RES power plants in conventional electrical systems do not undertake the role of guaranteed production units. On the contrary, the RES penetration simply aims to limit thermal generator operation, approaching to the maximum possible extent the above listed benefits. The requirement for uninterrupted alternative current production with specific, predefined nominal frequency and RMS value is ultimately based on the guaranteed power production from thermal generators. In conventional systems, thermal generators usually contribute the maximum percentage of the annual electricity production share, and essential tasks regarding the system's secure operation, such as the regulation of the grid's fundamental frequency and the system's reaction in case of a dynamic security contingency, are always taken over by them.

A hybrid power plant aims to swap the roles of guaranteed and nonguaranteed power production units. The ultimate scope of a hybrid power plant, which, as mentioned previously, is the maximization of RES penetration in an energy system, is not exhausted by techniques and measures that may lead only to a slight increase of the RES instant penetration percentage. In this case, thermal generators would still be the system's base units. Practically and theoretically, the operation philosophy of a hybrid power plant is totally different. To present it in a more perceivable way, let us assume a power plant with no thermal generators at all; consequently, there is no possibility for guaranteed power production. Under these conditions, a hybrid power plant should have the appropriate synthesis and operation algorithm in order to be able, by exploiting nonguaranteed units, to successfully meet the basic requirements of the electrical system, namely uninterrupted and secure power production with the minimum possible cost.

A reasonable question is how the nonguaranteed power production, imposed by the RES potential stochastic availability, could adequately face the electricity consumers' needs, as depicted on the power demand fluctuation versus time. Practically, what should be accomplished is that these two independent time-series, namely the available production from the nonguaranteed units and the power demand, must be adapted to each other. Taking for granted that the consumers' needs cannot change, the only way to approach the coincidence of these time-series is to adapt the stochastic nonguaranteed power production to the inelastic power demand. This can be achieved by combining RES power plants with storage units. The fundamental philosophy of the combined operation of RES and storage power plants is based on the following:

- Energy can be added to the storage units every time the available power production from the nonguaranteed power plants is higher than the power demand.
- The stored energy can be provided back to the consumers during time periods when the power demand is higher than the available power production from the nonguaranteed power plants.

This simplified approach of a hybrid power plant's operation is depicted in Figure 1.17 of Chapter 1.

Based on the above points, a hybrid power plant consists of at least two distinct components: nonguaranteed power production units (RES power plants) and storage units. Are these two components sufficient to ensure the uninterrupted and secure power production from the hybrid power plant? What will happen if, at a specific moment, the available RES power production and the stored energy in total are not sufficient to satisfy the current power demand? Such an eventuality is not improbable at all, because it depends on the power demand size and the hybrid power plant's dimensioning. In any case, to ensure the uninterrupted and secure power production from a hybrid power plant, concern should be taken to adequately face the above possibility, namely the insufficient power production from the RES units and the concurrent low charging level of the storage units. In this case, the only way to maintain uninterrupted power production is the availability of one or more guaranteed production units (e.g., thermal generators in the case of electrical systems).

Now we reveal the difference between conventional energy systems and hybrid power plants. In a hybrid power plant, power production is based, as a priority, on RES units. The thermal generators are employed only in the case of low power production from the RES units and low charging levels of the storage unit. Practically, the guaranteed units of the hybrid power plant are dispatched only after any possibility for power production from the RES or the storage units has been investigated and exhausted. The appropriate dimensioning of a hybrid power plant, accomplished on the basis of RES penetration criteria, will sensibly minimize the guaranteed units' production.

Conclusively, we may say that a hybrid power plant consists of three main and discrete components:

- The base units, which are nonguaranteed power production units, practically RES power plants. Base units constitute the principal production units of the hybrid plant. In electrical systems, these can be wind parks or photovoltaic (PV) stations. In thermal energy production systems, the base units' role is undertaken by a specific solar collector type, depending on the particular application.
- The storage power plant, aiming at the adaptation of the base units stochastic power production to the power demand. In electrical systems, the storage plant can be a pumped hydro

storage (PHS) system, electrochemical batteries, hydro production units, or compressed-air energy storage (CAES) systems. In thermal energy systems, the storage plant can be water storage tanks, concrete-solid materials, or pressurized water tanks.

- Backup units are dispatched exclusively when the production capacity from the base units and the storage plant is insufficient to cover the current power demand. In electrical systems, conventional thermal generators, most commonly diesel generators, are employed as backup units. In thermal energy systems, conventional oil or biomass heaters may be used.

This chapter examines small- and large-size hybrid power plants for electricity production, namely for decentralized and centralized production. The chapter can be summarized as follows:

- Thorough presentation of the available technologies employed as RES and storage units for electricity production hybrid power plants.
- Analysis of the fundamental layouts of small- and large-size hybrid power plants with their corresponding alternative operation algorithms.
- Based on the alternative operation algorithms, fundamental dimensioning methodologies will be presented, versus the employed RES and storage technologies. Optimized dimensioning of hybrid power plants can be executed based on energy criteria (maximization of the RES penetration) or economic criteria (optimization of the required investment's economic indices).
- Ultimately, characteristic examples of hybrid power plants' dimensioning and operation simulations will be presented as case studies.

3.2 CLASSIFICATION OF ELECTRICITY PRODUCTION HYBRID POWER PLANTS

Before proceeding to the main subjects' presentation of this chapter, it is necessary to introduce the size classifications for hybrid power plants. For different hybrid power plant sizes, the appropriate technologies for the base units and, more importantly, for the storage units, change, leading to different layouts and dimensioning procedures. Hence, hybrid power plants will be distinguished as small or large size, and a transition area among them will also be defined.

We consider hybrid power plants to be large-size if they aim to cover power demands higher than 5 MW. In the so-called developed world, such electrical systems are generally in small remote settlements or autonomous islands with populations in the range of 2,000–2,500. Consequently, large-size hybrid power plants aim to cover centrally the power demand configured by a number of distributed consumers (e.g., all the consumers of an autonomous insular grid).

Hybrid power plants focusing on the cover of power demands lower than 1 MW are considered small-size. Such electricity consumptions are met in remote settlements or islands with populations of less than 500 people. Depending on the overall topology of the electrical system, the power production from a small-size hybrid power plant can be considered either centralized or decentralized.

The potentially available RES and storage technologies for the formulation of a hybrid power plant are very specific and strongly depend on the plant's size. For example, a number of electrochemical battery strings, which seems to be the most sensible selection for a small hybrid power plant, will most probably be insufficient, from a technical and economic point of view, for a large hybrid power plant. On the other hand, a PHS, which is the identical solution for large-size projects, will not be cost-effective for small hybrid plants. Between these two main categories, there is a transition area, with power demand from 1–5 MW. For these transition-size hybrid power plants, the optimum base and storage unit technologies are not certainly defined beforehand. On the contrary, the choice of technologies depends on a series of parameters. For example, the availability of appropriate land morphology may favor the construction of a PHS, by keeping its setup cost low, even in cases of power demand lower than 3 MW. On the other hand, in cases of available cheap fossil fuel (e.g., US, Australia, Middle East, Africa), perhaps the selection of a CAES system could be more attractive, due to the fact that, as it will be explained in the next section of this chapter, the operation of this storage technology requires guaranteed thermal energy availability for heating the compressed air before its expansion. Additionally, an important parameter regarding the selection of the most feasible storage technology is also the daily operation schedule of the hybrid power plant. Generally, as the size and the daily operation time period of the hybrid power plant increase, the requirements for storage capacity increase too, and the use of electrochemical batteries tends to be less competitive.

Beyond the above listed categories, we could also define one more, for hybrid power plants of very small size, with nominal power production less than 100 kW. Electricity consumptions of such size are met in remote and non-interconnected consumers, such as farming units, desalination plants, biological treatment plants and remote settlements with populations fewer than 100 people. In such cases, the RES and storage unit technologies that can be potentially employed are definite, similar to the hybrid power plants of large size.

In the next sections, we present the most feasible and techno-economically mature and competitive technologies for RES and storage units that can be potentially employed based on the sizes and the operation algorithms of hybrid power plants.

3.3 TECHNOLOGIES FOR LARGE-SIZE HYBRID POWER PLANTS

Large hybrid power plants should be able to support energy production and storage of large size. The employed RES units must have low procurement and installation costs, low operation and maintenance costs, and high capacity factors. The capacity factor c_f of a power plant, calculated over a specific time period t, is defined as the ratio of the real energy production E from the power plant during this time period t versus the theoretically maximum energy that can be potentially produced by the specific power plant during the same time period t. This maximum theoretically produced energy equals the total energy produced under constant, nominal power production P_n from the power plant during the whole time period t:

$$c_f = \frac{E}{P_n \cdot t} \tag{3.1}$$

The capacity factors of wind parks and PV stations strongly depend on the available RES potential at their installation sites. Usually wind parks are installed at sites with annual capacity factors higher than 25%, while often sites have annual capacity factors higher even than 50%. Similarly, the annual capacity factor of a hydro power plant can exceed 80%, depending on the fluctuation of the available water flow rate during the year. The annual capacity factor of an electricity power plant supplied by a high-enthalpy geothermal field can reach values as high as 98%, configured mainly by the scheduled maintenance tasks of the power plant. Finally, typical capacity factors of PV stations installed in areas with annual incident solar irradiation at the range of 1,700 kWh/m^2 can be higher than 17%.

Additionally, the storage power plant must exhibit high storage capacity, capable of guaranteeing a long autonomous operation period for the hybrid power plant. The autonomy operation period of the hybrid power plant is defined as the maximum possible time period of power demand coverage exclusively by the storage plant, starting from fully charged level and without recharging it during this specific period. At the same time, given that the power production is supposed to be mainly based on the hybrid power plant, the storage units must have considerable flexibility regarding their charging and discharging process. The charging/discharging processes must exhibit high response rates and, depending on the operation algorithm of the hybrid power plant, they may also have to be executed simultaneously. Finally, the storage power plant must be able to adequately regulate the grid's fundamental frequency and to support the system's dynamic security.

Given the required specifications, the available technical solutions regarding the base and storage units of large-size hybrid power plants are presented in the next subsections.

3.3.1 Base Units

The most technically mature electricity production technologies from RES are wind parks and PV stations. These technologies can be theoretically used as base units in a hybrid power plant. After considerable reductions in setup costs of PV stations recorded after 2010, both technologies exhibit a final turnkey specific cost in the range of 1,000–1,200 $/kW. New emerging PV technologies may further reduce costs for PV station setups.

Regarding operation costs, generally wind turbines exhibit higher maintenance costs compared to PV stations. On the other hand, PV stations require much more land per unit of installed power; hence, PV stations exhibit higher land procurement or rent costs.

Finally, wind parks are typically installed in areas with annual capacity factors higher than 25%. The average monthly capacity factors of US wind parks are presented in Figure 3.1 [1]. The average annual values are calculated at 32.6% and 34.0% for the years 2013 and 2014, respectively. In central continental Europe, typical annual wind park capacity factors vary between 25% and 30%. In the western European coast, annual capacity factors higher than 40% are often met, while, in some specific windy territories (e.g., Aegean Sea islands), annual capacity factors higher than 50% are often calculated, based on captured wind potential measurements.

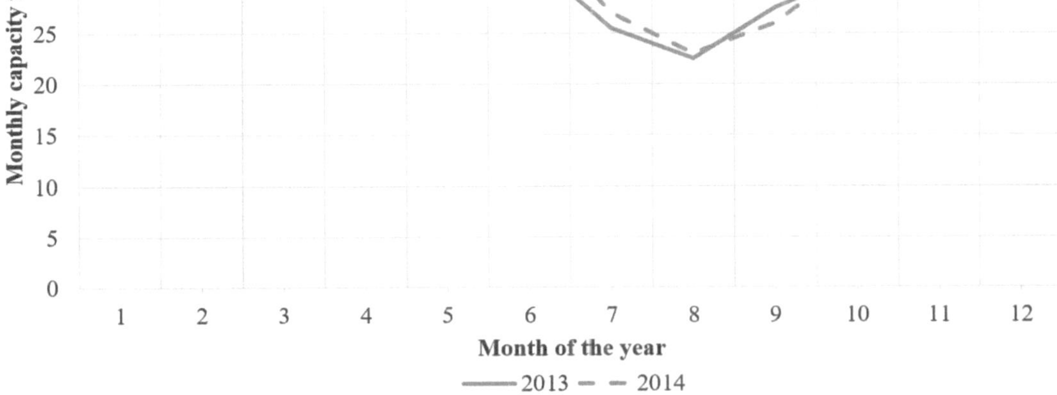

FIGURE 3.1 Variation of wind park's monthly average capacity factors in US [1].

On the other hand, PV stations installed in dry and sunny climates (geographical latitudes between 25°N and 35°N) can exhibit annual capacity factors of 17%–20%. Such areas include the Mediterranean basin, the southwestern US, the Middle East, Australia, and so on. These capacity values can be a bit higher (up to 23%) when the PV panels are installed on sun route trace trackers. For sites with higher geographical latitudes for PV installation, the expected annual capacity factor decreases considerably. These relatively low annual capacity factors are due to the fact that PV stations practically remain inactive for half of the year, namely during nighttime hours.

PV stations' relatively low capacity factors indicate that in order to produce the same electricity as a wind park, the installed PV power must be 1.5–2 times higher than the corresponding required nominal power of the wind park, with a corresponding increase of setup and production cost. This negative feature actually outweighs the PV stations' advantage of low maintenance costs, making wind parks the better solution for base units of hybrid power plants, on a technical and economic basis.

The above presented technical and economic comparison parameters are depicted on the final electricity production-specific cost. A wind park with an annual final (after losses) capacity factor of 40% exhibits a final production-specific cost of 0.03–0.05 $/kWh. A PV station, with annual final capacity factor of 20% exhibits a final production-specific cost of 0.07–0.09 $/kWh. These specific costs are calculated by accounting for operation costs, maintenance costs, and the amortization of the power plant setup costs.

Another crucial advantage of wind parks versus PV stations, as far as their introduction as base units in hybrid power plants is concerned, is their more uniform production profile during the year. This feature has a direct impact on the dimensioning of the hybrid power plant's storage units. A PV station produces power only during the daytime. Consequently, the storage capacity must be higher, in order to cover the total lack of energy production throughout the night. This conclusion is much more intensified, particularly in northern geographical areas, where PV production is mainly provided during the summer period. This fact is clearly depicted in Figure 2.36 of Chapter 2. Indeed, a PV station installed at a location of 35°N geographical latitude, with annual incident solar irradiation of 1,700 kWh/m^2, produces approximately around 65% of its annual total production from April 15 to October 15. Consequently, if the ultimate target of a hybrid power plant supported exclusively by PV panels is the maximization of the RES penetration in an electrical system over the entire year, the storage plant should be oversized in order to ensure enough energy availability during the winter, most possibly stored during summer. This option is not ideal. From a technical point of view, there will be extensive energy storage losses due to the long-term storage. On the economic side, the formulation of a storage plant with disproportionally increasing capacity will lead to a corresponding unfeasible increase of the hybrid power plant's setup cost.

Summarizing the above, given that the base unit should:

- be able to produce large amounts of electricity on an annual basis with high capacity factors,
- exhibit low setup and operation cost, and
- favor the minimization of the required storage capacity of the storage units,

it is concluded that the most appropriate base unit technology for large-size hybrid power plants is wind parks.

The above comparison is valid for geographical areas with availability of both high wind and solar potential, where annual capacity factors of 40% and 20% for wind parks and PV stations, respectively, are met. Put another way, the above conclusions stand for sites with equivalent wind and solar potential. Obviously, for geographical locations with, for instance, wind potential much higher than solar potential, or vice versa, the above conclusions are not any more valid and, certainly, the optimum RES technology will be the one with the highest potential. For example, despite the above general conclusions, it should be obvious that there is no call for vast installations of wind parks in areas such as the Middle East or Arabian Peninsula, because these locations have abundant solar radiation almost for the whole annual period and winds of relatively medium intensity. In such cases, the use of PV stations seems to be the better choice, while wind park installation should be restricted only to the areas with the highest available wind potential.

In any case, investigating sites with remarkable wind or solar potential and essential favorable features that facilitate the RES plant installation has its benefits. Researching the ideal sites may reduce project setup costs (e.g., close to the existing electricity grids, easy accessibility, cheap land, mild geographical terrain), and lay the groundwork for the successful design and implementation of a hybrid power plant, capable of ensuring high RES penetration share in the electricity system.

3.3.2 Storage Units

For large-size hybrid power plants, the storage unit should have high storage capacity. Based on studies and the dimensioning of several hybrid power plants, it is empirically estimated that the storage unit of a large-size hybrid power plant should exhibit a storage capacity of 1%–3% of the annual electricity production of the hybrid plant's wind park, depending on the operation algorithm of the hybrid power plant, and for a wind park with an annual capacity factor of 30%–40%. The discharging power of the storage unit should be at least equal to the annual peak power demand, in case the hybrid power plant aims to fully undertake the power demand coverage. The charging power should be a little lower than the nominal power of the hybrid plant's wind park. For example, for a hybrid plant's wind park of 50 MW nominal power and annual capacity factor of 30%, the storage capacity of the hybrid plant's storage unit should be around 1,300 MWh. To get an idea what this storage capacity means, in case of a PHS system, the above storage capacity can be provided with an upper reservoir with effective capacity of 1,700,000 m^3 and a geostatic head of 300 m.

Even if, theoretically, alternative storage technologies are available, the needs of such high storage and charging/discharging capacities can be met only by some very specific storage plants. Essentially, the available storage technologies for the management of large energy amounts and the adequate combined operation with a wind park or a PV station within the frame of a hybrid power plant are the following:

- compressed-air energy storage (CAES) systems
- pumped storage systems (PSS) or pumped hydro storage (PHS) systems.

These technologies are presented in the following sections.

3.3.2.1 Compressed-Air Energy Storage Systems

With CAES systems, energy is stored in the form of elastic energy, through air compression. CAES systems are classified in two categories: conventional and adiabatic CAES (AA-CAES). Two large-size CAES systems have been installed and are in operation so far, one in Neuen Huntorf, Germany, and one in McIntosh, Alabama (US) [2]. In the near future, perhaps the first AA-CAES systems will be constructed [3].

3.3.2.1.1 Conventional Compressed-Air Energy Storage

Figure 3.2 presents the layout and the operation of a conventional CAES system. Specifically, the layout presented in Figure 3.2 shows the implemented CAES system in Neuen Huntorf, Germany [4].

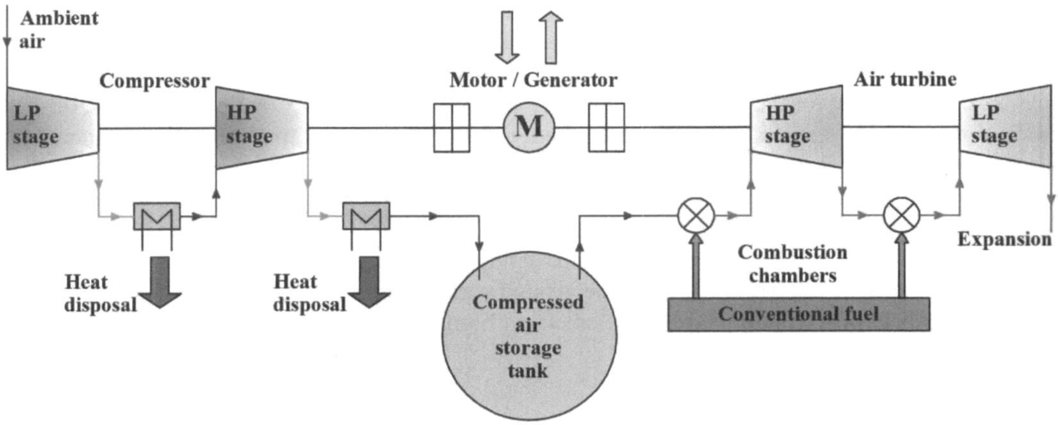

FIGURE 3.2 The layout of the conventional CAES in Neuen Huntorf, Germany [2].

The main operation concept of a conventional CAES is analyzed below:

- Power surplus available from the RES units of the hybrid power plant is provided to drive a two-stage compressor with intercooling of the compressed air between the two stages. The working medium is ambient air, which is eventually compressed in pressures between 40 and 70 bars. After the compressor's high-pressure stage, the compressed air is cooled again, in order to keep its temperature close to the ambient one. This aims to minimize the air density and, consequently, its volume, leading, eventually, to the final storage of the maximum possible air mass in the available storage tank.
- When power is required, compressed air with a specific flow rate, depending on the required power production, is released from the storage tank, heated in a combustion chamber to increase its specific enthalpy and, finally, led to a two-stage air turbine for power production, with intermediate heating between the turbine's two stages.

The other existing CAES system, installed in McIntosh, Alabama (US), uses a variation of the above presented conventional system, with the main improvement of the introduction of a heat recuperator, reducing fuel consumption for the air heating before its expansion in the air turbine's stages. Due to this alteration, presented in Figure 3.3, the compressed air, once released from the storage tank and before its expansion in the turbine's high-pressure stage, is preheated in the recuperator, by recovering heating from the expanded air after the turbine's low-pressure stage. The introduction of the recuperator has increased the efficiency of the system by 10%. On the other hand, a considerable drawback of this variation is the large size required for the recuperator and the corresponding increase in the system's setup cost [5].

The fundamental features of the two existing conventional CAES systems are presented in Table 3.1.

For the air compression, either axial compressors, achieving pressure ratios at the range of 20 with flow rates around 1.4 Mm^3/h, or radial compressors, with maximum flow rates up to 100,000 m^3/h and maximum compression pressure at 1,000 bar, may be used. With the available technology, the air compression is implemented in two stages with intercooling, in temperatures varying between 40°C and 200°C [6]. The compressed air is expanded in air turbines with pressure ratios up to 22 and maximum inlet air temperature at 1,230°C.

As already mentioned, air storage close to ambient temperatures enables a high density of stored air, thus reducing the storage reservoirs' required volume. Air storage in conventional large-size

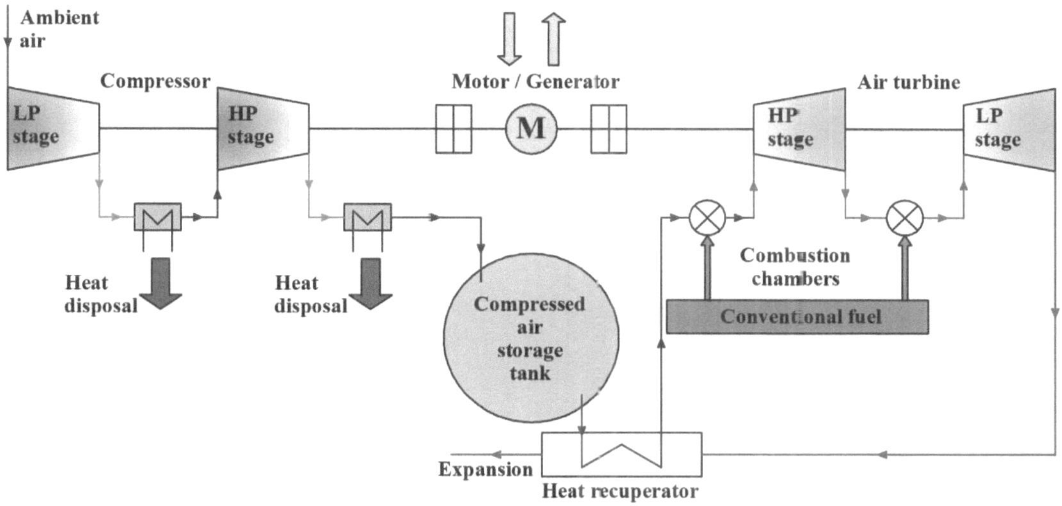

FIGURE 3.3 The operation of the existing conventional CAES system in McIntosh, Alabama (US), with heating recovery from the expanded air [2].

CAES systems is implemented in underground caverns with high-quality rocks regarding their pressure resistance and airtight integrity, the aquifer, depleted natural gas storage caves and salt domes with total storage volume between 300,000 and 600,000 m^3. Another feasible storage alternative is the use of underground pipelines of high pressure (20–100 bar).

3.3.2.1.2 Adiabatic Compressed-Air Energy Storage

In AA-CAES systems, instead of disposing the released heat from the air during compression in the ambient, the idea is to store it in a separate insulated heat tank. This fundamental change, presented in the AA-CAES basic layout in Figure 3.4, constitutes the main difference between the conventional and the adiabatic CAES systems. With the AA-CAES, the fuel consumption for the air preheating before its expansion is eliminated. This significant benefit forms the principal reason for the study and the development of AA-CAES.

TABLE 3.1
Fundamental Features of the Two Existing Conventional CAES Systems

Technical–Economic Features	Neuen Huntorf, Germany	McIntosh, Alabama, US
Power (MW)	321	110
Storage capacity (MWh)	1,160	2,640
Storage tank volume (m^3)	310,000 (2 caverns)	560,000
Maximum storage pressure (bar)	70	75
Turbines' total flow rate (kg/s)	416	154
Compressors' total flow rate (kg/s)	104	96
Setup cost ($ in year 2010)	167,000,000	65,000,000
Storage specific cost ($/kWh)	143,966	24,621
Setup specific cost ($/kW)	520.25	590.91

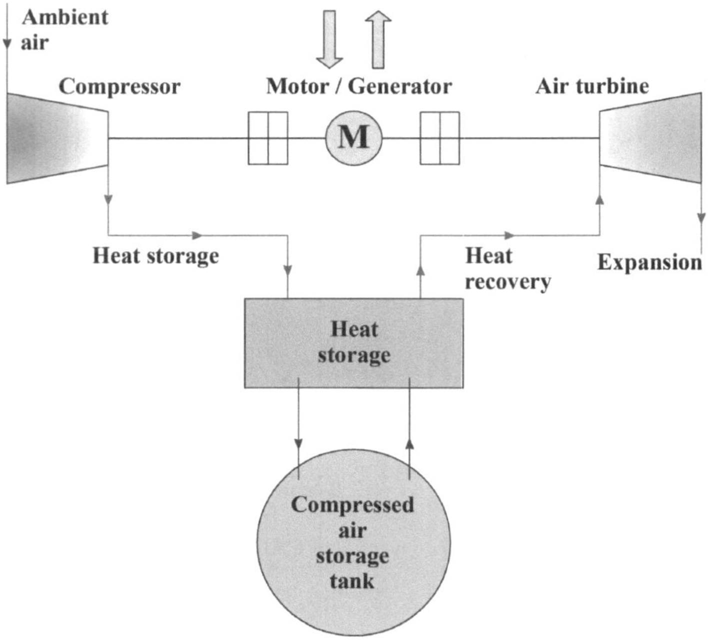

FIGURE 3.4 Layout of an AA-CAES [2].

The overall efficiency of the storage cycle may be up to 70% with AA-CAES [2, 3, 5], approaching, thus, the efficiency of a PHS system.

The basic operation layout of AA-CAES is analyzed below, classified in single-stage and two-stage CAES.

3.3.2.1.2.1 Single-Stage AA-CAES

- When power surplus is available from the hybrid power plant's RES unit, air is compressed in a single-stage compressor without cooling. The heat gained during compression is disposed in a separate heat storage tank, before the air is stored in the cool, compressed air storage tank.
- When power is required, the stored compressed air is heated up to 600°C before its expansion to the air turbine, by recovering the stored heat from the heat storage tank.

3.3.2.1.2.2 Two-Stage AA-CAES

- In two-stage AA-CAES, the heat gained during air compression is stored in two separate heat storage tanks.
- During the storage unit discharging, the heat stored in the separate tanks after the low and high compression's stage, is recovered by the compressed cool air before it is expanded in the turbine's low and high stage, respectively.

The main benefit achieved with two-stage AA-CAES is the increasing stored energy density (kWh of stored energy per m^3 of storage volume), compensating, in this way, for the more complicated construction of the whole storage plant (i.e., two heat storage tanks, air transfer pipelines).

The most important advantages of AA-CAES are:

- the elimination of the fuel consumption for the air preheating before its expansion in the turbine, and
- the air compression without cooling at any stage (intercooling or after the high-pressure stage), which enables the conservation of high air temperatures after compression and, consequently, large heat storage in the heat storage reservoir.

Yet, the development of AA-CAES will require new and more sophisticated designs with specific innovations regarding the system's basic components. In particular, the necessary upgrades on the AA-CAES components are summarized as follows:

- A special design for the required heat storage tanks with thermal storage capacity from 120 to 1,800 MWh_{th} will be required, in order to approach high heat transfer efficiencies during heat storage and recovery, minimizing, at the same time, all the potential heat losses from the tank to the ambient environment, for as long as the heat remains stored in the tank. The heat losses during the heat storage in and recovery from the heat tank must be also minimized [7–9].
- With conventional compressors, the high pressures and temperatures (100 bar/620°C for single-stage and 160 bar/450°C for two-stage systems) required for AA-CAES cannot be achieved, combined with the demand for compressors' quick response and high efficiency. Hence, the design of new compressor models is necessary, in order to approach the above operating specifications. A realistic approach is the construction of a compressor with three parts. The first will consist of an axial or radial compressor, as the low-stage unit, for high or low air flow rates, respectively. The second and the third part will consist of two radial compressors with common shaft, as the medium- and high-pressure stages, respectively.
- The turbine units must also be redesigned, in order to operate with high air inlet temperatures and varied air flow rates, without negatively affecting the overall system's efficiency. To approach the above standards, an innovative, nonconventional regulation stage should be designed, for improved management of the incoming air pressure and flow rates. The turbine's preheating seems also to contribute to the improvement of the turbine's response time [7–9].

3.3.2.2 Pumped Hydro Storage Systems

Electricity storage with PHS systems or pumped storage systems constitutes the most technologically mature and economically competitive technology for electrical systems of medium, large, and, under favorable conditions, even small size. Tens of PHS systems, already installed and operating worldwide under entirely different conditions, cover a wide power range from 5 MW to 2 GW and provide huge practical experience regarding their technical specifications and operation details. The basic layout and operation concept of a PHS system is presented in Figure 3.5.

A PHS system consists of two water reservoirs, constructed in neighboring locations, with adequate height difference (geostatic head) between them, usually at the range of some hundreds of meters. The water reservoirs' storage capacities can be from some hundreds of thousands to some millions of cubic meters. The contained water in these reservoirs can be transferred between them either through a single pipeline, used for both water pumping and falling, or through a double penstock, consisting of two separate pipelines, one used exclusively for water pumping and another used exclusively for water falling. The construction of a single or a double penstock for the connection of the PHS system's reservoirs depends on the operation algorithm of the hybrid power plant,

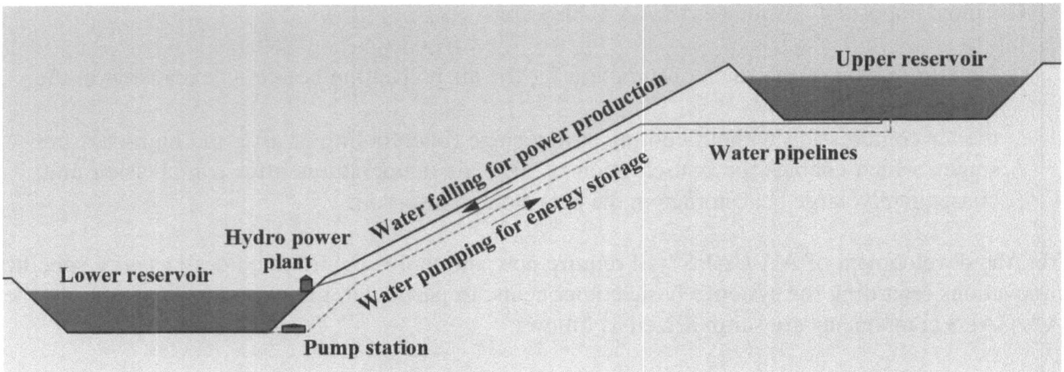

FIGURE 3.5 Basic layout and operation concept of a PHS system.

which determines the potential requirement for simultaneous water pumping and falling, namely, the need for simultaneous charging and discharging of the storage plant. The lower edges of the pipelines end to a pump station and a hydro power plant. When, during the hybrid power plant's operation, there is power surplus from the RES unit, this power is used to drive the PHS system's pumps, through which water is pumped and stored from the lower to the upper reservoir. In this way, the available energy surplus is stored in the upper reservoir in the form of gravitational energy. On the other hand, when power production is required, the water stored in the PHS system's upper reservoir is released, passes through the pipelines, and reaches the hydro turbines in the hydro power plant, providing, in this way, the requested power production.

3.3.2.2.1 Pumped Hydro Storage Systems for Power Peak Shaving

Tens of PHS systems have been constructed so far worldwide, introduced in conventional electrical systems, aiming for the so-called power peak shaving. Power peak shaving is achieved by storing electricity produced by the base thermal generators (steam turbines, diesel generators, combined cycles) with low specific cost during power demand low periods (in most cases the early morning hours), in order to return it back to the grid during power demand peak periods (at noon and in the afternoon, depending on the season), avoiding, thus, the power production from peak thermal power generators (e.g., gas turbines) with increased production-specific cost. This operation concept is presented in Figure 3.6.

Power peak shaving with the support of PHS systems is usually combined with nuclear power plants (NPP) of large size (nominal power at the range of GW), where the reduction of the total power production during low power demand periods is not possible, due to the operation of large

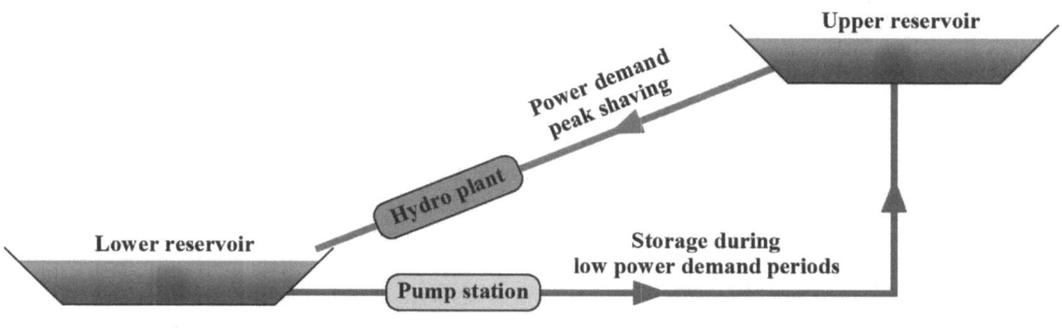

FIGURE 3.6 Power peak shaving with the support of a PHS system.

steam turbines with low response. For power peak shaving, the PHS systems are constructed with a single penstock, because simultaneous water pumping and falling, in other words simultaneous power storage and production, does not make sense.

The results of power peak shaving through the support of PHS systems on the daily power production synthesis on the island of Crete, Greece, is presented in Figure 3.7, based on a computational simulation of the electrical system's annual operation. In the upper power production synthesis graph (Figure 3.7a), the electrical system operates without the support of any PHS system (real state). The power production synthesis is formulated following the basic principles presented in Chapter 2, regarding the thermal generators dispatch order and the demand for spinning reverse.

In the lower power production synthesis graph (Figure 3.7b), it is assumed that a PHS system has been introduced in the insular autonomous grid, aiming at the power peak shaving on a 24-hour cycle. This means that on a daily basis energy is stored during the early morning, off-peak hours (e.g., from midnight until dawn), to be returned back to the grid during power demand peak periods (at noon and in the evening). Both graphs refer to the same 24-hour period with common power demand. Energy storing is performed by increasing the power production from either steam turbines or diesel generators, namely the base thermal generators with low production-specific cost. The stored energy is provided back to the grid during the day and night power demand periods, minimizing thus the gas turbine production, which is restricted only during the night power demand,

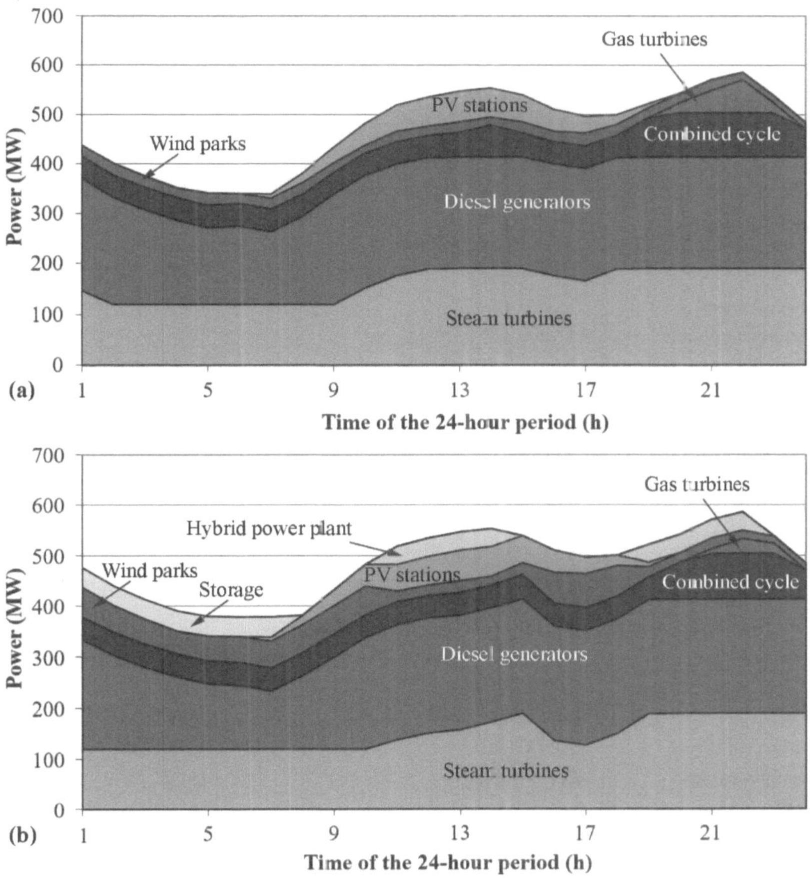

FIGURE 3.7 Graphs depicting power peak shaving in power production synthesis (**a**) without the support of a PHS system, and (**b**) with a PHS system in place.

due to the depletion of the stored energy. Another fact that leads to the gas turbines dispatch during night power demand peak is the lack of production from the available PV stations after sunset. During the day power demand peak, the introduction of a PHS system substitutes power production from the combined cycle.

3.3.2.2.2 Pumped Hydro Storage Systems and Renewable Energy Source Power Plants

The combined operation of PHS systems and RES power plants, practically wind parks and, more rarely PV stations, despite being one of the most popular subjects in the relevant academic literature for the past two decades, has so far been implemented only twice in practice. The first combined operation of a wind park and a PHS system was installed in 2016 on the Spanish island of El Hierro in the Canary Archipelago (eastern Atlantic Ocean), and the installation of the second one was completed in 2019 on the Greek island of Ikaria, located in the eastern Aegean Sea. In both projects the wind park–PHS hybrid power plant aims at the maximization of the RES (wind energy) penetration to the annual electricity production share. Both El Hierro and Ikaria are non-interconnected islands. The basic technical and economic features of these two hybrid power plants are presented in Table 3.2.

The ultimate target of an RES–PHS system is the maximization of the annual RES penetration in the electricity production annual share. This target implies fundamental modifications on the design and the operation algorithm of the PHS system, with regard to the power peak shaving concept. The major innovation of a PHS system combined with an RES unit within the frame of a hybrid power plant is the installation of a double penstock, through which simultaneous water falling and pumping is possible [10]. The necessity for the installation of a double penstock is

TABLE 3.2
Basic Technical and Economic Features of the Wind Park–PHS Hybrid Power Plants in El Hierro and Ikaria

Technical/Economic Features	El Hierro, Spain	Ikaria, Greece
Annual peak power demand (MW)	13.3	7.8
Annual electricity consumption (MWh)	41,000	27,600
Wind park (wind turbines number/nominal power)	5 × 2.3 MW = 11.5 MW	4 × 600 kW = 2.4 MW
Upper reservoir capacity (m^3)	380,000	900,000 (1st reservoir) 80,000 (2nd reservoir)
Lower reservoir capacity (m^3)	150,000	80,000
Total head (m)	655	724 (1st reservoir) 555 (2nd reservoir)
Storage capacity (MWh)	580	1,500
Pump station (units/nominal power)	2 × 1,500 kW + 6 × 500 kW = 6 MW	8 × 250 kW = 2 MW
Hydro power plant (units/nominal power)	4 Pelton × 2,830 kW = 11.32 MW	2 × 1,550 kW + 1,050 kW = 4.15 MW
Total setup cost (€)	64,700,000	26,000,000
PHS system's setup cost (€)	50,000,000	23,000,000
Setup specific cost per storage capacity (€/kWh)	86.21	15.33
PHS system's overall storage cycle efficiency (%)	65	69
Hybrid power plant annual penetration (%)	50	50

imposed by the maximum direct penetration percentage versus the power demand, applied in the RES unit, due to security and stability requirements. The installation of a double penstock enables the storage of the available wind power that cannot directly penetrate the electricity system, while, at the same time, water downfall flow is possible through the separate water falling pipeline, in order to cover the remaining power demand, after the wind power direct penetration. The installation of a double penstock also improves the flexibility of the hybrid power plant and its ability to react adequately after the grid's events and contingencies. For example, in case of a sudden power production drop from the wind park, the downfall pipeline enables direct power availability from the hydro turbines in order to undertake the appeared power production loss, regardless of any probable concurrent water pumping at the moment of the contingency's occurrence. An alternative reaction against the power production loss could be the decoupling of any probable pumps load from the grid.

The same flexibility can be also theoretically approached with the installation of the so-called hydraulic short-circuit, namely a hydraulic junction between the pump station and the hydro power plant, which enables the water direct transfer from the pumps to the hydro turbines (Figure 3.8). This technique provides the hybrid power plant with the same flexibility regarding the units' response against any abrupt power demand fluctuations, without the necessity for a double penstock installation. Practically, the hydro turbines, instead of being supplied with water flow from the upper reservoir, are directly supplied by the pumps through the hydraulic short-circuit. In this way, the stability of the system is no longer based on the stored water in the upper reservoir, but on the electronic control of the turbines and pump operation, as well as their features as electrical machines (generators and motors). It is conceivable that with this approach, the installation of a double penstock is no longer necessary, while the upper reservoir required storage capacity may also be reduced. Consequently, the hydraulic short-circuit contributes to a corresponding reduction of the PHS system's setup cost and to the increase of the storage cycle overall efficiency, due to the reduction of the water flow losses inside the pipelines. On the other hand, the stored energy in the upper reservoir reduces considerably too, affecting, thus, the PHS system's operation autonomy, namely its ability to take over completely the power production in the electrical system for a long time period, without any support from the RES unit (either direct penetration or storage). Additionally, in cases of small electrical systems, where the introduction of a hybrid power plant can focus on 100% RES penetration, the operation of a PHS system equipped with a double penstock and with the adequate storage capacity of the upper reservoir, is traditionally considered more secure than supporting the power plant's security and stability exclusively on a hydraulic short-circuit implementation.

The operation concept of a PHS system and a wind park, focusing on the maximization of the RES penetration, is given in Figure 3.9 [10].

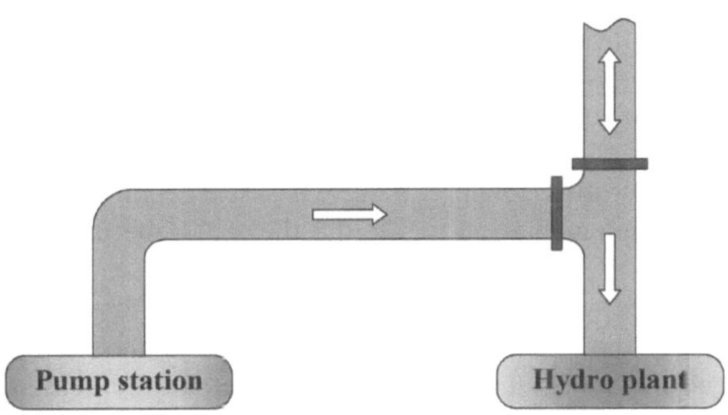

FIGURE 3.8 The fundamental design of the hydraulic short-circuit.

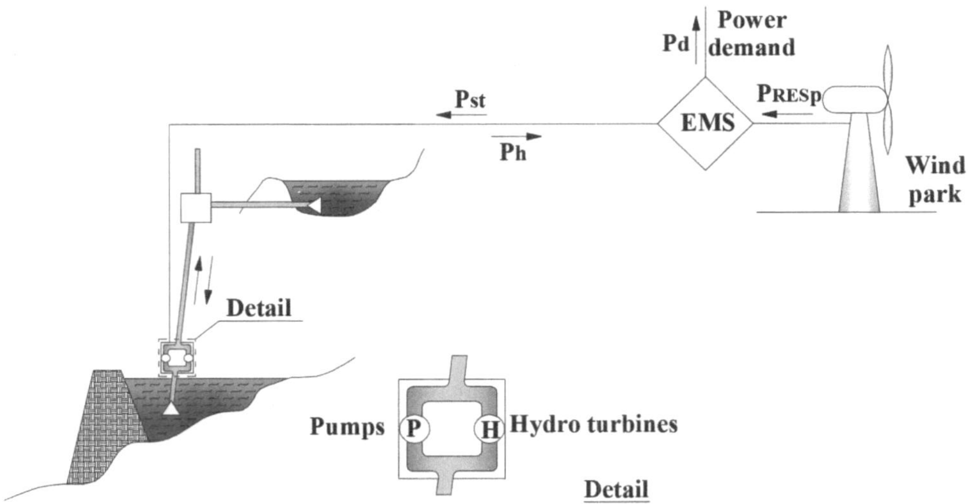

FIGURE 3.9 The operation concept and the main layout of a wind park–PHS hybrid power plant focusing on the maximization of the RES penetration [10].

Referring to Figure 3.9, it is assumed that at a specific time point the power demand in an electrical system is P_d and the wind power availability is P_w. The wind power direct penetration to the grid P_{wp} is not allowed to exceed a maximum percentage versus the current power demand $P_{wp} = a \cdot P_d$ $(0 < a < 1)$, in order to ensure the system's dynamic security. In case the available wind power exceeds the maximum possible direct penetration, the pump units of the PHS system are dispatched in order to absorb the excess wind power production and store the corresponding energy.

Two cases are distinguished:

1. If the stored water volume in the PHS system's upper reservoir is not enough, the remaining power demand, not covered by the wind power direct penetration, is provided by the hybrid power plant's backup units, whose power production P_{th} equals $P_{th} = P_d - P_{wp}$. The hydro turbines' power production is null: $P_h = 0$. The PHS system's pumps are provided with the wind park's power surplus $P_p = P_w - P_{wp}$, to be stored by pumping water from the lower to the upper reservoir.
2. If there is enough water stored volume in the PHS system's upper reservoir, the hydro turbines are dispatched to cover the remaining power demand: $P_h = P_d - P_{wp}$. At the same time, the possible wind power surplus $P_w - P_{wp}$ is again stored in the PHS system's upper reservoir, on the condition that this is not full. In the event of a fully charged upper reservoir, the excess wind power cannot be stored. This power surplus should be either rejected, by reducing the wind turbine's set point, or exploited in other alternative consumptions (e.g., complementary hydrogen production with electrolysis). The power production from the backup units will be null: $P_{th} = 0$.

From the above described operation concept, it is clear that RES units have a base production role in the hybrid power plant implementation, while the thermal generators are kept only as backup units.

A special category of PHS systems, of particular importance for geographical areas with low rainfalls, is seawater PHS systems. In these systems, the working medium is the seawater, pumped possibly directly from the sea, which, in this case, has the role of the PHS system's lower reservoir. The operation of a PHS system with seawater provides an invaluable alternative approach for geographical regions with low rainfalls, because it guarantees the availability of the working medium without affecting any possible limited water resources. Such cases can be met in areas close to the equator, in dry climates, in isolated insular territories, etc. Only one seawater PHS system has been installed operating for power peak shaving; it is on the island of Okinawa, Japan. This plant was constructed in 1999 and constitutes a valuable source of knowledge and experience for the development of similar projects [11–13].

It is self-evident that seawater PHS systems must be installed closed to the coastline [14]. The terrain morphology close to the coastline constitutes a crucial parameter for the project's technical feasibility and total setup cost. Small hills with absolute altitudes from 200 m to 600 m are considered ideal for the installation of such systems. The hills' mild slopes from the top towards the coastline, as well as the mild land morphology of both the slopes and the coastline (absence of abrupt cliffs, canyons, streams, etc.) contribute to the minimization of any required earth works for the installation of the penstock and the hydrodynamic machine stations. The final consequence of all the above is the reduction of the PHS system's setup cost. Abrupt terrain slopes, particularly, may impose the necessity for tunnel construction for the penstock installation. Such kind of works significantly increase the total setup cost and can, under specific conditions, affect considerably the economic feasibility for the construction of the project, especially in cases of small plants, such as the ones presented previously for the islands of El Hierro and Ikaria. In such small projects, the minimization of setup costs, with the terrain morphology being a principal parameter, constitutes a crucial prerequisite for the hybrid power plant's economic feasibility, due to the relatively low annual electricity consumption of the autonomous insular systems.

3.4 TECHNOLOGIES FOR SMALL-SIZE HYBRID POWER PLANTS

Hybrid power plants of small size aim at the electricity production starting from small annual contributions to 100% annual RES penetration in electricity systems with annual peak power demand up to 1 MW. The introduction of PHS or CAES systems as storage plants is not feasible in such small systems, as will be shown in the next sections, because their high setup cost cannot be justified by the low annual electricity consumption. The selection of RES units is also affected by the different storage unit's technologies as well.

3.4.1 Renewable Energy Source Units

The same parameters presented in the previous section regarding the selection of the appropriate RES technologies for the integration of large-size hybrid power plants are also applied for small-size hybrid systems. Wind parks, generally, exhibit higher capacity factors than PV stations, hence their electricity production is maximized, while the corresponding specific-production cost is minimized. Consequently, wind parks feature again as the most feasible technology, compared to PV stations, with regard to the formulation of hybrid power plants of small size.

Notwithstanding, for hybrid power plants of small size, wind park priority does not impose exclusivity, too, just as for large-size hybrid power plants. This is not because of the technical or economic specifications of the RES technologies themselves, but rather due to the supporting storage technologies. Specifically, in a large-size hybrid power plant, a properly dimensioned and sited PHS system with adequate storage capacity can guarantee on its own the electricity supply of the power demand for more than 1 week, starting from full charging state and without any intermediate storage from the RES units. In the short term, it can provide long autonomy operation periods, as defined in the previous section. Consequently, it is understood that in cases of large hybrid power plants, the storage technologies provide the possibility for high storage capacity, capable of supporting the secure operation of the plant for several days, by storing large energy amounts from the RES units whenever they are available. This feature is extremely important for the guaranteed electricity production from the hybrid power plant and the electricity security supply during periods of low RES potential availability.

For small hybrid power plants, the possibility for high storage capacity is usually not available, especially when electrochemical batteries are used as storage units. In particular, electrochemical batteries, several types of which will be presented in the following sections, exhibit specific technical features that do not permit the installation of storage systems with high storage capacity, such as:

- They exhibit high procurement cost and short service lives. For example, lead acid batteries are currently the most economically competitive electrochemical storage technology. Their procurement specific cost ranges from 50–150 $/kWh, much higher than the corresponding feature of a PHS system. Under ideal operation conditions (ambient temperature lower than 26°C, maximum discharge depth at 60%), lead acid batteries can operate satisfactorily, without significant reduction of their storage capacity, for 6–7 years. Consequently, during the lifetime period of a hybrid power plant, which for due diligence analyses is typically considered to be 20 years, the lead acid batteries must be replaced two or three times. This means that their overall procurement cost can exceed 400 $/kWh for a 20-year lifetime period of the hybrid power plant.
- Most electrochemical battery technologies exhibit a maximum possible discharge depth, which practically restricts analogously their effective storage capacity.
- Electrochemical batteries are characterized with limited flexibility. For example, it is possible to simultaneously charge and discharge a PHS or CAES system. On the contrary, simultaneous charging and discharging is not possible with a single electrochemical battery unit. To enable this possibility with electrochemical storage technologies, separate and independent battery groups must be developed, so-called battery strings. This feature, as it may be perceived, increases the required number of battery units and, consequently, the total battery procurement and replacement costs.
- Some specific technologically improved types of electrochemical batteries (e.g., redox floating batteries) exhibit high discharge depth (higher than 80%) and service lives up to 20 years, yet they still exhibit very high procurement cost, which may reach 1,000 $/kWh.
- A highly promising technology is lithium-ion batteries. They are characterized with high discharge depth (up to 80%), a service life of 10 years, and 100% efficiency. Yet their procurement cost still remains as high as 400–500 $/kWh, although rapidly decreasing. It is estimated that this cost will be around 100 $/kWh, or even lower, around 2030, boosting a technological revolution in small-scale storage technologies, including small-size hybrid power plants.

For all the above reasons, at least for the time being, the installation of storage plants with high storage capacity is not feasible for hybrid power plants of small size. The installed storage capacity is restricted by technical and economic constraints, leading to limited ability for long-term power demand coverage without any intermediate energy storage. Practically, the autonomy operation period usually achieved with electrochemical batteries in typical syntheses of small-size hybrid power plants can be from 1–2 days.

As a direct consequence of the above, in order to approach high RES penetration in small-size electrical systems, given the low storage capacity of the storage plant, the power production from the RES base units should be as consistent as possible, with the shortest possible time intervals of null production. This can be practically approached in two alternative ways, substantiated below.

The first way is to install the RES plant (usually the wind park) with very large nominal power, approximately up to twice higher than the annual peak power demand of the electrical system. Preferably, the total installed power should be distributed in more than one location, in order to maximize the probability for continuous RES potential availability. This approach may prolong the time periods with power production adequate to cover the power demand, with the support of short-term storage. Obviously, the continuous availability of even low RES potential constitutes a fundamental prerequisite.

Yet, as it is statistically almost certain, in case of installation of only one RES technology (e.g., wind park), some time periods during a year will experience inadequate availability of the corresponding RES potential (e.g., wind potential). During these periods, regardless of the installed RES plant's power, electricity production will not be possible from the RES base units. In case of a wind park, low RES potential actually implies periods with low wind. These periods, if met during summer, favor the increase of ambient temperatures and the appearance of heat waves, during which the annual peak power demands are usually recorded, due to the simultaneous and high power consumption from cooling active systems. The low, or even null, power production from the wind park and the low storage capacity of the storage plant result in an inability of the hybrid power plant to cover the power demand for more than 1 or 2 days, namely for more than the autonomy operation period offered by the electrochemical storage plant.

The more effective solution for this case is the installation of an alternative RES technology, which will operate complementarily to the basic one. A principal prerequisite is the availability of remarkable potential for this alternative RES plant during these time periods of inadequate potential of the primary RES technology. In the precedent example, the alternative RES plant, appropriate for power production during low wind potential periods, could by all means be a PV station. PV panels can successfully cover the power demand, within the operation frame of a hybrid power plant, during periods of null power production from the wind park in summer, when the solar radiation availability is, as a matter of fact, guaranteed and high for most regions in the world. Besides, the PV technology, as mentioned in the previous section, practically constitutes the only alternative RES technology (apart from wind parks), technically mature and economically competitive for hybrid plant implementations. The experience from the simulation and the dimensioning of small-size hybrid power plants proves that, usually, the installation of a PV station with a nominal power in the range of 80% of the electrical system's annual peak power demand is enough to guarantee annual RES penetration percentage higher than 90%, of course with the proper sizing of the wind park and the storage plant too, given an annual incident solar irradiation at 1,700 kWh/m^2.

The installation of a secondary RES technology, actually the PV station, is practically preferred to the alternative choice of an increasing storage unit's capacity, mainly due to economic reasons. Namely, the same result regarding the annual electricity production share from the hybrid plant can be achieved either with the installation of the secondary RES plant or the significant increase of the storage capacity, which, in the case of electrochemical batteries, implies a disproportional increase in the hybrid plant's setup cost.

Given the data presented above, we may come to the following conclusions on the available options for RES units:

- Wind parks still represent the principal RES unit option for small hybrid plants. The nominal power of the wind park is usually calculated from the simulation procedure up to twice higher than the peak power demand of the small electrical system.
- Complementarily, mainly for the peak power demand cover during the summer period, the installation of a PV station features as an ideal option. The PV station installed power is estimated at the range of 80% of the system's annual peak power demand.

Generally, as a rule of thumb, it may be stated that in cases of low storage capacity, the availability of power production from a secondary RES technology may significantly contribute to the maximization of the annual RES penetration in the electricity production share.

Finally, in cases of hybrid power plants of very small size, with a peak power demand lower than 100 kW (e.g., small insular communities or mountainous settlements), the small electricity consumption is rather easily faced with both approaches, namely the installation of two RES technologies and the increase of the storage capacity. The power demand and, consequently, the hybrid power plant's very small size, enables the implementation of each one of the available options, or maybe both of them, without the dramatic or prohibitory increase of the equipment's procurement cost. In such cases, the installation of higher storage capacity may be selected instead of the introduction of the secondary RES plant for several reasons, such as the simplification of the hybrid plant's layout, the low potential availability for the one of the available RES, or environmental or aesthetic constraints.

3.4.2 Storage Units

The available storage technologies for hybrid power plants of small and very small size are:

- several types of electrochemical batteries,
- small-size CAES systems, the so-called micro CAES, and
- fuel cells, which, practically, constitute a special electrochemical storage technology.

These technologies are analyzed in the following subsections.

3.4.2.1 Electrochemical Batteries

Electrochemical batteries are energy storage devices that:

- convert the stored chemical energy to electricity, during discharging operation mode, and
- store the provided electricity in the form of chemical energy, during charging operation mode.

The most popular types of electrochemical storage devices are presented in Figure 3.10.

Electrochemical batteries can be classified into low-temperature internal storage and high-temperature external storage units. The low-temperature batteries normally operate under common indoor space conditions. The basic difference between these two technologies is that in external storage batteries, the electricity production part is separated from the battery's storage part. This configuration enables the design and the dimensioning of the discharging and charging part independently. In external storage batteries, the electricity production and storage units are separate, yet connected to each other for exchanging electrochemical reactions during charging and discharging operation modes.

Lead acid (PbO_2) batteries, nickel cadmium (NiCd) batteries, lithium ion (Li-ion) batteries, sodium nickel chloride (NaNiCl) batteries, and nickel metal hydride (NiMH) batteries are the most common examples of internal storage technologies. On the other hand, the fundamental external storage technologies are sodium sulfur (NaS) batteries, sodium nickel chloride (NaNiCl) batteries, and vanadium redox flow batteries.

The basic features and the application fields of the most common electrochemical battery technologies are presented in Table 3.3 [15].

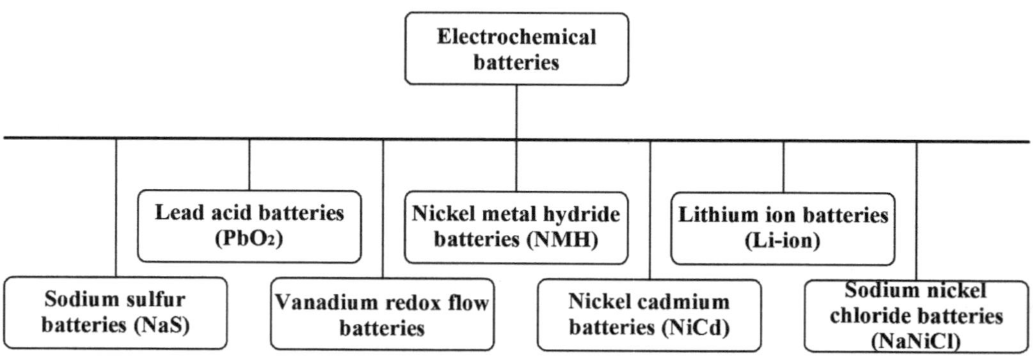

FIGURE 3.10 Different types of electrochemical storage devices [15].

TABLE 3.3
Basic Features and Application Fields of the Most Common Electrochemical Battery Technologies [15]

Technology	Maximum Discharging Rate/Storage Capacity	Installation Location/Usage	Features
Flooded lead acid (FLA)	10 MW/40 MWh	Chino, California (US)/Load management	• efficiency: 72%–78% • procurement cost: 50–150 $/kWh • service life: 1,000–2,000 charging cycles/ discharge depth: 70% • operating temperature: –5°C–40°C • storage density: 25 Wh/kg • self-discharge: 2%–5% per month • regular maintenance to replace water loss, heavy
Valve regulated lead acid (VRLA)	300 kW/580 kWh	Milwaukee, Wisconsin (US)/ Load management	• efficiency: 72%–78% • procurement cost: 50–150 $/kWh • service life: 200–300 charging cycles/ discharge depth: 80% • operating temperature: –5°C–40°C • storage density: 30–50 Wh/kg • self-discharge: 2%–5% per month • low required maintenance, portable, safe
Nickel cadmium (NiCd) batteries	27 MW/6.75 MWh	Alaska (US)/Load management	• efficiency: 72%–78% • procurement cost: 200–600 $/kWh • service life: 3,000 charging cycles/ discharge depth: 100% • operating temperature: –40°C–50°C • storage density: 45–80 Wh/kg • self-discharge: 5%–20% per month • high discharging rate, low required maintenance, poisoning risk from the NiCd cells
Sodium nickel chloride batteries (NaNiCl)			• efficiency: 85%–95% • procurement cost: 550–750 $/kWh • service life: 4,500 charging cycles/ discharge depth: 100% • operating temperature: 270°C–350°C • storage density: 100–120 Wh/kg • self-discharge: 0% • due to the ceramic electrolyte the battery has no electrochemical self-discharge
Sodium sulfur (NaS) batteries	9,6 MW/64 MWh	Tokyo, Japan/Load management	• efficiency: 89% (at 325°C) • procurement cost: 200–300 $/kWh • service life: 2,500 charging cycles/ discharge depth: 100% • operating temperature: 325°C • storage density: 100 Wh/kg • self-discharge: 5%–20% per month • must be heated during standby mode due to high required operation temperature, leading to overall efficiency reduction

(*Continued*)

TABLE 3.3 (Continued)

Technology	Maximum Discharging Rate/Storage Capacity	Installation Location/Usage	Features
Lithium ion (Li-ion) batteries			• efficiency: 100% • procurement cost: 400–700 $/kWh • service life: 3,000 charging cycles/ discharge depth: 80% • operating temperature: –30°C–60°C • storage density: 90–190 Wh/kg • self-discharge: 1% per month • high procurement cost due to special required packaging and internal overcharging protection
Vanadium redox flow batteries	1 MW/4 MWh	Japan/Power peak shaving	• efficiency: 85% • procurement cost: 360–1,000 $/kWh • service life: 10,000 charging cycles/ discharge depth: 75% • operating temperature: 0°C–40°C • storage density: 70 Wh/kg • self-discharge: 0% • high procurement cost, long life period

3.4.2.1.1 Lithium Ion (Li-ion) Batteries

Lithium ion batteries are mainly employed in both high- and low-power apparatus, in portable electronic devices, and in telecommunication equipment. Their main advantages are high efficiency and high storage density, which enables the construction of light batteries with high energy storage capacity [16]. Indeed, as presented in Table 3.3, Li-ion batteries exhibit the highest storage density among all the different electrochemical storage technologies, reaching values of 90–190 Wh/kg.

In Li-ion batteries, the anode is constructed with carbon graphite, while the cathode is constructed with lithiated metal oxide [17, 18]. The chemical energy storage medium consists of a mixture of lithium salts ($LiBF_4$, $LiClO_4$, $LiPF_6$) and organic carbonates (dimethyl carbonate or diethyl carbonate). During the discharging process, positive Lithium ions (Li^+) move from the negative electrode to the positive one, carrying electrical current. The reverse movement is executed during the charging process (Figure 3.11).

The realized electrochemical reactions are the following:
Chemical reaction in the positive electrode (charging):

$$LiCoCO_2 \leftrightarrow Li_{1-x}CoO_2 + xLi^+ + xe^- \tag{3.2}$$

Chemical reaction in the negative electrode (discharging):

$$xLi^+ + xe^- + C \leftrightarrow Li_nC \tag{3.3}$$

Modern Li-ion batteries exhibit service lives of maximum 3,000–4,000 charging cycles [17]. This means, approximately, a service life of a maximum 8–10 years, with 400–500 charging cycles annually. Other important technical features of Li-ion batteries are the high charging/discharging rate [18], the high discharge depth [20], and the almost negligible self-discharge in standby mode. Currently, the basic drawback of this technology is the high procurement cost. For this reason, Li-ion batteries have not yet been introduced in hybrid power plants.

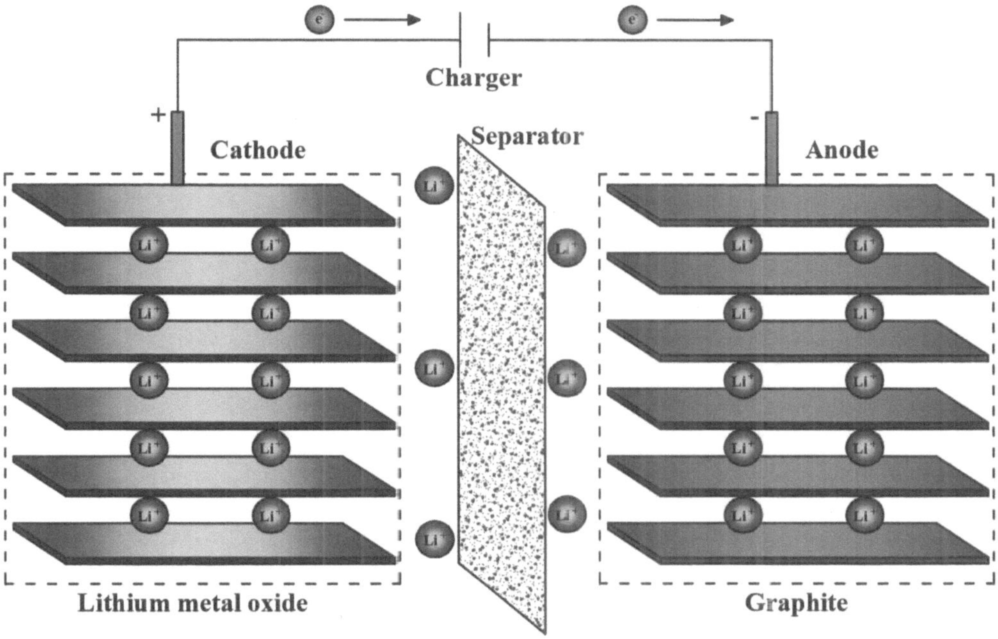

FIGURE 3.11 Operation concept of lithium ion batteries [19].

However, it must be noted that this cost is rapidly decreasing. In 2014, it was around 700 \$/kWh; just 2 years later it had dropped to 500 \$/kWh in the US. It is estimated that by 2025 this cost will have dropped to 200 \$/kWh, and below 100 \$/kWh after 2030. With Li-ion technology improvement and the anticipated decreases in their procurement costs, it is expected that this storage technology will dominate in small storage applications (small hybrid power plants, domestic storage, smart grids, and electrical vehicles) in the approximate future.

3.4.2.1.2 Sodium Sulfur (NaS) Batteries

The basic construction layout of sodium sulfur batteries is presented in Figure 3.12. They have high storage density, high discharge depth (reaching 100%), and high efficiency (almost 90%). Their procurement cost can be considered rather low, compared to the corresponding figure of other electrochemical technologies. This is mainly due to the inexpensive materials used for their construction, which can be recycled and reused. This electrochemical technology has been widely introduced in the US and Japan.

In NaS batteries, molten electrodes of sodium (negative) and sulfur (positive) are used. During the discharge process, sodium ions are conducted to knock-off free electrons, which in turn, while moving, generate electrical current. Electricity is eventually produced by the molten sodium electrode to the external load connected to the battery. The electrolyte used in a NaS battery is a beta-alumina solid electrolyte, referred to as BASE membrane. Apart from the power carrier task, the electrolyte also acts as a separator, permitting selectively only the sodium positive ions to pass through its mass and react with sulfur, producing eventually sodium polysulfide (Na_2S_4) [21] (Figure 3.12).

The operation of the NaS battery requires the maintenance of high temperatures, from 270°C to 300°C, in order to retain the execution of the chemical reactions. The required thermal energy for the maintenance of such temperatures is provided by the heat produced by the chemical reactions during the charge and discharge processes themselves, without the demand for any external heat source. However, in standby mode, an external heat source is necessary to keep the battery's

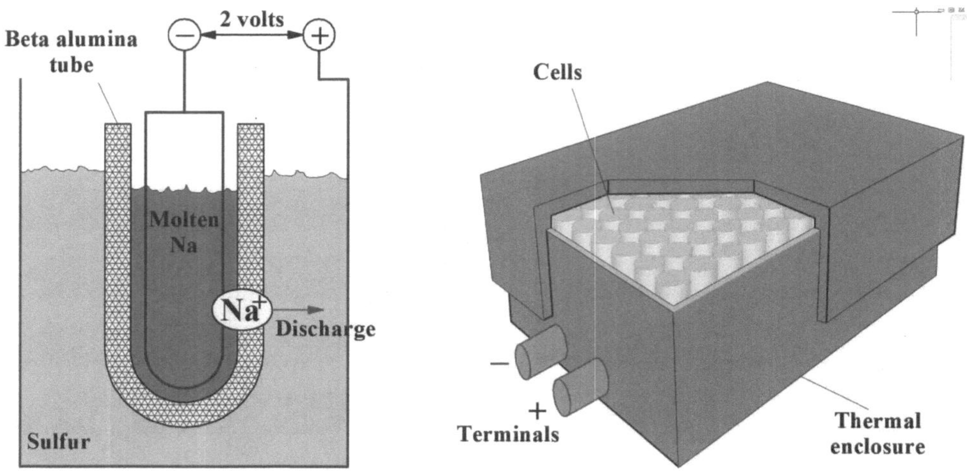

FIGURE 3.12 Layout and operation concept of sodium sulfur batteries [15].

temperature in the required level, resulting to a final decrease of the battery's overall efficiency. Regarding this drawback, some specific manufacturers have made efforts to develop an alternative technology of low-temperature NaS batteries.

The whole construction of an NaS battery must be integrated with an adequate thermal insulated and secure casing, in order to eliminate any thermal losses to the ambient environment and, thus, keep the efficiency at high levels, as well as to prevent any possible liquid leakage from the battery.

With a discharge depth of 100%, the service life of NaS batteries varies around 2,500 charging cycles, namely a period of 5–6 years with a normal use of 400–500 charging cycles per year. NaS batteries have already been employed in several applications in electricity systems, such as grid voltage and fundamental frequency regulation, power peak shaving, power quality support and power production curve smoothness, combined with wind parks. NaS batteries were used for the electricity supply in Presidio, New York, during the blackout that occurred in 2010 [22].

3.4.2.1.3 Lead Acid (PbO_2) Batteries

Lead acid batteries were the first type of electrochemical storage rechargeable battery constructed for either domestic or industrial use. The use of lead acid batteries in professional applications is rather restricted, due to the availability of alternative electrochemical storage technologies with improved technical features (higher efficiency and discharge depth, longer service life). Lead acid batteries are still in use, despite their significant technical and environmental drawbacks, mainly because they exhibit the lowest procurement cost of all the other available battery types. Additionally, due to the long use of this specific technology, they are characterized with reliable operation, mature technological level, and direct response to load fluctuations. Generally, they are used in cases in which the batteries' weight does not constitute a crucial parameter.

As already mentioned, lead acid batteries are the cheapest electrochemical storage technology, with a procurement-specific cost from 50 to 150 \$/kWh and relatively high efficiency, from 70% to 90%, depending on operating conditions (ambient temperature, charge/discharge rate, discharge depth [23]). The competitiveness of lead acid batteries against PHS or CAES systems for hybrid power plants of small size is configured mainly by these two features, namely their relatively low procurement cost and their high efficiency. One of their major disadvantages is their short service life, which, under ideal operating conditions may reach 2,000 charging cycles, namely a time period of 5–7 years. Nevertheless, it must be underlined that the lead acid batteries' service lives are strongly affected by ambient temperature, which must be kept normally below 26°C, and the discharge depth, which should not exceed 60%–70%, depending on the manufacturer.

There are two alternative implementations of lead acid batteries. Valve regulated lead acid (VRLA) are closed batteries with a pressure regulatory valve, as designated by their name, and the flooded lead acid (FLA) batteries, which are open type batteries. These two lead acid battery types exhibit similar characteristics regarding their operation principal concept; however, they differ from each other as far as procurement cost, maintenance requirements, and physical size. Compared to FLA batteries, VRLA batteries are more expensive, have shorter service lives, smaller size, and lower maintenance costs.

During charge mode, the provided electricity forces electron migration from the positive electrode to the negative one. A reverse process occurs during discharge mode, converting electrodes to lead sulfate, while the sulfuric acid amount decreases, producing water in large quantities. The regular maintenance required for the FLA batteries is mainly due to the need for distilled water removal.

The VRLA battery has a pressure regulatory valve, which enables automatic expansion of the internal pressure in case of overcharging.

3.4.2.1.4 Nickel Cadmium (NiCd) Batteries

Nickel cadmium (NiCd) batteries exhibit high efficiency (72%–78%), high energy storage density, acceptable service life, satisfactory response to electrical load fluctuations, and possible wide range of construction sizes. They are robust and, generally, characterized with features similar to lead acid batteries, apart from procurement cost.

A NiCd battery basically consists of a nickel oxide—hydroxide electrode, serving as the positive electrode, and a cadmium electrode, which serves as the negative one. Potassium hydroxide, an alkaline electrolyte, is employed as the battery's electrolyte. The whole construction is tightly sealed in a metal casing.

The chemical reactions during charge and discharge mode are the following:
Chemical reaction on cadmium electrode (negative electrode—charging):

$$Cd + 2OH^- \rightarrow Cd(OH)_2 + 2e^- \tag{3.4}$$

Chemical reaction on nickel electrode (positive electrode—charging):

$$2NiO(OH) + 2H_2O + 2e^- \rightarrow 2Ni(OH)_2 + 2(OH)^- \tag{3.5}$$

Total chemical reaction during discharge process:

$$2NiO(OH) + Cd + 2H_2O \rightarrow 2Ni(OH)_2 + Cd(OH)_2 \tag{3.6}$$

One of the most serious problems related to the use of NiCd batteries is their high construction cost, directly connected to the high procurement cost of the batteries' basic materials, namely cadmium and nickel. Another crucial issue is the potential environmental impacts, arisen from a possible leakage of the electrolyte to the environment. Nickel and cadmium are toxic heavy metals, with considerable effects on human health and on the natural environment. Another negative feature of this electrochemical technology is the so-called memory effect. Due to this fact, the battery must always be fully charged or discharged after charging or discharging respectively. In a different case, after partial charging or discharging, the battery's storage capacity will be further on defined by the last cycle's charge or discharge level respectively. This feature has proven to be particularly negative, because it strongly and gradually restricts the battery's flexibility as a part of a hybrid power plant. One more important shortcoming of this technology is the high self-discharge, which can reach up to 20% loss of total stored energy per month.

Despite the above drawbacks, NiCd batteries have been widely used in portable electronic devices, standby electric power systems, in aviation systems, in electrical vehicles, and in emergency lighting, because of some very specific advantages among the other available electrochemical storage technologies, such as:

- long service life (more than 3,500 charging cycles), combined with low maintenance requirements and total exploitation of their nominal storage capacity, through the 100% discharge depth,
- robust construction, making it less vulnerable to savageness, and
- operation with high current during discharging, implying high discharge rate.

3.4.2.1.5 Sodium Nickel Chloride (NaNiCl) Batteries

In this battery type, liquid sodium is used as the negative electrode. The role of the positive electrode is undertaken by nickel during the discharge process and nickel chloride during the charge process. The electrodes are separated by a beta-alumina ceramic wall that is conductive for sodium ions but an isolator for electrons (Figure 3.13). This beta-alumina ceramic acts as an electrolyte and enables the conduction of sodium ions between the anode and the cathode of the cells.

The battery temperature is kept between 270°C and 350°C to maintain the electrodes in a molten state. For this reason, independent heaters are part of the battery system.

This electrochemical storage technology was initially developed for electrical vehicles (EV) [25]. With the evolution achieved during the last years in EV technology, new storage devices were invented, more appropriate for such use. NaNiCl batteries have seen revived interest from researchers and manufacturers regarding the implementation of integrated systems with RES units, namely hybrid power plants of small size. They exhibit high storage capacity density.

The single battery size ranges from 4–25 kWh, suitable for a wide range of applications. Due to the ceramic electrolyte, the battery has no electrochemical self-discharge. Depending on operating conditions, the thermal loss is balanced by the internal electrical loss that is converted to heat, so that the overall efficiency is 80%–95%.

In contrast to other types of high-temperature batteries, $NaNiCl_2$ batteries have inherent overcharge capabilities and lower operating temperatures. Also, unlike other batteries, they may have a flexible power-to-energy ratio and can be cooled to ambient temperatures without component damage.

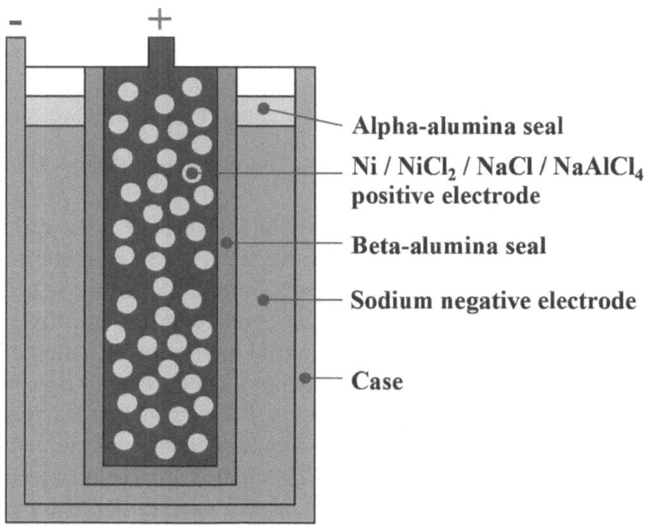

FIGURE 3.13 Layout and operation concept of sodium nickel chloride batteries [24].

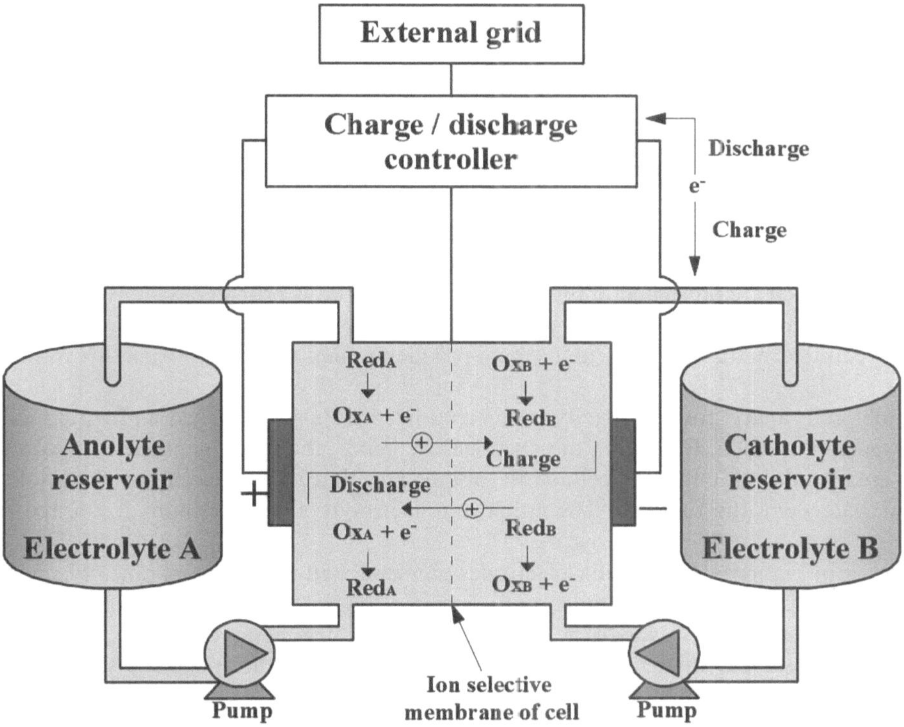

FIGURE 3.14 Layout and operation concept of vanadium redox flow batteries [15].

NaNiCl batteries are used in several applications, such as submarine power production, military applications, telecommunications infrastructure, and for the support of RES penetration in electricity grids.

3.4.2.1.6 Vanadium Redox Flow Batteries

The operation concept and the layout of vanadium redox flow batteries is presented in Figure 3.14. The overall implementation consists of two independent electrolyte storage reservoirs, separated from the electricity converter unit. The electricity conversion process is executed in the electrochemical cell, once the electrolytes have been transferred in this cell with a pump. The electrolytes are chemical active substances, flowing through an electrochemical cell, with a reversible ability to convert chemical energy to electricity.

Vanadium redox flow batteries are used for electricity production in remote settlements. This technology is ideal for the combined operation with RES units, precisely due to its fundamental design, which enables the independent operation of the charging and discharging parts of the battery. They have already been used for voltage and frequency regulation, RES penetration support in weak grids, and so on. They have the ability of 100% discharge level without being destroyed. Technically, they possess all the required specifications for cooperation with RES units. However, currently their procurement and maintenance costs remain too high to justify their economic feasibility.

3.4.2.2 Fuel Cells

Fuel cells constitute one of the main technologies for the exploitation of hydrogen in power systems. They can be considered as a special type of electrochemical storage. The main difference between fuel cells and the other electrochemical storage technologies, presented in the previous subsection, is that in fuel cells the consumption of some kind of fuel is required. This fuel can be pure hydrogen, which, yet, can be inserted in the fuel cell as a content of another substance or fuel, after appropriate processing. Such substances can be ammonia, natural gas, oil products, liquid propane, or biomass. Pure hydrogen can be also produced with water electrolysis, which may be performed with electricity produced by RES plants. In that case, it can be considered that the fuel cell's operation is totally based on RES and the corresponding electricity production imposes 100% RES penetration in the electrical grid.

In cases where hydrogen is stored in the fuel cell in order to ensure uninterrupted power production, the whole system is called a regenerative fuel cell (RFC). The layout and the operation concept of such an approach are presented in Figure 3.15. The fuel cell is provided with hydrogen produced by either an electrolysis device, or by a hydrocarbon dissolution process. The hydrogen is stored in the fuel cell to eventually react with oxygen, producing thus electricity, when this is required by the grid.

In a fuel cell, apart from electricity, water and heat are also produced as by-products of the electrochemical reaction. Following the chemical sequence, the reacting elements (hydrogen and oxygen) are introduced in the fuel cell, while, at the same time, the reaction's products (electricity, heat, and water) exit the fuel cell. The employed electrolyte always remains inside the fuel cell [26, 27].

As shown in Figure 3.15, the cell's electrodes are separated with a membrane, which serves as the electrolyte. Between this membrane and each electrode there is a catalyst layer. Hydrogen is concentrated in the fuel cell's anode (negative electrode), which, being in contact with the catalyst, is separated in positive hydrogen ions and electrons. The shape of the anode and the catalyst facilitate the uniform dispersion of the hydrogen ions. The released electrons are transmitted through an outer electrical circuit to the cathode producing, in this way, electrical current, given the fact that the membrane does not allow their passage through it. For this reason, the anode and the catalyst are constructed with conductive materials.

The hydrogen positive ions, practically isolated protons, pass through the membrane and react with oxygen, which is provided for the cathode (the positive electrode), to produce water. The homogenous dispersion of oxygen on the catalyst is also facilitated by the electrode's shape.

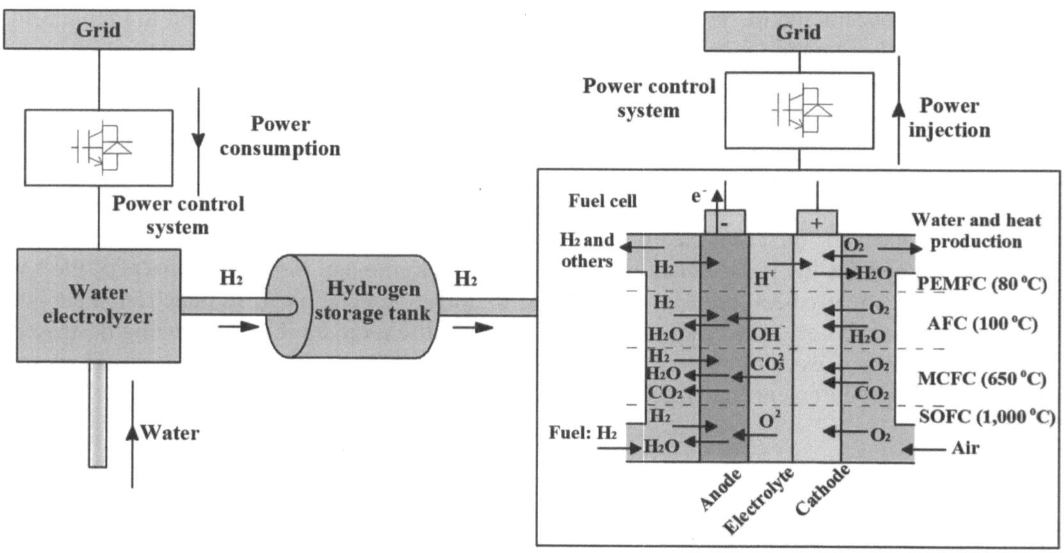

FIGURE 3.15 Operation and layout of a regenerative fuel cell [15].

The catalyst's role is to accelerate the water production Apart from the hydrogen and oxygen molecules, the electrons transmitted through the outer electrical circuit to the fuel cell's cathode also participate in this process.

The two layers of catalyst (one next to the anode and another one next to the cathode) serve for the acceleration of the hydrogen molecules dissolution reaction in the anode, as well as for the reaction between hydrogen and oxygen towards the water production in the cathode. Usually, a very thin layer of platinum (Pt) on a carbon surface serves as the catalyst. This layer is put on the catalyst's surface, which is in contact with the membrane. This layer is rough and porous, so as to maximize its total surface.

The above described process is expressed with the following chemical reactions:
Chemical reaction in the anode:

$$2H_2 \rightarrow 4H^+ + 4e^- \tag{3.7}$$

Chemical reaction in the cathode:

$$O_2 + 4H^+ + 4e^- \rightarrow 2H_2O \tag{3.8}$$

Overall chemical reaction:

$$2H_2 + O_2 \rightarrow 2H_2O. \tag{3.9}$$

With the above reactions, a regular fuel cell produces electricity with voltage around 0.7 Volts. In order to achieve higher and, practically, utilizable voltage, an adequate number of fuel cells connected in series (fuel cell stack) should be developed.

With the above presented implementation, the fuel cells perform similarly with a fuel tank, because the hydrogen is initially stored and then it can be used for electricity production according to the user's will. The fuel cell's operation is managed by automatic control systems and configured, firstly, by the availability of the primary energy source for hydrogen production (usually an RES in a hybrid power plant or hydrocarbons) and, secondly, by the required power production from the fuel cell, as it is imposed by the overall operation algorithm of the electrical system and the power demand.

A number of different fuel cell technologies have been developed in recent years, aiming at the substitution of internal combustion engines in electrical systems or in portable consumptions and as an auxiliary power source in electrical or hybrid vehicles. These alternative technologies are classified based on the ion exchange mechanism and the type of electrochemical reactions, affected by the type of the employed electrolytes and reacted elements. A classification of the so far developed fuel cells, depending on the employed electrolyte, follows:

- alkaline (AFC)
- proton exchange membrane (PEMFC)
- direct methanol (DMFC)
- direct ethanol (DEFC)
- phosphoric acid (PAFC)
- molten carbonate (MCFC)
- solid oxide (SOFC).

The major fuel cell technologies are summarized in Table 3.4, with their fundamental construction and operation features [28–30]. These technologies will be further presented in the following subsections.

TABLE 3.4
Technical Features of the Different Fuel Cell Technologies [15]

Fuel Cell Technology	Operating Temperature (°C)	Electrolyte	Charge Carrier	Catalyst Anode	Fuel for Cell	Electrical Efficiency (%)	Nominal Power of Commercial Models (kW)
Alkaline (AFC)	70–100	aqueous solution (liquid)	H^+	Ni	H_2	60–70	10–100
Proton exchange membrane (PEMFC)	50–100	Perfluor-sulfonated polymer (solid)	H^+	Pt (platinum)	H_2	30–50	0.1–500
Direct methanol (DMFC)	90–120	Perfluor-sulfonated polymer (solid)	H^+	Pt (platinum)	Methanol	20–30	100–1,000
Direct ethanol (DEFC)	90–120	Perfluor-sulfonated polymer (solid)	H^+	Pt (platinum)	Ethanol	20–30	100–1,000
Phosphoric acid (PAFC)	150–220	Phosphoric acid (immobilized liquid)	H^+	Pt (platinum)	H_2	40–55	5–10,000
Molten carbonate (MCFC)	650–700	Alkaline carbonate (immobilized liquid)	CO^{2-}	Ni	CO/H_2	50–60	100–300
Solid oxide (SOFC)	800–1,000	Yttria-stabilized zirconia (solid)	O^{2-}	Ni	$CO/H_2/CH_4$	50–60	0.5–100

The overall efficiency of a fuel cell ranges between 40%–65%, with an increase trend on the condition that the produced heat is exploited in parallel consumptions. A major shortcoming, which currently constitutes an inhibitory parameter for the wide commercial introduction of fuel cells in power systems, is their high procurement and installation cost, which ranges between 5,000 and 8,000 $/kW. This cost mainly comprises the cost of the required materials for the construction and the operation of the fuel cell.

3.4.2.2.1 Hydrogen Fuel Cells

Hydrogen constitutes, without doubt, the most commonly employed fuel in fuel cells for electricity production. Considerable research and pilot projects have been accomplished so far on hydrogen production from RES through electrolysis process. Yet, currently the hydrogen production from RES, mainly wind energy, solar radiation, and biomass, costs almost twice or three times more than hydrogen production from natural gas [31]. Consequently, the running research focuses on the improvement of the hydrogen production processes from RES, aiming, ultimately, at increasing the overall efficiency and reducing production costs [28].

Hydrogen fuel cells can be employed in a series of potential applications. One of the most popular is their usage in automobiles. Hydrogen fuel cells for vehicles (HFCVs) have become popular recently as a zero gas-emission solution.

Hydrogen fuel cells consist of three basic parts: the hydrogen storage chamber; the conversion unit, through which the hydrogen in transformed to electricity; and the electrolyzer unit, which serves the continuous production of hydrogen and the supply of the fuel cell.

The most common categories of hydrogen fuel cells are the proton exchange membrane, the alkaline, and the phosphoric acid cells. Hydrogen fuel cells, despite their advantages, exhibit a series of important drawbacks, such as high construction costs, relatively low cycle efficiency, and highly affected operation by a potential hydrogen's pollution.

3.4.2.2.2 Proton Exchange Membrane Fuel Cells

Proton exchange membrane fuel cells (PEMFC) constitute one of the most promising technologies for the development of fuel cells [32]. Their layout and operation concept is presented in Figure 3.15 and was thoroughly described above. The power production from a PEMFC can theoretically be from 10 to 500 kW, with an overall storage–production cycle efficiency of 30%–50%. PEMFC advantages include simple and robust construction, light weight, and the corresponding high power density. Up to now, PEMFC have been used in stationary power plants [33], for the support of uninterrupted power supply systems (UPS) [34], in portable electronic devices [35], in lightweight vehicles [36], in power bicycles [37], in hybrid buses [38], and in sailing yachts [39].

3.4.2.2.3 Molten Carbonate Fuel Cells

Molten carbonate fuel cells (MCFC) utilize lithium–sodium or lithium–potassium carbonate salts as the electrolyte at temperatures as high as 650°C. MCFCs exhibit high ionic conductivity. The fuel gas used in these fuel cells is a gasified mixture of H_2 and CO. Oxygen is provided in a mixture of O_2 and CO_2, which may contain water vapor. The operation pressure varies between 1 and 10 atm. Due to the high required operating temperature, a nonprecious metal (usually Ni) is selected to serve as the anode electrode, while, its oxide (in case of Ni, the NiO) is used as the cathode electrode. During the operation of a fuel cell, the salts contained in the cell melt because of the high prevailing temperature, producing, thus, carbonate ions on the cathode electrode $\left(CO_3^{2-}\right)$. The carbonate ions move towards the anode electrode, where they are combined with hydrogen to create water vapor, CO_2, heat, and electrons. The released electrons are conducted towards an externally connected electrical circuit, to eventually produce electricity, as presented in Figure 3.16.

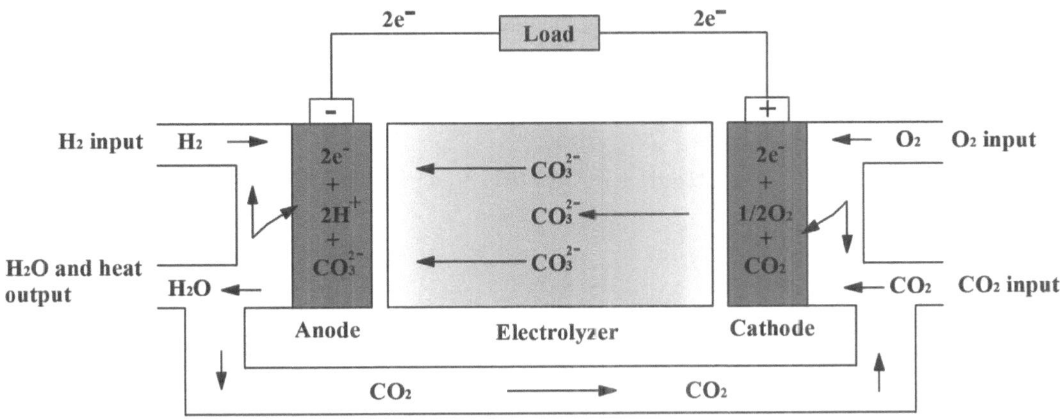

FIGURE 3.16 Layout and operation concept of a molten carbonate fuel cell [15]

MCFCs can operate with different fuels, unlike the other fuel cell technologies. This versatility is one of their major advantages. MCFCs are currently developed in Japan and the US with specific technological innovations, in order to enable their operation with synthetic gas (syngas), natural gas, and biogas for power plant use. Another important advantage is the possibility of replacing the platinum catalyst with a nickel one, leading to a considerable reduction of the fuel cell's cost. Due to their flexibility regarding the use of different fuels, MFCFs have been introduced in military and marine applications. Additionally, the required high operating temperature creates the prerequisites for their exploitation in combined electricity and heat cogeneration applications.

On the other hand, the most important drawbacks of MCFCs are the long time periods during which their temperature must be kept at high values, their low immunity to sulfur corrosive effect, the CO_2 emissions, and the difficulties with handling the liquid electrolytes. For the time being, MCFCs exhibit a rather low penetration in the market mainly due to technological immaturity.

3.4.2.2.4 Solid Oxide Fuel Cells

Solid oxide fuel cells (SOFC) feature as a particularly attractive technology due to their high efficiency, which theoretically can reach 60%. In this fuel cell type, solid oxide serves as the electrolyte, which facilitates the migration of negative oxide ions (O^{2-}) from the cathode to the anode, through a nickel or cobalt electrode, operating in considerably high temperatures (between 700°C and 1,000°C). The SOFC advantages include: relatively medium construction cost, high efficiency, relatively low CO_2 emissions, fast internal chemical reactions, and constant power production from a few Watts up to 2 MW. Currently, there are efforts towards the reduction of the required operating temperature. On the other hand, as in MCFC, the high operating temperature makes them suitable for heat and electricity cogeneration applications. Finally, due to the high operating temperature of these fuel cells, the use of expensive platinum as a catalyst is not required. Some of their most serious drawbacks are the corrosion of fuel cell components from sulfur and the low maturity level of the required technology.

3.4.2.2.5 Direct Methanol Fuel Cells

Direct methanol fuel cells (DMFC) are practically a subcategory of proton exchange fuel cells, in which methanol (CH_3OH) is directly employed as the fuel for the operation of the cell. DMFCs exhibit lower efficiency and lower operation temperature; yet, they have rapidly evolved into a reliable storage technology, due to their long life period and their high power density.

The DMFC fundamental operation concept is the hydrogen extraction from the methanol through an electro-oxidation process, leading to CO_2 production in the anode electrode. The hydrogen ions (H^+) are conducted towards the cathode, after passing through the proton exchange membrane, to eventually react with O_2 and produce water. The electrolytes employed in DMFCs are the same as those used in PEMFC.

The chemical reactions that take place in a DMFC are:
Chemical reaction in the anode:

$$CH_3OH + H_2O \rightarrow 6H^+ + 6e^- + CO_2 \tag{3.10}$$

Chemical reaction in the cathode:

$$\frac{3}{2}O_2 + 6H^+ + 6e^- \rightarrow 3H_2O \tag{3.11}$$

Total chemical reaction:

$$CH_3OH + \frac{3}{2}O_2 \rightarrow 2H_2O + CO_2 \tag{3.12}$$

The maximum power production capacity of DMFCs is lower than in hydrogen fuel cells. Nevertheless, DMFCs are characterized with usability and flexibility, based on the fact that methanol is a fuel much easier to transport and store, compared to hydrogen. DMFCs have already been used in portable electronic devices, such as phones, digital cameras, and portable computers of small size (notebooks). The most important disadvantage of DMFCs is their significant CO_2 emissions.

3.4.3 Storage Plant Selection for Small-Size Hybrid Power Plants

After the comprehensive presentation of the alternative storage technologies for hybrid power plants of small size, provided in the previous sections, a comparative analysis between them can now be performed, in order to determine the most appropriate ones for combined operation with RES power plants, within the frame of hybrid power plant integration. The storage unit of a hybrid power plant should be characterized by specific features, to facilitate the achievement of the hybrid plant's fundamental targets, namely the maximization of the RES secure penetration in the electrical system, ensuring the system's dynamic security and stability and with minimum possible setup and operation costs.

To accomplish these goals, the following features should be available by the storage unit:

- large storage capacity
- low procurement and maintenance cost
- quick response to power demand fluctuations.

From the precedent analysis, the high storage-specific cost of electrochemical batteries and the fuel cells has been revealed, compared to the corresponding figure of the PHS and CAES systems, employed for hybrid power plants of large size. Indeed, for the two existing large-size CAES, the storage-specific cost is calculated from 20 to 110 $/kWh (see Table 3.1), whereas this figure for the two existing PHS systems in the El Hierro and Ikaria islands is calculated from 15 to 85 €/kWh (see Table 3.2). On the contrary, for the electrochemical storage technologies, according to Table 3.3, the lowest storage-specific cost is configured at 50–150 $/kWh for lead acid batteries, while, in the worst case, this figure may reach the value of 1,000 $/kWh for vanadium redox flow batteries. Yet, these values quickly increase when we consider the required replacements of the electrochemical storage devices due to their short service lives. Specifically, assuming a life time period of a hybrid power plant of 20 years, and taking into account that the service life of lead acid batteries can be, under the most favorable conditions, around 6 years, it is concluded that during the overall hybrid plant's life, the batteries should be replaced three times. This implies that the total storage-specific cost of the storage plant, over a 20-year lifespan, will be four times the above mentioned procurement price (the initial procurement plus three replacements during the operation period). Fuel cells exhibit even higher storage-specific costs.

Electrochemical batteries, consequently, can become competitive compared to the large-size storage technologies (PHS or CAES), when the hybrid power plant focuses on small-scale electricity production. In such cases, the high infrastructure setup costs required for the construction of a PHS or a CAES system cannot be compensated by the low electricity production, because the large storage units practically remain unexploited and the corresponding high setup costs cannot be recouped.

For example, according to the results of several simulations of hybrid power plants' integration in small-size electrical systems, the setup cost of the storage unit for a hybrid power plant aiming at 100% RES penetration in an electrical system with annual power peak demand at 500–1,000 kW, is estimated:

- around $8,000,000, in case of a PHS system with reservoirs' capacities at 150,000 m^3 and a geostatic head height of 120 m
- around $2,000,000 totally, including the required replacements over the hybrid power plant's period, in case of lead acid batteries.

It is obvious that for such electrical systems of small size, the large-size storage plants seem to exhibit disproportionately high setup cost. This conclusion is a direct consequence of the fact that the setup cost of the large-size storage plants is configured by the construction of several infrastructure works (e.g., access roads, reservoirs, penstocks, buildings), which cannot be reduced proportionately with the power demand size. Hence, for small-size electrical systems, these projects have proven to be too large and expensive. As the size of the electrical system increases, yet, and, consequently, the electricity consumption as well, large storage plants become more and more competitive, compared to electrochemical batteries. For example, the previously mentioned PHS plant will be also able to support the 100% RES penetration for an electrical system with annual power demand peak at 1 MW, most probably only with a slight increase in setup costs, depending on the available land morphology and geological features. For this larger system (1 MW annual power peak demand instead of 500 kW) the electrochemical batteries' total procurement cost, including replacements, is estimated at $4,500,000. So, even for electrical systems with annual power peak demand as low as 1 MW, the difference between the storage plant's setup cost for a PHS system and electrochemical batteries is considerably decreased. Practically, for power plants with annual peak demand higher than 2 MW, the use of electrochemical batteries cannot be feasible, neither from a technical, nor from an economic point of view, given the currently existing technological and economic status of electrochemical storage, of course on the condition of availability of appropriate land morphology for the construction of a PHS or a CAES system.

At this point it must be noted that all the above conclusions stand on the ground that the hybrid power plant aims at the maximization of RES penetration in the electrical system, focusing, ultimately, on 100% annual electricity production from the RES units. This operation concept maximizes the annual electricity production from the hybrid plant and, respectively, the required storage capacity. If this is not the case, for example when the hybrid power plant is integrated in the electrical system for power peak shaving, hence the power production is restricted for specific time intervals during the 24-hour period, the employment of large storage plants can be economically feasible only for larger electrical systems. Generally, as the total electricity production from the hybrid power plant drops, the electrochemical storage technologies seem to be more competitive compared to the large storage plants, even for electrical systems of larger size.

Finally, the feasibility of electrochemical batteries is expected to raise, even for electrical systems with power demand higher than 2 MW, with the evolution of the relevant technology and the reduction of procurement costs at the range of 100 $/kW, which, as already mentioned, for Li-ion batteries, is expected to happen before 2030.

3.5 OPERATION ALGORITHMS OF LARGE-SIZE HYBRID POWER PLANTS

Large-size hybrid power plants can be integrated in electrical systems with the following operation concepts:

- Continuous power production follows the power demand during the whole 24-hour period, aiming at 100% annual electricity production from the RES units. So far, such projects have been proposed and studied for electrical systems with annual peak power demand from 3 to 50 MW, met, most commonly, in insular territories.
- Power production is guaranteed during specific time intervals over the 24-hour period. These time intervals usually coincide with the peak power demand periods. Practically, in these cases, the hybrid power plant is integrated in the electrical system for power peak demand shaving, by storing energy from the RES units anytime and provide it for the power demand coverage during the peak periods. Currently, such kinds of proposals and projects have been submitted for electrical systems with peak power demand higher than 100 MW.

The hybrid plant's operation algorithm changes versus its operation concept. For the above two alternative cases, the operation algorithms are presented in the following subsections, accompanied by all the math formulas required for the calculation of the involved magnitudes, depending on the employed storage technology (i.e., PHS or CAES system).

3.5.1 Hybrid Power Plants for 100% RES Penetration

Large-size hybrid power plants can be introduced in electrical systems in order to contribute to the 100% annual electricity production from RES [40]. The simulation process of the hybrid power plant is ultimately affected by the involved storage technology. In the following sections, the operation algorithms for hybrid power plants aiming at 100% RES penetration are presented, with PHS and CAES systems introduced as storage units.

3.5.1.1 PHS Systems as Storage Unit

The operation simulation of hybrid power plants with PHS systems as storage units is executed with the method of annual time series, with, most commonly, hourly time calculation steps, although time series with higher time discretization (e.g., of 30-min or 10-min intervals) can be used for higher accuracy. The operation algorithm of hybrid power plants supported by PHS systems and aiming at 100% annual RES penetration is analyzed in the following steps:

1. For every calculation time step, the currently available power production P_{RES} from the RES plant (most often a wind park) and the power demand P_d are introduced. Additionally, a maximum RES direct penetration percentage p_{max} versus the current power demand is defined, imposed by the system's security and stability requirements. As explained in Chapter 2, this maximum RES penetration percentage is a function of the system's size, the grid's topology, the currently dispatched thermal generators, the maintained spinning reserve, the climate conditions, etc.
2. Following the above definitions, the RES direct penetration P_{RESp} will be calculated as follows:
   i. if $P_{RES} \geq p_{max} \cdot P_d$, then $P_{RESp} = p_{max} \cdot P_d$
   ii. if $P_{RES} < p_{max} \cdot P_d$, then $P_{RESp} = P_{RES}$.
3. Once the RES direct penetration has been defined, there will be:
   i. a potential RES power production surplus: $P_{RES} - P_{RESp}$
   ii. a remaining power demand still uncovered: $P_d - P_{RESp}$.
4. The water volume V_p is then calculated, required to be pumped in the PHS upper reservoir, in order to store the power surplus $P_{RES} - P_{RESp}$, available for the duration t of the calculation step (H_p the available net pumping head, γ the water's specific weight, η_p the pump unit's average overall efficiency during the current calculation time step):

$$V_p = \frac{\left(P_{RES} - P_{RESp}\right) \cdot t \cdot \eta_p}{\gamma \cdot H_p} \tag{3.13}$$

5. Similarly, the water volume V_h is calculated, required to be removed from the PHS upper reservoir, so the remaining power demand $P_d - P_{RESp}$ will be produced by the hydro turbines for the duration t of the calculation step (H_T the available water falling net head):

$$V_h = \frac{\left(P_d - P_{RESp}\right) \cdot t}{\gamma \cdot H_T \cdot \eta_h} \tag{3.14}$$

6. The remaining stored water volume in the PHS upper reservoir after the end of the current calculation time step j will be:

$$V_{st}(j) = V_{st}(j-1) + V_p - V_h \tag{3.15}$$

7. The remaining water volume $V_{st}(j)$ in the PHS upper reservoir is checked whether or not it exceeds the reservoir's maximum storage capacity V_{max}:
   i. If $V_{st}(j) > V_{max}$, then:
      $P_h = P_d - P_{RESp}$
      $P_{th} = 0$
      $P_{st} = 0$
      $P_{rej} = P_{RES} - P_{RESp}$
      $V_{st}(j) = V_{st}(j-1) - V_h$.
   where P_h the power produced by the hydro turbines, P_{th} the power produced by the thermal generators, P_{st} the power absorbed by the pump units, and P_{rej} the RES units rejected power.
   ii. If $V_{st}(j) \leq V_{max}$, then we proceed to the following step.
8. Finally, the remaining water volume $V_{st}(j)$ in the PHS upper reservoir is checked whether or not it is lower than the minimum water stored volume V_{min}, which always remains stored in the reservoir, mainly due to constructive reasons (e.g., due to the position of the water's intake):
   iii. If $V_{st}(j) < V_{min}$, then:
      $P_h = 0$
      $P_{th} = P_d - P_{RESp}$
      $P_{st} = P_{RES} - P_{RESp}$
      $P_{rej} = 0$
      $V_{st}(j) = V_{st}(j-1) + V_p$.
   iv. If $V_{min} \leq V_{st}(j) \leq V_{max}$ then:
      $P_h = P_d - P_{RESp}$
      $P_{st} = P_{RES} - P_{RESp}$
      $P_{th} = 0$
      $P_{rej} = 0$
      $V_{st}(j) = V_{st}(j-1) + V_p - V_h$.

The above analysis shows that power production and storage from the PHS system may be executed simultaneously. The necessity for such an option is revealed by the hybrid plant's operation concept. Specifically, at a certain time point, because the RES unit direct penetration is restricted by a maximum percentage p_{max} defined versus the current power demand, the remaining power demand should be covered as a priority by the hydro turbines, of course on the condition of enough water volume stored in the PHS upper reservoir. However, at the same time, any potential power production surplus from the RES plant, remaining after the RES power direct penetration, should be stored. This means that the PHS plant should support concurrent power production and storage, which implies concurrent water pumping and falling. Simultaneous water falling and pumping can be achieved with the installation of a double penstock, namely of two separate and independent pipelines, one employed exclusively for water pumping and one for water falling.

With the computational execution of the above presented procedure for every time step and for the whole operation period (e.g., for a whole year) the fundamental involved magnitudes are calculated, such as the RES plant direct power penetration, the hydro turbines and thermal generator power production, the power absorbed by the pump units, the water volume stored in and removed from the PHS upper reservoir, and, finally, the remaining water volume in it. With the integration of the above calculations over the whole year, the corresponding annual magnitudes are eventually calculated, such as the annual electricity production and storage from the hybrid power plant's components and the charge level variation of the storage plant (i.e., the remaining water volume in the PHS upper reservoir). Having calculated the power production from the available generators, the annual power production synthesis can be depicted graphically. Finally, the maximum annual power production from the hydro turbines and the thermal generators, as well as the maximum annual power absorbed by the pump units will define the corresponding minimum required nominal power for each of these components.

The above described operation algorithm is graphically presented in Figure 3.17.

The required data for the above calculations are:

1. the power demand annual time series of, at least, mean hourly values
2. the power production annual time series from the RES plant of, at least, mean hourly values
3. the PHS system's upper and lower reservoir capacity
4. the hydro power plant's and the pump station's natural geostatic head
5. the efficiency variation curves versus the operating point of the hydro turbine and the pump selected models
6. the pipeline length between the two reservoirs.

The reservoirs' capacities, the water pumping and falling natural geostatic heads, and the pipeline length are imposed by the available land morphology at the PHS installation location and the final siting of the overall system. The calculation of the annual power production time series from the RES power plant is based on the available RES potential at the installation site, the total

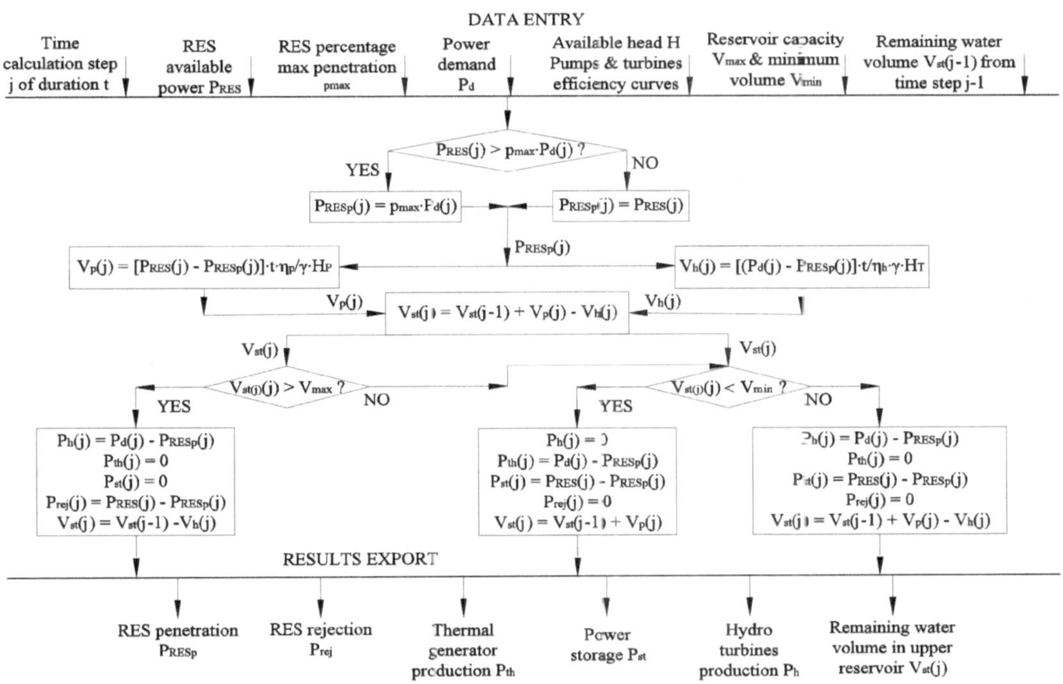

FIGURE 3.17 Operation algorithm graphical representation for a hybrid power plant with a PHS system as storage unit, aiming at 100% annual RES penetration [41].

installed RES plant's power, the power curve of the selected wind turbine (in case of wind park) and the wind turbines or PV panels micrositing at the installation site. However, the RES plant's nominal power is not known beforehand. On the contrary, it constitutes one of the main results of the dimensioning procedure. Consequently, in order to execute the above mentioned operation simulation algorithm, an initial assumption should be made for the required power of the RES plant. The operation simulation is then iteratively executed, with different assumed RES nominal power for each iterative loop. In this way, the hybrid power plant's annual operation is simulated again and again for different dimensioning scenarios, aiming to define the optimum dimensioning, based on either RES penetration maximization criteria, or on economic criteria, such as the optimization of the economic indices of the required investment. The iterative procedure for the dimensioning optimization of the hybrid power plant will be analytically presented in a following section.

A particular part of the dimensioning procedure of hybrid power plants with PHS systems refers to the hydraulic network calculation, mainly the dimensioning of the water transfer pipelines. The hydraulic network calculation will enable the estimation of the flow friction losses in the pipelines which, in turn, will lead to the calculation of the water falling and pumping net heads, H_T and H_P respectively, using the following general relationships (for upper and lower reservoirs open to atmospheric pressure):

$$H_T = H_{Tgeo} - \sum h_{fT} \tag{3.16}$$

$$H_P = H_{Pgeo} + \sum h_{fP} \tag{3.17}$$

where:

H_{Tgeo}: the natural geostatic head between the upper reservoir's surface and the hydro turbine's axis

H_{Pgeo}: the natural geostatic head between the upper and the lower reservoirs' free surfaces

$\sum h_{fT}$: the total flow losses through the hydraulic network during water falling

$\sum h_{fP}$: the total flow losses through the hydraulic network during water pumping.

The total flow losses in the hydraulic network are derived by the sum of the linear flow losses and the localized flow losses in several hydraulic components and instruments of the overall hydraulic installation:

$$\sum h_f = h_f + \sum h_{fk} \tag{3.18}$$

The localized flow losses in several hydraulic components and instruments are given by the general relationship:

$$\sum h_{fk} = \sum k \cdot \frac{u^2}{2 \cdot g} \tag{3.19}$$

where k is the localized flow losses coefficient, given separately for each hydraulic component in tables and graphs provided by the manufacturer or in the relevant literature and u the water flow velocity in the pipeline.

The linear flow losses for water falling and pumping are respectively given by the relationships:

$$h_{fT} = f \cdot \frac{L_T}{D_T^5} \cdot \frac{8 \cdot Q_T^2}{g \cdot \pi^2} \tag{3.20}$$

$$h_{fP} = f \cdot \frac{L_P}{D_P^5} \cdot \frac{8 \cdot Q_P^2}{g \cdot \pi^2} \tag{3.21}$$

where:

f:	the linear flow losses coefficient
D_T and D_P:	the inner diameters of the water falling and pumping pipelines respectively
L_T and L_P:	the length of the water falling and pumping pipelines respectively
Q_T and Q_P:	the water falling and pumping flows
g:	the acceleration of gravity.

The linear losses coefficient f depends on the water flow velocity, the tubes' material, and their inner surface roughness. Generally, the coefficient f is theoretically calculated by the Moody diagram or it may be given by the tubes' manufacturer, on the assumption of water flow velocity lower than 2 m/s. Especially for steel welded tubes, the f coefficient may be calculated by the empirical relationship of Nikuradse [42] (ε_s the tube's inner surface absolute roughness in mm):

$$\frac{1}{\sqrt{f}} = 2 \cdot \log\left(\frac{1}{\varepsilon_s}\right) + 1.14 \tag{3.22}$$

For the linear flow losses calculation, according to the Equations 3.19 and 3.20, the water flow inside the pipelines during power production and storage is required. These water flows are calculated by the relationships:

$$Q_T = \frac{P_{ht}}{\eta_T \cdot \rho \cdot g \cdot H_T} \tag{3.23}$$

$$Q_p = \frac{\eta_p \cdot P_p}{\rho \cdot g \cdot H_p} \tag{3.24}$$

In the above equations, for the water flow calculation, the net heads for water falling and pumping must be known, which, in turn, requires knowledge of the linear flow losses and the pipelines' inner diameter. However, these magnitudes are not known beforehand. Consequently, it is concluded that the calculation procedure requires prior knowledge of the magnitudes that are reasonably expected to be among the results of the calculations.

Such kind of problems are solved with the application of an iteration calculation, by assuming at the beginning some initial values for specific magnitudes involved in the calculation process and confirming them at the end of the calculation loop. For this particular case, the required iterative process can be analyzed in the following steps:

1. An initial value is assumed for the nominal power of the RES plant. The corresponding RES power production time series is then developed, based on the available RES potential, the final siting of the RES generators (wind turbines or PV panels), and the power curve of the selected wind turbine model. The RES plant's nominal power constitutes the independent parameter for the dimensioning optimization of the hybrid power plant.

2. Given that the pipelines' diameters are not yet known, the flow losses calculation is not possible. Hence, at a first approach, the flow losses are neglected and the total net heads are assumed equal to the corresponding geostatic heads:

$$H_T = H_{NT}$$

$$H_P = H_{NP}.$$

3. The computational simulation of the hybrid plant's operation is executed based on the above described algorithm. For every calculation time step, the hydro turbines' power production P_h and the power absorbed by the pump units P_P are calculated.
4. The water flow rates Q_T and Q_P during water falling and pumping respectively are calculated with Equations 3.22 and 3.23 for every calculation time step. Eventually, once this calculation has been integrated for the whole annual time period, the maximum annual water flow rates Q_{Tmax} and Q_{Pmax} are calculated.
5. By introducing the linear and the localized flow losses coefficients f and k respectively and the pipelines lengths, the total flow losses are calculated with Equations 3.18–3.20 separately for water falling and pumping for a range of potential inner diameters. The pipeline's inner diameter range is selected, according to the engineer's experience, between a minimum and a maximum value, with an increment step of 0.2–0.5 m, depending on the size of the transmitted water flows. With the above procedure, a final inner diameter is selected for the water falling and pumping pipelines, in order to achieve a maximum acceptable flow losses head. An empirical rule is to keep the overall flow losses head lower than 5% of the natural geostatic heads. Equivalently, the pipeline's inner diameter can be selected in order to keep the water flow maximum velocity lower than 2 m/s.

   The results of the above calculation can be presented in a manner similar to Table 3.5. Assuming that the maximum flow losses head should be kept below 20 m, the inner diameter for the water falling pipeline should be equal to 2 m, while for the water pumping pipeline, it should be equal to 2.40 m.

   The results presented in Table 3.5 are also graphically presented in Figure 3.18, regarding the linear flow losses head variation versus the pipeline's inner diameter.

TABLE 3.5
Inner Diameter Selection for the Pipelines of a PHS System

Pipeline's Inner Diameter (m)	Water Falling Flow Losses Head (m)	Water Pumping Flow Losses Head (m)
1.40	114.50	195.03
1.60	58.73	100.03
1.80	32.59	55.51
2.00	19.24	32.78
2.20	11.95	20.35
2.40	7.73	13.17
2.60	5.18	8.83
2.80	3.58	6.09
3.00	2.53	4.32

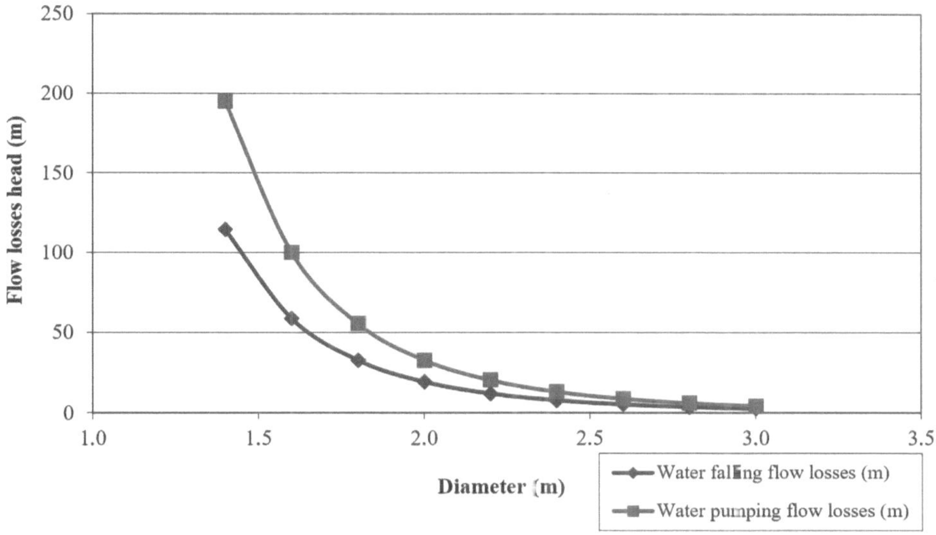

FIGURE 3.18 Linear flow losses head variation versus the pipelines inner diameter of a PHS system.

6. With the pipeline's inner diameter known, the flow losses calculation is now possible versus the transmitted water flow, practically, versus the hydro turbines' power production and the pump unit's absorbed power. The dimensioning procedure is now integrated with the following steps:
   6.1 Given the water flow calculated in the first calculation loop, the water flow losses are calculated with Equations 3.18–3.20.
   6.2 The water falling and pumping net heads are calculated with Equations 3.15 and 3.16, respectively.
   6.3 The annual operation simulation of the hybrid power plant is executed again and annual time series are developed for the hydro turbine power production and pump unit's absorbed power, based on the net heads calculated previously.
   6.4 The water falling and pumping flow rates are calculated again with Equations 3.22 and 3.23. If the new values differ considerably from the previously calculated ones, steps 6.1–6.3 are executed again.
   6.5 The procedure is iteratively executed until there is a satisfactory coincidence of the water flow rates between the consecutive calculation steps. Once this is achieved, the annual time series for the hydro power production, the power absorbed by the pump units, the RES plant direct penetration, and the water stored volume in the PHS upper reservoir constitute the fundamental results of the calculation process.
7. The above procedure is repeated for different nominal powers of the RES power plant. The final selected dimensioning is the one that fulfills the optimization criteria. These could purely be of energy nature, such as the achievement of annual RES penetration above a minimum percentage, or clearly economic, such as the optimization of the required investment's economic indices.

For example, an optimization criterion related to the hybrid plant's production could be the maximization of the annual electricity penetration of the RES unit E_{RESp}, yet without a corresponding significant increase of the RES annual rejected energy E_{rej}. As the RES plant's nominal power increases, the annual total electricity direct penetration and storage from the RES plant (RES penetration) $E_{RESp} + E_{st}$ obviously increases too. At the same time, the RES annually rejected energy E_{rej} increases as well, strongly depending on the storage capacity of the PHS system. The optimization criterion could be a RES annual penetration percentage higher than 90%, with simultaneous limitation of the annual RES rejection percentage lower than 10%. The achievement of this target is not always possible, because it depends on two major parameters:

- the available RES potential
- the available storage capacity of the PHS system.

The operation algorithm and the dimensioning procedure presented in this section for hybrid power plants with PHS systems as storage units is common for all hybrid power plants of this type, regardless their size.

3.5.1.2 CAES Systems as Storage Unit

Following a similar approach as the above section, the simulation of hybrid power plants with CAES systems as storage units is also executed with the method of annual time series, with, most commonly, an hourly calculation time step. The operation algorithm of hybrid power plants supported by CAES systems, aiming at the maximization of the annual RES penetration during the whole 24-hour period, is analyzed in the following steps:

1. For every calculation time step, the current available power production P_{RES} from the RES plant and the power demand P_d are introduced. Additionally, a maximum percentage p_{max} for the RES plant direct penetration versus the current power demand is defined, imposed by the system's security and stability requirements.
2. Following the above definitions, the direct RES penetration P_{RESp} is given as:
   i. if $P_{RES} \geq p_{max} \cdot P_d$, then $P_{RESp} = p_{max} \cdot P_d$
   ii. if $P_{RES} < p_{max} \cdot P_d$, then $P_{RESp} = P_{RES}$.
3. Once the RES direct penetration has been defined, there will be:
   i. a potential RES power production surplus: $P_{RES} - P_{RESp}$
   ii. a remaining power demand still uncovered: $P_d - P_{RESp}$.
4. The air mass m_{comp} is then calculated, which should be stored in the CAES air storage tank, in order to store the available power surplus $P_{RES} - P_{RESp}$ for the duration t of the time step ($\eta_{el,M}$, $\eta_{m,M}$, and $\eta_{m,comp}$ are the motor's electrical and mechanical and the compressor's mechanical average efficiency respectively during the current time step; Δh_c is the air's specific enthalpy difference in the compressor, assuming steady-state and adiabatic compression process):

$$m_{comp} = \frac{(P_{RES} - P_{RESp}) \cdot \eta_{el,M} \cdot \eta_{m,M} \cdot \eta_{m,comp} \cdot t}{\Delta h_c} \tag{3.25}$$

5. Similarly, we must calculate the air mass m_{turb} that should be removed from the CAES storage tank, so the remaining power demand $P_d - P_{RESp}$ will be produced by the air turbines for the duration t of the time step ($\eta_{el,G}$, $\eta_{m,G}$, and $\eta_{m,turb}$ are the generator's electrical and mechanical and the turbine's mechanical average efficiency respectively during the current calculation step; Δh_t is the air's specific enthalpy difference in the turbine, assuming steady-state and adiabatic expansion process):

$$m_{turb} = \frac{(P_d - P_{RESp}) \cdot t}{\Delta h_t \cdot \eta_{el,G} \cdot \eta_{m,G} \cdot \eta_{m,turb}} \tag{3.26}$$

6. The remaining air mass stored in the air tank after the end of the current calculation time step j will be:

$$m_{st}(j) = m_{st}(j-1) + m_{comp} - m_{turb}.$$

7. The new pressure in the air storage tank is calculated as:

$$p_{tank}(j) = \frac{m_{st}(j)}{V_{st}} \cdot R_m \cdot T_{tank}(j) \quad (3.27)$$

where V_{st} is the volume capacity of the storage tank and R_m is the atmospheric air constant, given by the following relationship versus the dry air constant R_d (R_d = 286.9 J/kg·K) and the ratio r of the humid versus the dry air density ρ_h and ρ_d respectively

$$R_m = R_d \cdot (1 + 0.61 \cdot r) \quad (3.28)$$

$$r = \frac{\rho_h}{\rho_d} \quad (3.29)$$

Typical fluctuation curves of the ratio r versus the ambient air temperature and for different atmospheric relative humidity are presented in Figure 3.19.

Given that the compressed air is cooled down after its compression, the air temperature inside the storage tank can be approximately considered equal to the current ambient temperature.

8. The air pressure $p_{tank}(j)$ in the CAES air storage tank is checked to see whether or not it exceeds the tank's maximum pressure p_{max}
   i. If $p_{tank}(j) > p_{max}$, then:
      turbine power production: $P_{turb} = P_d - P_{RESp}$
      thermal generators' power production: $P_{th} = 0$
      power storage: $P_{st} = 0$
      RES power rejection: $P_{rej} = P_{RES} - P_{RESp}$
      $m_{st}(j) = m_{st}(j-1) - m_{turb}$.
   ii. If $p_{tank}(j) \leq p_{max}$, we proceed to the next step.
9. The air pressure $p_{tank}(j)$ in the CAES air storage tank is checked to see whether or not it is lower than the minimum air pressure p_{min} in the tank, defined, commonly, by the atmospheric pressure or the minimum expansion rate of the involved air turbine model:
   i. If $p_{tank}(j) < p_{min}$, then:
      $P_{turb} = 0$
      $P_{th} = P_d - P_{RESp}$
      $P_{st} = P_{RES} - P_{RESp}$
      $P_{rej} = 0$
      $m_{st}(j) = m_{st}(j-1) + m_{comp}$.

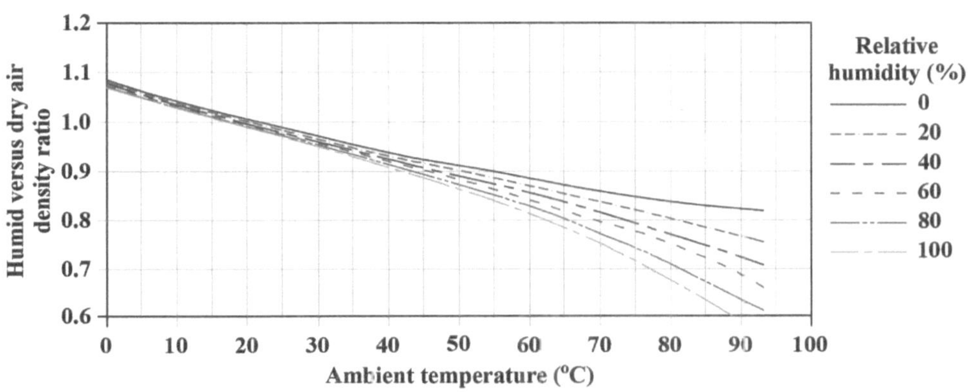

FIGURE 3.19 Typical fluctuation curves of the ratio r versus the ambient air temperature and for different atmospheric relative humidity [43].

ii. If $p_{min} \leq p_{tank}(j) \leq p_{max}$, then:
$P_{turb} = P_d - P_{RESp}$
$P_{th} = 0$
$P_{st} = P_{RES} - P_{RESp}$
$P_{rej} = 0$
$m_{st}(j) = m_{st}(j-1) + m_{comp} - m_{turb}$.

10. The pressure $p_{tank}(j)$ in the air storage tank at the end of the current calculation time step is calculated again by Equation 3.27, introducing the eventually remaining air mass $m_{st}(j)$ in the tank.

The above described operation algorithm is graphically presented in Figure 3.20.

The operation algorithm of a hybrid power plant supported by a CAES system is similar, as expected, to the corresponding algorithm of a hybrid power plant with a PHS system as a storage unit. The simultaneous power production and storage is also possible with the CAES hybrid power plant. This is ensured by the fundamental layout construction of a CAES system, because the air compression and expansion are two completely independent processes, implemented through separate pipeline networks.

With the execution of the above procedure, for every calculation time step the RES plant's direct penetration, the air turbine, and thermal generators (backup units) power production, and the power absorbed by the compressor during the storage process are calculated. Additionally, the air mass stored in and removed from the air storage tank and the remaining air mass, as well as the air pressure in the tank at the end of every time step, are calculated. With the integration of the operation algorithm over the annual time period, the annual electricity production and storage from the hybrid power plant's components are calculated, as well as the storage unit's charge level annual variation

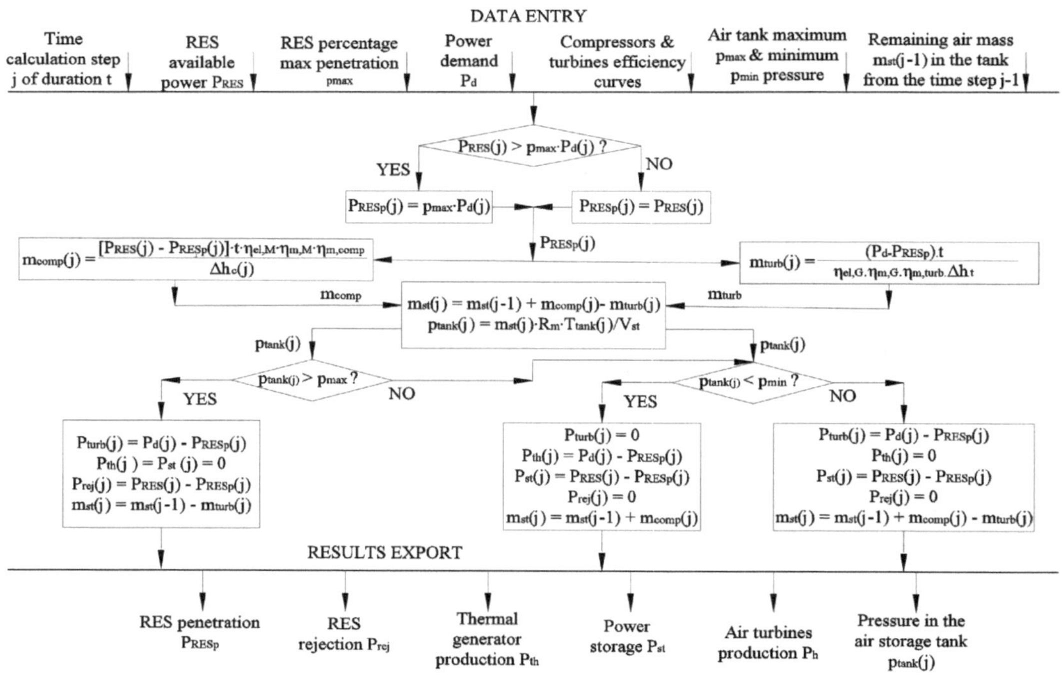

FIGURE 3.20 Operation algorithm graphical representation for a hybrid power plant with a CAES system as storage unit, aiming at 100% annual RES penetration.

(practically the air mass or the air pressure in the storage tank). With the availability of the power production time series, the annual power production synthesis graphs can be plotted. Finally, the maximum annual power production from the air turbines and the thermal generators, as well as the maximum annual power absorbed by the compressors, will define the minimum required values for the nominal power of the corresponding components.

In the above simulation, the air's specific enthalpy differences in the compressor and the turbine must be calculated. The theoretical background for the calculation of the involved air specific enthalpies is presented below.

- *Specific enthalpy difference in the compressor* Δh_c:

  The air's specific enthalpy difference in the compressor is equal to: $\Delta h_c = h_{c,out} - h_{c,in}$.

  The $h_{c,in}$ is the inlet ambient air specific enthalpy in the compressor. This specific enthalpy can be found on the ambient air's psychrometric chart, once the ambient air's conditions are known, e.g., the dry bulb temperature and the relative humidity.

  The air compression is approached by an isentropic process (reversible adiabatic), with the compression ratio β defined as:

$$\beta = \frac{p_{c,out}}{p_{c,in}} \tag{3.30}$$

where $p_{c,in}$ and $p_{c,out}$ are the inlet and outlet air pressure, respectively. Given that the inlet air pressure in the compressor is equal to the atmospheric pressure, the compression ratio must be at least equal to the existing pressure in the air storage tank from the previous calculation step. Practically the compression ratio will be defined by the operation point of the selected compressor's model with pressure ratio higher than the existing pressure in the air storage tank and the maximum possible efficiency.

Additionally, for an isentropic compression process, the following equation stands ($T_{c,in}$ the inlet air temperature in the compressor and $T^{is}_{c,out}$ the outlet air temperature from the compressor following an isentropic process):

$$\frac{T^{is}_{c,out}}{T_{c,in}} = \left(\frac{p_{c,out}}{p_{c,in}}\right)^{\frac{k-1}{k}} \tag{3.31}$$

where k is the ratio of the outlet air specific heat under constant pressure c_P and volume c_V:

$$k = \frac{c_p}{c_v} \tag{3.32}$$

The ratio k depends on the outlet air temperature and pressure. The values of the ratio k for temperatures from 250–1,000 K are presented in Table 3.6 for ideal gases and atmospheric pressure. Assuming atmospheric air is an ideal gas, the ratio k ranges from 1.400, for temperatures around 25°C, to 1.336, for temperatures around 725°C. For atmospheric conditions, the ratio k can be considered equal to 1.4 [44].

The ratio k of air versus temperature and pressure can be calculated with the c_p and c_v curves presented in Figure 3.21.

The real outlet temperature $T_{c,out}$ is calculated using the compressor's isentropic efficiency η_c, defined as:

$$\eta_c = \frac{T^{is}_{c,out} - T_{c,in}}{T_{c,out} - T_{c,in}} \Leftrightarrow T_{c,out} = \frac{T^{is}_{c,out} - T_{c,in}}{\eta_c} + T_{c,in} \tag{3.33}$$

TABLE 3.6
Atmospheric Air Properties (Assuming An Ideal Gas) Versus Its Temperature

Temperature (K)	Temperature (°C)	Specific Heat Under Constant Pressure c_P (kJ/kg·K)	Specific Heat Under Constant Volume c_V (kJ/kg·K)	Specific Heat Ratio $k = c_P/c_V$
250	–23	1.003	0.716	1.401
300	27	1.005	0.718	1.400
350	77	1.008	0.721	1.398
400	127	1.013	0.726	1.395
450	177	1.020	0.733	1.392
500	227	1.029	0.742	1.387
550	277	1.040	0.753	1.381
600	327	1.051	0.764	1.376
650	377	1.063	0.776	1.370
700	427	1.075	0.788	1.364
750	477	1.087	0.800	1.359
800	527	1.099	0.812	1.353
900	627	1.121	0.834	1.344
1,000	727	1.142	0.855	1.336

The compressor's isentropic efficiency is derived by its operation point defined on the basis of the existing pressure in the air storage tank from the previous calculation step and the maximum possible efficiency.

From Equations 3.31 and 3.33, with the ratio k for the outlet air temperature in the compressor calculated with Figure 3.21, and knowing the compression ratio, the real outlet air temperature $T_{c,out}$ from the compressor can be calculated.

The specific enthalpy $h_{c,out}$ of the compressed outlet air can be now calculated by the relationship:

$$h_{c,out} = h_{air} + h_w = c_{p,air} \cdot T_{c,out} + w \cdot \left(c_{p,w} \cdot T_{c,out} + h_{w,e}\right) \tag{3.34}$$

where:

$h_{air} = c_{p,air} \cdot T_{c,out}$:	the specific enthalpy of the compressed dry air
$h_w = w \cdot (c_{p,w} \cdot T_{c,out} + h_{w,e})$:	the specific enthalpy of the outlet water vapor from the compressor
$c_{p,air}$:	the specific heat under constant pressure of the compressed air, given in Table 3.7 versus its temperature $T_{c,out}$ and pressure $p_{c,out}$
w:	the humidity ratio (in kg of water vapor per kg of dry air) of the compressed air, which can be found on the ambient air psychrometric chart, for the inlet air stream conditions (e.g., dry bulb temperature and relative humidity)
$c_{p,w}$:	the specific heat of water, assumed equal to 4.184 kJ/(kg·K)
$h_{w,e}$:	the evaporation heat of water at 0°C, assumed equal to 2,501 kJ/kg
$T_{c,out}$:	the outlet air temperature from the compressor in K.

Once the specific enthalpies $h_{c,out}$ and $h_{c,in}$ are known, their difference Δh_c can be also calculated.

- *Specific enthalpies difference in the turbine Δh_t:*

  The specific enthalpy difference in the turbine is equal to: $\Delta h_t = h_{t,in} - h_{t,out}$.

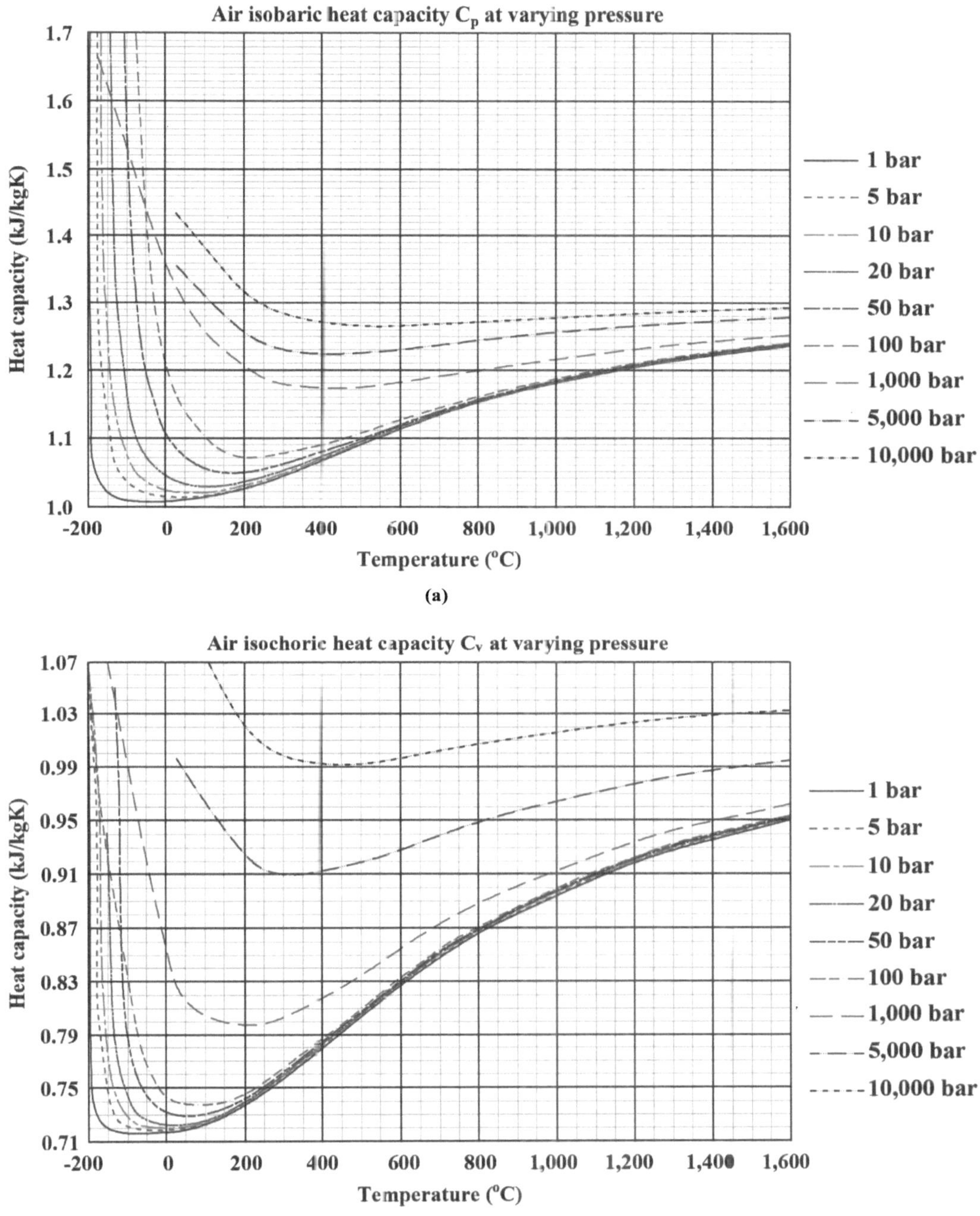

FIGURE 3.21 (**a**) Typical fluctuation curves of the air specific heat c_p versus temperature and pressure [45]. (**b**) Typical fluctuation curves of the air specific heat c_v versus temperature and pressure [45].

TABLE 3.7
Hydrogen's Specific Heat Under Constant Pressure c_p and Volume c_v, and their Ratio k versus Temperature

Temperature (K)	Temperature (°C)	Specific Heat Under Constant Pressure c_P (kJ/kg·K)	Specific Heat Under Constant Volume c_V (kJ/kg·K)	Specific Heat Ratio $k = c_P/c_V$
250	−23	14.051	9.927	1.416
300	27	14.307	10.183	1.405
350	77	14.427	10.302	1.400
400	127	14.476	10.352	1.398
450	177	14.501	10.377	1.398
500	227	14.513	10.389	1.397
550	277	14.530	10.405	1.396
600	327	14.546	10.422	1.396
650	377	14.571	10.447	1.395
700	427	14.604	10.480	1.394
750	477	14.645	10.521	1.392
800	527	14.695	10.570	1.390
900	627	14.822	10.698	1.385
1,000	727	14.983	10.859	1.380

Regardless of the use of an adiabatic or a conventional CAES system, the air, once released from the air storage tank, is heated either in a combustion chamber or through the heat recovery from the heat storage. Hence, in both alternative CAES layouts, the released air from the air tank is heated before it is led into the turbine, thus increasing its specific enthalpy.

The inlet air temperature $T_{t,in}$ in the turbine is defined according to the turbine's operation specifications. For a conventional CAES, usually it ranges between 1,000°C and 1,200°C. For an AA-CAES, the inlet air temperature is usually set at 600°C. A major parameter that affects the inlet air temperature in the turbine is the required power production. Depending on the available operation points, depicted in the selected turbine model's charts, in case of low required power production, the inlet air temperature may be even lower than the above presented values. Specifically, the required power production from the turbine is given by the following relationship, versus the inlet air temperature:

$$P_t = \dot{m}_{turb} \cdot c_{p,air} \cdot \left(T_{t,in} - T_{t,out}\right) \tag{3.35}$$

By assuming an inlet air temperature $T_{t,in}$, once the air mass flow rate $\dot{m}_{turb}$ and the real output air temperature $T_{t,out}$ have been calculated following the process described below, the turbine power production P_t can then be calculated. In case this power is higher than required, the inlet air temperature is reduced and the process is executed again. Eventually, once the inlet air temperature has been determined, its specific enthalpy $h_{t,in}$ can be calculated using Equation 3.34, by replacing $T_{c,out}$ with $T_{t,in}$. All of the other involved magnitudes are known (w, $c_{p,air}$, $h_{w,e}$, $c_{p,w}$). The humidity ratio w can be assumed equal to that of the inlet air stream in the compressor, considering that the whole CAES network is a closed system, without any water vapor interchange between it and the ambient environment. The inlet air specific heat $c_{p,air}$ can be found in Figure 3.21 versus its temperature $T_{t,in}$ and pressure $p_{t,in}$.

Regarding the outlet air specific enthalpy $h_{t,out}$ from the turbine, assuming an isentropic expansion process in the turbine, the following relationship stands ($T_{t,in}$ the inlet air temperature and $T_{t,out}^{is}$ the outlet air temperature after the isentropic process):

$$\frac{T_{t,in}}{T_{t,out}^{is}} = \left(\frac{p_{t,in}}{p_{t,out}}\right)^{\frac{k-1}{k}} \quad (3.36)$$

If the expansion ratio π is known:

$$\pi = \frac{p_{t,in}}{p_{t,out}} \quad (3.37)$$

then the outlet air temperature $T_{t,out}^{is}$ of the isentropic process can be calculated with Equation 3.36. Given that the output pressure from the turbine equals the atmospheric, the expansion ratio is practically defined by the existing pressure in the air storage tank and the available operation points of the selected turbine's model, which exhibit an expansion ratio lower than the air storage tank pressure.

The real outlet air temperature $T_{t,out}$ from the turbine can be now calculated from the isentropic temperature $T_{t,out}^{is}$ and the turbine's isentropic efficiency η_t, derived by the selected operation:

$$\eta_t = \frac{T_{t,in} - T_{t,out}}{T_{t,in} - T_{t,out}^{is}} \Leftrightarrow T_{t,out} = T_{t,in} - \eta_t \cdot \left(T_{t,in} - T_{t,out}^{is}\right) \quad (3.38)$$

Finally, the specific enthalpy $h_{t,out}$ of the outlet air stream from the turbine can be calculated with Equation 3.34, by replacing $T_{c,out}$ with $T_{t,out}$. Again, once the outlet air temperature $T_{t,out}$ has been calculated, all the involved magnitudes in the relationship are known. The specific heat $c_{p,air}$ can be found in Table 3.6 versus the outer air temperature $T_{t,out}$ for atmospheric pressure.

With the calculation of the air-specific enthalpy differences in the compressor and the turbine, the computational execution of the hybrid power plant's operation algorithm, as described previously, is now possible.

By dividing both parts of Equation 3.26 with the duration t of the calculation step, the required air mass flow rate $\dot{m}_{turb}$ is calculated. Once the air mass flow rate $\dot{m}_{turb}$ is known, the thermal energy required for the air heating before its expansion can be calculated. Specifically, the provided thermal power $\dot{Q}_t$ by the external heat source in the air stream is given by:

$$\dot{Q}_t = \dot{m}_{turb} \cdot c_{p,air} \cdot \left(T_{t,in} - T_{st}\right) \quad (3.39)$$

where T_{st} is the air temperature in the storage tank, roughly equal to the ambient temperature, assuming isothermal storage, and $T_{t,in}$ the inlet air temperature in the turbine, calculated as presented just above. The specific heat $c_{p\,air}$ is found in Figure 3.21 versus the stored air temperature T_{st} and pressure $p_{t,in}$.

By calculating the externally provided thermal power $\dot{Q}_t$ for the heating of the inlet air flow in the turbine for every time step and by integrating over the annual time period, the annually offered thermal energy is calculated. For a conventional CAES, the annually consumed fossil fuel's mass m_f or volume V_f for the air heating before its expansion in the turbine is then given by:

$$m_f \ (\text{or } V_f) = \frac{\sum\left(\dot{Q}_t\right) \cdot t}{H_u \cdot \eta_c \cdot \eta_{ex}} \quad (3.40)$$

where:

- H_u: the consumed fuel heat capacity in kWh/kg or in kWh/lt
- η_c: the combustion chamber efficiency
- η_{ex}: the heat exchanger efficiency for the heat transfer from the combustion chamber to the inlet air flow.

In general, in order to introduce the air pressure drop to the whole calculation process, an iterative procedure should be executed in the following stages:

1. The computational simulation of the plant's annual operation is executed, based on the above described operation algorithm. Among the results, the annual time series of the air mass flow rates from the compressor to the air storage tank and from the latter to the turbine are developed. For the air temperature in the storage tank, the air volume flow rates are calculated, because the air is transferred in the pipelines between the compressor, the turbine, and the tank under the prevailing conditions in the storage tank.
2. The total air pressure drop due to friction losses is calculated separately for the compressor's discharge and the turbine's supply pipelines. The air pressure drops in the pipelines to and from the air storage tank are calculated as follows:

$$\Delta p = \frac{450 \cdot q_c^{1.85} \cdot L}{p \cdot D^5} \tag{3.41}$$

where:

q_c: air volume flow rate in the pipelines under standard conditions (pressure 1 bar, temperature 20°C and 0% relative humidity) or, alternatively free air delivery (FAD) flow in lt/s
L: overall linear pipeline length in m
p: absolute inlet pressure in bar
D: pipeline inner diameter in mm.

Apart from the air pressure drop in the pipelines, there is also pressure drop in the air network components (valves, elbows, diaphragms, junctions, etc.). These localized pressure drops are approached by corresponding equivalent pipe lengths to each different component and, eventually, again using Equation 3.41. These equivalent pipe lengths can be provided by the components' manufacturers or in the form of tables contained in the relevant literature.

The total pressure drop in the involved compressed-air networks (from the compressor to the tank and from the tank to the turbine) comes from the sum of the total pressure drop in the pipelines and the contained components.

3. The selected pipeline's diameter should ensure a total pressure drop Δp below a maximum acceptable limit, for example 2% of the outlet air pressure from the compressor or the air tank, or, alternatively, a maximum absolute pressure drop of 0.1 bar.
4. Finally, the air transfer efficiencies to and from the storage tank are calculated, defined as percentages versus the initially available pressure at the beginning of each pipeline:

$$\eta_{tr,comp} = \frac{p_{c,out} - \Delta p_{comp}}{p_{c,out}} \tag{3.42}$$

$$\eta_{tr,turb} = \frac{p_{tank} - \Delta p_{turb}}{p_{tank}} \tag{3.43}$$

where:

$\eta_{tr,comp}$: air transfer efficiency from the compressor to the storage tank
$\eta_{tr,turb}$: air transfer efficiency from the storage tank to the turbine
$p_{c,out}$: outlet air pressure from the compressor
p_{tank}: air pressure in the storage tank.

The outlet air pressure $p_{c,out}$ from the compressor and the air pressure in the storage tank $p_{tank}(j)$ have been calculated through the process described previously.

5. The following magnitudes are calculated again:
   i. the air mass m_{comp} that should be stored in the storage tank in order to store the RES power surplus $P_{RES} - P_{RESp}$ during the calculation time step of duration t, recalculating the corrected specific enthalpy difference $\Delta h_{c,corr}$, with the updated pressure ratio, by subtracting the pressure drop Δp_{comp} from the final achieved air pressure in the tank p_{tank}:

$$m_{comp} = \frac{(P_{RES} - P_{RESp}) \cdot \eta_{el,M} \cdot \eta_{m,M} \cdot \eta_{m,comp} \cdot t}{\Delta h_{c,corr}} \tag{3.44}$$

   ii. the air mass m_{turb} that should be removed from the storage tank, in order to produce the remaining power of $P_d - P_{RESp}$ from the air turbines during the calculation time step of duration t, recalculating the corrected specific enthalpy difference $\Delta h_{t,corr}$, with the updated pressure ratio, by subtracting the pressure drop Δp_{turb} from the final air pressure inlet from the tank p_{tank}:

$$m_{turb} = \frac{(P_d - P_{RESp}) \cdot t}{\Delta h_{t,corr} \cdot \eta_{el,G} \cdot \eta_{m,G} \cdot \eta_{m,turb}} \tag{3.45}$$

6. Steps 1–4 of the above described dimensioning procedure are executed again, leading to updated results for the turbine power production, the compressor absorbed power, the air mass flow rate to and from the storage tank, and the final pressure in the tank. The results from the last iterative calculation loop are compared to the results from the previous loop. If there is significant divergence between them, then Steps 1–4 of the current optimization process are executed again.

Similarly to hybrid power plants with PHS systems, the above presented dimensioning methodology for hybrid plants with CAES systems is iteratively executed with different RES units power or different storage tank features (i.e., storage pressure and volume capacity), in order to determine the optimum dimensioning scenario, according to the predefined optimization criteria. As explained in the earlier section, these criteria can be

- of energy interest, such as the maximization of the hybrid power plant's annual penetration in the electrical system, keeping, at same time, the annual RES production surplus below a maximum acceptable level
- of economic interest, such as the minimization of the hybrid plant's life cycle cost, or the optimization of the investment's economic indices.

The above operation algorithm and dimensioning optimization process for hybrid power plants with CAES systems remains the same, regardless of the size of the examined power plant.

With the dimensioning procedure for hybrid power plants with CAES systems, the basic layout practically does not differ from the one presented in the previous section, for hybrid plants with PHS systems as storage units. Any differences between the two methodologies are traced on the secondary relationships involved in the stored energy calculation, rather than the fundamental dimensioning algorithm. This is due to the fact that in the first examined systems, energy is stored in the form of the gravitational energy of water, whereas, in the second case, energy is stored in the form of compressed air's elastic energy.

From the above analysis, it is perceivable that the presented methodologies are developed on the basis of the theoretical background of thermodynamics and turbomachinery. An alternative, accurate, and quite simplified approach is based on the use of the involved compressor's and turbine's charts for the determination of their operation points for each calculation step, according to the available operation conditions, practically the required power production, available power for storage, and existing pressure in the air storage tank.

Typical theoretical charts for a compressor unit with nominal power 1.2 MW are presented in Figure 3.22. As seen in this figure, the characteristic charts are provided in the form of the compressor's pressure ratio and the efficiency fluctuation curves versus the air mass flow rate and the levelized rotational speed. The levelized rotational speed is the ratio of the compressor's rotational speed over its nominal rotational speed.

Given the available power for storage and the existing pressure in the air storage tank, the compressor's operation point is determined with the following process:

1. All the operation points with pressure ratio higher than the existing pressure in the air storage tank are selected. Given that the inlet air pressure is equal to the atmospheric, the above condition practically means that each one of the selected operation points will give output air pressure higher than the existing pressure in the air storage tank.
2. The selected operation points are put in descending order versus their efficiency (starting at the highest and ending with the lowest efficiency).

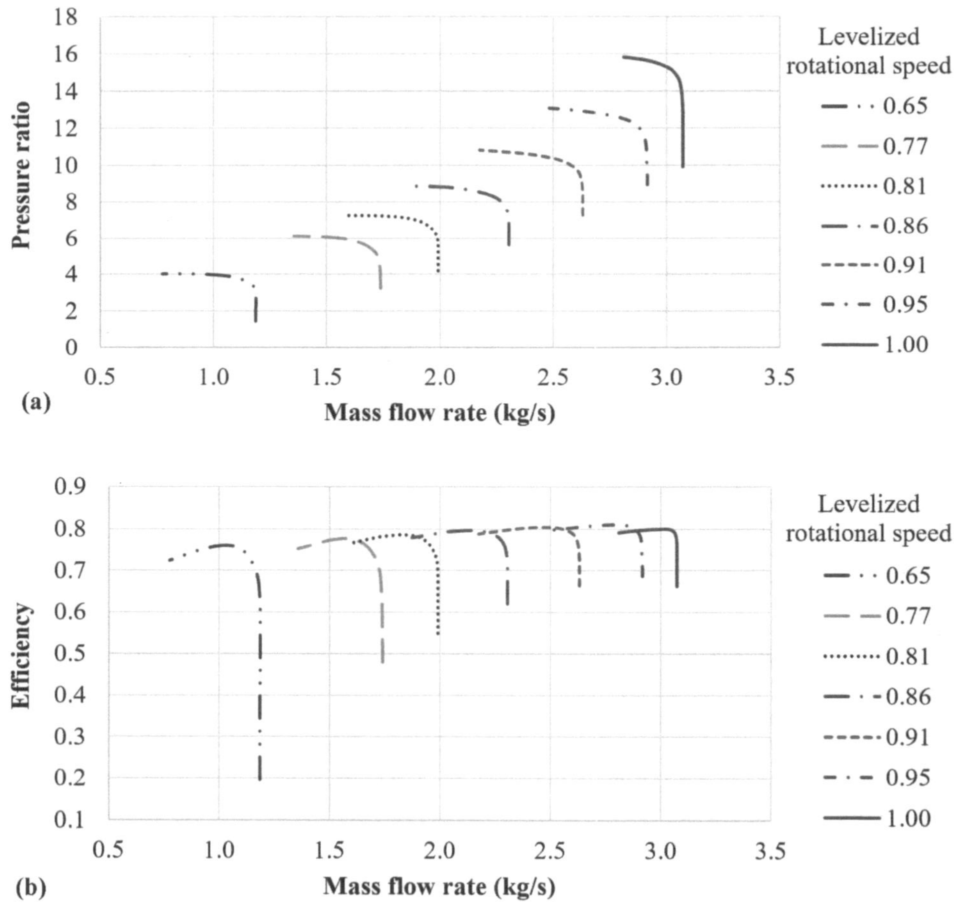

FIGURE 3.22 (a) Characteristic theoretical charts for a compressor unit of 1.2 MW nominal power: pressure ratio fluctuation versus air mass flow rate for different rotational speeds. (b) Characteristic theoretical charts for a compressor unit of 1.2 MW nominal power: efficiency fluctuation versus air mass flow rate for different rotational speeds.

3. For each of the selected points and following the descending order set in the previous step:
   a. First, the real outlet air temperature $T_{c,out}$ from the compressor is calculated, with the relationships 3.30 and 3.32, once the involved pressure ratio and the isentropic efficiency have been retrieved from each operation point
   b. Next, the power absorbed by the compressor is calculated with the following equation, until it becomes lower than the available power for storage (the air mass flow rate has been also retrieved from each operation point):

$$P_c = \dot{m}_{comp} \cdot c_{p,air} \cdot \left(T_{c,out} - T_{c,in}\right) \tag{3.46}$$

4. For better accuracy on the absorbed power calculation, a linear interpolation can be applied for the first operation point that fulfills the above set condition (power absorbed by the compressor lower than the available power for storage) and the previous one in the operation points' descending order.

The following special cases are distinguished:

1. If the available for storage power is very high with regard to the existing pressure in the air storage tank, then the absorbed power by the compressor can be lower than the available for storage power directly with the first operation point in the descending order (operation point with the maximum efficiency). In this case, the available for storage power is too high to be fully absorbed by the compressor. The eventually absorbed power will be equal to the maximum possible power that can be absorbed by the compressor.
2. If the available for storage power is very low with regard to the existing pressure in the air storage tank, such that none of the compressor's available operation points gives power absorption lower than this, then the available for storage power cannot be stored. The power storage in this case will be null.

Typical theoretical charts for a turbine unit with 1.0 MW nominal power output are presented in Figure 3.23 in the form of the turbine's expansion ratio and efficiency fluctuation curves versus the air mass flow rate and the turbine's levelized rotational speed.

Following an approach similar to the compressor case, the determination of the turbine's operation point is based on the following steps:

1. All the turbine's operation points are selected with an expansion ratio lower than or equal to the existing pressure in the air storage tank, given that the output air pressure from the tank equals the atmospheric pressure.
2. The selected points are put in descending order versus their efficiency (starting at the highest and ending at the lowest efficiency).
3. For each of the selected points, following their descending order:
   a. The real output air temperature $T_{t,out}$ is calculated with Equations 3.36 and 3.37, having retrieved the involved magnitudes (turbine's isentropic efficiency, expansion ratio) from the selected operation points.
   b. The final power production from the turbine is calculated with Equation 3.35, until it becomes higher than or equal to the required power production. The involved air mass flow rate in Equation 3.35 is also retrieved from the selected operation points.
4. For better accuracy on the turbine's power production calculation, a linear interpolation is applied between the first point that fulfills the above set condition (power production higher than the required) and the previous one in the operation points' descending order.

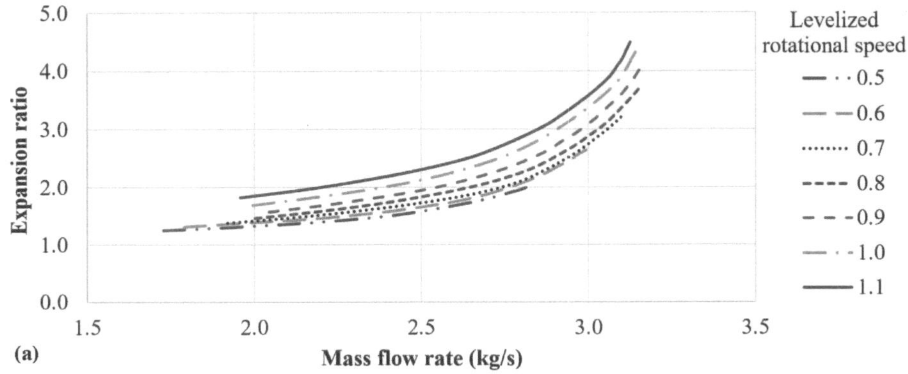

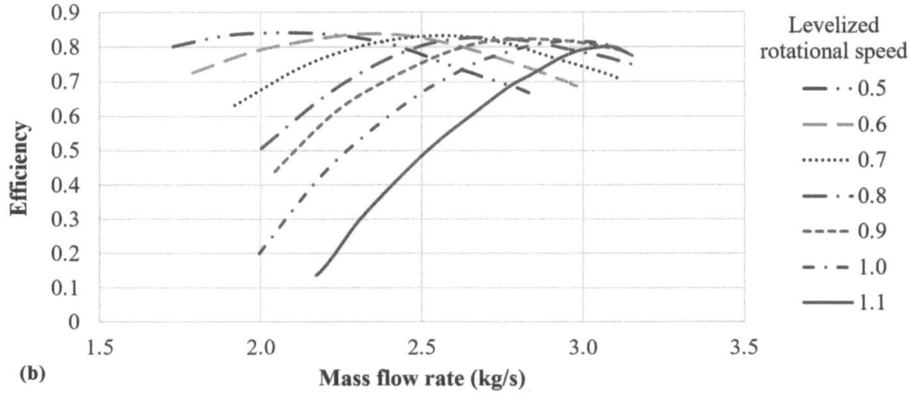

FIGURE 3.23 (**a**) Characteristic theoretical charts for a turbine of 1.0 MW nominal power output: expansion ratio fluctuation versus air mass flow rate for different rotational speeds. (**b**) Characteristic theoretical charts for a turbine of 1.0 MW nominal power output: efficiency fluctuation versus air mass flow rate for different rotational speeds.

The following special cases are distinguished:

1. If the required power production is very high with regard to the turbine's size or the existing pressure in the air storage tank, then the turbine's power production is set equal to the maximum possible, given the above mentioned technical parameters (turbine's size and existing pressure in the air storage tank).
2. If the required power production is very low with regard to the turbine's size, such that the power production from the turbine is higher than that required directly with the first operation point in the descending order, then the air inlet temperature in the turbine can be gradually reduced, until the turbine's power production becomes equal to the required. The minimum air inlet temperature in the turbine can be theoretically equal to the ambient temperature. If the turbine's power production still remains higher than the required for ambient inlet air temperature, then the inlet air mass flow rate can be alternatively gradually reduced.

Following the above processes, for each calculation step the operation points of the compressor and the turbine are selected and, in this way, all the involved magnitudes in the simulation process (air mass flow rate, pressure/expansion ratio, isentropic efficiency) are derived by the compressor's and turbine's charts. The operation power production and storage can then be calculated with Equations 3.35 and 3.46, respectively. With this approach, all the thermodynamic calculations are avoided.

3.5.2 Hybrid Power Plants for Power Peak Shaving

In large-size electrical systems, a single hybrid power plant is most possibly not able to support high RES penetration, with regard to the overall power demand, mainly due to economic and practical, constructive constraints. In such cases, the hybrid power plants are usually introduced is sensible sizes, usually much lower than the peak power demand, aiming at guaranteed power production during the power demand peak periods [46, 47], more specifically aiming at the so-called power peak shaving.

The hybrid plant's operation algorithm in this case obviously changes. Consequently, the dimensioning methodology is certainly different from the ones presented in the previous sections, aiming at high RES annual penetration percentages (higher than 70%). Generally, the operation algorithm of a hybrid power plant aiming at power peak shaving, regardless of the involved storage technology, is analyzed with the fundamental steps presented below.

Once again the procedure is accomplished using an annual time series of average values for a predefined time step duration (usually hourly). First of all, we must determine whether or not the current calculation time step falls into a power peak demand period. The following cases are distinguished:

A. *Guaranteed power production periods (peak power demand periods):*

1. During peak power demand periods, there should be enough energy stored in the hybrid plant's storage unit, in order to be able to satisfy the requirement for specific, usually predefined power production (guaranteed power production) for a certain time period. From the previous day, the available energy stored in the storage unit has been estimated and the guaranteed power production that can be provided for a specific time interval during the next day's peak periods has been determined accordingly.
2. For the current calculation step, the available power production P_{RES} from the RES unit and the predefined guaranteed power production from the hybrid power plant P_g are introduced. Additionally, a maximum RES direct penetration percentage p_{max} versus the predefined guaranteed power production P_g is set.
3. Following the above definitions, the direct RES contribution P_{RESp} on the guaranteed power production will be given as:
   i. if $P_{RES} \geq p_{max} \cdot P_g$, then $P_{RESp} = p_{max} \cdot P_g$
   ii. if $P_{RES} < p_{max} \cdot P_g$, then $P_{RESp} = P_{RES}$.
4. The guaranteed power production from the hybrid power plant's storage unit (hydro or air turbines) will be:

$$P_h \left(\text{or } P_{turb}\right) = P_g - P_{RESp}.$$

5. The removed water volume V_h from the PHS system's upper reservoir, or the removed air mass m_{turb} from the CAES system's air storage tank, is calculated according to Equations 3.14 and 3.45, respectively.

B. *Anytime during a 24-hour period:*

6. The RES power production surplus, available for storage, is given by the relationship:

$$P_{sur} = P_{RES} - P_{RESp}.$$

7. The water volume V_p that should be stored in the PHS system's upper reservoir or the air mass m_{comp} that should be stored in the CAES system's air storage tank, in order to store the RES power production surplus P_{sur} for a time step of duration t are calculated with Equations 3.13 and 3.44, respectively.
8. The remaining water volume or air mass in the corresponding storage reservoirs at the end of the current calculation step j will be:

$$V_{st}(j) = V_{st}(j-1) + V_p - V_h$$

$$m_{st}(j) = m_{st}(j-1) + m_{comp} - m_{turb}.$$

9. The remaining water volume $V_{st}(j)$ in the PHS upper reservoir or the air pressure $p_{tank}(j)$ in the CAES air storage tank is checked whether or not it exceeds the reservoir's maximum storage capacity V_{max} or pressure p_{max}, respectively:
   i. If $V_{st}(j) > V_{max}$ (or if $p_{tank}(j) > p_{max}$) then:
      P_h (or P_{turb}) $= P_g - P_{RESp}$
      $P_{th} = 0$
      $P_{st} = 0$
      $P_{rej} = P_{RES} - P_{RESp}$
      $V_{st}(j) = V_{st}(j-1) - V_h$ or $m_{st}(j) = m_{st}(j-1) - m_{turb}$.
   ii. If $V_{st}(j) \leq V_{max}$ (or if $p_{tank}(j) \leq p_{max}$) we proceed to the next step.
10. The remaining water volume $V_{st}(j)$ in the PHS upper reservoir or the air pressure $p_{tank}(j)$ in the CAES air storage tank is checked whether or not it is lower than the reservoir's minimum contained water volume V_{min} or the air storage tank's minimum pressure p_{min}, respectively:
   i. If $V_{st}(j) < V_{min}$ ($p_{tank}(j) < p_{min}$), then:
      P_h (or P_{turb}) $= 0$
      $P_{th} = P_g - P_{RESp}$
      $P_{st} = P_{RES} - P_{RESp}$
      $P_{rej} = 0$
      $V_{st}(j) = V_{st}(j-1) + V_p$ (or $m_{st}(j) = m_{st}(j-1) + m_{comp}$).
   ii. If $V_{min} \leq V_{st}(j) \leq V_{max}$ (or $p_{min} \leq p_{tank}(j) \leq p_{max}$), then:
      P_h (or P_{turb}) $= P_g - P_{RESp}$
      $P_{th} = 0$
      $P_{st} = P_{RES} - P_{RESp}$
      $P_{rej} = 0$
      $V_{st}(j) = V_{st}(j-1) + V_p - V_h$ (or $m_{st}(j) = m_{st}(j-1) + m_{comp} - m_{turb}$).

The iterative procedure for the calculation of the flow losses or the pressure drop calculation in the water or air transfer pipelines, described in the previous sections, should be also applied in this case.

The operation algorithm of a hybrid power plant for power peak shaving is presented in Figure 3.24.

From the above operation algorithm, it is concluded that the possibility of continuous energy storage in the hybrid power plant, even during the peak power demand periods, namely during the guaranteed power production period, still remains an aspiration. This implies that the hybrid power plant should be equipped with the ability of simultaneous energy storage and production. As mentioned in the earlier sections, this feature is definite for CAES systems, while for PHS systems it is provided by the installation of a double penstock.

However, it should be noted that the requirement of simultaneous energy storage and production in case of hybrid power plants for peak power shaving is not crucial, as with hybrid power plants

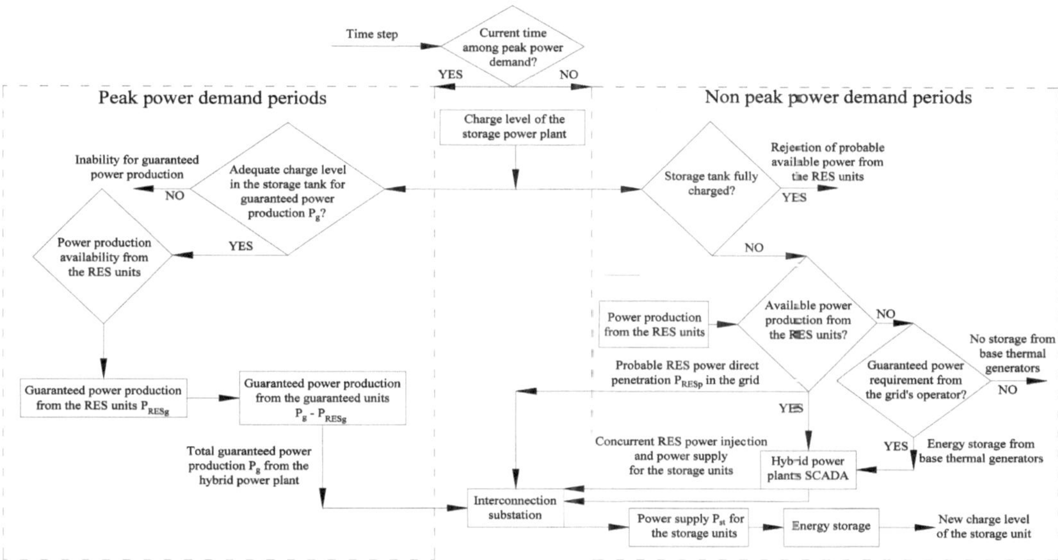

FIGURE 3.24 Operation algorithm of a hybrid power plant for power peak shaving with a PHS system as the storage unit.

aiming at RES penetration maximization. For example, in case of a PHS system with a single penstock, the hybrid power plant will not be able to store energy only during the guaranteed power production period. On the dilemma of which option is better, namely the construction of a double or a single penstock in a PHS system, there is not a definite answer. Practically, the installation of a single penstock will certainly lead to the reduction of the annual stored energy. However, ultimately, the crucial point is not the annual energy stored, but the annual guaranteed energy production. Potentially, the reduced annual energy stored can lead to a corresponding reduction of the annual guaranteed electricity production. Yet, this is not certain in advance, because it depends on the RES potential availability and the storage capacity of the storage plant. For example, in the case of a high RES potential availability and large storage capacity, the construction of a single penstock maybe will not impose any reduction at all on the annual guaranteed electricity production, due to the large amounts of stored energy available.

Yet, even if the installation of a single penstock of a PHS system leads to the reduction of the annual guaranteed electricity production, it should be checked whether this reduced electricity production causes a considerable reduction on the investment's anticipated revenues. In any case, the ultimate question should be if the additional setup cost of a double penstock, required to avoid this revenue reduction, could be compensated by a reasonable payback period caused by the anticipated increase on the investment's revenues. Conclusively, the installation of a double penstock should be investigated as a parameter of the overall dimensioning optimization of the hybrid power plant, based on specific predefined criteria. If the optimization criterion is the RES annual penetration maximization, then perhaps the installation of a double penstock will be a necessity, depending, of course, on the available RES potential and the storage capacity of the storage unit. If the optimization criterion is the optimization of the required investment's economic indices, then maybe the installation of a double penstock would not be the best option.

Finally, another parameter that can affect the decision of whether or not to install a double penstock in a PHS system integrated with a RES plant for power peak shaving, is the overall flexibility and the contribution of the hybrid plant to the dynamic security of the electrical system. This criterion, although always important, in cases of large electrical systems, will not perhaps be very crucial for hybrid power plants introduced for power peak shaving, specifically due to the relatively small size of the hybrid power plant compared to the size of the electrical system and the limited guaranteed power production periods only during the peak power demand periods.

3.6 OPERATION ALGORITHMS OF SMALL-SIZE HYBRID POWER PLANTS

A hybrid power plant of small size can be theoretically introduced in an electricity system either for power peak shaving or aiming at RES penetration maximization. However, practically, in most cases the small-size hybrid plants are introduced with the second operation concept, mainly for the following reasons:

- In small electricity systems, the low power consumption can be easily and continuously covered up to 100%, without the necessity of large-size, difficult, and expensive technical works.
- In small hybrid power plants, the storage unit's relatively high procurement cost can be compensated only with the maximum possible electricity production from the hybrid plant, approached with continuous power production for the whole 24-hour period and not only during the peak power demand hours.
- In small electricity systems, there is practically no difference between base and peak thermal generators. Most commonly in such systems the conventional thermal power plants are equipped exclusively with diesel generators. Consequently, given the absence of thermal generators with higher production cost (gas turbines), there is no difference on the electricity production cost between the low and the high power demand periods. Hence, there is no technical or economic expediency for power peak shaving.

Given the above list, the next section presents the operation algorithm of small-size hybrid power plants, with power demand from 100 kW to 1 MW, as well as for hybrid power plants of very small size, with power demand lower than 100 kW. The first case may refer to small non-interconnected islands or remote settlements with a permanent population of some hundreds of inhabitants, while the second case may refer to autonomous electricity consumptions, such as isolated residences, cottages, stock-farming units, desalination plants, lighthouses, and biological treatment facilities.

3.6.1 Hybrid Power Plants of Small Size

The general layout of a small-size hybrid power plant is presented in Figure 3.25. As explained in the previous sections, the RES units in such a hybrid plant can be a small wind park and a PV station. Electrochemical batteries seem to be the most feasible option as storage units (lead acid batteries, floating batteries, lithium ion batteries, etc.).

In Figure 3.25, an electrolysis unit is also involved as an alternative storage unit. This unit can be complementarily introduced in the hybrid power plant for hydrogen production and be utilized for energy storage whenever the basic storage units are fully charged.

All the power production units (RES and backup units), as well as the storage devices and the consumers, are connected to the same bus. This implies that the response of the power electronics (inverters) that determine the performance of the power production and storage units versus the power demand fluctuation is direct. As a result of the overall grid's topology, the RES unit's direct instant penetration can be up to 100%. Even in this case, the system can be considered secure, on the condition that there is always a maximum discharge depth kept for the storage units, which should not be higher than 60%. On this prerequisite, it is ensured that there will always be enough energy stored in the storage units, in order to guarantee the power demand coverage in case of a sudden power production loss, until the backup thermal generators are dispatched.

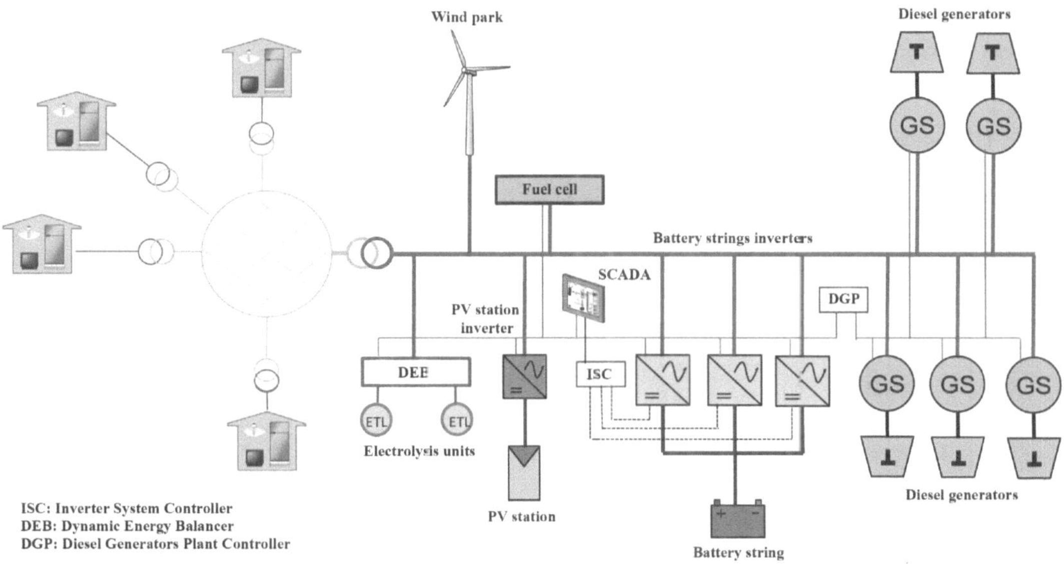

FIGURE 3.25 General layout of a small-size hybrid power plant [10].

Electrochemical batteries are grouped in strings or modules (depending on the specific technology). In case of a battery string with N units connected in series with nominal voltage V of each unit, the total string voltage V_{st} will be:

$$V_{st} = V \cdot N. \tag{3.47}$$

If I_{max} is the maximum permitted electrical current in the involved battery units, then the nominal charge/discharge rate of the battery string will be:

$$P_{bat} = V_{st} \cdot I_{max}. \tag{3.48}$$

The maximum electrical current I_{max} is determined by the battery manufacturer, imposed by specific technical parameters, directly affecting the battery's integrity and safety, such as the maximum temperature that can be born from the battery's constructive materials.

If the storage capacity of each battery unit C_n is given in Ah (e.g., for lead acid batteries), then the nominal capacity C_{bat} in Wh for each battery string will be:

$$C_{bat} = V_{st} \cdot C_n \tag{3.49}$$

where V_{st} is the battery string nominal voltage in volts.

If d_{dis} (discharge depth) is the maximum discharge percentage of each battery unit versus its nominal storage capacity, then the effective storage capacity of each battery string will be:

$$C_{bat.eff} = C_{bat} \cdot d_{dis}. \tag{3.50}$$

Finally, each battery string or module cannot be charged and discharged simultaneously. If the hybrid power plant should be provided with such an option, then the available battery units should be grouped in an appropriate number of strings (or modules), which definitely will be among the results of the power plant's operation simulation and its dimensioning process.

Following the above mentioned, referring to the operation algorithm of small-size hybrid power plants with electrochemical batteries as storage units aiming at RES penetration maximization, for every calculation time step j the following actions are executed:

1. The power production P_{RES} from the RES units is compared to the power demand P_d:
   i. If $P_{RES} < P_d$, then all the available power production from the RES units is directly absorbed by the grid for the power demand coverage (direct RES penetration): $P_{RESp} = P_{RES}$.
   ii. If $P_{RES} \geq P_d$, then the power demand is totally covered by the RES power production, namely: $P_{RESp} = P_d$ (100% RES penetration percentage).
2. The charge level $b_i(j-1)$ of each battery string (or module) i from the previous calculation time step j – 1 is checked. The following cases are distinguished:
   i. $P_{RES} < P_d$:

First subcase:

If there is enough energy stored in the battery strings, namely, if the following inequality is true:

$$\sum b_i(j-1) \cdot C_{bati} - t \cdot (P_d - P_{RESp}) \geq \sum (1 - d_{dis}) \cdot C_{bati}$$

where C_{bat} the total storage capacity of all the batteries strings, t the duration of the calculation time step, and C_{bati} the total storage capacity of the battery string i, then the remaining power demand will be covered by the batteries: $P_{bat} = P_d - P_{RESp}$.

The power production from the backup units will be null: $P_{th} = 0$.

The new charging level of each battery string will be:

$$b_i(j) = b_i(j-1) - P_{bati} \cdot t/(\eta_{disi} \cdot C_{bati})$$

where P_{bati} and η_{disi} the discharge rate and efficiency, respectively, of the battery string i during the discharge process.

Second subcase:

If there is not enough energy stored in the battery strings, then the battery strings will be fully discharged. In this case, the total power production from the storage units will be:

$$P_{bat} = \sum \left[b_i(j-1) \cdot C_{bati} - (1 - d_{dis}) \cdot C_{bati} \right] \cdot \eta_{disi} / t$$

where d_{dis} the maximum discharge depth, assumed common for every battery string.

The remaining power demand will be covered by the backup units:

$$P_{th} = P_d - P_{RES} - P_{bat}$$

The new charging level of the storage units will be equal to the minimum possible one, namely:

$$b_i(j) = 1 - d_{dis}.$$

For the above subcases, the stored power in the storage units P_{st} will be null, as well as the available power for the electrolysis unit P_{el}.

ii. $P_{RES} > P_d$:
First subcase:
If there is enough storage space in the battery strings to store the RES power production surplus, namely if the following inequality is true:

$$\sum b_i (j-1) \cdot C_{bati} + t \cdot (P_{RES} - P_{RESp}) \leq \sum C_{bati}$$

then all the RES power production surplus $P_{RES} - P_{RESp}$ will be stored:

$$P_{st} = P_{RES} - P_{RESp}.$$

In this case there will not be any available power for the electrolysis unit.
The power production surplus can be stored only in the battery strings not currently employed for power production. These strings will be available for energy storage units. The new charge level for each one of these batteries strings will be:

$$b_i (j) = b_i (j-1) + \eta_{chi} \cdot P_{sti} \cdot t/C_{bati}$$

where P_{sti} and η_{chi} the charge rate and efficiency of the string i during the charge process.
Second subcase:
If there is not enough storage space in the battery strings to store all the RES power production surplus, then the absorbed power by the battery strings will be determined by their maximum storage capacity, as expressed by the following relationship:

$$P_{st} = \sum \left[C_{bati} - b_i (j-1) \cdot C_{bati} \right] / (\eta_{chi} \cdot t)$$

In this case, the remaining non-stored power production from the RES units will be:

$$P_{RESav} = P_{RES} - P_{RESp} - P_{st}.$$

If the nominal power of the electrolysis units is P_{el}, then the absorbed power P_{RESel} by this unit will be:

$$P_{RESel} = P_{RESav}, \text{ if } P_{RESav} \leq P_{el}$$

$$P_{RESel} = P_{el}., \text{ if } P_{RESav} > P_{el}$$

Finally, the rejected power production from the RES units will in any case be:

$$P_{RESrej} = P_{RES} - P_{RESp} - P_{st} - P_{RESel}.$$

The new charge level of the battery strings employed for the charge process, namely excluding those employed for power production at the same time step, will be equal to 100%, namely the involved battery strings will be fully charged.

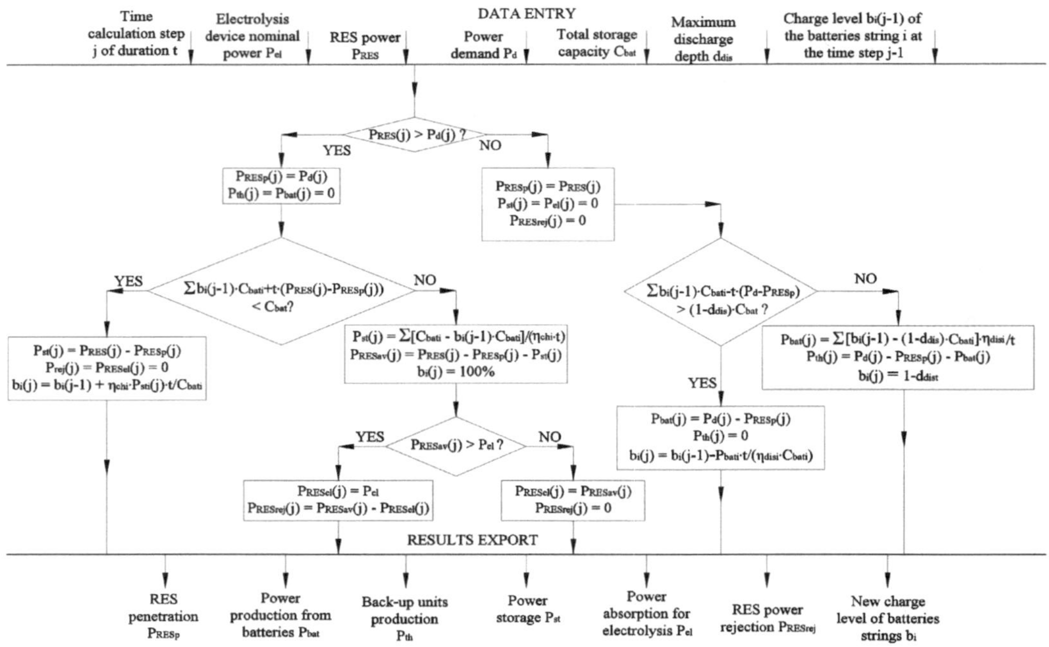

FIGURE 3.26 Operation algorithm of a small-size hybrid power plant with electrochemical batteries as storage units.

The above described operation algorithm is depicted in Figure 3.26.

The above analyzed operation algorithm exhibits two principal differences, compared to the ones presented in the previous sections, for large-size hybrid power plants with PHS or CAES systems as storage units. Specifically:

- The possibility for RES units' direct penetration up to 100%, based on the direct response of the storage units' inverters versus any potential power demand fluctuation.
- The fact that the same battery string cannot be charged and discharged at the same time. This feature affects the dimensioning of the hybrid power plant, because, in order to ensure the simultaneous charge and discharge process, the required battery units should be properly grouped in a number of battery strings. Sensibly, the battery distribution in a number of parallel and independent strings leads to increases in the storage plant size and the corresponding procurement cost.

Finally, it is obvious that the dimensioning of the storage plant and the calculation of the electricity transfer from and to the battery strings do not require an extensive analytical calculation process, as in the cases of PHS or CAES systems. On the contrary, the hybrid power plant operation simulation with electrochemical batteries as storage units can be simply accomplished once the charge/discharge efficiency curves are known for the employed battery model.

Indicative charge efficiency curves are presented for valve regulated lead acid batteries in Figure 3.27. The charge efficiency of this battery model is affected by the charging current, namely the absorbed power, and the battery's charge level. These particular curves refer to a battery model with a nominal storage capacity of 3,000 Ah and a maximum charge/discharge current of 600 A. The charge efficiency remains higher than 80% for charge levels lower than 90%, regardless of the charge current. Finally, the charge efficiency drops as the charging current or the existing charge level increases.

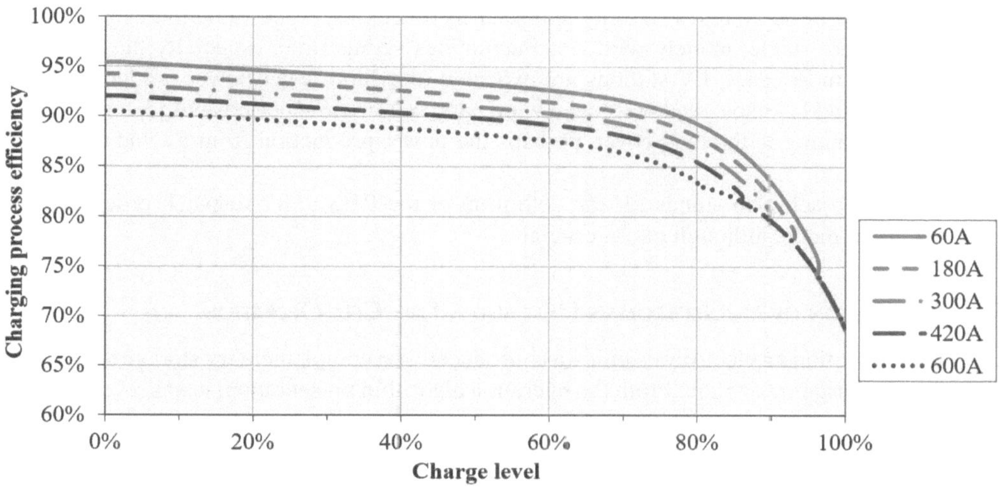

FIGURE 3.27 Characteristic charge efficiency curves for valve regulated lead acid batteries.

Similarly, characteristic discharge efficiency curves for valve regulated lead acid batteries are presented in Figure 3.28. These curves also refer to the same battery model as above, with 3,000 Ah nominal storage capacity and maximum charging/discharging current of 600 A. The discharge efficiency depends on the discharging current, namely the provided power from the battery, and the existing charge level. Regardless of the operation conditions, the discharge efficiency remains higher than 93%. Finally, the discharge efficiency drops with the increase of the discharging current and the decrease of the battery's charge level.

A last issue worthy of mentioning for the case of small-size hybrid power plants comes from the potential cooperation of two different RES technologies, namely a wind park and a PV station. This issue has to do with the dispatch priority between these two available technologies, either during RES direct penetration or power storage operation mode. This priority is not defined in advance. Referring to the direct RES penetration in the electrical grid, it can be stated as a general rule that the dispatch priority will be given to the RES technology with the highest reliability and the

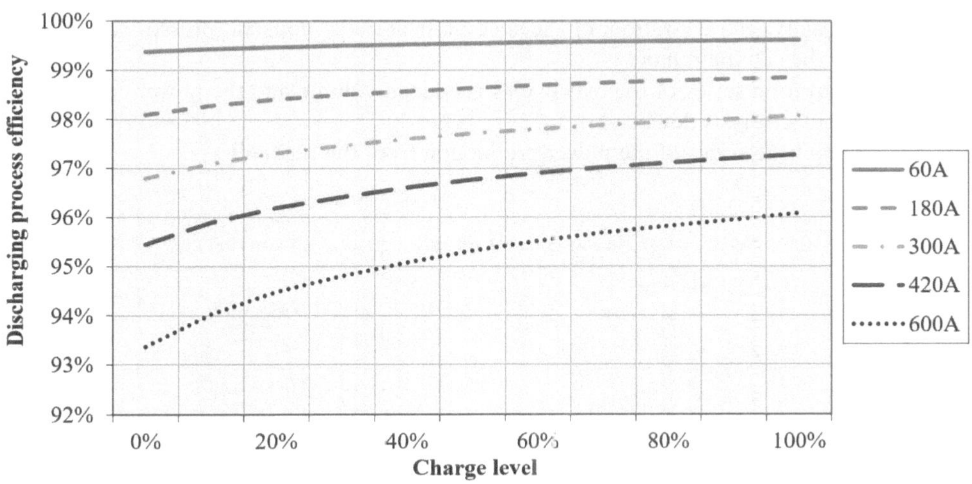

FIGURE 3.28 Characteristic discharging efficiency curves for valve regulated lead acid batteries.

minimum effects on the system's stability and security. Generally, solar radiation can certainly be considered as more stable, namely with less fluctuations versus time, especially during summer. Consequently, in most cases, PV stations are expected to exhibit fewer power production fluctuations, so they should be dispatched before the wind park. Yet, in time periods with intensive periods of sunshine alternating with cloud cover, perhaps the power production from a wind park may be more constant.

The same approach also stands for the definition of the RES unit's dispatch order during the storage operation mode, although not so crucial.

3.6.2 Simulation of an Electrolysis Unit and a Fuel Cell Operation

In the previous section an electrolysis unit was introduced as a complementary storage technology in the examined hybrid power plant. From the operation algorithm presentation, it was revealed that the electrolysis unit is employed as an alternative storage technology. Energy is stored in the electrolysis device through hydrogen production, any time the basic storage units (electrochemical batteries) are fully charged and there is still power available for storage from the RES units. The produced and stored hydrogen can then be used in a fuel cell for electricity production, operating as a backup unit instead of the conventional thermal generators. The low dispatch priority of the electrolysis unit is imposed by the high procurement cost of the required equipment and the relatively low efficiency of the storage and production cycle compared to the conventional electrochemical batteries.

The general layout of an electrolysis unit fuel cell system is presented in Figure 3.29. The produced hydrogen is compressed in a compressor, cooled down in a heat exchanger, and stored in a tank, from which it is eventually provided for the fuel cell for electricity production whenever it is required.

The dimensioning and the annual simulation of an electrolysis unit fuel cell system require the following data:

- the annual time series of the power available for the electrolysis unit
- the volume and the maximum pressure of the hydrogen storage tank
- the initial pressure in the hydrogen storage tank at the beginning of the calculation process
- the nominal power and the hydrogen production efficiency for the introduced electrolysis device model, versus the consumed power
- the nominal power and the efficiency curve of the fuel cell device
- the hydrogen fundamental properties, such as the specific heat under constant pressure (c_p), the heat capacity, and the ratio k of the specific heat under constant pressure and volume and aims at the calculation of:
    - the annual time series of the hydrogen's produced volume and the power consumption from the electrolysis device
    - the annual time series of the power production from the fuel cell

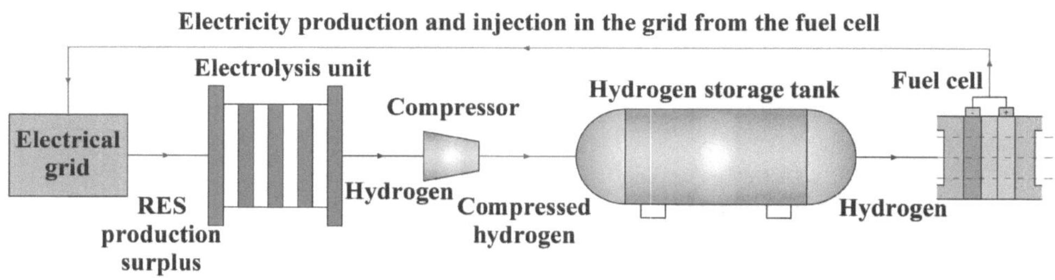

FIGURE 3.29 The layout of an electrolysis unit fuel cell system for electricity production.

- the annual time series of the power consumed by the compressor
- the annual time series of the stored hydrogen's mass and pressure in the storage tank.

The simulation procedure of an electrolysis fuel cell system is presented below. For every calculation time step:

1. If P_{RESav} is the available power for the electrolysis unit and P_{el} is the electrolysis device nominal power, then:
   i. if $P_{RESav} > P_{el}$, the absorbed power from the electrolysis device will be: $P_{RESel} = P_{el}$
   ii. if $P_{RESav} \leq P_{el}$, the absorbed power from the electrolysis device will be: $P_{RESel} = P_{RESav}$.
2. The specific electricity consumption per produced hydrogen volume curve versus the consumed power is introduced as provided by the manufacturer of the employed electrolysis device model. A characteristic curve for a commercial electrolysis device model with a nominal power of 250 kW is presented in Figure 3.30. Figure 3.30 shows that the specific electricity consumption depends on the consumed power by the electrolysis unit, yet, this variation is not intensive.
3. From the introduced curve, the finally produced hydrogen volume V_{el} by the electrolysis device is estimated, given the absorbed power P_{RESel}. The hydrogen's volume V_{el} is estimated in Nm^3 (normal cubic meter), namely for 0°C temperature and for absolute pressure of 1 atm (101,325 Pa). Given the hydrogen's density for the above temperature and pressure conditions, equal to 0.0899 kg/Nm^3, the produced hydrogen mass from the electrolysis unit m_{el} can be calculated.
4. The required power production from the fuel cell must be also calculated, based on the requirements and the operating schedule of the fuel cell. For example, if the fuel cell is employed as a backup unit, within the operation of a hybrid power plant, instead of conventional thermal generators, then the fuel cell annual power production time series is determined by the requirements for power production by the hybrid plant's backup units. Let's assume that the required power production from the fuel cell at the current calculation time step is P_{fc}.

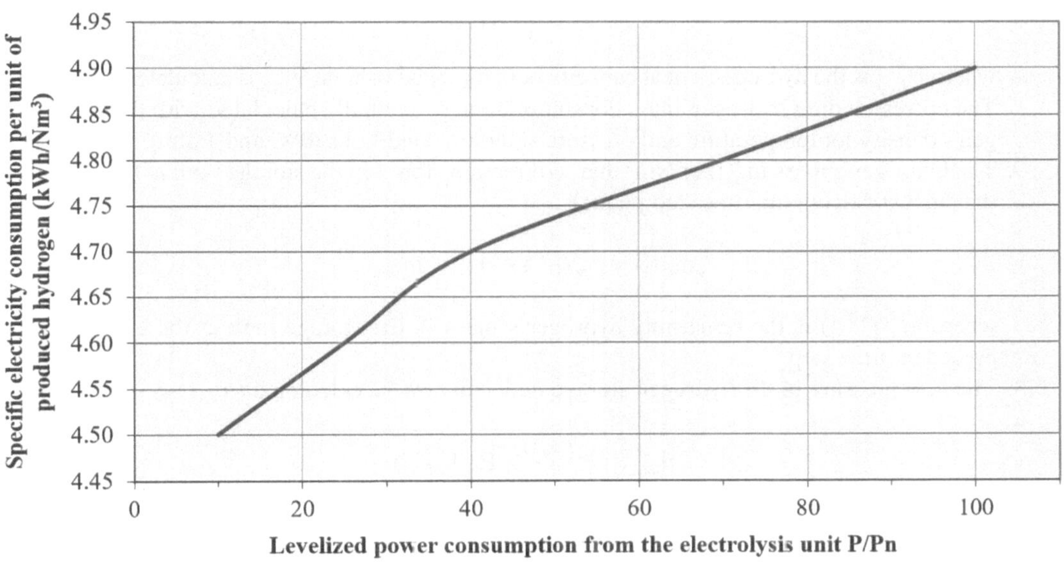

FIGURE 3.30 The specific electricity consumption per produced hydrogen volume curve versus the consumed power for a commercial electrolysis device of 250 kW nominal power.

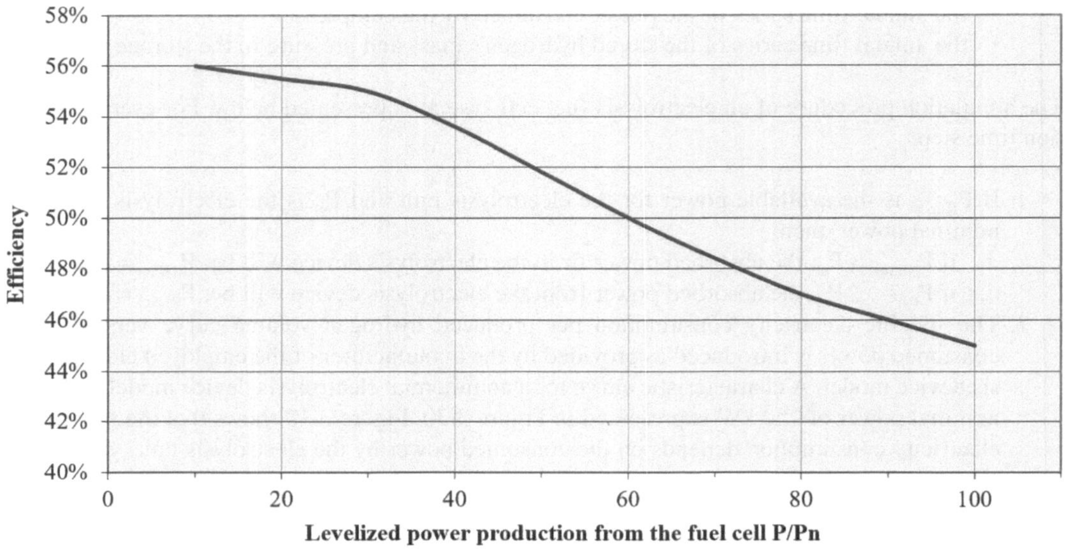

FIGURE 3.31 Power production efficiency characteristic curve for a commercial proton exchange membrane fuel cell device of 200 kW nominal power.

5. The power production efficiency curve for the employed fuel cell device is introduced. Such a curve for a commercial proton exchange membrane fuel cell model with a nominal power of 200 kW is presented in Figure 3.31. Using this curve, the fuel cell's efficiency η_{fc} is estimated, given the power production requirement P_{fc} from the fuel cell.
6. The hydrogen's consumption volume V_{cons} required for power production equal to P_{fc} from the fuel cell for the duration t of the time step is calculated with the relationship:

$$V_{cons} = \frac{P_{fc} \cdot t}{\eta_{fc} \cdot H_{hydr}} \tag{3.51}$$

where H_{hydr} is the hydrogen's heat capacity per Nm^3. The volume V_{cons} is calculated in Nm^3. The corresponding hydrogen mass consumption m_{cons} is finally calculated with the hydrogen's density for temperature and pressure standard conditions (0°C and 1 atm).
7. The hydrogen's mass $m_{st}(j)$ in Nm^3 that will remain stored in the storage tank at the end of the current calculation time step j equals:

$$m_{st}(j) = m_{st}(j-1) + m_{el} - m_{cons}$$

where $m_{st}(j - 1)$ is the remaining hydrogen's mass in the storage tank at the end of the precedent time step j – 1.
8. The new pressure in the hydrogen storage tank can now be calculated as:

$$p_{tank}(j) = \frac{m_{st}(j)}{V_{st}} \cdot R \cdot T_{tank}(j) \tag{3.52}$$

where V_{st} is the volume storage capacity of the storage tank and $T_{tank}(j)$ the hydrogen temperature in the tank, which can be considered equal to the ambient temperature, given that the compressed hydrogen has been cooled down before being stored. R is the hydrogen's ideal gas constant, equal to 4,124.2 J/(kgK).

9. It is examined whether the new pressure in the hydrogen tank is lower than the minimum p_{min} (e.g., 1 bar) or higher than the maximum p_{max} possible value. The following cases are distinguished:
   i. If $p_{tank}(j) < p_{min}$, then:
      $V_{cons} = 0$
      $P_{fc} = 0$
      $V_{st}(j) = V_{st}(j-1) + V_{el}$.
   ii. If $p_{tank}(j) > p_{max}$, then:
      $V_{el} = 0$
      $P_{RESel} = 0$
      $V_{st}(j) = V_{st}(j-1) - V_{cons}$.
   iii. If $p_{min} \leq p(j) \leq p_{max}$, then the magnitudes V_{cons}, V_{el}, $V_{st}(j)$, P_{fc} and P_{RESel} are calculated with the relationships presented above.
10. The actual power consumption from the compressor is given by the following relationship:

$$p_{comp} = \eta_c \cdot p_{in} \cdot \dot{V}_{el} \cdot \frac{\gamma}{\gamma-1} \cdot \left[\left(\frac{p_{out}}{p_{in}} \right)^{\frac{\gamma-1}{\gamma}} - 1 \right] = \eta_c \cdot p_{in} \cdot \dot{V}_{el} \cdot \frac{\gamma}{\gamma-1} \cdot \left[\beta^{\frac{\gamma-1}{\gamma}} - 1 \right] \quad (3.53)$$

where:

η_c: compressor's isentropic efficiency
p_{in}: hydrogen's inlet pressure in the compressor (practically equal to the atmospheric one)
$\dot{V}_{el}$: hydrogen's volume flow rate during compression
k: ratio $k = c_p/c_v$ of the hydrogen's specific heat under constant pressure c_p and volume c_V
p_{out}: hydrogen's outlet pressure from the compressor
β: compression ratio.

The hydrogen's specific heat under constant pressure c_p and volume c_v, as well as their ratio k for atmospheric pressure are presented versus the hydrogen's temperature in Table 3.7 [44].

With the presentation of the operation simulation of the electrolysis fuel cell system, the analysis of the dimensioning methodology and the simulation of small-size hybrid power plants has been integrated.

3.6.3 Hybrid Power Plants of Very Small Size

The study on the operation algorithms and the computational simulation processes of hybrid power plants is integrated in this section with hybrid power plants of very small size, with power demands up to 100–200 kW. Such hybrid power plants are usually developed for remote (non-interconnected with the utility grid) electricity consumptions of corresponding size, such as cottages, lighthouses, stock-farming units, and desalination plants. The size of these hybrid plants and the fact that they are supposed to support a non-interconnected consumption impose a different design and operation concept. The hybrid power plant should obviously be able to securely support up to 100% of the non-interconnected power system. Additionally, in such small-size applications, the hybrid plant's procurement and operation cost constitute fundamental parameters for the optimum design, dimensioning, and configuration of the operation algorithm, because these costs are not distributed among a large number of users or consumers, but, on the contrary, they are afforded by a small number of individuals or even a single user. For the above reasons, the operation algorithm and the dimensioning process of hybrid power plants of very small size differ from the previously presented ones for larger energy systems.

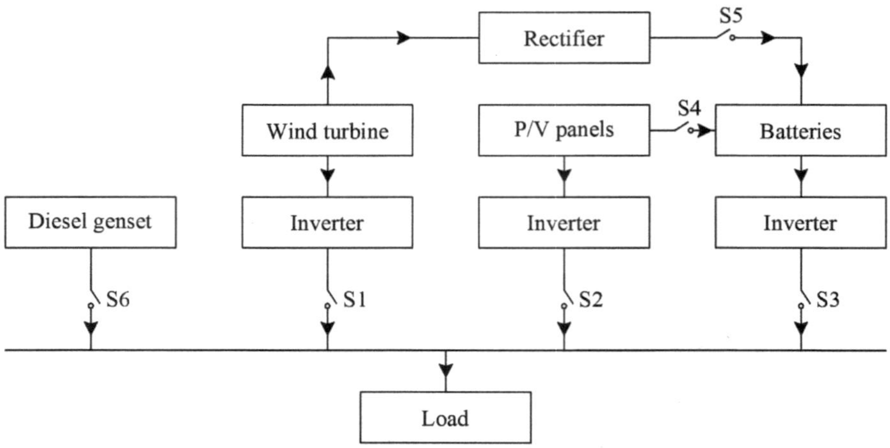

FIGURE 3.32 Main structure of a hybrid power plant of very small size.

In a hybrid power plant of very small size, the following technologies can be used:

- RES units: small wind turbines, with nominal power up to 20–50 kW and PV panels
- storage units: electrochemical batteries and mainly, due to the currently low procurement cost, lead acid batteries; however, in the future most probably other technologies too (e.g., lithium ion batteries)
- backup units: usually a diesel generator set (diesel genset).

The main structure of a hybrid power plant of very small size is presented in Figure 3.32.

According to Figure 3.32, all the power production and storage units (wind turbines, PV panels, batteries, and diesel genset) are connected at the same bus with the load. The RES and storage units provide the produced power through their inverters, so the predefined standards for the produced alternative current are fulfilled. Also, the RES units are directly connected with the storage units of the hybrid power plant. The produced alternative current from the wind turbines must first be rectified before being stored in the batteries. Finally, switches are installed in all the connecting cables between the RES, the storage units, and the load, as well as between the RES and storage units.

The hybrid power plant's ultimate target is the guaranteed power production following the power demand fluctuations, based on the maximization of the RES units' production, which, in turn, implies the minimization of the electricity production from the diesel genset and the subsequent minimization of the plant's operation cost, configured mainly by the fuel's consumption cost. For the above targets:

- a minimum charge level in the batteries should be continuously maintained, adequate to ensure the power production from the hybrid power plant at emergency events, minimizing the power production from the diesel genset and ensuring the security of the weak electrical system
- the available energy for storage will be provided only by the RES units and the energy storage will be accomplished during periods of excess RES electricity production
- the direct penetration of the RES units will be possible during periods with high batteries' charge level
- the produced electricity by the diesel genset will be provided directly for the power demand only when there is low power production from the RES units and inadequate charge level in the batteries
- no energy storage will be performed by the diesel genset.

Based on the above general operation principles, the fundamental criterion for the operation mode of the plant is the batteries' charge level. Based on this parameter, the operation algorithm of this power plant can be approached as described below.

Let's designate as b_{min} and b_{max} the minimum and the maximum respectively possible charge level of the batteries and C_{max} their maximum storage capacity. Let's also assume that at a certain time point j the batteries' current charge level is b(j). At the same time point j let the power production from the wind turbine and the PV panels be P_W and P_{PV} respectively and the power demand (load) P_L. Finally, the maximum charging rate (power) of the batteries is assumed equal to P_{ch}. The following cases are distinguished:

1. $b(j-1) = b_{max}$ and $P_W + P_{PV} \geq P_L$:

   In this case obviously there is no possibility for energy storage at the already fully charged batteries from the previous time step j – 1. Also the available power production from the RES units overgrows the power demand, which will be, consequently, totally covered by the RES units. The power produced by the RES units is regulated by their inverters. Conclusively:

   i. RES power direct penetration: $P_{RESp} = P_L$
   ii. power storage in batteries: $P_s = 0$
   iii. power production from the batteries: $P_{bat} = 0$
   iv. batteries' new charge level: $b(j) = b_{max}$
   v. power production from the diesel genset: $P_{dg} = 0$.

   The above operation is achieved with the switches of the electrical circuit as follows:

   - S_1, S_2: closed
   - S_3, S_4, S_5, S_6: open.

2. $b = b_{max}$ and $P_W + P_{PV} < P_L$:

   In this case there is not any possibility for energy storage in the batteries either, because, on the one hand, they are already fully charged from the previous time step j – 1, and, on the other, there is no power production surplus from the RES units available for storage. The RES power production is not enough for the total power demand coverage, which will be covered by both the RES units and the batteries. The power produced by the RES and the battery units is regulated by their inverters. Conclusively:

   i. RES power direct penetration: $P_{RESp} = P_W + P_{PV}$
   ii. power storage in batteries: $P_{st} = 0$
   iii. power production from the batteries: $P_{bat} = P_L - P_{RES\delta}$
      Namely, it is equal to the RES power production shortage compared to the power demand.
   iv. batteries' new charge level: $b(j) = b_{max} \cdot C_{max} - P_{bat} \cdot t/\eta_{dis}$
      where η_{dis} is the batteries' discharge efficiency and t the duration of the calculation time step
   v. power production from the diesel genset: $P_{dg} = 0$.

   The above operation is achieved with the switches of the electrical circuit as follows:

   - S_1, S_2, S_3: closed
   - S_4, S_5, S_6: open.

3. $b = b_{min}$ and $P_W + P_{PV} < P_d$:

   In this case the batteries' charge level is the minimum one. Additionally, the total power production from the RES units is lower than the power demand. The power production will be covered by the RES units and the diesel genset. No energy storage can be performed. Conclusively:

   i. RES power direct penetration: $P_{RESp} = P_W + P_{PV}$
   ii. power storage in batteries: $P_{st} = 0$
   iii. power production from the batteries: $P_{bat} = 0$

iv. batteries' new charge level: $b(j) = b_{min}$
v. power production from the diesel genset: $P_{dg} = P_L - P_{RESp}$.

The above operation is achieved with the switches of the electrical circuit as follows:

- S_1, S_2, S_6: closed
- S_3, S_4, S_5: open.

4. $b = b_{min}$ and $P_W + P_{PV} \geq P_L$:

In this case the batteries' charge level is the minimum one. The power demand will be totally covered by the power production from the RES units. The excess RES power production will be stored in the battery units. The power storage will be equal to their maximum charging capacity P_{ch}, if the RES power surplus P_{RESav} is higher than P_{ch}, or equal to P_{RESav} in the opposite case. Conclusively:

i. RES power direct penetration: $P_{RESp} = P_L$
ii. power storage in batteries:
if $P_{RESav} = P_{RES} - P_{RESp} \geq P_{ch}$, then $P_{st} = P_{ch}$
if $P_{RESav} = P_{RES} - P_{RESp} < P_{ch}$, then $P_{st} = P_{RESav}$
iii. power production from the batteries: $P_{bat} = 0$
iv. batteries' new charge level: $b(j) = b_{min} \cdot C_{max} + P_{st} \cdot t \cdot \eta_{ch}$
where η_{ch} is the batteries' charging efficiency
v. power production from the diesel genset: $P_{dg} = 0$.

The above operation is achieved with the switches of the electrical circuit as follows:

- S_1, S_2, S_4, S_5: closed
- S_3, S_6: open.

5. $b_{min} \leq b \leq b_{max}$:

This is the general case with the batteries' charge level between their minimum and maximum value. The power demand will be covered by the RES units and perhaps the battery units and the diesel genset. At the same time, power storage can be performed if there is an RES power production surplus. Conclusively:

i. RES power direct penetration:
if $P_W + P_{PV} \geq P_L$, then: $P_{RESp} = P_L$
else: $P_{RESp} = P_W + P_{PV}$
ii. power storage in batteries: $P_{st} = P_W + P_{PV} - P_{RESp}$
iii. power production from the batteries: $P_{bat} = P_L - P_{RESp}$
iv. batteries' new charge level: $b(j) = b(j-1) \cdot C_{max} + P_{st} \cdot t \cdot \eta_{ch} - P_{bat} \cdot t/\eta_{dis}$
v. power production from the diesel genset: $P_{dg} = P_L - P_{RESp} - P_{bat}$.

In this case all the switches of the electrical circuit are closed. Practically, the hybrid power plant exhibits the same operation philosophy with the plants presented in the previous section.

The operation algorithm presented above is not unique. Depending on several involved parameters, such as the kind and the sensitivity of the power demand, the available RES potential, the battery technology, and the climate conditions, different operation approaches may be developed. A completely different version of a very small-size hybrid power plant is presented in Figure 3.33.

As seen in this figure, all the power production units, namely both the RES and the backup units, are used for energy storage. The power demand is exclusively supported by the batteries. Priority regarding the batteries' charging is given to RES units. The diesel genset is employed for batteries charging only if there is low available RES potential and the batteries' charge level is close to the minimum value. It is understood that the overall efficiency of such a hybrid power plant is strongly reduced, because all the primarily produced electricity is first stored in the batteries and then provided for consumption. On the other hand, this approach is simpler as far as its implementation, design, and operation are concerned. Moreover, given that the power demand is exclusively supplied by the batteries, the power quality of the produced alternative current is ensured and the system's security is improved. Perhaps for hybrid power plants of very small size, such as cottages, with

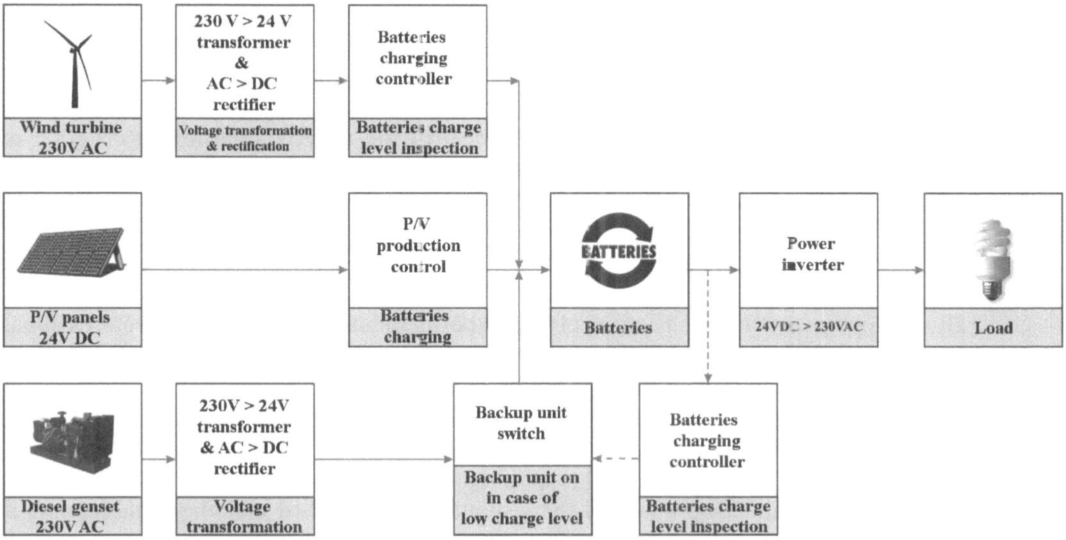

FIGURE 3.33 Alternative layout of a hybrid power plant of very small size [48].

limited use per year, this approach is the optimum option from both a technical and economic point of view, because it offers simplicity and operation reliability, while the relatively increased operation cost may not constitute a crucial issue, due to the short operation time periods.

3.7 OPTIMIZATION CRITERIA FOR THE DIMENSIONING OF HYBRID POWER PLANTS

It has been explained in the earlier sections that hybrid power plant dimensioning is optimized on the basis of iterative simulation executions of their operation algorithm, with varying values of their essential features, such as the RES units nominal power and the storage plants' capacity. Usually an independent design variable is selected, while the rest are computationally calculated through the execution of the operation algorithm. In the dimensioning methods presented in this chapter, the independent variable was the RES plant's nominal power. For each different examined dimensioning, a particular magnitude was selected to be calculated and employed as a common comparison index. This magnitude can refer to energy or economic features and constitutes the optimization criterion of the dimensioning process. The hybrid plant's dimensioning is integrated for the specific design point at which the predefined criterion is optimized.

3.7.1 Optimization of Hybrid Power Plants Based on Energy Criteria

The optimization of a hybrid power plant's dimensioning based on energy criteria can be approached in several ways. For example, if E_{th} is the annual electricity production from the hybrid power plant's thermal generators, E_d the annual electricity demand, E_{RES} the annual electricity production from the RES units, and E_{rej} the annual energy rejection of the RES units, then an optimization criterion could be used that requires the following ratios should be both lower than a maximum acceptable percentage:

- E_{th}/E_d
- E_{rej}/E_{RES}.

Indicatively, under favorable conditions (high RES available potential and high available storage capacity of the storage plant), these ratios can be in the range of 10% or even lower. For higher RES unit's nominal power, the thermal generator's electricity production share versus the consumption (first ratio), drops. However, at the same time, the rejected RES energy will most probably increase, hence, the second ratio too. So the fulfillment of the above dimensioning criterion is practically deduced to the task to maintain a balance between the RES penetration and rejection, which strongly depends on specific design parameters, such as the available RES potential and its seasonal variation, the total storage capacity of the storage units, the seasonal variation of the power demand, and the possible coincidence of the peak power demand periods with the availability of high RES potential.

Generally, the optimization dimensioning of hybrid power plants based on energy features aims, obviously, at the maximization of the annual electricity production share from the hybrid plant versus the electricity demand, while keeping, at the same time, the RES annual rejection below acceptable limits. The high energy rejection from the RES units implies the installation of too high RES nominal power, without being supported by a storage plant with a corresponding storage capacity. Alternatively, the high RES energy rejection may also imply that the initial target for the maximization of the annual RES production share should be approached with the installation of a second RES technology.

3.7.2 Dimensioning Optimization of Hybrid Power Plants Based on Economic Criteria

The dimensioning of hybrid power plants may be optimized based on economic criteria with two different approaches.

3.7.2.1 Optimization of the Investment's Economic Indices

The most commonly economic indices used for investment evaluation are the internal rate of return (IRR), the net present value (NPV), the undiscounted and the discounted payback period, the return on investment (ROI), and the return on equity (ROE). The dimensioning of a hybrid power plant can be executed on the basis of the economic indices optimization of the corresponding investment, if the whole project's principal target is the economic efficiency maximization of the invested capital.

The calculation of the investment's economic indices is executed versus the net profit cash flows. To this end, the commonly applied procedure is briefly presented below:

1. First, the investment's annual revenues are calculated, which usually come from the produced electricity selling to the grid's operator or directly to final consumers, in the frame of electricity wholesale markets. In some cases, revenues may also come from the availability of the hybrid power plant's guaranteed power. The prices for both the selling electricity and the guaranteed power availability are defined in the existing electricity production and distribution national legislation regimes. Generally, the most common pricing policies are:
   - the feed-in-tariff regime, in which a fixed selling price is predefined for each different RES electricity production plant, configured by specific parameters, such as the specific RES technology (including hybrid plants), the interconnection or not of the electricity grid with other mainland grids, and the existing electricity production-specific cost in case of hybrid power plants introduced in non-interconnected grids
   - liberalized electricity market, in which a number of electricity producers, providers, and grid operators collaborate towards the electricity provision for the final consumers, with the prices configured according to the fundamental market rules or particular introduced pricing policies (more on liberalized electricity markets is presented in Chapter 6).

2. Second, the annual expense cash flows are estimated, usually consisting of:
   - public rates for the local municipalities or prefectures, which usually, according to local regulations, are defined between 3% and 5% versus the annual revenues
   - maintenance costs
   - annual load payments which, in most cases, have been received from commercial banks for the implementation of the project
   - equipment insurance cost
   - staff salaries
   - land and premises rent
   - several other minor costs
   - taxes.
3. The annual net profit cash flows are calculated by subtracting the annual expense cash flows from the annual revenue cash flows.
4. All the above calculations are executed on an annual basis and for the overall lifetime of the investment. For the purpose of economic evaluation, the investment's life period is usually assumed from 20 to 25 years, even if the real life period of the project or of some of its components (e.g., the PHS system of a hybrid power plant) can be considerably longer.
5. The investment's economic indices are calculated for the net profit cash flows and for the whole assumed investment's life period. The calculation formulas for each of the previously mentioned economic indices are given below:
   - Net present value (NPV):

$$\mathrm{NPV} = \sum_{t=1}^{N} \left(\frac{P_t}{(1+i)^t} \right) - C \tag{3.54}$$

where P_t is the project's net profit at the year t, i is the discount rate, N is the investment's life period expressed in years, and C is the investment's total setup cost. The NPV provides a figure for the cumulative net profits during the project's total life period, deduced in the present value. The higher the NPV, the more efficient the investment will be. However, in order to compare two different investments, the calculation of the NPV should be accompanied with the knowledge of the investments' setup costs.

The NPV can also be calculated versus the investment's equities. In that case, the corresponding amount should be introduced in the above relationship at the position of the magnitude C, instead of the total setup cost.

   - Internal rate of return (IRR):

The IRR is the discount rate i that nullifies the NPV:

$$\mathrm{NPV} = \sum_{t=1}^{N} \left(\frac{P_t}{(1+\mathrm{IRR})^t} \right) - C = 0 \tag{3.55}$$

Because the discount rate is an index expressing the overall economic situation in a wide geographical region, configured, among others, by the bank deposits' rates, the IRR practically expresses a theoretical deposit rate for which the investment's capital would have the same economic efficiency with the examined project. For example, an investment's IRR of 10% means that the invested capital will exhibit an economic efficiency as in the case of a bank deposit with a rate of 10%. It is perceivable that the IRR constitutes a characteristic index of an investment, which can provide integrating information on its economic efficiency, without the necessity for the calculation or the knowledge of any additional figures.

As for the NPV, the IRR can also be calculated versus the investment's equities. This can be achieved if in the previous relationship the corresponding amount is introduced instead of the project's total setup cost at the position of the magnitude C.

- Payback period and discounted payback period:
The payback period t_p (or undiscounted payback period) is the time period from the beginning of the project's operation to the end of which the sum of the so far earned net profits equals the investment's setup cost. Consequently the investment's payback period t_p is calculated by solving the following equation versus t_p:

$$\sum_{t=0}^{t_p} P_t - C = 0 \tag{3.56}$$

The discounted payback period t_p is the time period from the beginning of the project's operation to the end of which the sum of the so far earned net profits, deduced in the present value, equals the investment's setup cost. Consequently the investment's discounted payback period t_p is calculated by solving the following equation versus t_p:

$$\sum_{t=1}^{t_p} \left(\frac{P_t}{(1+i)^t} \right) - C = 0 \tag{3.57}$$

If the investment's equities E are introduced in the above two relationships, instead of the total setup cost C, the payback periods t_p (discounted and undiscounted) are defined and calculated versus the investment's equities.

The payback periods constitute characteristic economic indices for the evaluation of an economic investment, giving concrete information on the investment's efficiency.

- Return on investment (ROI):
The ROI is defined with the following relationship:

$$ROI = \frac{\sum_{t=1}^{N} \left(\frac{P_t}{(1+i)^t} \right)}{C} \tag{3.58}$$

The ROI practically expresses the investment's economic efficiency percentage over the investment's total setup cost. For example, an ROI equal to 120% means that every invested dollar of the investment's total setup cost will return back $1.20.

- Return on equities (ROE):
The ROE is defined by the following relationship:

$$ROE = \frac{\sum_{t=1}^{N} \left(\frac{P_t}{(1+i)^t} \right)}{E} \tag{3.59}$$

The ROE practically expresses the investment's economic efficiency percentage over the investment's equities. For example, an ROE equal to 250% means that every invested dollar of the investment's equities will return back $2.50.

The dimensioning iterative procedure of the hybrid power plant is integrated with the synthesis of its components, which leads to the maximization of the NPV, IRR, ROI, ROE, and the minimization of the payback periods.

3.7.2.2 Minimization of Setup and Operation Costs

With this approach, an integrated index most frequently used is the life cycle cost (LCC), which includes both the setup cost of the hybrid power plant (e.g., equipment procurement cost, installation, civil engineering works) and the operation cost, for the whole life period of the project. This method is applied when the implementation of the hybrid power plant does not exhibit a strict investment character, but generally aims to cover a specific electricity demand with the minimum possible cost. Such cases are hybrid power plants of small size, introduced in isolated grids and remote consumptions. For example, a farmer wants to cover the electricity needs of a stock-farming unit by installing a hybrid power plant with the minimum possible setup and operation cost.

For hybrid power plants, the LCC can be expressed as presented in Equation 3.60 [48]. An estimation for all the components involved in this relationship will be given below for hybrid power plants of small size, with electrochemical batteries employed as storage units.

$$LCC = C_{RES} + C_{inst.} + C_{maint.} + C_B + C_{inv.} + C_{DG.} - S \quad (3.60)$$

- C_{RES}:

  The RES unit procurement and installation cost. This cost can be determined with relative accuracy, based on the specific setup cost per installed unit of power. For example, for small wind turbines and PV panels, the following setup-specific costs may be used:

small wind turbines with nominal power lower than 5 kW:	2,000–2,500 \$/kW
small wind turbines with nominal power higher than 5 kW:	1,500 \$/kW
PV panels:	1,000–1,200 \$/kW

  However, a relevant market research may lead to figures that differ greatly from these, especially in cases of small wind turbines, because many alternative technologies developed by manufacturers from different locations on earth may lead to substantially reduced setup costs.

- $C_{inst.}$:

  The hybrid power plant's installation cost. Because this cost is mainly configured by the installation of the RES units, it may be estimated as a percentage of the procurement cost of the RES equipment. This percentage can be assumed between 5% and 8%. Consequently, the installation cost of the hybrid power plant is given by the relationship:

$$C_{inst.} = a_{inst} \cdot C_{RES}, \; a_{inst} = 5\text{–}8\%. \quad (3.61)$$

- $C_{maint.}$:

  The total maintenance cost of the hybrid power plant for its whole life period, deduced in present value. Because this cost is mainly configured by the maintenance of the RES units (mainly the wind turbines), it may be estimated as a percentage of the procurement cost of the RES equipment. This percentage is usually assumed around 2%. The annual maintenance cost $C_{maint,an}$ will be:

$$C_{maint.an} = a_{main} \cdot C_{RES}, \; a_{maint} = 2\%. \quad (3.62)$$

The sum of the annual maintenance cost for the hybrid plant's total life period, deduced in present value, is given by the relationship:

$$C_{maint.} = \sum_{t=1}^{N} \frac{a_{maint} \cdot C_{RES}}{(1+i)^t} = a_{maint} \cdot C_{RES} \sum_{t=1}^{N} (1+i)^{-t} \tag{3.63}$$

- C_B:

  The batteries' procurement cost. This cost is found after relevant market research. Based on 2017 prices, VRLA batteries with 3,000 Ah storage capacity cost around €1,000, while a lithium ion battery with 14 kWh storage capacity costs \$5,000–\$7,000.

  The service life of a battery most commonly is shorter than the hybrid power plant's typical life period of 20 years. For example, the service life of a lead acid battery can reach 6–7 years, under favorable operating conditions, while the life period of a lithium ion battery is around 10 years. Consequently, the battery procurement costs should also include the battery replacement costs throughout the 20-year period, calculated when they take place and deduced in present value. If C_{B0} is the battery's first procurement cost, namely at the time of the hybrid power plant's installation, its total cost C_B for the hybrid plant's whole life period will be:

$$C_B = C_{B0} \cdot \left(1 + \sum_{t=1}^{n} (1+i)^{-N_t} \right) \tag{3.64}$$

  where n is the battery replacements number during the hybrid power plant's life period and N_t the years of replacements.

- C_{inv}:

  The inverters' procurement cost which depends on their size, type, and manufacturer. It should result from a relevant market search. Indicative prices for some characteristic sizes are provided below (2017 prices):

inverter 3 kVA, 24 V/220 V	\$3,000
inverter 8 kVA, 24 V/220 V	\$4,500
inverter 30 kVA, 24 V/ 20 V	\$10,000

- C_{DG}:

  This cost component includes the procurement C_{DG0} and operation C_{FC} cost of the diesel genset for the whole life period of the hybrid plant. The latter is practically configured by the consumed fuel cost:

$$C_{DG} = C_{DG0} + C_{FC}. \tag{3.65}$$

  Similar to the inverters, the diesel genset procurement cost depends on the size, the type, and the manufacturer. It should also be found via relevant market search. Some indicative prices are given below (2017 prices):

diesel genset 10 kVA, 220 V, 50 Hz	€4,500
diesel genset 100 kVA, 220 V, 50 Hz	€20,000

The consumed fuel total cost C_{FC} for the whole life period of the hybrid power plant, deduced in present value, is calculated by the relationship:

$$C_{FC} = V_{FC} . c_{FC} . \sum_{t=1}^{N} \frac{(1+FPVR(t))^{t}}{(1+i)^{t}} \tag{3.66}$$

where

V_{FC}: annual consumed fuel volume in lt (in most cases diesel fuel)

c_{FC}: fuel price at the installation time of the hybrid power plant in $/lt (or in €/lt, etc.)

FPVR(t): fuel price variation rate, namely the annual variation percentage of the fuel price for each year of the hybrid plant's life period

i: discount rate

N: life period of the hybrid power plant.

From the above equation, we see that the LCC calculation requires the knowledge of the annual consumed fuel for the diesel genset operation, which may be gained only with the computational simulation of the hybrid power plant's annual operation and the calculation of the annual electricity production from the backup unit (diesel genset). Consequently, the hybrid power plant's annual operation and the contribution of the diesel genset to the annual electricity production, which imposes the calculation of the annual electricity production from the RES units, are involved in the calculation of the LCC through the component C_{FC}.

- S:

  The residual value of the hybrid power plant at the end of its life period. This value is usually estimated as a small percentage, in the range of 10%, of the initial procurement cost of the hybrid plant's equipment (RES units, inverters, batteries, and backup units). To keep a conservative calculation, this value is often assumed equal to zero.

From the LCC components, the C_{RES}, $C_{inst.}$, C_B, C_{inv}, and C_{DG0} refer to the initial procurement and installation cost of the hybrid plant's equipment, including also the batteries' replacement cost during the operation of the hybrid plant, whereas the components C_{maint} and C_{FC} refer to the hybrid plant's annual operation cost.

The LCC is iteratively calculated for different syntheses and sizes of the hybrid power plant components. As mentioned previously, independent parameters may be the RES and storage unit technologies and their nominal power or storage capacity. The iterative process is executed within a predefined range for these independent design parameters. With the integration of the iterative calculations, the optimum hybrid power plant's synthesis and dimensioning is the one with the minimum LCC. The dimensioning optimization process, based on the minimization of the LCC, is shown in Figure 3.34.

The above presentation shows that the LCC constitutes an integrated criterion for the economic evaluation of hybrid power plants, because it includes both the setup and the operation cost of the hybrid plant for its entire life period. Practically, it expresses the electricity production total cost from the hybrid power plant for its whole life period, deduced in present value. Hence, it constitutes an objective criterion for the dimensioning and the economic evaluation of hybrid power plants, when the ultimate goal is coverage of specific electricity demands with minimum possible costs.

Depending on the hybrid power plant's synthesis and operation concept, perhaps additional cost components can be involved in the LCC calculation. Therefore, the above approach for the LCC calculation of a hybrid power plant should be viewed as indicative and not definitive.

An alternative approach is the dimensioning optimization of a hybrid power plant based on the minimization of the electricity production specific cost, instead of the hybrid plant's LCC. The electricity production-specific cost is defined as the quotient of the total annual production cost

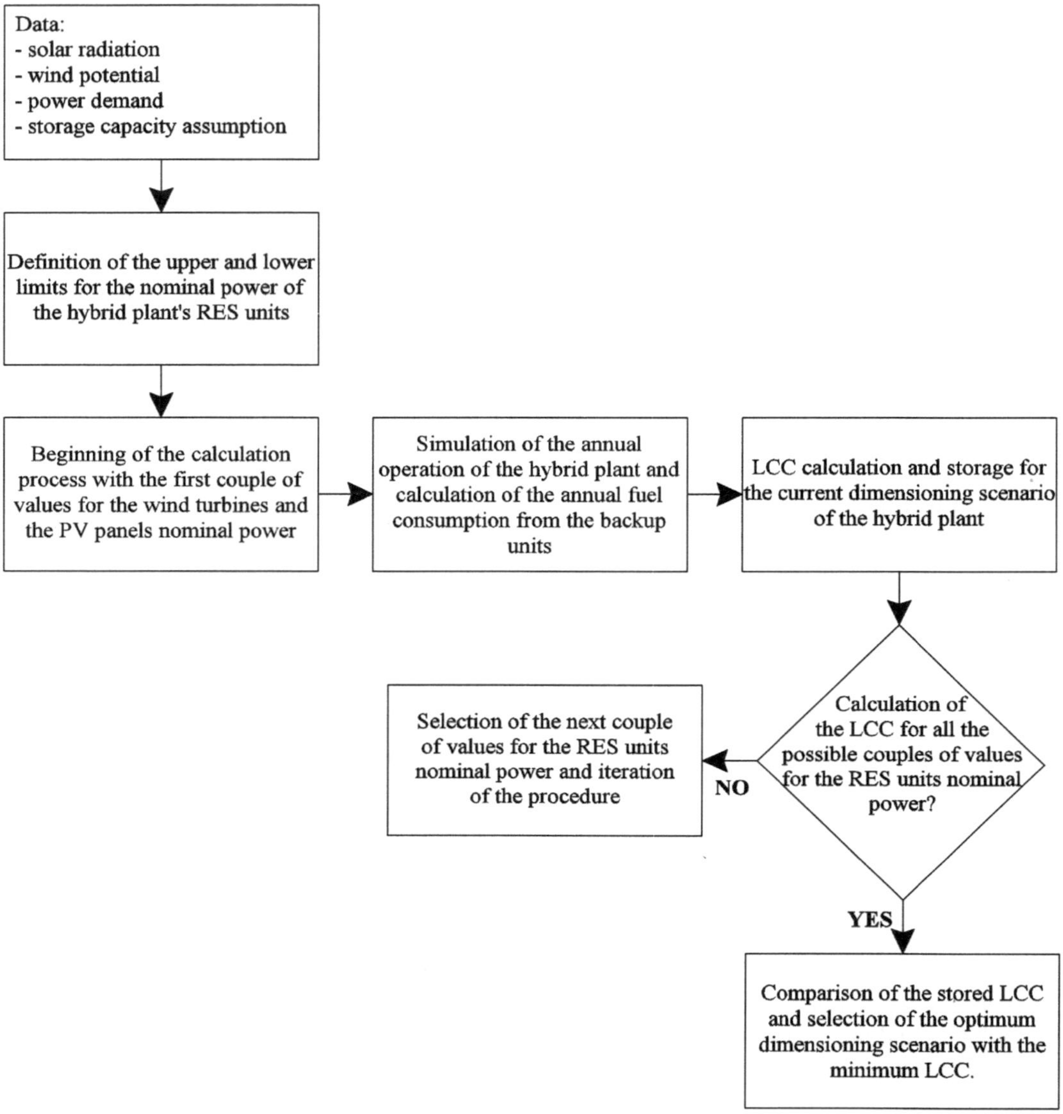

FIGURE 3.34 Iterative process algorithm for the dimensioning optimization of small-size hybrid power plants, based on LCC minimization.

over the annual electricity production. For the calculation of the annual production cost, apart from the obvious annual operation cost, the contribution of the total setup cost must be also considered. The simplest way to do this is to divide the total setup cost with the total life period of the hybrid plant expressed in years. In this way, the total setup cost is evenly allocated over the hybrid plant's whole life period, calculating, thus, its contribution to the annual production cost configuration. Consequently, the electricity production annual specific cost can be calculated with the relationship:

$$c_p = \frac{\frac{C}{N} + O.C.}{E_{el}} \tag{3.67}$$

where C is the total setup cost, N the hybrid power plant's life period in years, OC the annual operation cost (maintenance and fuel consumption), and E_{el} the annual electricity production. If the annual operation cost varies from year to year, the average value for the whole life period should be introduced in the above relationship.

3.8 HYBRID POWER PLANT CASE STUDIES

Having presented all the alternative technologies and approaches for the integration and operation of hybrid power plants in electrical systems, in this section four case studies on the dimensioning and the operation of hybrid power plants of different technologies, syntheses, and sizes will be presented. Three of these case studies refer to non-interconnected insular systems. This is because autonomous electrical systems often constitute characteristic cases for hybrid power plant introduction, aiming at the transition from fossil fuel consumption to energy independency through the exploitation of clean energy resources. Given the lack of support from any neighboring electrical grids, the only way to achieve the above targets is the introduction of hybrid power plants. Additionally, the restricted size of insular, autonomous systems, even of large size, offers an excellent option for the simulation and study of hybrid power plants for demonstrative or educational purposes.

The fourth investigated case study refers to an isolated building with a corresponding power consumption of very small size.

The four hybrid power plants that will be presented as case studies in this last section of this chapter will be:

- A wind park–PHS system for the insular, autonomous system of the Faroe Islands. The Faroe Islands' electrical system exhibits an annual power peak demand higher than 70 MW, classifying it as a medium-size system. Abundant wind potential, rich rainfalls available for ten months annually, and intensive land morphology create a favorable environment for the introduction of this specific hybrid power plant.
- A wind park–seawater PHS plant for the insular, autonomous system of the small Aegean Sea island of Sifnos, Greece. The annual power peak demand of this insular system is 6.5 MW, classifying it as a small-size system. The lack of adequate rainfalls in the restricted insular territory imposes the use of seawater as the working medium of the PHS system. This example is chosen mainly to demonstrate the efforts of the local energy cooperative company (Sifnos Island Cooperative–SIC) towards energy independency in the insular system. Additionally, it is chosen as a characteristic case for the introduction of PHS systems even in cases of small-size electrical systems, under favorable conditions, such as high available wind potential and appropriate land morphology. With the proper dimensioning and siting of the system's components, the introduction of a wind park–PHS system for the island of Sifnos can guarantee secure 100% RES penetration in the local insular grid.
- The last case study for an insular system will be for the autonomous system of the island of Kastelorizo, Greece. The annual peak power demand of the system is lower than 1 MW, classifying it as a small-size system. For this case, a hybrid power plant supported alternatively by a CAES system or electrochemical batteries will be presented.
- Finally, a small hybrid power plant for an isolated residential building will be investigated, consisting of small wind turbines, PV panels, and electrochemical batteries. The optimization of the dimensioning of this hybrid power plant will be based on the minimization of the LCC.

3.8.1 A Hybrid Power Plant for the Faroe Islands

3.8.1.1 The Aim of the Dimensioning

The Faroe Islands complex consists of 18 islands, located in the North Atlantic Ocean, between Scotland and Iceland, with a permanent population of approximately 50,000 inhabitants. A map of the Faroe Islands and their location in Europe is provided in Figure 3.35.

The local utility in the Faroe Islands (SEV) has emerged the initiative for 100% RES penetration in the local electrical system (see http://secure.interreg-npa.eu/news/show/faroe-islands-100-renewable-generation-of-electricity-by-2030/). Despite the abundant RES potential available in the insular complex of the Faroe Islands, mainly wind and hydro, as it will be shown below, the achievement of this target will not be easy. Following this energy independency program of SEV, the dimensioning of the introduced hybrid power plant will be accomplished focusing on 100% RES penetration in the electrical grid.

3.8.1.2 Independent Parameters of the Dimensioning

The dimensioning of the hybrid power plant will be executed on the basis of two basic, independent parameters:

- the installed power of the RES units, mainly wind parks and supplementary PV stations
- the storage capacity of the storage unit, which, given the site of the power demand and the intensive land morphology, will sensibly be a PHS system.

The dimensioning process will be executed iteratively, based on the annual operation simulation of the hybrid power plant, employing annual time series with average hourly values. For each simulation execution, alternative syntheses and dimensioning scenarios will be investigated, aiming to achieve 100% RES penetration in the local electrical system, ensuring the system's security and keeping the hybrid power plant's setup cost as low as possible.

3.8.1.3 Required Data

The required data for the execution of the simulation process are:

1. the power demand annual time series of average hourly values
2. the wind velocity and solar radiation annual time series of average hourly values, as measured by existing meteorological stations and wind parks
3. the storage capacity of the storage plant, practically, for a PHS system, the effective capacity of the upper reservoir versus the available geostatic head (absolute altitude difference from the lower reservoir)
4. characteristic efficiency curves of the involved pumps and hydro turbine models and the power curve of the introduced wind turbine
5. digitized map of the PHS system installation site for the siting of the system and the calculation of fundamental features regarding the PHS system, such as the volume of the upper reservoir and the length of the penstock route.

3.8.1.4 Building the Power Demand Annual Time Series

For the dimensioning of the introduced hybrid power plant, a future projection of the power demand will be attempted, considering, also, apart from the current electrical loads, the transition of heating from oil to heat pumps and the introduction of electrical vehicles. The annual power demand in the Faroe Islands in 2016 is given in Figure 3.36. The lower demand during summer reflects industry holidays and less need for light and heating. The maximum power demand is measured equal to 55.14 MW at the 8,298th hour of the year, namely between 5 p.m. and 6 p.m. on December 12.

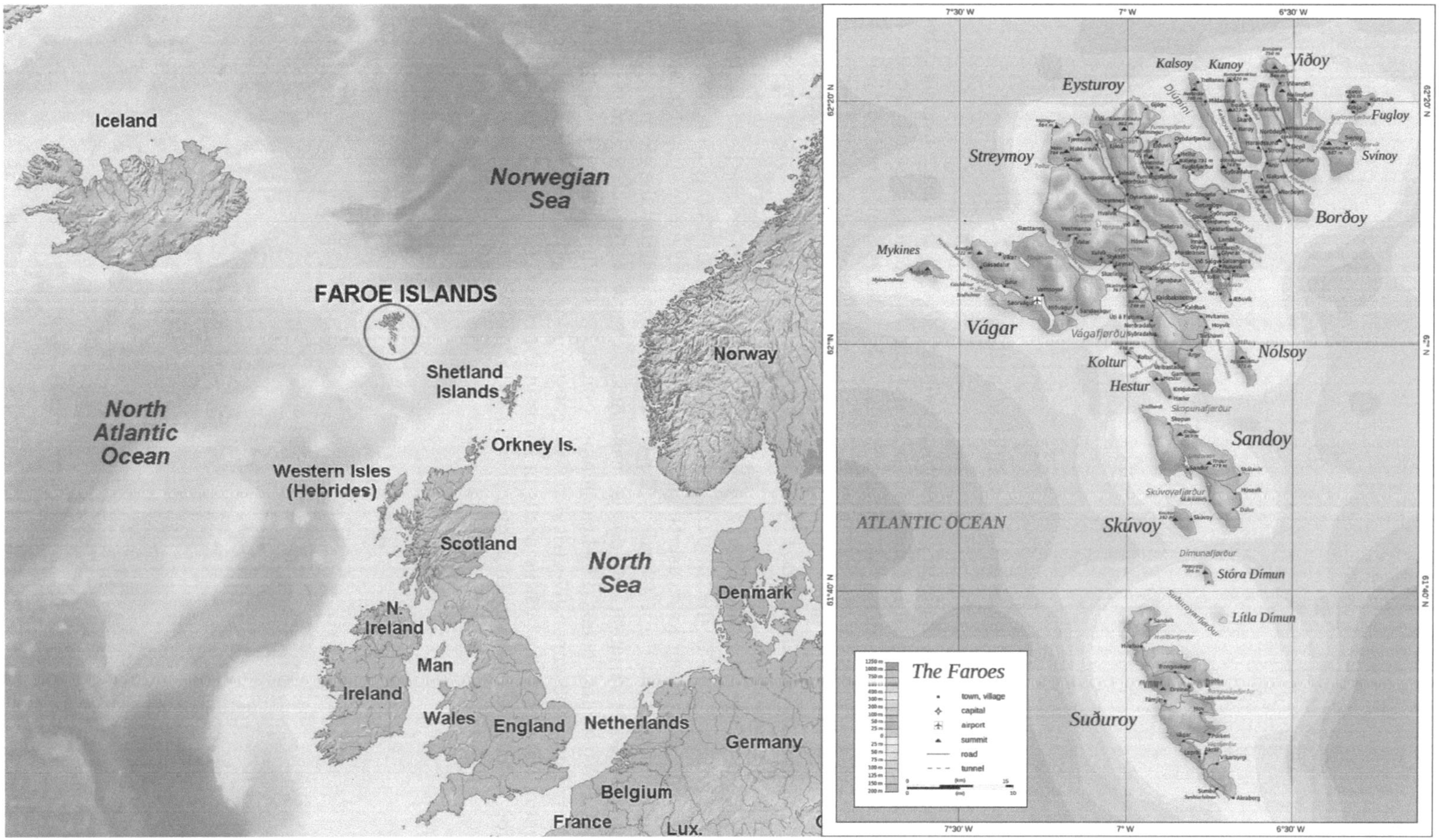

FIGURE 3.35 Map of the Faroe Islands and their location in Europe [49].

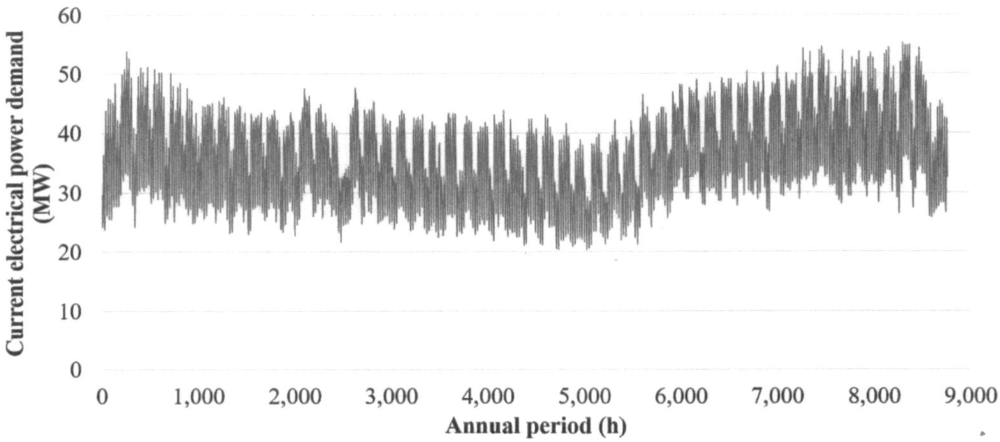

FIGURE 3.36 Existing annual power demand variation in the Faroe Islands [49].

The Islands are located at the high geographical latitude of 62°N and space heating is required throughout the year. Traditional heating from peat and coal was replaced by oil burners around 50 years ago and ever since, almost all heating needs in the Islands are covered by imported oil. More than 20,000 oil burners are in operation in 17,500 households, offices, and industry buildings. However, due to increasing oil prices, alternative heating options have garnered attention in recent years and a number of households have made a turn towards heat pumps. This conversion is also encouraged by the authorities.

In 2016, around 24% of the total 236.475 tons of oil import was used for heating. Taking into account the efficiency loss in the oil burner, the oil used for heating corresponds to 520 GWh of heating needs. According to measurements by the Energy Directorate (unpublished data) an electrical heat pump in the Faroe Islands will have a COP around 3 and the electric demand will have a distinct seasonal variation with double magnitude during winter, with regard to summer. To meet the 520 GWh heating demand by heat pumps, around 175 GWh of electricity will be needed. Figure 3.37 presents the annual electrical power consumption for indoor space heating, arisen from the entire transfer of the heating needs in Faroe in the electrical grid. The annual electricity consumption, calculated by integrating this curve, is 180.126 GWh.

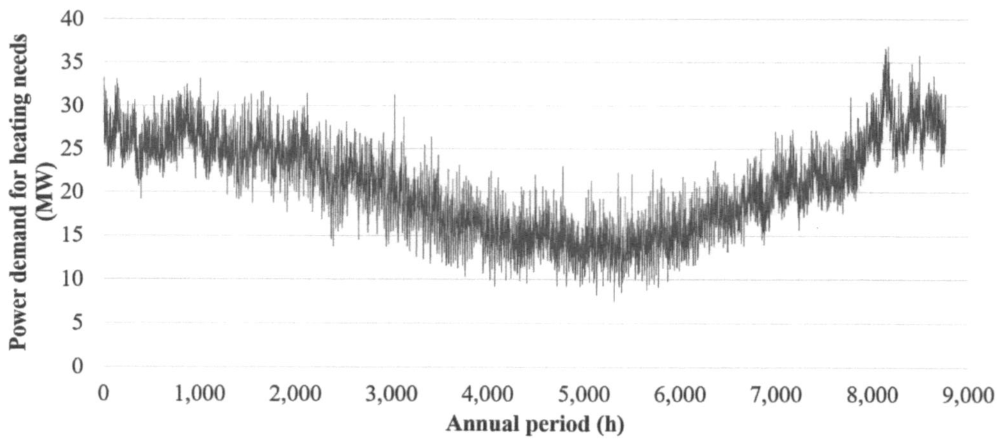

FIGURE 3.37 Estimated annual electrical power demand for the heating needs in the Faroe Islands [49].

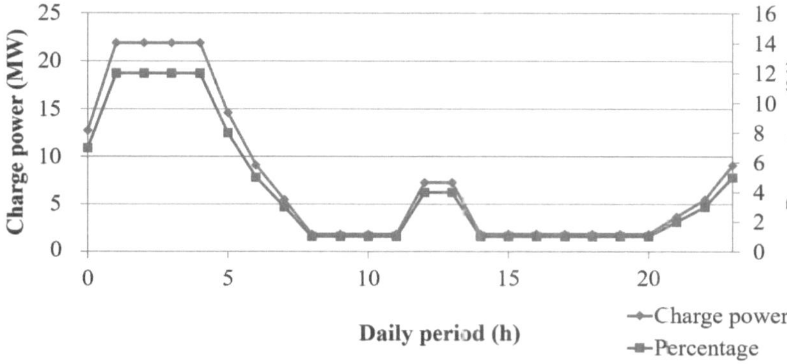

FIGURE 3.38 Introduced electrical vehicle charge daily profile [49].

The electricity consumption for the transportation sector is approached based on statistical data. A typical levelized energy consumption of 20 kWh per 100 km (0.20 kWh/km) is adopted for electrical vehicles of average size. Additionally, the average distance covered by a typical vehicle during a whole yearly period in the Faroe Islands is estimated at 15,000 km. This implies an average daily covered distance of 41 km. Furthermore, the above assumptions impose annual electricity production per vehicle at 3,000 kWh. Finally, the number of personal cars registered in the Faroe Islands is 22,200, leading to an overall electricity consumption for the onshore transportation sector of 66,600 MWh.

The final step is to develop a typical charge daily profile for all the cars in the Islands. The main charging process will be certainly executed at night, during which the daily peak for the corresponding power demand will appear. A second, minor peak is also introduced in the early afternoon hours. The final developed daily charge profile is presented in Figure 3.38.

By adding the above time series, we get the overall power demand time series, combining the current electricity consumption, the indoor space heating needs, and the onshore transportation needs. Yet, the Faroe Islands are blessed with remarkably high hydro potential. Annual rainfalls higher than 3,000 mm are measured in several locations in the country. Sensibly, hydro electricity has been a fundamental production technology for the Islands. In 2016, the annual electricity production from all the hydro power plants reached 106,348 MWh, a figure that corresponds to 34% of the annual electricity consumption. The maximum annual hydro power production was at 33.41 MW in 2016. This hydro production will be maintained in the new examined system, contributing to the target of 100% RES penetration. The net overall power demand will be derived by subtracting the hydro power production time series from the overall power demand time series. This subtraction is depicted graphically in Figure 3.39. The net annual electricity consumption is calculated at 455,234 MWh and the annual peak net power demand is calculated at 81.72 MW.

3.8.1.5 Available RES Potential

Solar radiation in the Faroe Islands is not high, as sensibly expected. Solar radiation measurements since 2008 indicate total annual incident solar irradiation on horizontal planes at 780 kWh/m^2. A typical annual time series of the levelized electrical power production from a PV station per installed kWp is presented in Figure 3.40. The total annual electricity production is calculated at 630kWh/kWp. In this figure it is seen that, despite the low power production during winter, the PV station can significantly contribute during summer, with power production as high as 900 kW/kWp, a figure that can be considerably crucial for the successful achievement of the 100% energy independency target in the Islands, given the fact of the relatively low wind potential during the same season, as documented below.

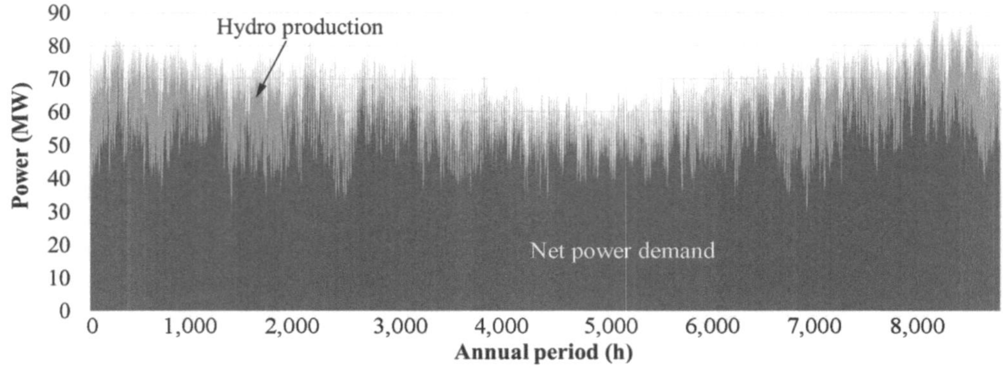

FIGURE 3.39 Hydro production and overall net power demand [50].

The Faroe Islands are blessed with world record wind potential. The official strategy of the local state utility company (SEV) is to focus on large-scale wind power development. Fast track development of wind parks is essential when expecting considerable increases in electricity production to supply the needs of heating and onshore transportation. Wind energy is especially well-suited as supply for heating demand, because both have coinciding profiles throughout the year.

Remarkably high wind potential is often met in Faroe Islands, with average annual wind velocity above 10 m/s. For the purpose of this study, the annual wind velocity time series depicted in Figure 3.41 was employed. According to the captured wind potential measurements, the annual average wind velocity is calculated at 10.8 m/s.

However, the wind potential in the Faroe Islands exhibits two major drawbacks:

1. It is the result of strong ocean winds, with regular extreme values (wind gusts), which are not exploitable, because they fall out of the wind velocity operating range of the wind turbines (usually 25 m/s). As seen in Figure 3.42 (dotted curve), these values range from 2.5% to 4% during the winter period, imposing significant hysteresis losses for the wind turbines.

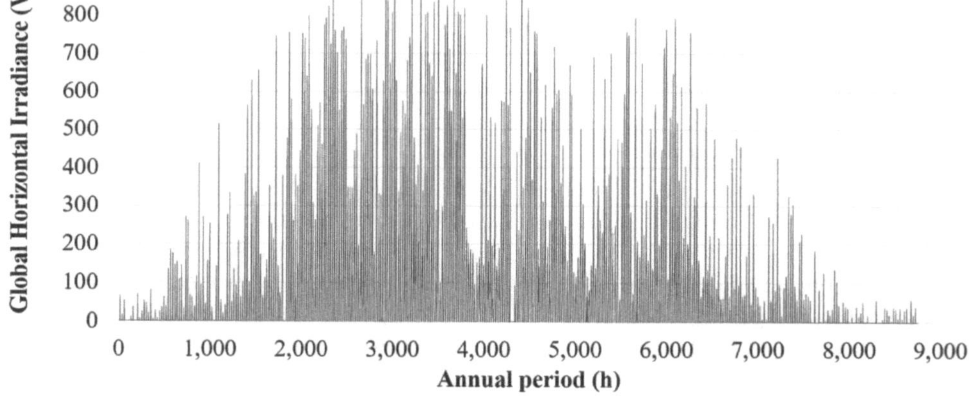

FIGURE 3.40 Annual variation of the levelized power production from PV stations in the Faroe Islands [49].

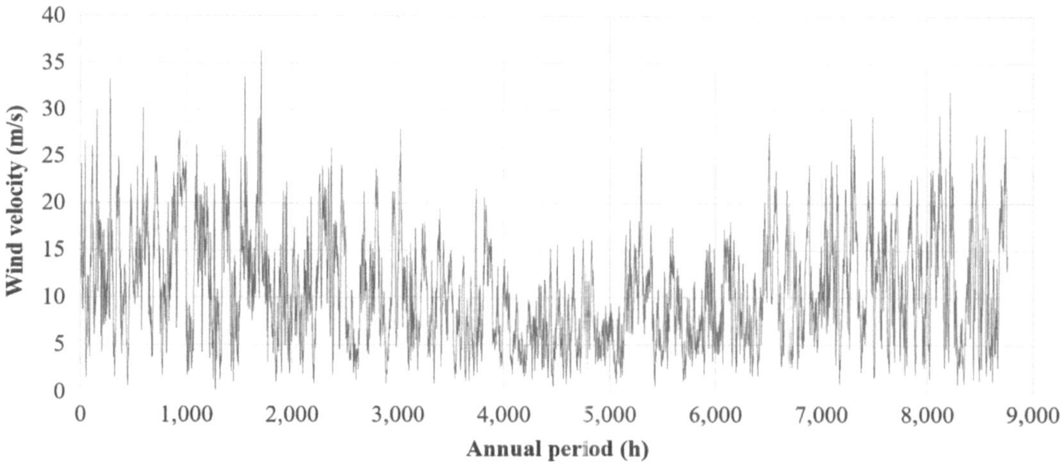

FIGURE 3.41 Annual wind velocity variation at a specific site available for wind park installation.

2. It is concentrated mainly during the winter period. This fact is also depicted in Figure 3.42 (continuous curve), where the monthly average wind velocity values are plotted. In this figure it is seen that the average wind velocity remains above 11 m/s from October to April; it is still adequately high for August (8.8 m/s), September (9.5 m/s), and May (10.0 m/s); but for June and July it drops to 7.6 m/s and 6.7 m/s, respectively. Hence, for a period of two months the available wind potential exhibits a remarkable fall, a fact that will be crucial for the dimensioning of the hybrid power plant, as will be shown below.

3.8.1.6 The Proposed Hybrid Power Plant

Energy independency in the Faroe Islands will certainly be based on wind energy and solar radiation, the most commonly used primary energy sources in insular systems, along with the already exploited hydrodynamic potential. Particularly in the Faroe Islands, energy autonomy will be mainly

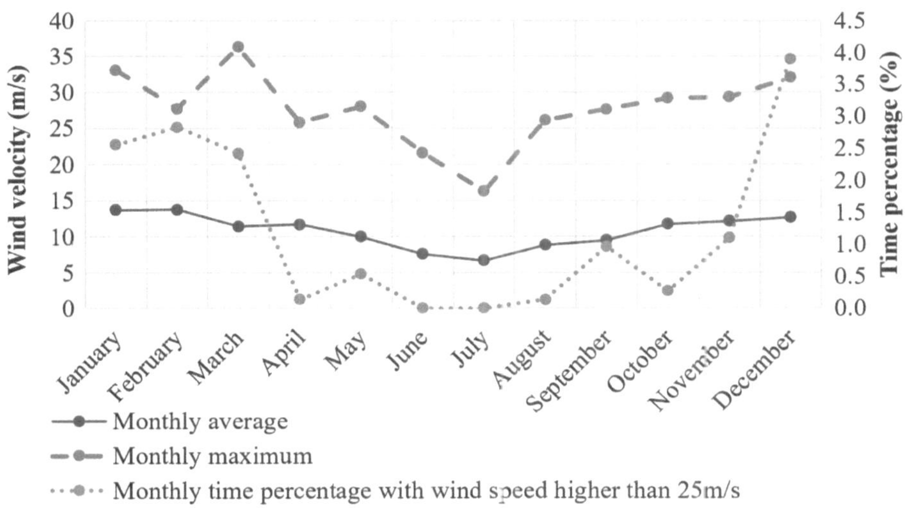

FIGURE 3.42 Monthly average and maximum wind velocity at a specific site available for wind park installation in Faroe Islands and monthly time percentage with wind velocity values higher than 25 m/s [49].

based on wind parks, given the remarkably high wind potential for 9 months during the year. PV stations will also be examined as complementary RES power plants, substantially during summer, when the available wind potential is considerably reduced. The RES power plants will be supported by a PHS system. The proposed hybrid power plant aims at 100% RES annual electricity production. To approach this target, the adequate dimensioning of the proposed plant is achieved with the iterative execution of the computational simulation of the system's annual operation, following the operating algorithm described in Section 3.5.1.1.

With regard to the appropriateness of the available land morphology, the Faroe Islands seem to constitute an ideal field for PHS installations. The insular terrain generally is configured from mild mountain slopes, valleys, and plateaus with altitude differences between 300 and 500 m often met, namely ideal sites for PHS installations. What is more important is that this excellent land morphology is also combined with reach annual rainfalls, making it very easy to gather water and form technical reservoirs for the integration of the PHS system.

In this study, a particular site is selected, given the recommendations of the Energy Department of the Environment Agency and SEV, for the installation of the PHS. This site is located in the area of Vestmanna, at the north side of the island Streymoy, which also hosts the capital of the country Tórshavn. In this particular site, there are already two existing reservoirs, a lower one with a capacity of 2,100,000 m^3 and its free surface at the absolute altitude of 107 m, and an upper one with a capacity of 4,100,000 m^3 and its free surface at the absolute altitude of 346 m. The absolute height difference between the two reservoirs as they exist today is 239 m. The lower reservoir is formulated with an arch dam, while the upper reservoir with a gravity dam. The overall area with the two reservoirs is presented in three-dimensional (3D) satellite view and in digitized map in Figure 3.43.

Neither of the existing reservoirs' capacities is adequate to cover the requirements for the energy storage plant of the Faroe Islands. Both of them should be enlarged. After essential on-site geotechnical assessment, especially on the lower reservoir, it is estimated that a capacity around 8,700,000 m^3 for the lower reservoir can be technically feasible. Hence, in this simulation, the capacities of both reservoirs will be considered equal to the above presented maximum constructive limit. Eventually, the new capacities for the lower and the upper reservoirs are precisely calculated, through the digitization of the land terrain and the computational volume measuring at 8,980,300 m^3 and 9,442,400 m^3, respectively. The new absolute height difference of the reservoirs' free surfaces is 222 m. Given the size of the PHS system and the intensive slope close to the lower reservoir, the penstock should be rather constructed underground, as a tunnel, directly with the appropriate diameter, which is also facilitated by the relatively low height difference and the corresponding low maximum hydrostatic pressures. The length of the penstock should be in the range of 2,000 m.

3.8.1.7 Results

The proposed system is computationally simulated following the adopted operation algorithm presented in Section 3.5.1.1. Various executions were performed for different dimensioning scenarios, with regard to the reservoirs' capacities and the nominal power of the wind park and the PV stations. The wind power production was estimated on the basis of the available wind velocity time series and the introduced power curve of the above mentioned wind turbine model. No precise siting of the wind turbines was accomplished and no wind map was developed. The aerodynamic shading losses and the hysteresis losses were empirically considered equal to 5% and 3.5% respectively, while the availability of the wind turbines was assumed at 95%. For both technologies and for the PHS system, the electricity transportation losses were assumed 3%.

The results from the executed simulations regarding the RES annual penetration are summarized, among the essential differences between the final presented scenarios, in Table 3.8.

From the results presented in Table 3.8 we come to the following conclusions:

- Annual RES penetration higher than 90% or even 95% can be relatively easily achieved, maintaining also the economic feasibility of the required investment, as documented below.

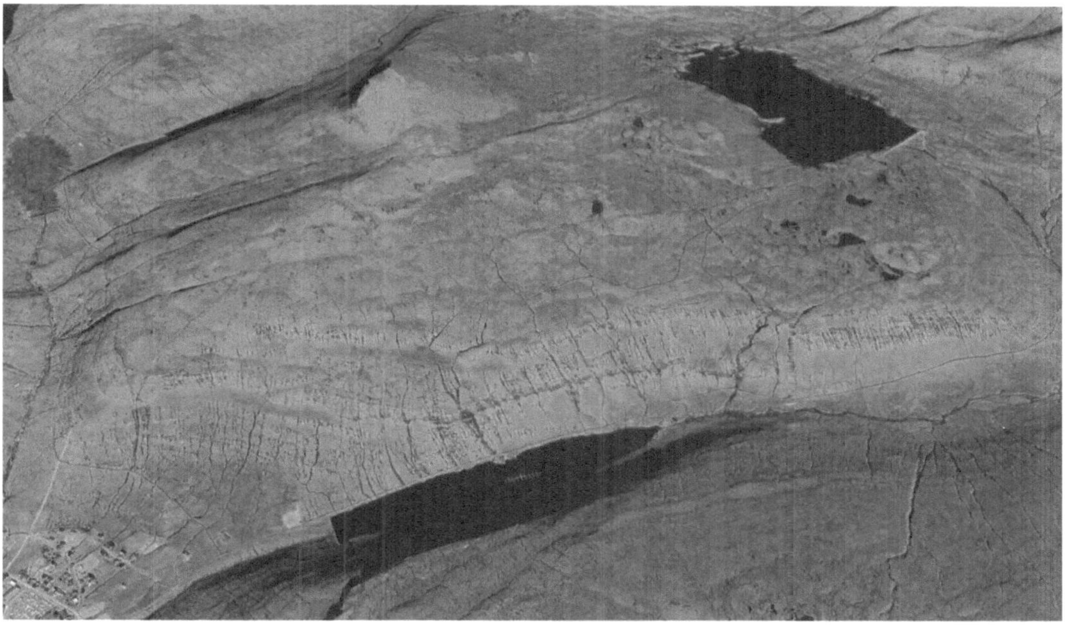

(a)

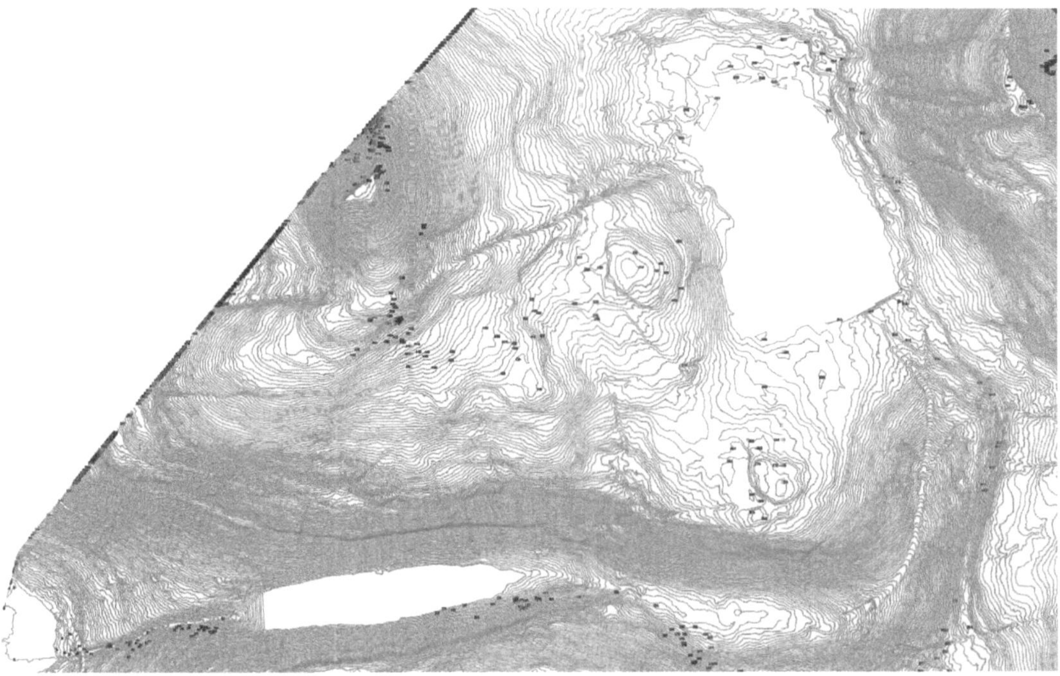

(b)

FIGURE 3.43 (a) The area of Vestmanna in 3D satellite view [49]. (b) The Vestmanna site top view in digitized map [49].

TABLE 3.8
Dimensioning and RES Penetration Main Results [49]

Dimensioning Scenario	1	2	3	4	5	6
			Dimensioning			
Reservoirs' nominal capacity ($10^6 \cdot m^3$)	8.9	8.9	8.9	4.5	4.5	4.5
Wind park nominal power (MW)	184.0	230.0	184.0	184.0	230.0	184.0
PV nominal power (MW)	0.0	0.0	50.0	0.0	0.0	50.0
Hydro turbine maximum power (MW)	74.0	74.0	74.0	74.0	74.0	74.0
Pump maximum power (MW)	143.5	184.4	176.7	143.5	184.4	176.7
			RES Penetration			
RES direct penetration (%)	43.8	45.2	46.0	43.8	45.2	46.0
Hydro turbine production (%)	46.2	49.2	47.4	43.5	45.2	44.1
Total RES penetration (%)	90.0	94.4	93.4	87.3	90.4	90.1
Total RES rejection (%)	16.8	29.5	18.0	20.2	33.3	21.8
PHS storage cycle efficiency (%)	63.8	63.0	63.8	64.1	63.4	64.1

- The major obstacle towards 100% energy independence, as expected, is the low wind potential availability during the summer period. This is also justified with the annual fluctuation of the water stored volume in the PHS system's upper reservoir, depicted in the graphs shown in Figure 3.44. From these figures we can see that the low RES availability during summer causes the water stored volume in the reservoir to drop to the minimum level, imposing the introduction of diesel generators in the production mixture.

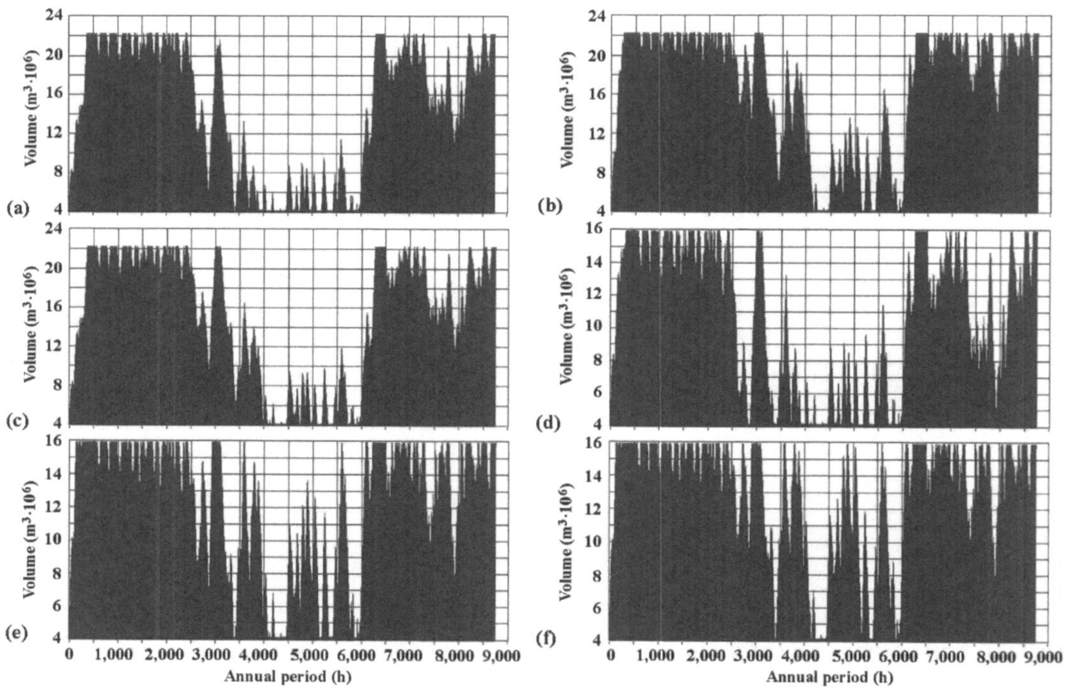

FIGURE 3.44 Fluctuation of the water stored volume in the upper reservoir for the various investigated scenarios [49].

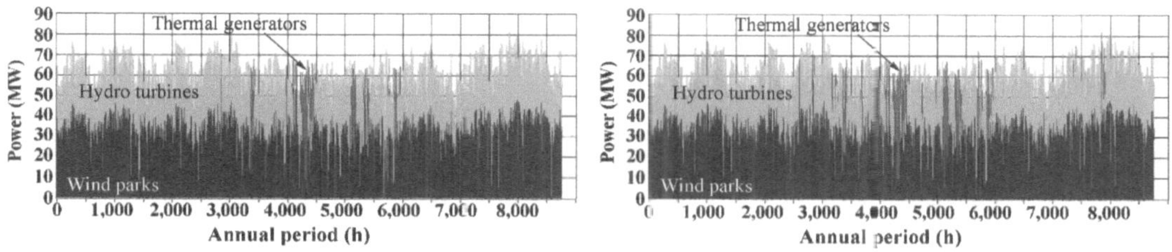

FIGURE 3.45 Annual power production synthesis graphs for the above investigated scenarios 3 and 6 (graph (a) and (b) respectively) [49].

- The addition of PV stations in the hybrid power plant with nominal power up to 50 MW in order to maintain the economic feasibility of the investment, does not seem to constitute an effective solution. We can come to this conclusion by observing that roughly the same annual RES penetration can be achieved with the introduction of either 50 MW of PVs or additional 46 MW of wind turbines.
- The storage capacity of the reservoirs seems to be high enough for the purposes of the power plant, namely the energy independence of the Faroe Islands. We come to this conclusion by observing that the reduction of the nominal capacity of the reservoirs from 8.9 to 4.5 million m^3 practically does not affect the achieved RES annual penetration.

Figure 3.45 presents the annual power production synthesis for the above scenarios 3 and 6. It is clearly seen that thermal generator production is exclusively required during summer.

The setup cost of the wind park and the PVs is calculated on the basis of an adopted specific setup cost, set equal to 1,200 €/kW for the wind park and 1,000 €/kWp for the PVs. The setup of the PHS system, estimated following the experience from similar previous projects, is presented in Table 3.9. The PHS setup cost ranges versus the different scenarios for the reservoir sizes and the pumps' required power, as calculated from the executed simulations. The funding scheme assumed 50% equities and 50% banking loan with 1.50% loan rate and a 15-year payback period.

The selling price of the produced electricity from the hybrid power plant to the final consumers was adopted equal to 0.15 €/kWh. The results presented in Table 3.10 prove the economic feasibility of the investigated hybrid power plant. The presented economic indices refer to the investment's equities.

TABLE 3.9
PHS System Setup Cost Calculation [49]

No.	Setup Cost Component	Cost (10^6·€)
1	Hydro turbines and generators (hydro power plant)	44.4
2	Pumps and motors (pump station)	101.5–133.0
3	Penstocks	21.3
4	Upper reservoir	0.0–13.3
5	Lower reservoir	5.0–17.8
6	New utility network	7.5
7	Several infrastructure works	2.0
8	SCADA	1.5
9	Consultant fees	2.0
10	Several other costs	2.0
	Total PSS setup cost	**187.2–244.8**

TABLE 3.10
Economic Evaluation Results [49]

Dimensioning Scenario	1	2	3	4	5	6
Setup cost (10^6·€)	431.4	518.8	511.6	406.0	492.7	485.5
Equities payback period (years)	6.4	8.4	7.8	6.2	8.5	7.7
IRR on equities (%)	11.8	7.8	8.8	12.4	7.8	9.0

Summarizing the results of the executed study, it seems that the achievement of 100% energy independence in the remote insular system of the Faroe Islands is a real challenge. The topography of the Faroe Island is truly blessed with abundant wind and hydrodynamic potential and excellent sites for PHS installations. Yet, the low wind potential availability during summer constitutes a significant obstacle, for a clear, 100% exclusive energy production in Faroe from RES. To this end, a highly promising solution can be the integration of the proposed hybrid power plant with parallel energy saving actions in the existing and new electricity consumptions, with heating being, of course, the main sector for potential energy saving applications.

3.8.2 A Hybrid Power Plant in the Island of Sifnos, Greece

3.8.2.1 The Aim of the Dimensioning

Sifnos is a small island located in the Western Aegean Sea, with a permanent population of 2,500 inhabitants. Administratively, it belongs to the Cyclades Complex. A map of the island with its location in Greece is given in Figure 3.46.

Sifnos today constitutes a non-interconnected insular system with an annual power peak demand of 6.5 MW and an annual electricity consumption of 18,500 MWh. It is considered a small-size electrical system. Currently the electricity production in Sifnos is based exclusively on imported diesel oil, consumed in 8 diesel generators of 1 MW nominal power each, installed in the local

FIGURE 3.46 Map of the island of Sifnos and its location in Greece [41].

thermal power plant. The electricity production overall specific cost ranges above 0.25 €/kWh. Despite the considerably high electricity production cost, the small size of the insular system and the distance of Sifnos from the nearby islands make the perspective of its interconnection economically rather unfeasible.

In 2013, a local energy cooperative was founded in the island, with the brand name Sifnos Island Cooperative (SIC). The main target of SIC is to achieve energy independency of the island by substituting the currently consumed imported diesel oil for electricity and the transportation sector with the locally available RES, mainly wind energy and solar radiation. A major challenge for SIC is to approach the above goal with the optimum and most favorable economic terms, in order to maximize the economic and social benefits for the SIC and the local community as well. This is because the ultimate target of SIC is to claim a feasible and sustainable development of the insular community by exploiting the expecting revenues from the operation of the required RES projects to create the essential technical infrastructure that will facilitate the creation of new professional activities in the island, apart from tourism, on which more than 95% of the annual community's income is currently based.

Given the fact of the autonomous insular system and the rather improbable option for a future electrical interconnection, the only way to achieve energy independency in Sifnos is through the development of a hybrid power plant. After extensive research, SIC concluded on the optimum synthesis of the hybrid power plant: a wind park, probably a PV station, and a PHS system operating with seawater, given the low availability of rainfall. A cluster of additional actions, such as the introduction of electrical vehicles and demand-side management techniques, integrate the overall effort towards energy independency. However, in this section, we focus only on the dimensioning, siting, and simulation of the centralized hybrid power plant, which will be the foundations of the overall effort.

3.8.2.2 Independent Parameters of the Dimensioning

As in the case of the Faroe Islands, the scope of the dimensioning process will be the development of a hybrid power plant capable of undertaking 100% electricity production in the island from RES. The dimensioning of the hybrid power plant will be executed on the basis of two basic, independent parameters:

- the RES unit required power, most probably a wind park and perhaps a PV station
- the storage capacity of the storage units, namely the PHS system.

The dimensioning process will be executed iteratively, based on the annual operation simulation of the hybrid power plant, employing annual time series with average hourly values. For each simulation execution, alternative synthesis and dimensioning scenario will be investigated, aiming to achieve 100% RES annual penetration in the local electrical system, ensuring the system's security and keeping the hybrid power plant's setup cost as low as possible.

3.8.2.3 Required Data

Similarly to the previous case study, the required data for the execution of the simulation process are:

1. the power demand annual time series of average hourly values
2. the wind velocity and solar radiation annual time series of average hourly values
3. the storage capacity of the storage plant, practically of the PHS system, the effective capacity of the upper reservoir versus the available absolute altitude difference from the sea
4. characteristic efficiency curves for the involved pump and hydro turbine models and the power curve of the employed wind turbine
5. digitized map of the PHS system installation site for the siting of the system and the calculation of fundamental features regarding the PHS system, such as the volume of the upper reservoir and the length of the penstock route.

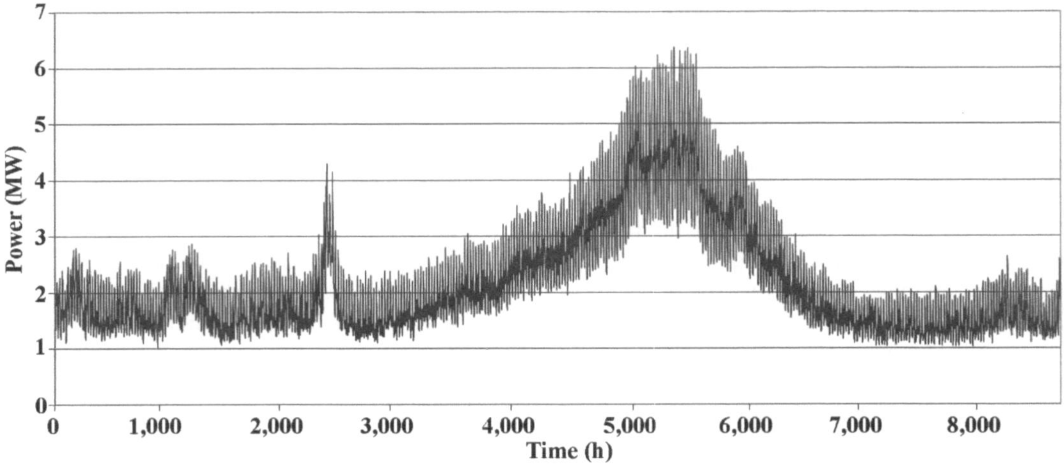

FIGURE 3.47 Annual power demand variation in Sifnos Island [41].

The annual power demand in Sifnos is given in Figure 3.47. The main features of this curve are the maximization of the power demand in summer due to tourist activities, the minimization during autumn and spring because of the mild weather, and the slight increase during winter due to heating needs. Actually, for the case of Sifnos, the power demand in summer is more than three times higher than in autumn, revealing the almost exclusive dependence of the local economy on seasonal tourism. This form of power demand, as indicated in Chapter 2, constitutes a characteristic form for the most autonomous insular systems on earth, with considerable tourist economic activities in summer. Another occasional increase of power demand is also observed in spring due to the Easter period.

The maximum power demand is measured equal to 6.37 MW close to August 15. On the other hand, the minimum power demand is measured at 1.03 MW. In general, such low power demand values are met for almost all of October through May, mainly during the early morning hours. The annual electricity consumption is calculated by integrating this curve at 18,857 MWh.

The annual wind velocity time series is depicted in Figure 3.48. The wind potential was evaluated at the specific site where the wind park is planned to be erected with an 11 m height meteorological

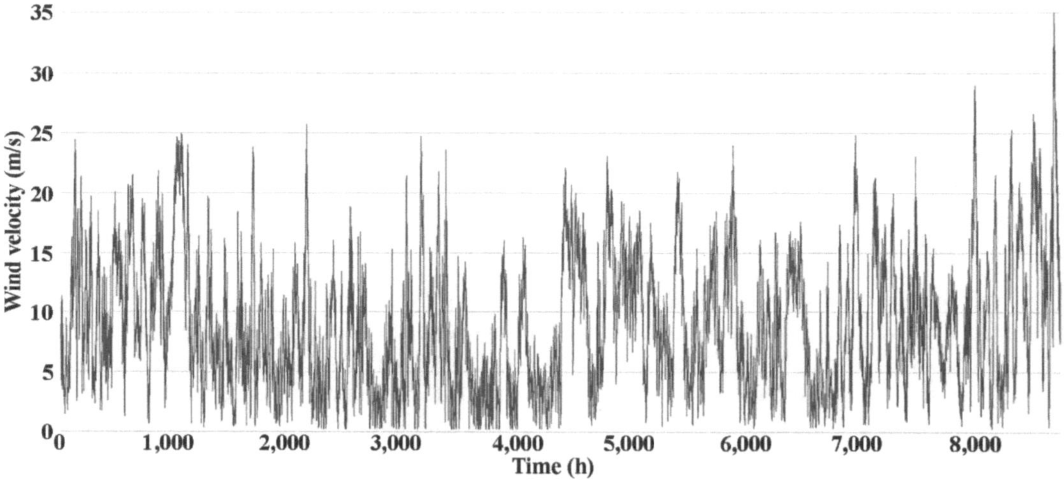

FIGURE 3.48 Annual wind velocity variation at the specific site selected for the wind park installation on the island of Sifnos [41].

mast, installed exclusively for this specific purpose. According to the captured measurements, the annual average wind velocity is calculated at 9.0 m/s, revealing the high available wind potential. What is most important, however, is the coincidence of high wind velocity availability with the peak power demand period. Specifically, during July 2016, the average monthly wind velocity was measured at 11.8 m/s, while the same feature was measured at 9.1 m/s during August 2016. The corresponding values for the same months in 2017 are 10.1 m/s and 12.0 m/s respectively, proving that the availability of high wind potential during this particular period was not incidental during the first summer period of the wind measurement process. Additionally, Figure 3.48 also shows that the wind velocity remains almost constantly lower than 25 m/s, apart from a period during the last weeks of the year. This implies that the high annual average wind velocity is configured by wind potential totally exploitable by the wind turbine operating range. Furthermore, the lack of extreme wind velocity favors the safe operation of the wind turbines and contributes to the security of the electrical system.

Another important feature of the available wind potential is the stability of the prevailing wind blowing direction. As seen in the wind rose presented in Figure 3.49, for more than 74% annually wind is constantly blowing from north, northeast. Especially during summer, namely the peak power demand period, this blowing direction remains constant for more than 95% of the particular time period. This characteristic is also depicted in the high value of the Weibull k parameter, also presented in Figure 3.49 (k=1.73). The high value of this parameter implies low variation of the wind velocity. This feature is very important with regard to any potential impacts of the introduced hybrid power plant on the autonomous system's dynamic security. The constant north, northeast wind blowing direction, especially during summer, contributes towards the improvement of the power quality injected in the local grid by the wind turbines.

Given the high available wind potential at the wind park's installation site, in combination with its coincidence with the peak power demand period and its high quality (low wind velocity variation with regard to its measure and direction), most probably there will be no need for the introduction of a PV station. Nevertheless, this will be also checked through the iterative dimensioning procedure. To this end, an annual solar radiation time series must be introduced too. Such a time series, captured in the Aegean Sea island of Samos (Easter Aegean Sea), is presented in Figure 3.50.

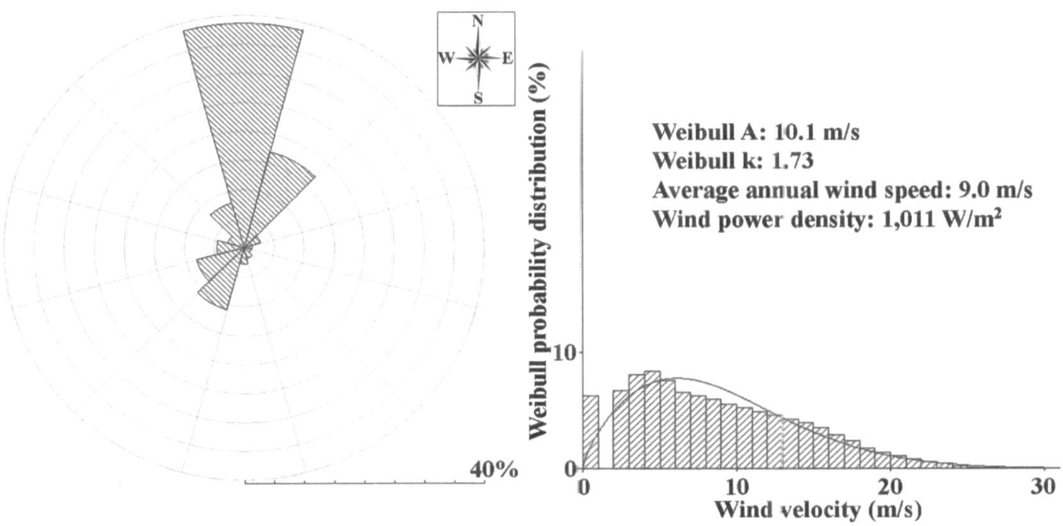

FIGURE 3.49 Wind rose and Weibull distribution based on the annual wind potential measurements at the wind park's installation site [41].

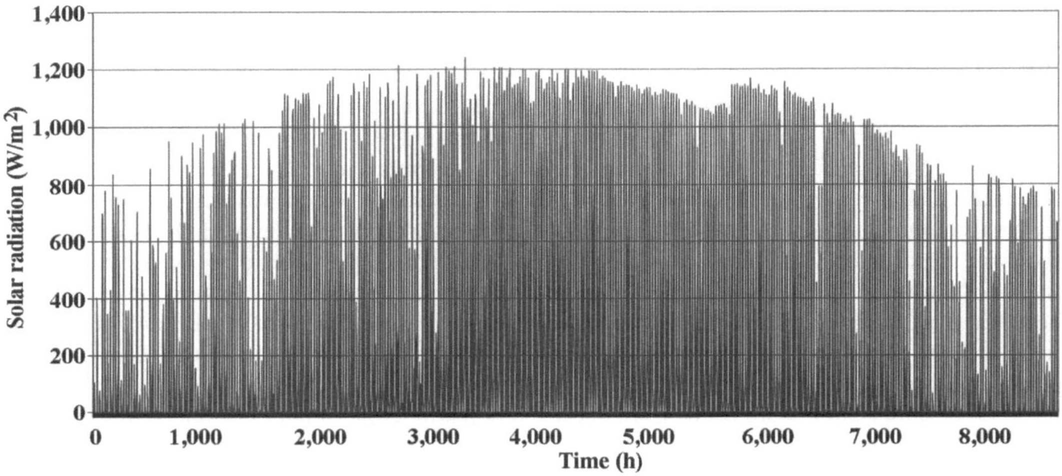

FIGURE 3.50 Annual solar radiation time series from the island of Samos, employed also for the simulation procedure for the island of Sifnos.

Finally, with regard to the available land morphology, the intensively varied terrain in the island offers a number of alternative locations for the installation of the PHS system. Eventually, it was decided to exploit the selected site for the installation of the wind park for the construction of the PHS upper reservoir too. The specific site, located at the northeastern coastline of the island, apart from the high wind potential, also exhibits all the essential favorable characteristics for the installation of a seawater PHS, which are:

- proximity to the coastline
- absolute altitude at the location for the construction of the upper reservoir at 320 m, which is considered an excellent option for the required size of the hybrid power plant
- mild land morphology at the coastline, facilitating the construction of the pump station and the hydro power plant
- mild terrain slopes from the upper reservoir location to the coastline, which also contribute to the reduction of the penstock installation cost.

The construction of both the wind park and the PHS system at the same site exhibits a number of positive features, such as the minimization of the project's setup cost, through the restriction of the required infrastructure works (access roads, interconnection grid, land rent, etc.), the minimization of the electricity transportation losses from the RES to the storage units, and more.

The overall siting of the PHS system is presented in Figure 3.51 on a vertical cross-section view. The available plateau at the top of the hill offers the possibility for the construction of an upper reservoir with an effective capacity of more than 1,000,000 m^3 with absolute altitude at 320 m. A double penstock is constructed, with a total length of 1,380 m.

Given the digitized land terrain at the hybrid power plant's installation site and the available wind potential measurements, the wind potential map was developed with a relevant software application. Five wind turbines of 2.3 MW nominal power each were sited. This means that the maximum possible installed wind park's power is 5×2.3 MW = 11.5 MW. The siting of the wind turbines is presented on a 3D wind potential background map in Figure 3.52. The siting of the overall hybrid power plant (wind park, upper reservoir, penstock, and hydrodynamic machine stations) is presented in Figure 3.53.

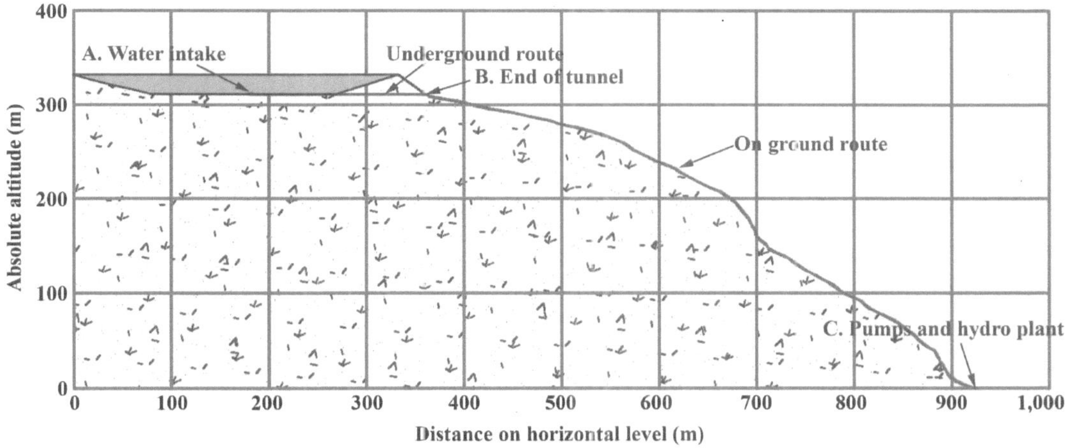

FIGURE 3.51 Vertical cross-section view of the PHS system siting in Sifnos Island [41].

3.8.2.4 Dimensioning Procedure

The dimensioning procedure follows the hybrid power plant's operation algorithm presented in Section 3.5.1.1 and the dimensioning optimization process described in Section 3.7.1. More specifically, the simulation of the hybrid power plant's operation is executed iteratively for annual time periods and with varied wind park and PV station nominal power. For each iteration, the following magnitudes are calculated:

1. the maximum hourly average power production and consumption of the hydro turbines and the pumps respectively
2. the maximum direct penetration of the wind park, given that the maximum direct penetration percentage is defined at 15% versus the current power demand
3. the pipeline minimum required diameter for the minimization of water flow losses

FIGURE 3.52 The siting of the wind turbines on a 3D wind potential background map [41].

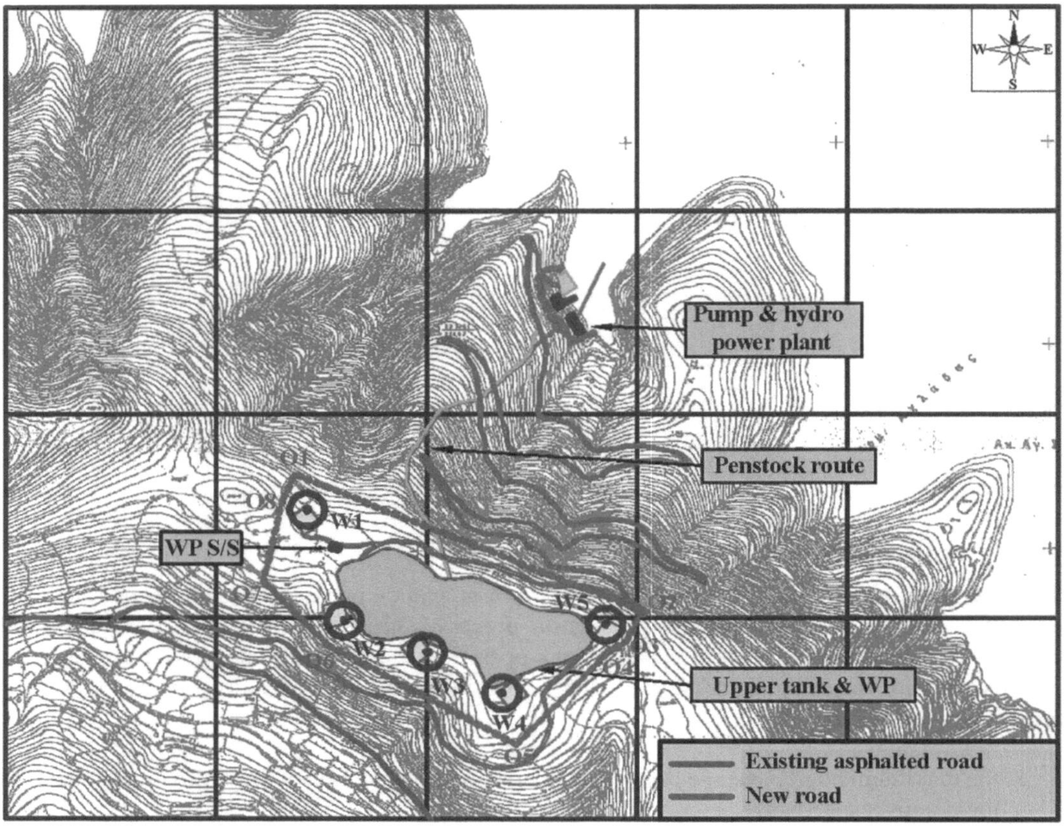

FIGURE 3.53 Top view of the overall hybrid power plant's siting [41].

4. the annual electricity production and storage by integrating the corresponding annual time series of the wind park, the hydro turbines, the thermal generators, and the pump time series
5. the economic features of the projects, such as the setup and the operation cost, the annual revenues, the electricity production-specific cost, and, eventually, the investment's economic indices.

For this specific hybrid power plant, the dimensioning procedure was executed with the ultimate target of achieving 100% RES annual penetration in the insular system.

3.8.2.5 Results

The results of the iterative execution of the hybrid power plant operation simulation are summarized in Table 3.11. For all the investigating scenarios, the upper reservoir's capacity is 1,100,000 m^3 with a geostatic head of 320 m. The economic results presented in this table are based on the assumption of a funding scheme equities/banking loan of 12,75%/87,25% respectively, with loan payback period of 15 years and a loan rate of 6%. The produced electricity selling net price (excluding VAT, taxes, etc.) was set at 0.22 €/kWh, while there is also a fee of 188 €/kW a year for the availability of the guaranteed power of the hybrid power plant. The IRR is calculated on the investment's equities.

TABLE 3.11
Dimensioning Procedure Results [41]

	Dimensioning Scenario					
Magnitude	**1**	**2**	**3**	**4**	**5**	**6**
Wind turbine number	2	3	3	4	4	5
Wind park (MW)	4.6	6.9	6.9	9.2	9.2	11.5
PV station (MW)	2	0	2	0	0.5	0
RES penetration (%)	86.13	94.07	100.00	99.60	100.00	100.00
Wind energy surplus (%)	6.62	21.04	25.87	38.39	39.56	51.04
Wind energy surplus (MWh)	1,440	5,935	8,170	14,595	15,345	24,360
Setup cost (million €)	32.165	32.195	34.695	34.725	35.225	37.255
IRR (%)	11.37	17.32	17.59	16.06	15.48	10.89

From the above table we can state:

- The ultimate target of SIC, namely the 100% RES penetration in the autonomous insular system, is alternatively achieved with:
  - 6.9 MW of wind park and 2 MW of PV station
  - 9.2 MW of wind park and 0.5 MW of PV station
  - 11.5 MW of wind park.
- The economically optimum scenario is the installation of 6.9 MW wind park and 2 MW PV station. This scenario also exhibits the minimum setup cost from the options that can satisfy the 100% RES penetration in the insular system.
- The last scenario, namely the installation of 11.5 MW, exhibits the worst economic efficiency.
- The last scenario exhibits the maximum wind energy production surplus, which, although at first glance it may be considered a drawback, given the overall approach of SIC regarding the future development of the insular community and the possible introduction of additional consumptions, such as electrical vehicles, desalination plants, and new economic activities, it may feature as the unique scenario that can support a generalized development approach, without being restricted exclusively to the coverage of the currently existing power demand. Furthermore, taking into account any potential exploitation of the wind energy surplus available mainly from the dimensioning scenario 5 and 6 in these additional electrical loads, the economic efficiency of the overall investment may be considerably improved.

The inadequacy of the wind park to fully support the power demand with three or four wind turbines is due to a considerable reduction in the available wind potential recorded mainly in June. This is depicted in Figure 3.54, where the monthly average wind velocity values have been plotted, based on the captured wind potential measurements at the wind park's installation site. Although the monthly average wind velocity remains almost higher than 7.5 m/s for 10 months of the year, there is a considerable drop in June, when the average wind velocity drops to 4.8 m/s. This value was measured in June 2017, while for the same month in 2016 the average monthly value was 6.3 m/s, which was again the lowest value measured during the year. The probability for this considerable wind potential drop in June dramatically affects the ability of the wind park to take over the full support of the power demand.

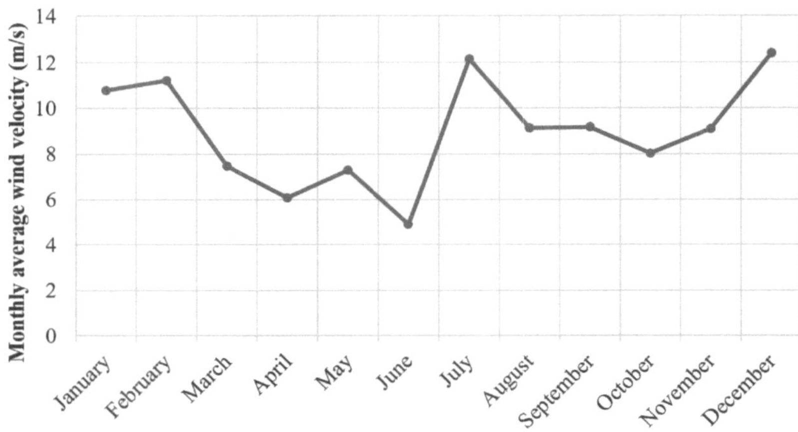

FIGURE 3.54 Monthly average wind velocity variation at the wind park's installation site [41].

All the above are characteristically depicted in Figure 3.55, where the annual variation of the water stored volume in the PHS system's upper reservoir is plotted for scenarios 2–6 of the above table.

It is clearly seen that the stored water volume in the upper reservoir is fully removed after a period of 20 days with low wind potential during June. In scenario 2 (no PV available) the thermal generators are employed to compensate for the power production drop. In scenario 3, the introduction of 2 MW of PV panels, given the availability of high solar radiation during the summer period in Greece, adequately supports the hybrid power plant and the operation of the thermal generators is avoided. Scenario 4 presents that 100% RES is not achieved only for a slight time period. This is corrected with the introduction of only 500 kW of PV station in scenario 5 (see the corresponding

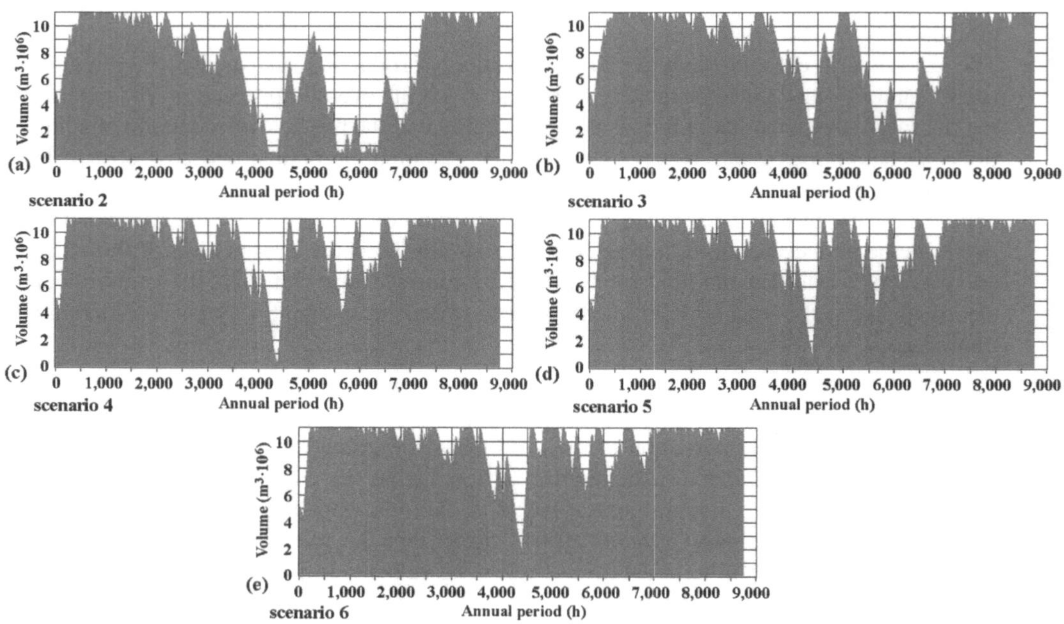

FIGURE 3.55 Annual variation of the water stored volume in the PHS system's upper reservoir for the different investigating dimensioning scenarios [41].

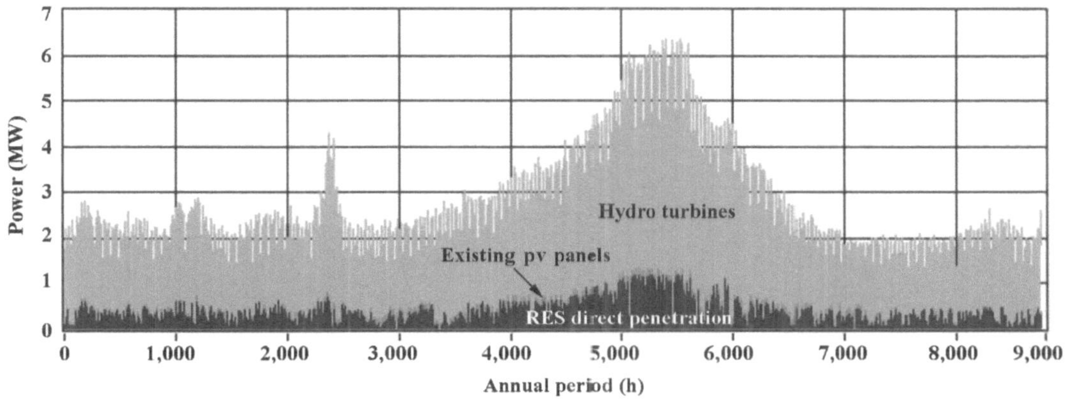

FIGURE 3.56 Annual power production synthesis in the island of Sifnos with the introduced hybrid power plant, following the above investigated scenario 5.

annual power production synthesis graph in Figure 3.55). Finally, in scenario 6 the operation of the system is further ensured, however with considerable wind energy surplus for long time periods during the year.

From the above analysis what is also highlighted is the strong effect of the available RES potential and its variability during the year and from year to year. For autonomous systems and hybrid power plants aiming at 100% RES penetration, this variability may be of high importance and may significantly affect the dimensioning, the operation and, eventually, the economic feasibility of the hybrid power plant. Another crucial parameter is the wind power direct penetration percentage, which was set for the above simulations at 15% versus the current power demand, a value that is doubtlessly rather low. Just to provide an example, if this percentage is raised to 30% for the above simulations, the RES annual penetration in scenario 4, namely with the production of four wind turbines and without the installation of a PV station, becomes 100%.

3.8.3 A Hybrid Power Plant for the Island of Kastelorizo, Greece

3.8.3.1 Objective of the Case Study

In this case study, alternative hybrid power plant configurations will be studied for another Greek island, Kastelorizo. Kastelorizo is a very small island. with 500 permanent inhabitants and a total area of 9.1 km^2. It is the eastern most Greek island, located 72 n.m. to the east of Rhodes and 1.25 n.m. to the south of the Turkish coast (Figure 3.57). The objective of this case study is to indicate a hybrid power plant synthesis, adequate to enable high RES annual penetration in the weak electrical system.

3.8.3.2 Hybrid Power Plant Components

Due to the small size of power demand, the introduction of a PHS system will not be economically feasible, given the high imposed setup cost. Hence, for electricity storage a micro-CAES system and lead acid batteries will be alternatively investigated. Conclusively, the hybrid power plant's components will be:

- RES units: a wind park and a PV station
- storage units: a micro-CAES system or lead acid batteries.

Lead acid batteries are selected among the other available electrochemical storage technologies because they currently exhibit the lower procurement cost.

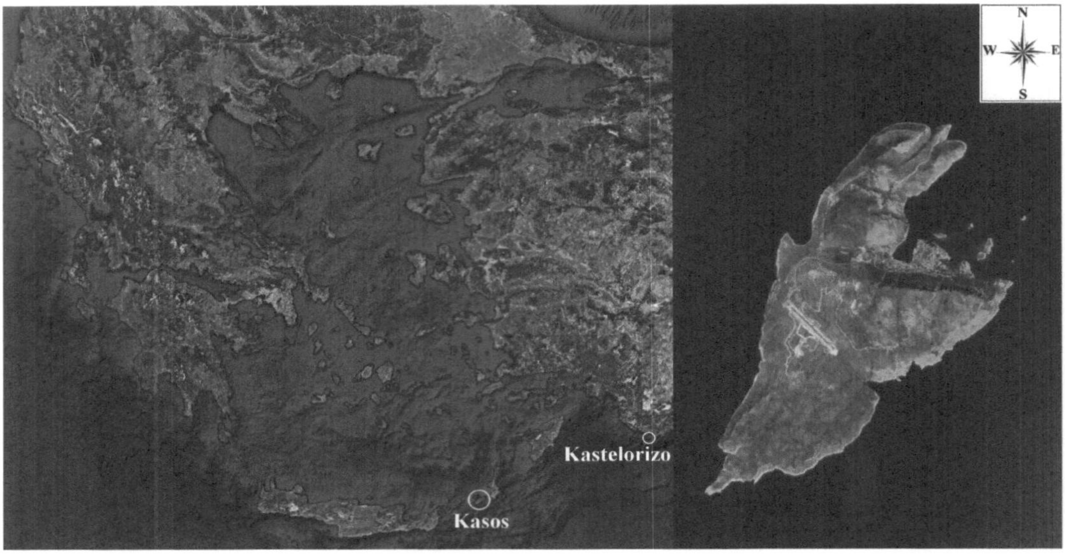

FIGURE 3.57 Map of the island of Kastelorizo and its location in Greece.

3.8.3.3 Dimensioning Parameters and Required Data

Similar to the above case studies, the dimensioning process will be executed on the computational simulation of the investigated systems' annual operation, according to the algorithms presented in Sections 3.5.1.2 and 3.6.1 for the hybrid power plant supported by the CAES system and electrochemical batteries, respectively. The simulations will be iteratively executed, with independent parameters the wind park and the PV panel nominal power and the storage capacity of the involved storage devices. The target of the dimensioning process will be the maximization of the RES annual penetration. To this end, the following data are employed:

1. the annual power demand time series of average hourly values
2. the annual wind velocity time series of average hourly values
3. the annual incident solar radiation time series of average hourly values
4. the annual ambient temperature time series of average hourly values
5. the charge/discharge efficiency curves of the employed lead acid batteries
6. characteristic compressor and turbine operation maps.

The annual power demand time series for the island of Kastelorizo for the year 2019 is presented in Figure 3.58. Characteristic features of the annual electricity demand in the island are presented in Table 3.12.

With regard to the other required annual time series:

- Given the lack of wind potential measurements for the island of Kastelorizo, the annual wind velocity time series presented in Figure 3.59 will be employed, captured in the island of Kasos, located in the southeast Aegean Sea, 137 n.m. southwest from Kastelorizo. The location of Kasos is also depicted in Figure 3.57. The employed wind velocity time series exhibits an annual average value of 11.61 m/s, with considerable concentration of the wind blowing direction from the north–northwest, as depicted in the wind rose presented in Figure 3.60 and in the impressively high k parameter of the Weibull distribution, calculated equal to 2.85.

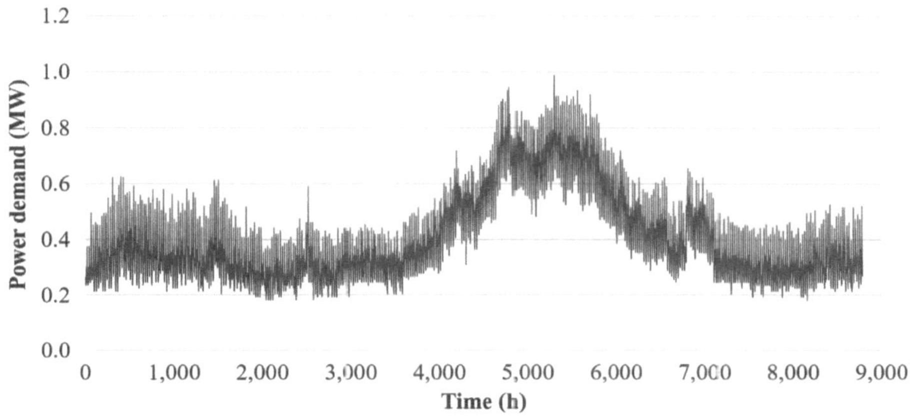

FIGURE 3.58 Annual power demand fluctuation in Kastelorizo for 2019.

TABLE 3.12
Characteristic Features of the Annual Electricity Demand in Kastelorizo in 2019

Annual peak power demand (MW)	0.99
Minimum annual power demand (MW)	0.18
Annual electricity consumption (MWh)	3,510.2
Daily average electricity consumption (MWh)	9.6

- With regard to the required solar radiation data, the annual time series employed also for the previous case study for the island of Sifnos is used in this case (Figure 3.50).
- The annual time series of the ambient temperature is also required, for the calculation of the heat transfer from the PV panels to the ambient environment. A typical annual time series also captured in Samos is employed, presented in Figure 3.61.
- The charge/discharge efficiency curves presented in Figures 3.27 and 3.28 for lead acid batteries will be employed for the current case study. Similarly, the compressor and turbine operation maps presented in Figures 3.22 and 3.23 will be employed for the corresponding CAES simulation.

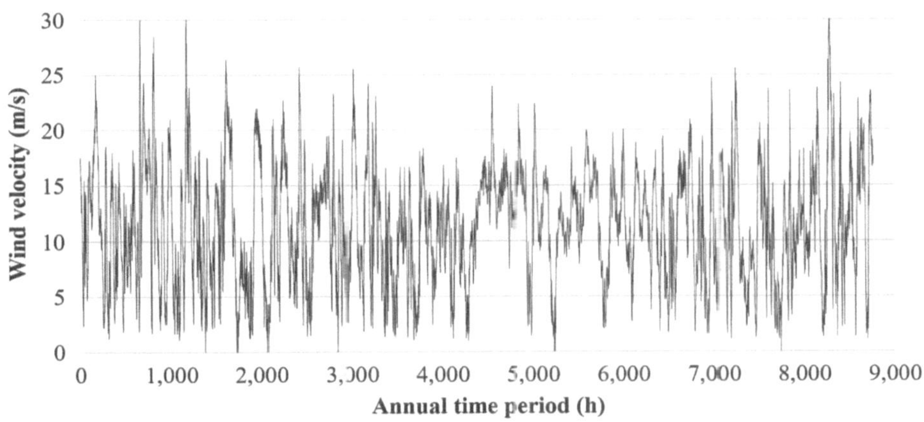

FIGURE 3.59 The employed annual wind velocity time series, captured on the island of Kasos.

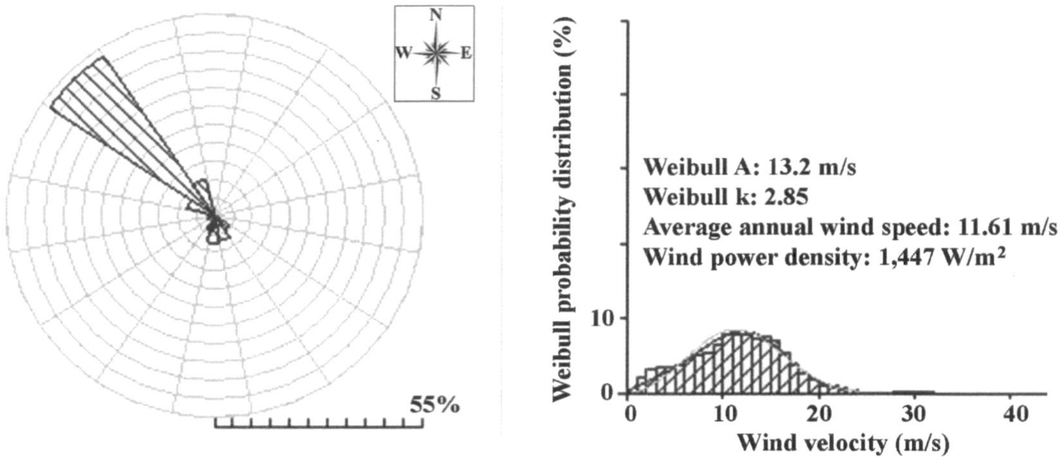

FIGURE 3.60 Wind rose and Weibull distribution based on the annual wind potential measurements from the island of Kasos, employed for the case study for the island of Kastelorizo.

3.8.3.4 Dimensioning of the Hybrid Power Plant Supported with CAES

The dimensioning of the hybrid power plant with the CAES system was based on the presented operation algorithm and the simulation process presented in Section 3.5.1.2. The crucial parameter was the dimensioning of the compressed-air storage tank features, namely its volume and nominal pressure. Given the small size of the system under consideration, the goal was to keep as low as possible both the above constructive features, in order to facilitate the construction of the compressed-air storage tank with the minimum possible setup cost, approaching, at the same time, as high annual RES penetration percentage, as possible.

After several iterative executions of the simulation of the hybrid power plant's annual operation, the conclusion is the construction of a storage tank with 20,000 m^3 volume capacity and a nominal pressure of 10 bar. Given this size of storage tank, several dimensioning scenarios were investigated, regarding the nominal power of the wind park and the PV station. For the iterative simulations, a wind turbine model with 330 kW nominal power was introduced. The goal was to achieve annual RES penetration percentage close to 80% with regard to the annual electricity consumption.

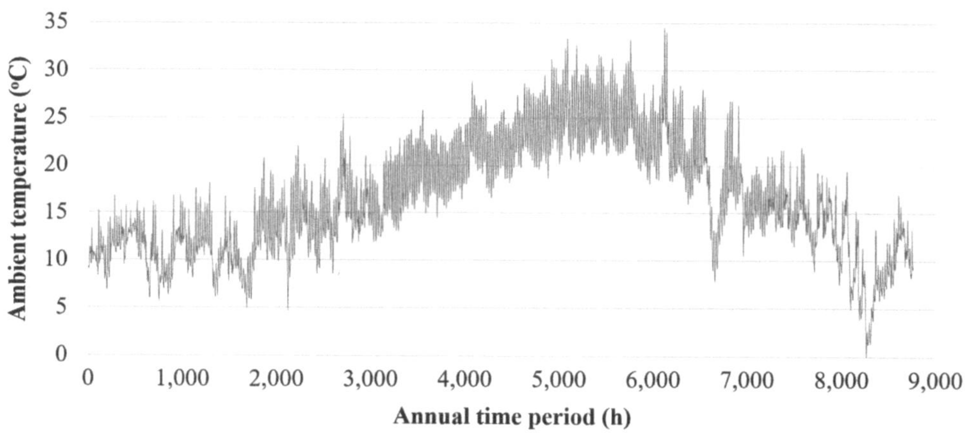

FIGURE 3.61 Annual ambient temperature time series employed for the under consideration case study.

TABLE 3.13
Dimensioning Procedure Results

	Dimensioning Scenario					
Magnitude	**1**	**2**	**3**	**4**	**5**	**6**
Wind turbine number	4	4	4	5	5	5
Wind park power (MW)	1.32	1.32	1.32	1.65	1.65	1.65
PV station power (MW)	0	1.0	1.0	0	0.5	0.5
Compressor units	1	1	2	1	1	2
Turbine units	1	1	1	1	1	1
Annual Energy Results						
RES direct penetration (MWh)	1,159	1,274	1,274	1,188	1,279	1,278
Turbine production (MWh)	1,190	1,513	1,615	1,441	1,540	1,554
Backup units production (MWh)	1,161	723	621	881	691	678
Total production (MWh)	3,510	3,510	3,510	3,510	3,510	3,510
Electricity storage (MWh)	4,191	5,230	4,125	5,117	5,497	4,266
RES rejection (MWh)	152	395	1,474	573	788	2,018
RES penetration (%)	66.9	79.4	82.3	74.9	80.3	80.7
RES production surplus (%)	2.7	5.7	21.4	8.3	10.4	26.7

The direct RES maximum penetration percentage was set at 40% versus the current power demand. The overall dimensioning results are presented in Table 3.13.

In Figure 3.62, the annual power production synthesis is presented for the sixth investigated dimensioning scenario. The direct RES penetration is always restricted below 40% versus power demand. Thermal power production is required whenever the existing pressure in tank becomes lower than the minimum expansion ratio, required for the power production from the air turbines. Whenever this happens, the air turbine power production is set equal to zero. In this figure it is also seen (as in Table 3.13) that air turbines exhibit the highest contribution to the annual electricity

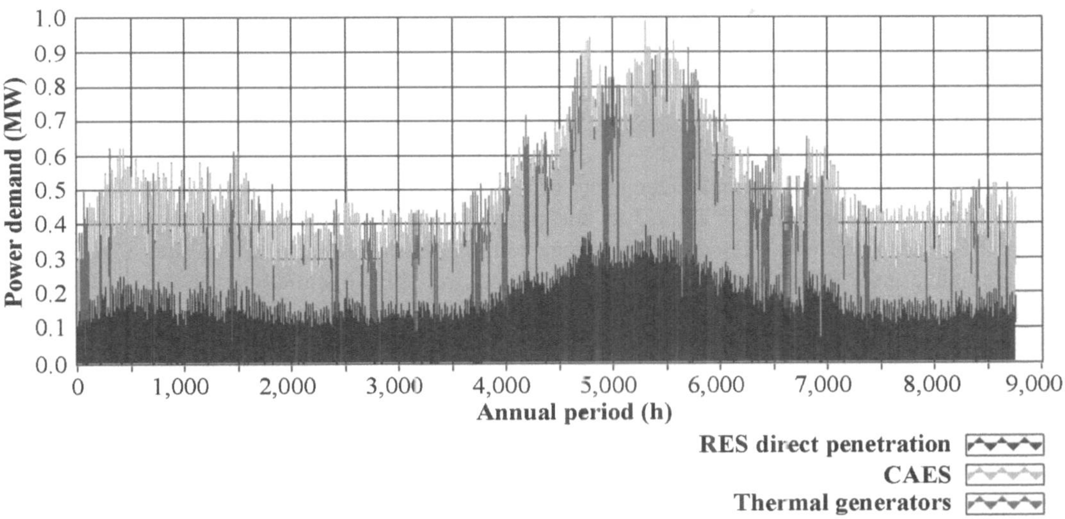

FIGURE 3.62 Annual power production synthesis of the under study hybrid power plant with the CAES system.

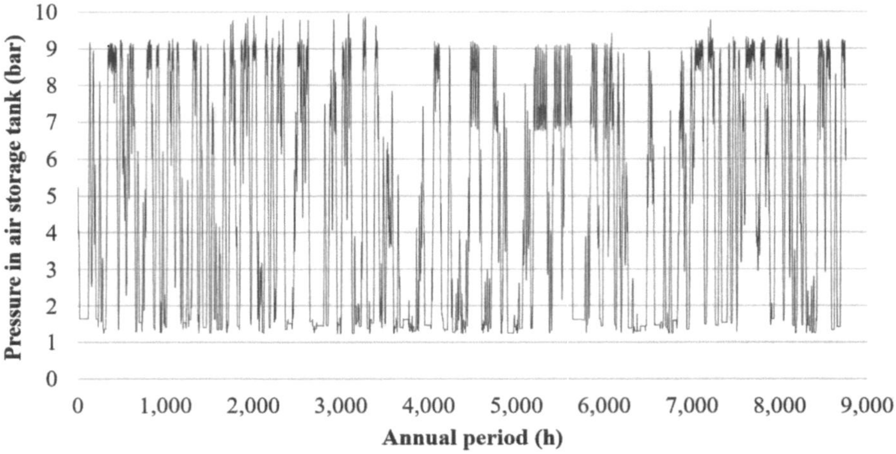

FIGURE 3.63 Pressure annual fluctuation in the air storage tank.

production share. The annual fluctuation of the pressure in the air storage tank is depicted in Figure 3.63, again for the sixth investigated scenario.

3.8.3.5 Dimensioning of the Hybrid Power Plant Supported with Electrochemical Storage

The dimensioning of the hybrid power plant supported by lead acid batteries is presented in Section 3.6.1 and the corresponding layout is given in Figure 3.25. Every calculation time step, the battery string with the minimum charge level is charged first. Respectively, power production from battery strings starts with the string with the maximum charge level.

The computational simulation is iteratively executed versus the RES unit nominal power. Given the size of the power demand, the iterative simulation starts with the assumption of a single 900 kW wind turbine. The nominal power of the PV station is increased with a step of 100 kW. Additionally, the installation of at least two battery strings is selected, in order to provide the capability of concurrent charge and discharge process. The charge/discharge rate should be configured according to the peak power demand size. For each combination of RES installed power and battery strings size and number, the computational simulation of the hybrid power plant's annual operation is executed, according to the operation algorithm presented in Section 3.6.1. Eventually, the final annual electricity production and storage are calculated. The goal of the dimensioning process is to approach annual RES penetration higher than 80%. The results from the above described dimensioning process are summarized in Table 3.14.

The results regarding the annual electricity production and storage are presented in Table 3.15.

Given the results presented in Table 3.15, the target of annual RES penetration percentage higher than 80% is achieved with the concluded dimensioning. Nevertheless, it is also seen that there is considerable annual RES rejection, due to the high required RES installed power and the low storage capacity of the involved storage units. The RES nominal power remains practically unexploited during the low demand period.

The annual fluctuation of the battery strings charge level is presented in Figure 3.64. Due to the adopted priority regarding the battery string charge and discharge order, the charge level remains almost the same among the involved battery strings throughout the overall annual period.

The annual power production synthesis is presented in Figure 3.65. The thermal generators are involved in the production process mainly during summer, namely during the peak power demand period.

Finally, in Figure 3.66 the power production synthesis from August 1–10 is depicted, namely during peak power demand period.

TABLE 3.14
Dimensioning Results of the Hybrid Power Plant with Lead Acid Batteries

Wind park nominal power (kW)	900
PV station nominal power (kW)	900
Total nominal charge/discharge rate of the battery strings (kW)	1,440
Total storage capacity of the battery strings (MWh)	7.20
Battery strings' maximum discharge depth (%)	60
Total effective storage capacity (MWh)	4.32
Battery Strings	
Battery strings number	4
Battery number per string	300
Nominal voltage per battery unit (V)	2
Nominal capacity per battery unit (Ah)	3,000
Battery type	OPzV
Charge/discharge rate per battery string (kW)	360
Storage capacity per battery string (MWh)	1.80

TABLE 3.15
Annual Electricity Production and Storage with the Hybrid Power Plant Supported by Lead Acid Batteries, for the Island of Kastelorizo

Initial Production from the RES Units	
Wind park production (MWh)	3,037
PV station production (MWh)	1,303
RES Unit Direct Penetration	
Wind park direct penetration (MWh)	1,495
PV station direct penetration (MWh)	1,144
Electricity Production and Storage from the Storage Units	
Storage units production (MWh)	375
Electricity storage (MWh)	386
Total electricity consumed for storage (MWh)	561
Storage units initial charge level (MWh)	5.04
Storage units final charge level (MWh)	7.20
Annual average charge cycle efficiency (%)	66.47
Thermal Generators Production	
Annual electricity production from thermal generators (MWh)	498
Annual production percentage from thermal generators (%)	14.2
RES Production Surplus	
RES production surplus (MWh)	1,140
RES production surplus with regard to the initial production (MWh)	26.3
Summary	
RES total annual direct penetration (MWh)	2,639
Storage units annual production (MWh)	375
Thermal generators annual production (MWh)	498
Total annual electricity production (MWh)	3,512
Annual RES penetration percentage (%)	85.8

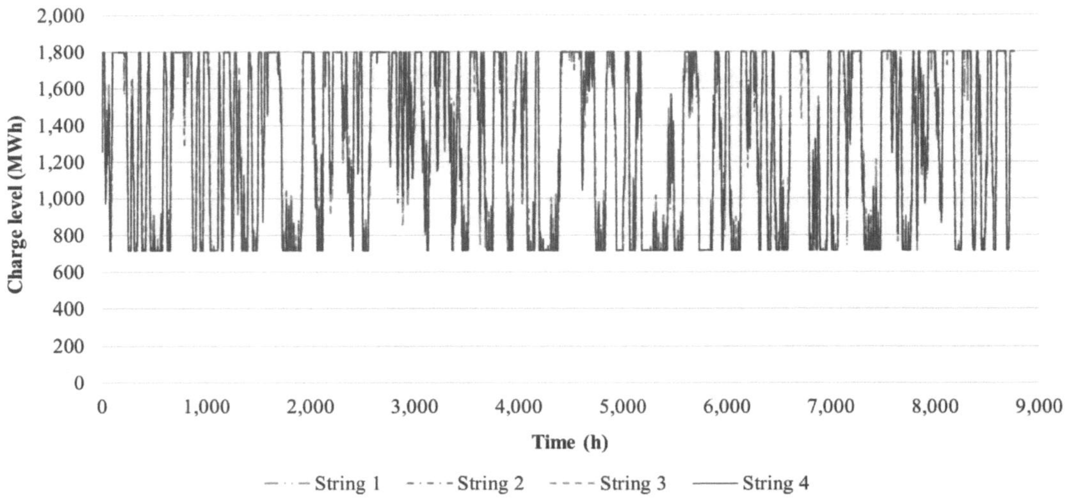

FIGURE 3.64 Charge level annual fluctuation of the battery strings.

3.8.4 A Hybrid Power Plant for a Remote Cottage

The last presented case study refers to a hybrid power plant of very small size, introduced for a non-interconnected, typical cottage. The parameters, the requirements, and the available data for this case are completely different from the previously examined examples. The essential differences between the current and the previously investigated case studies can be classified as explained below:

1. For very small autonomous consumptions in buildings, most probably there are no power demand data available beforehand, because, normally, apart from the overall electricity consumption versus a regular time period (e.g., monthly), no other power consumption data

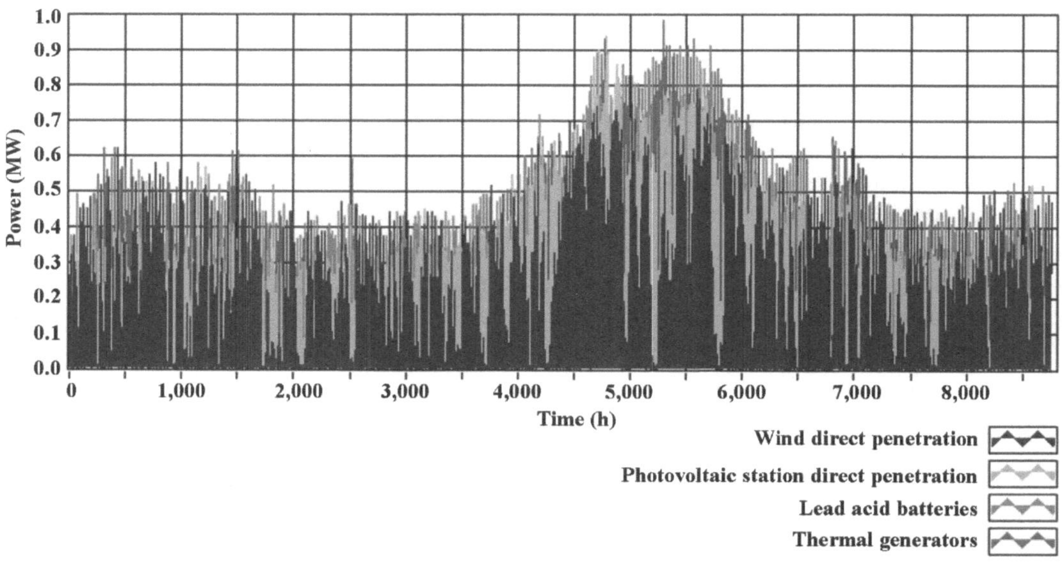

FIGURE 3.65 Annual power production synthesis of the under study hybrid power plant with the support of the lead acid batteries.

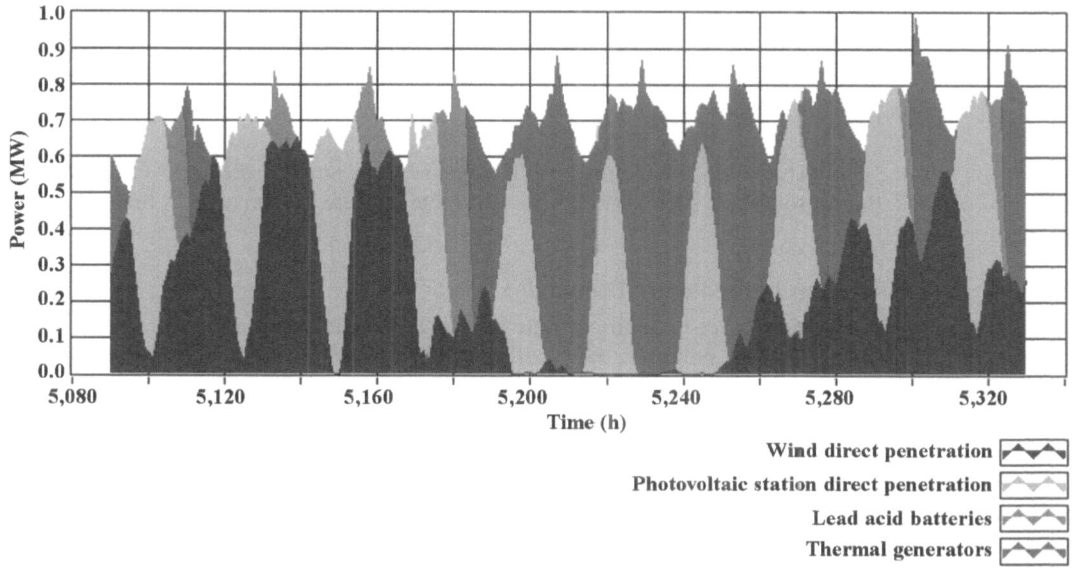

FIGURE 3.66 Power production synthesis of the under study hybrid power plant with the support of lead acid batteries, from August 1–10.

are usually recorded. So, in order to proceed with the dimensioning of a hybrid power plant in such cases, the consultant must, first of all, try to estimate the existing or the expected (for a new building) power demand and to develop the corresponding time series.

2. Even remote and alienated buildings are usually erected in areas that can be inhabited, namely with normally not as high wind potential as in top of hills or mountains, selected usually for the installation of large wind parks. This means that in cases of hybrid power plants for autonomous electricity consumptions in buildings and other remote human activities, the wind potential is not expected considerably high, such as for centralized hybrid power plants. This, in turn, implies that the installation of PV panels becomes highly competitive against the installation of small wind turbines, with regard to the expected electricity production. If we also take into account the aesthetics, the noise emission, the required space, and other parameters that can be crucial for installations close to inhabited areas, the introduction of PV panels in such hybrid power plants becomes more attractive.
3. In cases of hybrid power plants for decentralized power production in buildings and other types of autonomous electricity consumptions, the economics of the overall project emerge as a major parameter. This means that, once the reliability of the power production has been ensured, the owner/user of the hybrid power plant is above all interested in the coverage of the power demand with the minimum possible cost. To this end, the dimensioning of the hybrid power plant is optimized versus the minimization of the overall setup and operation cost, expressed by the project's LCC, as also explained in a former section.
4. The aim of the dimensioning is 100% electricity production from the hybrid power plant. Because the remote infrastructure is considered non-interconnected to the local grid, apart from the hybrid power plant there is no other alternative power supply source available. However, this does not mean that the ultimate target of the dimensioning is 100% RES penetration, which, most probably, would require an over-dimensioning of the introduced hybrid power plant with a subsequent negative impact on the economics of the project

through a disproportional raise of the setup cost. On the contrary, because the ultimate target of the dimensioning is the minimization of the overall setup and operation cost, the contribution to the power production of the backup thermal generators and the corresponding consumption of a relatively limited amount of diesel oil may also be required.

Given the above peculiarities met for hybrid power plants of very small size, the overall simulation and dimensioning process of such a system for a remote cottage is analyzed in the following paragraphs.

3.8.4.1 The Estimation of the Power Demand

As stated previously, in most cases of small, remote buildings or other remote infrastructures (e.g., desalination plants, lighthouses), normally there will not be any prior power demand data available. So, the consultant must develop the required power demand time series for the simulation of the system. A simple procedure is provided in this section. Specifically:

- First, all the electrical and electronic devices, practically all the apparatus that consumes electricity, are written down. In such systems, a wise approach is to try to avoid the use of electrical devices of very high electricity consumption, mainly devices related to the transformation of electricity to thermal power, such as electrical ovens and hotplates, electrical heating devices, etc. The corresponding needs should be covered with alternative primary energy sources, such as oil (for heating) and liquefied petroleum gas (for cooking). The introduction of such energy-consuming devices will lead to the formulation of an oversized hybrid power plant, with considerably increased setup costs.
- With the support of the owner/user of the building, typical 24-hour period power consumption profiles must be developed for every different and characteristic period of the year. For example, the different climate and natural lighting conditions imply the construction of a typical 24-hour power demand profile for the summer period, another one for the winter period, and another common one for autumn and spring. If there are any other special parameters, apart from the ambient conditions, that may affect the user's attribute and the subsequent electricity consumption, then additional corresponding power demand profiles should be introduced for the specific periods of the year. For this particular case study, the three following 24-hour power demand profiles have been developed, for winter, summer, and spring–autumn respectively, presented in Figure 3.67. These profiles have been developed with hourly time steps. For each hourly time step, the average power demand in W of each device has been estimated, leading, practically, to the corresponding hourly electricity consumption in Wh.
- For each developed power demand profile, the daily average power consumption is calculated.
- Finally, by introducing the above calculated daily average power consumptions for each month and for a whole annual period, the annual power demand is developed with monthly average values. For this particular case study, the developed power demand with monthly average values is presented in Table 3.16.

TABLE 3.16
Monthly Average Power Consumption for the Investigated Case Study

Month	Jan	Feb	Mar	Apr	May	Jun	Jul	Aug	Sep	Oct	Nov	Dec
Power (W)	0.409	0.409	0.361	0.361	0.361	0.316	0.316	0.316	0.361	0.361	0.361	0.409

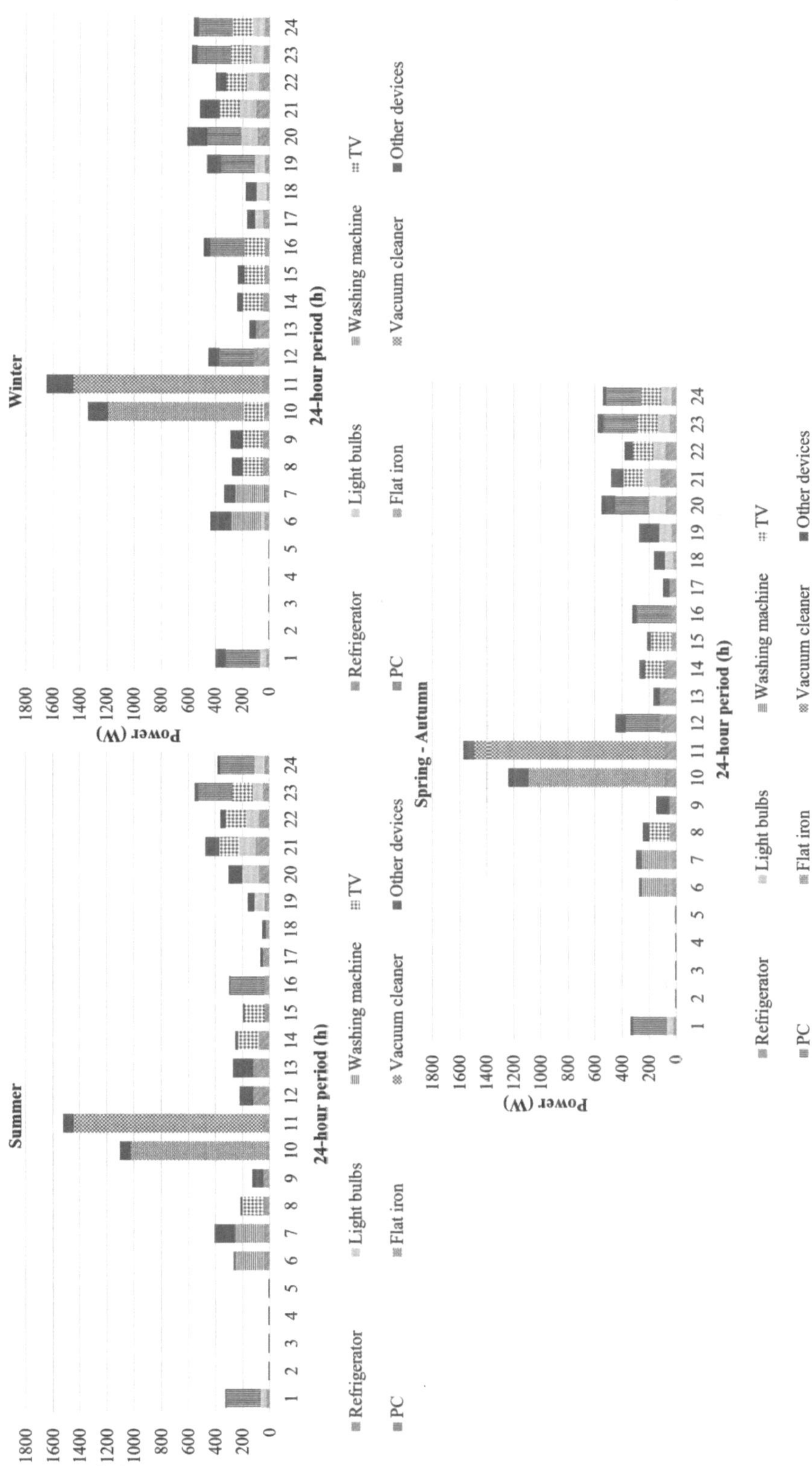

FIGURE 3.67 24-hour power demand profile for the winter, summer and spring–autumn period.

3.8.4.2 The Estimation of the Available RES Potential

The accurate knowledge of the available RES potential, namely the wind potential and the solar radiation, constitutes another crucial parameter for the valid dimensioning of the hybrid power plant. First of all, because the required power demand has been approached through average monthly values, it would be adequate to estimate the available RES potential with the same time discrimination. We will investigate the two alternative primary energy sources separately:

- *Wind potential*

  Obviously it makes no sense to proceed with the installation of a wind mast for the precise estimation of the available wind potential. The cost of such a task maybe be higher than the cost of the hybrid power plant itself, not to mention the time-consuming procedure. So, the only way to proceed to the estimation of the wind potential is through the approach of existing wind measurements from neighboring wind masts and the development of wind maps with the use of relevant, valid, and reliable software applications. The consultant should request the provision of the required measurements from academic institutes and laboratories or specialized technical and consulting offices. Because the final required piece of information is the monthly average wind power production, the knowledge of the monthly Weibull A and k parameters for the installation site would be sufficient.
- *Solar radiation*

  Access to solar radiation data is usually easier and, in most cases, it is expected to be free of charge. Because the variability of solar radiation is not as intensive as the wind potential, and there are plenty of relevant measurements and data available online from valid and acclaimed institutes, national meteorological services, etc., it is almost certain that the consultant will not have any difficulties finding reliable solar radiation data from a representative location for the installation site of the hybrid power plant.

  Along with the solar radiation data, the ambient temperature annual variation should be also known, because it is required for the calculation of the expected power production from the PV panels.

3.8.4.3 Power Production Calculation from the RES Units

Given the possession of wind potential and solar radiation data, the calculation of the expected power production from the installed wind turbines and PV panels is easy and straightforward. Of course the knowledge of the exact values of the wind turbine and PV panel nominal power is not available beforehand, because, both of them are among the main results of the dimensioning procedure. However, the power production is calculated by assuming different values for the nominal power of both RES technologies iteratively, as explained in the dimensioning procedure presented in Section 3.7.2.

Apart from the RES potential data, the power curve of the involved wind turbine model is also required. Indicatively, in Figure 3.68 the power curve of a 5 kW wind turbine model is presented. This wind turbine is fully passively controlled and has been developed in the Wind Energy & Power Plants Synthesis Laboratory of the Technological Educational Institute of Crete, Greece.

The results of the power production calculation from a 1 kW wind turbine and 1.5 kW PV panel are presented in Table 3.17, together with the employed RES potential data. The presented results in Table 3.17 represent average monthly values, that is, they refer to the whole monthly time interval.

According to the RES potential data presented in Table 3.17, the investigated area exhibits a wind potential with annual wind velocity 4.95 m/s and annual solar irradiation of 1,775 kWh/m^2, which represents a typical value for the western US, southern Europe, etc.

A summary of the available RES potential is also provided graphically in Figure 3.69.

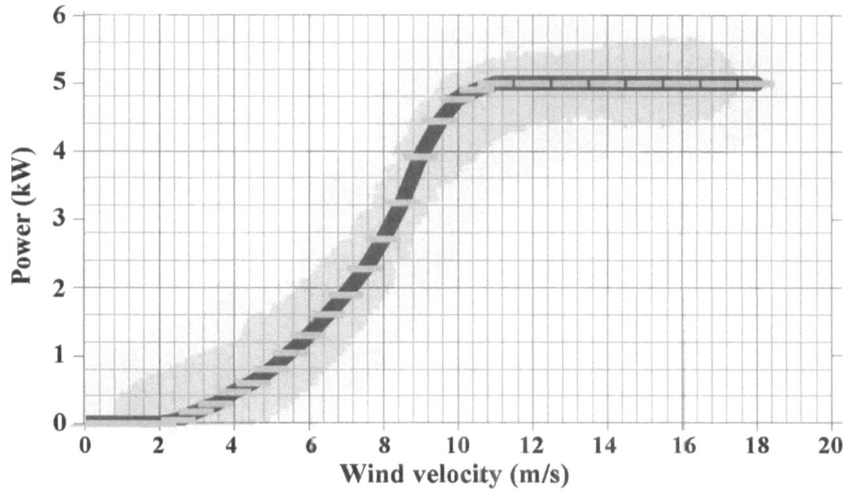

FIGURE 3.68 Characteristic power curve of a 5 kW small wind turbine.

3.8.4.4 Power Production from the Thermal Generators and Fuel Consumption

Having estimated the expected power demand and the power production from the RES units, the final step before the calculation of the LCC is the estimation of the required power production from the thermal generators and the subsequent fuel consumption. The required power production from the thermal generators (hybrid power plant's backup units—practically a diesel genset) depends on the fundamental operation algorithm of the hybrid power plant. As explained in Section 3.6.3, in cases of hybrid power plants of very small size, a common policy is to approach

TABLE 3.17
Available RES Potential Data and Power Production from the RES Units

Month	Weibull C Parameter (m/s)	Weibull k Parameter	Average Wind Velocity (m/s)	Solar Irradiation (kWh/m²)	Ambient Temperature (°C)	Production from 1 kW Wind Turbine (kW)	Production from 1.5 kW PV Panels (kW)
Jan	6.34	2.08	5.30	66.86	11.3	0.232	0.120
Feb	6.62	1.99	5.61	81.56	11.3	0.255	0.160
Mar	7.27	1.95	6.00	118.31	12.7	0.296	0.207
Apr	6.26	1.59	5.57	159.89	15.5	0.226	0.282
May	4.92	1.48	3.94	209.53	19.3	0.149	0.346
Jun	5.03	1.93	4.44	236.94	23.5	0.131	0.396
Jul	6.70	2.46	6.10	250.33	25.5	0.257	0.404
Aug	5.35	1.80	4.79	228.03	25.1	0.163	0.369
Sep	4.64	1.76	4.09	177.03	22.8	0.113	0.301
Oct	5.02	1.69	4.00	109.67	19.4	0.146	0.186
Nov	5.41	1.93	4.68	75.56	15.5	0.161	0.136
Dec	5.79	1.76	4.96	61.64	12.8	0.197	0.110

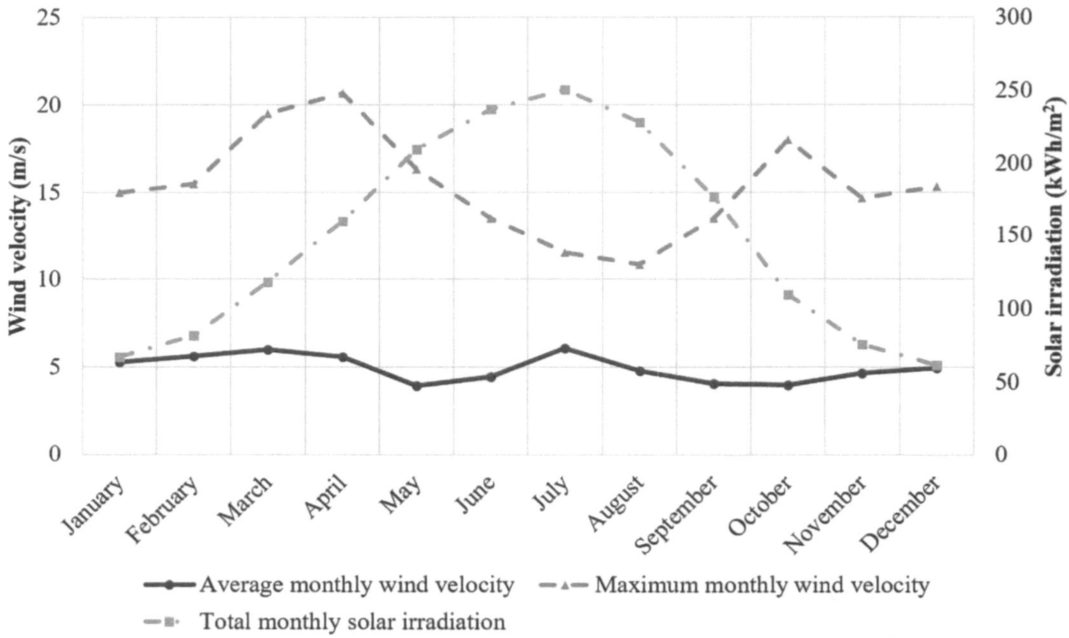

FIGURE 3.69 Average and maximum monthly wind velocity and total monthly solar irradiation.

the operation of the hybrid power plant through the algorithm presented in Figure 3.34. According to this figure, the power consumption is always covered by the batteries. Any produced power by the RES or the backup units is first stored in the batteries. This means that the final provided power by the batteries must be calculated by multiplying the available power for storage with the charge–discharge cycle efficiency. Consequently, if η_b is the overall charge–discharge cycle efficiency, the power production P_{dg} from the diesel genset is calculated according to the following algorithm:

- if $(P_{wt} + P_{pv})\cdot\eta_b \geq P_d$, then $P_{dg} = 0$
- if $(P_{wt} + P_{pv})\cdot\eta_b < P_d$, then $P_{dg} = P_d - (P_{wt} + P_{pv})$

where P_{wt} and P_{pv} are the power production from the wind turbines and the PV panels respectively and P_d the current power demand.

After the calculation of the power production from the diesel genset, the subsequent fuel consumption is calculated based on the average efficiency of the thermal generator and the heat capacity of the consumed fuel. The total results from the application of the above described process for the presented case study are provided in Table 3.18.

3.8.4.5 Calculation of the LCC Dimensioning Optimization

For the above first dimensioning, the LCC is calculated following the described methodology in Section 3.7.2 and the provided costs in this current section. The life period of the hybrid power plant for the calculation of the LCC was considered 20 years. The calculation is repeated for several combinations for the installed power of the wind turbine and the PV panels.

The batteries' total storage capacity is selected at 10,000 Ah, capable to guarantee a power production autonomy for a period of 4–6 days (average daily consumption at 7–9 kWh, depending on the season), starting from a fully charged stage and with a maximum discharge depth of 60%.

TABLE 3.18
Power Production from Diesel Genset and Fuel Consumption

Month	Power Demand (kW)	Production from 1 kW Wind Turbine (kW)	Production from 1.5 kW PV Panels (kW)	Power Production from the Diesel Genset	Fuel Consumption (lt)
Jan	0.409	0.232	0.120	0.057	10.548
Feb	0.409	0.255	0.160	0.000	0.000
Mar	0.361	0.296	0.207	0.000	0.000
Apr	0.361	0.226	0.282	0.000	0.000
May	0.361	0.149	0.346	0.000	0.000
Jun	0.316	0.131	0.396	0.000	0.000
Jul	0.316	0.257	0.404	0.000	0.000
Aug	0.316	0.163	0.369	0.000	0.000
Sep	0.361	0.113	0.301	0.000	0.000
Oct	0.361	0.146	0.186	0.028	5.215
Nov	0.361	0.161	0.136	0.063	11.348
Dec	0.409	0.197	0.110	0.102	18.845
Total diesel oil consumption (lt):					**45.955**

Two battery strings of lead acid batteries will be developed, consisting of five battery units each with 1,000 Ah nominal storage capacity (5,000 Ah each battery string). The nominal voltage for each battery string will be 20 V and the nominal charge–discharge capacity will be 4 kW. The nominal storage capacity for each battery string will be 20 kWh.

The batteries' total setup cost for one installation was estimated approximately at €4,000, while the cost of the required battery inverter was found to be €1,500. The lead acid batteries will be replaced every 7 years, namely in total twice during the 20-year life period of the hybrid power plant, at the 7th and 14th year.

Based on relevant market research, the procurement and setup costs for various types of RES units are presented in Table 3.19.

The diesel genset procurement price was found to be €730 and the current diesel oil price was assumed at 1.35 €/lt. A constant annual increase percentage of 1% was introduced for the diesel oil price, leading to a final price after 20 years of 1.65 €/lt. The discount rate was considered at 3%. Finally all the inverters and the diesel genset will be replaced once at the 10th year of the hybrid plant's life period.

Given the above assumptions and data, the operation of the system was executed for several combinations of the wind turbine and PV panel nominal power and the corresponding LCC was calculated. The calculation follows the procedure thoroughly described in Section 3.7.2, given the assumptions and the prices presented in this current section. The results are presented in Table 3.20. In this table the following notation for the LCC components has been adopted:

- $C_{W/T}$: wind turbine procurement cost
- $C_{P/V}$: PV panels procurement cost
- C_{inv}: inverters procurement cost, including replacement
- $C_{D/G}$: diesel genset procurement and replacement cost
- C_{inst}: installation cost
- C_{maint}: maintenance cost for the whole life period
- C_{diesel}: fuel consumption cost for the whole life period
- C_B: batteries total procurement cost, including replacements

TABLE 3.19
Indicative Procurement and Setup Costs for RES Units in the European Market

Wind Turbine (kW)	Price (€)	Inverter (kVA)	Price (€)	Total (€)
1	1,544.00	1	750.50	2,294.50
2	3,069.00	2	1,240.56	4,309.56
3	6,058.00	3	1,532.52	7,590.52
4	8,010.00	4	1,955.50	9,965.50
5	9,951.00	5	2,450.76	12,401.76
PV Panels (kW)	**Price (€)**	**Inverter (kVA)**	**Price (€)**	**Total (€)**
0.5	467.00	0.5	725.55	1,192.55
1.0	934.00	1.0	750.50	1,684.50
1.5	1,401.00	1.5	775.51	2,176.51
2.0	1,868.00	2.0	1,240.56	3,108.56
2.5	2,335.00	2.5	1,240.56	3,575.56
3.0	2,802.00	3.0	1,532.52	4,334.52
3.5	3,269.00	3.5	1,532.52	4,801.52
4.0	3,736.00	4.0	1,955.50	5,691.50
5.0	4,670.00	5.0	2,450.76	7,120.76

A relevant figure with the variation of the calculated LCC versus the investigated dimensioning scenario is also provided in Figure 3.70.

From the last table and figure it is seen that the minimum LCC is calculated for 1 kW wind turbine and 1.5 kW PV panels. The introduction of the RES technologies leads approximately to a 15% drop of the LCC compared to the system's operation with the support of solely the diesel genset. By introducing a 2 kW wind turbine, the fuel consumption is eliminated. Hence, the increase of the PV panel nominal power above 1 kW, given the existence of a 2 kW wind turbine, actually does not offer anything but raising the hybrid plant's setup cost.

TABLE 3.20
Analysis of the LCC Calculation for All the Examined Combinations of the Hybrid Power Plant's Synthesis

	Wind Turbine–PV Panels Installed Nominal Power (kW)						
LCC Component	**0–0**	**1.0–1.0**	**1.0–1.5**	**1.0–2.0**	**2.0–1.0**	**2.0–1.5**	**2.0–2.0**
	Cost (€)						
$C_{W/T}$	0	1.544	1.544	1.544	3.069	3.069	3.069
$C_{P/V}$	0	934	1.400	1.868	934	1.400	1.868
C_{inv}	0	5.236	5.279	6.090	6.090	6.134	6.945
$C_{D/G}$	1.273	1.273	1.273	1.273	1.273	1.273	1.273
C_{inst}	500	500	500	500	500	500	500
C_{maint}	3.176	2.721	2.869	3.018	3.457	3.625	3.773
C_{diesel}	23.500	2.863	1.372	553	0	0	0
C_B	0	9.900	9.900	9.900	9.900	9.900	9.900
LCC (€)	28.449	24.971	**24.138**	24.746	25.223	25.901	27.329

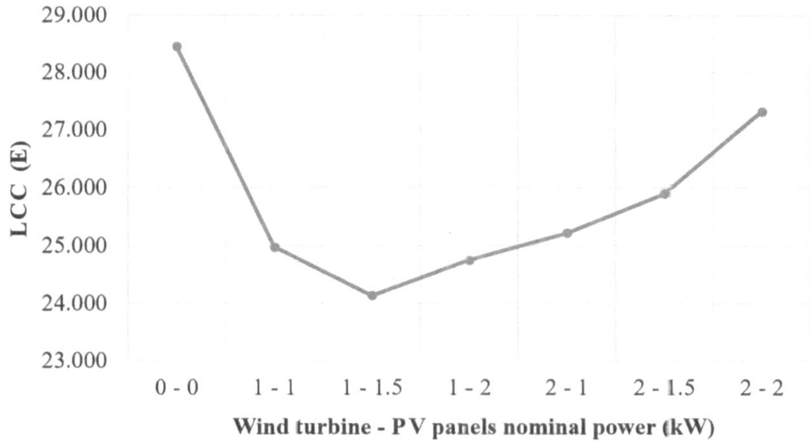

FIGURE 3.70 The variation of the LCC versus the nominal power of the wind turbine and the PV panels.

REFERENCES

1. Electric Power Monthly—U.S. Energy Information Administration. www.eia.gov/electricity/monthly/ (Accessed on September 2019).
2. Karellas, S., & Tzouganatos, N. (2014). Comparison of the performance of compressed-air and hydrogen energy storage systems: Karpathos island case study. *Renewable and Sustainable Energy Reviews*, *29*, 865–882.
3. Meyer, F. (2007). Integration von regenerativen Stromerzeugern. Druckluft-Speicherk- raftwerke, Projektinfo.
4. Crotogino, F. (2006). KBB Underground Technologies GmbH, Hannover, Compressed Air Storage, Internationale Konferenz "Energieautonomie durch Speicherung Erneuerbarer Energien," 30-31.
5. Ibrahim, H., Ilinca, A., & Perron, J. (2008). Energy storage systems–characteristics and comparisons. *Renewable and Sustainable Energy Reviews*, *12*, 1221–1250.
6. Nölke, M. (2006). Compressed Air Energy Storage (CAES)–Eine sinnvolle Ergänzung zur Energieversorgung? Promotionsvortrag.
7. Jakiel, C. (2005). Entwicklung von Großdampfturbinen, Wärmespeichern und Hochtemperatur–Kompressoren für adiabate Druckluftspeicherkraftwerke. 5. dena-EnergieForum "Druckluftspeicherkraftwerke." Berlin 8 September 2005.
8. Zunft, S., Tamme, R., Nowi, A., & Jakiel, C. (2005). Adiabate Druckluftspeicherkraftwerke: Ein Element zur netzkonformen Integration von Windenergie. Energiewirtschaf–tliche Tagesfragen, 55 Jg 2005, Heft 7.
9. Nowi, A., Jakiel, C., Moser, P., & Zunft, S. (2006). Adiabate Druckluftspeicherkraftwerke zur netzvertäglichen Windstrominegration VDI-GET Fachtagung "Fortschrittliche Energiewandlung undanwendung. Strom–und Wärmeerzeugung. Kommunale und industrielle Energieanwendungen," Leverkusen, 09–10 Mai 2006.
10. Katsaprakakis, D.A. (2016). Hybrid power plants in non-interconnected insular systems. *Applied Energy*, *164*, 268–283.
11. Hiratsuka, A., Arai, T., & Yoshimura, T. (1993). Seawater pumped-storage power plant in Okinawa island, Japan. *Engineering Geology*, *35*, 237–246.
12. Japan Commission on Large Dams. http://web.archive.org/web/20030430004611/http://www.jcold.or.jp/Eng/Seawater/Summary.htm (Accessed on September 2019).
13. Fujihara, T., Imano, H., & Oshima, K. (1997). Development of pump turbine for seawater pumped - storage power plant. *Hitachi Review*, *47*(5).
14. Katsaprakakis, D. A., Christakis, D. G., Stefanakis, I., Spanos, P., & Stefanakis, N. (2013). Technical details regarding the design, the construction and the operation of seawater pumped storage systems. *Energy*, *55*, 619–630.

15. Yekini Suberu, M., Wazir Mustafa, M., & Bashir, N. (2014). Energy storage systems for renewable energy power sector integration and mitigation of intermittency. *Renewable and Sustainable Energy Reviews*, *35*, 499–514.
16. Hall, P. J., & Bain, E. J. (2008). Energy-storage technologies and electricity generation. *Energy Policy*, *36*, 4352–4355.
17. Leadbetter, J., & Swan, L. G. (2012). Selection of battery technology to support grid integrated renewable electricity. *Journal of Power Sources*, *216*, 376–386.
18. Zaghib, K., Dontignya, M., Guerfi, A., Charest, P., Rodrigues, I., Mauger, A., et al. (2011). Safe and fast-charging Li-ion battery with long shelf life for power applications. *Journal of Power Sources*, *196*, 3949–3954.
19. Wegayehu, T., Mulushoa, Y., Murali, N., Vikram Babu, B., & Arunamani, T. (2017). Brief review of solid electrolyte for lithium ion batteries in particular to garnet-structured $Li_7La_3Zr_2O_{12}$ solid-state electrolyte. *International Journal of Advanced Research*, *5*, 1657–1663.
20. Wakihara, M. (2001). Recent developments in lithium ion batteries. *Mater Sci Eng*, *33*, 109–134.
21. Chen, H., Cong, T. N., Yang, W., Tan, C., Li, T., & Ding, Y. (2009). Progress in electrical energy storage system: a critical review. *Progress in Natural Science*, *19*, 291–312.
22. Justin Coleman Shannon Bragg-Sitton. Eric Dufek, Sam Johnson Joshua Rhodes. Todd Davidson. Michael E. Webber. An evaluation of energy storage options for nuclear power. Idaho National Laboratory. June 2017. https://www.osti.gov/servlets/purl/1372488 (Accessed on February 2020).
23. Mc Dowall, J. (2005). Integrating energy storage with wind power in weak electricity grids. *Journal of Power Sources*, *162*, 959–964.
24. Commons Wikimedia–File: Sodium nickel chloride cell. https://commons.wikimedia.org/wiki/File:Sodium-nickel-chloride_cell.svg (Accessed on September 2019).
25. Broussely, M., & Pistoia, G. (2007). *Industrial applications of batteries: From cars to aerospace and energy storage*. Amsterdam, London: Elsevier BV.
26. Brett, D. L. J., Aguiar, P., & Brandon, N. P. (2006). System modelling and integration of an intermediate temperature solid oxide fuel cell and ZEBRA battery for automotive applications. *Journal of Power Sources*, *163*, 514–522.
27. Sudworth, J. L. (1994). Zebrabatteries. *Journal of Power Sources*, *51*, 105–114.
28. Hotza, D. & Costa, J. C. D. (2008). Fuel cells development and hydrogen production from renewable sources in Brazil. *International Journal of Hydrogen Energy*, *33*, 4915–4935.
29. Song, C. Fuel processing for low-temperature and high temperature fuel cells–challenges and opportunities for sustainable development in the 21st century. (2002). *Catalysis Today*, *77*, 17–49.
30. Dicks, A. L., Costa, J. C. D., Simpson, A., & McLellan, B. (2004). Fuel cells, hydrogen and energy supply in Australia. *Journal of Power Sources*, *131*, 1–12.
31. European Commission. (2003). *Hydrogen Energy and Fuel Cells–A Vision of Our Future*. Luxembourg: Office for Official Publications of the European Communities.
32. Peihgambardoust, S. J., Rowshanzamir, S., & Amjadi, M. (2010). Review of the proton exchange membranes for fuel cell applications. *International Journal of Hydrogen Energy*, *35*, 9349–9384.
33. Wang, C., Mao, Z., Bao, F., Li, X., & Xie, X. (2005). Development and performance of 5kW proton exchange membrane fuel cell stationary power system. *International Journal of Hydrogen Energy*, *30*, 1031–1034.
34. Lin, M., Cheng, Y., & Yen, S. (2005). Evaluation of PEMFC power systems for UPS base station applications. *Journal of Power Sources*, *140*, 346–349.
35. Tuber, K., Zobel, M., Schmidt, H., & Hebling, C. (2003). A polymer electrolyte membrane fuel cell system for powering portable computers. *Journal of Power Sources*, *122*, 1–8.
36. Hwang, J. J., Wang, D. Y., & Shih, N.C. (2005). Development of a light weight fuel cell vehicle. *Journal of Power Sources*, *141*, 108–115.
37. Hwang, J. J., Wang, D. Y., Shih, N. C., Lai, D. Y., & Chen, C. K. (2004). Development of fuel-cell-powered electric bicycle. *Journal of Power Sources*, *133*, 223–228.
38. Folkesson, A., Andersson, C., Alvofors, P., Alakula, M., & Overgaard, L. (2003). Real life testing of a hybrid PEM Fuel cell bus. *Journal of Power Sources*, *118*, 349–357.
39. Beckhaus, P., Dokupil, M., Heinzel, A., Souzani, S., & Spitta, C. (2005). On-board fuel cell power supply for sailing yachts. *Journal of Power Sources*, *145*, 639–643.
40. Katsaprakakis, D. A., & Christakis, D.G. (2006). A wind parks, pumped storage and diesel engines power system for the electric power production in Astypalaia. In *European Wind Energy Conference and Exhibition 2006, EWEC*, *1*, 621–636.

41. Katsaprakakis, D. A., & Voumvoulakis, M. (2018). A hybrid power plant towards 100% energy autonomy for the island of Sifnos, Greece. Perspectives created from energy cooperatives. *Energy*, *161*, 680–698.
42. Papantonis, D. (2008). *Small Hydro Power Plants*. Athens: Symeon Editions, ISBN-13: 978-960-7888-23-5.
43. The Engineering ToolBox: Density of Moist Humid Air. Available online at: https://www.engineeringtoolbox.com/density-air-d_680.html (Accessed on September 2019).
44. Hilsenrath, J., et al. Tables of Thermal Properties of Gases. U.S. Department of Commerce, National Bureau of Standards Circular 564. November 1955.
45. The Engineering ToolBox: Air-Specific Heat at Constant Pressure and Varying Temperature. Available online at: https://www.engineeringtoolbox.com/air-specific-heat-capacity-d_705.html (Accessed on 19 March 2019).
46. Katsaprakakis, D. A., Christakis, D. G., Zervos, A., Papantonis, D., & Voutsinas, S. (2008). Pumped storage systems introduction in isolated power production systems. *Renewable Energy*, *33*, 467–490.
47. Katsaprakakis, D. A., Christakis, D. G., Pavlopoylos, K., Stamataki, S., Dimitrelou, I., Stefanakis, I., & Spanos, P. (2012). Introduction of a wind powered pumped storage system in the isolated insular power system of Karpathos–Kasos. *Applied Energy*, *97*, 38–48.
48. Katsaprakakis, D. A., Papadakis, N., Kozirakis, G., Minadakis, Y., Christakis, D., & Kondaxakis, K. (2009). Electricity supply on the island of Dia based on renewable energy sources (RES). *Applied Energy*, *86*, 516–527.
49. Katsaprakakis, D. A., Thomsen, B., Dakanali, I., & Tzirakis, K. (2019). Faroe Islands: Towards 100% RES penetration. *Renewable Energy*, *135*, 473–484.

4 Hybrid Plants for Thermal Energy Production

4.1 INTRODUCTION

In this chapter, we will examine subjects related to hybrid plants for thermal energy production. Obviously, the final product is now thermal energy, instead of electricity. Similar to the hybrid power plants for electricity production investigated in the previous chapter, thermal hybrid plants aim to meet the guaranteed thermal power production, determined by the necessity to cover a specific thermal power demand, based on nonguaranteed thermal power units.

As in the case of electricity hybrid power plants, a thermal hybrid plant consists of the following discrete components:

- Base units, which are nonguaranteed power production units. For thermal hybrid plants, solar radiation constitutes the primary energy source, because this is the only form of nonguaranteed renewable energy source (RES) that can be directly transformed to thermal energy. This implies that the base units in a thermal hybrid plant can only be various types of solar thermal collectors. Just for clarity, it should be underlined that thermal power can be also produced from other RES technologies, such as biomass burners or geothermal plants. However, the availability of these primary energy resources is guaranteed, so there is no sense considering these technologies as the base units of a thermal hybrid plant.
- Storage units. In active thermal power plants, thermal storage can be achieved through thermal energy transfer from the nonguaranteed base unit to a medium with high heat capacity. The typical medium for such a task is water. Thermal energy storage technologies usually consist of insulated water tanks. Advanced thermal energy storage technologies can also be based on solid materials, such as rocks or concrete, molten salts, or pressurized water tanks.
- Backup units. The backup units in a thermal hybrid plant can be common central heating burners and, in more specialized applications, small compact steam turbines.

An essential difference between electricity and thermal hybrid plants is that the latter directly provide the final product (thermal energy) for the consumers, namely the produced thermal energy does not have to be transformed to another form of energy before consumption. Because thermal energy constitutes one of the major final energy forms (more than 60% of the final energy consumption in residential and commercial buildings corresponds to thermal energy for indoor space conditioning and hot water production [1]), the significance of thermal hybrid plants is high. Additionally, thermal energy is sensible and desirable to be produced close to the final consumption's location, mainly due to the fact that its transportation is not as easy as with electricity, requiring significant and expensive infrastructure and exhibiting considerable losses. Consequently, thermal hybrid plants should be installed close to the point where thermal energy will be consumed. This peculiarity determines both the size and the type of thermal hybrid plants. Given the above analysis, it follows that thermal hybrid plants are mainly decentralized units. Each thermal hybrid plant aims to cover an autonomous thermal power demand of a single consumer, or, in more advanced cases, of a small group of consumers (e.g., a block of offices or a small settlement). Due to the nature of thermal energy, the development of thermal hybrid plants of large size is not technically or economically feasible. Conclusively, in this chapter we will examine issues on decentralized hybrid plants for thermal energy production.

The use of thermal hybrid plants extends to any application and activity with required final thermal power consumption. Hence, thermal hybrid plants can be introduced for the heating of commercial or residential buildings, the production of hot water, the heating of swimming pools, the heating production for industrial processes, the drying of agricultural products, etc. Due to the usually high setup cost of thermal hybrid plants, configured mainly by the base and storage units' procurement cost, their introduction is feasible in cases of considerable final thermal energy needs (e.g., swimming pools, industries, agricultural processes, blocks of buildings). In such cases, compared with the high heating production cost of conventional technologies (e.g., oil burners), the economic feasibility of thermal hybrid plants is maximized.

The examined topics in this chapter are:

- base unit major technologies for thermal hybrid plants and energy production analysis,
- major technologies for thermal energy storage, and
- operation simulation and dimensioning of the operation of thermal hybrid plants.

4.2 SOLAR COLLECTORS

Solar collectors constitute the technology for the transformation of the primary solar radiation to the final thermal energy, undertaking the role of the base units in thermal hybrid plants. The four major solar collector types are:

- uncovered solar collectors
- flat-plate solar collectors: black painted, semi-selective, selective coating collectors
- vacuum tube solar collectors
- concentrating solar collectors
- photovoltaic (PV) thermal hybrid solar collectors.

These solar collector types are presented in the following sections.

4.2.1 Uncovered Solar Collectors

Uncovered solar collectors are presented in Figure 4.1. Their construction and operation concept is simple. The uncovered solar collector consists of a set of plastic, dark colored (usually black) pipelines, aiming to minimize the thermal losses from the heated medium to the ambient environment and to maximize the collector's absorptance coefficient. The material used for the construction of uncovered solar collectors is an organic polymer (e.g., ethylene propylene diene M-class—EPDM). The incident solar radiation is absorbed in the form of thermal energy by the flowing medium inside the collector's pipelines. The captured thermal energy is then exploited as required.

According to the above description and Figure 4.1, it is conceivable that the uncovered solar collectors is a simple and economic technology, ideal for cases of low temperature heating demand, when a solar collector's operation at high efficiency is not a critical parameter. For these cases, given the low procurement cost of the required equipment and the simple required hydraulic installation, this particular technology is the optimum one.

The most common application of uncovered solar collectors is swimming pool heating, particularly during summer periods, when the available solar radiation maximizes. A simplified layout diagram of such an application is presented in Figure 4.2.

As seen in Figure 4.2, the swimming pool water itself flows through the solar collector and, once heated by the sun, is directly dispensed back into the swimming pool. Hence, this system does not

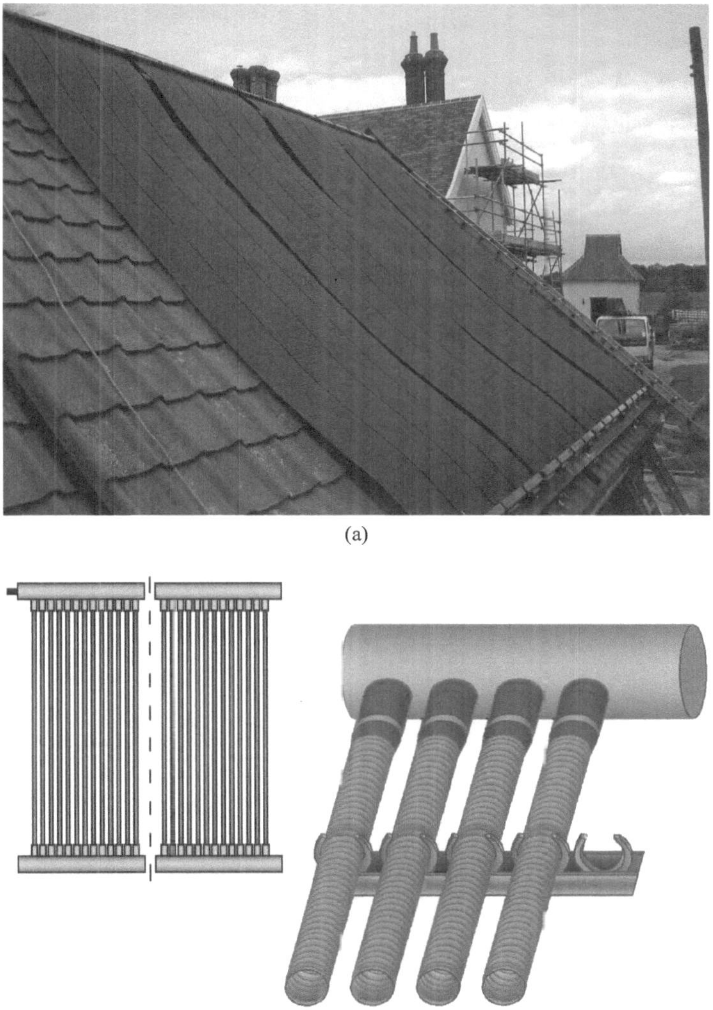

(a)

(b)

FIGURE 4.1 Uncovered solar collectors. (a) Installation of uncovered solar collectors on the roof of a residential building. (b) The construction concept of uncovered solar collectors.

require the installation of a heat exchanger for the thermal energy transfer from a primary hydraulic loop (solar collector loop) to a secondary one (swimming pool loop), a fact that considerably reduces the overall setup cost, while also maximizing its efficiency. The system's operation is supervised and determined by a control unit, which, by receiving temperature signals from the solar collectors' outlet and the swimming pool, begins to circulate the water.

The use of swimming pools in relatively warm weather conditions can be prolonged from April to October with the introduction of uncovered solar collectors (annual Global Horizontal Irradiance higher than 1,600 kWh/m^2). The computational simulation of such systems leads to a total required solar collectors' area equal to the 80% of the heated swimming pool's surface.

The lack of a heat exchanger and the direct water flow from the swimming pool to the solar collector and backwards, a feature mentioned previously as an advantage of such systems due to the simplified installation and the reduced setup cost, constitutes, at the same time, a potential drawback, due to the effect of the pool's chlorine on the collector's technical and thermal features. The

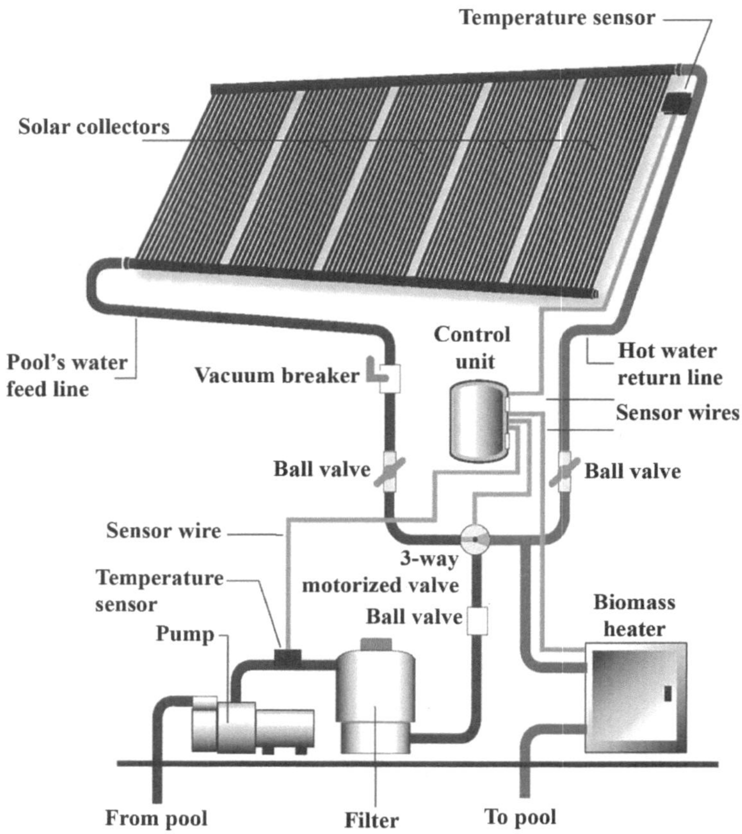

FIGURE 4.2 Layout of swimming pool heating system based on uncovered solar collectors [2].

gradual cover of the collector pipeline's inner surface with solid precipitators from the swimming pool's water will lead to increasing inner surface roughness, raise of the water flow viscosity losses, and a concurrent reduction of the thermal energy transmittance coefficient from the pipeline plastic material to the flowing water. The above effects will lead, with time, to the reduction of the collector's thermal efficiency and the increase of the energy consumed by the pump for the circulation of the water in the hydraulic loop. To avoid the above consequences, the solar collector should be regularly cleaned, ideally once before the beginning and another one at the end of the swimming pool usage period, by injecting under high pressure a special solution inside the pipeline for the disintegration of the solid precipitators.

4.2.2 Flat-Plate Solar Collectors

Flat-plate solar collectors are the most widely used technology for the exploitation of solar radiation to produce thermal energy. A graphical representation of a flat-plate solar collector is provided in Figure 4.3. It consists of an absorber plate, attached inside an insulated metallic frame (usually an aluminum alloy). The construction is closed from above with the placement of one or more toughened glass frames.

The absorber plate can be flat, grooved, or pressurized, with the working medium pipelines attached on its surface (Figure 4.4). As the solar radiation falls on to the solar collector through the glass cover, it is converted into thermal energy at the absorber plate, characterized by high absorptance coefficient. The largest part of the delivered thermal energy is transferred with conductivity

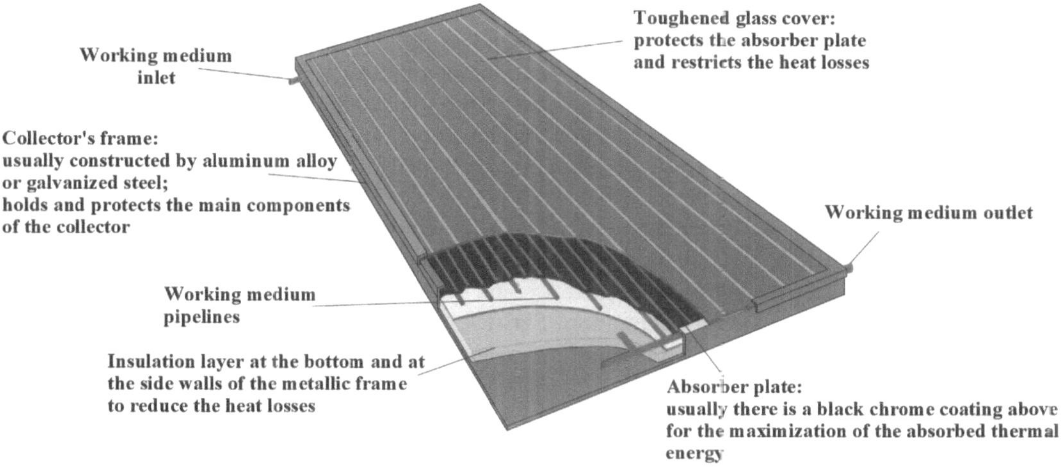

FIGURE 4.3 Constructive features of a flat-plate solar collector (under due diligence of greenspec.co.uk) [3].

and convection from the absorber plate to the flowing working medium inside the attached collector's pipelines. The thermal energy is eventually provided with the working medium for use or storage.

The absorptance coefficient of the absorber plate determines the type of the flat-plate solar collector. Generally, flat-plate solar collectors are distinguished in two types, depending on the absorptance coefficient:

- Black painted or semi-selective coating collectors. In these collectors the absorber plate is covered with a black semi-selective painting with absorptance coefficients at the range of 80%. The flat-plate collectors are usually used for water and indoor space heating.
- Selective coating collectors, with the absorber plate painted with a selective coating, raising the absorptance coefficient at 90%–95% and the reflectance coefficient at 5%–15%. With selective coating solar collectors, higher efficiencies and higher water temperatures are achieved. Selective coatings are also used for indoor space heating and hot water production, however under unfavorable conditions (e.g., lower available solar radiation) or aiming at higher solar radiation penetration.

The transparent cover of the solar collector exhibits high transparency coefficient at shortwave solar radiation and low transmittance coefficient at longwave thermal energy. It should, at the same time,

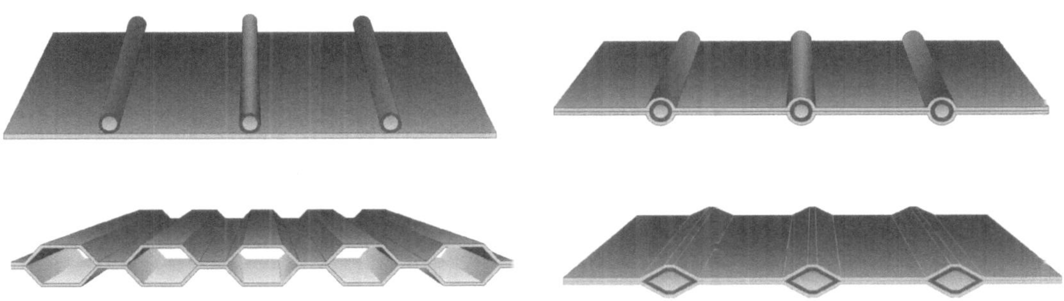

FIGURE 4.4 Different integration types of the working medium pipelines on the absorber plate of a solar collector.

be resilient against adverse weather conditions (e.g., hail) and inexpensive. Usually toughened glass is used with low iron content, shortwave radiation transmittance coefficient at 0.9, while for the longwave thermal radiation the transmittance coefficient is almost zero. Antireflective coating on the glass surface can also contribute to the increase of the solar radiation transmittance coefficient. Additionally, the transparent cover exhibits low thermal conductivity, for the minimization of thermal losses through conductivity from the solar collector's interior to the ambient environment, due to overheating of the absorber plate.

Apart from glass, the use of plastic materials for the construction of the transparent cover is also possible. Generally, only a few plastic materials are able to withstand the solar radiation fluctuations during the year, the subsequent diastoles and systoles, and the wear on the plastic material caused by the solar radiation penetration. Plastics, usually, exhibit higher thermal radiation transmittance coefficient, up to 0.4. On the other hand, plastics are lighter, less fragile, and more easily installed.

Another important constructive parameter of flat-plate solar collectors is the route followed by the pipelines on the absorber plate. Some pipeline routes are presented in Figure 4.5. The route presented in Figure 4.5a is a single-line route, crossing the collector's overall area. This installation implies maximization of the outlet temperature. If the achievement of high temperatures of the working medium at the collectors' outlet constitutes an essential requirement of an application, then the single-line route should be selected.

The parallel pipelines route presented in Figure 4.5b results in the maximization of the working medium flow, while the water outlet temperature remains at lower levels. This implementation is preferable in applications in which the basic requirement is the achievement of high efficiency for the solar collectors, while the increase of the water temperature is not a crucial parameter (e.g., warm water production for residential use). Finally, the implementation presented in Figure 4.5c constitutes a combination of the previous two alternatives. From the above it is concluded that the correct choice of the optimum flat-plate solar collector type, with regard to the pipeline route, is a function of the available climatic data (solar radiation) and the requirements of the final thermal energy demand.

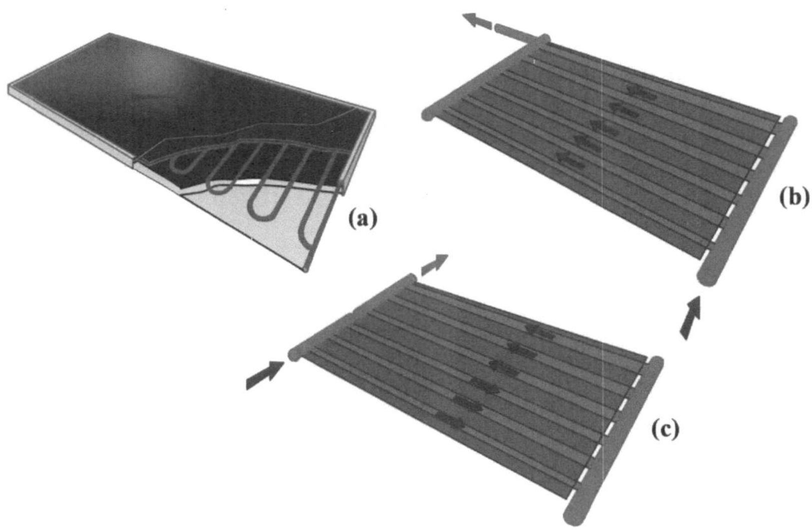

FIGURE 4.5 Pipeline routes in flat-plate collectors: (a) single-line route, (b) parallel pipeline route, (c) combination route.

FIGURE 4.6 Image of a vacuum tube solar collector (under due diligence) [4].

4.2.3 Vacuum Tube Solar Collectors

The constructive layout of vacuum tube solar collectors is based on the configuration of a vacuum space between two co-axis transparent tubes, where the absorber surface is introduced (Figure 4.6). The main objective of this layout is to minimize the heat transfer losses with convection and radiation from the collector to the ambient environment. The advantages of vacuum tube solar collectors emerge mainly under unfavorable climate conditions (e.g., cloudy skies, low solar radiation, low ambient temperatures, high wind velocities). The efficiency of flat-plate solar collectors is considerably reduced in such climates. On the contrary, the efficiency of vacuum tube solar collectors is maintained at high levels even under unfavorable weather conditions. Hence, they feature as an ideal option for indoor space heating and hot water production in cold climates and for geographical regions with low available solar radiation. Additionally, the working medium's temperature with vacuum tube solar collectors can be increased up to 300°C, making them suitable for special applications with high required temperatures (e.g., industrial processes, agricultural products drying).

The operation concept of vacuum tube solar collectors is significantly different from flat-plate collectors (Figure 4.7). They are constructed with a series of couples of parallel glass tubes. Each

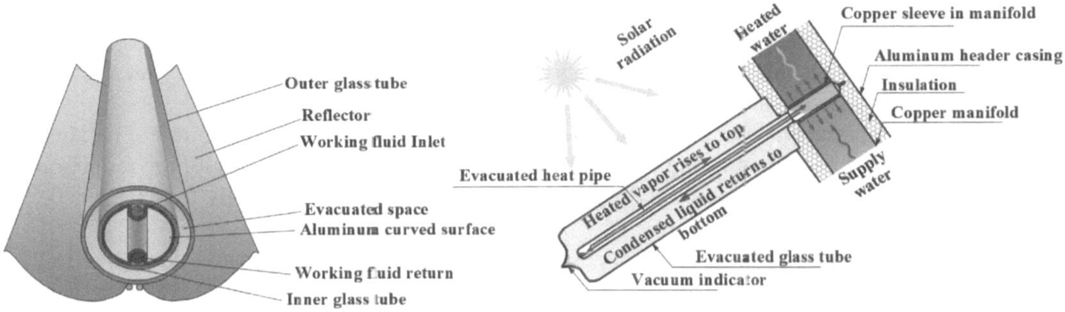

FIGURE 4.7 Construction details of vacuum tube solar collectors.

couple consists of two glass tubes, placed one inside the other, with vacuum space between them. The outer tube is made of bio-silicic glass of high hardness and resistance (it can withstand the strike of 25 mm diameter hail falling with a speed of 100 km/h). Inside each tube there is a flat or curved aluminum surface connected with a metallic tube, usually made of copper. This surface is painted with a thermal-absorbing material, of low reflectance, leading to a total transformation of more than 92% of the incident solar radiation to thermal energy. The working medium inside the copper tube can be an inorganic, volatile fluid (boiling point at 25°C). The working medium is converted to superheated steam and lifted to the tube's upper side, located inside a heat exchanger, through which the thermal energy is transmitted to the heated water. The fluid is condensed, after transmitting the thermal energy to water, and drops to the lower part of the copper tube, to start a new thermal transmission cycle. Some types of vacuum tube solar collectors have outer reflectors behind the vacuum tubes or inside the glass tubes. The outer reflectors increase the radiation collected by the collector, by leading the radiation that escapes through the vacuum space back to the absorber.

The high efficiency of vacuum tube solar collectors is achieved with the combined use of aluminum reflective sheets, of the nonconductive vacuum space due to which the thermal losses with conductivity and convection are minimized, and of the absorbing coating on the aluminum surface. The vacuum environment minimizes the thermal losses and contributes to the increase of the collector efficiency and the achieved water temperature. The tubular construction of the vacuum tube collectors enables the increase of the produced thermal power even for low solar radiation incident angles and low solar radiation, prolonging the effective operation of the collector during the whole daytime period.

4.2.4 Concentrating Solar Collectors

The temperatures developed in the working medium of solar collectors can be significantly increased in the following two ways:

- By reducing the heat losses surface from the collector to the ambient environment.
- By concentrating the incident solar radiation on a small collecting surface of the collector. This is achieved by introducing an optical device between the radiation source, namely the sun, and the absorber surface of the collector.

Concentrating solar collectors exhibit the following advantages compared to conventional solar collectors:

- With concentrating solar collectors, higher temperatures can be achieved for the working medium, compared to flat-plate solar collectors with the same absorber surface.
- The efficiency of concentrating solar collectors is higher than in flat-plate collectors, due to the smaller heat losses surfaces.
- The reflective surfaces of concentrating solar collectors require lower quantities of constructive materials and are constructively simpler than in flat-plate collectors. As a result, the cost per unit of installed area is lower for the concentrating collectors.
- Due to the relatively low required surface of the solar radiation receiver per unit of incident solar radiation in concentrating collectors, the application of technologies for the reduction of heat losses aiming to increase the collector's efficiency is economically feasible.

On the other hand, concentrating solar collectors exhibit the following drawbacks compared to flat-plate solar collectors:

- The diffused solar radiation absorbed by the concentrating solar collectors is lower.

- The installation of a tracking base is required for the concentrating solar collectors, to follow the sun's orbit in the horizon.
- The reflectance of the reflective surface of the concentrating solar collectors normally reduces with time. For this reason, the reflective surfaces must be regularly cleaned or replaced after specific time periods.

Concentrating solar collectors can be of two types:

- cylindrical concentrating solar collectors (line-focus solar collectors)
- spherical or parabolic solar collectors.

The operation of both cylindrical and spherical or parabolic concentrating solar collectors is based on the reflection of solar radiation from two-dimensional (2D) spherical and cylindrical reflectors, as presented in Figure 4.8. The technical features of these two reflector types with regard to the concentration of the incident solar radiation parallel rays, falling perpendicularly on the reflector's surface, are:

- all the parallel rays reflected by a cylindrical reflector go through a line crossing the circular projection's center of the spherical reflector on the vertical level, parallel to the incident rays
- all the parallel rays reflected by a spherical or parabolic reflector intersect at a particular point when they are parallel to the parabolic reflector's axis of symmetry.

Additionally, a circular reflector exhibits full symmetry with regard to any rotational movement around its center. As a result, if the solar rays are not perpendicular to the reflector's opening (aperture), the pattern of the reflected solar rays remains the same, but twisted with regard to the reflector's axis. On the other hand, a parabolic reflector does not exhibit a symmetric attribute with regard to its focal point, as seen in Figure 4.9. For the accurate concentration of the solar rays in the focal point of a parabolic reflector, the sun's orbit should be precisely traced, so the axis or level of symmetry remains always parallel with the incident solar rays.

The new term *concentration ratio* is introduced for the concentrating solar collectors. Concentration ratio is used to describe the concentrated light energy achieved with a concentrating collector. Two different definitions are generally available.

- Optical concentration ratio
  It is defined as the ratio of the average solar radiation I_r integrated over the reflector's surface A_r, versus the incident solar radiation on the collector's aperture I_a:

$$CR_o = \frac{\frac{1}{A_r} \cdot \int I_r \cdot dA_r}{I_a} \tag{4.1}$$

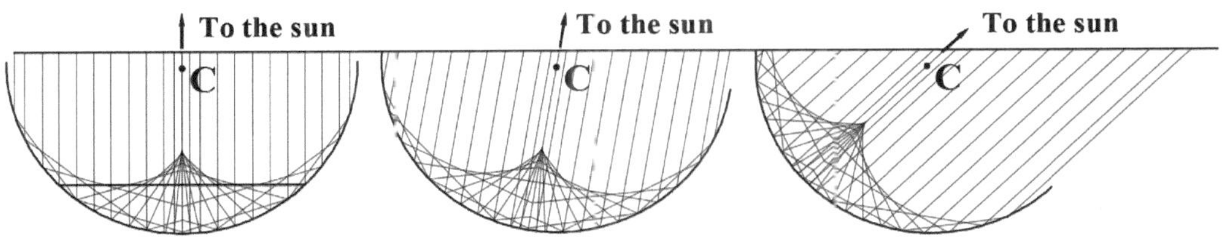

FIGURE 4.8 Solar radiation reflection and concentration on spherical or cylindrical reflectors [5].

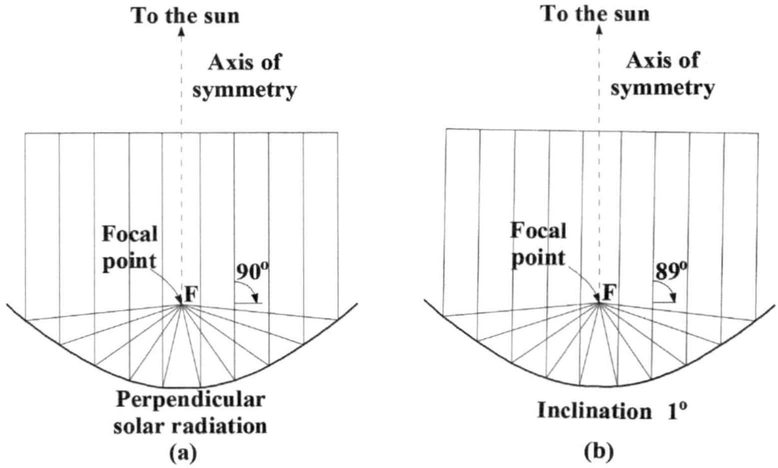

FIGURE 4.9 Characteristics of parabolic optics for (a) solar rays parallel with the axis of symmetry and (b) solar rays direction angle 1° with the axis of symmetry [5].

- Area concentration ratio
  It is defined as the ratio of the collector's aperture area A_a over the receiver's area A_r:

$$CR_s = \frac{A_a}{A_r} \tag{4.2}$$

4.2.4.1 Line-Focus Concentrating Solar Collectors

Line-focus concentrating solar collectors collect high-density solar radiation on a focus line. The cylindrical parabolic solar collectors belong to this category. To create a parabolic or cylindrical concave reflector, the 2D reflectors presented in Figure 4.8 must be expanded perpendicularly to the level of curvature, as shown in Figure 4.10.

A parabolic concave reflector exhibits also a focus line and must be equipped with a tracking mechanism, so it can maintain the solar radiation concentration on the focus line. The appropriate tracking angle is defined by the reflector's orientation with regard to the position of the sun in the horizon. The scope of the tracking mechanism is to always maintain a perpendicular incidence of the solar rays on the reflector's aperture.

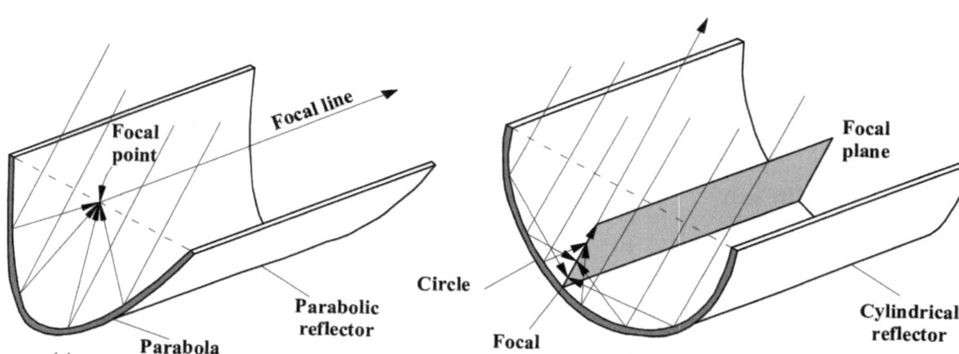

FIGURE 4.10 (a) Parabolic and (b) cylindrical concave reflectors [5].

4.2.4.2 Spherical Concentrating Collectors

If the 2D reflectors presented in Figure 4.8 twist around their axes, instead of being expanded, the result will be spherical concentrating collectors. A spherical concentrating collector should be equipped with two axes of rotation, in order to keep the incident solar radiation perpendicular to the reflector's aperture.

Paraboloid spherical concentrating solar collectors belong to this category, which, generally, are able to gather higher density solar radiation around a focal point.

4.2.4.3 Compound Parabolic Collectors

Compound parabolic concentrators (CPC) are collectors of average temperature and combine elements from both concentrating and flat-plate solar collectors. Specifically, the CPC do not only concentrate the solar radiation from the reflector to the receiver, but they also exploit the diffused solar radiation, like the flat-plate collectors. The solar radiation is exploited within a wide range of incident angles, so the requirement for solar tracking is respectively restricted. The achieved concentration ratios are relatively low. The CPC with two axes of rotation exhibit higher concentration ratios, yet their maintenance requirements affect their economic feasibility. With regard to heat losses, these can be up to 32% lower compared to flat-plate collectors.

The most usual form of CPC is the 2D collector of trough type, presented in Figure 4.11. As seen in this figure, each side of the collector follows the pattern of a parabola. Both parabolas are extended until they are parallel to the axis of symmetry of the CPC. The angle configured by the CPC axis and the line connecting the focus of the one parabola with the opposite edge of the collector's aperture is called acceptance half angle θ_c. If the parabolic reflective surfaces are optically perfect, then the incident solar radiation inside the collector's aperture with an angle θ: $-\theta_c < \theta < \theta_c$ is reflected towards the receiver, located at the base of the CPC. For $\theta < -\theta_c$ or $\theta > \theta_c$ the solar radiation cannot reach the receiver. The angle range $[-\theta_c, \theta_c]$ is called *acceptance angle* and determines the incident angles for which the solar radiation is exploited.

The concentration ratio for an optically perfect elongated CPC is given by the relationship [5]:

$$C_i = \frac{1}{\sin(\theta_c)} \tag{4.3}$$

FIGURE 4.11 Linear, concentrating solar collectors.

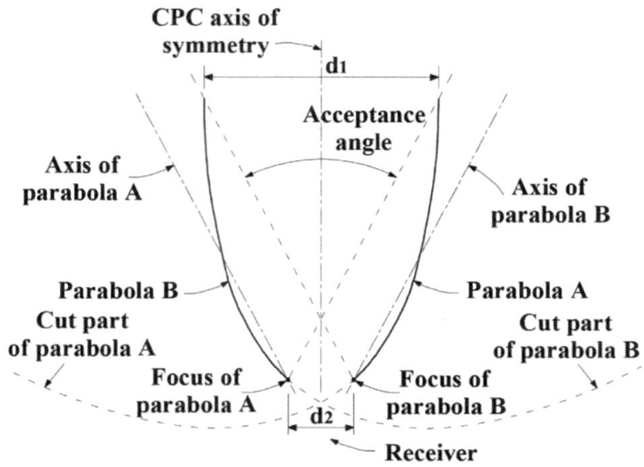

FIGURE 4.12 Geometry of a CPC [6, 7].

The upper edges of the parabolic sides of a CPC have low contribution to the reflection of the solar radiation towards the receiver; hence, they can be cut off, leading to a subsequent reduction of the CPC height, with a negligible reduction of its efficiency and a significant reduction of the manufacturing cost.

Figure 4.12 provides an example of a CPC with a cylindrical receiver, with a graphical depiction of the geometrical shape development of the reflective surface required by the receiver. Specifically, the perpendicular line on the random point P of the surface CE (continuous short line in the figure), should split in half the angle configured by the tangential line PT on the absorber plate and by the line that crosses the point P and forms an angle θ_c with the CPC axis of symmetry. The above method for the development of the reflective surface geometry can be used for the design of the shape of any bent receiver, creating thus collectors of CPC type that are not necessarily parabolic.

Apart from the above presented type of CPC, there is also a series of alternative implementations, such as the fixed-mirror solar collector (FMSC), the moving reflector stationary receiver (SLATS), the fixed-mirror distributed focus (FMDF), and the Fresnel collector. The extensive presentation of these collector types is beyond the scope of this chapter. More can be found in the provided literature [7, 8] and in Chapter 6.

From a mechanical point of view, the concentrating solar collectors exhibit specific problems compared to the flat-plate collectors. First, they must by founded on a solar tracking system, a fact that increases the complexity and the setup cost of the whole installation. Additionally, their maintenance requirements are considerably increased, mainly due to the necessity to keep the solar radiation concentrating optical systems in good condition, taking into account the existence of dust, the climate conditions, and the atmospheric corrosion environment. For the above reasons, the use of concentrating solar collectors is restricted in very special applications, for which the achievement of very high temperatures and the production of high thermal power is required, such as solar thermal power plants, examined in Chapter 6.

4.2.5 Photovoltaic Thermal Hybrid Solar Collectors

Photovoltaic thermal (PVT) hybrid solar collectors are formulated as a combination of PV cells with thermal solar collectors, aiming at the concurrent production of electricity and thermal energy.

The necessity for the design and the development of PVT systems comes from mainly two major parameters:

- The fact that the efficiency of a pure PV panel decreases with the increase of the PV cell's temperature. Indeed, it has been proved that a 0.4%–0.5% drop of the PV cell's efficiency occurs for every 1°C temperature rise for silicon cells [9]. This imposes the requirement for its adequate cooling, in order to maintain its temperature close to standard thermal conditions (25°C) and the electricity production efficiency among acceptable limits. This, in turn, creates the prerequisites for the installation of a proper cooling system and the exploitation of the removed heat from the PV cell for the coverage of parallel heating needs. As a result, the PVT can exhibit an overall annual average efficiency in the range of 60%–80% [10], considerably higher than the corresponding feature of a pure PV panel or a solar collector, which normally can be in the range of 15% and 40%–50%, respectively.
- The installation of a common panel for both electricity and thermal energy production implies less required space and improved aesthetics, compared to the installation of separate PV panels and solar collectors, next to each other.

Based on the above motives, the study of PVT systems began in the 1970s with the investigation of different layouts and materials, aiming, in any case, at the maximization of the electrical and thermal efficiency, with the minimization of the manufacturing cost and the prolongation of the PVT panels' service life. The different configurations are mainly formulated by the involved cooling system and the introduced constructive layout. Specifically, the following alternative PVT systems have been proposed:

- PVT collector with single-pass air heat, with or without fins
- PVT collector with double-pass air heat with or without fins
- PVT collector with double-pass air heat with compound parabolic concentrators (CPC) and fins
- PVT collector with single-pass air heater and V-groove or rectangular shape absorber plate
- PVT collector with water heater
- PVT collector with air and water combination heater.

The fundamental layouts of an air-based and a water-based PVT system are presented in Figures 4.13 and 4.14 respectively. As seen in these figures, the main concept is simple and does

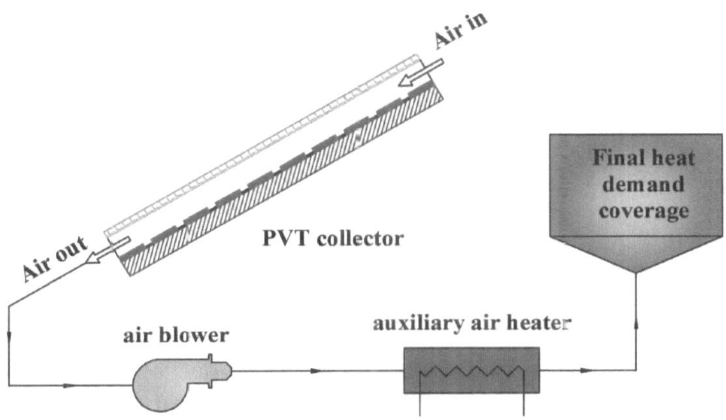

FIGURE 4.13 Essential operation principle of an air-based PVT system.

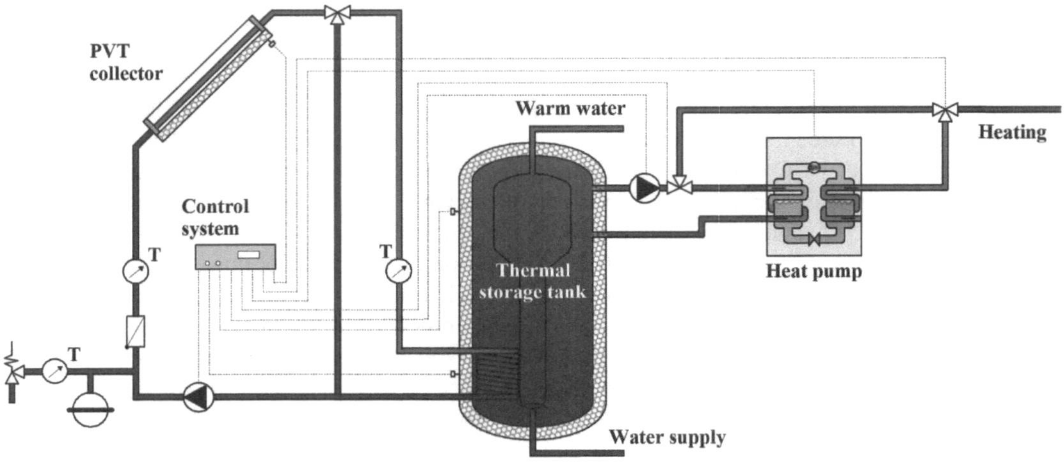

FIGURE 4.14 Essential operation principle of a water-based PVT system.

not differ substantially from the corresponding operation concepts of conventional solar collector systems. In both cases, the produced electricity is injected in the grid, as with common PV systems. In case of air-based PVT systems, atmospheric air is forced with inverter-regulated fans to pass through the PVT panel and the captured heat, by the air stream, is led by an air blower to an auxiliary heater, if additional heating is required, and, eventually, the hot air stream is utilized for the coverage of the final heating demand (e.g., indoor space heating). In case of water-based PVT panels, the arisen heat in the panel is captured through a closed hydraulic network, with its flow controlled and regulated by an inverter operated circulator. The captured heat is led to a thermal storage tank and then to the primary loop of a water-to-water heat pump. The produced heating from the heat pump is finally exploited according to the final users' needs.

Figure 4.15a presents the main configuration of air-based PVT panels, with regard to the air stream flow, the positioning of the PV panel–absorber plate, and the main constructive features [11]. It can be seen that the available configurations can be distinguished as single- and double-pass air flow. The PV cells and the absorber plate can be located below, above, or between the single-pass air flow and always between the double-pass air flow. For double-pass air flow or for single-pass air flow below or between the panel and the plate, metallic fins can be adapted underneath the plate, to facilitate the heat removal factor, by increasing the heat transfer surface, as shown in Figure 4.15b [12]. The PV cells are always adjusted at the top side of the absorber plate. As shown in the latter figure, this configuration is formulated by the top, transparent cover, the metallic, insulated frame that forms the casing of the collector, and the absorber plate, with the PV panels adjusted on its top and the fins on its bottom side.

A variation of the above PVT layouts is the double-pass air-based systems, with fins and CPCs. This layout is shown in Figure 4.16. Apart from the PV panels and the fins, the CPCs have also been introduced above the absorber plate. The PV cells are placed on the CPCs. In this way, the CPCs concentrate the incident solar radiation on the PV cells, increasing the captured solar radiation and the produced electricity. The inlet air stream enters the collector above the plate, flows between the CPCs, is heated by absorbing the produced thermal energy in the cells, cools at the same time the PV cells, and exits the PVT collector through an opposite flow direction below the plate.

A major drawback of the above two configurations is the housing of the overall system inside the casing of the collector and the placement of the PV cells below a transparent top cover, which leads to considerable reduction of the final electricity production from the cell, which can be as high as 50%, compared to the corresponding production of an uncovered PV cell. To overcome this inadequacy, the alternative configurations of the V-groove and the rectangular shape solar collectors,

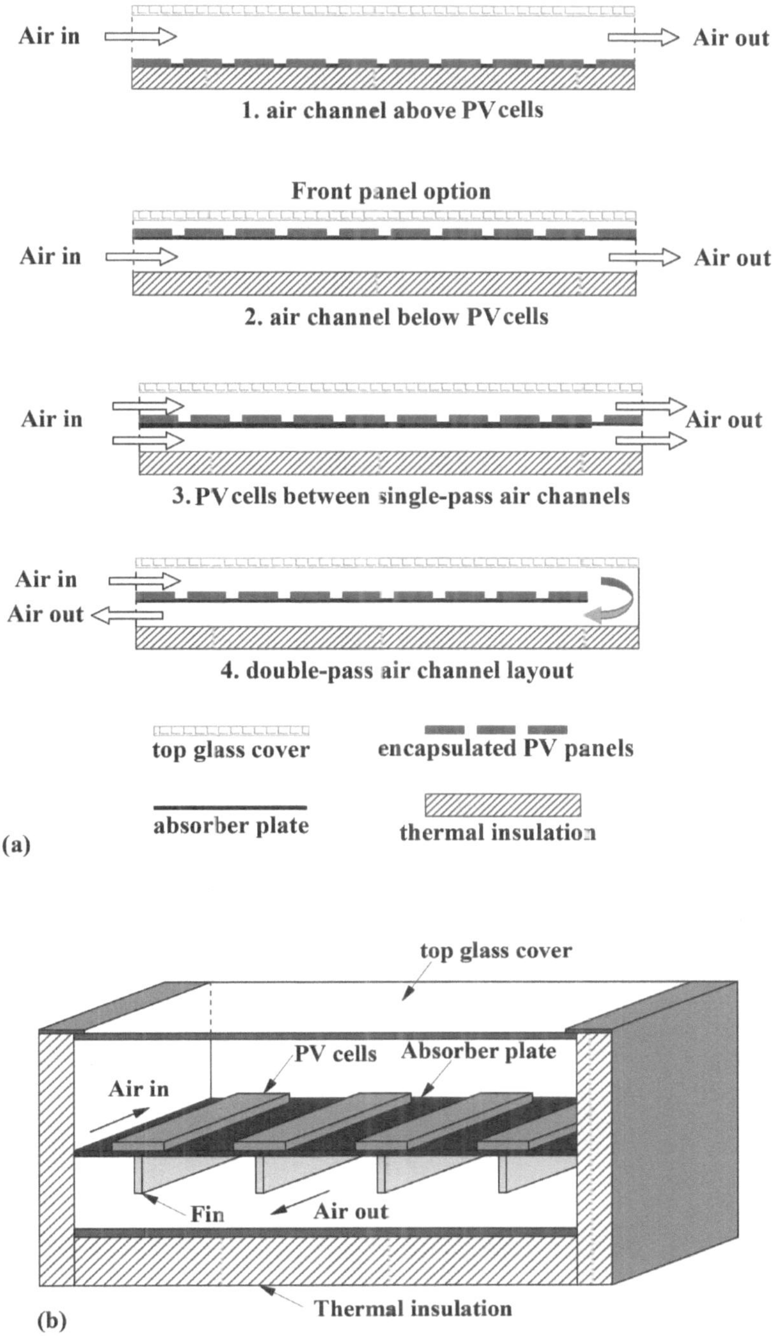

FIGURE 4.15 The operation principle (a) and constructive layout (b) of air-based PVT panels, with single- or double-pass air flow [11, 12].

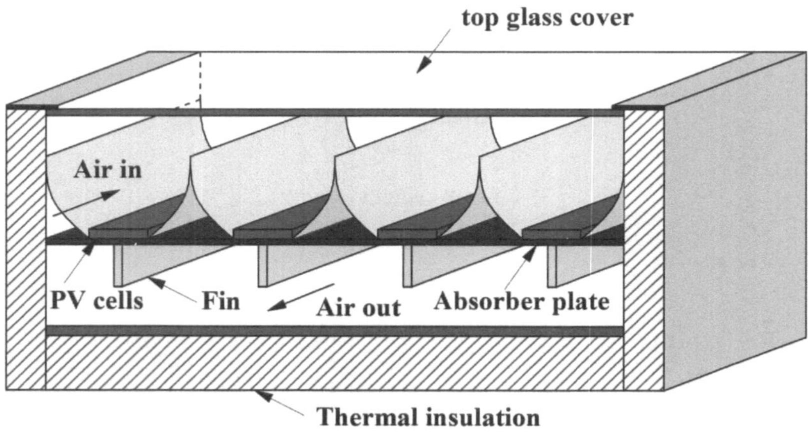

FIGURE 4.16 The constructive layout and the operation principle of air-based PVT panels, with double-pass air flow, CPC, and fins [12].

presented in Figures 4.17 and 4.18 respectively have been formulated. The configurations are simple. The PV panels are exposed uncovered to the sun, just like in conventional PV applications. The heat production on the PV panels is transferred to the adjusted on their bottom side V-groove or rectangular absorber plate. The single-pass air stream enters the collector in the gaps of the absorber plate, absorbing the gathered heat and cooling the PV cells.

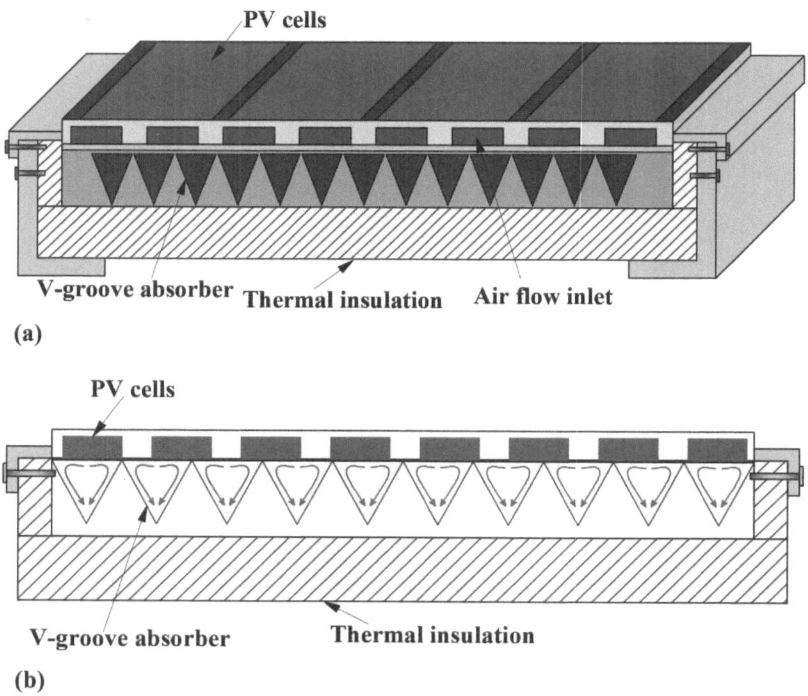

FIGURE 4.17 The constructive layout (a) and the operation principle (b) of air-based PVT panels, with single-pass air flow and a V-groove absorber plate [12].

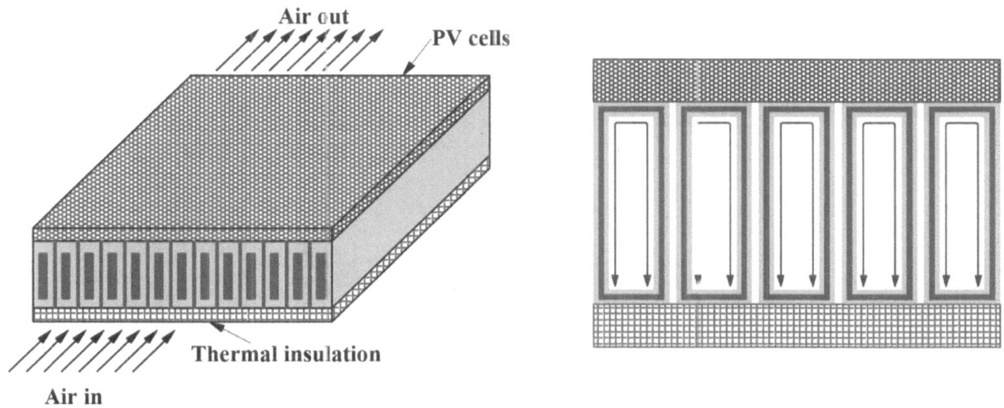

FIGURE 4.18 The constructive layout and the operation principle of air-based PVT panels, with single-pass air flow and a rectangular shape absorber plate [14].

Water-based configurations have been developed similarly to the above presented air-based PVT systems. They are shown in Figure 4.19. As seen in this figure, the water-based PVT can be with or without a top transparent cover. The water flow can be realized inside parallel tubular pipelines or rectangular channels, adjusted underneath the absorber plate, which, in turn, in any case, is adjusted below the PV panels. An alternative configuration is with the water flow inside channels above or below the absorber plate. Additionally, instead of linear pipelines, the spiral flow collector, presented in Figure 4.20a, has been proposed. The spiral PVT water-based system was indicated as the optimum configuration among seven other alternative investigated PVT water-based systems, according to simulations on their operation executed under a controlled incident solar radiation of 600 W/m^2 and a fluid mass flow rate of 0.01 kg/s. A thermal efficiency of 50.1% and an electrical efficiency of 12.0% was measured for this particular layout [13].

Finally, in Figure 4.20b, a combined air- and water-based PVT system is presented. Specifically, the V-groove air-based configuration has been adapted respectively, so that underneath the absorber plate a spiral shape water pipeline is adjusted. In this way, heat is removed from the PV panel and the absorber plate both with the air and the water flow. The heat removal rate is higher than in the previous systems, leading to higher electrical efficiency for the PV panels too. Additionally, the availability of thermal power concurrently with two different forms offers a larger variety for potential applications for the exploitation of the captured heat. This layout, exploiting the combined cooling of the PVT panel both with water and air stream, is known as a PVT-combi system.

4.3 ENERGY ANALYSIS OF A FLAT-PLATE SOLAR COLLECTOR

The following energy flows take place in a flat-plate solar collector:

- heat transfer through the solar radiation from the environment to the absorber plate of the solar collector
- heat transfer from the absorber plate towards the collector's inner space and the transparent cover
- heat transfer from the base, the peripheral side surfaces, and the transparent cover of the collector to the ambient environment, through radiation, conductivity, and convection.

These heat flows are depicted in Figure 4.21.

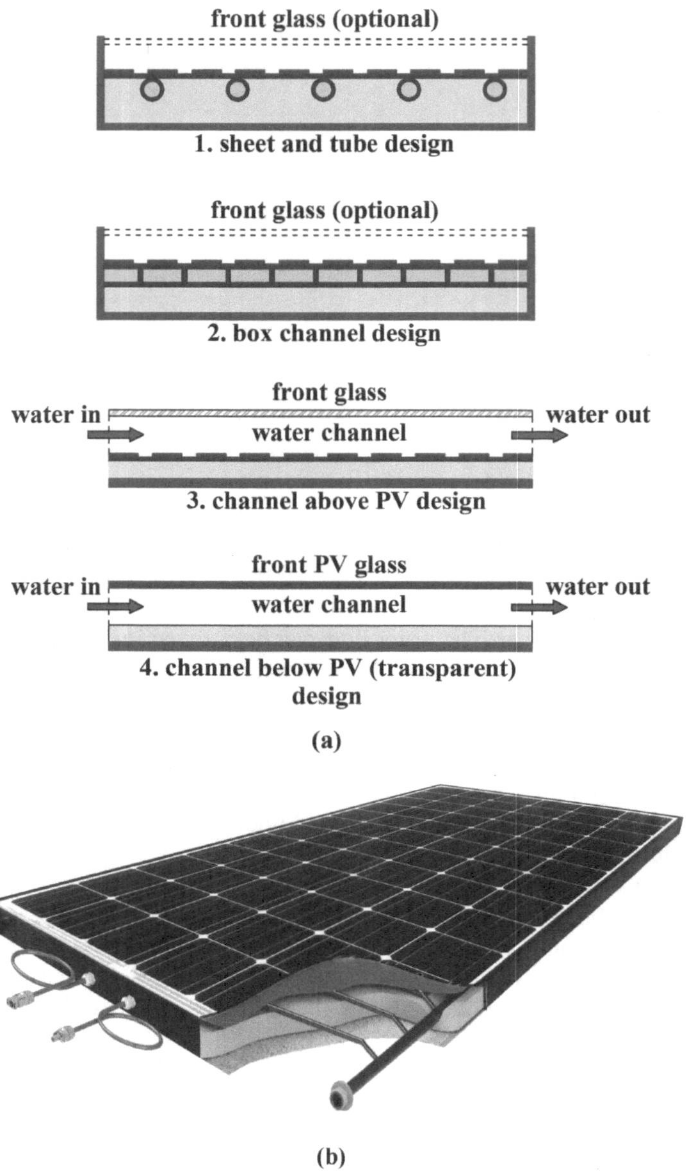

FIGURE 4.19 The operation principle (a) and the constructive layout (b) of water-based PVT panels (under due diligence) [11, 15].

The thermal power production from flat-plate solar collectors is given by Equation 4.4 [6]:

$$\dot{Q} = A_c \cdot F_R \cdot \left[G_t \cdot (\tau \cdot \alpha) - U_L \cdot (T_{fi} - T_a)\right] \tag{4.4}$$

where:

A_c: the solar collector's effective area in m^2

F_R: the heat removal factor (or correction factor) from the collector to the environment, defined below

G_t: the total incident solar radiation in the collector in W/m^2

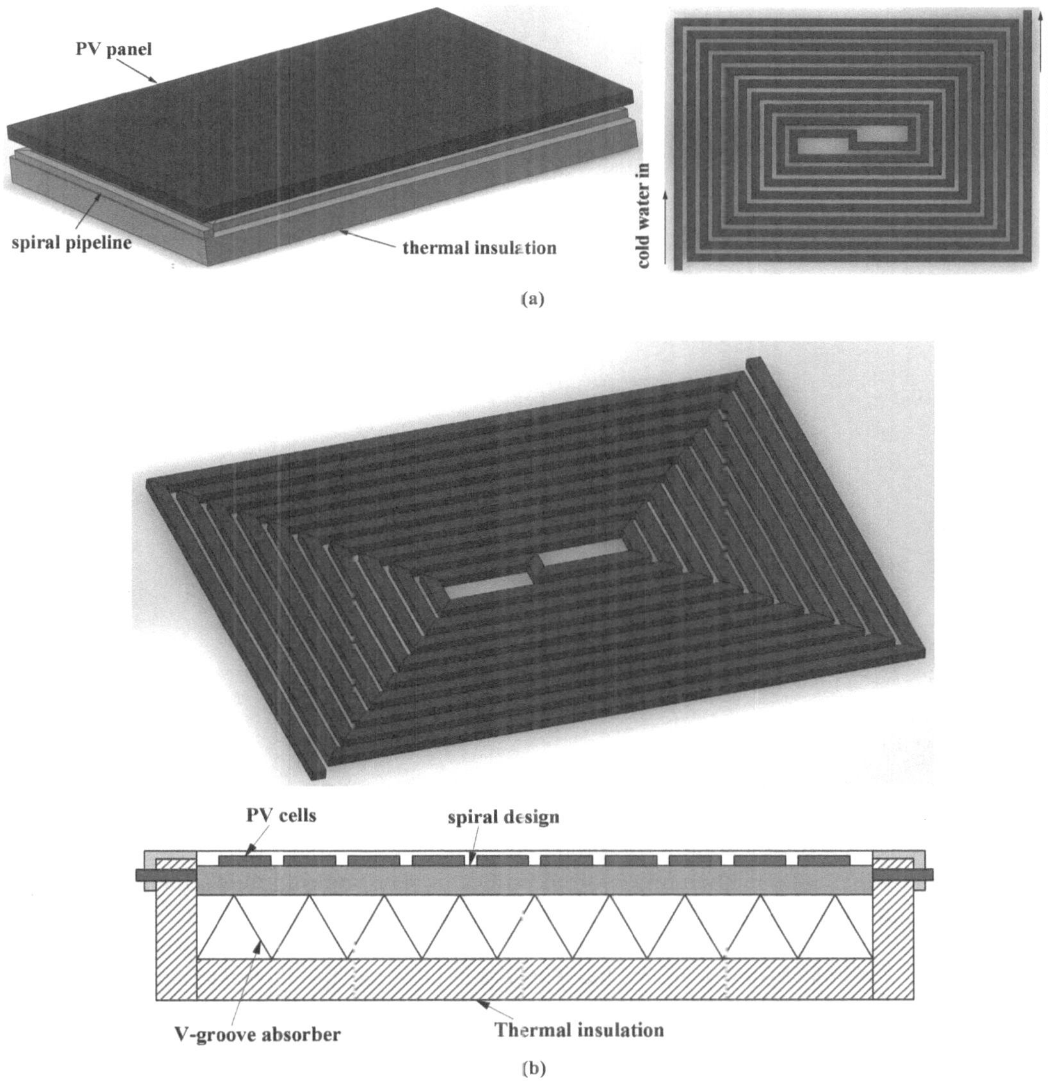

FIGURE 4.20 The constructive layout (a) and the operation principle (b) of water-based PVT panels, with a spiral-shape pipeline and the formulation of the PVT-combi system [16].

- $\tau \cdot \alpha$: the transmittance–absorptance product; the transmittance refers to the transparent cover of the solar collector and the absorptance refers to the collector's absorber plate
- $G_t \cdot (\tau \cdot \alpha)$: the final, net solar radiation absorbed by the solar collector in W/m^2
- U_L: the thermal transmittance factor for the heat transfer from the collector to the environment in $W/(m^2 \cdot K)$
- T_{fi}: the working medium inlet temperature in the collector in °C
- T_a: the ambient temperature in °C.

The calculation procedures for the parameters involved in the thermal power production calculation from a flat-plate collector, according to the previous relationship, are presented in the following sections.

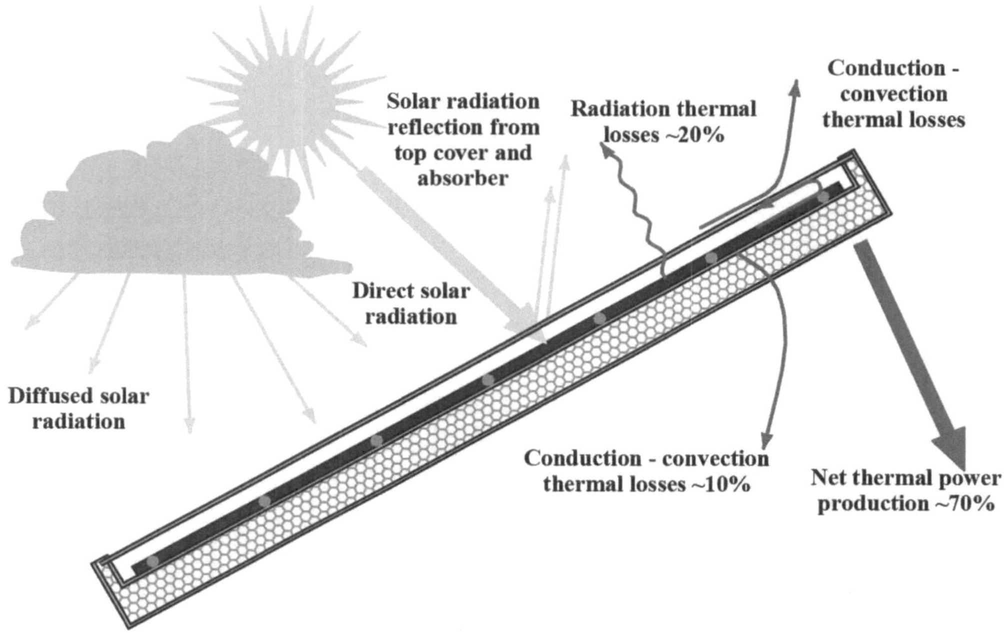

FIGURE 4.21 Energy flows to and from a flat-plate solar collector.

4.3.1 Heat Removal Factor F_R

The heat removal factor is defined as the ratio of the real thermal power production from the solar collector over the theoretical thermal power that would be produced if the working fluid's temperature remained constant and uniform through the collector pipelines' whole length, and equal to the working fluid's inlet temperature T_i in the collector. The working fluid's inlet temperature in the collector depends on the technical features and the operation mode of the overall thermal power production system and the thermal power demand. The heat removal factor depends on the solar collector's technical specifications, the working fluid's type and flow rate, according to the following [6]:

$$F_R = \frac{\dot{m} \cdot c_p}{A_c \cdot U_L} \cdot \left[1 - \exp\left(-\frac{A_c \cdot U_L \cdot F'}{\dot{m} \cdot c_p}\right)\right] \tag{4.5}$$

where:

$\dot{m}$: the working fluid's mass flow rate in the collector in kg/s
c_p: the working fluid's specific heat in kJ/(kg·K)
F': the collector's efficiency factor.

The collector's efficiency factor is defined as the ratio of the real thermal power production over the thermal power that would be produced if the collector's absorber plate temperature were equal to the working medium temperature in the collector [6]. The collector's efficiency factor, for uncovered solar collectors, is given by:

$$F' = \frac{\frac{1}{U_L}}{\frac{1}{U_L} + \frac{D_o}{D_i \cdot h_f} + \frac{D_o}{2 \cdot k} \cdot \ln\left(\frac{D_o}{D_i}\right)} \tag{4.6}$$

where:

- D_i: the inner diameter of the collector's pipelines in m
- D_o: the outer diameter of the collector's pipelines in m
- k: the thermal conductivity factor of the solar collector's material in W/(m·K)
- h_f: the thermal convection factor for the heat transfer from the collector's pipelines to the working fluid in W/(m²·K).

For flat-plate solar collectors the efficiency factor F′ is given by the relationship [6]:

$$F' = \frac{\frac{1}{U_L}}{W \cdot \left[\frac{1}{U_L \cdot [D + (W - D) \cdot F]} + \frac{1}{C_b} + \frac{1}{\pi \cdot D_i \cdot h_{fi}}\right]} \tag{4.7}$$

where:

- W: the distance between two consecutive pipelines of the solar collector in m (see Figure 4.22)
- D: the outer diameter of the collector's pipelines in m (see Figure 4.22)
- D_i: the inner diameter of the collector's pipelines in m (see Figure 4.22)
- F: the standard fin efficiency, defined below
- C_b: the bond conductance of the collector's pipelines with the absorber plate, analyzed below in W/(m·K)
- h_{fi}: the thermal convection factor for the heat transfer from the working fluid to the collector's pipelines in W/(m²·K).

The standard fin efficiency F is given by the following relationship [6]:

$$F = \frac{\tanh\left[\frac{m \cdot (W - D)}{2}\right]}{\frac{m\ (W - D)}{2}} \tag{4.8}$$

where the parameter m is defined versus the absorber plate's thermal conductivity factor k and thickness δ (Figure 4.22) from the relationship [5]:

$$m = \sqrt{\frac{U_L}{k \cdot \delta}} \tag{4.9}$$

The bond conductance between the collector's pipelines and the absorber plate is calculated by the relationship [6]:

$$C_b = \frac{k_b \cdot b}{\gamma} \tag{4.10}$$

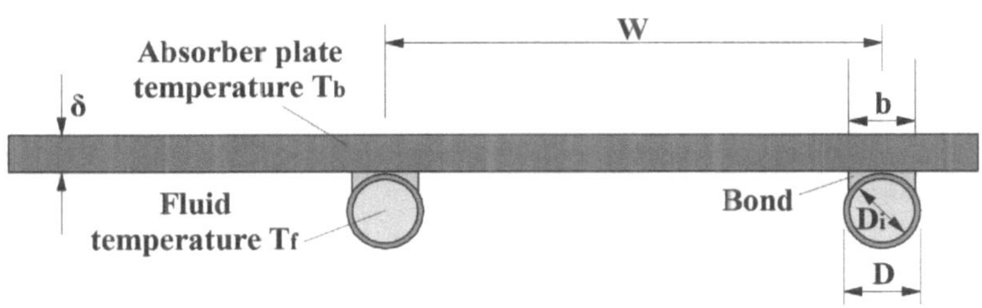

FIGURE 4.22 Inner structural layout of a flat-plate solar collector.

where:

- k_b: the thermal conductivity factor for the bond material in W/(m·K)
- b: the bond width in m
- γ: the bond thickness in m.

4.3.2 Thermal Transmittance Factor U_L

Regarding the thermal transmittance factor U_L for the thermal losses from the solar collector to the ambient, in case of uncovered solar collectors this is approached by the following empirical relationship [17]:

$$U_L = 4.15 + 2.05 \cdot V + 0.655 \cdot \overline{T}^{0.25} + 2 \cdot \sigma \cdot \overline{T}_s^3 \cdot \varepsilon + 2 \cdot \sigma \cdot \overline{T}_r^3 \cdot \varepsilon \qquad (4.11)$$

where:

- V: the average wind velocity for the under consideration time period in m/s
- $\overline{T}_r$: the average temperature difference between the collector and the ambient air in °C
- $\overline{T}_s$: the average temperature difference between the collector and the installation surface °C
- σ: the Stefan–Boltzmann constant, referring to thermal transmission with radiation, equal to $5.67 \cdot 10^{-8}$ W/(m²·K⁴)
- ε: the emissivity for the collector's pipeline material.

For flat-plate solar collectors, the overall thermal transmittance factor U_L constitutes a complex function of the constructive features and operation conditions of the collector. In general, it is given by the following relationship:

$$U_L = U_t + U_b + U_e \qquad (4.12)$$

where the subscripts t, b, and e designate thermal power losses from the top (through the transparent cover), the bottom, and the side edges of the collector, respectively.

The thermal transmittance factor from the collector's top cover is given by the following relationship [18]:

$$U_t = \frac{1}{\dfrac{N}{\dfrac{C}{T_b} \cdot \left(\dfrac{T_{pm} - T_a}{N + f}\right)^e} + \dfrac{1}{h_f}} + \frac{\sigma \cdot \left(T_{pm} + T_a\right) \cdot \left(T_{pm}^2 + T_a^2\right)}{\dfrac{1}{\varepsilon_p + 0.00591 \cdot N \cdot h_t} + \dfrac{2 \cdot N + f - 1 + 0.133 \cdot \varepsilon_p}{\varepsilon_g} - N} \qquad (4.13)$$

where:

- N: the number of the transparent protective covers of the solar collector
- h_t: the thermal transition factor referring to the heat transfer above the top side of the solar collector in W/(m²·K)
- T_{pm}: the average temperature of the collector's absorber plate in K
- T_a: the average ambient air temperature in K
- ε_g: the emissivity of the collector's transparent cover
- ε_p: the emissivity of the collector's absorber plate
- C, f, e: empirical parameters presented below.

The parameter C is given by the relationship [18]:

$$C = 520 \cdot \left(1 - 0.000051 \cdot \beta^2\right) \qquad (4.14)$$

where β is the collector's installation angle versus the horizontal plane. The above relationship is valid for inclinations from 0° to 70° with regard to the horizontal plane. If the solar collector's installation angle is higher than 70°, the value of the C parameter is kept at β = 70°.

The parameter f is given by the relationship [19]:

$$f=\left(1+0.089\cdot h_t-0.1165\cdot h_t\cdot \varepsilon_p\right)\cdot\left(1+0.07866\cdot N\right) \tag{4.15}$$

The parameter e is given by the relationship [18]:

$$e=0.430\cdot\left(1-\frac{100}{T_{pm}}\right) \tag{4.16}$$

As seen in the above presented relationship for the calculation of the thermal transmittance factor U_t, the average temperature T_{pm} of the absorber plate is involved. This temperature can be calculated by the relationship:

$$T_{pm}=T_{fi}+\frac{\dot{Q}/A_c}{F_R\cdot U_L}\cdot\left(1-F_R\right) \tag{4.17}$$

where T_{fi} is the working medium inlet temperature in the solar collector.

The thermal transmittance factors for the heat transfer from the collector's base and side edges are given by introducing the fundamental relationships for the heat transfer from solid materials to the ambient through conductivity, convection, and radiation. In general, these factors can be considered constant and equal to (the subscripts e and b in the following equation designate its applicability for the heat transfer calculation from both the collector's side edges and bottom, respectively):

$$U_{e,b}=\frac{1}{\frac{t_{e,b}}{k_{e,b}}+\frac{1}{h_{e,b}}} \tag{4.18}$$

where:

- $t_{e,b}$: the insulation layer thickness applied at the side edges and the bottom of the collectors respectively, in m
- $k_{e,b}$: the thermal conductivity factor for the insulation material applied at the collector's side edges and bottom respectively in W/(m·K)
- $h_{e,b}$: the thermal transition factor referring to the heat transfer from the collector's side edges and bottom respectively to the ambient environment in W/(m²·K).

Due to the thermal insulation applied at the solar collector's side edges and bottom, the thermal power losses from these surfaces usually remain lower than 10% of the total thermal power losses from the solar collector to the ambient environment.

From the above relationships it is seen that, for the calculation of the parameters and the magnitudes involved in the heat removal factor F_R calculation, some magnitudes that normally should be expected to be among the results of the overall calculation process, actually must be known in advance. For example, for the calculation of the final net thermal power production from the solar collector $\dot{Q}$, the average temperature T_{pm} of the absorber plate is required (involved in the F_R calculation). However in the corresponding Equation 4.17 for the calculation of the temperature T_{pm}, it is seen that the net thermal power production $\dot{Q}$ from the solar collector is introduced as a known magnitude. In these cases the calculation is approached through an iterative procedure, starting with initial assumptions for some of the involved magnitudes and verifying them at the end of each calculation loop. The analytical calculation and dimensioning procedure of a solar collectors' field will be presented in a later section.

4.3.3 Transmittance–Absorptance Product ($\tau \cdot \alpha$)

The last parameter required for the calculation of the final thermal power production from the solar collector is the product $G_t \cdot (\tau \cdot \alpha)$, which gives the final, net solar radiation absorbed by the collector in W/m^2. The term G_t refers to the total incident solar radiation on the solar collector. This term is a function of the Global Horizontal Irradiance (GHI) at the collector's installation position, measured with pyranometers, the clarity of the atmosphere, the geographical location, the surrounding surfaces of the installation position, and the orientation of the solar collector. It can be accurately calculated, following the analytical procedures and methodologies defined and based on the fundamentals of solar geometry and solar radiation [19].

The transmittance–absorptance product is considered an indivisible magnitude, referring to the combination of the collector's transparent cover and absorber plate, rather than the product of two different and independent parameters. If we consider that the collector's transparent cover exhibits a transmittance τ for a specific incident angle of solar radiation, and the absorber plate exhibits an absorptance α, then a part of the total global incident solar radiation G_t on the collector, equal to $G_t \cdot \tau$, penetrates through the transparent cover, and another part of the penetrated radiation is absorbed by the absorber plate, equal to $G_t \cdot \tau \cdot \alpha$. The amount $\tau \cdot (1 - \alpha)$ of the penetrated but not absorbed solar radiation returns back to the transparent cover. Yet, another part, equal $\tau \cdot (1 - \alpha) \cdot \rho_d$, from the solar radiation returning back from the absorber plate does not penetrate the transparent cover to escape to the ambient, but it is reflected from the cover back to the absorber plate. A part from this reflected radiation will be absorbed by the absorber plate and the rest will return back to the transparent cover, and so on. The magnitude ρ_d is the reflectance of the solar collector's transparent cover, referring to the diffused solar radiation reflected from the absorber plate towards the inner side of the cover. The above described iterative process is graphically depicted in Figure 4.23.

The consecutive solar radiation reflection from the absorber plate to the transparent cover, and vice versa, implies that the final transmittance–absorptance product is eventually configured by the following equation [20]:

$$(\tau\alpha) = \tau \cdot \alpha \cdot \sum_{n=0}^{\infty} \left[(1-\alpha) \cdot \rho_d\right]^n = \frac{\tau \cdot \alpha}{1-(1-\alpha) \cdot \rho_d} \tag{4.19}$$

The transmittance of the transparent cover and the absorptance of the absorber plate of a solar collector are not constant magnitudes. Both of them basically depend on the solar radiation incident angle, which is measured with regard to the perpendicular line on the collector's surface (Figure 4.24).

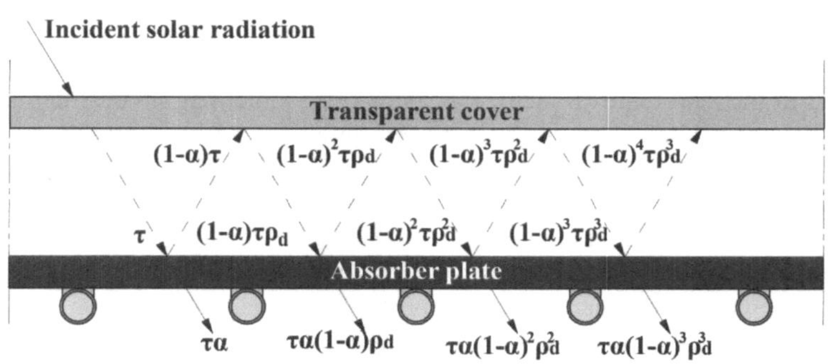

FIGURE 4.23 Solar radiation absorption from an absorber plate below a transparent cover [6].

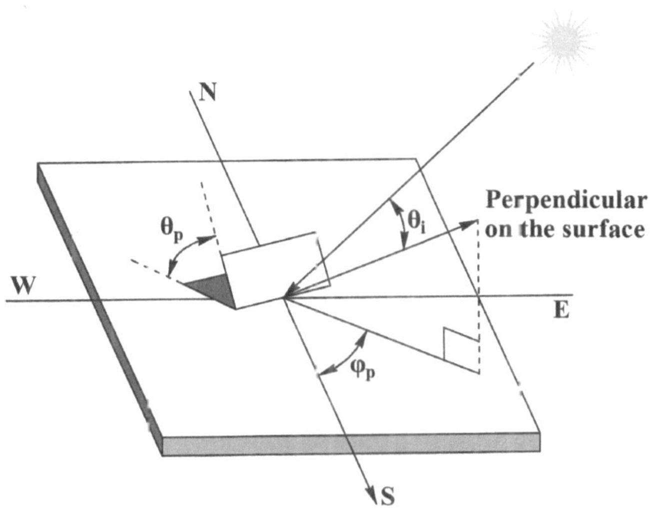

FIGURE 4.24 Incident angle θ_i of solar radiation on an inclined surface with regard to the horizontal plane.

The fluctuation of the product $(\tau\alpha)$ versus the solar radiation incident angle on a solar collector is given in Figure 4.25 [20]. Specifically, in this figure the product $(\tau\alpha)$ is levelized over the product $(\tau\alpha)_n$ which refers to perpendicular incidence of solar radiation on the collector's surface (incident angle 0°).

The transmittance and absorptance of the solar radiation for zero incident angle (perpendicular incidence) are given by the solar collector's manufacturer.

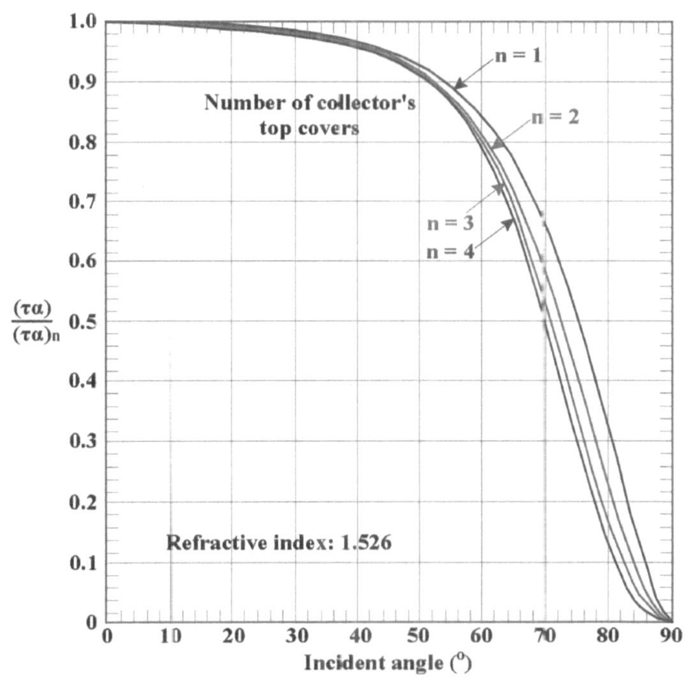

FIGURE 4.25 Fluctuation of the transmittance–absorptance product of solar collectors with different numbers of top covers, versus the solar radiation incident angle θ [20].

TABLE 4.1
Typical Refractive Indices at Solar Spectrum for Characteristic Materials Used for the Construction of Solar Collector Covers [21]

Cover Material	Refractive Index	Solar Transmittance 0.2–4 μm	Solar Transmittance 3–50 μm
Float glass	1.518	0.840	0.02
Polymethyl methacrylate	1.510	0.875	0.02
Polyvinyl fluoride	1.49	0.896	0.02
Polyfluorinated ethylene propylene	1.64–1.67	0.869	0.178
Polytetrafluoroethylene	1.586	0.726	0.02
Polycarbonate	1.46	0.922	0.207

Additionally, the curves presented in Figure 4.25 are valid for collector transparent covers of pure glass, with a refractive index at solar spectrum of 1.526. Nevertheless, it has been experimentally proved that Figure 4.25 can be used for all the materials that can be possibly used for collector transparent cover manufacturing, on the condition they exhibit a refractive index close to that of glass. Table 4.1 presents typical solar spectrum refractive indices for some characteristic materials used for the construction of solar collector covers [21].

It must be underlined that the above relationship and figure for the calculation of the transmittance–absorptance product are based on the assumption that there is no solar radiation absorptance from the transparent cover. Practically this is not absolutely valid, because there is always a small amount of solar radiation absorbed by the collector's cover. To correct this error, the magnitude of the effective transmittance–absorptance product is introduced, which will be designated with the subscript e, namely as $(\tau\alpha)_e$. The absorbed solar radiation by the collector's cover is not lost. On the contrary, it is later provided back to the absorber plate. As a result, the final effective transmittance–absorptance product is approximately 1%–2% higher than the theoretical product, as defined previously. There is an empirical methodology developed for the accurate calculation of the product $(\tau\alpha)_e$, the presentation of which is beyond the scope of this chapter. Besides, the difference in the final arithmetic result is rather low. Conclusively, for the calculation of the final thermal power production from a flat-plate solar collector, the effective transmittance–absorptance product should be used, which can be considered equal to:

$$(\tau \cdot \alpha)_e \cong 1.02 \cdot (\tau \cdot \alpha) \quad (4.20)$$

for solar collectors with glass covers (high absorptance), and equal to:

$$(\tau \cdot \alpha)_e \cong 1.01 \cdot (\tau \cdot \alpha) \quad (4.21)$$

for solar collectors with covers made of material with low absorptance [20].

The product $(\tau\alpha)_n$ and the factors F_R and U_L vary between specific limits, depending on the collector's type. Consequently, for fast, approximate calculations, averaged values can be used for these magnitudes, retrieved by statistical data from different solar collector manufacturers and solar collector types. In Table 4.2 indicative values for the products $F_R \cdot (\tau\alpha)_n$ and $F_R \cdot U_L$ are given for different solar collector types and under typical characteristic environmental conditions (i.e., mild weather conditions, medium wind velocities, global solar radiation 600–800 W/m^2).

TABLE 4.2
Indicative Values for the Parameters Involved in the Thermal Power Production Calculation Process from Solar Collectors

Type	Description	$F_R \cdot (\tau\alpha)_n$	$F_R \cdot U_L$ (W/m²·K)
I	Black color, one cover flat-plate collector	0.82	7.50
II	Black color, two covers or selective coating with one cover flat-plate collector	0.75	5.00
III	Vacuum tube solar collector	0.45	1.25
IV	Uncovered solar collector (wind speed at 2 m/s)	0.86	21.50

4.3.4 Efficiency of Flat-Plate Solar Collector

The efficiency of a solar collector is given by the ratio of the final thermal power production $\dot{Q}$ from the collector over the total incident solar radiation G_t at the collector's effective area A_c:

$$\eta = \frac{\dot{Q}}{G_t \cdot A_c} = F_R \cdot \left[(\tau \cdot \alpha)_e - \frac{U_L \cdot (T_{fi} - T_a)}{G_t} \right] \tag{4.22}$$

By definition, the efficiency of the solar collector is also given by the relationship:

$$\eta = \frac{\dot{m} \cdot c_p \cdot (T_{fo} - T_{fi})}{G_t \cdot A_c} \tag{4.23}$$

where T_{fo} and T_{fi} are the working fluid's outlet and inlet temperatures respectively from and in the solar collector.

As seen in Figure 4.25, for solar radiation incident angles below 35°, the transmittance–absorptance product remains almost constant; hence, the collector's efficiency fluctuation versus the term $(T_{fi} - T_a)/G_t$ can be considered approximately linear, as the thermal transmittance factor U_L remains constant. In Figure 4.26, a typical variation curve of a flat-plate collector's efficiency versus the term $(T_{fi} - T_a)/G_t$ is presented. The almost linear relationship of these two magnitudes is observed. The term $(T_{fi} - T_a)/G_t$ is called *heat loss parameter.*

The efficiency curve presented in Figure 4.26 constitutes the result of the linear interpolation based on real efficiency experimental measurements. Such curves are provided by certified laboratories in the frame of the certification of commercial solar collectors. The efficiencies that correspond to the executed measurements have been also depicted in the graph. The efficiency curve intersects the vertical axis at the point $F_R \cdot (\tau\alpha)_e$ for $(T_{fi} - T_a)/G_t = 0$, namely for $T_{fi} = T_a$, namely, when the working fluid's inlet temperature in the collector equals the ambient temperature. As the heat loss parameter $(T_{fi} - T_a)/G_t$ increases, practically, as the working fluid's inlet temperature T_{fi} increases with regard to the ambient temperature T_a, the collector's efficiency drops, following a linear pattern with slope $-F_R \cdot U_L$.

In practice, the collector's thermal transmittance factor U_L is not constant. On the contrary, it depends on the fluid's temperature in the collector and the ambient temperature. In general, this relationship follows a linear pattern, hence we can write:

$$F_R \cdot U_L = c_1 + c_2 \cdot (T_{fi} - T_a) \tag{4.24}$$

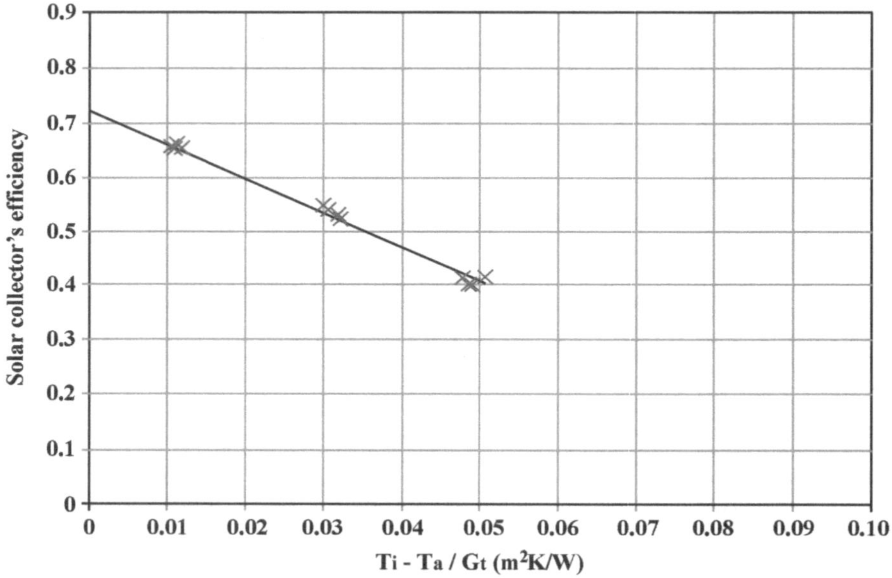

FIGURE 4.26 Variation of the collector's efficiency versus the heat loss parameter for a typical, commercial flat-plate collector.

From Equation 4.24, the relationship from Equation 4.22 can be written as:

$$\eta = F_R \cdot (\tau \cdot \alpha)_e - c_1 \cdot \frac{(T_{fi} - T_a)}{G_t} - c_2 \cdot \frac{(T_{fi} - T_a)^2}{G_t} \tag{4.25}$$

The above equations include all the parameters that can potentially affect the efficiency of a flat-plate solar collector, apart from the fluid's flow rate inside the collector's pipelines and the solar radiation incident angle. The fluid's flow rate certainly affects the collector's efficiency. Low flow rate imposes a corresponding increase on the fluid's temperature; hence, according to Equation 4.22, the collector's efficiency drops. Hence, low fluid flow rates inside the collector imply low thermal power absorptance from the absorber plate, which will cause a corresponding increase on the plate's temperature. This, in turn, will cause an increase of the thermal power losses from the collector to the environment. The reverse results will be caused by an increase of the fluid's flow rate inside the collector.

The integrated calculation of the efficiency of a solar collector requires the introduction of two more magnitudes:

- the solar radiation incidence angle modifier, with which the optical losses due to the incident angle are also introduced in the calculation
- the collector time constant, which introduces in the calculation the collector's heat capacity, determined by the materials used for the collector's manufacturing and its mass.

4.3.4.1 Incidence Angle Modifier

The equations for the collector's efficiency and the thermal power production presented above are valid on the condition of perpendicular incident solar radiation on the collector's surface (incident angle 0°). This assumption, of course, given the fact that the flat-plate solar collectors are usually installed in constant bases, practically is not valid, perhaps apart from some very short time intervals during the annual period. The solar radiation incident angle affects the reflected solar radiation on the collector's top cover, a fact which, in turn, affects the transmittance–absorptance product.

In order to approach an estimation of the effect of the solar radiation incident angle on the transmittance–absorptance product ($\tau\alpha$), the incidence angle modifier $K_{\tau\alpha}$ is defined, as the ratio of the transmittance–absorptance product $(\tau\alpha)_\theta$ for incident angle θ, over the corresponding value of the product for perpendicular solar radiation $(\tau\alpha)_n$:

$$K_{\tau\alpha}(\theta) = \frac{(\tau\alpha)_\theta}{(\tau\alpha)_n} \tag{4.26}$$

With the definition of the solar radiation incidence angle modifier, Equation 4.22 can be written as:

$$\eta = F_R \cdot \left[K_{\tau\alpha}(\theta) \cdot (\tau \cdot \alpha)_e - \frac{U_L \cdot (T_{fi} - T_a)}{G_t} \right] \tag{4.27}$$

For the calculation of the solar radiation incident angle modifier, the following empirical relationship has been introduced [22] for the case of flat-plate collectors:

$$K_{\tau\alpha}(\theta) = 1 - b_0 \cdot \left(\frac{1}{\cos(\theta)} - 1 \right) - b_1 \cdot \left(\frac{1}{\cos(\theta)} - 1 \right)^2 \tag{4.28}$$

The above relationship can be used for solar radiation incident angle θ lower than 60°. For solar collectors with one cover, it is $b_1 = 0$. The factor b_0 ranges around 0 1.

An alternative calculation relationship for the factor $K_{\tau\alpha}(\theta)$ is given by the following [23]:

$$K_{\tau\alpha}(\theta) = 1 - b_0 \cdot \left(\frac{1}{\cos(\theta)} - 1 \right)^n \tag{4.29}$$

which can be applied for incident angles higher than 60° with the appropriate definition of the factors b_0 and n. The factor b_0 again ranges around 0.1.

For the calculation of the solar radiation incidence angle modifier of flat-plate solar collectors, or the factors b_0, b_1, and n, ASHRAE suggests the execution of experimental measurements for the calculation of the $(\tau\alpha)_\theta$ product for different incident angles of the solar radiation. A characteristic variation curve of the incidence angle modifier is presented in Figure 4.27 for a typical single-cover solar collector. The curve is plotted using Equation 4.28 and setting $b_0 = 0.136$ and $b_1 = 0$.

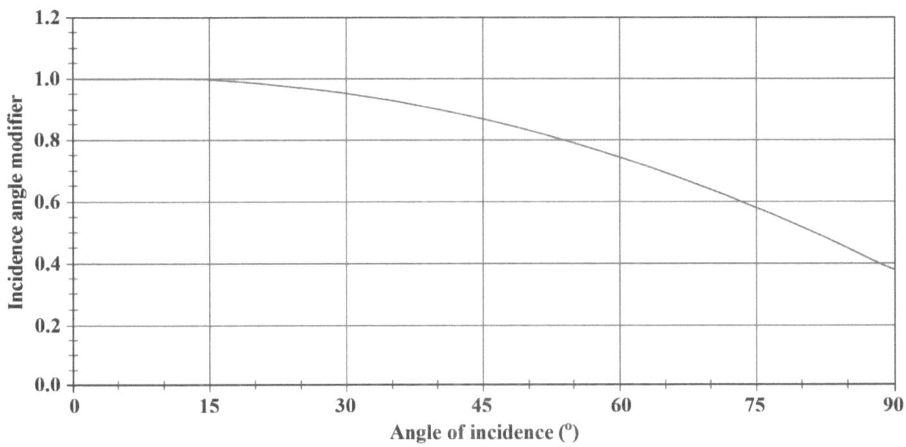

FIGURE 4.27 Variation of the incidence angle modifier versus the solar radiation incident angle.

4.3.4.2 Collector's Time Constant

The last item that affects the efficiency of a solar collector is its thermal capacity. This effect is taken into account by introducing the collector's time constant, defined as the time required for the outlet fluid's temperature to drop, after the fluid's circulation stoppage inside the collector, to a percentage equal to:

$$1-\frac{1}{e}=0.632 \tag{4.30}$$

where e = 2.72 the Euler's number.

The time constant is calculated experimentally as described below. The solar collector operates at constant conditions, preferably with the fluid's inlet temperature close to the ambient one. Suddenly the incident solar radiation is interrupted, e.g., by placing an opaque cover above the collector's top cover. With the collector's closed-loop circulator still operating, the fluid's outlet temperature is monitored. The time constant is the time required for the following relationship to be satisfied:

$$\frac{T_{o,t}-T_{fi}}{T_{o,init}-T_{fi}}=\frac{1}{e}=0.368 \tag{4.31}$$

where:

$T_{o,t}$: the fluid's outlet temperature from the collector after time t from the blocking of the incident solar radiation

$T_{o,init}$: the fluid's outlet temperature from the collector precisely when the incident solar radiation is blocked

T_{fi}: the fluid's inlet temperature in the solar collector.

In Figure 4.28 a typical variation curve of the fluid's outlet temperature from the solar collector is presented during the execution of the time constant measurement experiment, with air employed as the working medium and with a double top cover [24].

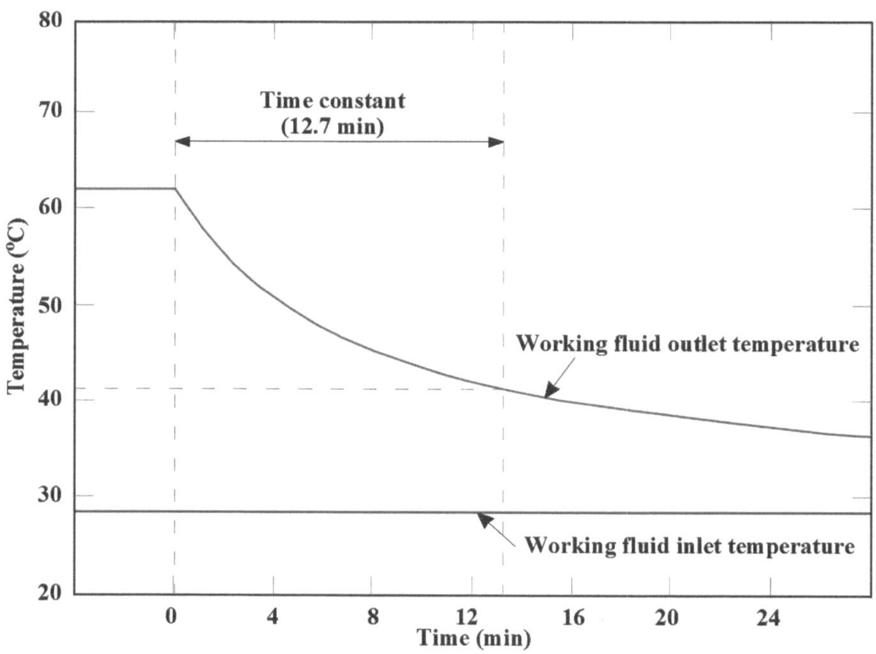

FIGURE 4.28 Fluid's outlet temperature variation from a solar collector after the abrupt stoppage of the incident solar radiation (with due diligence of US Department of Energy) [24].

For a fluid's inlet temperature at 28°C and initial outlet temperature from the solar collector at 63°C, the collector's time constant will be the time required after the interruption of the incident solar radiation, for the fluid's outlet temperature to drop at:

$$\frac{T_{o,t} - T_{fi}}{T_{o,init} - T_{fi}} = 0.368 \Leftrightarrow T_{o,t} = \left(T_{o,init} - T_{fi}\right) \cdot 0.368 + T_{fi} \Rightarrow T_{o,t} = (63 - 28) \cdot 0.368 + 28 \Leftrightarrow T_{o,t} = 40.88°C$$

In this case, the time constant is calculated at 12.7 min from Figure 4.28.

4.3.5 Calculation Procedure of the Thermal Power Production from Flat-Plate Collectors

In this section, we describe the calculation procedure of the thermal power production from a flat-plate solar collector, based on the equations and graphs presented in the earlier sections. The current section aims to unify all the previously provided information, in order to distinguish the required data, the assumptions, and the expected results from the thermal power production calculation procedure from flat-plate solar collectors, which is analytically presented step-by-step below.

In an actual problem regarding the calculation of the thermal power production from flat-plate collectors for a specific time period (e.g., a day, a month or a year), the total time period is discretized in a number of calculation time steps with certain duration, e.g., hourly. For every calculation time step the following methodology is executed.

1. *Introduction of fundamental calculation data*
   According to Equation 4.1, the fundamental data required for the calculation of the thermal power production from a flat-plate solar collector are:
   - The total installed solar collector's effective area A_c in m^2. The term *effective area* is used to describe the solar collector's net surface interacting with the incident solar radiation and not the total one, which also includes the collector's frame. The collector's effective area coincides with the absorber's plate surface.
   - The total incident solar radiation G_t σε W/m^2, which is given by the sum of the direct, the diffused, and the reflected from the neighboring surfaces solar radiation. The total incident solar radiation is calculated according to specific methodology, based on the fundamentals of solar geometry and radiation [19].
   - The ambient temperature T_a, which must be available in the form of annual time series with the above mentioned time discrimination, based on available measurements from meteorological stations installed in the under consideration geographical area.
   - The specific heat capacity of the working fluid c_p in kJ/(kg·K). In case of a water–glycol solution, the specific heat capacity can be practically assumed equal to the one of pure water, namely 4.184 kJ/(kg·K). In case of air, the specific heat capacity under constant pressure is used, namely $c_p = 1.005$ kJ/(kg·K).
   - Constructive features and technical specifications of the collector's materials, such as:
     - the number N of the collector's protective transparent top covers
     - the collector's insulation thickness t_e and t_b in the side edges and in the bottom of the solar collector respectively, in m
     - the distance W in m between two consecutive pipelines of the solar collector (see Figure 4.22)
     - the collector's pipeline outer diameter D in m
     - the collector's pipeline inner diameter D_i in m
     - the thermal conductivity factors of the collector's insulation k_e and k_b for the side edges and for the bottom of the collector respectively, in W/(m·K)
     - the emissivity ε_g of the collector's transparent cover

- the emissivity ε_p of the collector's absorber plate
- the thermal conductivity factor k of the absorber plate in W/(m·K)
- the thickness δ of the absorber plate in m
- the thermal conductivity factor k_b of the bond's material of the collector's pipelines and the absorber plate in W/(m·K)
- the bond's thickness b of the collector's pipelines and the absorber plate in m
- the bond's thickness γ of the collector's pipelines and the absorber plate in m.

• Data regarding the particular collector's installation, such as:
- the collector's installation angle β in degrees
- the thermal convection factor h_{fi} for the heat transfer from the working fluid to the collector's pipelines in W/(m²·K), retrieved by tables regarding the heat transfer with convection for water or air flows inside cylindrical tubes, versus the flow's velocity
- the thermal transition factors h_e and h_b for the heat transfer from the collector's side edges and bottom respectively to the ambient, in W/(m²·K)
- the mass flow rate $\dot{m}$ of the working fluid inside the collector in kg/s.

2. *Assumptions*

For the execution of the solar collector's thermal power production calculation procedure, an initial assumption of the absorber's plate average temperature T_{pm} is required. For flat-plate collectors with liquid working fluid and a mass flow rate at 0.01 – 0.02 kg/(m²·s), a sensible assumption for the absorber's plate average temperature is:

$$T_{pm} = T_{fi} + 10°C \tag{4.32}$$

where T_{fi} is the fluid's inlet temperature in the solar collector. In case of a solar collector with air employed as working fluid, a logical approach is:

$$T_{pm} = T_{fi} + 20°C \tag{4.33}$$

The fluid's inlet temperature in the solar collector T_{fi} depends on the operation mode of the collector. In case of an open loop, where the collectors are constantly provided with water pumped from a water tank (e.g., a thermal tank), the inlet temperature T_{fi} will be equal to the temperature of the water stored in the tank. In case of a closed loop, where the produced thermal power is delivered from the primary hydraulic loop to the thermal storage tank through a heat exchanger (Figure 4.29), the fluid's inlet temperature in the solar collectors will be equal to the return temperature of the working fluid in the primary loop.

In this case, if $\dot{Q}_{tot}$ is the total delivered thermal power from the primary solar collectors' closed loop to the thermal tank's heat exchanger, including also the heat losses during the fluid's flow inside the pipelines, T_{fo} is the outlet temperature of the working fluid from the solar collectors and $\dot{m}$ the mass flow rate of the working fluid in the primary closed loop, then the following relationship stands:

$$\dot{Q}_{tot} = \dot{m} \cdot c_p \cdot (T_{fo} - T_{fi}) \tag{4.34}$$

More on the calculation of the working fluid's return temperature will be presented in the following section, which is dedicated to thermal energy storage.

3. *Calculation of thermal transmittance factor U_L*

The total thermal transmittance factor U_L is given by Equation 4.12:

$$U_L = U_t + U_b + U_e$$

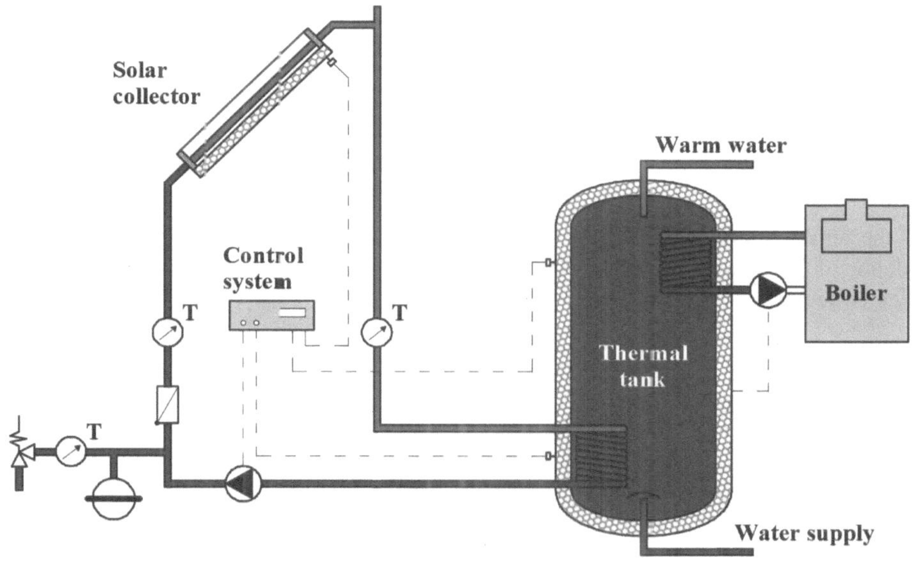

FIGURE 4.29 Hydraulic connection diagram of solar collectors with a thermal storage tank, through an independent closed hydraulic loop.

The thermal transmittance factors U_b and U_e from the collector's bottom and side edges respectively are given by Equation 4.18:

$$U_{e,b} = \frac{1}{\frac{l_{e,b}}{k_{e,b}} + \frac{1}{h_{e,b}}}$$

The thermal transmittance factor U_t from the top of the collector is given by Equation 4.13:

$$U_t = \frac{1}{\frac{N}{\frac{C}{T_b}\cdot\left(\frac{T_{pm}-T_a}{N+f}\right)^e} + \frac{1}{h_f}} + \frac{\sigma\cdot\left(T_{pm}+T_a\right)\cdot\left(T_{pm}^2+T_a^2\right)}{\frac{1}{\varepsilon_p + 0.00591\cdot N\cdot h_t} + \frac{2\cdot N + f - 1 + 0.133\cdot\varepsilon_p}{\varepsilon_g} - N}$$

In the above equation:

- The parameter C is calculated with Equation 4.14 versus the collector's installation angle β.
- The parameter f is given by Equation 4 15 versus the number N of the collector's transparent top covers, the absorber's plate emissivity ε_p, and the thermal transition factor h_t referring to the heat transfer from the collector to the ambient.
- The parameter e is given by Equation 4.16 versus the average temperature of the absorber plate T_{pm}.
- The parameter σ is the Stefan–Boltzmann constant, equal to $5.67\cdot10^{-8}$ W/(m$^2\cdot$K^4).

4. *Calculation of the heat removal factor F_R*
   The heat removal factor F_R is calculated from Equation 4.5:

$$F_R = \frac{\dot{m} \cdot c_p}{A_c \cdot U_L} \cdot \left[1 - \exp\left(-\frac{A_c \cdot U_L \cdot F'}{\dot{m} \cdot c_p}\right)\right]$$

   In the above relationship, all the involved magnitudes are known. The parameter F′ is calculated from Equation 4.6 for uncovered solar collectors and from Equation 4.7 for flat-plate collectors:

$$F' = \frac{\frac{1}{U_L}}{W \cdot \left[\frac{1}{U_L \cdot [D + (W - D) \cdot F]} + \frac{1}{C_b} + \frac{1}{\pi \cdot D_i \cdot h_{fi}}\right]}$$

   In the above relationship, the standard fin efficiency F is calculated by Equation 4.8, versus the collector's dimensions W and D and a new parameter m. This new parameter m is given by Equation 4.9, versus the total heat transmittance factor U_L, the absorber's plate thermal conductivity factor k, and thickness δ. The bond conductance C_b is calculated by Equation 4.10 versus the thermal conductivity factor k_b of the bond's material, the bond's width b, and thickness γ.
5. *Calculation of the transmittance – absorptance product*
   The transmittance–absorptance product should be calculated for every calculation time step versus the solar radiation incident angle. For this purpose, the variation of the product versus the incident angle should be known, which is provided by the collector's manufacturer in the form of a diagram or table sheet (see Figure 4.22). The transmittance–absorptance product $(\tau\alpha)_n$ for perpendicular incidence of the solar radiation on the solar collector must also be known.

   For higher accuracy, the effective transmittance–absorptance product can eventually be used, which is calculated, versus the type of the collector's top cover, from Equation 4.20 or 4.21.
6. *Calculation of the thermal power production from the solar collector*
   The thermal power production from the solar collector is eventually given by Equation 4.4. All the involved magnitudes are known, either by assumptions or by calculations. Consequently, the calculation of the thermal power production from the solar collector is now possible.
7. *Confirmation of the initial assumption*
   Having calculated the thermal power production from the solar collectors, the initial assumption for the absorber's plate average temperature T_{pm} must be confirmed through Equation 4.17. If there is significant divergence between the initial assumption for the temperature T_{pm} and the final calculated value, then Steps 3–6 are executed again by introducing the new calculated temperature T_{pm}. The iterative procedure is repeated again, until there is convergence between the initial and the final calculated values for the temperature T_{pm}.

   The above presented calculation procedure is graphically presented in Figure 4.30.

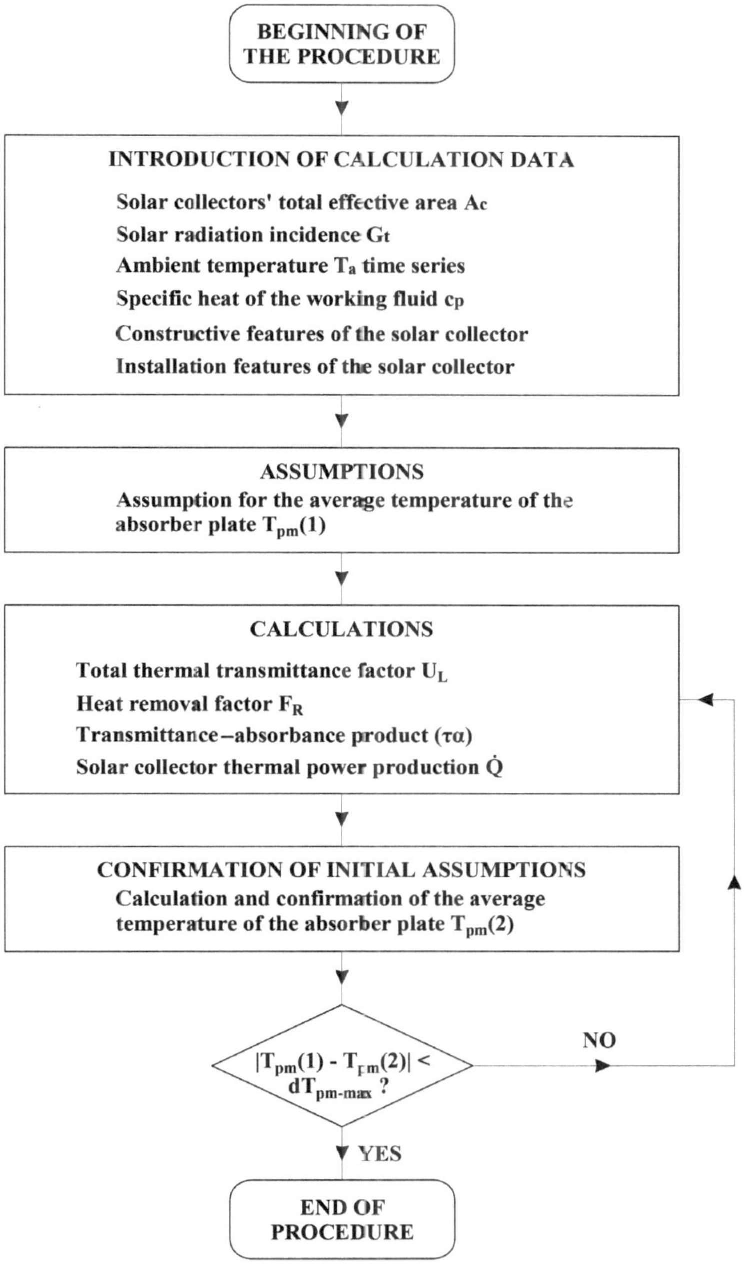

FIGURE 4.30 Flow diagram of the calculation procedure for the thermal power production from a flat-plate solar collector.

Example 4.1 Calculation of the Thermal Power Production from a Flat-Plate Solar Collector

Calculate the thermal power production and the efficiency of a flat-plate solar collector with the technical features and for the operating conditions presented below:

Total installed effective area A_c:	2 m^2
Total incident solar radiation G_t:	700 W/m^2
Ambient temperature T_a:	22°C
Working fluid:	water–glycol solution
Working medium supply:	water supply network
Number N of the collector's top covers:	1
Insulation thickness t_e and t_b for the collector's side edges and bottom respectively:	0.0075 m
Thermal conductivity factors k_e and k_b for the collector's insulation layer at the side edges and the bottom respectively:	0.025 W/(m·K)
Distance W between two consecutive pipelines on the absorber plate:	0.116 m
Outer diameter D of the collector's pipelines:	0.011 m
Inner diameter D_i of the collector's pipelines:	0.010 m
Emissivity ε_p of the collector's absorber plate:	0.95
Emissivity ε_g of the collector's top cover:	0.90
Thermal conductivity factor k of the absorber plate:	350 W/(m·K)
Thickness δ of the absorber plate:	0.0005 m
Thermal conductivity factor k_b of the bond between the collector's pipelines and the absorber plate:	1,000 W/(m·K)
Width b of the bond between the collector's pipelines and the absorber plate:	0.120 m
Thickness γ of the bond between the collector's pipelines and the absorber plate:	0.060 m
Collector's installation angle β:	35°
Thermal convection factor h_{fi} from the working fluid towards the collector's pipelines:	3,000 W/(m^2·K)
Thermal transition factors h_t, h_e, and h_b for the heat transfer from the collector's top, side edges, and bottom respectively to the ambient:	10 W/(m^2·K)
Mass flow rate of the working medium inside the collector's pipelines:	0.025 kg/(s·m^2)
Refractive index of the collector's top cover:	1.526
Top cover's transmittance for perpendicular solar radiation incidence:	0.94
Absorber plate's absorptance for perpendicular solar radiation incidence:	0.90
Solar radiation incident angle:	20°

SOLUTION

Following the methodology described above, we have:

1. *Initial assumption for the absorber plate's average temperature*
   Given that the fluid inlet temperature T_{fi} in the collector coincides with the temperature of the water supply network, which can be assumed at 16°C, the initial assumption for the average temperature of the absorber plate, given by Equation 4.32, is:

$$T_{pm} = T_{fi} + 10°C \Rightarrow T_{pm} = 16 + 10°C \Leftrightarrow T_{pm} = 26°C = 299 \text{ K}$$

2. *Calculation of the thermal transmittance factor U_L*
   The total thermal transmittance factor U_L is given by the relationship:

$$U_L = U_t + U_b + U_e$$

The thermal transmittance factors U_b and U_e from the collector's bottom and side edges respectively are calculated by the relationship:

$$U_{e,b} = \frac{1}{\frac{t_{e,b}}{k_{e,b}} + \frac{1}{h_{e,b}}} \Rightarrow \frac{1}{\frac{0.0075m}{0.025W/(m\cdot K)} + \frac{1}{10W/(m^2\cdot K)}} \Leftrightarrow U_{e,b} = 2.500\ W/(m^2\cdot K)$$

The thermal transmittance factor U_t from the collector's top is given by Equation 4.13, where:

$$C = 520\cdot(1-0.000051\cdot \beta^2) \Rightarrow C = 520\cdot(1-0.000051\cdot 35^2) \Leftrightarrow C = 487.513 \quad \text{(Equation 4.14)}$$

$$f = (1+0.089\cdot h_t - 0.1166\cdot h_t\cdot \varepsilon_p)\cdot(1+0.07866\cdot N) \Rightarrow$$
$$f = (1+0.089\cdot 10 - 0.1166\cdot 10\cdot 0.95)\cdot(1+0.07866\cdot 1) \Leftrightarrow f = 0.844 \quad \text{(Equation 4.15)}$$

$$e = 0.430\cdot\left(1-\frac{100}{T_{pm}}\right) \Rightarrow e = 0.430\cdot\left(1-\frac{100}{299}\right) \Leftrightarrow e = 0.286 \quad \text{(Equation 4.16)}$$

The thermal transmittance factor from the collector's top is calculated with Equation 4.13:

$$U_t = \frac{1}{\frac{N}{\frac{C}{T_b}\cdot\left(\frac{T_{pm}-T_a}{N+f}\right)^e} + \frac{1}{h_f}} + \frac{\sigma\cdot(T_{pm}+T_a)\cdot(T_{pm}^2+T_a^2)}{\frac{1}{\varepsilon_p + 0.00591\cdot N\cdot h_t} + \frac{2\cdot N + f - 1 + 0.133\cdot\varepsilon_p}{\varepsilon_g} - N} \Rightarrow$$

$$U_t = \frac{1}{\frac{1}{\frac{487.513}{299}\cdot\left(\frac{299-293}{1+0.844}\right)^{0.286}} + \frac{1}{10}} + \frac{5.67\cdot 10^{-8}\cdot(299+293)\cdot(299^2+293^2)}{\frac{1}{0.95+0.00591\cdot 1\cdot 10} + \frac{2\cdot 1 + 0.844 - 1 + 0.133\cdot 0.95}{0.88} - 1} \Leftrightarrow$$

$$U_t = 4.559\ W/(m^2\cdot K)$$

Finally, the overall thermal transmittance factor is given by Equation 4.12:

$$U_L = U_t + U_b + U_e \Rightarrow U_L = 4.599 + 2.500 + 2.500 \Leftrightarrow U_L = 9.599\ W/(m^2\cdot K)$$

3. *Calculation of the heat removal factor F_R*
The parameter F′ is calculated by Equation 4.7 for flat-plate collectors:

$$F' = \frac{\frac{1}{U_L}}{W\cdot\left[\frac{1}{U_L\cdot[D+(W-D)\cdot F]} + \frac{1}{C_b} + \frac{1}{\pi\cdot D_i\cdot h_{fi}}\right]}$$

where the standard fin efficiency F is given by the relationship:

$$F = \frac{\tanh\left[\frac{m\cdot(W-D)}{2}\right]}{\frac{m\cdot(W-D)}{2}} \quad \text{(Equation 4.8)}$$

with the parameter m being defined versus the absorber plate's thermal conductivity factor k and thickness δ from the relationship:

$$m = \sqrt{\frac{U_L}{k \cdot \delta}} \Rightarrow m = \sqrt{\frac{9.559}{350 \cdot 0.0005}} \Leftrightarrow m = 23.371 \frac{1}{m} \qquad \text{(Equation 4.9)}$$

The standard fin efficiency F can be now calculated with the relationship:

$$F = \frac{\tanh\left[\frac{m \cdot (W-D)}{2}\right]}{\frac{m \cdot (W-D)}{2}} \Rightarrow F = \frac{\tanh\left[\frac{23.371 \cdot (0.116-0.011)}{2}\right]}{\frac{23.371 \cdot (0.116-0.011)}{2}} \Leftrightarrow F = 0.686 \qquad \text{(Equation 4.8)}$$

The conductance of the pipeline's bond with the absorber plate is calculated with the relationship:

$$C_b = \frac{k_b \cdot b}{\gamma} \Rightarrow C_b = \frac{1{,}000 \cdot 0.120}{0.060} \Leftrightarrow C_b = 2{,}000 \frac{W}{m \cdot K} \qquad \text{(Equation 4.10)}$$

The parameter F′ can now be calculated with Equation 4.7:

$$F' = \frac{\frac{1}{U_L}}{W \cdot \left[\frac{1}{U_L \cdot [D + (W-D) \cdot F]} + \frac{1}{C_b} + \frac{1}{\pi \cdot D_i \cdot h_{fi}}\right]} \Rightarrow$$

$$F' = \frac{\frac{1}{9.559}}{0.116 \cdot \left[\frac{1}{9.559 \cdot [0.011 + (0.116-0.011) \cdot 0.686]} + \frac{1}{2{,}000} + \frac{1}{\pi \cdot 0.010 \cdot 3{,}000}\right]} \Leftrightarrow F' = 0.710$$

Finally, the calculation of the heat removal factor F_R is eventually possible. The fluid's total mass flow rate for a collector's effective surface of 2 m^2 is calculated at 0.05 kg/s.

$$F_R = \frac{\dot{m} \cdot c_p}{A_c \cdot U_L} \cdot \left[1 - \exp\left(-\frac{A_c \cdot U_L \cdot F'}{\dot{m} \cdot c_p}\right)\right] \Rightarrow$$

$$F_R = \frac{0.05 \cdot 4.184}{2 \cdot 9.559} \cdot \left[1 - \exp\left(-\frac{2 \cdot 9.559 \cdot 0.710}{0.05 \cdot 4.184}\right)\right] \Leftrightarrow F_R = 0.687 \qquad \text{(Equation 4.5)}$$

4. *Calculation of the transmittance–absorptance product (τα)*
According to the provided data, for perpendicular solar radiation, the top cover's transmittance is given equal to 0.94 and the absorber plate's absorptance is given equal to 0.90. Consequently, the transmittance–absorptance product for perpendicular solar radiation is calculated equal to:

$$(\tau\alpha)_n = 0.94 \cdot 0.90 = 0.846.$$

Given the top cover's refractive index (1.526), from Figure 4.25, for solar radiation incident angle 20° and one top cover, the ratio $(\tau\alpha)/(\tau\alpha)_n$ is found equal to 0.99. Hence, the product (τα) for 20° incident angle of solar radiation is calculated:

$$(\tau\alpha) = (\tau\alpha)/(\tau\alpha)_n \cdot (\tau\alpha)_n = 0.99 \cdot 0.846 = 0.838.$$

TABLE 4.3
Major Results of the Iterative Procedure for the Calculation of the Thermal Power Production from the Flat-Plate Solar Collector

	First Loop	Second Loop	Third Loop	Fourth Loop
U_t (W/m²·K)	4.559	5.156	5.139	5.140
U_e (W/m²·K)	2.500	2.500	2.500	2.500
U_b (W/m²·K)	2.500	2.500	2.500	2.500
U_L (W/m²·K)	9.559	10.156	10.139	10.140
F_R	0.687	0.675	0.676	0.676
Q (W)	874.194	862.478	862.806	862.797
T_{pm} (°C)	36.834	36.420	36.431	36.431

Finally, in order to take into account the effect of the solar radiation absorptance from the collector's top cover, the transmittance–absorptance effective product is finally calculated for a single top cover from the relationship:

$$(\tau\cdot\alpha)_e \cong 1.02\cdot(\tau\cdot\alpha) \Rightarrow (\tau\cdot\alpha)_e \cong 1.02\cdot 0.838 \Leftrightarrow (\tau\cdot\alpha)_e \cong 0.854 \quad \text{(Equation 4.20)}$$

5. *Thermal power production calculation from the solar collector*
   Calculation of the thermal power production from the solar collector is now possible:

$$\dot{Q} = A_c\cdot F_R\cdot\left[G_t\cdot(\tau\cdot\alpha) - U_L\cdot(T_{fi} - T_a)\right] \Rightarrow \dot{Q} = 2\cdot 0{,}687\cdot\left[700\cdot 0.854 - 9.559\cdot(16-22)\right] \Leftrightarrow \dot{Q} = 874.19\text{W}$$

6. *Verification of the initial assumption on the absorber plate's average temperature*
   After the calculation of the thermal power production from the solar collector, the initial assumed average temperature of the absorber plate must be verified from the relationship:

$$T_{pm} = T_{fi} + \frac{\dot{Q}/A_c}{F_R\cdot U_L}\cdot(1-F_R) \Rightarrow T_{pm} = 16 + \frac{874.19/2}{0.687\cdot 9.559}\cdot(1-0.687) \Leftrightarrow T_{pm} = 36.83°\text{C}$$

7. Re-execution of the calculation procedure
   It is seen that there is significant divergence between the initial assumption and the calculated value of the absorber plate's average temperature. Consequently, the calculation procedure should be executed again, by introducing, this time, the new calculated value of the plate's average temperature. The results of the iterative calculation procedure are presented in Table 4.3.
   It is seen that in the fourth loop the calculated average temperature of the absorber plate matches the value from the previous loop, while the difference of the calculated thermal power production between the third and the fourth loop is lower than 0.01 W.
8. *Solar collector's efficiency*
   The solar collector's efficiency can now be calculated with the relationship:

$$\eta = \frac{\dot{Q}}{G_t\cdot A_c} \Rightarrow \eta = \frac{862.797}{700\cdot 2} \Leftrightarrow \eta = 61.63\%$$

4.3.6 Operation Features of Flat-Plate Solar Collectors

Solar collectors are tested in certified laboratories and institutes, to confirm their performance under specific operation conditions. The evaluation processes are defined in specific international certification standards (e.g., EN 12975, EN ISO 9806). One of the essential evaluation procedures

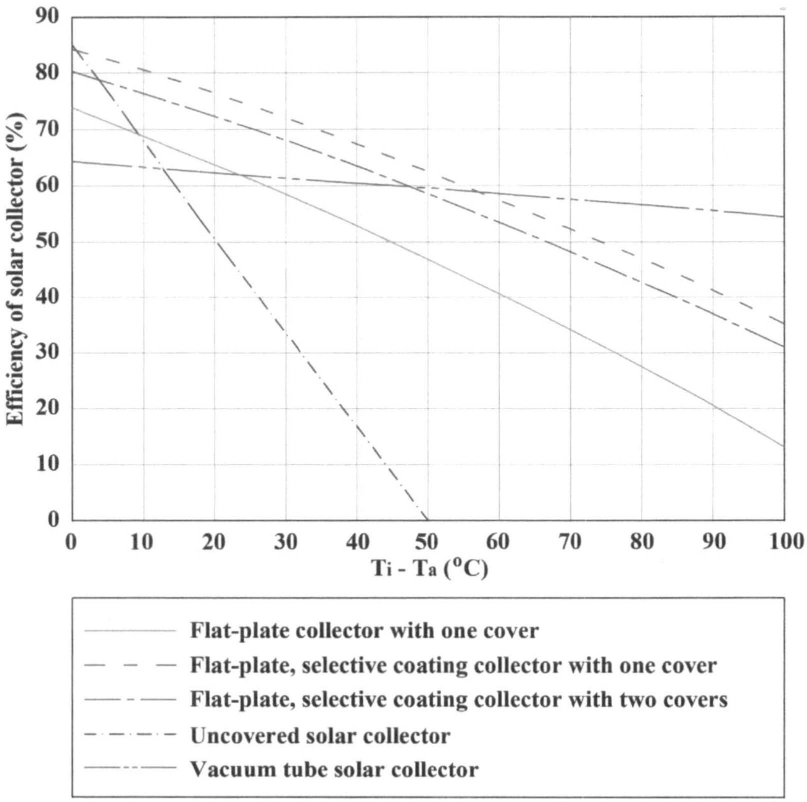

FIGURE 4.31 Efficiency variation curves for different types of solar collectors versus the inlet fluid and the ambient environment temperature difference.

is the measurement of the collector's instant efficiency under different conditions regarding the incident solar radiation, the ambient temperature, and the inlet fluid temperature in the collector. For this purpose, the incident solar radiation and the thermal power received by the working fluid inside the collector's pipelines are continuously inspected and measured under constant (or almost constant) operation conditions. This procedure is automatically executed and controlled by an electronic measurement recording system.

Figure 4.31 shows the efficiency variation curves for different types of solar collectors versus the temperature difference between the inlet fluid in the collector and the ambient environment, as retrieved from the corresponding certification procedures of the specific solar collector types [25]. The relatively confined operation range of the uncovered solar collectors is observed, as well as the almost constant efficiency of the vacuum tube solar collectors, regardless of the previously mentioned temperature difference.

Further tests are executed with regard to the measurement of the heat losses from the collector to the ambient, the estimation of the incident solar radiation angle impact on the thermal power production from the collector, the estimation of the collector's time constant, etc.

Figure 4.31 also demonstrates that as the temperature difference between the inlet fluid and the ambient environment increases, the collector's efficiency decreases and vice versa. This observation constitutes one of the most characteristic operational features of solar collectors. Thus, in a solar collector's closed hydraulic circuit, the collector's efficiency will be high during the first hours after dawn, with the beginning of the fluid's circulation inside the closed hydraulic circuit, for as long as its temperature remains close to the ambient one. Contrarily, with the passage of time and the

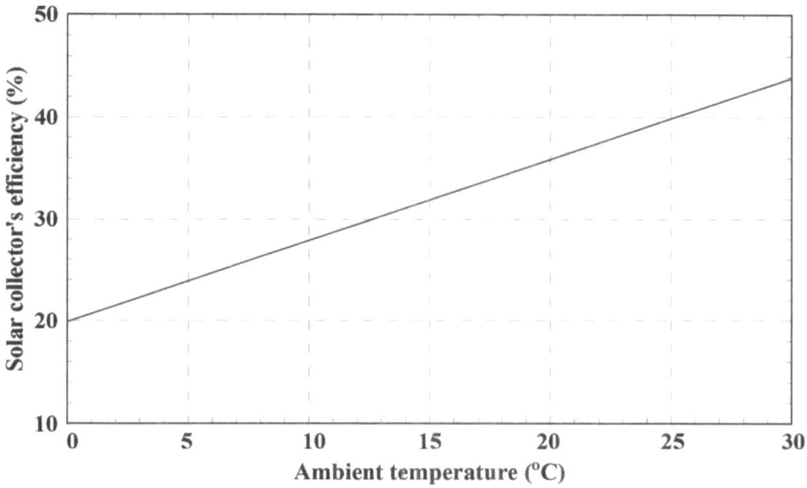

FIGURE 4.32 Fluctuation of the efficiency of a solar collector versus the ambient temperature [21].

increase of the working fluid's temperature inside the closed circuit, consequently with the concurrent increase of the temperature difference between the working fluid and the ambient too, the collector's efficiency drops.

A direct consequence of the above is the increase of the collector's efficiency with the ambient temperature, because higher ambient temperature T_a implies reduction of the temperature difference $T_i - T_a$, as the working fluid's temperature T_i increases (Figure 4.32) [21, 26, 27].

Another direct consequence of the above operational feature is the solar collector's efficiency increase with the increase of the working fluid's flow rate inside the collector's pipelines (Figure 4.33) [21]. This arises from the fact that increased fluid flow rate within the collector will lead to a reduction of the achieved fluid's outlet temperature T_o, so, in a closed hydraulic circuit, the fluid's inlet temperature will be respectively reduced too.

Solar collector efficiency drops as the wind velocity increases, because this implies higher thermal transition factor for the heat transfer from the collector to the ambient and, hence, higher

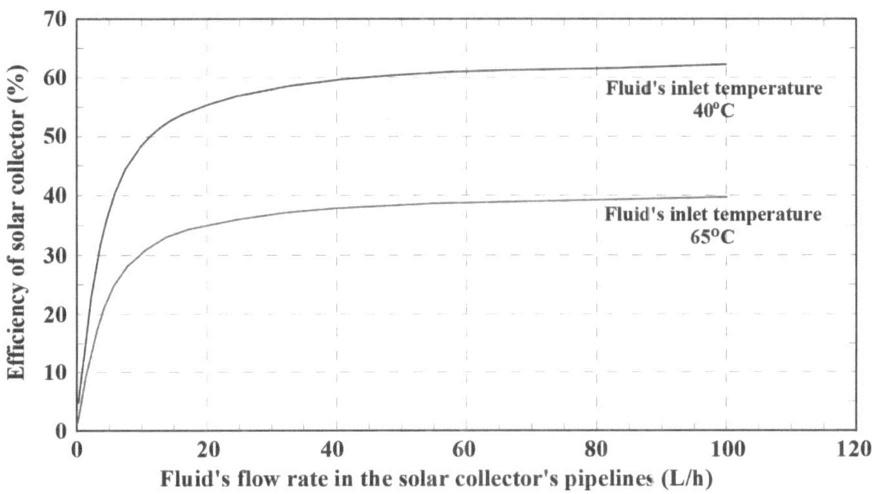

FIGURE 4.33 Fluctuation of solar collector's efficiency versus the fluid's flow rate inside the collector [21].

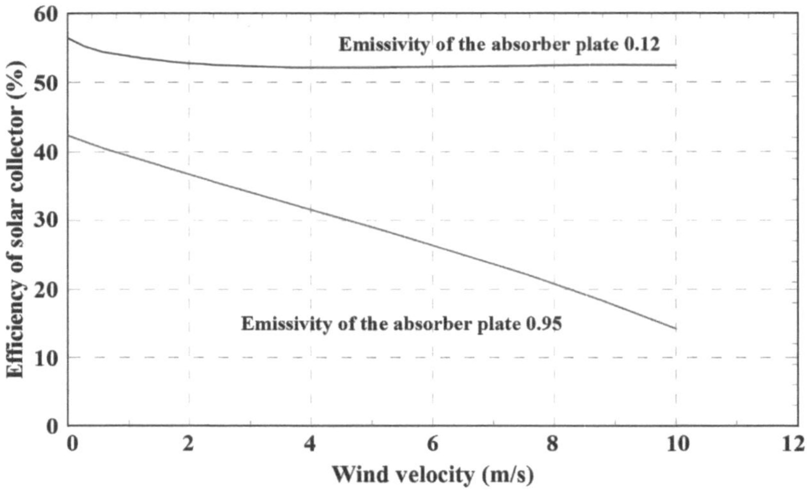

FIGURE 4.34 Fluctuation of the solar collector efficiency versus the wind velocity [21].

heat losses. This effect is more intensive for absorber plates with high emissivity. For absorber plates with low emissivity (at the range of 0.10), the wind velocity effect on the collector's efficiency is restricted (Figure 4.34) [21].

The effect of solar radiation is also important on solar collectors' efficiency (Figure 4.35) [21]. With the increase of the incident solar radiation, the collector's efficiency also increases following an almost exponential pattern, while, for incident solar radiation below a lower threshold, the collector's efficiency is null.

4.3.7 Optimum Installation Angle

Beyond the above presented parameters, one of the most essential issues, which maybe more than any other involved factor affects the thermal power production from a solar collector, is the solar collector's installation angle with regard to the horizontal plane.

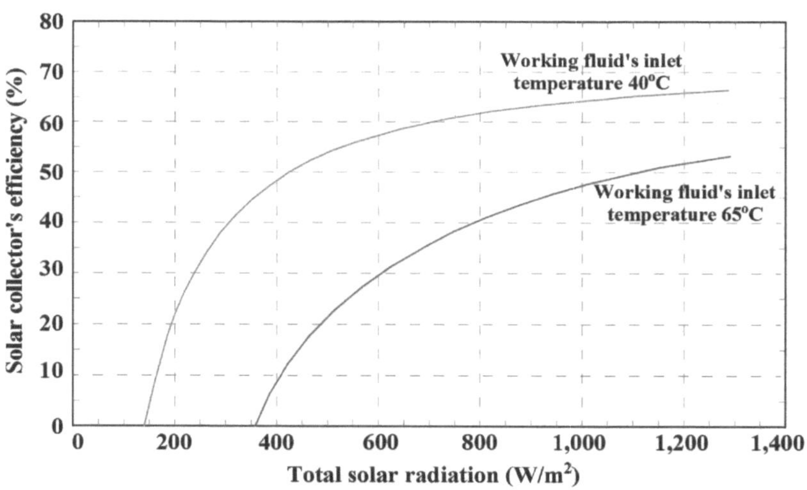

FIGURE 4.35 Fluctuation of solar collector's efficiency versus the incident solar radiation [21].

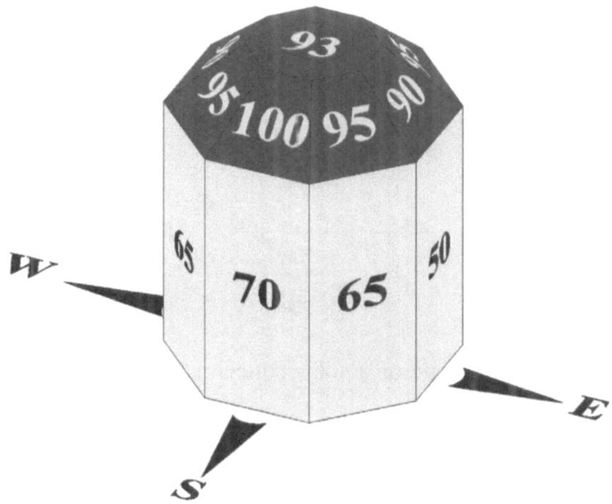

FIGURE 4.36 Fluctuation of the solar radiation incidence on a flat surface versus its orientation and installation angle (with due diligence of Centre of Renewable Energy Sources of Greece).

Of course, before the installation angle, another crucial parameter to be considered is the solar collector's orientation. It is obvious that solar collectors should be oriented towards the south, for geographical locations on the north hemisphere, namely the collector's top surface azimuth should be 0°. Likewise, for locations on the south hemisphere, solar collectors should be oriented towards the north. Any other installation orientation imposes a relevant reduction on the incident solar radiation, which becomes higher as the collector's orientation diverges from the south or north direction. Indicatively, in Figure 4.36, the solar radiation incidence reduction for different installation orientations and inclinations is presented as a percentage over the maximum possible solar radiation incidence achieved for south orientation and optimum installation angle (percentage 100%).

With regard to the installation angle versus the horizontal plane, the optimum options are not certain or obvious beforehand. The optimum inclination of a solar collector depends on its usage, which usually determines the necessity for the maximization of the solar radiation incidence during specific seasons throughout the year. For example, the installation of a solar collector field for warm water production in a tourist hotel operating exclusively during the summer period, implies that the solar radiation incidence should be maximized during summer, namely when the solar elevation angles are high. Given that the solar radiation is maximized for perpendicular incidence, for high solar elevation angles at the range of 70°–80°, the installation angle should be between 10°–20° versus the horizontal plane. On the contrary, for a solar collector field installed for the heating of a building's indoor space, the solar radiation incidence should be maximized during the winter season, when the solar elevation angles are low (at the range of 30°–40° for geographical latitudes between 30°–40°). Consequently, to approach again perpendicular incident solar radiation during the winter season, the installation angle should be raised. The above conclusions are graphically summarized in Figure 4.37.

In Figure 4.38, the effect of a solar collector's installation angle on its efficiency is presented [21]. The figure refers to a geographical area with 35° geographical latitude. It is observed that during the summer solstice (high solar elevation angle), the collector's efficiency is maximized for low installation angles (at the range of 20°), while during the vernal equinox, the collector's efficiency is maximized for installation angles at the range of 40°.

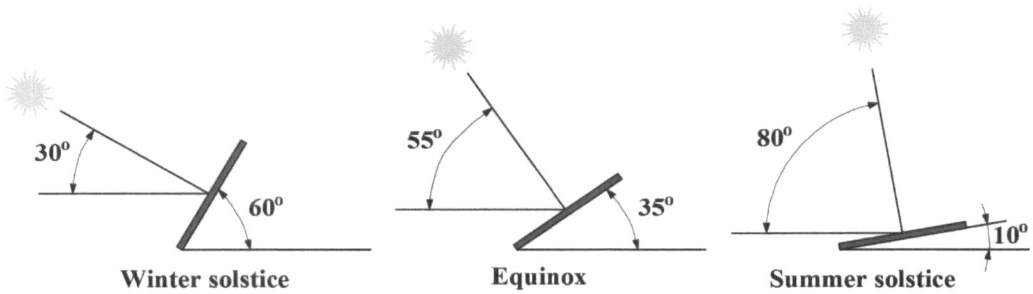

FIGURE 4.37 Optimum installation angle of a solar collector versus the horizontal plane aiming to the maximization of the solar radiation incidence.

In Table 4.4 the thermal power production results versus the collector's installation angle are presented, obtained by a computational simulation of a solar collector field executed by the Solar Thermal Department of the Centre for Renewable Energy Sources (CRES), for a location in the proximity of Athens, Greece (geographical latitude close to 38°) [28].

From Table 4.4 we may state the following:

- The final annual thermal energy production per unit of installed solar collector effective area is maximized for installation angle equal to 30°.
- The final thermal energy production from April to October is maximized with the collector's installation angle at 20°.
- The final thermal energy production from November to March is maximized with the collector's installation angle at 50°–55°.
- Particularly during the summer months, namely from June to August, the thermal energy production is maximized with the collector's installation angle at 10°.
- Particularly during the winter months, namely from December to February, the thermal energy production is maximized with the collector's installation angle at 55°–60°.

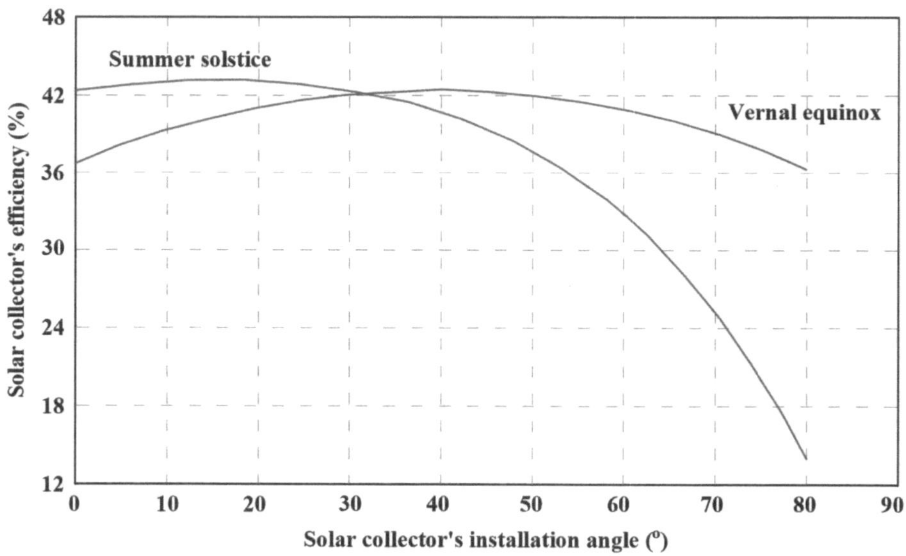

FIGURE 4.38 Fluctuation of a solar collector's efficiency versus its installation angle with regard to the horizontal plane [21].

TABLE 4.4
Thermal Power Production Per Unit of Effective Solar Collector Area, Installed in the Proximity of Athens, Greece, versus the Installation Angle (with due diligence of the Centre for Renewable Energy Sources of Greece) [28]

Time Period	Collector's Installation Angle with Regard to the Horizontal Plane (°)																
	0	10	20	25	30	35	40	45	50	55	60	65	70	75	80	85	90
	Final Thermal Power Production (kWh/m²)																
January	66	80	91	96	100	104	107	109	111	112	113	112	111	109	107	104	100
February	75	84	91	93	96	97	99	99	99	99	98	96	94	91	88	84	80
March	104	112	116	118	119	119	119	118	116	114	111	108	104	99	94	89	83
April	146	151	152	152	151	149	147	143	139	134	129	123	116	108	101	92	84
May	182	183	181	178	175	170	165	159	153	145	137	128	119	109	100	90	79
June	200	200	195	191	185	180	173	166	158	149	139	128	118	108	96	85	75
July	213	214	210	205	199	194	187	180	171	162	151	139	128	117	105	91	80
August	200	206	206	204	202	199	194	188	182	174	165	155	144	132	121	109	96
September	156	168	176	179	180	181	180	178	175	171	166	161	154	146	138	128	118
October	106	120	130	134	138	140	142	143	142	142	140	137	134	130	125	119	113
November	66	77	86	90	94	96	99	100	101	102	102	101	99	97	95	92	88
December	53	63	72	76	79	82	85	87	88	89	89	89	88	87	85	83	90
Annual Production	**1,567**	**1,658**	**1,706**	**1,716**	**1,718**	**1,711**	**1,697**	**1,670**	**1,635**	**1,593**	**1,540**	**1,477**	**1,409**	**1,333**	**1,255**	**1,166**	**1,086**
April–October production	1,203	1,242	**1,250**	1,243	1,230	1,213	1,188	1,157	1,120	1,077	1,027	971	913	850	786	714	645
November–March production	364	416	456	473	488	498	509	513	**515**	**516**	513	506	496	483	469	452	441
December–February production	194	227	254	265	275	283	291	295	298	**300**	**300**	297	293	287	280	271	270
June–August production	613	**620**	611	600	586	573	554	534	511	485	455	422	390	357	322	285	251

The measurements of Table 4.4 reveal the optimum installation angle of a solar collector field with regard to the horizontal plane, in order to achieve maximization thermal energy production during specific periods of the year. The optimum installation angle can vary from 10° to 60°, depending on the scope of the solar collector field, and particularly if thermal energy production is more crucial to be maximized during the winter, the summer, or the whole year.

It must be emphasized that the above measurements and values are valid for geographical locations with geographical latitudes at 37°–38°. With the results presented in Table 4.4, the following empirical rules are confirmed:

- The optimum installation angle of a flat surface in order to maximize the solar radiation incidence on an annual basis should be approximately the same as the geographical latitude of the installation site.
- If maximization of the thermal energy production during the summer period is more crucial, the optimum installation angle should be approximately 10°–15° lower than the installation site's geographical latitude.
- If maximization of the thermal energy production during the winter period is more crucial, the optimum installation angle should be approximately 10°–15° higher than the installation site's geographical latitude.
- If the installation site is surrounded with surfaces with high reflectance (e.g., snow), the installation angle can be further lifted to increase the collection of the reflected solar radiation.

4.3.8 Application for Water-Based Photovoltaic Hybrid Thermal Collectors

The set of equations and relationships presented above can also be used for water-based PV hybrid thermal (PVT) collectors, with slight modifications. The structure and the energy flows in a water-based PVT collector are presented in Figure 4.39. As seen in this figure, the PV panel is placed below a top, transparent cover, with air gap in the space between them. Underneath the PV panel, the absorber plate is attached with an adhesive layer between them. The water pipelines are adjusted below the absorber plate, evenly arranged throughout the plate's width. A proportion of the incident solar radiation is absorbed by the transparent top cover and transferred to the PV panel. A part of it is utilized for electrical power production and the rest is transferred through conduction as thermal power towards the collector's absorber plate. If $(\tau\alpha)_p$ is the effective transmittance–absorptance product of the top cover–PV panel, then the absorbed thermal power $\dot{Q}_p$ from the PV panel is given by the relationship:

$$\dot{Q}_p = G_t \cdot (\tau \cdot \alpha)_p - P_{el} \tag{4.35}$$

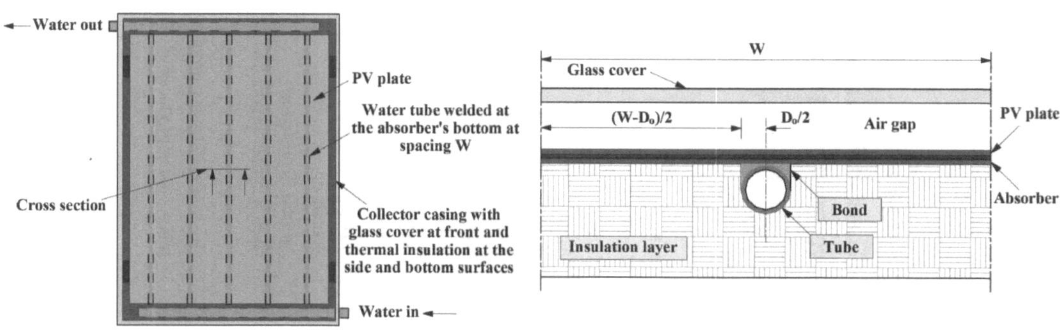

FIGURE 4.39 Construction details and energy flows for a water-based PVT collector with top cover [29].

where G_t is the incident solar radiation on the collector's surface and P_{el} is the electrical power output from the PV panel. The effective transmittance–absorptance product $(\tau\alpha)_p$ can be calculated with the relationship:

$$(\tau \cdot \alpha)_p = \frac{\tau_\alpha \cdot \tau_\rho \cdot \alpha_p}{1 - (1 - \alpha_p) \cdot r} \tag{4.36}$$

where:

- τ_α: the top cover absorptance, considering only absorption losses
- τ_ρ: the top cover absorptance, considering only reflection losses
- α_p: the absorptance of the PV plate
- r: the reflectance of glass

The thermal power production from the PVT collector can be calculated with a relationship similar to Equation 4.4, by substituting the term $G_t \cdot (\tau\alpha)$, which designates the captured solar radiation in the form of thermal energy by the flat-plate collector, with the captured thermal power $\dot{Q}_p$ by the PVT collector:

$$\dot{Q} = A_c \cdot F_R \cdot \left[\dot{Q}_p - U_L \cdot (T_{fi} - T_a)\right] \tag{4.37}$$

From this point on, the process presented above for a typical flat-plate collector can be applied for the calculation of the thermal power production from a PVT collector. The eventually calculated temperature of the absorber plate T_{pm} can be assumed equal to the temperature of the PV cell, required for the calculation of the electrical power output.

4.4 ENERGY ANALYSIS FOR A CONCENTRATING SOLAR COLLECTOR

The calculation of the thermal power production from concentrating solar collectors generally follows the same concept as flat-plate solar collectors. The major difference is that with concentrating solar collectors, first the solar radiation entering the reflector's aperture S must be calculated, which, in turn, will be the input for the calculation of the solar radiation concentration in the collector's receiver, given the optical properties of the reflector and the receiver. Having calculated the thermal power available in the receiver, the next step is the estimation of the total thermal transmittance factor U_L, which determines the heat losses from the receiver to the ambient. Finally, the heat removal factor F_R is also introduced for the case of the concentrating solar collectors, which will enable the calculation of the heat losses, given the temperatures of the working fluid and the ambient. With the factors U_L and F_R known, the thermal power production calculation from the solar collector is now possible with a relationship similar to the one used for the flat-plate collectors (Equation 4.4). Yet, another essential difference between flat-plate and concentrating collectors arises from the high appearing temperatures in the concentrating collector's receiver, which impose significant heat losses to the ambient. This, is turn, means that the thermal transmittance factor U_L is affected and does not remain constant, because it depends on the working fluid's temperature.

4.4.1 Total Thermal Transmittance Factor U_L for the Heat Losses from the Receiver

The heat losses from the receiver of a concentrating solar collector follow the same physical laws and processes as flat-plate collectors. The receiver may be equipped with a transparent cover. The transmittance–absorptance product is also introduced for the concentrating solar collectors, in order to take into account the effect of the cover's transmittance and the absorber's absorptance. Due to the high solar radiation concentration from the collector to the receiver, the material used for the manufacturing of the receiver's cover should exhibit considerably low absorptance, in order to avoid its wear due to absorbed solar radiation.

Generally the thermal transmittance factor U_L depends on the geometry and the technical specifications of the concentrating solar collector, exactly as with flat-plate collectors. Analytical mathematical expressions can be developed by applying the essential laws of heat transfer. For example, for a cylindrical receiver without cover, the heat loss rate to the ambient, determined by the heat transfer through conductivity, convection, and radiation, is given by [6]:

$$\frac{\dot{Q}_{loss}}{A_r} = h_w \cdot (T_r - T_a) + \varepsilon \cdot \sigma \cdot \left(T_r^4 - T_{sky}^4\right) + U_{cond} \cdot (T_r - T_a) = (h_w + h_r + U_{cond}) \cdot (T_r - T_a) \Leftrightarrow \frac{\dot{Q}_{loss}}{A_r} = U_L \cdot (T_r - T_a) \tag{4.38}$$

where:

$\dot{Q}_{loss}$: the heat loss rate (power) from the receiver to the ambient
A_r: the receiver's total outer cylindrical area
h_w: the convection factor for the heat transfer from the receiver to the ambient
T_r: the working fluid temperature in the receiver
T_a: the ambient temperature close to the receiver
ε: the emissivity of the receiver's material
T_{sky}: the ambient temperature far from the receiver
U_{cond}: the thermal transmittance factor for the heat transfer through conductivity.

According to the above, we conclude the following relationship for the total thermal transmittance factor:

$$U_L = h_w + h_r + U_{cond} \tag{4.39}$$

where the factor h_r stands for the heat transfer with radiation and is given from the relationship:

$$h_r = \frac{\varepsilon \cdot \sigma \cdot \left(T_r^4 - T_{sky}^4\right)}{T_r - T_a} \tag{4.40}$$

If the cylindrical receiver of a concentrating solar collector is protected with an outer transparent cover, the heat losses:

1. first, from the receiver of temperature T_r to the space inside the protective cover, of temperature T_{ci}
2. next, from the cover's inner surface, of temperature T_{ci}, to its outer surface, of temperature T_{co}
3. eventually, from the cover's outer surface to the ambient, of temperature T_a (close to the receiver) and T_{sky} (far from the receiver)

are respectively given from the following relationships:

$$\dot{Q}_{loss} = \frac{2 \cdot \pi \cdot k_{eff} \cdot L}{\ln(D_{ci}/D_r)} \cdot (T_r - T_{ci}) + \frac{\pi \cdot D_r \cdot L \cdot \sigma \cdot \left(T_r^4 - T_{ci}^4\right)}{\frac{1}{\varepsilon_r} + \frac{1 - \varepsilon_c}{\varepsilon_c} \cdot \frac{D_r}{D_{ci}}} \tag{4.41}$$

$$\dot{Q}_{loss} = \frac{2 \cdot \pi \cdot k_c \cdot L \cdot (T_{ci} - T_{co})}{\ln(D_{co}/D_{ci})} \tag{4.42}$$

$$\dot{Q}_{loss} = \pi \cdot D_{co} \cdot L \cdot h_w \cdot (T_{co} - T_a) - \varepsilon_c \cdot \pi \cdot D_{co} \cdot L \cdot \sigma \cdot \left(T_{co}^4 - T_{sky}^4\right) \tag{4.43}$$

where:

D_r: the receiver's outer diameter
D_{ci}: the inner diameter of the receiver's cover
D_{co}: the outer diameter of the receiver's cover
L: the length of the cylindrical receiver
k_c: the receiver's thermal conductivity factor
k_{eff}: the thermal conductivity factor for the space between receiver and cover
ε_r: the emissivity of the receiver's material
ε_c: the emissivity of the receiver's cover material.

For a cylindrical receiver, the thermal convection factor h_w is calculated as described below. Initially, the Reynolds number for the air flow outside the receiver's cover is calculated:

$$Re = \frac{u \cdot D_{co}}{\nu} \tag{4.44}$$

where ν is the kinematic viscosity of air. Next, the Nusselt number for the heat transfer from the receiver to the ambient is calculated from the following empirical relationships, versus the Reynolds number [30]:

$$Nu = 0.40 + 0.54 \cdot Re^{0.52} \quad \text{for } 0.10 < Re < 1{,}000 \tag{4.45}$$

$$Nu = 0.30 \cdot Re^{0.60} \quad \text{for } 1{,}000 < Re < 50{,}000 \tag{4.46}$$

The above relationships are valid for air flow around a cylindrical tube and for atmospheric conditions. Having calculated the Nusselt number, the thermal convection factor h_w can now be calculated from the definition relationship of the Nusselt number:

$$Nu = \frac{h_w \cdot D_{co}}{k} \tag{4.47}$$

where k is the thermal conductivity factor of the tube's material that is directly exposed to the ambient (cover or receiver tube, depending on the existence of protective cover or not respectively).

With the above available relationships, the following procedure is followed for the calculation of the heat losses rate from the receiver to the ambient:

1. The required data are the ambient temperatures T_a and T_{sky}, the magnitudes referring to the geometry of the receiver L, D_r, D_{ci}, D_{co}, and the thermal conductivity factors k_c and k_{eff}, the convection factor h_w, and the emissivity ε_r and ε_c.
2. A first assumption is introduced for the temperature of the cover's outer surface T_{co}. This temperature is usually selected close to the ambient temperature T_a, instead of the fluid's temperature inside the receiver tube T_r.
3. From Equation 4.43, a first calculation for the heat losses rate $\dot{Q}_{loss}$ is estimated.
4. Next, given the calculated $\dot{Q}_{loss}$, the temperature T_{ci} of the cover's inner surface is calculated with Equation 4.42.
5. From Equation 4.38 the receiver's average temperature T_r is calculated with trials.
6. Finally, Equation 4.41 is used to confirm the heat losses rate of the collector, based on the initial assumption for the temperature T_{co}. In case there is significant divergence between the two alternative calculations, the procedure is executed again, by introducing a new temperature T_{co} as calculated by Equation 4.43, given the calculated $\dot{Q}_{loss}$.

The above described process will be outlined in the following example.

Example 4.2 Calculation of Heat Losses Rate from Concentrating Solar Collector

Calculate the total heat losses rate from the cylindrical receiver of a concentrating solar collector to the ambient. The receiver's outer diameter is 65 mm and the working fluid's temperature is 270°C. The emissivity of the receiver's material is 0.25. The receiver's cover has an outer diameter of 100 mm and a thickness of 5 mm. The space between the receiver and its cover is vacuum. The wind velocity is 4 m/s and the ambient temperatures close to and far from the receiver are 15°C and 5°C respectively. The thermal conductivity of the cover's material should be considered equal to 0.022 W/(m·K). The emissivity of the cover's material is 0.90 and the receiver's thermal conductivity is 1.45 W/(m·K). The heat losses rate calculation should be performed per unit of receiver's length.

SOLUTION

First of all, the heat convection factor for the heat transfer from the receiver to the ambient will be calculated. The kinematic viscosity of air with ambient temperature 15°C is found equal to $1.456 \cdot 10^{-5}$ m²/s. The Reynolds number for the air flow outside the receiver's cover is:

$$Re = \frac{u \cdot D_{co}}{\nu} \Rightarrow Re = \frac{4 \cdot 0.10}{1.456 \cdot 10^{-5}} \Leftrightarrow Re = 27{,}473$$

From Equation 4.46 the Nusselt number is calculated:

$$Nu = 0.30 \cdot Re^{0.60} \Rightarrow Nu = 0.30 \cdot 27{,}473^{0.60} \Leftrightarrow Nu = 138.18$$

Finally, from Equation 4.47 the heat convection factor h_w is calculated for the heat transfer from the outer cover's surface to the ambient:

$$Nu = \frac{h_w \cdot D_{co}}{k} \Leftrightarrow h_w = \frac{Nu \cdot k}{D_{co}} \Rightarrow h_w = \frac{138.18 \cdot 0.022}{0.10} \Leftrightarrow h_w = 30.40 \frac{W}{m^2 \cdot K}$$

The first estimation for the heat losses rate can now be calculated with Equation 4.43. For this purpose, given that the ambient temperature is given 15°C = 288 K, the temperature of the cover's outer surface is assumed equal to 20°C = 293 K. The application of Equation 4.43 gives, taking into account that the calculation is executed per unit of receiver's length:

$$\dot{Q}_{loss} = \pi \cdot D_{co} \cdot L \cdot h_w \cdot (T_{co} - T_a) + \varepsilon_c \cdot \pi \cdot D_{co} \cdot L \cdot \sigma \cdot (T_{co}^4 - T_{sky}^4) \Rightarrow$$

$$\frac{\dot{Q}_{loss}}{L} = \pi \cdot 0.10 \cdot 30.40 \cdot (293 - 288) + 0.90 \cdot \pi \cdot 0.10 \cdot 5.67 \cdot 10^{-8} \cdot (293^4 - 278^4) \Leftrightarrow \frac{\dot{Q}_{loss}}{L} = 70.15 \frac{W}{m}$$

The temperature on the cover's inner surface is now calculated with Equation 4.42:

$$\dot{Q}_{loss} = \frac{2 \cdot \pi \cdot k_c \cdot L \cdot (T_{ci} - T_{co})}{\ln(D_{co}/D_{ci})} \Leftrightarrow T_{ci} = \frac{\dot{Q}_{loss}}{2 \cdot \pi \cdot k_c \cdot L} \cdot \ln\left(\frac{D_{co}}{D_{ci}}\right) + T_{co} \Rightarrow$$

$$T_{ci} = \frac{70.15}{2 \cdot \pi \cdot 1.45} \cdot \ln\left(\frac{0.10}{0.90}\right) + 293 \Leftrightarrow T_{ci} = 294\ K$$

TABLE 4.5
Results from the Iterative Procedure for the Calculation of the Heat Losses Rate

	First Loop	Second Loop	Third Loop	Fourth Loop
Temperature T_{co} (°C/K)	20/293	34/307	33.3/306.6	33.36/306.36
Initial estimation of heat losses rate (W/m)	70.153	228.115	220.135	220.819
Temperature T_{ci} (°C/K)	21.82/293.82	36.64/309.64	35.88/308.85	35.92/308.92
Final estimation of heat losses rate (W/m)	228.115	220.610	220.876	220.853

Finally, Equation 4.41 is used for the verification of the initial assumption for the temperature T_{co}. Because it is given that the space between the receiver and its cover is vacuum, the thermal conductivity factor k_{eff} is considered null.

$$\frac{\dot{Q}_{loss}}{L} = \frac{2 \cdot \pi \cdot k_{eff}}{\ln(D_{ci}/D_r)} \cdot (T_r - T_{ci}) + \frac{\pi \cdot D_r \cdot \sigma \cdot (T_r^4 - T_{ci}^4)}{\frac{1}{\varepsilon_r} + \frac{1-\varepsilon_c}{\varepsilon_c} \cdot \frac{D_r}{D_{ci}}} \Rightarrow$$

$$\frac{\dot{Q}_{loss}}{L} = 0 + \frac{\pi \cdot 0.065 \cdot 5.67 \cdot 10^{-8} \cdot (543^4 - 294^4)}{\frac{1}{0.25} + \frac{1-0.90}{0.90} \cdot \frac{0.065}{0.090}} \Leftrightarrow \frac{\dot{Q}_{loss}}{L} = 228.11 \frac{W}{m}$$

There is a significant divergence between the initial and the final calculation of the heat losses rate. Hence, the calculation procedure is repeated with a new assumption for the temperature T_{co}. From Equation 4.43, it is found with trials that for heat losses rate of 225.55 W, the temperature T_{co} should equal 34°C. With this assumption, the calculation procedure is executed once again. The overall results from the iterative calculation are presented in Table 4.5.

With the heat losses rate calculated at 220.85 W, the total thermal transmittance factor can be finally calculated with Equation 4.38:

$$\frac{\dot{Q}_{loss}}{A_r} = U_L \cdot (T_r - T_a) \Leftrightarrow U_L = \frac{Q_{loss}}{A_r \cdot (T_r - T_a)} \Rightarrow U_L = \frac{\dot{Q}_{loss}}{\pi \cdot D_r \cdot L \cdot (T_r - T_a)}$$

$$\Rightarrow U_L = \frac{220.85}{\pi \cdot 0.065 \cdot (543 - 288)} \Leftrightarrow U_L = 4.241 \frac{W}{m^2 \cdot K}$$

In the above example, it was assumed that the air has been removed from the space between the receiver and the cover, creating vacuum. In this case, there is no heat transfer with convection through this space. However, in the general case, with air contained in this particular space, the heat transfer with convection cannot be ignored. The thermal conductivity factor k_{eff} for the space between the receiver and its cover is retrieved from air properties tables, or, in most generally, from properties tables or graphs of the contained in this space medium.

Additionally, in the above example, only one value was calculated for the total thermal transmittance factor U_L. In Figure 4.40, the fluctuation of the thermal transmittance factor U_L versus the fluid's temperature in the receiver is presented [6], which is considerable; hence, it cannot be ignored. In case the receiver's temperature is not constant versus its length, a fact that imposes different thermal transmittance factors for different parts of the receiver, the receiver should be discretized in separate parts and for each one of them a different thermal transmittance factor should be applied.

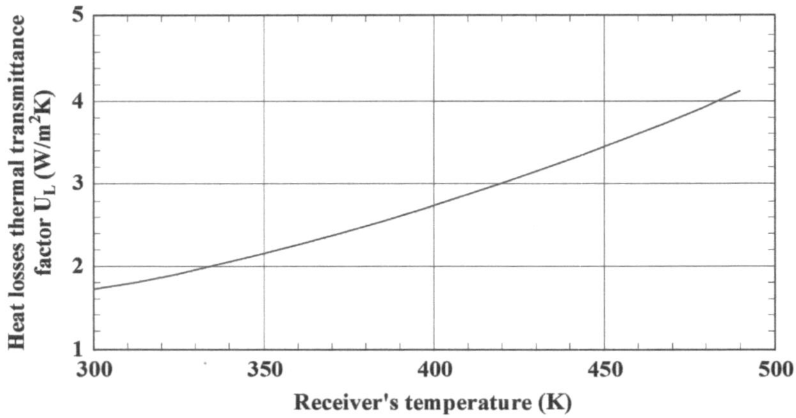

FIGURE 4.40 Fluctuation of the total thermal transmittance factor for the heat transfer from the receiver to the ambient, versus the receiver's temperature.

4.4.2 Thermal Power Production from Concentrating Solar Collectors

The total thermal transmittance factor U_o for the heat transfer from the receiver to the working fluid inside the receiver's pipelines is given by the relationship [6]:

$$U_o = \left(\frac{1}{U_L} + \frac{D_o}{h_{fi} \cdot D_i} + \frac{D_o}{2 \cdot k} \cdot \ln\left(\frac{D_o}{D_i} \right) \right)^{-1} \tag{4.48}$$

where:

- D_o: the receiver's tube outer diameter in m
- D_i: the receiver's tube inner diameter in m
- h_{fi}: the thermal transition factor for the heat transfer from the tube to the working fluid in W/(m^2·K)
- k: the thermal conductivity factor of the receiver's tube material in W/(m·K).

The specific thermal power production per unit of the receiver's length is given by the relationship [6]:

$$q'_u = \frac{A_a \cdot S}{L} - \frac{A_r \cdot U_L}{L} \cdot (T_r - T_a) \tag{4.49}$$

where:

- A_a: the nonshaded area for the collector's aperture in m^2
- A_r: the receiver's outer area in m^2, which, for cylindrical receiver, is $\pi \cdot D_o \cdot L$
- L: the receiver's length in m
- S: the incident solar radiation in W/m^2.

The thermal power production from the concentrating solar collector expressed as a function of the working fluid's temperature T_f inside the receiver's tube is written as presented below:

$$q'_u = \frac{(A_r/L) \cdot (T_r - T_f)}{\frac{D_o}{h_{fi} \cdot D_i} + \frac{D_o}{2 \cdot k} \cdot \ln\left(\frac{D_o}{D_i} \right)} \tag{4.50}$$

By eliminating the receiver's temperature T_r from the above two relationships we get:

$$q'_u = F' \cdot \frac{A_a}{L} \cdot \left[S - \frac{A_r}{A_a} \cdot U_L \cdot (T_f - T_a)\right] \tag{4.51}$$

where the efficiency factor F′ of the concentrating solar collector is:

$$F' = \frac{1/U_L}{\frac{1}{U_L} + \frac{D_o}{h_{fi} \cdot D_i} + \frac{D_o}{2 \cdot k} \cdot \ln\left(\frac{D_o}{D_i}\right)} \tag{4.52}$$

or alternatively:

$$F' = \frac{U_o}{U_L} \tag{4.53}$$

Similarly with Equation 4.4, the thermal power production from the concentrating solar collector can be written as:

$$\dot{Q}_u = A_a \cdot F_R \cdot \left[S - \frac{A_r}{A_a} \cdot U_L \cdot (T_i - T_a)\right] \tag{4.54}$$

where T_i is the receiver's temperature and T_a the ambient temperature.

Also, similarly with Equation 4.5, the collector's heat removal factor F_R is given by the relationship:

$$F_R = \frac{\dot{m} \cdot c_p}{A_r \cdot U_L} \cdot \left[1 - \exp\left(-\frac{A_r \cdot U_L \cdot F'}{\dot{m} \cdot c_p}\right)\right] \tag{4.55}$$

The above relationships are applied for receivers either with or without covers. The differences between these two alternatives are introduced with the magnitudes S and U_L.

Example 4.3 Final Thermal Power Production from Concentrating Solar Collector

The concentrating solar collector examined in the previous example exhibits heat losses total thermal transmittance factor at 4.50 W/(m²·K). The aperture of the concentrating collector has a width of 2.5 m and a length of 12 m. The incident solar radiation at a specific moment is 550 W/m². Inside the receiver's tube, the flowing fluid exhibits a specific heat capacity of 3.500 kJ/(kg·K). The working fluid enters the receiver with a flow rate of 0.05 kg/s and an inlet temperature of 140°C. The thermal convection factor for the heat transfer from the receiver to the working fluid is 400 W/(m²·K). The receiver's tube has been constructed with stainless steel, with a thermal conductivity factor of 16 W/(m·K) and a thickness of 5 mm. If the ambient temperature is 15°C, calculate the final thermal power production from the collector and the working fluid's outlet temperature from the receiver. The geometrical dimensions of the cover and the receiver have been given in Example 4.2, namely the receiver's outer diameter is 65 mm, the cover's outer diameter is 100 mm, and its thickness is 5 mm. It is reminded that the average temperature of the receiver is 270°C.

SOLUTION

The receiver's inner area is:

$$A_r = \pi \cdot D_r \cdot L \Rightarrow A_r = \pi \cdot 0.065 \cdot 12 \Leftrightarrow A_r = 2.45\ m^2$$

Taking into account the shading of the collector from the receiver, the collector's effective area will be:

$$A_a = (W - D_r) \cdot L \Rightarrow A_a = (2.50 - 0.10) \cdot 12 \Leftrightarrow A_a = 28.80\ m^2$$

The F′ factor is then calculated from Equation 4.52:

$$F' = \frac{1/U_L}{\dfrac{1}{U_L} + \dfrac{D_o}{h_{fi} \cdot D_i} + \dfrac{D_o}{2 \cdot k} \cdot \ln\left(\dfrac{D_o}{D_i}\right)} \Rightarrow F' = \frac{1/4.50}{\dfrac{1}{4.50} + \dfrac{0.065}{400 \cdot 0.055} + \dfrac{0.065}{2 \cdot 16} \cdot \ln\left(\dfrac{0.065}{0.055}\right)} \Rightarrow F' = 0.985$$

The heat removal factor F_R is then calculated by Equation 4.55:

$$F_R = \frac{\dot{m} \cdot c_p}{A_r \cdot U_L} \cdot \left[1 - \exp\left(-\frac{A_r \cdot U_L \cdot F'}{\dot{m} \cdot c_p}\right)\right] \Rightarrow$$

$$F_R = \frac{0.05 \cdot 3{,}500}{2.45 \cdot 4.50} \cdot \left[1 - \exp\left(-\frac{2.45 \cdot 4.50 \cdot 0.985}{0.05 \cdot 3{,}500}\right)\right] \Leftrightarrow F_R = 0.940$$

From Equation 4.54, the final thermal power production from the concentrating solar collector is calculated:

$$\dot{Q}_u = A_a \cdot F_R \cdot \left[S - \frac{A_r}{A_a} \cdot U_L \cdot (T_i - T_a)\right] \Rightarrow$$

$$\dot{Q}_u = 28.80 \cdot 0.940 \cdot \left[550 - \frac{2.45}{28.80} \cdot 4.50 \cdot (270 - 15)\right] \Leftrightarrow \dot{Q}_u = 12{,}447.4\ W$$

The fluid's temperature inside the receiver and the fluid's outlet temperature from the receiver are calculated from the relationship:

$$\dot{Q}_u = \dot{m} \cdot c_p \cdot (T_{fo} - T_{fi}) \Leftrightarrow T_{fo} - T_{fi} = \frac{\dot{Q}_u}{\dot{m} \cdot c_p} \Rightarrow$$

$$T_{fo} - T_{fi} = \frac{12{,}447.4}{0.05 \cdot 3{,}500} \Leftrightarrow T_{fo} - T_{fi} = 71.13°C \Leftrightarrow T_{fo} = 71.13°C + T_{fi}$$

$$\Rightarrow T_{fo} = 71.13°C + 140°C \Leftrightarrow T_{fo} = 211.13°C$$

The temperature drop from the receiver's outer surface to the working fluid can be calculated from the relationship:

$$\dot{Q}_u = \frac{\overline{T}_{ro} - \overline{T}_{fi}}{\dfrac{1}{\pi \cdot D_{ri} \cdot L \cdot h_{fi}} + \dfrac{1}{2 \cdot \pi \cdot k \cdot L} \cdot \ln\left(\dfrac{D_{ro}}{D_{ri}}\right)} \Leftrightarrow \overline{T}_{ro} - \overline{T}_{fi} = \dot{Q}_u \cdot \left[\frac{1}{\pi \cdot D_{ri} \cdot L \cdot h_{fi}} + \frac{1}{2 \cdot \pi \cdot k \cdot L} \cdot \ln\left(\frac{D_{ro}}{D_{ri}}\right)\right] \Rightarrow$$

$$\overline{T}_{ro} - \overline{T}_{fi} = 12{,}447.4 \cdot \left[\frac{1}{\pi \cdot 0.055 \cdot 12 \cdot 400} + \frac{1}{2 \cdot \pi \cdot 16 \cdot 12} \cdot \ln\left(\frac{0.065}{0.055}\right)\right] \Leftrightarrow \overline{T}_{ro} - \overline{T}_{fi} = 16.73°C$$

4.4.3 Solar Radiation Absorptance from Concentrating Solar Collectors

The solar radiation absorptance from the concentrating solar collector generally depends strongly on the geometry and the optical features of the collector. A general relationship for the calculation of the absorbed solar radiation S by the nonshaded aperture of the concentrating solar collector is [6]:

$$S = I_b \cdot \rho \cdot (\gamma \cdot \tau \cdot \alpha)_n \cdot K_{\gamma\tau\alpha} \tag{4.56}$$

In the above relationship, I_b is the incident solar radiation on the nonshaded surface of the concentrating solar collector, consisting mainly of the direct solar radiation, apart from solar collectors with concentration ratio lower than 10, for which an amount of diffused radiation should also be taken into account for the calculation of the total incident solar radiation, depending on the orientation angle of the solar collector with respect to the south.

The factor ρ is the specular reflectance of the solar collector. In case of combined use of diffuse reflectors with cylindrical absorbers, the factor ρ will be the diffuse reflectance. If the solar collector is a refractor, then the factor ρ coincides with the transmittance of the refractor.

The diffused reflected solar radiation is propagated evenly in all directions, while the reflected radiation is catoptrical towards the opposite direction with the same incident angle θ. The reflection depends on the sheen of the reflector's surface. In practice, there is no reflection that can be considered absolutely catoptrical or diffused. Actually, reflections are formed rather as a combination of the two forms of solar radiation, which is called *general reflection*. In general reflection, the size of the intensity of the reflected solar radiation towards a specific direction and for a particular reflective surface is a function of the incident solar radiation wave length and spatial distribution. The alternative forms of reflected solar radiation mentioned above are presented in Figure 4.41 [6].

The next three factors γ, τ, and α are functions of the solar radiation incident angle.

The factor γ is the intercept factor, defined as the percentage of the solar radiation which, after being reflected from the reflector, hits the receiver. The basic target for the development of concentrating solar collectors is the minimization of the heat losses from the collector to the ambient, achieved due to the reduction of the receiver's area. Most concentrating solar collectors are constructed with receiver areas large enough to retrieve a considerable amount of the reflected solar radiation from the collector. Usually the intercept factor γ ranges close to 0.90.

As with flat-plate solar collectors, τ is the transmittance of the collector's cover and α the absorptance of the absorber's material. The determination of these two magnitude may not be easy, due to the fluctuation of the solar radiation incidence angle from the collector to the receiver.

In Equation 4.56 the product of the γ, τ, and α factor refers to perpendicular solar radiation incidence on the receiver. In order to account for the effect of the solar radiation incidence angle, the incidence angle modifier $K_{\gamma\tau\alpha}$ is introduced. In general, as the solar radiation incident angle on the reflector increases, the reflected solar radiation to the receiver decreases.

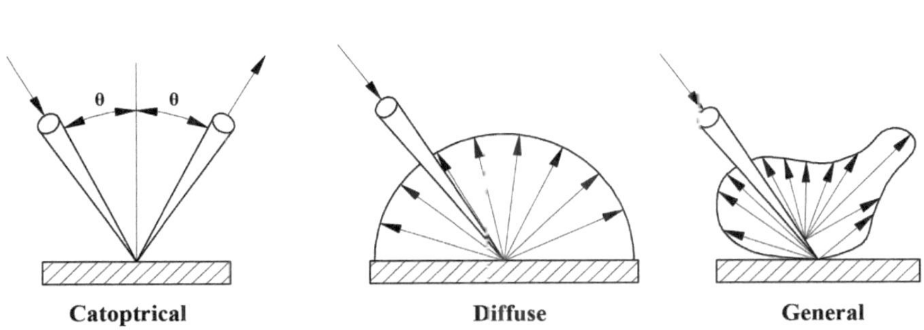

FIGURE 4.41 Catoptrical, diffuse, and general solar radiation reflection [6].

4.4.4 Solar Radiation Absorbed from Compound Parabolic Collectors

Compound parabolic collectors (CPCs) were presented in Section 4.2.4.3. In this section it was mentioned that the collector receives solar radiation only if the incident angle θ ranges between the limits $-\theta_c < \theta < \theta_c$, where θ_c is the acceptance half angle of the CPC (see Figure 4.12). In this case, the absorbed solar radiation from the CPC is given by the relationship [6]:

$$S = C\cdot\left(G_{b,CPC}\cdot\tau_{c,b}\cdot\tau_{CPC,b}\cdot\alpha_b + G_{d,CPC}\cdot\tau_{c,d}\cdot\tau_{CPC,d}\cdot\alpha_d + G_{g,CPC}\cdot\tau_{c,g}\cdot\tau_{CPC,g}\cdot\alpha_g\right) \tag{4.57}$$

where:

$$G_{b,CPC} = F\cdot G_{bn}\cdot\cos(\theta) \tag{4.58}$$

$$G_{d,CPC} = \begin{cases} \dfrac{G_d}{C} & \text{for} \quad \beta + \theta_c < 90° \quad (4.59) \\ \dfrac{G_d}{2}\cdot\left(\dfrac{1}{C} + \cos(\beta)\right) & \text{for} \quad \beta + \theta_c > 90° \quad (4.60) \end{cases}$$

$$G_{g,CPC} = \begin{cases} 0 & \text{for} \quad \beta + \theta_c < 90° \quad (4.61) \\ \dfrac{G_d}{2}\cdot\left(\dfrac{1}{C} - \cos(\beta)\right) & \text{for} \quad \beta + \theta_c > 90° \quad (4.62) \end{cases}$$

The first term in Equation 4.57 expresses the contribution of the direct solar radiation to the total solar radiation absorbed by the solar collector. Respectively, the second and the third term express the contribution of the diffuse and the reflected from the ground solar radiation to the absorbed by the collector solar radiation.

In the first term of Equation 4.57:

- the magnitude $G_{b,CPC}$ is the direct solar radiation falling on the reflector's aperture, within the acceptable limits of the incident angle
- the magnitude $\tau_{c,b}$ is the transmittance to the direct solar radiation of the receiver's cover
- the magnitude α_b is the absorptance to the direct solar radiation of the receiver
- the factor $\tau_{CPC,b}$ expresses the effect on the solar collector's transmittance of the consecutive solar radiation reflections on the collector's reflector, before the solar radiation is eventually absorbed by the receiver, and it is a function of the number of reflections.

The magnitudes appearing in the second and the third term of Equation 4.57 for the diffuse and the reflected solar radiation respectively are similar to the ones described just previously for direct solar radiation.

The magnitude G_{bn} in Equation 4.58 designates the total available solar radiation on the horizontal plane at the solar collector's installation site. The magnitude G_d in Equations 4.59, 4.60, and 4.62 designates the diffuse solar radiation at the collector's installation site. The magnitude C in Equations 4.59, 4.60, and 4.62 is the solar collector's concentration ratio, while the magnitude θ in Equation 4.58 is the incident angle of the solar radiation on the collector's aperture.

According to Equations 4.61 and 4.62, the reflected from the ground solar radiation can be significant only if $\beta + \theta_c > 90°$, a condition that implies that the reflector of the collector has a direct optical contact with the ground. It is reminded that, in the above relationships, the angle β is the solar collector's installation angle with regard to the horizontal plane. The angles involved in the above relationships are graphically explained in Figure 4.42.

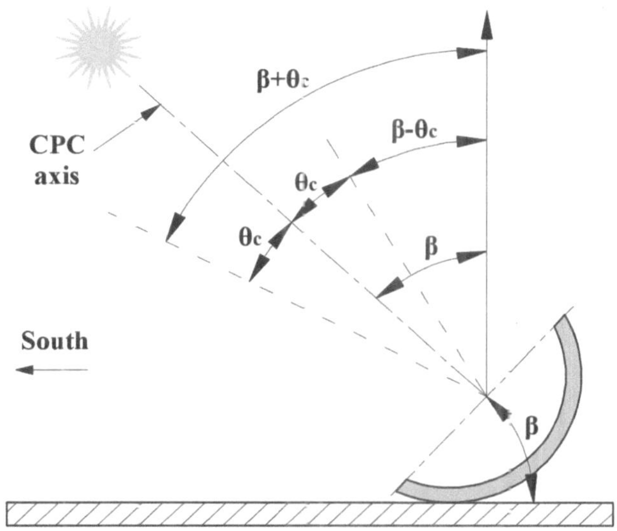

FIGURE 4.42 Solar radiation acceptable incident angle range on a CPC, with south orientation (for the north hemisphere), versus the installation angle.

Additionally, it is proved that direct solar radiation can exhibit a considerable contribution to the solar radiation absorptance by the collector, only if the following relationship is satisfied [31], where θ_z and φ_s are the sun's zenith and azimuth angle respectively:

$$\beta - \theta_c \leq \tan^{-1}\left(\tan(\theta_z)\cdot\cos(\varphi_s)\right) \leq \beta + \theta_c \quad (4.63)$$

The criterion expressed with Equation 4.63 is introduced in the calculation of the direct solar radiation through Equation 4.58 with the control function F, which takes the value 1, if the above criterion is satisfied, and the value 0 in the opposite case.

The transmittance factors $\tau_{c,d}$ and $\tau_{c,g}$ express the transmittance of the collector's cover for the diffuse and the reflected from the ground solar radiation respectively.

With regard to the diffuse solar radiation, only a percentage of the total available in the ambient diffuse radiation enters the reflector's aperture. This percentage depends on the acceptance half angle θ_c. If the diffuse solar radiation in the ambient environment can be considered isotropic, then the relationship between the angle θ_c and the incoming diffuse solar radiation in the reflector can be expressed as a function of a new magnitude, the so-called equivalent angle of incidence θ_e of the isotropic diffuse radiation. The cover's transmittance for the diffuse radiation will be eventually determined by the equivalent angle of incidence θ_e. This relationship depends on the cover's material. For materials with refractive indices from 1.34 to 1.526 (glass), the following polynomial approach can be developed for the relationship between the angles θ_c and θ_e [6]:

$$\theta_e = 44.86 - 0.0716\cdot\theta_c + 0.00512\cdot\theta_c^2 - 0.00002798\cdot\theta_c^3 \quad (4.64)$$

For example, if the acceptance half angle for a concentrating solar collector is 25°, then the equivalent angle of incidence of the isotropic diffuse solar radiation is calculated with the above relationship equal to 45.83°. In this case, the transmittance of the receiver's cover for the diffuse radiation will be determined for incidence angle of 45.83°. Similarly, the equivalent angle of incidence of the isotropic diffuse radiation is given graphically versus the acceptance half angle from the diagram of Figure 4.43 [32].

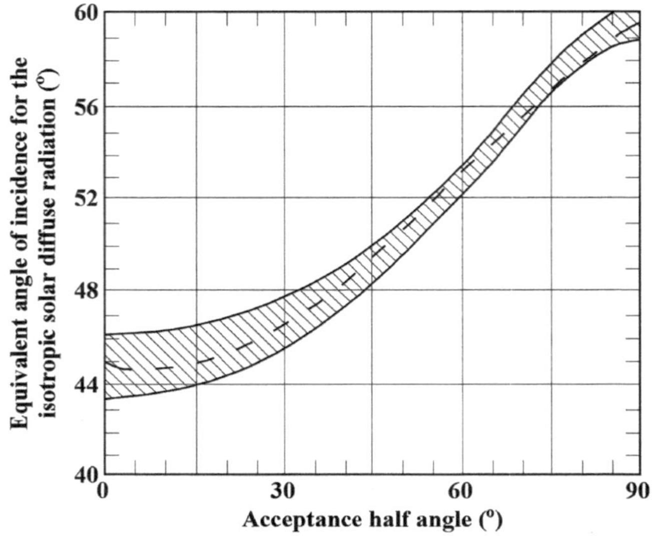

FIGURE 4.43 Fluctuation of the equivalent angle of incidence of the isotropic solar radiation on a CPC versus the acceptance half angle [32].

The transmittance factors for the direct and the diffuse solar radiation can be estimated from Figure 4.44, versus the incidence angle [6]. In case of diffuse radiation, the same figure can be used with the equivalent angle of incidence, as defined previously.

In Figure 4.44, apart from the solar radiation angle of incidence, the transmittance factors are given also versus the product K·L. The magnitudes involved in this product are the thickness L of the receiver's cover in m and a new parameter K, named as *extinction coefficient*, measured in m^{-1}. The extinction coefficient depends exclusively on the cover's material, so for the same material it remains unchanged. For glass it varies from 4 m^{-1}, for so-called white glass, namely for glass that appears to be white when seen from its edge, to 32 m^{-1} for glass with high iron content, which appears to be green when seen from its edge. Hence, the product K·L constitutes a characteristic feature of the cover, taking into account both its thickness and its material. It should be also stated that Figure 4.44 can be generally used for the estimation of the transmittance factors for the direct and the diffuse solar radiation for either flat-plate or concentrating solar collectors.

The magnitudes $\tau_{CPC,b}$, $\tau_{CPC,d}$, and $\tau_{CPC,g}$ in Equation 4.57 express losses of the absorbed solar radiation from the collector's receiver due to the consecutive catoptrical reflections on the collector's reflector, before the radiation is eventually absorbed by the receiver. These three terms are usually considered equal between them and are calculated versus the number n_i of the catoptrical reflections of the solar radiation and the reflectance ρ of the reflector from the relationship [6]:

$$\tau_{CPC} = \rho^{n_i} \quad (4.65)$$

In Figure 4.45 the design concept of a CPC is presented. Each side of the collector follows the form of a parabola. The parabolas and their axes are presented for the two sides of the collector. The collector's height h is determined by the intersection of the line parallel with the axis of the one parabola (left or right parabola) that passes from the lowest point of the corresponding part of the collector (left or right part), with the extension of the parabola from the opposite part of the collector. The acceptance half angle θ_c is also depicted in the same figure, defined by the collector's vertical axis and the line parallel with the axis of any one of the collector's parabolas that connects the lowest point of the one side of the collector with the highest point of the opposite side of the collector.

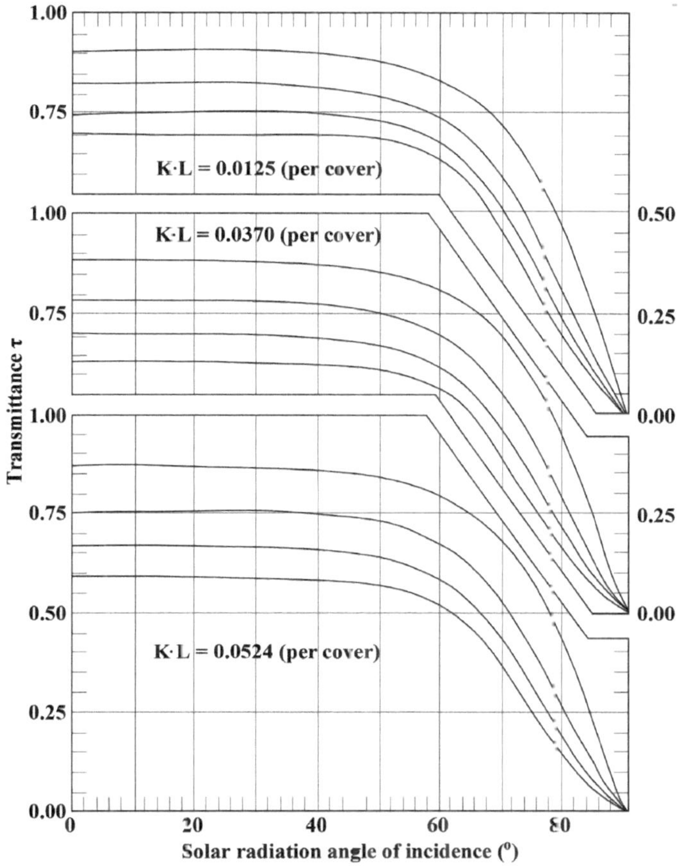

FIGURE 4.44 Transmittance factors for different types of protective cover versus the solar radiation angle of incidence [6].

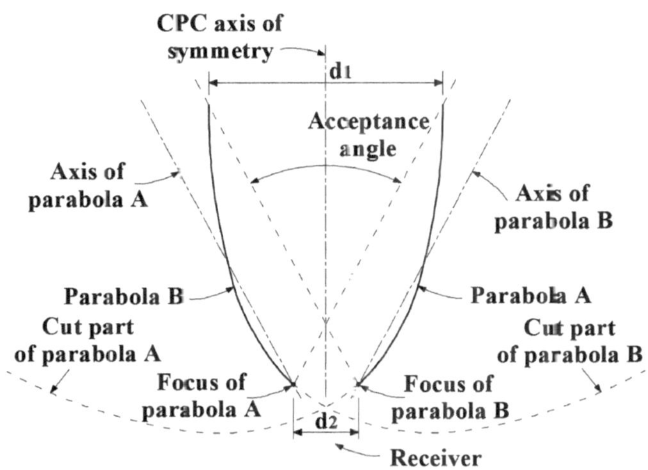

FIGURE 4.45 Geometrical design concept of a compound parabolic collector [33].

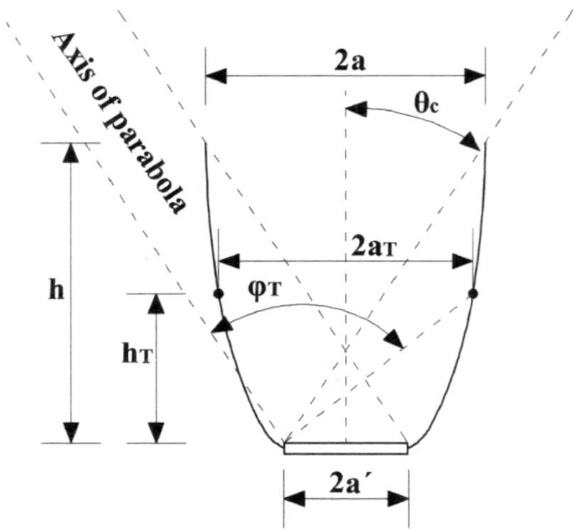

FIGURE 4.46 Geometrical features of a CPC [33].

Usually CPCs are constructed with height h_T lower than the height h, as defined above. The reason for this height reduction arises from the fact that with this modification a significant reduction of the manufacturing cost is achieved with only a low drop of the absorbed solar radiation by the collector. In this case, the collector is called a truncated solar collector and its geometry is given in Figure 4.46 [33].

According to Figure 4.46:

- h is the collector's theoretical height, according to the previously described design concept
- h_T is the collector's actual construction height
- a′ is the half width of the collector's receiver
- a is the reflector's half aperture for collector height h
- a_T is the reflector's half aperture for collector height h_T
- θ_c is the collector's acceptance half angle
- ϕ_T is the angle formed between the parabola's axis for one of the collector's sides and the line connecting the intersection point of the same parabola's axis with the receiver and the opposite side of the reflector at height h_T.

Given the receiver's half width a′, the acceptance half angle θ_c, and the angle ϕ_T (or alternatively the ratio h_T/h), the analytical calculation of the average reflections number n_i is possible with the following relationships [33]:

$$f = a' \cdot (1 + \sin\theta_c) \tag{4.66}$$

$$a = \frac{a'}{\sin\theta_c} \tag{4.67}$$

$$h = \frac{f \cdot \cos\theta_c}{\sin^2\theta_c} \tag{4.68}$$

$$a_T = \frac{f \cdot \sin(\varphi_T - \theta_c)}{\sin^2 \frac{\varphi_T}{2}} - a' \tag{4.69}$$

$$h_T = \frac{f \cdot \cos(\varphi_T - \theta_c)}{\sin^2 \frac{\varphi_T}{2}} \tag{4.70}$$

$$C = \frac{a_T}{a'} \tag{4.71}$$

$$\text{or, for } h_T = h,\ C = \frac{a}{a'} = \frac{1}{\sin\theta_c} \tag{4.72}$$

$$\frac{A_{RT}}{1 \cdot a_T} = \frac{f}{2} \cdot \left[\frac{\cos\frac{\varphi}{2}}{\sin^2\frac{\varphi}{2}} + \ln\left(\cot\frac{\varphi}{4}\right) \right] \Bigg|_{\varphi = \theta_c + \frac{\pi}{2}}^{\varphi = \varphi_T} \tag{4.73}$$

$$x = \frac{1 + \sin\theta}{\cos\theta} \cdot \left[-\sin\theta + \sqrt{1 + \frac{h_T}{h}\cot^2\theta} \right] \tag{4.74}$$

$$n_i = \max\left[C \cdot \frac{A_{RT}}{4 \cdot a_T} - \frac{x^2 - \cos^2\theta}{2 \cdot (1 + \sin\theta)},\ 1 - \frac{1}{C} \right] \tag{4.75}$$

In the above relationships the magnitude A_{RT} is the reflection area per unit of depth of the truncated collector. In case $h = h_T$, then $\varphi_T = 2 \cdot \theta_c$ and $A_{RT} = A_R$. Also, the solar radiation angle of incidence on the collector's aperture is designated as θ.

Equivalently, instead of using the above relationships, the average reflections number of the solar radiation on the collector's reflector can be estimated from the diagram of Figure 4.47 versus the collector's concentration ratio and the acceptance half angle.

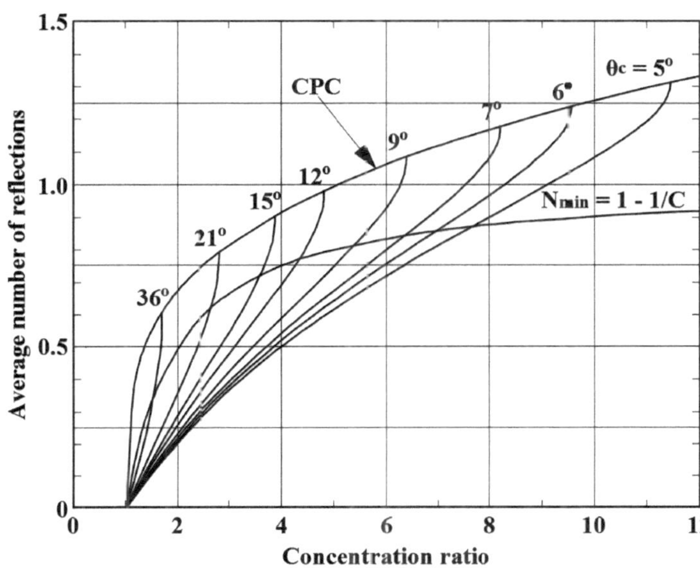

FIGURE 4.47 Fluctuation of the average reflections number of solar radiation on a CPC versus the concentration ratio C and the collector's acceptance half angle θ_c [6, 33].

Example 4.4 Calculation of Solar Radiation Incidence on a CPC

Calculate the total solar radiation incidence on a truncated CPC, installed in a location with geographical latitude 35°N and with installation angle $\beta = 25°$ with regard to the horizontal plane, at a specific time moment when the global available solar radiation is $G_{bn} = 800$ W/m², the diffuse solar radiation is $G_d = 350$ W/m², the sun's zenith and azimuth angles are $\theta_z = 32°$ and $\varphi_s = -65°$ respectively, while the solar radiation angle of incidence is $\theta = 30°$. The acceptance half angle is $\theta_c = 12°$ and the collector's concentration ratio is C = 4. The product K·L for the collector's cover is 0.0125. The solar collector's reflectance and absorptance are given as $\rho = 0.88$ and $\alpha = 0.95$ respectively.

SOLUTION

First of all, the criterion expressed with Equation 4.63 should be checked:

$$\beta - \theta_c \le \tan^{-1}\left(\tan(\theta_z)\cdot\cos(\varphi_s)\right) \le \beta + \theta_c \Rightarrow$$

$$25 - 12 \le \tan^{-1}\left(\tan(32)\cdot\cos(-65)\right) \le 25 + 12 \Leftrightarrow 13 \le 14.79 \le 37$$

Hence, Equation 4.63 is satisfied and the control function F in Equation 4.58 takes the value 1. The direct solar radiation on the solar collector is calculated with Equation 4.58:

$$G_{b,CPC} = F\cdot G_{bn}\cdot\cos(\theta) \Rightarrow G_{b,CPC} = 1\cdot 800\cdot\cos(30) \Leftrightarrow G_{b,CPC} = 692.8\ \frac{W}{m^2}$$

Because the sum $\beta + \theta_c$ equals to 37°, namely it is lower than 90°, the incident diffuse solar radiation on the solar collector will be given by the first part of Equation 4.59:

$$G_{d,CPC} = \frac{G_d}{C} \Rightarrow G_{d,CPC} = \frac{350}{4} \Leftrightarrow G_{d,CPC} = 87.5\frac{W}{m^2}$$

Additionally, from Equation 4.61, for $\beta + \theta_c = 37° < 90°$, the reflected from the ground solar radiation $G_{g,CPC}$ is taken equal to 0.

For angle of incidence 30° and K·L equal to 0.0125, from Figure 4.44 the transmittance of the collector's cover for the direct radiation is estimated $\tau_{c,b} = 0.90$. Additionally, for angle of incidence 30° from Figure 4.43, the equivalent angle of incidence for the isotropic diffuse solar radiation is estimated 46°. Then, again from Figure 4.44, for equivalent angle of incidence 46° of the isotropic solar radiation, the transmittance of the collector's cover for the diffuse and the reflected solar radiation are estimated $\tau_{c,d} = \tau_{c,g} = 0.89$. Finally, from Figure 4.47, for concentration ratio equal to 4 and acceptance half angle 12°, the average reflections number is estimated 0.68. Given the above estimations, the calculation of the absorbed solar radiation by the solar collector is now possible with Equation 4.57:

$$S = C\cdot\left(G_{b,CPC}\cdot\tau_{c,b}\cdot\tau_{CPC,b}\cdot\alpha_b + G_{d,CPC}\cdot\tau_{c,d}\cdot\tau_{CPC,d}\cdot\alpha_d + G_{g,CPC}\cdot\tau_{c,g}\cdot\tau_{CPC,g}\cdot\alpha_g\right) \Rightarrow$$

$$S = C\cdot\left(G_{b,CPC}\cdot\tau_{c,b}\cdot\rho^{n_i}\cdot\alpha_b + G_{d,CPC}\cdot\tau_{c,d}\cdot\rho^{n_i}\cdot\alpha_d + G_{g,CPC}\cdot\tau_{c,g}\cdot\rho^{n_i}\cdot\alpha_g\right) \Rightarrow$$

$$S = 4\cdot\left(692.8\cdot 0.90\cdot 0.88^{0.68}\cdot 0.95 + 84.5\cdot 0.89\cdot 0.88^{0.68}\cdot 0.95 + 0\cdot 0.89\cdot 0.88^{0.68}\cdot 0.95\right) \Leftrightarrow$$

$$S = 2{,}443.5\ \frac{W}{m^2}$$

4.5 THERMAL ENERGY STORAGE

As with hybrid power plants for electricity production, examined in Chapter 3, another necessary component for thermal hybrid plants is the storage unit. After the presentation of the basic technologies for thermal power production from RES, namely the alternative solar collector types, covered in the previous sections of this chapter, in this section there will be a discussion about the basic technologies for thermal energy storage. Particularly, the basic analytical relationships and methodologies will be presented for the computational simulation of energy storage processes in thermal storage tanks utilizing either water or solid materials, because these energy storage technologies are the most basic and mature applicable technologies for the most common applications of active thermal power production systems.

Thermal energy storage can be achieved in the form of either sensible or latent heat, increasing the temperature or changing the phase respectively of the thermal storage medium. The choice of the storage technology depends on the type and the requirements of the particular application. For example, for the production of warm water for domestic use, it is logical to store thermal energy at the same medium in which it is intended to be heated up, so typical water thermal storage tanks are commonly used. Water, due to its high heat capacity and its, generally, availability with relatively low cost, as well as a number of additional attractive properties (e.g., odorless, nonflammable, nonvolatile), constitutes an ideal medium for thermal energy storage in cases of typical heating applications for domestic or commercial buildings and warm water production for domestic use. For air-heated solar collectors, thermal energy storage can also be performed in the form of sensible or latent heat with solid or liquid means respectively. A classic example of the thermal energy storage is a pebble or rock bed. Special concrete materials with high heat capacity or vessels with pressurized water can be employed for thermal energy storage in high temperatures (up to 300°C) and exploitation for electricity and thermal energy cogeneration. Finally, in cases of passive solar systems, thermal energy is stored as sensible heat at the structural elements of a building.

The basic features of a thermal energy storage system are:

- the heat capacity per unit of volume of the storage medium, known as specific heat capacity
- the temperature range of the medium employed for thermal energy storage
- the working medium with which the heat transfer is executed from and to the storage medium
- the temperature distribution inside the working medium employed for the thermal energy storage
- the thermal energy storage system procurement and operation cost.

The final choice of the employed thermal energy storage system is eventually configured, together with the type of the application, by the above technical and economic features.

A fundamental peculiarity of thermal hybrid plants is based on the physical heat transfer process, which is always executed from a high to a low temperature medium (from high to low potential). This implies that heat transfer can be achieved only if the temperature of the working fluid from the hybrid power plant's base unit, e.g., the water–glycol solution temperature from the solar collector's primary hydraulic circuit, is higher than the temperature of the thermal energy storage medium in the thermal tank, e.g., the temperature of the contained water in a water thermal tank. If this condition is not satisfied, any potential thermal power production surplus from the hybrid thermal plant's base unit will not be possible to be stored.

In reality, the temperature of the working fluid in the primary hydraulic loop while reaching the storage tank should be expected to be slightly lower than the outlet temperature from the solar collectors, due to heat losses in the transfer pipelines from the base units (solar collectors) to the storage unit. Consequently, the criterion that will determine whether the heat transfer will be feasible or not, is if the inlet temperature of the working fluid in the heat exchanger of the storage unit is higher or not than the temperature of the contained medium in the storage tank. This limitation, as seen in a next section, is attempted to be overcome with the introduction of stratification storage in thermal tanks, namely the achievement of different temperatures for different layers of the working medium inside the storage tank.

4.5.1 Thermal Energy Storage in Water Tanks

The use of water tanks constitutes the ideal, most popular technology for thermal energy storage for usual applications of active thermal systems. The fundamental philosophy is presented in Figure 4.48. Specifically, in Figure 4.48a the working medium in the solar collectors coincides with the medium in the thermal storage tank. After being heated in the solar collector, it is led to the storage tank, where it is stored until its final disposal to the consumption. In contrast, in Figure 4.48b, the working medium in the solar collectors flows inside an autonomous, closed circuit, independent from the medium in the thermal storage tank. The two mediums are totally separated from each other and do not come into contact at any point in the hydraulic network. The heat transfer from the primary circuit to the storage tank is executed with a heat exchanger, through which the two independent circuits (and not the mediums) are brought into contact.

The open-loop system of Figure 4.48a is simpler, and, consequently, exhibits lower cost. Additionally, the thermal energy storage is direct. Hence, by eliminating the heat transfer losses from the primary closed loop to the thermal storage tank through the heat exchanger, the efficiency of the thermal production–storage system is maximized. The alternative implementation presented in Figure 4.48b is safer, firstly for the system itself, because water–glycol solution can be used as the working fluid with lower freezing temperature, hence the freezing probability of the working fluid inside the collector's pipelines during periods with low ambient temperatures is reduced, restricting the contingency for serious damages in the collectors. Additionally, the circulation of the collectors' fluid inside a closed loop minimizes the requirements for the refilling of the closed circuit and, subsequently, the wear of the pipelines due to chemical corrosion and solid precipitators on the pipelines' inner side. Finally, the system is safer for the final user as well, because the warm water with which the final user may come into contact (e.g., in case of warm water production), is totally independent from the working medium inside the solar collectors.

If the medium in the thermal storage tank exhibits a specific heat capacity of c_p and the delivered thermal power causes a uniform temperature rise of ΔT_s of the medium in the thermal storage tank after a time period t, then the stored thermal power will be $\dot{Q}_s$:

$$t \cdot \dot{Q}_s = m \cdot c_p \cdot \Delta T_s \tag{4.76}$$

where m is the total mass of the storage medium in the thermal tank, assumed to be evenly heated.

The thermal power $\dot{Q}_s$ that will be eventually stored in the thermal tank, corresponds only to a part of the initially available thermal power $\dot{Q}_u$ from the hybrid plant's base unit. An amount $\dot{Q}_d$

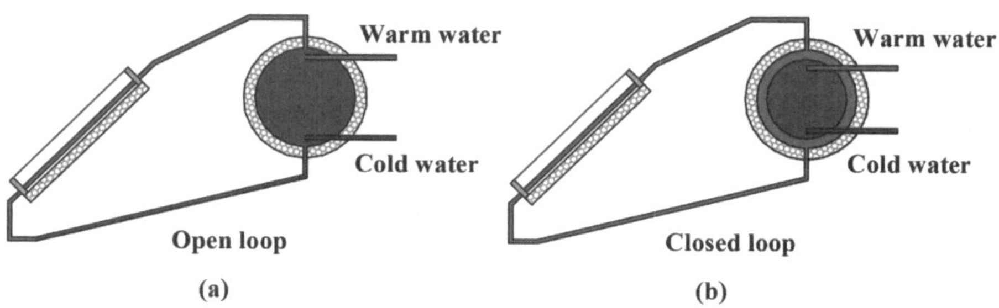

FIGURE 4.48 Thermal energy storage from the solar collector to the thermal storage tank, with an (a) open-loop and a (b) closed-loop system.

from the initial available power $\dot{Q}_u$ will be provided directly for the coverage of the concurrent thermal power demand, while another part $\dot{Q}_L$ will be lost to the ambient with the form of heat losses from the tank storage itself. The thermal power balance during the storage process can be written as:

$$t \cdot \dot{Q}_s = m \cdot c_p \cdot \Delta T_s = t \cdot \left[\dot{Q}_u - \dot{Q}_d - U_s \cdot A_s \cdot (T_s - T_a) \right] \quad (4.77)$$

where:

U_s: the total thermal transmittance factor, for the heat transfer from the thermal storage tank to the ambient

A_s: the total heat transfer area between the thermal storage tank and the ambient

T_s: the uniform temperature of the storage medium inside the thermal tank storage

T_a: the ambient environment temperature.

It must be underlined that the above relationships are valid for thermal storage tanks without stratification storage, namely with a uniform temperature of the storage medium in the storage tank. The above thermal energy balance is graphically presented in Figure 4.49.

Because in the above relationship all the involved magnitudes are known, the problem is deduced to the calculation of the temperature of the storage medium in the thermal storage tank after the storage of thermal power $\dot{Q}_s$. The calculation procedure includes the discretization of the total time interval in discrete time calculation steps, usually hourly. Assuming an initial temperature $T_s(0)$ of the storage medium in the thermal storage tank at the beginning of the calculation process, the temperature of the storage medium is calculated at the end of each calculation step i with the relationship:

$$m \cdot c_p \cdot \Delta T_s = m \cdot c_p \cdot \left[T_s(i) - T_s(i-1) \right] = t \cdot \left[\dot{Q}_u - \dot{Q}_d - U_s \cdot A_s \cdot \left(T_s(i-1) - T_a(i) \right) \right] \Leftrightarrow$$
$$T_s(i) = \frac{t \cdot \left[\dot{Q}_u - \dot{Q}_d - U_s \cdot A_s \cdot \left(T_s(i-1) - T_a(i) \right) \right]}{m \cdot c_p} + T_s(i-1) \quad (4.78)$$

The thermal power demand $\dot{Q}_d$ should be available in the form of a time series with the same time discretization. Similarly, the thermal power production $\dot{Q}_u$ from the solar collectors should be calculated by applying the procedures described in Sections 4.3 and 4.4, depending on the collector

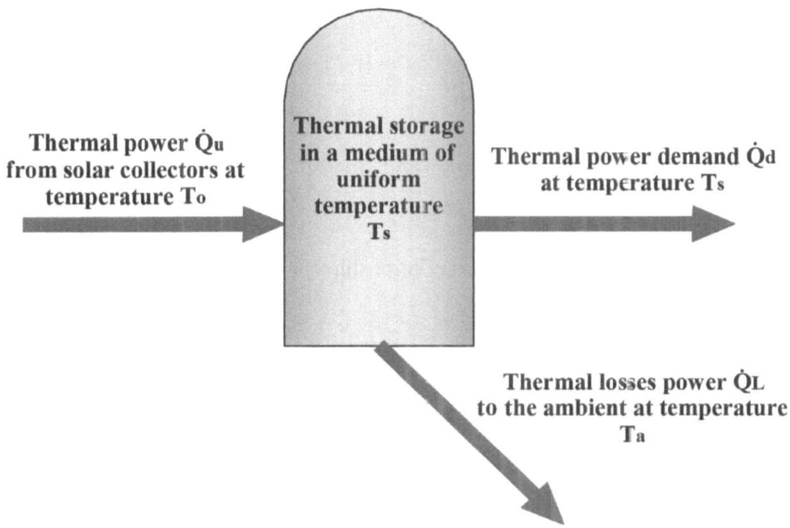

FIGURE 4.49 Heat transfer flows during the thermal energy storage process in a thermal storage tank.

type. The calculation of the storage medium's temperature at the end of each calculation time step constitutes a basic prerequisite for the operation simulation of the overall thermal hybrid plant.

Example 4.5 Calculation of the Storage Medium's Temperature in a Thermal Storage Tank

In a thermal energy production system from solar collectors and a thermal storage tank, the evolution versus time of the thermal power production and demand is presented in Table 4.6 for the first 12 hours of a 24-hour period.

The thermal storage tank has a total volume of 1,000 L and is characterized with a thermal transmittance factor and heat transfer area product of $U_s{\cdot}A_s = 10.5$ W/K. At the beginning of the under examination time period, the storage medium's temperature inside the thermal storage tank has been configured at 50°C, while the ambient temperature remains constant during the whole examined time period at 20°C. Calculate the working medium's temperature inside the thermal storage tank at the end of each time step for the whole time period of the 12 hours. Consider the working medium's temperature uniform for the whole volume of the storage tank. The storage medium's density can be assumed equal to 1,000 kg/m^3 and its specific heat capacity is given as 4.184 kJ/(kg·K).

SOLUTION

The procedure for the solution of this example is based on the repeating application of Equation 4.78 for the calculation of the storage medium's temperature at the end of each calculation time step. All the magnitudes involved in Equation 4.78 are known. For the first time step, namely for the first hour of the examined period, the temperature $T_s(i-1)$ is given equal to 50°C.

The total mass of the thermal storage medium contained in the tank is:

$$m = V{\cdot}\rho = 1{,}000 \text{ kg}$$

where V = 1,000 L the volume capacity of the storage tank and $\rho = 1{,}000$ kg/m^3 the density of the storage medium in the tank. The duration of the time step is:

$$t = 1 \text{ h} = 3{,}600 \text{ s}$$

Indicatively, the calculation of the temperature for the first time step is presented below:

$$T_s(i) = \frac{t{\cdot}\left[\dot{Q}_u - \dot{Q}_d - U_s{\cdot}A_s{\cdot}\left(T_s(i-1) - T_a(i)\right)\right]}{m{\cdot}c_p} + T_s(i-1) \Rightarrow$$

$$T_s(i) = \frac{3{,}600\cdot\left[0 - 3{,}000 - 10.5\text{W/K}{\cdot}(50-20)\text{K}\right]}{1{,}000 \text{ kg}{\cdot}4{,}184 \text{ J/kgK}} + 50°\text{C} \Rightarrow T_s(i) = 47.15°\text{C}$$

The results from the application of the above relationship for the total examined time period are presented in Table 4.7.

TABLE 4.6
Thermal Power Production and Demand Evolution versus Time in a Thermal Power Production System

Time (h)	1	2	3	4	5	6	7	8	9	10	11	12
Thermal Power Production (kW)	0	0	0	0	0	0	0	0	7	13	17	21
Thermal Power Demand (kW)	3	3	2	2	5	3	4	4	5	6	7	6

TABLE 4.7
Calculation Results for the Storage Medium's Temperature in the Thermal Storage Tank

Time (h)	Thermal Power Production Q_u (kW)	Thermal Power Demand Q_d (kW)	Thermal Storage Tank Temperature $T_s(i-1)$	Thermal Storage Tank Temperature $T_s(i)$
1	0	3	50.00	47.15
2	0	3	47.15	44.32
3	0	2	44.32	42.38
4	0	2	42.38	40.46
5	0	5	40.46	35.97
6	0	3	35.97	33.25
7	0	4	33.25	29.68
8	0	4	29.68	26.15
9	7	5	26.15	27.82
10	13	6	27.82	33.77
11	17	7	33.77	42.25
12	21	6	42.25	54.96

4.5.2 Stratification Thermal Storage in Water Tanks

It was mentioned in the earlier section that the thermal energy storage from solar collectors to water tanks can be maximized via stratification storage. With stratification thermal storage, a discrete temperature distribution is achieved in the storage medium, so the low temperatures at the tank's bottom gradually increase as we approach the higher layers of the water tank. Stratification thermal storage is naturally approached, with the recirculation of the warmer and lighter water masses from the lower towards the upper layers of the thermal tank, exclusively with physical flow.

The result of stratification thermal energy storage in a water thermal tank is presented graphically in Figure 4.50, where the fluctuation of the storage medium's temperature in the thermal tank is depicted, starting from low temperatures, close to the tank's base, and ending at high temperatures, close to the tank's top. Heating demand is covered with water flow always supplied from the upper layer of the thermal tank. In this way, thermal power with the maximum available temperature is

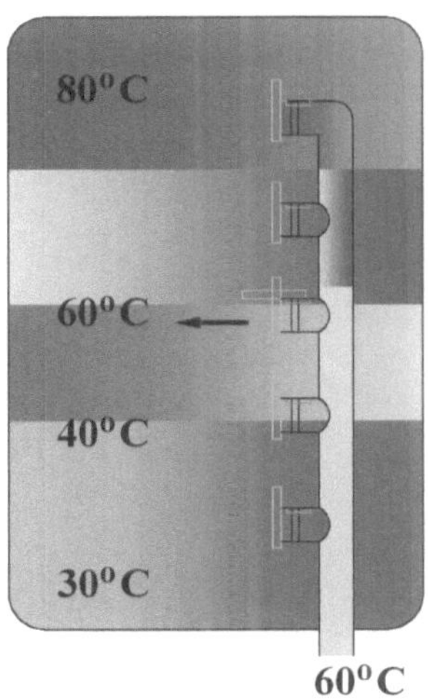

FIGURE 4.50 Stratification thermal energy storage in a water tank [34].

provided for the heating needs coverage. Furthermore, the offered thermal power from the solar collectors is transferred through a heat exchanger, connected with the thermal storage tank with multiple inlets. With temperature sensors, the thermal power is automatically injected inside the thermal storage tank in the inlet with the maximum temperature, which remains lower than the solar collectors' output temperature. In this way, the storage of the available thermal energy is maximized. The final result is the achievement of stratification thermal energy storage in the water tank, as shown in Figure 4.50.

It is conceivable that stratification thermal energy storage facilitates the thermal power storage from the solar collectors, because this can be feasible even for relatively low temperatures of the working fluid in the solar collectors' primary loop.

Figure 4.50 also presents a technique for the facilitation of the achievement of stratification thermal storage, commonly applied in water thermal tanks. The depicted thermal tank is equipped with five alternative thermal power inlet entries, located at different heights from the thermal tank's base. By measuring the working fluid's temperature in the primary loop (in this figure this is assumed to be 60°C) and the temperature distribution versus the thermal tank's height, the thermal power entry in the thermal tank is selected to be the one with the highest temperature in the tank that still remains lower than the working fluid's temperature in the primary loop.

The simulation of stratification thermal energy storage in a water tank is a relatively simple procedure with regard to its essential concept. However, the solution of the problem is feasible only with the computational execution of the required calculations. The discretization of a thermal storage tank in three calculation nodes is presented in Figure 4.51 [6]. Each node is considered discrete and independent from its neighboring, while the mass of the storage medium contained inside the borders of each node has a uniform and constant for each calculation time step temperature. The storage medium's temperature in the inferior node is the lowest one, while the temperature of the storage medium in the upper node is the highest one.

In case of an open loop, the working medium leaves the thermal storage tank for the solar collectors always from the lowest node with the lowest temperature, with a mass flow rate of $\dot{m}_c$. Additionally, the thermal power provided for the consumers is always removed from the upper node of the storage tank with the highest temperature, with a mass flow rate of $\dot{m}_L$. The returning working fluid from the solar collectors or from the thermal power distribution network will be led to the node with the highest temperature from the other nodes, yet lower than the temperature of the returning fluid.

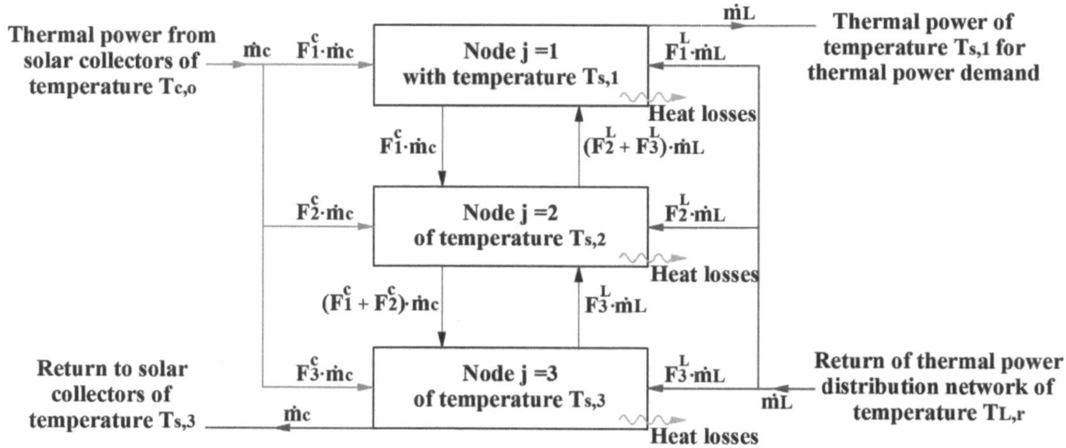

FIGURE 4.51 Discretization in three nodes of a thermal storage tank with stratification thermal storage [6].

The thermal power storage process in the water tank is simulated with the introduction of a control function F^c, which describes the node j at which the thermal power available for storage from the solar collectors will be stored. It is defined as:

$$F_j^c = \begin{cases} 1 & \text{if } j=1 \text{ and } T_{c,o} > T_{s,1} & (4.79) \\ 1 & \text{if } j=2 \text{ or } 3 \text{ and } T_{s,j-1} \geq T_{c,o} > T_{s,j} & (4.80) \\ 0 & \text{for any other case} & (4.81) \end{cases}$$

where $T_{c,o}$ is the temperature of the working fluid from the solar collectors.

The above definition ensures that in case there is thermal power production from the solar collectors, the control function will have a nonzero value for only one node. For example, if the fluid's temperature in the three nodes is 40°C, 50°C, and 60°C, respectively and the fluid's temperature from the solar collectors is $T_{c,o} = 55°C$, then:

- for j = 1: $T_{c,o} = 55°C < T_{s,1} = 60°C$, hence $F_1^c = 0$
- for j = 2: $T_{s,j-1} = T_{s,1} = 60°C > T_{c,o} = 55°C > T_{s,j} = T_{s,2} = 55°C$, hence $F_2^c = 1$
- for j = 3: $T_{s,j-1} = T_{s,2} = 50°C < T_{c,o} = 55°C$, hence $F_3^c = 0$.

Consequently, as it is sensible and expected, the available for storage thermal power from the solar collectors will be stored in node 2, namely in the node with the highest temperature in the tank, yet lower than the fluid's temperature from the collectors.

Likewise, the node selection process for the return of the working fluid from the thermal power distribution network to the storage tank is simulated with a similar approach. Another control function F^L is introduced, defined as presented below, for the general case of the discretization of the water thermal tank in N nodes:

$$F_j^L = \begin{cases} 1 & \text{if } j=N \text{ and } T_{L,r} < T_{s,N} & (4.82) \\ 1 & \text{if } j<N \text{ and } T_{s,j-1} \geq T_{L,r} > T_{s,j} & (4.83) \\ 0 & \text{for any other case} & (4.84) \end{cases}$$

where $T_{L,r}$ is the temperature of the returning working fluid in the thermal storage tank, after the thermal power distribution for the final consumers.

Beyond the thermal power storage from the solar collectors and the thermal power supply for the final distribution network, there is also thermal energy flow between the nodes of the storage tank. This thermal energy transfer is implemented with the storage medium's mass flow upwards or downwards, depending on the mass flow rates $\dot{m}_c$ and $\dot{m}_L$ from the collectors to the tank and from the tank to the final consumers respectively. To simulate the mass flow rate between the nodes of the water tank, a new, cumulative mass flow rate $\dot{m}_m$ is defined from the node j to the node j – 1, which constitutes the algebraic sum of the mass flow rates $\dot{m}_c$ and $\dot{m}_L$, from the precedent or the following nodes:

$$\dot{m}_{m,1} = 0 \tag{4.85}$$

$$\dot{m}_{m,j} = \dot{m}_c \cdot \sum_{n=1}^{j-1} F_n^c - \dot{m}_L \cdot \sum_{n=j+1}^{N} F_n^L \tag{4.86}$$

In the above equations, the mass flow rates from node 1 to node N are considered with a positive sign, while the mass flow rates towards the opposite direction, namely from node N to node 1, are considered with a negative sign.

After the definition of the above control functions and magnitudes, the energy balance for the node j is written as:

$$\dot{m}_j \cdot c_p \cdot \left[T_{s,j}(i) - T_{s,j}(i-1)\right] = \dot{Q}_{s,j} = F_i^c \cdot \dot{m}_c \cdot c_p \cdot \left[T_{c,o}(i) - T_{s,j}(i-1)\right]$$

$$+F_i^L \cdot \dot{m}_L \cdot c_p \cdot \left[T_{L,r}(i) - T_{s,j}(i-1)\right] - U_{s,j} \cdot A_{s,j} \cdot \left[T_{s,j}(i) - T_a(i)\right]$$

$$\begin{cases} \dot{m}_{m,j} \cdot \left[T_{s,j-1}(i-1) - T_{s,j}(i-1)\right], \text{ if } \dot{m}_{m,j} > 0 & (4.87) \\ \dot{m}_{m,j+1} \cdot \left[T_{s,j}(i-1) - T_{s,j+1}(i-1)\right], \text{ if } \dot{m}_{m,j+1} < 0 & (4.88) \end{cases}$$

The application of the above equation in water tanks with a discretization of the stratification thermal storage process in three to four nodes is capable to describe the problem adequately.

In the above equation, two important procedures have not been considered, actually present in practice and affecting the final temperature distribution in the water thermal tank:

- In reality, the stratification thermal storage is more intensive and remarkable for as long as the thermal energy storage and supply processes last. In case of no thermal power transfer from and to the thermal storage tank for a considerable time period, e.g., for a commercial building during a weekend, the different temperatures of the storage medium initially met at different heights of the tank tend to be equated, through the heat transfer with thermal conductivity inside the storage medium's mass. Consequently, the stratification thermal storage gradually faints.
- Most thermal storage tanks combine two or three alternative heat sources, apart from the basic one. These thermal storage tanks usually combine the possibility for thermal energy storage from a nonguaranteed power unit, e.g., solar collectors, and another one or two additional heat sources from guaranteed power units, such as a conventional oil or biomass burner and/or an electrical resistance. These additional heat sources are connected to the tank's upper layers, with the highest storage medium's temperatures, in order to directly contribute to the availability of peak thermal power production for immediate thermal power demand coverage. In this case, in the above thermal energy balance, another term should be introduced, for the simulation of the thermal power storage in the thermal tank from the guaranteed power units. This new term should be introduced for the calculation relationship of the node $j = 1$.

The solution of the above equations, even computationally, requires the development of rather complicated algorithms and software applications. Usually, the solution of such problems is approached with various arithmetic methodologies and finite-element software applications.

4.6 OPERATION SIMULATION OF THERMAL HYBRID POWER PLANTS

Having integrated the presentation of the essential relationships and calculation methodologies for the thermal power production from solar collectors and for the thermal energy storage in thermal storage tanks, it is now possible to proceed to the analysis of the simulation procedure of the operation of thermal hybrid plants. The fundamental layout of a thermal hybrid plant is presented in Figure 4.52.

The thermal hybrid plant presented in Figure 4.52 aims to cover the heating load of a building's indoor space. The following discrete components are distinguished:

- flat-plate or vacuum tube solar collectors, as base units
- water insulated tanks, as thermal storage units

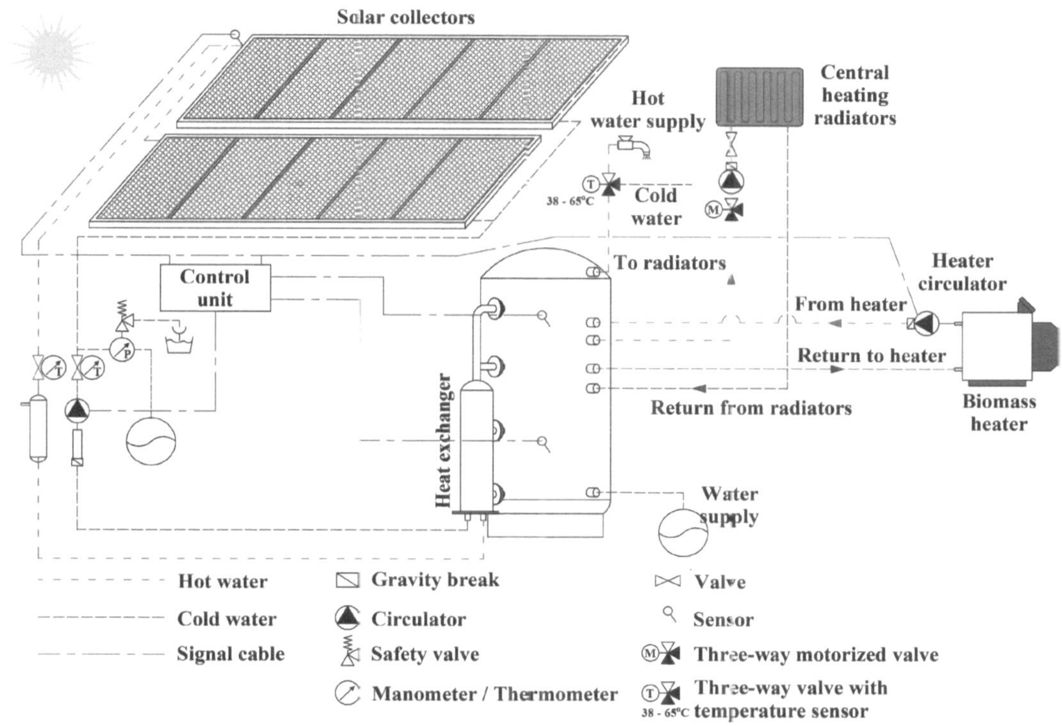

FIGURE 4.52 Connection and operation layout of a thermal hybrid plant with a thermal storage tank with two heat entries [34].

- a conventional central heater, as the backup unit
- a control unit, for the control and the management of the system's operation
- hydraulic devices and equipment aiming to ensure the secure circulation of the working fluid in the hydraulic network and to realize the automatic operation of the hybrid thermal plant.

The central control unit executes the following required checks, which determine the operation mode of the hybrid plant:

- *Comparison of the working fluid's outlet temperature $T_{c,o}$ from the solar collectors and temperatures $T_{s,j}$ of the storage medium at different heights inside the storage tank.*
  If the temperature $T_{c,o}$ is higher than any of the temperatures $T_{s,j}$ of the working medium in the storage tank's discrete nodes j, then thermal energy storage from the solar collectors to the storage tank is executed. The thermal power is stored in the node j of the thermal tank with the highest temperature in the tank, yet lower than the temperature $T_{c,o}$. It must be underlined that the temperature $T_{c,o}$ is not measured at the solar collectors' outlet, but in the heat exchanger employed for the heat transfer from the collectors' primary loop to the storage tank, so the heat losses at the pipelines are also taken into account.
- *Requirement for thermal power production and maximum temperature $T_{s,1}$ of the storage medium in the storage tank.*
  The thermal power production requirement can be declared in two ways:
  - either with the existing temperatures of the building's indoor space, in case of automatic operation of the central heating system with thermostats
  - or with the operation of the circulator of the heating distribution network, in case of manual operation of the central heating system.

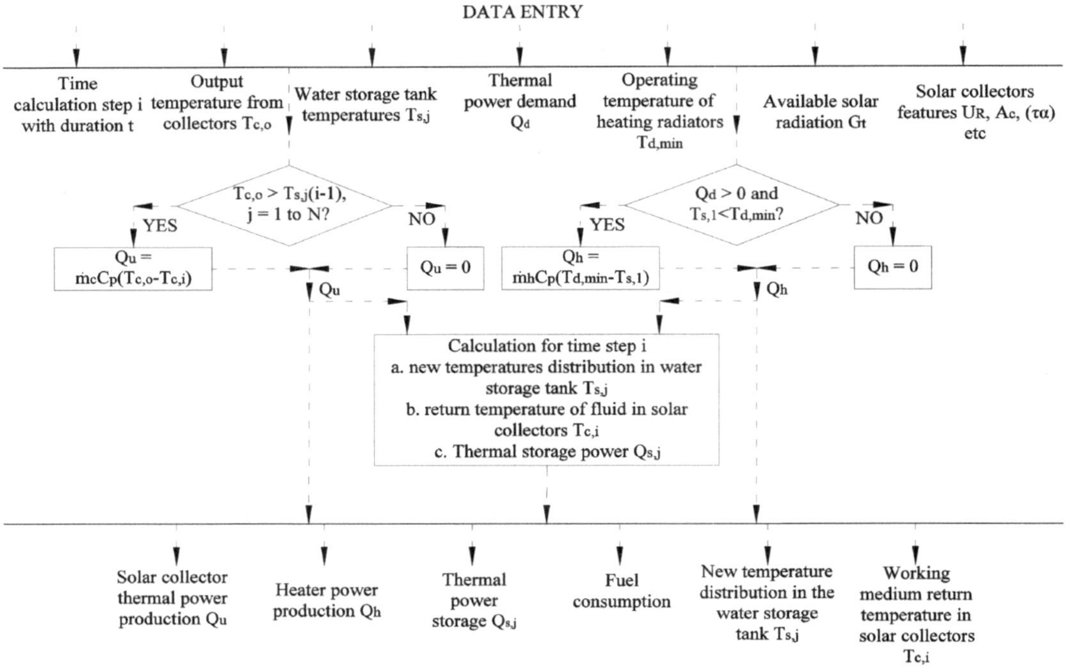

FIGURE 4.53 Operation algorithm of a thermal hybrid plant.

In case there is a current thermal power demand and the maximum temperature $T_{s,1}$ of the storage medium in the storage tank is lower than the minimum required temperature $T_{d,min}$ for the operation of the thermal distribution terminals, then the backup unit is turned on, so the storage medium's temperature at the storage tank's highest layers becomes higher than the minimum required value.

The operation algorithm of the thermal hybrid power plant is presented in Figure 4.53.

The required relationships for the calculation of the involved magnitudes in the above described procedure are summarized below.

For the calculation of thermal power production from the solar collectors, Equation 4.4 is used for flat-plate collectors and Equation 4.54 for concentrating solar collectors. These relationships can be written in the following general form:

$$\dot{Q} = A_c \cdot F_R \cdot \left[S - U_L \cdot \left(T_{fi} - T_a\right)\right] \tag{4.89}$$

The above thermal power can be exploited for storage if $T_{c,o} > T_{s,j}$, for any one of the discrete nodes $j = 1$ to N.

The same thermal power production is also given by the relationship:

$$\dot{Q}_u = \dot{m}_c \cdot c_p \cdot \left(T_{c,o} - T_{c,i}\right) \tag{4.90}$$

where $\dot{m}_c$ is the mass flow rate of the working fluid in the solar collectors' primary loop, $T_{c,o}$ and $T_{c,i}$ the outlet and inlet temperatures of the working fluid from and in the solar collectors.

The thermal power delivered by the central heating burner in case of inadequate temperature of the storage medium in the storage tank is given by the relationship:

$$\dot{Q}_h = \dot{m}_h \cdot c_p \cdot \left(T_{d,min} - T_{s,i}\right) \tag{4.91}$$

where $\dot{m}_h$ is the working fluid's mass flow rate from the storage tank to the thermal terminals.

Finally, the thermal power storage in the storage tank is given from Equation 4.76 for a uniform temperature of the storage medium in the tank and from Equations 4.87 and 4.88 for stratification thermal storage.

In addition to the above, some other important parameters affect the thermal hybrid plant's operation, which should be taken into account in the operation simulation. These are:

- the existence of the heat exchanger for the heat transfer from the solar collectors' primary loop to the storage tank
- the heat losses appeared during the working fluid's flow in the pipelines from the solar collectors to the storage tank and from the storage tank to the thermal distribution terminals
- the thermal power production in a real installation is accomplished by a number of solar collectors connected in series and in parallel, and not only from one single solar collector
- the thermal energy storage in a real installation can be implemented in a number of thermal storage tanks, depending on the size of the system.

The above subjects are investigated in the following sections.

4.6.1 Heat Exchanger Factor

The heat exchanger is involved in a hybrid thermal plant for the heat transfer from the solar collectors' primary loop to the thermal storage tank and, potentially, also in other points of the hydraulic network. Relevant schemes have been presented in Figures 4.29 and 4.52. The introduction of the heat exchanger in the heat transfer circuit from the collectors to the storage tank practically implies a reduction on the heat transfer overall efficiency, expressed with a new efficiency ε.

In general, the transferred heat through the heat exchanger is given by the relationship [30]:

$$\dot{Q}_{he} = \varepsilon \cdot \left(\dot{m} \cdot c_p\right)_{min} \cdot \left(T_{c,o} - T_i\right) \tag{4.92}$$

where:

$\left(\dot{m} \cdot c_p\right)_{min}$: the minimum from the fluid capacitance rates of the working fluid in the collectors' primary loop $\left(\dot{m} \cdot c_p\right)_c$ and the thermal storage medium $\left(\dot{m} \cdot c_p\right)_t$ in the thermal tank

$T_{c,o}$: the inlet temperature of the working fluid from the solar collectors' loop in the heat exchanger

T_i: the inlet temperature of the storage medium (water) from the thermal storage tank in the heat exchanger

ε: the efficiency of the heat exchanger.

In case of a counter-flow heat exchanger, its efficiency is given by the relationship:

$$\varepsilon = \begin{cases} \dfrac{1 - e^{-NTU \cdot \left(1 - C^*\right)}}{1 - C^* \cdot e^{-NTU \cdot \left(1 - C^*\right)}} & \text{if } C^* \neq 1 \quad (4.93) \\[2ex] \dfrac{NTU}{1 + NTU} & \text{if } C^* = 1 \quad (4.94) \end{cases}$$

where the magnitude NTU is the number of transfer units as defined with the relationship:

$$NTU = \frac{U \cdot A}{\left(\dot{m} \cdot c_p\right)_{min}} \tag{4.95}$$

where U is the thermal transmittance factor and A is the heat transfer area of the heat exchanger. The dimensionless fluid capacitance rate C* is given by the relationship:

$$C^* = \frac{(\dot{m} \cdot c_p)_{min}}{(\dot{m} \cdot c_p)_{max}} \tag{4.96}$$

If a heat exchanger is used for the heat transfer from the solar collectors' primary loop to the storage tank, Equation 4.4 is still utilized for the calculation of the produced thermal power from the collectors, yet instead of the heat removal factor F_R a new factor F_R' can be used:

$$\dot{Q} = A_c \cdot F_R' \cdot \left[S - U_L \cdot (T_{fi} - T_a)\right] \tag{4.97}$$

where the new factor F_R' is introduced to described the effect of the heat exchanger on the final thermal power production from the collector. This new factor is given by the relationship:

$$\frac{F_R'}{F_R} = \left[1 + \frac{A_c \cdot F_R \cdot U_L}{(\dot{m} \cdot c_p)_c} \cdot \left(\frac{(\dot{m} \cdot c_p)_c}{\varepsilon \cdot (\dot{m} \cdot c_p)_{min}} - 1\right)\right]^{-1} \tag{4.98}$$

The heat exchanger practically imposes that the solar collectors should deliver thermal power in higher temperature, due to the reduction of the overall heat transfer efficiency from the primary loop to the thermal storage tank, compared to the system's operation without heat exchanger. This, as has been mentioned previously, leads to a reduction of the efficiency of the solar collectors. This effect is expressed by the ratio F_R'/F_R.

Example 4.6 Calculation of Essential Features of a Heat Exchanger

A heat exchanger is used for the heat transfer from a solar collector to a thermal storage tank. A water–glycol solution is used for the collector's primary loop with specific heat capacity 3,850 J/(kg·K). The mass flow rate of the working fluid in the collector's loop is 1.3 kg/s. The storage medium in the storage tank is water with specific heat capacity 4,184 J/(kg·K). The mass flow rate in the heat exchanger is 0.90 kg/s. The product of the thermal transmittance factor with the heat transfer effective area of the heat exchanger is 6,700 W/K. Calculate the transferred thermal power through the heat exchanger. The inlet temperature of the water–glycol solution in the heat exchanger is given $T_{c,o}$ = 60°C and the outlet temperature of the water from the thermal tank is T_i = 30°C. Additionally, calculate the outlet temperatures of the water–glycol solution and the water from the heat exchanger.

If the product $F_R \cdot U_L$ is 3.80 W/(m²·K) and the collector's effective area is A_c = 2 m², calculate, finally, the ratio F_R'/F_R.

SOLUTION

The fluid capacitance rates of the two circuits are calculated as shown below (the subscripts c and t are used for the collector's and the tank's circuits, respectively):

$$C_c = \dot{m}_c \cdot c_{p,c} = 1.30 \frac{kg}{s} \cdot 3{,}850 \frac{J}{kg \cdot K} \Leftrightarrow C_c = 5{,}005 \frac{W}{K}$$

$$C_t = \dot{m}_t \cdot c_{p,t} = 0.90 \frac{kg}{s} \cdot 4{,}184 \frac{J}{kg \cdot K} \Leftrightarrow C_t = 3{,}766 \frac{W}{K}$$

The minimum from the above fluid capacitance rates is the one of the water from the storage tank. The dimensionless fluid capacitance rate is calculated as presented below:

$$C^* = \frac{(\dot{m}\cdot c_p)_{min}}{(\dot{m}\cdot c_p)_{max}} \Rightarrow C^* = \frac{3{,}766\ W/K}{5{,}005\ W/K} \Leftrightarrow C^* = 0.75$$

The magnitude NTU is calculated from the relationship:

$$NTU = \frac{J\cdot A}{(\dot{m}\cdot c_p)_{min}} \Rightarrow NTU = \frac{6{,}700\ W/K}{3{,}766\ W/K} \Leftrightarrow NTU = 1.78$$

Then, the efficiency of the heat exchanger is calculated, for $C^* \neq 1$, from the relationship:

$$\varepsilon = \frac{1 - e^{-NTU\cdot(1-C^*)}}{1 - C^*\cdot e^{-NTU\cdot(1-C^*)}} \Rightarrow \varepsilon = \frac{1 - e^{-1.78\cdot(1-0.75)}}{1 - 0.75\cdot e^{-1.78\cdot(1-0.75)}} \Leftrightarrow \varepsilon = 0.69$$

The transferred thermal power through the heat exchanger is given by the relationship:

$$\dot{Q}_{he} = \varepsilon\cdot(\dot{m}\cdot c_p)_{min}\cdot(T_{c,o} - T_i) \Rightarrow \dot{Q}_{he} = 0.69\cdot 3{,}766\cdot(60 - 30) \Leftrightarrow \dot{Q}_{he} = 77.9\ kW$$

The outlet temperature of the water–glycol solution from the heat exchanger is given by the relationship:

$$\dot{Q}_{he} = \dot{m}_c\cdot c_{p,c}\cdot(T_{ci} - T_{co}) \Rightarrow T_{co} = T_{ci} - \frac{\dot{Q}_{he}}{\dot{m}_c\cdot c_{p,c}} \Rightarrow T_{co} = 60°C - \frac{77.9\ kW}{1.30\frac{kg}{s}\cdot 3{,}850\frac{J}{kg\cdot K}} \Leftrightarrow T_{co} = 44.4°C$$

Similarly, the outlet temperature of the water from the heat exchanger is given by the relationship:

$$\dot{Q}_{he} = \dot{m}_t\cdot c_{p,t}\cdot(T_{to} - T_{ti}) \Leftrightarrow T_{to} = T_{ti} + \frac{\dot{Q}_{he}}{\dot{m}_t\cdot c_{p,t}} \Leftrightarrow T_{to} = 30°C + \frac{77.9\ kW}{0.90\frac{kg}{s}\cdot 4{,}184\frac{J}{kg\cdot K}} \Leftrightarrow T_{to} = 50.7°C$$

The ratio F_R'/F_R is finally calculated as shown below:

$$\frac{F_R'}{F_R} = \left[1 + \frac{A_c\cdot F_R\cdot U_L}{(\dot{m}\cdot c_p)_c}\cdot\left(\frac{(\dot{m}\cdot c_p)_c}{\varepsilon\cdot(\dot{m}\cdot c_p)_{min}} - 1\right)\right]^{-1} \Rightarrow \frac{F_R'}{F_R}$$

$$= \left[1 + \frac{2\ m^2\cdot 3.80\ W/m^2\cdot K}{5{,}005\ W/K}\cdot\left(\frac{5{,}005\ W/K}{0.69\cdot 3{,}766\ W/K} - 1\right)\right]^{-1} \Leftrightarrow \frac{F_R'}{F_R} = 0.99$$

4.6.2 Heat Losses from the Hydraulic Network

The working fluid's flow in the hydraulic network from the solar collectors to the heat exchanger or from the thermal storage tank to the final thermal distribution network imposes heat losses, due to which the working fluid:

- is finally delivered to the heat exchanger or to the final thermal distribution terminals with lower temperature
- returns to the solar collectors or to the thermal storage tank also with lower temperature.

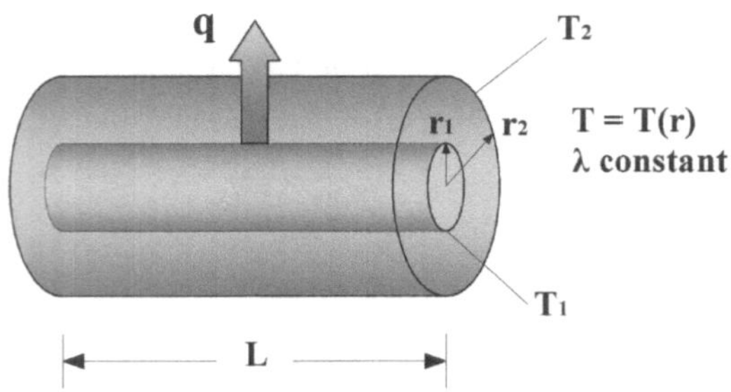

FIGURE 4.54 Radial heat transfer flow with thermal conductivity and convection through the walls of a single cylindrical pipeline.

The heat losses during the working fluid's flow in the pipelines usually cannot be ignored, apart perhaps from very particular cases, referring to transportation hydraulic networks of very short length, with adequate insulation and with relatively high ambient temperatures. When the above conditions are not present, the heat losses during the thermal power transfer should be calculated.

For the calculation of the heat losses from thermal power transportation networks, the essential relationships and concepts from the heat transfer theory are applied for pipelines of circular cross section area. The heat losses from cylindrical pipelines can be ideally approached as presented in Figure 4.54 with a radial heat flow through the pipeline.

The problem in this case is the calculation of the temperature's distribution and the heat transfer rate through a long, concave cylinder of length L, given that the temperatures inside and outside the cylinder are T_1 and T_2 respectively and with the assumption that there is no thermal power production inside the cylinder. Given that the temperatures at the inner and outer side of the cylinder are considered constant and the temperature distribution through the cylinder does not vary with time, the Fourier relationship for the heat transfer with conductivity will be:

$$q = -\lambda \cdot A \cdot \frac{dT}{dr} = \lambda \cdot 2 \cdot \pi \cdot r \cdot L \cdot \frac{dT}{dr} \tag{4.99}$$

where λ is the thermal conductivity factor of the cylindrical pipeline's material.

By integrating Equation 4.99 from the inner radius r_1 to the outer radius r_2 (or, respectively, from the inner diameter d_1 to the outer diameter d_2) and for the corresponding temperatures T_1 and T_2, we reach the following relationship for the heat flow through the pipeline:

$$q = \frac{2 \cdot \lambda \cdot \pi \cdot L}{\ln\left(\frac{r_2}{r_1}\right)} \cdot (T_1 - T_2) \Leftrightarrow q = \frac{T_1 - T_2}{\frac{1}{2 \cdot \lambda \cdot \pi \cdot L} \cdot \ln\left(\frac{r_2}{r_1}\right)} \tag{4.100}$$

The transferred heat flow per unit of length L of the pipeline is:

$$\frac{q}{L} = \frac{2 \cdot \lambda \cdot \pi}{\ln\left(\frac{r_2}{r_1}\right)} \cdot (T_1 - T_2) \Leftrightarrow \frac{q}{L} = \frac{T_1 - T_2}{\frac{1}{2 \cdot \lambda \cdot \pi} \cdot \ln\left(\frac{r_2}{r_1}\right)} \tag{4.101}$$

The above equations can be equivalently written versus the diameters d_1 and d_2:

$$q = \frac{T_1 - T_2}{\frac{1}{2 \cdot \lambda \cdot \pi \cdot L} \cdot \ln\left(\frac{d_2}{d_1}\right)} \Leftrightarrow \frac{q}{L} = \frac{T_1 - T_2}{\frac{1}{2 \cdot \lambda \cdot \pi} \cdot \ln\left(\frac{d_2}{d_1}\right)} \tag{4.102}$$

The term $R = \frac{1}{2 \cdot \lambda \cdot \pi} \cdot \ln\left(\frac{d_2}{d_1}\right)$ is called *thermal resistance factor of cylindrical layer*, measured in mK/W, while the term $R_\theta = \frac{1}{2 \cdot \lambda \cdot \pi \cdot L} \cdot \ln\left(\frac{d_2}{d_1}\right)$ is called *thermal resistance of cylindrical layer*, measured in K/W.

Finally, in case there is a composite cylindrical wall with n layers from different materials, with constant thermal conductivity factors λ_1, λ_2 … λ_n and under the condition that the different layers are firmly attached to each other, leaving no gaps between them, the transferred thermal power per unit of length is given by the relationship [30]:

$$\frac{q}{L} = \frac{T_1 - T_2}{\frac{1}{2 \cdot \lambda_1 \cdot \pi} \cdot \ln\left(\frac{d_2}{d_1}\right) + \frac{1}{2 \cdot \lambda_2 \cdot \pi} \cdot \ln\left(\frac{d_3}{d_2}\right) + \cdots + \frac{1}{2 \cdot \lambda_n \cdot \pi} \cdot \ln\left(\frac{d_{n+1}}{d_n}\right)} \tag{4.103}$$

where d_1 is the inner diameter of the cylindrical wall, d_2 the outer diameter of the first layer (e.g., the main pipeline material), d_3 the outer diameter of the second layer (e.g. insulation), and d_{n+1} the outer diameter of the nth layer, while T_1 is the temperature of the pipeline's inner side (refers to the side of diameter d_1) and T_2 is the temperature of the outer diameter of nth layer (refers to the diameter d_{n+1}).

Example 4.7 Heat Losses Calculation from Cylindrical Pipeline

A pipeline has an inner diameter d_1 = 3.81 cm and an outer diameter d_2 = 4.83 cm. The temperatures at the pipeline's inner and outer sides are respectively equal to 97°C and 87°C. The pipeline's length is 3 m and its thermal conductivity factor is $\lambda = 42.9$ W/mK. Calculate the thermal heat rate (q) and the thermal heat rate per length of pipeline (q/L) for the specific pipeline.

SOLUTION

The heat losses transfer rate is calculated with Equation 4.100:

$$q = \frac{2 \cdot \lambda \cdot \pi \cdot L}{\ln\left(\frac{d_2}{d_1}\right)} \cdot (T_1 - T_2) \Rightarrow q = \frac{2 \cdot 42.9 \cdot \pi \cdot 3}{\ln\left(\frac{4.83}{3.81}\right)} \cdot (97 - 87) \Leftrightarrow q = 34.1 \text{ kW}$$

Hence, the heat losses per unit of length is calculated as shown below:

$$\frac{q}{L} = \frac{34.1 \text{ kW}}{3 \text{ m}} = 11.4 \frac{\text{kW}}{\text{m}}$$

If we also take into account that:

$$A_1 = \pi \cdot d_1 \cdot L = \pi \cdot 0.0381 \cdot 3 = 0.359 \text{ m}^2$$

and

$$A_2 = \pi \cdot d_2 \cdot L = \pi \cdot 0.0483 \cdot 3 = 0.455 \text{ m}^2$$

the heat losses per unit of heat transfer area are (heat flow density):

$$\frac{q}{A_1} = \frac{34.1 \text{ kW}}{0.359 \text{ m}^2} = 94.9 \frac{\text{kW}}{\text{m}^2}$$

$$\frac{q}{A_2} = \frac{34.1 \text{ kW}}{0.455 \text{ m}^2} = 74.9 \frac{\text{kW}}{\text{m}^2}$$

4.6.3 Connection of Solar Collectors In-Parallel and In-Series

In most cases of thermal hybrid plants, there is the need for the installation of multiple solar collectors, connected to each other either in parallel or in series or with a combination of these two basic connection layouts, depending on the requirements of the thermal power demand, as far as both the consumed thermal power and the temperature of the working fluid are concerned. In Figure 4.55 the in-series and in-parallel connections of two solar collectors are presented. When two solar collectors are connected in-parallel, the total flow rate of the working fluid is divided between them. When the collectors are connected to each other in-series, the working fluid's flow rate is common for all of them and equal to the inlet flow rate of the working fluid in the solar collectors' field. Additionally, for the in-series connection, the flow outlet of the precedent collector constitutes the flow inlet for the next one. It is so conceivable that with the in-series connection the achieved outlet temperature from all the collectors is higher than in-parallel connection, because the same flow rate of the working fluid passes through all the solar collectors, gaining an additional temperature rise from each of them, contrary to the in-parallel connection, where each one of the divided fluid flow rates passes through only one solar collector.

Additionally, according to what has been mentioned in the previous sections, the operating point and the efficiency of two solar collectors connected in-parallel are expected to be the same, or almost the same, given that the working fluid's flow rate and inlet temperature remain the same for both collectors and on the conditions that both collectors receive the same solar radiation, have the same orientation and installation angle versus the horizontal plane, and are similarly shaded from the surrounding technical or physical obstacles. On the contrary, two solar collectors connected in-series, even though the above conditions may be valid, will never operate at the same operating

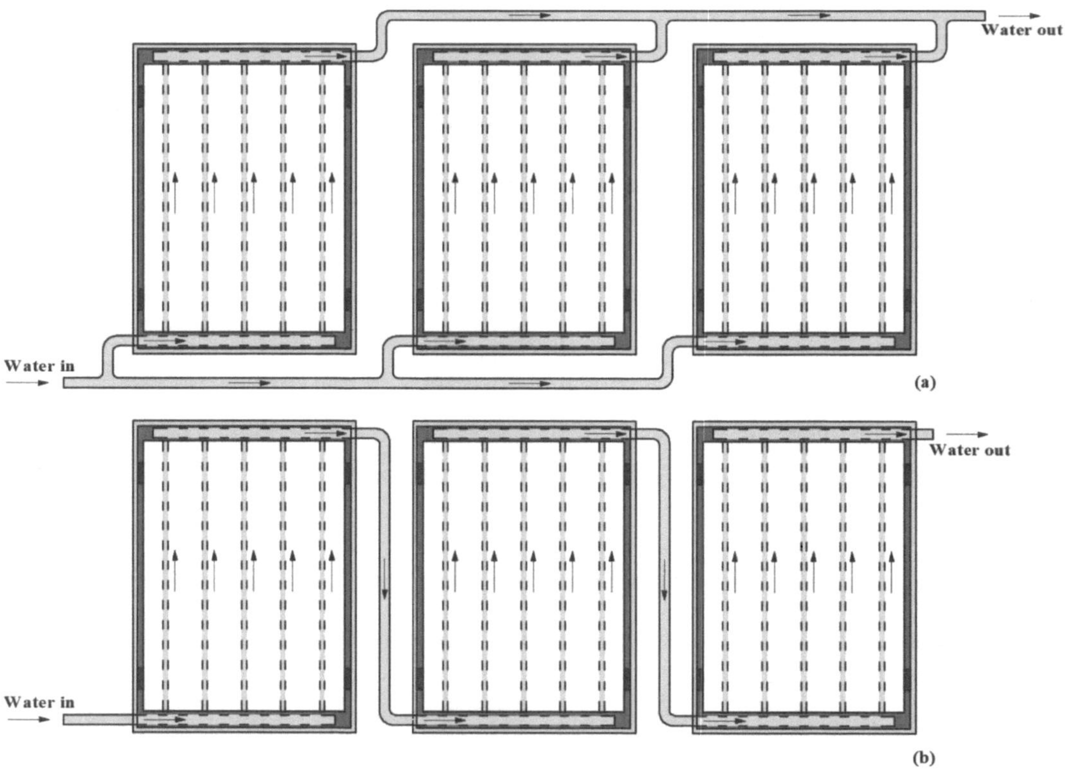

FIGURE 4.55 Solar connection essential connection modes (a) in-parallel and (b) in-series.

point and will never exhibit the same efficiency, because the working fluid's inlet temperature in the second collector will be higher than the inlet temperature in the first collector.

In most real applications, the basic connection modes of solar collectors are combined in order to approach the desirable result, with regard to the produced thermal power, the overall efficiency, and the achieved final outlet temperature. Such a case is presented in Figure 4.56, where the solar collectors are divided into five groups of six collectors each and another group of five collectors. The collectors included in each group are connected in-series, while all six groups are connected in-parallel. The above imply that

- the first solar collector from each group receives the working fluid as it is provided from the thermal storage tank
- the last solar collector from each group provides the working fluid for the common hydraulic network through which the produced thermal power will be provided for the thermal storage tank.

In Figure 4.56, which depicts an actual installation of a solar collectors' field in a building located in the north hemisphere, the siting concept of the solar collectors on the building's roof is also observed. The solar collectors are oriented towards the south and are installed in two parallel lines. The solar collectors' distance on the east–west direction is short, so as to minimize the required lengths of the hydraulic pipelines, which will lead to a reduction of the system's setup and operation cost, through the reduction of the flow and heat losses, but, also, to optimize the exploitation of the available space on the building's roof, to ensure that it will be enough for the installation of the required number of collectors. On the other hand, there is a considerable distance between the two parallel lines of the solar collectors on the south–north direction, approximately equal to three times the length of the collector's projection on the horizontal plane, so as to minimize the shading losses from the southern installation line to the northern one. It must be also noted that the solar collectors presented in Figure 4.56 are drawn taking into account their installation angle versus the horizontal plane. Hence, what is actually depicted in this figure is not the actual length of the collectors but the length of their projection on the horizontal plane.

For solar collectors connected in-parallel, the operation conditions are common for each one of them; hence, for the calculation of the total thermal power production from the total solar collectors'

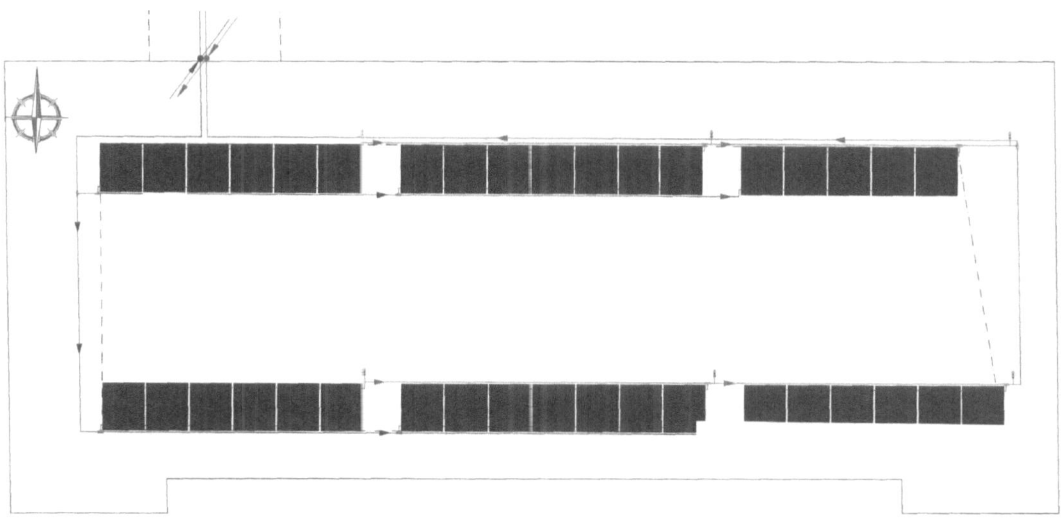

FIGURE 4.56 Combined connection of solar collectors in-series and in-parallel.

field, the methodologies presented in the previous sections are applied once for a typical solar collector from the overall installation. The calculated thermal power production from the one solar collector, multiplied by the total number of in-parallel connected collectors, will give the total thermal power production of the overall solar collectors' installation. The above approach, of course, can be applied on the conditions that all the solar collectors operate exactly with the same shading, are installed with the same orientation versus the south and with the same angle with regard to the horizontal plane, and receive the same solar radiation.

For an in-series connection, as already explained, the operation conditions for each solar collector are different. Consequently, the thermal power production from each solar collector will not be the same. At a first stage, let's assume a solar collectors field of two solar collectors connected in-series. The total thermal power production, according to the precedent sections, will be given by the following equation:

$$\dot{Q}_1 + \dot{Q}_2 = A_{c1} \cdot F_{R1} \cdot \left[G_t \cdot (\tau \cdot \alpha)_1 - U_{L1} \cdot (T_{fi1} - T_a)\right] + A_{c2} \cdot F_{R2} \cdot \left[G_t \cdot (\tau \cdot \alpha)_2 - U_{L2} \cdot (T_{fo1} - T_a)\right] \quad (4.104)$$

where T_{fi1} is the inlet temperature of the working fluid in the first collector and T_{fo1} is the outlet temperature of the working fluid from the first collector, which coincides with the inlet temperature in the second collector. This temperature, as presented, can be calculated from equation:

$$\dot{Q}_1 = \dot{m}_f \cdot c_p \cdot (T_{fo1} - T_{fi1}) \Leftrightarrow T_{fo1} = T_{fi1} + \frac{\dot{Q}_1}{\dot{m}_f \cdot c_p} \quad (4.105)$$

By eliminating the temperature T_{fo1} from the above two equations, we get [6]:

$$\dot{Q}_1 + \dot{Q}_2 = \left[A_{c1} \cdot F_{R1} \cdot (\tau \cdot \alpha)_1 \cdot (1-K) + A_{c2} \cdot F_{R2} \cdot (\tau \cdot \alpha)_2\right] \cdot G_t - \left[A_{c1} \cdot F_{R1} \cdot U_{L1} \cdot (1-K) + A_{c2} \cdot F_{R2} \cdot U_{L2}\right] \cdot (T_{fi1} - T_a) \quad (4.106)$$

where the magnitude K is equal to:

$$K = \frac{A_{c2} \cdot F_{R2} \cdot U_{L2}}{\dot{m} \cdot c_p} \quad (4.107)$$

The form of Equation 4.106, which is similar to the form of Equation 4.4, leads to the conclusion that the two solar collectors can be considered as one equivalent solar collector with the following characteristics [6]:

$$A_c = A_{c1} + A_{c2} \quad (4.108)$$

$$F_R \cdot (\tau \cdot \alpha) = \frac{A_{c1} \cdot F_{R1} \cdot (\tau \cdot \alpha)_1 \cdot (1-K) + A_{c2} \cdot F_{R2} \cdot (\tau \cdot \alpha)_2}{A_c} \quad (4.109)$$

$$F_R \cdot U_L = \frac{A_{c1} \cdot F_{R1} \cdot U_{L1} \cdot (1-K) + A_{c2} \cdot F_{R2} \cdot U_{L2}}{A_c} \quad (4.110)$$

In case of three solar collectors connected in-series, the above relationships can be applied initially for the first two collectors, leading to a new, equivalent collector. At a second stage, these relationships can be applied again between the equivalent collector and the third one. This procedure can be iteratively applied for more collectors connected in-series, leading, eventually, to an equivalent collector for all the in-series connected collectors.

In case of a total number of N similar solar collectors connected in-series, it is proved that for the equivalent collector, the following relationships stand [6]:

$$F_R\cdot(\tau\cdot\alpha)=F_{R1}\cdot(\tau\cdot\alpha)_1\cdot\frac{1-(1-K)^N}{N\cdot K} \quad (4.111)$$

$$F_R\cdot U_L=F_{R1}\cdot U_{L1}\cdot\frac{1-(1-K)^N}{N\cdot K} \quad (4.112)$$

Example 4.8 Calculation of Equivalent Solar Collector

Calculate the magnitudes $F_R\cdot(\tau\alpha)$ and $F_R\cdot U_L$ for the equivalent solar collector of four similar solar collectors connected in-series. The working fluid's common mass flow rate is 0.080 kg/s. For this particular mass flow rate, the characteristic technical features for the installed solar collector model are $F_R\cdot(\tau\alpha)=0.70$ and $F_R\cdot U_L=3.8$ W/(m²·K). The dimensions of the employed collector are 2 m × 1.2 m, while the specific heat capacity of the working fluid can be considered equal to $c_p=4.184$ J/(kg·K).

SOLUTION

Initially the parameter K is calculated from Equation 4.107:

$$K=\frac{A_c\cdot F_R\cdot U_L}{\dot{m}\cdot c_p}\Rightarrow K=\frac{2\text{m}\cdot 1.2\text{m}\cdot 3.8\text{W}/(\text{m}^2\cdot\text{K})}{0.080\text{kg/s}\cdot 4.184\text{J}/(\text{kg}\cdot\text{K})}\Leftrightarrow K=0.0272$$

Then the characteristic features of the equivalent collector are calculated from Equations 4.111 and 4.112:

$$F_R\cdot(\tau\cdot\alpha)=F_{R1}\cdot(\tau\cdot\alpha)_1\cdot\frac{1-(1-K)^N}{N\cdot K}\Rightarrow F_R\cdot(\tau\cdot\alpha)=0.70\cdot\frac{1-(1-0.0272)^4}{4\cdot 0.0272}\Leftrightarrow F_R\cdot(\tau\cdot\alpha)=0.6719$$

$$F_R\cdot U_L=F_{R1}\cdot U_{L1}\cdot\frac{1-(1-K)^N}{N\cdot K}\Rightarrow F_R\cdot U_L=3.8\cdot\frac{1-(1-0.0272)^4}{4\cdot 0.0272}\frac{\text{W}}{\text{m}^2\cdot\text{K}}\Leftrightarrow F_R\cdot U_L=3.65\frac{\text{W}}{\text{m}^2\cdot\text{K}}$$

4.6.4 Thermal Energy Storage in Multiple Storage Tanks

For thermal hybrid plants of large size, the required total thermal storage capacity will most probably be higher than the maximum capacity of a commercially available thermal storage tank. In such cases, in order to approach the required total thermal capacity of the thermal storage plant, multiple thermal storage tanks should be installed. The installation of more than one thermal storage tank increases the complexity of the thermal hybrid plant, with regard to its operation algorithm, its operation simulation, the connectivity of the involved components, and the alternative operation modes. The major question of course is which thermal storage tank should be employed each time for the storage of the potentially available thermal energy.

The operation of a hybrid thermal plant with more than one thermal storage tank will be analyzed with the support of an example. In Figure 4.57, a vertical hydraulic diagram is presented for a thermal hybrid plant aiming at the heating of a school building with 1,000 m² covered area, with the support of three thermal storage tanks [34].

The figure shows three involved thermal storage tanks are at the same time in-series and in-parallel connected between them and with regard to the solar collectors respectively. The in-series connection

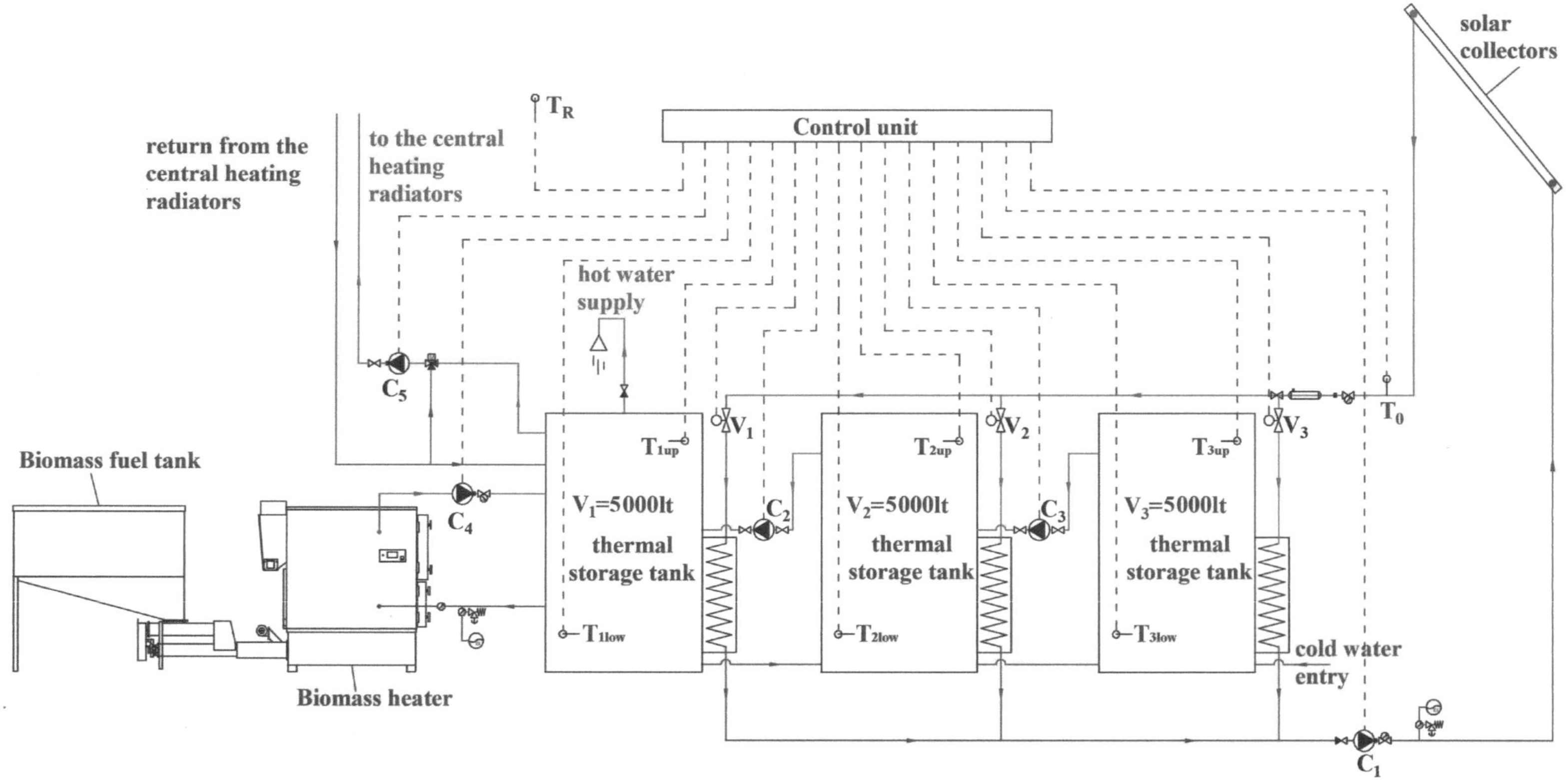

FIGURE 4.57 Vertical hydraulic diagram of a thermal hybrid power plant with three thermal storage tanks [34].

implies that the storage medium (water) can be transferred from one thermal tank to the other, while the in-parallel connection versus the solar collectors ensures that the thermal power produced by the collectors can be stored in any one of the installed tanks, without affecting the existing conditions in the others. The produced thermal power is transferred from the collectors to the thermal tanks through closed hydraulic circuits, at which the working fluid's flow is powered by the circulator C_1.

The central heating burner (backup unit) is connected exclusively to the first thermal tank, which constitutes the peak thermal tank. The same thermal tank is also used for the heat supply for the thermal distribution network, as well as for the produced warm water supply for domestic use. The produced thermal power from the conventional burner is transferred to the first thermal tank through a closed hydraulic loop, at which the working fluid's flow is powered by the circulator C_4.

The thermal energy flow between the storage tanks is also possible through the storage medium transfer, ensured with the circulator C_3, from the third to the second tank, and with the circulator C_2, from the second to the first tank.

The operation of the thermal hybrid plant is managed and supervised automatically by a central electronic control unit. The thermal power storage algorithm should focus on:

- the approach of the maximum possible temperature in the peak thermal tank (first tank), in order to minimize the operation of the conventional central heating burner
- the maximization of the exploitation of the supplied thermal power by the solar collectors.

Both the above targets are approached if we attempt to establish stratification thermal energy storage, starting from the lowest temperature in the third tank and ending with the highest possible temperature in the first, peak tank. This target can be achieved with the appropriate installation of circulators and motor-driven valves in the hybrid plant's hydraulic network, operating automatically according to the commands received by the central control unit. Nevertheless, the commands to be executed will be defined beforehand according to a number of temperature signals received by the central control unit from specific points of the network. Specifically, the signals received by the control unit for every time point are:

- temperature T_R of the indoor conditioned space,
- outlet temperature from the solar collectors T_0,
- temperature T_{1low} close to the base of thermal tank 1,
- temperature T_{1up} close to the top of thermal tank 1,
- temperature T_{2low} close to the base of thermal tank 2,
- temperature T_{2up} close to the top of thermal tank 2,
- temperature T_{3low} close to the base of thermal tank 3,
- temperature T_{3up} close to the top of thermal tank 3.

On the other hand, the control unit defines through commands the operation of the following devices:

- the circulator of the solar collector's primary closed loop C_1
- the circulator between the thermal storage tanks 1 and 2 C_2
- the circulator between the thermal storage tanks 2 and 3 C_3
- the circulator of the central heating burner closed loop C_4
- the motor-driven valve V_1, which controls the thermal power storage in tank 1
- the motor-driven valve V_2, which controls the thermal power storage in tank 2
- the motor-driven valve V_3, which controls the thermal power storage in tank 3.

The operation algorithm practically aims to determine which thermal energy storage tank should be used each time for the storage of the available for storage thermal energy from the solar collectors and whether the conventional burner should be used. The command for the burner's operation is given if there is a current thermal power demand Q_d, which is imposed in case the thermal comfort temperature T_{TC} is higher than the measured indoor space temperature T_R, and the maximum temperature T_{1up} in the highest level of the peak thermal tank is lower than the minimum temperature $T_{d,min}$ required for the operation of the thermal distribution terminals. The required control and the corresponding command are:

- if $T_{1up} < T_{d,min}$ and $T_R < T_{TC}$, then: C_4: ON, C_5: ON else C_4: OFF, C_5: OFF.

In order to ensure that the thermal energy stored in the thermal tanks will be the maximum possible one and, at the same time, the temperature distribution in the thermal tanks will increase from the third to the first tank, it should be checked every time whether thermal energy storage is possible, first of all, in the first thermal tank, then in the second and, finally, in the third tank. Practically, if the outlet temperature from the solar collectors is higher than the existing temperature in the peak thermal tank, then the available for storage thermal energy is stored in this first tank. If not, then the thermal energy will be stored in the thermal tank with the maximum temperature that remains lower than the outlet temperature from the solar collectors, with the second tank exhibiting priority versus the third one. The storage algorithm described above is implemented with the following controls and the corresponding commands:

- if $T_0 > T_{1low}$, then C_1: ON, V_1: open, V_2: close, V_3: close, else:
- if $T_0 < T_{1low}$ and $T_0 > T_{2low}$, then C_1: ON, V_1: close, V_2: open, V_3: close, else:
- if $T_0 < T_{1low}$ and $T_0 < T_{2low}$ and $T_0 > T_{3low}$, then C_1: ON, V_1: close, V_2: close, V_3: open, else:
- if $T_0 < T_{1low}$ and $T_0 < T_{2low}$ and $T_0 < T_{3low}$, then C_1: OFF, V_1: close, V_2: close, V_3: close.

The last command ensures that there will not be any flow of the working fluid from the solar collectors to any one of the available thermal storage tanks, if there is no temperature detected, in any thermal tank, lower than the outlet temperature from the solar collectors.

Finally, if during the thermal energy storage in the third or in the second thermal tank, its temperature becomes higher than the temperature in the second or the first tank respectively, then thermal power should be transferred between the tanks, specifically, from the third to the second tank or from the second to the first tank, in order to maintain the desirable temperature stratification between the tanks. This operation is ensured with the following control and commands:

- if $T_{2up} > T_{1low}$, then C_2: ON
- if $T_{3up} > T_{2low}$, then C_3: ON.

With the above example, we see that thermal energy storage in multiple thermal storage tanks constitutes a complex task, which should be designed and accomplished appropriately, given the conditions and the requirements met at each different application, in order to ensure maximum effectiveness of the thermal hybrid plant. The above presented algorithm, although obvious and rational, can be only characterized as indicative. Different requirements of the final thermal energy needs may impose different operation conditions for the thermal hybrid plant, which, in turn, may lead to modifications on the adopted algorithm. In any case, the involvement of multiple thermal storage tanks in thermal hybrid plants should be treated adequately with regard to both the operation algorithm of the thermal hybrid plant and its realization, with the formulation of the appropriate hydraulic network and the introduction of the required automations.

4.7 SOLAR THERMAL POWER PLANTS

The basic operation concept of solar thermal power plants is the concentration of solar radiation onto a receiver, for the heating of a working fluid in temperatures adequate for mechanical power production. For this purpose, concentrating solar collectors are employed. The working fluid is circulated either in a closed, primary loop, or inside a single loop. In the first case, through heat exchangers, the contained thermal power can be either stored in thermal tanks or transferred directly to a secondary loop, where it is utilized for steam production. The produced steam is then exhausted in a steam turbine for the production of mechanical power. In the second case, the working fluid is a water–steam mixture, which, after the absorption of the concentrated solar radiation, can be led either directly to a steam turbine, for direct power production, or to thermal tanks, for thermal energy storage. In most implementations, there is also a conventional backup heater, which, by consuming fossil fuels, is utilized for supplemental thermal power production during periods of low available solar radiation (cloudy days and nights). The overall operation concept is graphically presented in Figure 4.58.

According to this figure, a solar thermal power plant consists of the following discrete components:

- the concentrating solar collectors
- the solar radiation receiver
- the thermal energy storage tank
- the conventional heater
- the working fluid transportation hydraulic network
- the steam turbine(s).

The optimum configuration of a solar thermal power plant imposes adequate surface for the installed solar collectors, for the collection and the concentration of the incident solar radiation onto the

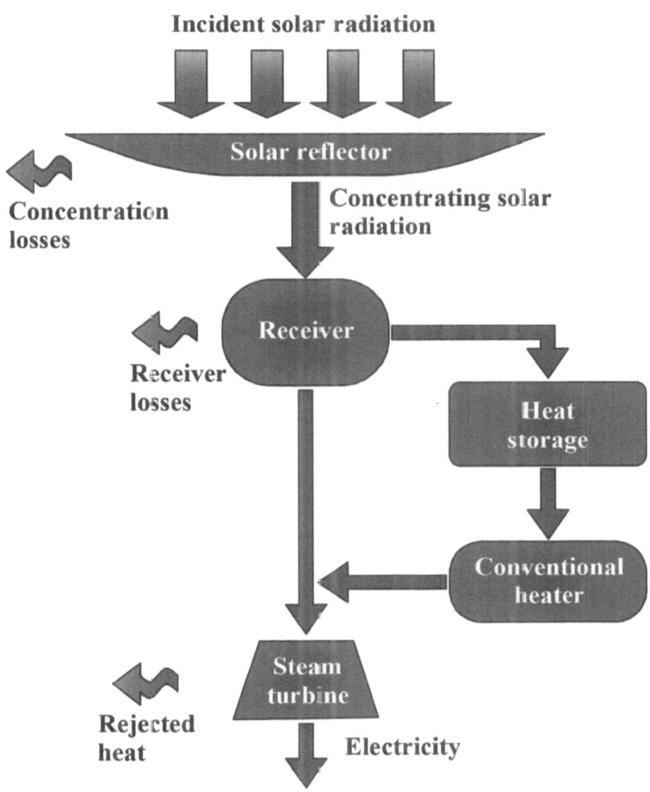

FIGURE 4.58 Operation concept of solar thermal power plants.

receiver. The solar receiver is characterized with high absorptance, approaching the performance of a black body. The concentrated solar radiation in the receiver is transferred to a working medium in the form of thermal energy at high temperatures (depending on the plant's technology, from 350°C to temperatures higher than 550°C). The absorbed thermal energy by the working medium is eventually utilized for power production.

Although the concept of the concentration of solar radiation has been generally known for more than a century, it gained attention after the first oil crisis in 1973. During the 1980s and until 1990, the first nine solar thermal power plants were constructed in the Mojave Desert, in California, US, with a total installed nominal power of 354 MW. Named Solar Electric Generating Systems (SEGS), they constituted in total the largest solar thermal power plant in the world at that time.

After the installation of these first solar thermal projects, there was a long period of close 20 years without any more construction of similar projects. The main reasons were the domination of cheap fossil fuels and the high setup costs of the required solar thermal technology, which reduced the attractiveness of the corresponding investments. Yet, with technology's evolution and the experience gained from the operation of the SEGS project, combined with the gradual increase in fossil fuel prices and the global trend towards the promotion of renewables, the interest in solar thermal power plants began to rise again. The production cost of the required equipment was gradually reduced, leading to a subsequent drop of the solar thermal power plants' operation cost. By the beginning of the twenty-first century, the operation cost of the SEGS plant dropped one-third compared to the first years of its operation [35]. These conditions led to the beginning of the construction of new solar thermal power plants after 2005. The first of them were integrated from 2009 to 2010. Today (2019), after nearly a decade of resurgence in solar thermal power plant installations, more than 50 different projects have been in total installed worldwide, specifically in the US, Spain, India, South Africa, UAE, and Morocco, with nominal power from 50 MW to 392 MW. The largest solar thermal power plant project is the Ivanpah Solar Electric Generating System (Figure 4.59), with a nominal power of 392 MW, completed in February 2014, also installed in the Mojave Desert, like the first SEGS plants. The total installed power globally of solar thermal power plants reached 4.9 GW by the end of 2017.

The installed solar thermal power plants have been erected in geographical locations with high available solar radiation, namely with global horizontal irradiance higher than 800–900 W/m^2, especially during summer, and annual cumulative solar irradiation from 1,600–2,800 kWh/m^2.

FIGURE 4.59 The Ivanpah solar thermal power plant in the Mojave Desert in the US (under due diligence) [36].

These figures imply time periods of full load operation at the range of 2,000–3,500 hours annually, or, equivalently, annual capacity factors from 22%–40%.

Solar thermal power plants today, after the experience gained from the installation and the operation of the above projects, exhibit the following positive features:

- tested operation and proven performances, characterized with high availability and setup and operation cost percentage reductions close to 50%, compared with the first installations in California, during the 1980s
- modular construction, which means that they can be easily expanded and adapted according to the needs of the power system or the economic incentives of the investors
- they can fully support the electricity production for autonomous grids, ensuring the stability and the security of the electrical systems
- they seem to be ideal for areas with high available solar radiation and restricted possibilities for alternative exploitation, like desert areas in the US, Africa, and Middle East
- they offer an alternative approach for increased RES penetration in areas with low wind potential availability and inappropriate land morphology conditions for the installation of pumped storage.

On the other hand, solar thermal power plants exhibit certain drawbacks as well, such as:

- they require the availability of considerably high solar radiation, which, practically makes them infeasible for geographical latitudes higher than 35°
- they can never be exclusively renewable, in the sense that they require the consumption of a type of fossil fuel to ensure continuous operation during periods of low available solar radiation
- most solar thermal power plant technologies require considerable amounts of land, which ranges close to 40,000 m^2 per MW of nominal power [37]
- despite the achieved reductions in setup costs, solar thermal power plants still remain a rather expensive technology, with setup cost, most commonly not lower than 3,500 $/kW$_e$, while, depending on the technology, the size and the installation area of the project, this figure may reach 7,500 $/kW$_e$.

4.7.1 Solar Thermal Power Plant Alternative Technologies

Solar thermal power plants are distinguished based on the types of the involved solar collectors and receivers. For the solar radiation concentration, either point focus or linear focus solar collectors can be employed. With linear focus collectors, optical concentration ratios from 70 to 100 and working medium's temperatures in the receiver up to 550°C can be achieved. On the other hand, the optical concentration ratios achieved with point focus collectors can reach 1,000, leading to temperatures of the working medium in the receiver higher than 1,000°C. The most technically mature solar thermal power plant technologies are:

- the parabolic trough collector (PTC) systems
- the linear Fresnel reflector (LFR)systems
- the central receiver (CRS) or solar power tower (SPT) systems
- the disk engine (DE) or parabolic disk collector (PDC) systems.

The first two of the above technologies are linear focus systems, while the last two are point focus systems.

4.7.1.1 Parabolic Trough Collector Systems

Compound parabolic trough collectors are the most commonly utilized types of concentrating solar collectors in solar thermal power plants. The system is composed of linear focus concentrating solar collectors, installed in parallel lines (Figure 4.60). The trough collectors' parallel lines are placed along the east–west axis, so the solar collectors are oriented to the north–south axis. The collectors are rotated around the horizontal axis, tracing the sun's orbit on the horizon from the east to the west (practically the changes of the solar height). There is no need for rotation of the solar collectors around the vertical axis, to trace the seasonal changes of the sun's orbit, because these changes are performed in parallel with the focal line, hence the solar radiation concentration is displaced only along the receiver's direction.

So far, in commercial applications, the compound parabolic trough collectors are constructed with thermoplastic glass of 4 mm thickness, which is both heavy and expensive. Modern trends aim to reduce the collectors' weight with the introduction of new techniques and materials, such as the use of glazed aluminum, instead of glass. The reflected incident solar radiation is concentrated at the focus of the parabola, where the absorber tubes (receiver) with the flowing working fluid are located. The receivers are designed and constructed in order to exhibit high incident solar radiation absorptance. For this purpose, the absorber tubes are manufactured from steel with a special black coating on the outer surface, aiming to maximize the absorbed solar radiation and minimize the thermal losses via radiation to the environment. The steel tube's absorptance exceeds 90%. The steel tube is placed inside a protective glass tube. The space between the steel and the glass tubes is vacuum, aiming to reduce the thermal losses from the absorber tube to the ambient environment (Figure 4.61). Typical inner diameters for the steel and the glass tubes of the receiver can be 70 mm and 115 mm respectively. The glass tube outer surface can be also covered with an antireflective coating, to facilitate the transmittance of the incident solar radiation.

The heat transfer fluid (HTF) in the receiver, usually a synthetic oil with high thermal conductivity, through its circulation in a closed loop, gathers the absorbed thermal energy from the solar collectors and transfers it to a series of heat exchangers (steam generators), where the transferred heat is utilized for the production of superheated steam, normally at temperatures of 370°C–400°C and pressures of 90–100 bars. The HTF inlet and outlet temperatures in and from the solar collectors field is at the range of 290°C and 390°C respectively. The produced steam is then led to the steam

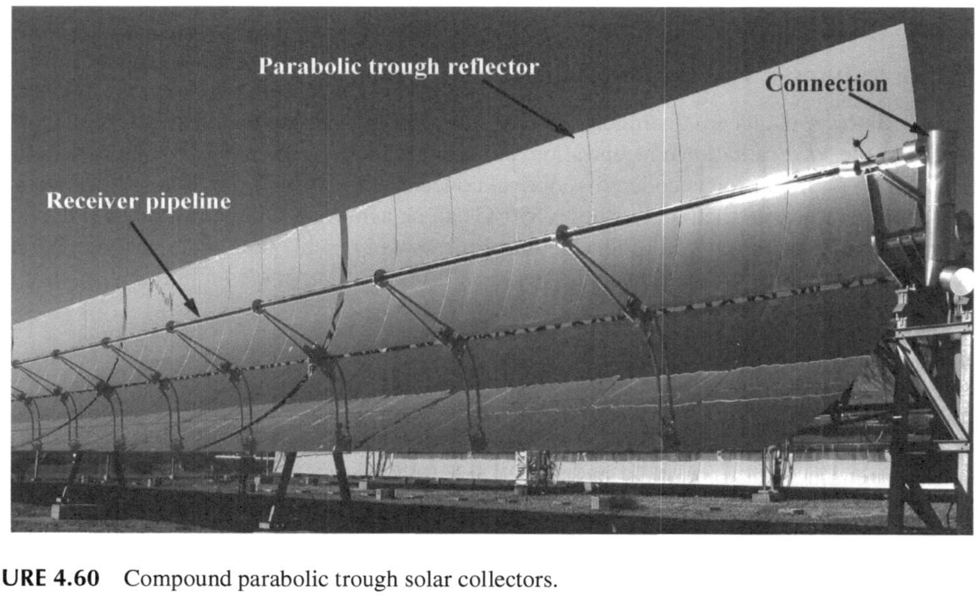

FIGURE 4.60 Compound parabolic trough solar collectors.

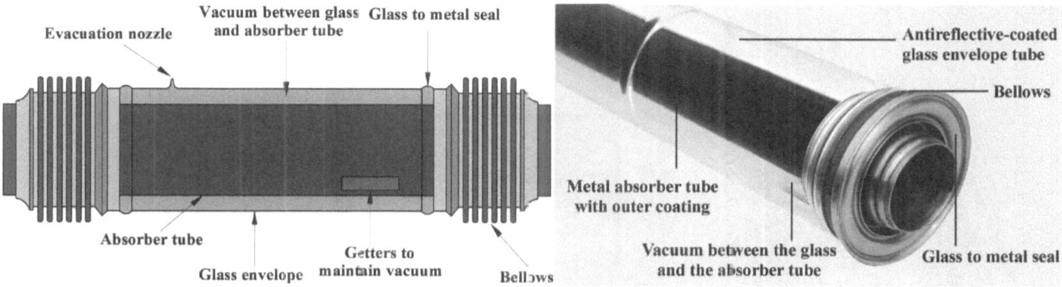

FIGURE 4.61 The fundamental constructive concept of the solar radiation receiver in compound parabolic trough collectors [38].

turbines for the production of mechanical power. The synthetic oil usually used is a eutectic mixture of 73.5% diphenyl ether oxide and 26.5% diphenyl ether. The main operational problem with this material is its high solidification temperature, around 12°C, due to which an auxiliary heat source is required during periods of low ambient temperature and solar radiation availability. Additionally, the boiling point of this employed synthetic oil under pressure of 1.013 bars is 257°C; hence, the synthetic oil inside the receiver's loop should be under pressure higher than the atmospheric one, together with an inert gas, such as nitrogen, argon, etc., in order to avoid its evaporation when its temperature becomes higher than the above mentioned boiling point.

The synthetic oil's maximum possible temperature of 400°C puts an upper limit for the steam turbine efficiency, which increases with the HTF temperature. For this reason, currently there is extended research on the use of advanced HTFs in the receiver, such as molten salts, or on the direct use of steam. The use of molten salts enables the increase of the HTF temperature at the level of 560°C, leading to the subsequent increase of the power production process efficiency. On the other hand, the direct steam generation (DSG) in the receiver eliminates the use of heat exchangers, enabling the reduction of the system's setup cost and the increase of the overall efficiency due to the elimination of heat transfer losses in the heat exchangers.

In a typical solar thermal power plant with compound parabolic trough collectors, a number of solar collectors are connected in-series to form a line and a number of lines are connected in-parallel to reach the required thermal power production at the nominal operation point. The number of the in-series connected solar collectors at each line depends on the HTF's temperature required increase during a single circulation in the receiver. The receiver's tubes between the collectors' lines should be connected with flexible joints, to enable the collectors' rotation. Additionally, the same joints should be able to treat the receiver tubes' linear thermal expansion, during their temperature rise. For these reasons, two main techniques are available: flex hoses and bole joint assemblies. Flex hoses were used in the first solar thermal power plants projects in the Mojave Desert. Today, bole joint assemblies are mainly used due to certain advantages regarding their reliability, their maintenance requirements, and the reduced subsequent cost.

The structural layout of a solar thermal power plant with PTCs is presented in Figure 4.62.

If the solar thermal power plant is equipped with a thermal storage plant, the thermal energy from the HTF primary loop can be alternatively disposed in the thermal tank. More about available thermal storage technologies is presented in a later section.

The operation of the whole system is usually facilitated with the introduction of a conventional heat production unit, such as a fossil fuel burner, which aims:

- at the improvement of the system's efficiency, through the increase of the produced steam temperature–pressure (in other words, the increase of the steam's exergy)
- at the prolongation of the solar thermal power plant's operation periods after sunset and during periods of low available solar radiation

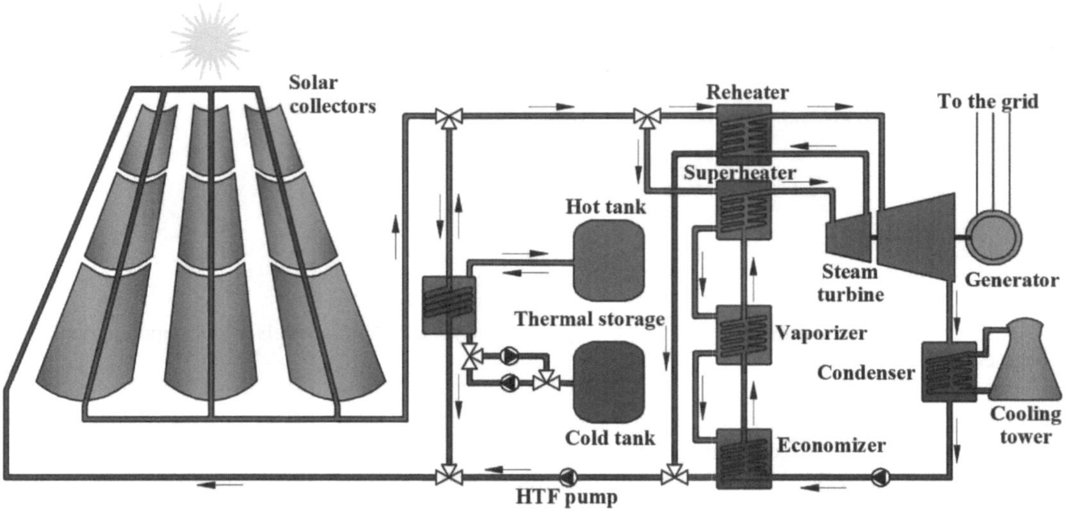

FIGURE 4.62 Structural layout of a solar thermal power plant with parabolic trough collectors, supported with thermal storage and conventional heater [39].

- to ensure constant and secure power production, crucial in cases of high percentage penetration of the solar thermal power plant in the electrical grid or in cases of weak, autonomous electrical systems
- to maintain the functionality of solar thermal power plant components during the winter period, when the possible low ambient temperatures can be destructive for the synthetic oil in the receiver or for the molten salts in the thermal storage tank [40].

The annual average efficiency of a solar thermal power plant with PTCs is in the range of 14%–16%, calculated as the ratio of the final produced electrical power over the incident solar radiation density.

Today (2019), most solar thermal power plants installed globally are constructed with PTCs. The total installed power exceeds 1 GW. Among the installed stations, the first SEGS plants of 354 MW in the Mojave Desert, US, and the first solar thermal power plants in Europe, installed in Granada, Spain, in 2008, 2009, and 2011 with a nominal power of 150 MW, are included.

4.7.1.2 Linear Fresnel Reflector Systems

The linear Fresnel reflector (LFR) system is the second linear focus technology of solar thermal power plants. These systems are integrated with long and narrow lines of flat or slightly curved reflectors, with their reflected surfaces made with glass, through which the solar radiation is concentrated to one or more linear receivers, placed above the reflectors (Figure 4.63). The receiver is placed along the focal line of the Fresnel reflectors and, most commonly, consists of a metal tube, coated with a selective coating, which extends along the entire receivers' line. A secondary parabolic reflector is usually installed above the receiver for further solar radiation concentration, by reflecting back the radiation that does not directly strike the receiver. The metal absorber can be also enclosed inside a glass tube, as in parabolic trough collectors, for further reduction of the heat losses from the absorber to the ambient. LFR systems exhibit lower optical performance and thermal power production, yet these drawbacks are compensated with lower setup and maintenance costs.

The main operating feature of the LFR systems is the use of Fresnel reflectors. The performance of these reflectors is based on the exploitation of the Fresnel lens phenomenon, which enables the construction of concentrating reflectors with high aperture and low focal length, leading to the reduction of the required materials quantities for the construction of the reflectors and the

FIGURE 4.63 The fundamental operation concept of linear Fresnel reflector systems (right photo under due diligence of U.S.A. Department of Energy) [41, 42].

subsequent drop of the manufacturing and procurement cost. Additionally, in LFR systems, the solar radiation is concentrated by a number of Fresnel reflectors at the same receiver, unlike the parabolic collectors, where each collector is equipped with its own receiver. At the same time, the one-axis tracing system is also maintained in this technology, as in parabolic trough collectors. The Fresnel reflector lines are installed along the east–west direction, so the reflectors are oriented to the north–south axis.

The receiver in LFR systems is placed in a stable and constant position, so there is no need for flexible hydraulic joints. The type of the receiver depends on the employed working fluid for the heat transfer. The receiver most commonly consists of parallel, black-coated absorber tubes, made of high-temperature metal alloys. This construction is used with water–steam mixture or molten salt as the working fluid. In cases of air, volumetric absorbers are employed, which use a highly porous ceramic structure to absorb the reflected solar radiation and heat the air flowing through them.

A major difficulty often occurred with LFR systems is the blocking of the incident solar radiation and the optical shading from neighboring reflectors. These malfunctions can be treated by increasing the height of the installation position of the receiver tube and by increasing its size, which will impose longer distances between the Fresnel collectors' lines and reduction of the shading effects. Yet, this solution will also imply higher required installation land. This inconvenience can be overcome with the use of appropriate focus techniques, such as reflectors focusing on different receivers for different time periods during the day. With this approach, the density of the installed reflectors on the available land can increase, leading to a subsequent drop of the required occupied land. This alternative will be further analyzed below.

As with the parabolic solar collectors, thermal energy is absorbed in the receiver by a working fluid, which can be a synthetic oil, a water–steam mixture, or a molten salt. The most popular application is with the use of a water–steam mixture. The water–steam mixture is heated up under pressure of 50–60 bars and temperature 280°C for the production of saturated or superheated steam, which is led directly to the steam turbine. The operation layout of an LFR system with direct steam production is depicted in Figure 4.64.

A compact variation of this technology is the so-called concentrating LFR or compact LFR; in both cases it is abbreviated as CLFR. In a simple LFR system there is only one linear receiver placed in a single linear construction. Hence, each Fresnel reflector is oriented towards this unique receiver, without any other alternative choices. On the contrary, with a CLFR implementation, there are different receivers placed inside the reflectors' installation area. Hence, if the receivers are installed close enough to each other, each single reflector will have the option to choose at least between two alternative receivers for the concentration of the reflected solar radiation. This option constitutes the

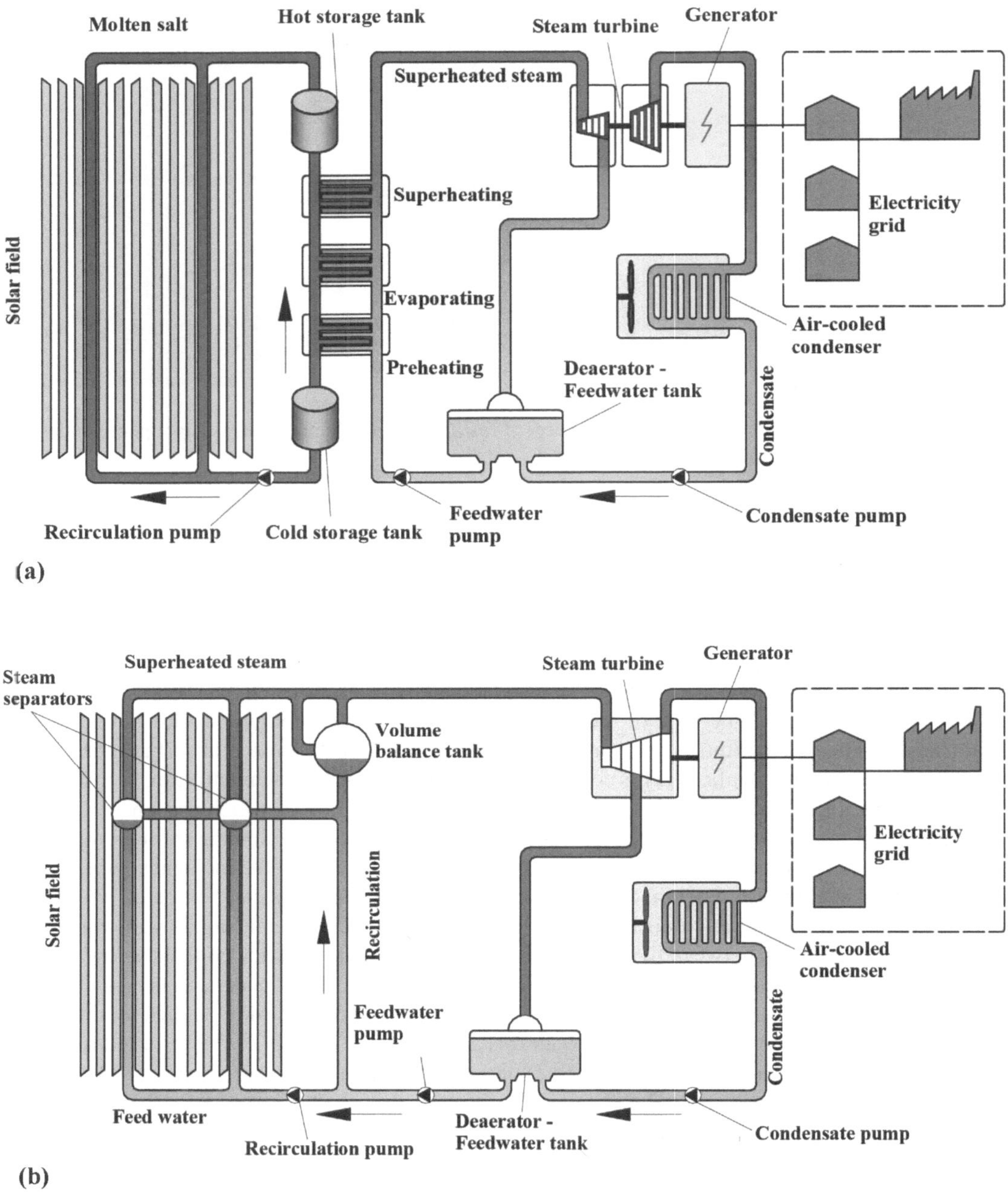

FIGURE 4.64 Operation layout of an LFR system alternatively with (a) heat storage and (b) direct steam production [43].

solution for the above mentioned shading and blocking problem of the incident solar radiation from neighboring reflectors, offering the possibility for thicker siting of the Fresnel reflector lines. This is achieved by adjusting the inclination for every reflector line differently, in combination with the neighboring lines, so each line is focused on a different receiver, avoiding, thus, the solar radiation blocking or shading from neighboring reflector lines. Additionally, the thicker siting of the Fresnel reflectors imposes lower total length of the absorber tubes and higher installed power per occupied land unit, which, in turn, contributes to the reduction of the thermal energy losses to the ambient and the total setup cost of the solar thermal power plant.

The first LFR power plant was constructed in 2008 in California, US, with a nominal power of 5 MW_e. It was the first solar thermal power plant after the SEGS systems installed in the same geographical region during the 1980s. The employed technology is CLFR. The second LFR system and the first in Europe was installed 1 year later in Spain, with a final electrical power output of 1.4 MW_e. Today (2019), around 50 MW_e of LFR systems have been installed worldwide.

The main advantages of the LFR systems are:

- Lower setup and operation costs than the other solar thermal power plant technologies.
- Considerably reduced land occupation with regard to the other solar thermal power plant technologies. While the required land for the other solar thermal power plant technologies is in the range of 40,000 m^2/MW_e, the same feature for the LFR systems is less than 20,000 m^2/MW_e, namely it appears more than 50% reduced.
- Simplified installations, restricted maintenance requirements.

The most important drawbacks of this technology are:

- The so far relatively few implemented and operating relevant projects and the corresponding restricted gained experience.
- The relatively low efficiency of the overall electricity production process, which normally ranges from 7% to 10%, defined as the ratio of the final produced electrical power over the incident solar radiation density [44].

4.7.1.3 Power Tower/Central Receiver Systems

Solar thermal power plants of power tower or central receiver systems (CRS) technology consist of individual, multiple tracking reflectors, called heliostats, with reflecting surfaces for each one from 1 m^2 to 140 m^2, placed in a circular layout around a central, external receiver, installed at the top of a tower. The heliostats are constructed with a flat or slightly curved glass surface. The reflected solar radiation from the heliostats is concentrated at the central receiver, which constitutes the focal point, through the appropriate orientation of the heliostats (Figure 4.65). For this reason, each heliostat is equipped with a two-axis tracking system, in order to alter its orientation continuously, relative to the sun's position in the horizon. Rotary or linear drivers, or, in modern applications, hydraulic actuators, can be used for the tracking system. The central tower can be as high as 200 m, depending on the size of the solar thermal power plant. From the above, we see that solar power tower plants are classified as point focus systems.

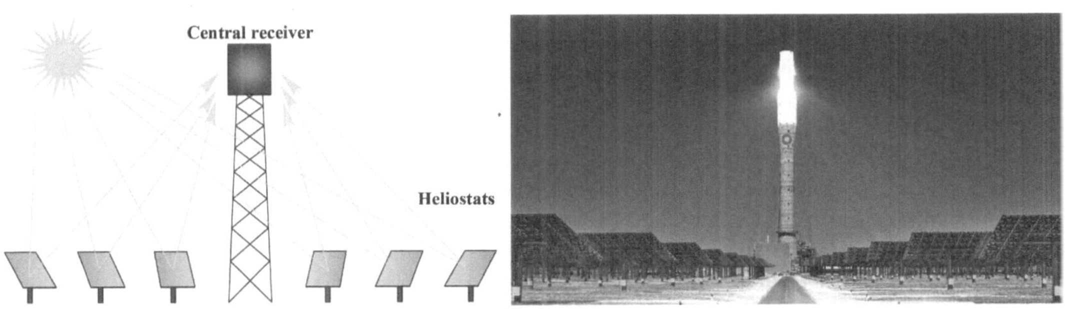

FIGURE 4.65 The fundamental operation concept of solar central receiver systems [41, 45] (right photo under due diligence of Torresol Energy).

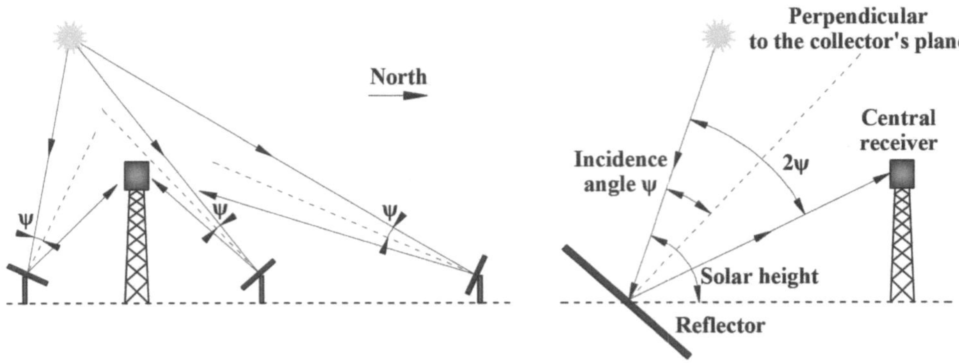

FIGURE 4.66 Incident angle of solar radiation on a flat surface.

The siting of the heliostats with regard to the location of the power tower depends on the geographical latitude of the solar thermal power plant's installation site. The goal is always to maximize the reflected solar radiation towards the central receiver. The reflected solar radiation depends on the radiation's incident angle on the heliostat. According to the fundamental theory of solar radiation/solar geometry, the total incident solar radiation on a flat surface is analogous to the cosine of the direct solar radiation incident angle θ_i [46]. This is clearly depicted in Figure 4.66. As shown in this figure, for a solar thermal power plant installed in the north hemisphere, the cos θ_i can take values close to 0.9, for the heliostats installed at the north side of the power tower, while for the heliostats installed at the south side of the power tower, the cos θ_i is considerably reduced. Hence, for plants installed at the north hemisphere, the solar field should be installed at the north side of the power tower, and vice versa. For solar plants installed close to the equator, the solar field should surround the power tower.

The concentrated solar radiation at the central receiver is transferred in the form of thermal energy towards the working fluid, similarly to the parabolic trough systems. The absorbed thermal energy from the working fluid is then utilized for the production of steam, which, in turn, is exhausted in a steam turbine for the production of mechanical power. The working fluid can be a molten salt, air, or water-steam mixture. In the last case, the produced steam is led directly to the steam turbines—there are no heat exchangers involved in the thermal power transportation process. There is also the option of thermal energy storage in a thermal storage tank. As in LFR systems, the receiver can consist of black-coated parallel absorber tubes, made of high-temperature metal alloys, if the working fluid is water–steam mixture or molten salt, or volumetric absorbers with highly porous ceramic structure, if the working fluid is air.

The optical concentration ratios with CRS solar thermal power plants can be from 600 to 1,000, hence the temperature of the working medium and, eventually, the produced steam can be strongly higher than with parabolic trough collectors. This is because the temperature of the molten salt, employed as the working medium in the receiver, can increase more, compared to the synthetic oil employed in parabolic collectors, with which the maximum achieved temperature is lower than 400°C, reducing, respectively, the contained exergy of the produced steam.

The temperature of the working fluid inside the absorber can be up to 565°C, according to measurements captured during commercial operation, while it has been experimentally observed that this feature can approach values in the range of 1,000°C. Due to the high possible temperatures in the central receiver, the CRS power plants can also be combined, apart from steam turbines, with gas turbines and combined cycles as well. Indeed, a pressurized gas at temperatures close to 1,000°C, approaches the properties of the gases exhausted from a gas turbine. Moreover, if the employed gas turbines are integrated in the frame of a combined cycle, then the overall efficiency

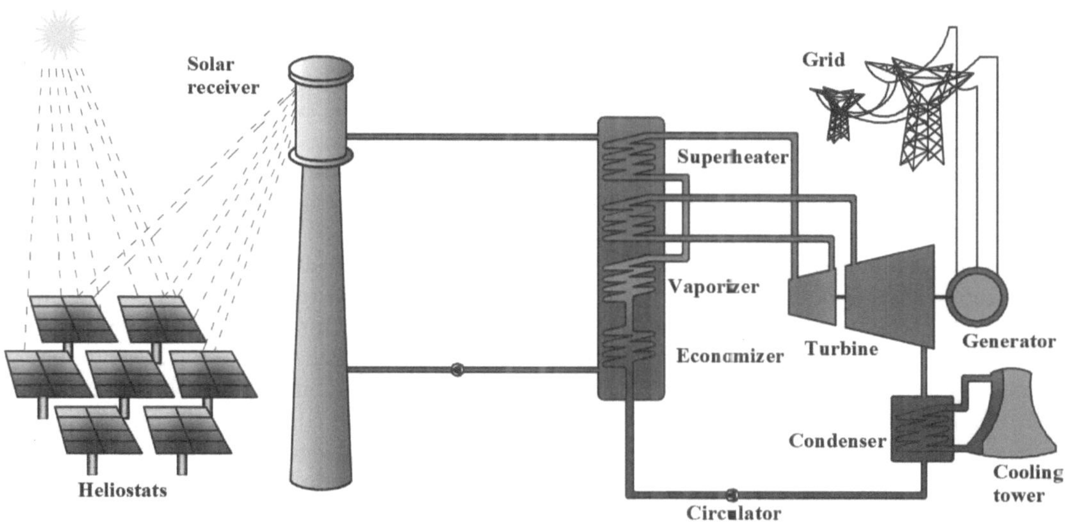

FIGURE 4.67 Operation layout of a CRS plant with a steam turbine [47].

of the solar thermal power plant can reach values as high as 35%, at nominal operation, and average annual efficiencies up to 25%.

In CRS solar thermal power plants, as in the case of parabolic trough collector plants, the whole system can be optionally integrated with the installation of a conventional thermal power production unit and a thermal energy storage tank. Most commonly, the working medium employed in the thermal energy tank is molten salts. These components offer to the overall plant similar advantages and flexibilities as for parabolic trough collector plants.

The capacity factor of a CRS plant can be increased, for a certain steam turbine nominal power, obviously by increasing the heliostats' installed effective surface, as well as with the increase of the thermal storage capacity, the increase of the central receiver's size, and the increase of the tower's height.

The operation layouts of a CRS plant with a steam turbine or a combined cycle are depicted in Figures 4.67 and 4.68.

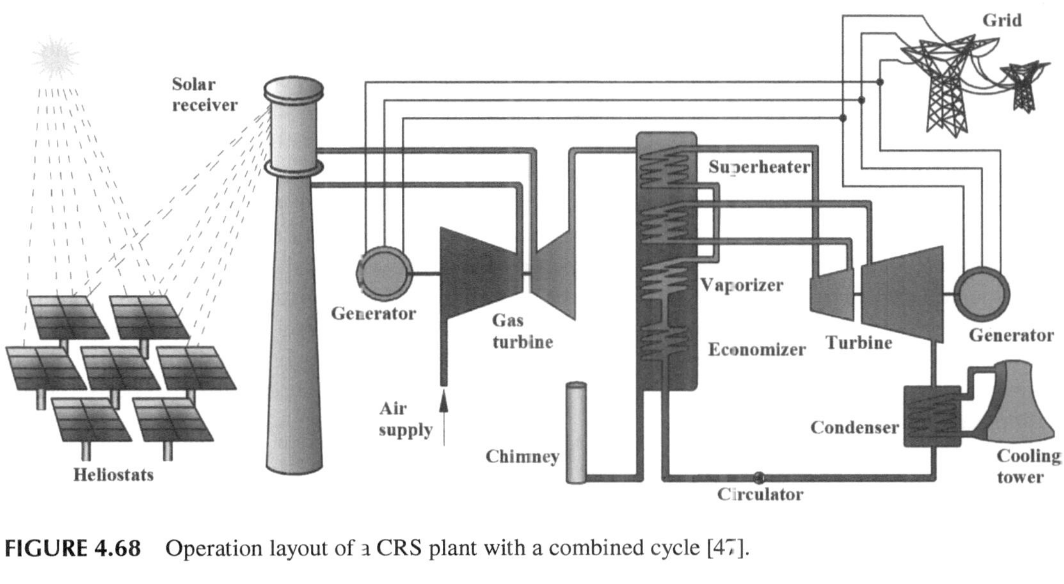

FIGURE 4.68 Operation layout of a CRS plant with a combined cycle [47].

The first commercial CRS plant was constructed in 2007 in Sanlucar la Mayor, close to Seville, Spain, with a nominal power of 11 MW_e. The solar field consists of 624 heliostats that cover a total surface of 600,000 m^2. The central power tower's height is 115 m. The produced thermal energy is stored in the form of pressurized steam in thermal tanks, offering the ability for the power plant for 1–2 hours of additional operation after sunset. Additionally, there is a supplementary conventional natural gas heater. With the annual solar irradiation at 2,100 kWh/m^2, the specific CRS solar thermal power plant produces annually 24.3 GWh of electricity, a figure that gives an annual capacity factor of 25%. The overall, annual average efficiency is configured at 17% [48]. The first CRS plant with molten salt employed as the working medium in the central receiver was constructed in 2011 at a location Fuentes de Andalucia, close to Seville, Spain again. The nominal power of this station is 19.9 MW_e. This station is equipped with thermal storage tanks with storage capacity that offers 15 hours additional operation at nominal power without any solar radiation available. This feature practically enables continuous operation of the power plant during summer. The solar field consists of 2,480 heliostats, with an effective surface of 300,000 m^2 and total covered land of 1,420,000 m^2. However, the most impressive CRS plant currently installed worldwide is the Ivanpah solar thermal power plant, mentioned in an earlier section, which constitutes the largest solar thermal power plant globally at the time of this book's printing. The plant's installation was completed in 2014. It has a nominal power of 392 MW_e, with 173,500 heliostats, covering total land of 16,000,000 m^2. It consists of three separate power plants, with three individual power towers. The annual net final electricity production in 2015 reached 650 GWh, configuring, thus, an annual average capacity factor of 18.9%. The annual average overall efficiency is estimated at 15.8%. Finally, the overall setup cost reached $2.2 billion, leading to a setup-specific cost of 5,600 €/kW_e. The total installed power of CRS plants globally is estimated at 570 MW_e.

Attempting a comparison between solar thermal power plants with parabolic trough collectors and power towers, the major advantage of the latter, without doubt, is the potentially higher temperatures in the central receiver. Thermal energy at higher temperatures can be converted to electricity or stored as thermal energy easier for further use. Another important advantage of the CRS plants is that, exactly due to the high achieved temperatures, the installation of an air-cooling system is more easily applicable, instead of a water-cooling system. This feature is also of particular importance given the fact that solar thermal power plants are usually installed in geographical areas with high available solar radiation and relatively low water resources. Additionally, CRS plants can also be installed in inclined land surfaces, such as a mountain slope, because the only requirement is the orientation of the heliostats towards the central receiver, where all the required hydraulic network is gathered.

On the other hand, although flat heliostats are cheaper than parabolic collectors, each heliostat must have its own two-axis tracking system, in order to continuously focus the reflected solar radiation at the central receiver, while at parabolic trough collectors only one single-axis tracking system is required for each collectors' line. Additionally, the reflected solar radiation from the heliostats does not entirely reach the receiver, because a portion of it is diffused and lost in the atmosphere, reducing the efficiency and the performance of the power plant. This phenomenon is known as atmospheric attenuation. Yet, currently, the most important drawback of the CRS plants is the limited experience and technical knowledge gathered from their operation, due to the relatively few implemented and operating plants. However, the simplicity of CRS construction and the heliostats' low procurement lay the groundwork for the rapid increase of corresponding implemented projects within the next decades. It is estimated that by 2020 the electricity production cost from CRS plants will be configured at 5.47 US cents/kWh_e and at 6.21 US cents/kWh_e for solar thermal power plants with parabolic trough collectors [49].

4.7.1.4 Parabolic Disk Systems

Parabolic disk systems utilize a large, reflective, parabolic disk for the solar radiation concentration on a focal point, where the receiver is placed (Figure 4.69). The concentrated solar radiation

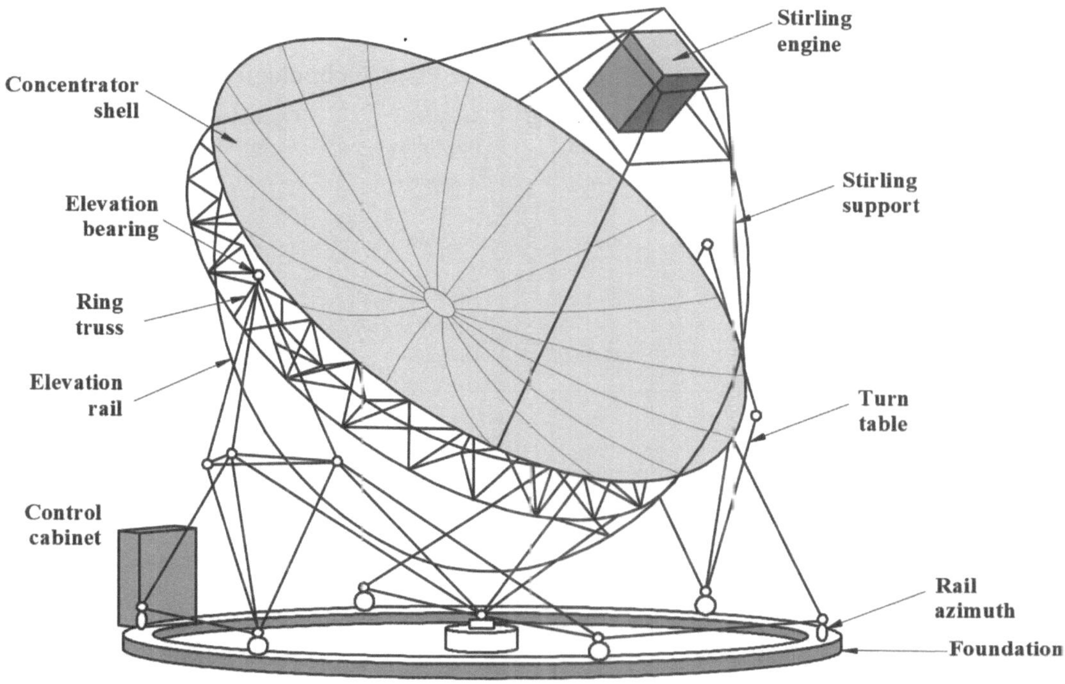

FIGURE 4.69 Fundamental structure of a parabolic disk system [50].

at the receiver is absorbed by the working fluid, a liquid or a gas, increasing its temperature up to 650°C–750°C. The heated working fluid is led to a thermal generator for the production of mechanical power. For parabolic disk systems, Stirling engines are most commonly used, exhibiting overall efficiencies at the range of 30%. Alternatively, gas turbines of small size can also be used.

The diameter of a typical parabolic disk employed in a solar thermal power plant can be from 5 m to 10 m. The disk's surface usually ranges from 40 m^2 to 120 m^2, although there are disks constructed with surfaces up to 400 m^2. The nominal electrical power output of a parabolic disk with 10 m diameter is at the range of 25 kW. The parabolic disk is composed of a number of circular or rectangular reflectors, assembled on a metallic truss. The reflective surfaces of the reflectors are constructed with glass or glazed metal. The receiver is placed on the extension of the central line of the parabolic disk, opposite the reflectors. Each parabolic disk system constitutes an individual power unity, with its own, separate thermal engine and electrical generator. This means that the parabolic disk system is a modular technology, adapted easily to the requirements or the design conditions of each specific project, by simply adding or removing systems. This feature makes this particular technology quite attractive for decentralized power production and for remote, autonomous electrical systems.

As in the case of CRSs with a power tower, the parabolic disk system should be equipped with a two-axis tracking system, to trace the sun's orbit in the horizon and approach the maximum possible solar radiation concentration on the receiver. Given the requirement for a two-axis rotational system and the heavy construction of the disk engine, the tracking system exhibits considerably increased setup cost.

As mentioned just above, the receiver is placed opposite the reflectors. There are two main receiver types for parabolic disk systems, the Brayton and the Stirling receivers. The thermal energy rate with a Brayton receiver is constant. The working medium is air, under relatively low pressure. In most advanced Brayton receivers, the concept of volumetric absorbers is used,

TABLE 4.8
Comparison Between Different Solar Thermal Power Plant Technologies [51]

CSP Technology	Operating Temperature (°C)	Concentration Ratio	Storage Possibility	Occupied Land (m^2/MW_e)	Setup Cost ($/$MW_e$)
Parabolic trough collector (PTC)	20–400	15–45	Possible	40,000	5,000–8,000
Linear Fresnel reflector (LFR)	50–300	10–40	Possible	20,000	3,500–5,000
Central receiver system (CRS)	300–1,000	150–1,500	Highly possible with low cost	40,000	5,000–7,000
Parabolic disk	120–1,500	100–1,000	Difficult	40,000	7,000–10,000

as in the case of the CRS receivers operating with air. A porous material is again used for the solar radiation absorption. The thermal energy is then transferred to the flowing air through the porous material.

There are two general types of Stirling receivers, direct illumination receivers (DIR) and indirect illumination receivers. With DIR Stirling receivers, the heating pipelines of the Stirling motor are utilized for the concentrated solar radiation absorption. The concentrated solar radiation is transferred in the form of thermal energy to a high-pressure gas, usually hellion or hydrogen, which is utilized for the mechanical power production in the Stirling engine. In this way, namely with the direct exploitation for power production of the heated working fluid from the solar radiation, the efficiency of the power production process is maximized and the manufacturing cost of the Stirling engine is minimized, due to the elimination of any heat exchangers and the construction of a simpler engine. With the use of high-pressure hellion or hydrogen, the particular receiver can absorb high solar radiation rates, approximately 75 W/cm^2. In indirect Stirling receivers, the thermal energy is transferred through a heat exchanger from the primary to the secondary loop, which is eventually utilized for power production.

The first commercial parabolic disk system was installed in March 2010, in the city of Peoria, Arizona, US. It consists of 60 parabolic disk systems with a total nominal electrical power output of 1.5 MW.

The main advantages of parabolic disk systems is their modular attribute and the relatively high efficiency (at the range of 30%), mainly due to the utilization of Stirling engines and the high achieved temperatures in the receiver. On the other hand, their major drawbacks are the relatively low power output (up to 25 kW_e for each disk system), low maturity, and restricted experience from implemented projects, which imposes reliability issues with regard to potential impacts on electrical systems' stability and dynamic security.

Depending on the power plant's installation site and size, the occupied land is approximately 40,000 m^2/MW_e. The required setup cost for a parabolic disk of 10 kW_e nominal power can be from 7,000 to 10,000 $/$kW_e$.

Table 4.8 summarizes the above four technologies of solar thermal power plants with their main features [51].

4.7.2 Thermal Energy Storage Systems

Solar thermal power plants must be able to follow a fluctuating power demand. Given the unstable availability of the primary solar radiation, just like in hybrid power plants, this can only be achieved with the introduction of a thermal energy storage system. Hence, if a solar thermal power plant is integrated with a thermal energy storage system, it can be considered a special case of a hybrid power plant, aiming to cover a certain power demand. With thermal storage, thermal energy is stored in an appropriate storage medium right after sunrise, in order to:

- ensure a stable and constant power production, according to the predefined power production schedule, regardless of the solar radiation fluctuations [52]
- prolong the power production during periods with low or no solar radiation available at all, e.g., during night periods.

Apart from the above flexibilities, thermal energy storage offers the obvious option to the solar thermal power plant operator to shift the power production to periods with higher electricity selling prices, improving, in this way, the economic efficiency of the corresponding investment.

In solar thermal power plants integrated with a thermal energy storage system, the dimensioning of the solar collectors' field is oversized, so the solar collectors during daytime periods are able to maintain their nominal power output for the electrical grid, while, at the same time, to store thermal energy in the storage system, in order to support the prolongation of the power production for a certain time period after sunset.

The alternative solar thermal power plant integrations impose different thermal energy storage systems, adapted to each solar technology involved. The parameters that prescribe the employed thermal energy technology are the type of working fluid in the receiver and, particularly, its temperature and pressure; the size of the solar thermal power plant; the operation algorithm of the power plant; and the size of the covered power demand. Thermal energy storage technologies are classified into two main categories, the direct and the indirect systems.

With indirect thermal storage systems, the thermal storage medium is not heated directly by the solar collectors. On the contrary, the concentrated solar radiation is first transferred to the working medium in the receiver (synthetic oil, air), which then is led to heat exchangers to transfer the thermal energy to the storage tank. The employed thermal storage materials must have high thermal capacity, such as pressurized steam, concrete, and several phase change materials (PCMs), particularly molten salts, such as sodium nitrate ($NaNO_3$) and potassium nitrate (KNO_3). Indirect storage is usually introduced when the thermal energy transfer medium is expensive (e.g., synthetic oil) or it is difficult to store (e.g., unpressurized steam). The main advantage of indirect storage is the low cost of the thermal storage medium, while the two main disadvantages are the heat transfer losses through the heat exchanger, as well as high setup cost.

With direct thermal energy storage, the thermal energy transfer medium is directly stored in the employed thermal storage tanks, being also the thermal storage medium. For the application of such a system, the thermal energy transfer medium should exhibit low procurement cost and high thermal capacity. These two prerequisites will enable the reduction of the thermal storage tank's required size.

Thermal energy storage (TES) can be achieved with a wide variety of working mediums. All of them must exhibit very specific thermophysical properties, such as appropriate melting point, depending on the executed thermal process, high latent and specific heat, and high thermal conductivity. Additionally, they must be nonflammable, nontoxic, and available in the market with low procurement cost. Furthermore, they must exhibit thermal and chemical stability, low expansion factor, so as their volume does not change considerably during heating, low super cooling, and low vapor pressure [53]. TES systems can be classified into three main categories with regard to the involved working medium:

- sensible heat storage systems
- latent heat storage systems
- thermochemical heat storage systems.

Sensible and latent heat storage systems are currently applied in thermal storage processes, including solar thermal power plants. Thermochemical heat storage systems still remain at laboratory stage. Extensive coverage on the above TES systems is given in the following sections.

4.7.2.1 Sensible Heat Storage Systems

TES in the form of sensible heat is based on the specific heat capacity of the working medium and its temperature rise during the thermal energy transfer process. When a mass m of a working medium with specific heat capacity c_p is heated, causing a uniform temperature rise of ΔT, the stored thermal energy is given by the relationship:

$$Q = m \cdot c_p \cdot \Delta T \tag{4.113}$$

During the thermal process, the transferred thermal energy in the form of latent heat is null, namely there is no phase change for the working mediums. Their temperature rise is the only change they face. The most commonly used materials with sensible heat TES systems are water or steam, synthetic oils, molten salts, liquid metals, concrete blocks, and earth materials.

- *Thermal energy storage with water or steam*

  Generally, water exhibits some very attractive features with regard to its use as thermal storage material, such as high specific heat (4.184 kJ/kg.K), high availability and low cost, nonflammability, and nontoxicity. On the other hand, the major drawbacks of water as a TES material are its high vapor pressure and its corrosiveness. The use of water in liquid phase as thermal storage material, namely under atmospheric pressure for temperatures lower than 100°C, is extensively presented in Chapter 4, dedicated to thermal hybrid plants. Apart from liquid phase, water can be also used as ice, for cooling applications, and steam. For solar thermal power plants, water is used in the form of steam, most commonly exploited in direct steam generation (DSG) systems [54].

  TES with steam can be characterized as a conventional method. The thermal energy is stored in thermal insulated tanks, filled with pressurized steam with pressures from 50 to 70 bars and temperatures at the range of 280°C. The steam follows a closed loop, from the receiver to the thermal tank and from the thermal tank to the steam turbines, or directly from the receiver to the steam turbines. During storage under high pressure, the steam is condensed, to be vaporized again with its pressure drop once released from the thermal tank. The storage capacity of the thermal tank is rather limited, due to the high imposed cost of large-size thermal tanks. Due to this limitation, the optimum use of this technology is for short-term storage applications, e.g., to undertake the peak power demand for a short time period, in case of inadequate power production from the solar collectors.

- *Synthetic oils as TES material*

  Synthetic oils constitute another popular medium employed for TES processes. They can be used as both heat transfer fluid and TES material. Currently they are organic fluids with appropriate thermophysical properties, such as high thermal conductivity and specific heat capacity. Some of the most common synthetic organic oils are listed in Table 4.9 [51], along with characteristic features. Nonedible vegetable oils have also been considered for use as TES materials [55].

  Synthetic oils are usually colorless liquids under atmospheric conditions. Their main advantage against water is that they remain in liquid phase for temperatures up to 250°C, under atmospheric pressure, due to their low vapor pressure. As mentioned in Section 4.2.1, synthetic oils exhibit a temperature operating range between 12°C and 400°C. This high temperature margin implies higher possibilities for thermal energy storage, according to Equation 4.113. Another positive feature of synthetic oils is their low vapor pressure. For example, Dow Chemical Company's Dowtherm A oil exhibits a vapor pressure of

TABLE 4.9
Most Common Synthetic Oils for TES with Their Main Properties [51]

Commercial Brand Name	Therminol® VP-1	Xceltherm 600	Syltherm XLT	Dowtherm® A	Vegetable oil
Manufacturer	Eastman Chemical Company	Radco Industries	Dow Chemical Company	Dow Chemical Company	[55]
Chemical Composition	Diphenyl ether Oxide/diphenyl ether	Paraffinic mineral oil	Dimethyl polysiloxane	Diphenyl ether Oxide/diphenyl ether	Triglycerides/free fatty acids
Maximum Operating Temperature (°C)	400	316	260	400	—
Melting Point (°C)	12	—	−111	12	—
Boiling Point under Atmospheric Pressure (°C)	257	301	200	257	295
Kinematic Viscosity at 40°C (mm²/s)	2.48	15.5	1.1	2.56	30
Density at 40°C (kg/m³)	1,068	841	834	1,043	926
Density at 210°C (kg/m³)	904	736	660	897	0.11
Thermal Conductivity at 210°C (W/mK)	0.11	0.12	0.06	0.1083	0.11
Specific Heat Capacity at 210°C (kJ/kgK)	2.075	2.643	2.171	1.630	2.509
Thermal Storage Capacity at 210°C (kJ/Km³)	1.876	1.945	1.433	1.462	2.012
Procurement Cost (€/t)	25,000	—	29,400	—	835
Energy Storage Cost for ΔT = 100°C (€/kWh)	464	—	573	—	12

7.6 bars at 374°C. This means that an ambient pressure of 8 bars would be enough to maintain it in liquid phase at this temperature. By comparison, the vapor pressure for water at the same temperature is 221 bars. In other words, the low vapor pressure of the synthetic oils implies reduced requirements for the hydraulic network through which the synthetic oil flows or is stored (reduced pipeline and storage tank thickness), which, in turn, contributes to the reduction of the project's budget. Additionally, the synthetic oil has low viscosity, leading to reduced flow losses and low setup and operating costs, through the installation of small-size circulators. The solidification temperature of synthetic oils, although relatively high (at 12°C), is higher than that of molten salts; hence, they do not require demanding antifreezing protection such as a TES with molten salts.

On the other hand, synthetic oils exhibit rather mediocre heat transfer features. As shown in Table 4.9, synthetic oils exhibit considerable low thermal conductivities, at the range of 0.1 W/m·K and roughly 50% reduced specific heat capacity, compared to water [56]. However, these moderate thermophysical properties can be improved with the addition to the synthetic oil of nano-additives, such as graphene, graphite, and metal oxides [57]. Yet, perhaps the most important drawback is their considerably high procurement cost, as shown in Table 4.9.

With regard to their potential environmental impacts, synthetic oils can exhibit low toxicity and can bio-concentrate in living organisms. If heated above their operating temperatures, as indicated in Table 4.9, they can react with air (oxidation) and produce acids, like carbolic acid and peroxide compounds [55], which may contribute to the corrosion of the metallic surfaces they come into contact with (pipelines, tanks, etc.). When evaporated they become flammable, especially when mixed with atmospheric air. For all these reasons, as well as for its high procurement cost, it is crucial to prevent any oil leakage from the hydraulic network to the ambient environment. Finally, synthetic oils degrade with aging, with negative impacts on their thermophysical impacts.

- *Thermal storage with molten salts*
Molten or liquefied salts feature today (2019) as the most appropriate medium, both for TES and transfer. Molten salts can be used mainly with indirect thermal storage systems, yet, because under atmospheric pressure they are met in liquid phase, they can also be utilized in direct thermal storage systems integrated with CRS plants (solar thermal power plants with power tower). Molten salts are characterized with high TES efficiency, due to their high volumetric heat capacity, which, combined with their low procurement cost, leads to low storage cost. They also exhibit high boiling point, very high thermal stability, and vapor pressure close to zero. The achieved storage temperatures (up to 565°C) [54] make them appropriate for high-pressure and high-temperature steam turbines, while they are neither toxic nor flammable. Their higher boiling point, compared to synthetic oil or water, and their high thermal stability impose a stable operation under high temperatures, a feature that improves the steam's Rankine cycle. Additionally, the higher boiling point implies higher temperature rise ΔT in Equation 4.113, hence higher potential TES. Molten salts are also readily available, nontoxic, and nonflammable. At the end of the service life of a solar thermal power plant, the employed molten salts in the thermal storage system can be crystallized and removed in solid phase to be used elsewhere, such as agriculture. Finally, there is considerable experience in the use of such material in thermal processes performed in chemical and metallurgical industries.

On the other hand, molten salts exhibit high melting point, usually above 200°C, due to which they may freeze in the pipelines, in the absence of any heat source. The optimum properties would be a melting point close to the ambient temperature and a boiling point as high as possible, so that the entire operating range of the heat transfer fluid could be exploited for energy storage. Practically, eutectic mixtures of two or more salts are integrated to bring down their melting point, without affecting their high boiling point. Furthermore, molten salts are characterized with high viscosity, compared to water and synthetic oils, which imply high flow losses in the pipelines and increased pumping or circulating costs. Their corrosive impact is also significant, further exacerbated at high temperatures.

With regard to their thermal performance, they exhibit rather mediocre properties. Just like synthetic oils, their thermal conductivity is remarkably low (roughly 0.5 W/m·K) and their specific heat capacity is roughly 60% lower than that of water. The improvement of these thermophysical features is examined with the addition of nanofluids [58].

The most popular molten salt mixtures used in thermal processes are given in Table 4.10. Among them, the HITEC and the HITEC XL salt mixture, with melting points 142°C and 120°C and boiling points 535°C and 500°C respectively have gained attention. Yet, until recently, the most widely used molten salt was a mixture of sodium nitrate ($NaNO_3$) and potassium nitrate (KNO_3), known as solar salt. While pure $NaNO_3$ and KNO_3 have melting points of 307°C and 334°C respectively, the melting point of solar salt drops to 220°C.

An indicative application of molten salts in TESs is found in the Andasol Solar Power Station, in Granada, Spain. The employed technology is PTCs. The power plant consists of three separate power plants of 50 MWe each (in total 150 MW_e), implemented in 2008, 2009, and 2010. The solar field for each station consists of 624 PTCs, connected in 156 loops with a total effective area of 510,120 m^2 and total occupied land of 2,000,000 m^2. The annual electricity production from each station is estimated at 179 GWh. For each station, 28,500 tn of molten salts are used, stored in two different thermal storage tanks, one cool and one hot, with 36 m diameter and 14 m height each (Figure 4.70) [59]. During

TABLE 4.10
Molten Salt Eutectic Mixtures and Their Fundamental Properties [51]

Salt/Eutectic	Hitec $NaNO_3$-KNO_3-$NaNO_2$ (7-53-40)	Hitec XL $NaNO_3$-KNO_3-$Ca(NO_3)_2$ (7-45-48)	Solar Salt $NaNO_3$-KNO_3 (50-50)	$LiNO_3$
Melting Point (°C)	142	120	220	250
Maximum Operating Temperature (°C)	535	500	600	600
Specific Heat (kJ/kgK)	1.561	1.447	1.5	—
Density (kg/m³)	1,640	1,992	1,899	2,380
Thermal Conductivity (W/mK)	0.60	0.519	0.55	—
Sensible Heat Storage Capacity (MJ/m³K)	2.56	2.9	2.8	—
Procurement Cost ($/kg)	—	1.19	—	—

daytime periods, the molten salt from the cool thermal tank, where it is stored at the temperature of 291°C, is heated up in the heat exchangers by the synthetic oil from the solar receivers, to be led and stored in the hot tank at approximately 384°C for later use. The adequately insulated thermal tanks retain the molten salt's temperature even for several weeks (the temperature drop is not higher than 1°C–2°C).

- *Liquid metals*
  Table 4.11 summarizes the most appropriate metals and alloys for use in thermal processes. As seen in this table, the sodium–potassium alloy, for example, exhibits a melting point of –12.6°C and a boiling point of 785°C. The high difference between its melting and boiling point creates a respectively high temperature margin ΔT while remaining in liquid phase, increasing the potential sensible heat storage, according to Equation 4.113. Additionally, the low melting point practically eliminates the solidification risk, under low solar radiation availability conditions. Finally, the executed thermal process under high

FIGURE 4.70 The thermal storage system of the Andasol Solar Power Station in Granada, Spain [59].

TABLE 4.11
Liquid Metals Appropriate for Use as HTF or TES Materials [51]

Metal/Alloy	Sodium (Na)	Sodium (22.2%)–Potassium (77.8%) Eutectic (NaK)	Lead (44.5%)–Bismuth (55.5%) Eutectic (LBE)
Melting Point (°C)	98	−12.6	125
Boiling Point (°C)	883	785	1,533
Specific Heat (kJ/kgK)	1.3	0.89	0.14
Density (kg/m³)	1,042	780	10,300
Thermal Conductivity (W/mK)	64.9	26.3	14.9
Sensible Heat Storage Capacity (MJ/m³K)	1.354	0.694	1.44
Cost ($/kg)	2	2	13

temperature leads to higher efficiency of the thermodynamic cycle. With these characteristics, the sodium–potassium alloy seems to be ideal for use as a heat transfer fluid (HTF).

Importantly, liquid metals exhibit excellent thermophysical properties. For example, the thermal conductivity of sodium–potassium alloy is 26.3 W/m·K, while the same feature of pure sodium (Na) is 64.9 W/mK. Due to these high thermal conductivity values, the wall temperature on solar receivers will remain low, because thermal energy is transferred quickly from the receiver's wall to the liquid metal, instead of concentrating on receiver walls, resulting in lower heat losses to the environment. Conclusively, the use of liquid metals as HTF has a positive impact on the receiver's efficiency.

On the other hand, among the major drawbacks of liquid metals as HTFs, their high procurement cost and their sensitivity to corrosion can be mentioned. Finally, lead–bismuth eutectic is a toxic material and Na is flammable.

- *Concrete blocks*
  The introduction of concrete mixtures as a thermal storage medium is still under experimental research with several projects of small size currently running or implemented, exhibiting positive results. Concrete can be used as a TES material, because it exhibits some attractive features, such as low cost, good mechanical properties, nontoxicity, and nonflammability. Its relatively low thermal conductivity and specific heat, close to 0.5 W/mK and 0.7 kJ/kgK respectively, can be increased up to 0.719 W/mK and 1.05 kJ/kgK with specific additives, like silane-coated silica fume. Normally, thermal storage can be executed with cement with temperatures up to 400°C, while with an admixture like blast furnace slag a maximum temperature of 550°C can be reached [60]. Currently the setup-specific cost per unit of thermal storage capacity is configured at 30 $\$/kWh_{th}$; however, the target is to reduce this cost to 15 $\$/kWh_{th}$. The first commercial generation of thermal storage systems with concrete slabs is available with nominal thermal storage capacities at 300–400 kWh_{th}.
- *Earth materials*
  Earth materials, like rocks, gravel, and sand, offer an alternative option for TES, due to a number of favorable features, such as nonflammability, nontoxicity, low cost and high local availability, relatively high specific heat capacity, and thermal conductivity. The most popular earth materials for thermal storage are presented with their mechanical, chemical, and thermophysical features in Table 4.12. Apart from the above properties, other desirable features can be also high surface strength to withstand wear from abrasion due to repeating heat compression and expansion, low porosity to prevent oil infiltration, high mechanical strength, etc.

Another crucial parameter, with regard to the appropriateness of a particular earth material as a thermal storage medium, is its chemical composition. For example, granite is composed of minerals containing hydroxyl bonds, which break under temperatures higher than 350°C, leading to a weight loss of the initial material close to 3% [61]. A similar effect is also observed with marble, mainly composed of $CaCO_3$, due to CO_2 escape during heating. Unlike marble and granite, other rock types, like quartzite, basalt, and hornfels, seem to be thermally stable and unaffected when heated in temperatures up to 400°C. These temperature values practically set the upper temperature limits for thermal storage processes with earth materials.

Regardless of whether they are available in small grains (e.g., sand) or in large pieces (e.g., rocks) they are usually laid in layers to form a packed bed structure inside a thermal tank. Rocks of relatively large sizes (e.g., stones, pebbles, small gravel) are used for packed bed fillers. On the other hand, materials with smaller granule size, like sand, are used in fluidized beds. This structure offers the option to pass the HTF pipelines through the packed bed and achieve thermal energy transfer from the HTF to the thermal storage material through direct contact conduction. In other words, the earth materials act both as the thermal storage medium and the thermal transfer surface, eliminating, thus, the requirement for the introduction of heat exchangers and contributing to the further reduction of the plant's setup cost. This structure also leads to the increase of the thermal transfer surface, contributing to the maximization of the storage process efficiency. Another positive impact can be the reduction of the required quantity of the employed HTF, which may be up to 80% [62]. For large solar thermal power plants, HTF is the most appropriate material for thermal energy transfer to earth materials. Yet, in small-scale applications, like residential building heating, air is also likely to be used.

In Table 4.12, it is seen that quartzite exhibits the highest thermal conductivity (7.7 W/mK) and thermal storage capacity (3,822 kJ/m^{3}K) of the other available earth materials for thermal storage. This fact, along with its high potential operation temperature (400°C) makes this particular mineral the most appropriate for TES. This conclusion is also transferred to sand, because it is mainly composed of quartz mineral with a silica content (SiO_2) over 90%.

TABLE 4.12
Most Commonly Used Types of Earth Materials for Thermal Storage with Their Properties [51]

Rock	Granite	Quartzite	Marble	Basalt	Hornfels
Type	Igneous	Metamorphic	Metamorphic	Igneous	Metamorphic
Porosity (%)	1.02–2.87	0.22–22.1	0.40–0.65	0.65–0.81	0.8–2.3
Density (kg/m^3)	2,530–2,620	2,210–2,770	2,510–2,860	2,610–2,670	2,400–2,800
Uniaxial Compressive Strength (MPa)	100–300	100–350	150–300	50–200	100–200
Specific Heat (kJ/kgK)	0.6–1.2	0.8–0.9	0.7–1	1.47	0.7–0.9
Thermal Conductivity (W/mK)	2.8	7.7	2	3.2	1.5
Sensible Heat Storage Capacity (kJ/m^{3}K)	1,440–2,880	3,822	1,750–2,500	1,680–2,520	2,560–2,880
Hardness	High	Very high	Low–Medium	Medium–High	Medium
Major Components	SiO_2 (69%) Al_2O_3 (14%)	SiO_2 (94%)	CaO (54%)	SiO_2 (47%) Al_2O_3 (17%) Fe_2O_3 (10%) CaO (13%)	SiO_2 (63%) Al_2O_3 (21%)

Sensible heat storage is usually executed with cheap and widely available materials, except for the use of liquid metals and synthetic oils. Sensible heat storage systems are the most mature. They are widely used in solar thermal power plants (e.g., in most plants in Spain and the US). On the other hand, the major drawback of sensible heat storage is the relatively low energy density, which ranges between 60 kWh_{th}/m^3 for sand (storage temperatures range 200°C–300°C), rock, mineral oil, and 150 kWh_{th}/m^3 for cast iron (storage temperature range 200°C–400°C). This restricted energy density imposes the construction particularly large storage units. Compared to latent heat thermal storage, the specific heat capacity of the materials used for sensible heat storage is 50 to 100 times smaller and, consequently, the available sensible thermal storage potential is proportionally reduced.

4.7.2.2 Latent Heat Storage Systems

This is another TES technology under development, regarding its potential integration in solar thermal power plants. The phase change materials (PCMs) can ensure more effective thermal storage processes. They can be either organic or inorganic chemical compounds, with high melting point. The basic concept of this thermal energy technology is that considerable thermal energy amounts are absorbed by the PCMs during their transition from the solid to the liquid phase, and vice versa.

In these systems, the working fluid from the solar receiver flows through a heat exchanger embodied inside the PCM mass, transferring thermal energy from the receiver's loop to the thermal storage material. The method has been so far tested in experimental and pilot applications, but not in commercial projects. The main advantage of this technology is the high volumetric storage density and the low procurement cost of the storage PCMs. On the other hand, the main obstacle towards the commercial development of this technology is the low thermal conductivity of the involved PCMs, which does not facilitate the thermal storage process.

Latent heat storage is based on the phase change of the employed material during a thermal process under constant temperature. The most usually executed phase change is from solid to liquid. Solid to solid phase change can also be used, although the specific latent heat is less. Yet, solid to solid phase change exhibits certain advantages, like no leakage risk, no need for enclosing the whole construction, etc. Finally, the liquid to gas phase change has the highest specific latent heat [63]. However, the liquid to gas phase change is accompanied by a huge increase of the fluid's volume, which requires respectively large encapsulations. For this reason, this phase change is generally not applied in practical thermal storage applications.

TES through latent heat is expressed with the following relationship:

$$Q = m \cdot L \tag{4.114}$$

where m is the mass of the phase changed material and L is its specific latent heat (in kJ/kg).

Latent heat thermal storage capacity can be 50 to 100 times larger than sensible heat. However, the main drawback of latent heat storage materials is the low thermal conductivity when organic or salt materials are used. All PCMs in general are nontoxic, yet organic PCMs are flammable. Long-chain fatty acids and esters cannot be stored and transported in plastic containers, because they are highly lipophilic. Additionally inorganic PCMs are corrosive to metal containers.

Latent heat TES is performed with either organic or inorganic materials. A short presentation is provided in the next sections.

4.7.2.2.1 Organic Materials

Organic materials exhibit a variety of favorable properties, with regard to latent heat thermal storage. They are widely available in nature, nontoxic, noncorrosive, and chemically stable in low temperatures. Additionally, their melting point, or, in other words, their solid to liquid change temperature, is very close to human thermal comfort conditions, namely in the range of 18°C–30°C.

For these reasons, they are most commonly used in residential applications. Among their major drawbacks, their low thermal conductivity and their chemical instability at high temperatures can be mentioned. Organic materials can be classified in the following categories:

- *Paraffin*
  Paraffin constitutes the most commonly used organic material for thermal storage applications. It is generally represented with the formula ($CH_3 - (CH_2)_{(n-2)} - CH_3$), with n designating the number of carbon atoms. Paraffin's melting point increases with the number of carbon atoms in the molecule. For thermal storage uses, usually paraffin with a melting point from 10°C (n = 10) to 65°C (n = 30) is employed. Under ambient conditions, paraffin with atom number lower than n = 16 usually exists in liquid phase, while for n greater than 16, paraffin is met in solid phase (like wax). Pure paraffin requires extended refinement, leading to subsequent increase of the procurement cost. For this reason, a cheaper technical paraffin wax, produced as a by-product of oil refining, is usually used as PCM in thermal storage applications. Paraffin is odorless, exhibits low super cooling, chemically stable, and compatible with metal containers. However, the large volume expansions (around 10%) during phase change, its low density, and thermal conductivity constitute considerable disadvantages [64].
- *Fatty acids*
  Fatty acids are represented with the general formula R-COOH where R designates an alkyl group. Fatty acids can be produced from natural oil and they exhibit a series of positive features, such as low cost and chemical stability. Certain fatty acids (e.g., lauric acid) can be extracted as a by-product from agricultural processes; hence, they can be considered renewable. This approach also contributes to the reduction of the production cost. On the other hand, they have odor, low density, and thermal conductivity and they exhibit large expansion (roughly 10%) during phase change. Fatty acids have sharp melting points that increase with the number of carbon atoms in their molecule. Usually, the saturated fatty acids are considered as TES materials, such as caprylic acid (n = 8, melting point at 16°C) or stearic acid (n = 18, melting point 69°C) [65]. Unsaturated acids are not suitable as TES, mainly due to their low phase change temperature, with the exemption of some of them, like oleic acid, with a melting point of 14°C. Expansion of the phase change temperature range can be achieved with the introduction of fatty acid eutectics [66].
- *Esters*
  Fatty acid esters have the general formula R-COO-R^1 where R and R^1 designate alkyl groups. They are produced from carboxylic acids and alcohols with a balanced, reversible, catalytic reaction called esterification [66]. Esterification is an alternative method to improve the thermophysical properties of fatty acids. For example, the high phase temperature of certain fatty acids, like stearic acid (69°C) and palmitic acid (62°C) is reduced with the esterification procedure and the production of methyl stearate (39°C) and methyl palmitate (29°C) respectively, which exhibit lower phase change temperature, closer to human thermal comfort conditions. Esters are characterized with low super cooling and chemical stability. Just like fatty acids, they have odor, low density, and thermal conductivity, and they are expensive.
- *Alcohols*
  Sugar alcohols exhibit the highest melting points and the highest latent heat from all the organic materials considered as TES. Given their phase change temperatures (20°C–165°C), alcohols are appropriate for thermal storage processes executed under medium temperatures (90°C–250°C), such as thermal storage from solar collectors or waste heat recovery in cogeneration plants [67]. They are cheap and nontoxic. Alcohols exhibit polymorphic attributes, meaning they can exist in two or more crystalline states.

Different polymorphs can have significant differences in physiochemical and thermophysical properties [66]. During TES, polymorphic changes can reduce the effective latent heat by releasing heat suddenly. In addition, polymorphism can affect the cycling stability of the alcohol. Consequently, before using a sugar alcohol, cycling and chemical stabilities must be examined.

4.7.2.2.2 Inorganic Materials

Inorganic materials can also be used as PCMs for TES processes at high temperatures, under which organic materials would have thermally decomposed. The most commonly used inorganic materials as PCMs are salts, salts eutectics, metals, and metal alloys.

- *Salts and salt eutectics*

  Generally, salts with high melting points are employed as PCMs for thermal storage processes, offering the option of thermal storage at high temperatures. Additionally, latent heat TES by using salts as PCMs with high melting point can offer high volumetric thermal storage capacity when the melting point falls into the operational temperature range of the thermal storage process. For sensible heat storage, salts and salt eutectics of relatively lower melting points are employed. In these cases, thermal energy is stored by exploiting the sensible heat capacity in liquid phase of the employed material. This implies a relatively restricted volumetric thermal storage capacity, imposed by the difference between the operating temperature of the thermal process and the melting point of the employed salt. For example, if the operating temperature of a thermal process is 400°C, by using solar salt for sensible heat storage, with a melting point of 220°C, a specific heat of 1.5 kJ/kg·K and a density of 1,899 kg/m³, according to Equation 4.113, the maximum volumetric thermal storage capacity is calculated at:

$$\frac{Q}{m} = c_p \cdot \Delta T \Rightarrow Q = 1.5 \frac{kJ}{kg \cdot K} \cdot (400 - 220)K \Leftrightarrow \frac{Q}{m} = 270 \frac{kJ}{kg} \Leftrightarrow \frac{Q}{V} = 270 \frac{kJ}{1/1{,}899 m^3} \Leftrightarrow \frac{Q}{V} = 512.7 \frac{MJ}{m^3}$$

  However, by choosing KNO_3, with a melting point at 335°C, a specific latent heat of 266 kJ/kg, a specific heat capacity at 0.940 kJ/kgK, and a density at 2,109 kg/m³, by exploiting both sensible heat and latent heat, the volumetric thermal storage capacity can be, according to Equations 4.113 and 4.114:

$$\frac{Q}{m} = c_p \cdot \Delta T + L \Rightarrow \frac{Q}{m} = 0.940 \frac{kJ}{kg \cdot K} \cdot (400 - 335)K + 266 \frac{kJ}{kg} \Leftrightarrow \frac{Q}{m} = 327.1 \frac{kJ}{kg} \Leftrightarrow \frac{Q}{V}$$

$$= 327.1 \frac{kJ}{1/2{,}109 m^3} \Leftrightarrow Q = 689.9 \frac{MJ}{m^3}$$

  Generally, pure salts and salt eutectics with melting points higher than 250°C are suitable as PCMs for latent heat storage. Latent heat thermal storage involves the PCM's heating up process while being in the solid phase, until its melting point. During this heating process, thermal energy is transferred through the PCM's mass only by conduction. Given the generally low thermal conductivity of inorganic salts, commonly between 0.5–1 W/mK, it is derived that this preliminary heating up process is performed with low efficiency and prolongs the thermal storage cycle time. For this reason, the increase of the PCM's thermal conductivity while being in solid phase is crucial [68]. This is intended with the dispersion of highly conductive fillers in the PCM. Such fillers can be:

- carbon-based fillers, like nanofiber, graphene, carbon nanotube, and expanded graphite
- metallic fillers, like copper nanowires, silver nanowires, and aluminum nanowires
- ceramic fillers, like alumina, boron nitride, silicon carbide, and zinc oxide.

  There are several types of inorganic salts, such as hydrates, nitrates, carbonates, chlorides, sulfates, fluorides, and hydroxides. Nitrate salts exhibit the lowest melting points. For this reason they are currently the most commonly employed salts as PCMs in solar thermal power plants. Hydroxides exhibit medium melting points, from 250°C to 600°C. The other salt types, namely the carbonates, chlorides, sulfates, and fluorides have melting temperatures higher than 600°C.

  While salts exhibit certain thermophysical properties, salt eutectics offer the option to adapt the thermophysical properties of pure salts according to the executed thermal process. Given the large number of available pure salts, the potential number of salt eutectics can be even larger. With salt eutectics, the operating temperature range of PCMs can be expanded, while the melting point of a pure salt can be reduced, to avoid the solidification risk inside a pipeline, etc.
- *Metals and alloys*

  Metals and alloys have the highest thermal conductivity and the highest volumetric thermal storage capacity among the alternative PCMs. On the other hand, they are expensive. Yet, if the occupied space is a decision parameter, they can constitute a feasible choice. Apart from their high cost, there are also some additional issues that should be considered if metals are used as PCMs. First of all, after repeated thermal cycles, they can undergo changes in their microstructure, because of precipitation, oxidation, segregation, etc. This can result in property changes, including melting points and latent heat. To prevent oxidation, an inert atmosphere is required, yet, metals can absorb these inert gases during melting and solidification cycles, affecting, in turn, their thermophysical properties. Another issue is the potential strains and ruptures that can be caused at the metal or the ceramic containers where metals are contained, due to their repeated thermal expansion.

Table 4.13 summarizes the above PCMs with their fundamental properties.

4.7.3.3 Thermochemical Heat Storage Systems

Thermochemical heat energy storage processes are based on reversible chemical reactions. Solar radiation is employed to drive an endothermic chemical reaction. The absorbed solar radiation by the reactants is then stored in the form of chemical potential. During discharging, the stored thermal energy is recovered by the reversed exothermic reaction, often executed with the support of a catalyst. The thermochemical heat storage process relies on two major advantages:

- it exhibits high energy density, up to 10 times greater than latent storage [69]
- thermal energy storage can have indefinitely long duration at ambient temperature.

Consequently, thermochemical heat storage constitutes an attractive and economically competitive option. Thermochemical heat storage is performed at medium or high temperatures (300°C–1,000°C) with metallic hydrides, carbonates systems, hydroxides systems, redox systems, ammonia systems, and organic systems [69].

Some thermochemical reactions are characterized with incomplete irreversibility. This constitutes a major drawback, because it contributes to the gradual reduction of the storage capacity, with repeated incomplete charging–discharging cycles.

The most promising reactions, exhibiting high reversibility, are presented in Table 4.14.

Thermochemical heat storage still remains in the development stage. Several critical issues still must be solved. For example, if the products of the chemical reaction are gases, such as in the above

TABLE 4.13
Popular Phase Change Materials for Thermal Storage Processes with Their Properties

Type	Thermal Storage Material	Phase Change Temperature (°C)	Latent Heat (kJ/kg)	Density (kg/m³)	Thermal Conductivity (W/mK)	Latent Heat Storage Capacity (MJ/m³)
Organic	Paraffin	−10–80	150–255	760 (L)–880 (S)	0.15 (L)–0.36 (S)	10–770
	Fatty acids	16–70	150–210	850 (L) –1,000 (S)	0.149 (L)–0.162 (L)	145–196
	Esters	11–63	100–215			
	Alcohols	23–165	220–345	1,450–1,520		164–500
	Glycols	5–70	117–175	1,128		133
	Organic eutectics	21–52	143–182			
Inorganic/ Salts	Salt hydrates	30–117	115–280	1,450 (L)–2,070 (S)	0.490 (L)–1.225 (S)	225–550
	Nitrate salts	300–560	172–260	2,100–2,260	0.5	300–561
	Carbonate salts	730–1,330	140–500	1,972 (L)–2,930 (S)		410–1,074
	Chloride salts	192–770	75–710	452 (L)–2,900 (S)		218–910
	Sulfate salts	850–1,460	85–210	2,000 (L)–2,680 (S)		186–560
	Fluoride salts	850–1,410	390–1,045	1,800 (L)–3,200 (S)		1,100–2,750
	Hydroxides	320–460	150–870	1,460 (S)–2,100 (S)		300–1,475
	Salt eutectics	13–770	95–790	1,800 (S)–3,000 (S)	0.5 (S)–1.0 (L)	155–960
Inorganic/ Metals	Copper	1,084	208	8,020 (L)–8,960 (S)	401	1,864
	Zinc	419	113	7,140 (S)	116	806,8
	Aluminum	660	397	2,375 (L)–2,707 (S)	204	1,074.7
	Alloys	340–950	90–760	1,380 (S)–8,670 (S)		370–2,100

reactions (H_2, SO_3, and NH_3) a major technical issue is the storage of these products, due to their increased volume. A potential solution could be the compression and the storage of these products in large tanks, leading to higher installation and operation costs. Alternative solutions can be the storage of H_2 with metal hydrides at low temperatures or by adsorption, yet, without avoiding additional costs.

The hydration/dehydration of CaO constitutes another promising reaction with several positive features, such as good reversibility, low operating pressure, low price, nontoxicity of the products, and large experimental feedback. Nevertheless, in this case other technical problems, like low thermal conductivity, agglomeration, and sintering still require remedy. Low thermal conductivity constitutes a common drawback of materials used for thermochemical storage, leading to slow heat transfer rates. Additionally, low permeability is another common drawback, which reduces the mass transfer and decelerates the chemical reaction's rate. Research efforts are focused on the treatment of these specific technical problems, namely they aim to enhance the mass transfer during direct reaction and to enhance the heat transfer during the reverse reaction.

4.7.4.4 Thermal Energy Storage Integration

The TES alternative integration concepts can be classified into active and passive systems, with regard to the motion state of the employed thermal storage medium. In active storage systems, the storage medium itself flows to absorb (charge) or dispose (discharge) thermal energy by forced convection. On the contrary, in passive storage systems, the storage medium (most commonly in solid phase) remains still while heated or cooled by the circulation of the HTF. Moreover, active systems can be either direct or indirect systems. In active direct systems, there is only one working fluid, being both the HTF and the storage medium. The common working fluid circulates between the solar collectors and the thermal storage tanks. During storage operation mode, the working fluid

TABLE 4.14
Main Reactions with High Reversibility Used in High-Temperature Thermochemical Heat Storage [70]

Reaction/Phase	Charge/Discharge Temperature and Pressure	Reaction Enthalpy and Energy Density	Advantages	Disadvantages
$2Co_3O_{4(s)} + \Delta H \leftrightarrow 6CoO_{(s)} + O_{2(g)}$ Solid-Gas	1416°C–1446°C 0–1 bar	205 kJ/mol 295 kWh/m^3	– Good reversibility – High reaction enthalpy – No catalyst needed – No by-product	– Cost of CH_4 – Toxicity – O_2 storage – Little experimental feedback
$SO_{3(l)} + \Delta H \leftrightarrow SO_{2(g)} + ½.O_{2(g)}$ Liquid-Gas	1346°C–1546°C/ 1046°C–1146°C 1–5 bars	98 kJ/mol 646 kWh/m^3	– High energy density – Industrial feedback	– Corrosive product – Toxicity – Catalyst needed – O_2 storage
$TiH_{2(s)} + \Delta H \leftrightarrow Ti_{(s)} + H_{2(g)}$	923°C–973°C 1–3 bars	150 kJ/mol 813 kWh/m^3 (experimental) 3,331 kWh/m^3 (theoretical)	– Good reversibility – High energy density – No catalyst needed – No by-product – Low cost	– H_2 storage – Little experimental feedback
$CaCO_{3(s)} + \Delta H \leftrightarrow CaO_{(s)} + CO_{2(g)}$	1406°C/1429°C 0–1 bar	178 kJ/mol 692 kWh/m^3	– High energy density – Availability and low cost of materials – No catalyst needed – No by-product	– Agglomeration and sintering – Important volume change – CO_2 storage

absorbs the concentrated solar radiation on the receiver and directly stores it at the hot storage tank, connected at the output of the solar field (Figure 4.71a). During thermal discharge operation mode, the hot stored medium is pumped from the hot tank and, after releasing its contained thermal energy in a heat exchanger or a boiler for steam production, it is stored in the cold tank, connected at the solar field's inlet. From this tank, the working medium will be led into the solar collectors to collect further concentrated solar radiation and act again as HTF. It is conceivable that with this integration the necessity of heat exchangers for the thermal energy transfer between the HTF and the storage medium is eliminated. The employed working fluid must exhibit appropriate properties to act both as HTF and as storage medium. Steam and molten salts with wide operating temperature range for sensible heat storage seem to be the most favorable materials for such systems. The main advantages of the direct active systems is the elimination of the heat exchanger and the separate storage of the cold and hot storage materials. The main disadvantages include the low thermal energy density of steam, the high cost of molten salts, and the possible freezing of molten salts inside the hydraulic network due to their high melting point.

Unlike direct systems, in active indirect systems, two different fluids are employed as HTF and storage medium. A typical active indirect layout is presented in Figure 4.71b. It consists of two separate thermal storage tanks, one hot and one cold. The concept has been described above, in the sensible heat storage with molten salts. During the thermal storage phase, the working fluid is pumped from the cold tank and led through a heat exchanger where it is heated by the HTF, which flows from the solar receivers. The hot thermal storage medium, after passing through the heat exchanger, is stored in the hot tank. During the discharge operation mode, the flow of the thermal storage medium is reversed. It is now pumped from the hot tank and, while passing through the heat exchanger, it releases thermal energy at the returning HTF from the steam generator (boiler). In this way, the hot thermal tank substitutes the thermal power production from the solar collectors, when it is not available or not enough to support the predefined operation of the power plant. Indirect active systems

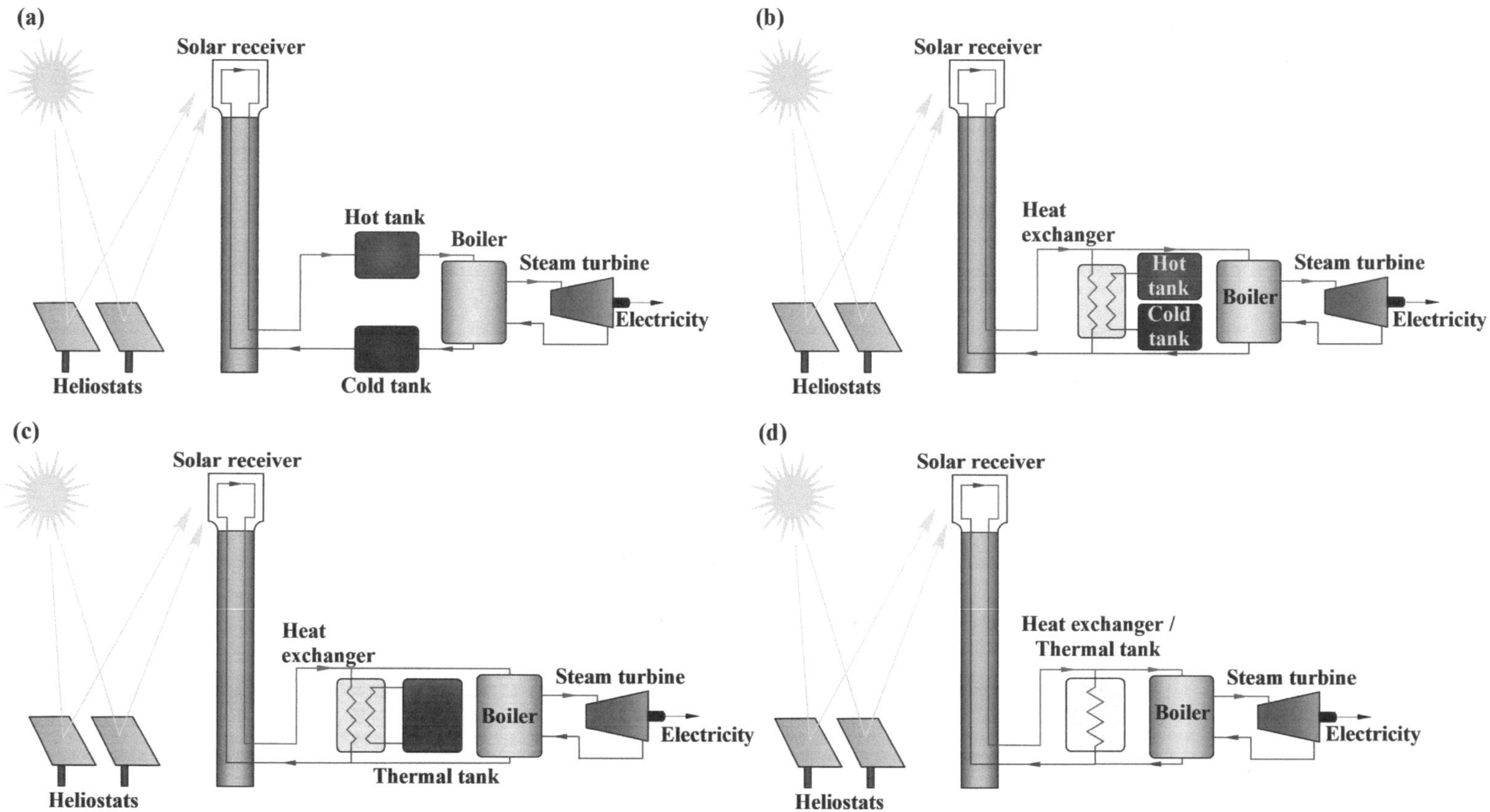

FIGURE 4.71 Alternative thermal energy storage integration concepts [70].

with two tanks exhibit the advantage of separate storage of cold and hot materials, while they also have the main disadvantages of direct systems (molten salts' high cost and possible freezing of them).

An alternative implementation of active indirect systems is presented in Figure 4.71c. It consists only of one thermal storage tank, where the hot and cold storage medium are separated based on the achieved temperature physical stratification, namely the cold material is concentrated at the bottom while the hot material is concentrated at the top of the storage tank, just like the water thermal storage tanks presented earlier in this chapter. The zone between the hot and cold fluids is known as the thermocline. The thermocline effect and the temperature stratification inside the unique thermal tank can be strengthened with the addition of specific filler materials, like quartzite rocks, sand, and concrete pieces in the storage medium. In this way, the required quantity of the storage medium is also reduced. With the construction of one thermal tank instead of two, a setup cost percentage drop of the thermal storage plant up to 35% can be achieved [71]. The achievement of temperature stratification can be also approached with the performance of an automatic, controlled procedure, executed with the combination of a central electronic device, a set of thermometers and the appropriate hydraulic network's layout. A relevant example is provided in Section 4.6.4, where the thermal storage in multiple water thermal tanks is investigated. A similar approach can also be applied in solar thermal power plants with a single storage tank. Another key factor is the configuration of potential fillers, especially in cases of solid filling materials. For example, it is found that in the case of concrete fillers, the packed-bed structure seems to have better performance with regard to the achieved temperature stratification, compared to channel-embedded structure, parallel-plate structure, and rod-bundle structure [72]. The main advantage of indirect systems with one tank is their reduced setup cost. On the other hand, yet, there is the difficulty for maintaining the temperature stratification in the thermal tank, a task which requires the design and the construction of a complex control system.

As mentioned above, in passive storage systems the thermal storage medium remains still inside the thermal tank, while the HTF circulates through the thermal medium's mass to dispose the concentrated thermal energy from the solar receivers or to absorb the stored thermal energy in the thermal tank. In case of sensible heat storage, the energy transfer from the HTF towards the thermal tank is performed with a tubular heat exchanger, embodied in the mass of the thermal material. In this way, the heat storage device and the heat exchanger form a compact unit (Figure 4.71d). The high thermal conductivity of the involved materials, combined with the large thermal energy exchange surface and the good contact between the heat exchanger and the storage medium maximize the thermal energy transfer rates. Passive storage systems are characterized with low material costs, simple and compact storage units, and possible high heat transfer rates. However, the cold and hot materials are not separated and, possibly, the discharge temperature can vary.

The alternative concepts of TES integration in solar thermal power plants are summarized in Table 4.15 [70].

TABLE 4.15
Thermal Energy Storage Alternatives in Solar Thermal Power Plants [70]

Layout	Principle	Storage Materials
Active direct	A common working fluid is used both as HTF and as storage material	– Molten salts – Steam
Active indirect	HTF transfers thermal energy via a heat exchanger to the thermal tank	– Molten salts – Steam – Solids
Passive	The storage medium remains still while HTF circulates through its mass to transfer heat	– Concrete – Earth materials – PCMs – Thermochemical materials

4.8 CHARACTERISTIC CASE STUDIES

The current chapter includes two characteristic case studies regarding the design, the operation simulation, and the dimensioning of two thermal hybrid plants for the heating of a swimming pool center and a school building. Both case studies are accomplished for geographical locations with annual solar irradiation above 1,500 kWh/m^2, an essential prerequisite to ensure the technical and economic feasibility of the introduced systems. This can be understood if we consider that heating is obviously required mainly during the winter period, which means that considerable solar radiation should be also available during winter, a fact that, in turn, implies annual solar irradiation higher than the above mentioned figure. Practically, the introduced systems can be applied for locations with geographical latitudes lower than 38°N or S. For northern or southern geographical regions, the annual results will not be as good as in the presented case studies below. In both case studies presented below, meteorological data were retrieved for the city of Palma, in the island of Mallorca, in the Balearic Archipelago, Spain. Specifically, the annual time series for the ambient temperature, the wind velocity, and the Global Horizontal Irradiance (GHI) presented in Figures 4.72, 4.73, and 4.74 respectively were introduced. These data were employed for the calculation of the heating loads and the solar heat gains for the facilities under consideration.

4.8.1 Thermal Hybrid Plant for Swimming Pool Heating

Heating of swimming pools constitutes one of the most energy consuming applications, configured by the usually large volume of the heated water, its high heat capacity, the operation timetable of the swimming pool, and, for colder locations, the cold climate conditions and the subsequent increasing heating needs. In many instances, the cost for heating a swimming pool is barely affordable by the responsible operator (e.g., local municipalities, sport clubs, private companies) and affects respectively the operation cost and, eventually, the fees for the use of the swimming pool.

The high primary energy consumption required for swimming pool heating and the subsequent costs have caused a relevant interest for the introduction and the application of alternative technologies, based on the exploitation of RESs and the introduction of energy saving measures, aiming to reduce swimming pool heating loads and to restrict the consumption of expensive fossil fuels. Starting from fundamental passive systems, such as the swimming pools' housing and the use of floating insulating covers on their surface whenever the pools are not in use, the proposed measures also expand to the introduction of active systems, such as the substitution of the existing oil burners

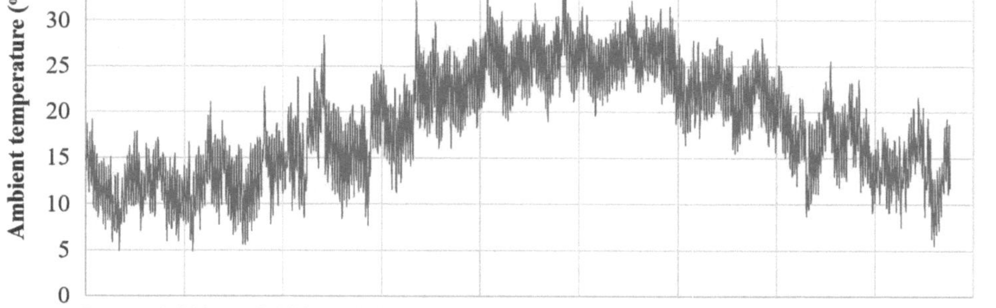

FIGURE 4.72 The annual ambient temperature time series at the city of Palma, Mallorca.

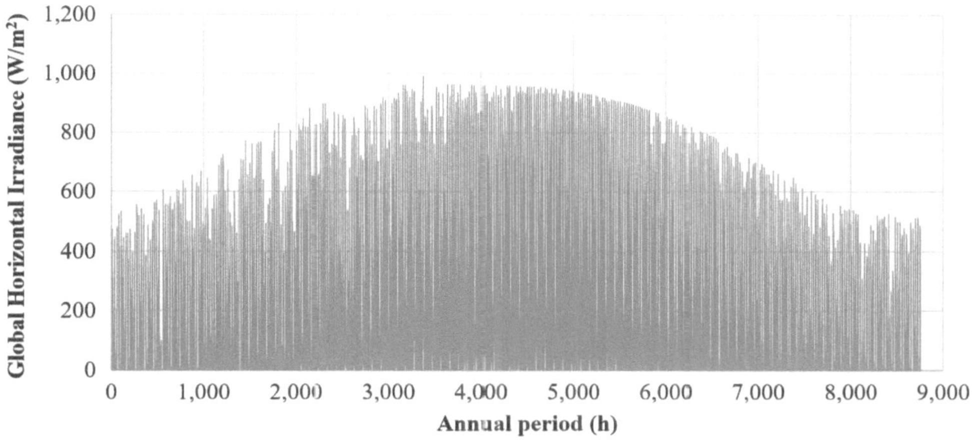

FIGURE 4.73 The annual Global Horizontal Irradiance time series at the city of Palma, Mallorca.

with biomass heaters, the installation of geothermal heat exchangers combined with geothermal heat pumps, or the development of a combined system of solar collectors, a biomass heater, and thermal storage tanks, namely a thermal hybrid plant as defined in this chapter, most commonly known in the technical and academic literature as a *solar-combi system*.

In this first case study we will focus on the introduction of a combi-solar system for a swimming pool center, consisting of two swimming pools, one of Olympic size (50 × 20 m free surface) and a second, smaller pool, for training use (25 × 6 m free surface).

a. *Calculation of the swimming pools' heating loads*
   The examined swimming pool center is housed in a polycarbonate massive telescopic enclosure, founded on an aluminum frame structure, with the ability to be removed/opened during summer months, in order to avoid high indoor space temperature, which, combined with high humidity, would make the use of the swimming pools rather impossible. Additionally, a floating insulating cover is rolled out on each pool's surface whenever it is not used, to restrict the heat losses from the pool to the indoor space. These are the passive measures introduced for the reduction of the swimming pools' heating loads.

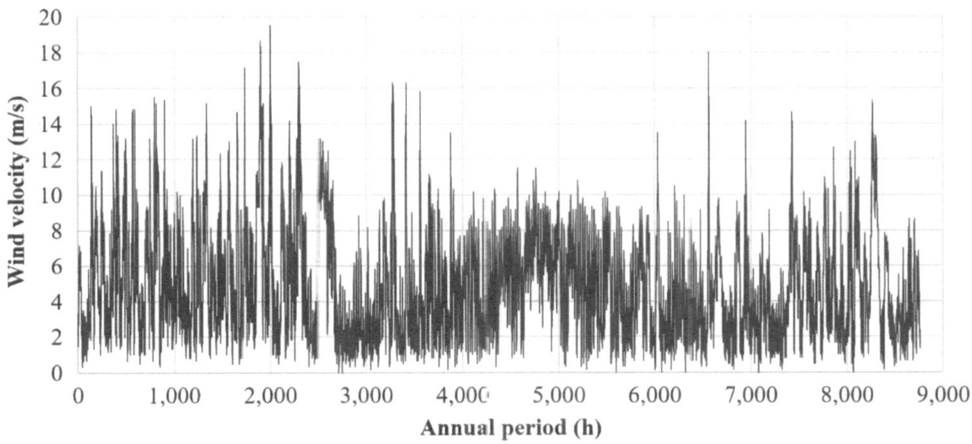

FIGURE 4.74 The annual wind velocity time series at the city of Palma, Mallorca.

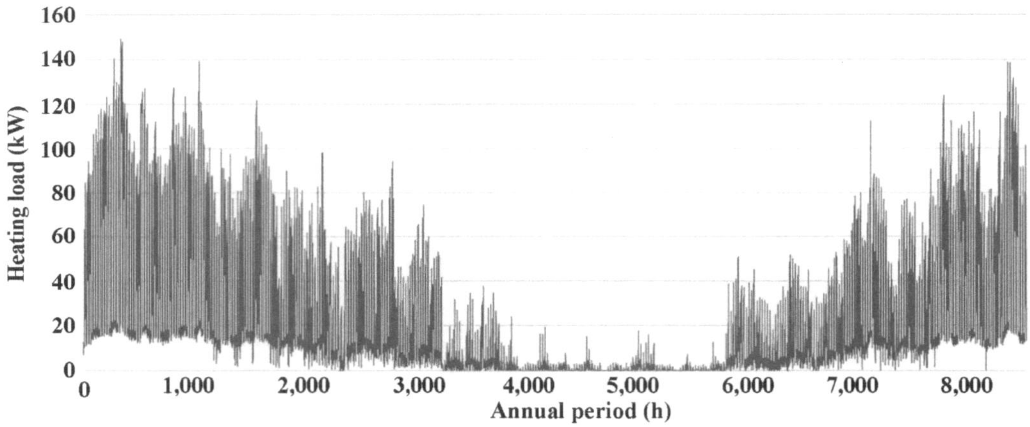

FIGURE 4.75 Annual time series of the examined swimming pools' heating loads [74].

The calculation of the swimming pools' heating loads was based on the meteorological data presented in Figures 4.72, 4.73, and 4.74. First, the heat losses from the indoor space to the ambient were calculated and the corresponding annual indoor space temperature time series was developed. At a second stage, the swimming pools' heating loads were calculated following the fundamental heat transfer theory from the swimming pools' free surface to the space above them [73], with convection and radiation, introducing the required factors for heat convection and radiation above the water free surface and the insulating cover and following the swimming pools' operation timetable. All this procedure is beyond the scope of this book, so it is omitted. The swimming pools' calculated heating loads annual time series (final required thermal power) is presented in Figure 4.75.

Figure 4.75 shows that the maximum heating loads, obviously during winter time, vary between 120 and 140 kW. It is also seen that there is a low heating load even during summer, which always remains below 20 kW. Additionally, the parameters involved in the heating load calculation and the annual final thermal energy consumption are summarized in Table 4.16. In the same table, the annual diesel oil consumption and the corresponding primary energy are also presented, on the assumption that the annual calculated swimming pools' heating loads are totally covered by a conventional oil burner.

b. *The layout of the introduced thermal hybrid plant*

In case of thermal hybrid plants for swimming pool heating, the swimming pool can play a double role:

- It obviously constitutes the ultimate destination of the produced thermal energy, namely the medium that is intended to be heated.
- It can be also exploited as the TES tank. This can be achieved by increasing the swimming pool's water temperature one or two degrees Celsius above the required temperature for its normal operation, in cases of excess solar radiation availability. Having increased the pool's water temperature, there is the margin to avoid the operation of the oil or the biomass heater until the water temperature drops again to the required level, when the available solar radiation will be inadequate to achieve the pool's heating load.

The overall design and layout of the introduced solar-combi system is presented in Figure 4.76. To exploit the available solar radiation to the maximum possible extent,

TABLE 4.16
Involved Parameters and Fundamental Results of the Swimming Pools' Heating Loads Calculation [74]

Parameters	
Water emissivity ε_w	0.957
Stefan-Boltzmann constant (W/m^2·K^4)	5.67·10^{-8}
Upper surface of Olympic size swimming pool (m^2)	1,000
Upper surface of training swimming pool (m^2)	150
Operation schedule	12:00–22:00/6 days per week
Required water temperature in the pools (°C)	26
Material of floating insulating cover	polyethylene
Emissivity of floating insulating cover ε_c	0.550
Convection factor for heat transfer from horizontal surface to indoor space (W/m^2·K)	2.5
Conduction factor of the floating insulating cover (W/m·K)	0.025
Thickness of floating insulating cover (m)	0.020
Conduction factor of massive polycarbonate panels (W/m·K)	0.200
Thickness of polycarbonate panels (m)	0.004
Solar gains' factor of massive polycarbonate panels	0.84
Required fresh air ventilation (m^3/(h·m^2 of indoor space covered area))	33.75

Results

	Annual Final Thermal Energy Consumption (kWh)	Annual Diesel Oil Consumption (klt)	Annual Primary Energy Consumption (kWh)	CO_2 Emissions (tn)
Olympic pool	137,832	18.250	202,154	53.369
Training pool	20,675	2.737	30,323	8.005
Total	158,507	20.987	232,477	61.374

given also the fact that the main thermal energy demand is expected apparently during winter, selective coating flat-plate collectors are introduced, instead of uncovered solar collectors, which are used for swimming pools heated mainly during summer. In order to eliminate the corrosion risk of the copper absorbers used in selective coating solar collectors from the swimming pool water, the required hydraulic network is separated in two independent loops, as shown in Figure 4.76. The thermal power produced by the solar collectors is transferred through the primary closed loop to a plate heat exchanger. Through the secondary open loop, the thermal energy is finally disposed in the swimming pools. The introduction of the heat exchanger and the separation of the system into two independent loops constitute the main differences between the introduced system in this case study and the one presented in Figure 4.2, with uncovered solar collectors.

The system's whole operation is automated with the support of a control unit, which receives two main temperature input signals: the solar collectors' outlet temperature and the pool water temperature. If the solar collectors' outlet temperature is at least 4°C higher than the required water temperature in the pool, then the circulators in the primary and secondary loops are activated and the three-way motorized valve opens. If the water temperature in the pools remains lower than the required, then the biomass heater is activated, supplied with the preheated water from the solar collectors, reducing, in this way, the consumed biomass fuel.

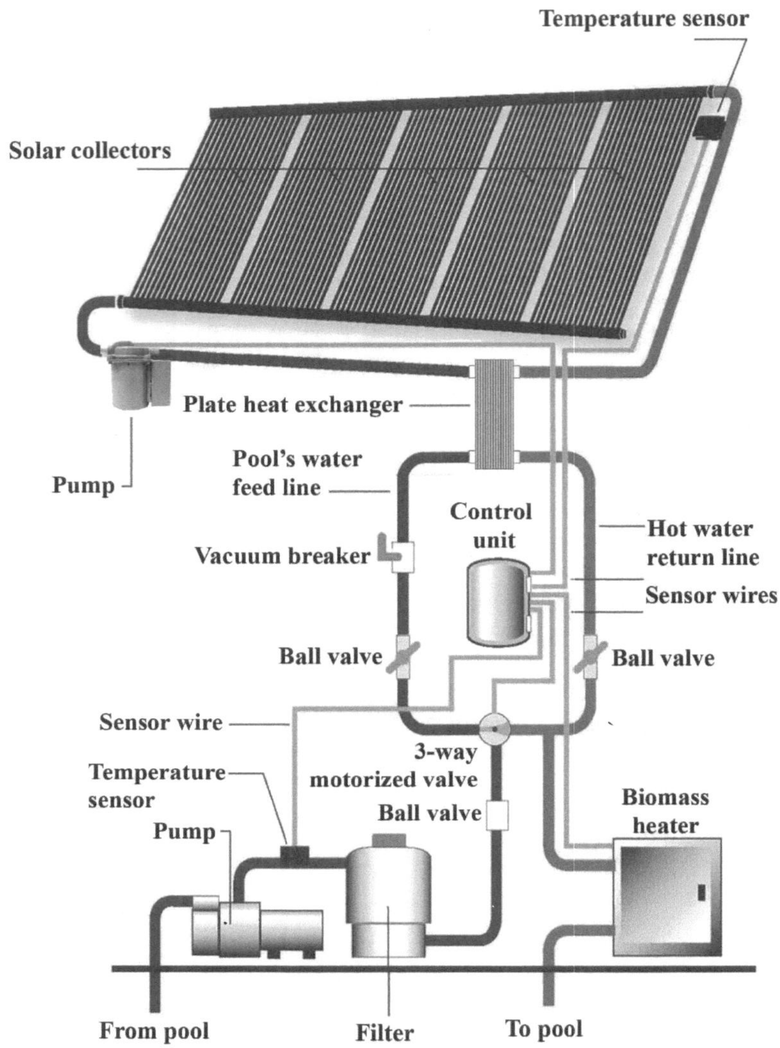

FIGURE 4.76 Typical design of a solar collector and biomass heater combi system for swimming pool heating [74].

c. *The dimensioning algorithm*

The dimensioning process of this solar-combi system is executed on the basis of iterative computational simulations of the system's annual operation, introducing as optimization criterion the minimization of the final thermal energy production-specific cost. It can be analyzed in the discrete steps presented below:

- The annual time series with average hourly values are introduced for all the involved magnitudes in the operation simulation, which are the ambient environmental and indoor space temperature, the incident solar radiation, and the swimming pools' heating loads.
- Following the process described in Section 4.3.5, the thermal transmittance coefficient U_L, the heat removal factor F_R, and, eventually, the produced thermal power from the solar collectors Q_{SC} are calculated for every time step.

- The efficiency of the plate heat exchanger and the transmitted power Q_s can be also calculated following the procedure described in Section 4.6.1.
- Once the swimming pools' heating load Q_L and the thermal power production from the solar collectors Q_s have been calculated, the following cases are distinguished:
  i. If $Q_L \geq Q_s$, then the produced thermal power from the solar collectors is not enough to fully undertake the swimming pools' current heating load. In this case, obviously the supplied power Q_s is fully absorbed for the swimming pools' heating. The remaining uncovered heating load, due to the insufficient power production from the solar collectors, should be covered by the backup unit, namely the biomass heater, which will produce power Q_b equal to $Q_b = Q_L - Q_s$.
  ii. If $Q_L < Q_s$, then the produced thermal power from the solar collectors is higher than the current heating load, which is fully covered by them. Yet, aiming at the maximization of the exploitation of the thermal energy produced by the solar collectors, additional thermal power is absorbed by them, allowing a maximum rise of the water temperature in the pools up to 28°C. In this way, thermal energy is stored in order to be utilized at another time point, with insufficient thermal power production from the solar collectors, compared to the heating load. The additionally stored thermal power in the current step will be subtracted from the required biomass heater thermal power production in this next step.
- The above steps are executed for every calculation time step and for the overall annual period. With the integration of the process on an annual basis, the following magnitudes are calculated:
  - the total annual thermal energy production from the solar collectors and the biomass heater
  - the required biomass fuel annual consumption and procurement cost, assuming alternatively the use of olive kernel or biomass pellets.
- The system is eventually evaluated with the calculation of the optimization index, which is the specific production cost c_{th} of the final produced thermal energy, defined with the following relationship:

$$S.C. = \frac{\frac{I.C.}{N} + \frac{\sum_{n=1}^{N} \frac{A.O.C.}{(1+i)^n}}{N}}{E_{th}} \quad (4.115)$$

where:

S.C.: the annual average, thermal energy production-specific cost (in €/kWh$_{th}$)
I.C.: the initial cost (setup cost) of the solar-combi system (in €)
A.O.C.: the overall annual operation and maintenance cost (in €/year)
i: the discount rate, assumed equal to 3%
N: the total life period of the solar-combi system, assumed equal to 20 years
n: the number of the current year of the system's operation
E_{th}: the annual final thermal energy production of the solar-combi system (kWh$_{th}$).

The annual operation cost equals the sum of the fuel consumption cost C_{fuel} and the annual maintenance cost C_{maint}. The latter is practically determined by the biomass heater required maintenance, which is estimated at €500 per year.

- The annual operation of the investigated solar-combi system is simulated iteratively, introducing each time a different quantity of solar collectors. The optimum dimensioning is obviously the one with the minimum production-specific cost.

d. *The dimensioning results*

With the previously described methodology, the optimum dimensioning scenario, based on economic criteria, will be formulated for the investigated solar-combi system. The optimization methodology is executed iteratively, increasing each time the solar collectors' installed surface. Taking into account that a typical surface of a selective coating flat-plate solar collector is 2 m^2, the calculation is executed for a solar collectors' surface range from 0 m^2 to 300 m^2, with an increment step of 2 m^2.

The solar collectors are installed in groups of maximum five solar collectors connected in series. The outlet temperature of a solar collector will be the inlet temperature for the next solar collector connected in series at the same group, as indicated in Section 4.6.3. The arisen groups are then connected together in parallel. This means that the power production and the outlet temperature from one group of five solar collectors will be the same for the other groups with the same number of solar collectors too.

The operation of the biomass heater is investigated for two alternative types of biomass fuels, namely for olive kernel with a procurement price of 80 €/tn and a heat capacity of 4.054 kWh/kg and for biomass pellets, with a procurement price of 350 €/tn and a heat capacity of 5.792 kWh/kg.

The final thermal production-specific cost fluctuation (introduced as the dimensioning optimization criterion) versus the solar collectors' total effective surface is presented in Figure 4.77.

The results regarding the optimum dimensioning are presented in Table 4.17. Among them, the thermal energy production from the solar collectors and the biomass heater, the biomass fuel, and the corresponding primary energy consumption, with the biomass fuel procurement cost, are included.

In Table 4.17 the effect of the biomass fuel price on the optimum dimensioning of the thermal hybrid power plant is clearly depicted, executed on the basis of economic criteria, such as the minimization of the overall thermal energy production-specific cost. In case of olive kernel use, the low procurement price has led to a rather limited installation of solar collectors (98 m^2) and a corresponding reduced annual thermal energy penetration from the solar collectors (33%). The required thermal energy production for the swimming pools' heating is mainly based on the biomass fuel consumption, which exceeds 35 tn annually, yet with a rather low cost, due to the low procurement

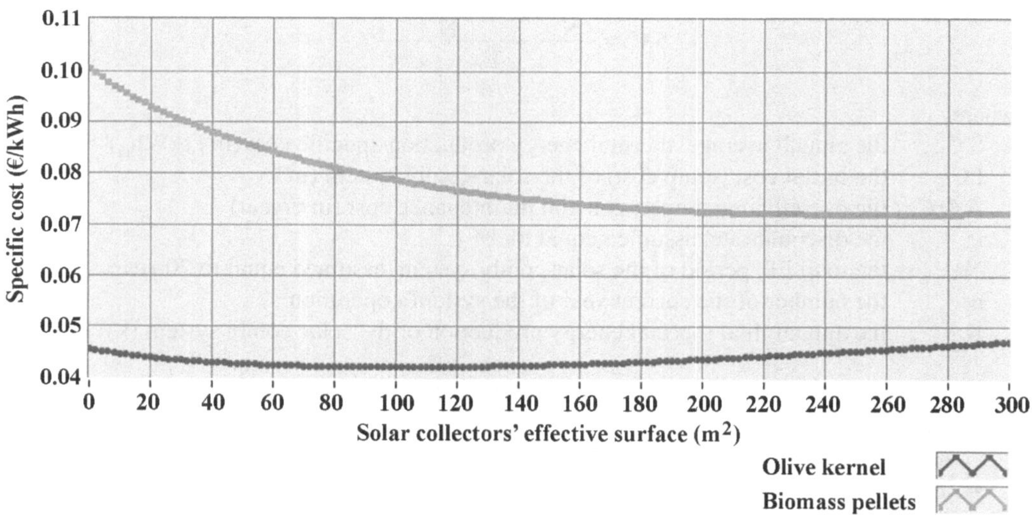

FIGURE 4.77 The fluctuation of the final thermal energy production-specific cost (optimization index) versus the solar collectors' total effective surface [74].

TABLE 4.17
Characteristic Features of the Optimum Synthesis for the Biomass Heater–Solar Collectors Combi System [74]

	Consumed Fuel	
Magnitude	**Olive Kernel**	**Biomass Pellets**
Total effective surface of solar collectors (m^2)	98.00	262.00
Solar collectors final annual thermal energy production (kWh)	52,346	82,536
Biomass heater final annual thermal energy production (kWh)	106,161	75,971
Total final annual thermal energy production (kWh)	158,507	158,507
Solar collectors' annual thermal energy production percentage (%)	33.02	52.07
Biomass heater annual thermal energy production percentage (%)	66.98	47.93
Biomass fuel annual consumption (tn)	35.388	17.725
Primary energy annual consumption (kWh)	143,461	102,664
Biomass fuel annual cost (€)	2,831.00	6,203.79

price (80 €/tn). On the other hand, in case of expensive biomass pellets use, with a procurement price more than four times higher than that of olive kernels, the optimum dimensioning synthesis implies a considerable increase of the required solar collectors' effective surface (262 m^2), leading to a subsequent rise of their final thermal energy annual production share (52%). The annual biomass consumption drops at 18 tn, yet with higher procurement cost, due to the corresponding increased price.

Table 4.18 is a comprehensive synopsis of the essential economic features of the examined systems. Assuming a procurement price of 0.95 €/lt for diesel oil, first the annual diesel oil consumption cost is calculated, on the theoretical assumption that the swimming pools' heating is covered exclusively by a diesel oil heater, given the annual calculated diesel oil consumption of 20,987 lt presented in Table 4.16. Then, the annual reduction of the fuel consumption cost is calculated for the two alternative biomass fuels. Given also an approximate estimation of the setup cost of the examined systems, a rough calculation of the anticipated payback period is eventually calculated. The effect of the biomass fuel price on the economic efficiency of the examined combi-solar systems is also depicted on the calculated payback periods.

Finally, in Figures 4.78 and 4.79 the annual final thermal power production synthesis from the biomass heater and the solar collectors is presented, showing the consumption of olive kernel and biomass pellets, respectively. From these figures it is seen that the operation of the biomass heater with expensive biomass pellet consumption is restricted to the winter season only, namely

TABLE 4.18
Fundamental Economic Features and Indexes of the Examined Swimming Pools' Heating Systems

	Total Setup Cost (€)	System's Annual Operating Cost (€)	Annual Operating Cost Reduction Compared to the Existing One (€)	Annual Operating Cost Reduction Compared to the Existing One (%)	Payback Time Period (years)
Diesel oil system	—	19,937.65	—	—	—
Combi solar-heater system burning olive kernels	54,700	2,831.00	17 106.65	85.80	3.20
Combi solar-heater system burning biomass pellets	120,300	6,203.79	13 733.86	68.88	8.76

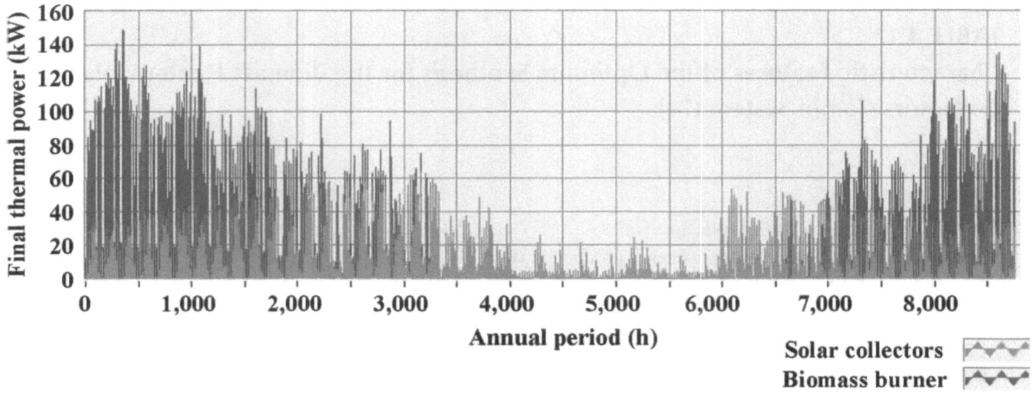

FIGURE 4.78 Annual final thermal power production synthesis from biomass heater and solar collectors with the use of olive kernel [74].

from December to February, when the swimming pool heating loads are maximized. On the contrary, the operation of the biomass burner with the consumption of olive kernel is considerably prolonged.

4.8.2 Thermal Hybrid Plant for School Building Heating

In this second case study, a solar-combi system consisting of flat-plate solar collectors, thermal storage water tanks, and a biomass heater will be studied for the indoor space heating of a school building. The school building extends to three floors (basement, ground floor, and first floor) with a total covered area of conditioned space at 1,050 m^2 and of nonconditioned space at 850 m^2. A general three-dimensional (3D) view of the school building is presented in Figure 4.80.

The calculation of the building's heating loads was accomplished by assuming appropriate insulation of the opaque surfaces with U-factors below 1 W/m^{2}K, openings with double glazing and aluminum frame with thermal break with U-factors at 2.7 W/m^{2}K, and thermal transition factors for the heat transfer with convection and radiation from the building's envelope to the ambient environment at 10 W/m^{2}K and 7.7 W/m^{2}K for air flows above horizontal and next to vertical surfaces, respectively. The desirable indoor space temperature and relative humidity, according to thermal

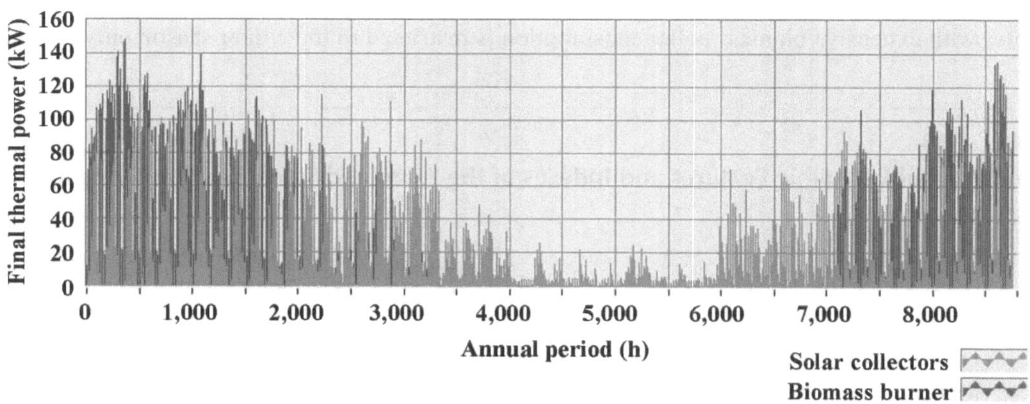

FIGURE 4.79 Annual final thermal power production synthesis from biomass heater and solar collectors with the use of biomass pellets [74].

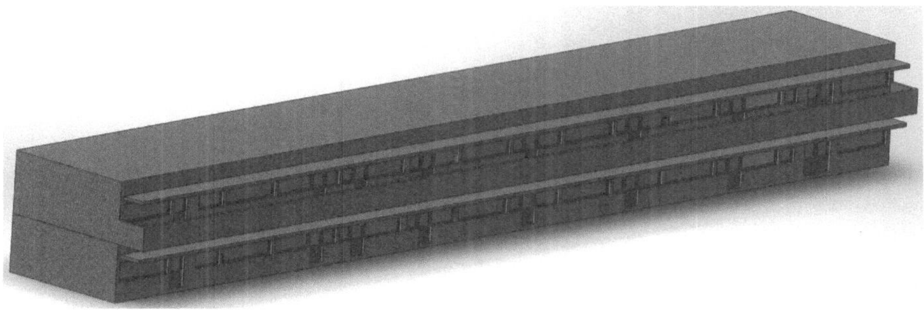

FIGURE 4.80 General 3D view of the case study school building.

comfort conditions, were set at 22°C/26°C for winter/summer and 50%, respectively. The methodology introduced by ASHRAE was applied, along with any other required parameters [19, 75]. Finally, the school building's operation schedule was also considered. Specifically, a daily operation program from September 10 to June 15 was adopted, from Monday to Friday, from 8 a.m. to 5 p.m. Based on the above overview of the facts and assumptions, the heating and cooling loads' annual fluctuation for the examined building is presented in Figure 4.81.

By integrating this time-series, the annual final thermal energy demand is calculated at 32,356 kWh. If this thermal energy demand is undertaken by a conventional diesel oil heater, by assuming average efficiency of 80% for the heater, 88% for the heat distribution network, and 92% for the heating radiators, and by adopting the value of 10.25 kWh/L as the heat capacity of diesel oil, the annual diesel oil consumption is calculated as:

$$32{,}356 \text{ kWh}/(0.80 \times 0.88 \times 0.92 \times 10.25 \text{ kWh/L}) = 4{,}874 \text{ L}$$

Furthermore, by assuming a diesel oil procurement price at 0.95 €/L for the Mediterranean basin, the corresponding annual procurement cost equals €4,630.

Alternatively, for the coverage of the above heating load, the solar-combi system's layout presented and analyzed in Figure 4.82 will be adopted. It consists of a number of thermal storage tanks, flat-plate solar collectors with selective coating, and a biomass heater, supported by the appropriate hydraulic network and components. The operation algorithm presented in Section 4.6.4 will also be

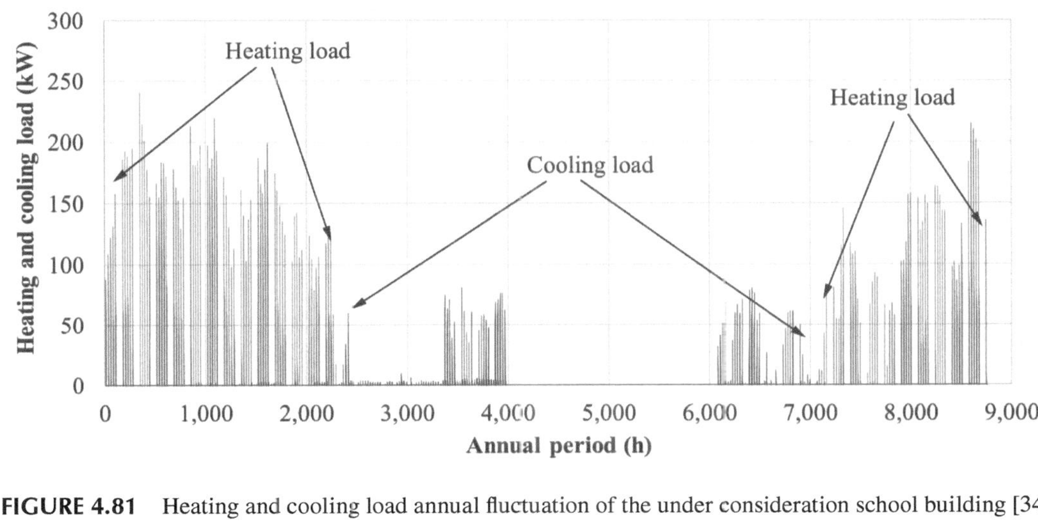

FIGURE 4.81 Heating and cooling load annual fluctuation of the under consideration school building [34].

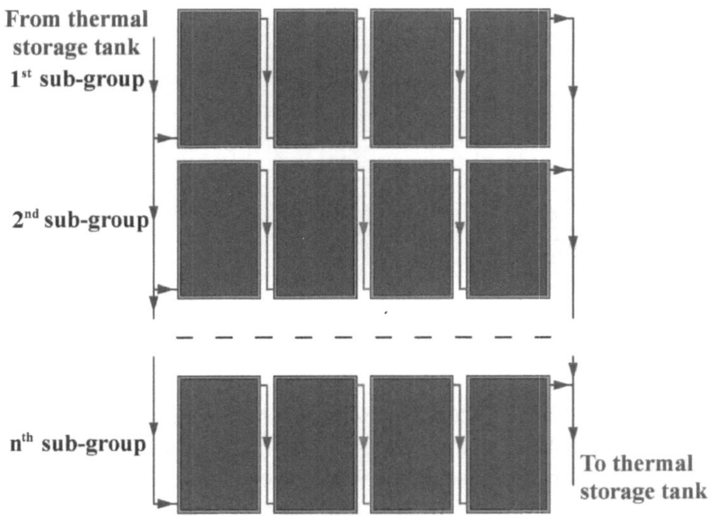

FIGURE 4.82 Connection layout of the solar collectors' field [34].

applied, realized by a central control unit, integrated with the required measurement instruments (thermometers), the motor vanes, and the circulators as placed in Figure 4.82.

The solar collectors' field is formulated in parallel groups of four in-series connected collectors, as depicted in Figure 4.82. The total number of these groups of four in-series connected collectors will be the result of the dimensioning process. The optimization criterion set for this dimensioning is the minimization of the final thermal energy production-specific cost, as in the previously presented case study (Equation 4.113). For each in-series connected group, the total thermal power production is first calculated, according to the process described in Section 4.6.3. The total thermal power production from the overall solar collectors' field will then be the thermal power produced by one in-series connected group multiplied by the number of groups.

A relevant software application was developed for the simulation of the examined solar-combi system annual operation and the computational realization of the proposed algorithm. The simulation was executed on an annual basis with hourly calculation steps. For every hourly calculation step, the thermal power demand P_{td} and the total thermal power production from the solar collectors P_{sc} are introduced. The direct thermal power penetration P_{sp} from the solar collectors for the thermal power demand coverage will be:

- if $P_{td} \geq P_{sc}$ then $P_{sp} = P_{sc}$;
- if $P_{td} < P_{sc}$ then $P_{sp} = P_{td}$.

In any case, the thermal power storage P_{sta} from the solar collectors will be:

$$P_{sta} = P_{sc} - P_{sp}.$$

The remaining thermal power demand P_{tdr}, after the direct penetration from the solar collectors will be:

$$P_{tdr} = P_{td} - P_{sp}.$$

Given the interrupted operation of the school building, to facilitate the simulation process, the remaining thermal energy demand and the total TES from the solar collectors are calculated for

TABLE 4.19
Results of the Iterative Dimensioning Procedure of the Solar-Combi System [34]

Number of solar collectors	36	40	44	48	52
Solar collectors' total surface (m^2)	82.8	92	101.2	110.4	119.6
Solar collectors' production from 15/10–15/4 (kWh)	17,072	18,774	20,480	22,097	23,801
Solar collectors' direct thermal energy penetration (kWh)	4,711	5,093	5,448	5,803	6,157
Solar collectors' thermal energy available for storage (kWh)	12,361	13,680	15,031	16,294	17,644
Solar collectors' thermal energy eventually stored (kWh)	7677	8354	9068	9721	10,292
Biomass heat thermal energy production (kWh)	19,968	18,909	17,840	16,833	15,907
Total thermal energy production (kWh)	32,356	32,356	32,356	32,356	32,356
Required thermal storage capacity (kWh_{th})	198.98	220.25	241.81	262.76	283.82
Required water tank capacity (kg)	11,370	12,586	13,818	15,015	16,219
Solar collectors' annual percentage coverage (%)	38.29	41.56	44.86	47.98	50.84
Solar collectors' thermal energy annual percentage surplus (%)	27.43	28.37	29.12	29.75	30.89

every 24-hour period, by integrating the corresponding 24-hour thermal power time series. In case of higher TES than the remaining thermal energy demand during a specific 24-hour period, this will be used for the thermal energy demand coverage for the next 24-hour period or periods. In cases of lower TES than the remaining thermal energy demand for a specific 24-hour period, the corresponding thermal energy shortage is undertaken by the biomass heater. With the above simulation approach, the thermal storage capacity is determined by the maximum required TES over a 24-hour period, over the year.

Indicatively, the results of five different runs of the simulation process are presented in Table 4.19 [34]. For these runs, a solar collector model with 2.3 m^2 of effective surface was introduced. As seen in this table, the total annual contribution of the solar collectors for the thermal energy demand coverage ranges from 38% to 50%. The thermal storage contribution to this thermal energy production is calculated between 27% and 28%. There is also significant annual thermal energy surplus, due to the seasonal operation of the solar-combi system, defined by the heating needs of the school building's indoor space.

The final dimensioning of the solar-combi system will be based on the minimization of the final thermal energy production-specific cost. To this end, essential economic setup and operation figures should be introduced. Specifically, the following costs and prices have been adopted:

- procurement and installation cost of the biomass heater and its accessories: €30,150
- procurement cost of one solar collector with selective coating: €220
- procurement cost of one water thermal storage tank of 5,000 lt capacity: €10,000
- procurement and installation cost of secondary hydraulic and electronic equipment (circulator, motor-valves, control system): €8,000
- procurement price for biomass pellets (in Mallorca): 350 €/tn
- annual average maintenance and operation cost configured by the consumed biomass pellets procurement cost and the biomass heater annual maintenance cost, on average: €200.

Given the above assumptions, the final thermal energy production-specific cost is analyzed for alternative combinations of solar collectors and thermal storage tank numbers in Table 4.20. The annual biomass consumption is calculated by introducing the specific heat capacity of biomass pellets equal to 5.2 kWh/kg and the average efficiency of the new biomass heat at 85%. In Table 4.20, it is seen that the installation of 48 solar collectors, or equivalently the installation of 110.4 m^2 of solar collectors' effective area, combined with the installation of three thermal storage tanks gives the minimum specific production cost; hence, it is selected as the optimum setup. The divergence

TABLE 4.20
Economic Analysis and Calculation of the Final Thermal Energy Production-Specific Cost [34]

	Investigating Scenario (Number of Solar Collectors/Thermal Storage Tanks)					
Cost Component	36/3	40/3	44/3	48/3	52/4	52/3
Biomass heater cost (€)	30,150	30,150	30,150	30,150	30,150	30,150
Solar collectors cost (€)	7,920	8,800	9,680	10,560	11,440	11,440
Thermal storage tanks cost (€)	30,000	30,000	30,000	30,000	40,000	30,000
Other equipment cost (€)	8,000	8,000	8,000	8,000	8,000	8,000
Total setup cost (€)	76,070	76,950	77,830	78,710	89,590	79,590
Biomass pellets annual consumption (tn)	3.840	3.636	3.431	3.237	3.059	3.096
Average annual maintenance and operation cost (€)	1,200	1,147	1,093	1,043	996	1,006
Thermal energy production levelized cost (€/kWh$_{th}$)	0.1546	0.1544	0.1541	0.1539	0.1692	0.1541

between the calculated values of the optimization criterion for the investigated dimensioning scenarios is rather small. This is because the only varying parameters among the alternatively configured scenarios is the quantity of solar collectors, which has a limited contribution to the system's total setup cost. Of course, the annual biomass pellet consumption is also affected, yet with only minimal impact on the final production-specific cost [34].

In Figure 4.83 the annual final thermal power production synthesis is presented. Two focused thermal power production synthesis graphs are given in Figure 4.84. The first one refers to the first 105 days of the year (from January 1 to April 15) and the second one refers to the last 76 days of the year (from October 15 to the December 31). From these graphs, the following observations are worth mentioning:

- the TES during the weekends is enough to undertake a considerable thermal energy demand of the first days from the next week
- during the last days of October and while approaching the last days of March, namely at the beginning and at the end of the heating season, the heating load is fully undertaken by the solar collectors and the thermal storage tanks.

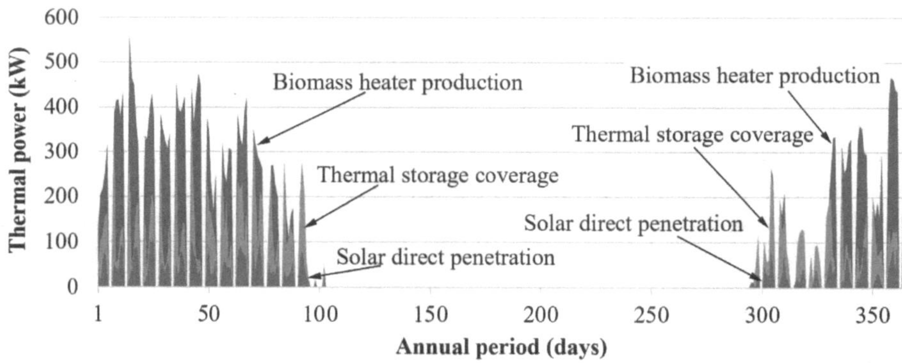

FIGURE 4.83 Annual thermal power production synthesis for the heating of the school building.

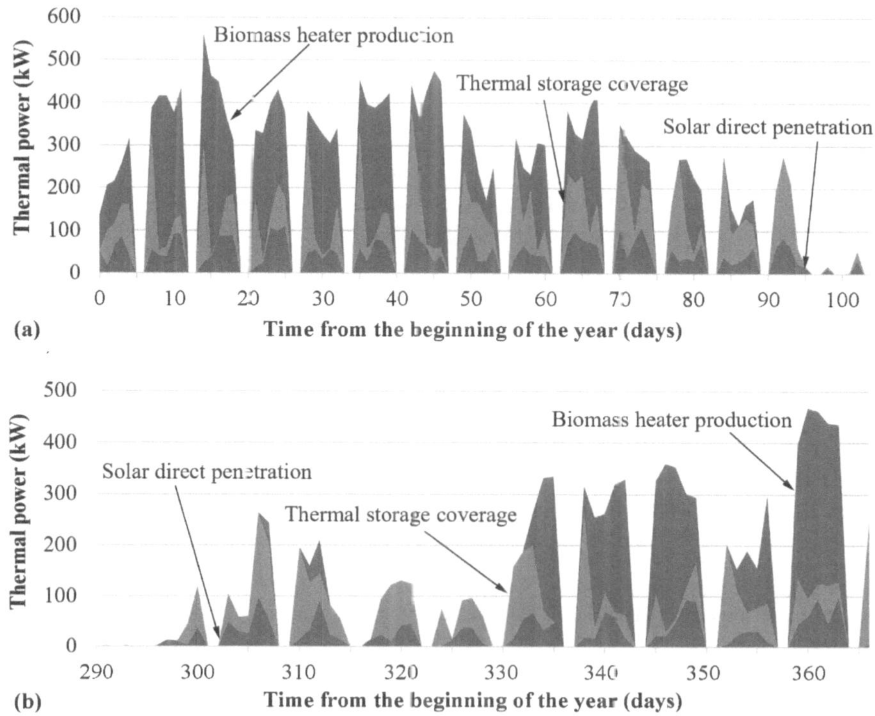

FIGURE 4.84 Thermal power production synthesis for the heating of the school building for the periods (a) from January 1 to April 15, and (b) from October 15 to December 31 [34].

REFERENCES

1. Pérez-Lombard, L., José Ortiz, J., & Christine Pout, C. (2008). A review on buildings energy consumption information. *Energy and Buildings*, *40*, 394–398.
2. Katsaprakakis, D. A. (2015). Comparison of swimming pools alternative passive and active heating systems based on renewable energy sources in Southern Europe. *Energy*, *81*, 738–753.
3. GreenSpec: Solar hot water collectors. http://www.greenspec.co.uk/building-design/solar-collectors/ (Accessed on September 2019).
4. Centrometal: Heating Technique. CVSKC-10 (vacuum tube solar collector). https://www.centrometal.hr/en/portfolio/cvskc-10-vacuum-tube-solar-collector_en/ (Accessed on December 2019).
5. Stine, W. B., Geyer, M., & Stine, S. R. Power from the sun.net. Chapter 8: Concentrator optics. http://www.powerfromthesun.net/Book/chapter08/chapter08.html (Accessed on October 2019).
6. Duffie, J. A., & Beckman, W. A. (2016). *Solar Engineering of Thermal Processes* (4th ed.). New Jersey: John Wiley & Sons.
7. Rabl, A. (1976). Optical and thermal properties of compound parabolic concentrators. *Solar Energy*, *18*, 497–511.
8. Pranesh, V., Velraj, R., Christopher, S., & Kumaresan, V. (2019). A 50 year review of basic and applied research in compound parabolic concentrating solar thermal collector for domestic and industrial applications. *Solar Energy*, *187*, 293–340.
9. Kincaid, N., Mungas, G., Kramer, N., Wagner, M., & Zhu, G. (2018). An optical performance comparison of three concentrating solar power collector designs in linear Fresnel, parabolic trough, and central receiver. *Applied Energy*, *231*, 1109–1121.
10. Sajjad, U., Amer, M., Muhammad Ali, H., Dahiya, A., & Abbas, N. (2019). Cost effective cooling of photovoltaic modules to improve efficiency. *Case Studies in Thermal Engineering*, *14*, 100420.
11. Chow, T. T. (2010). A review on photovoltaic/thermal hybrid solar technology. *Applied Energy*, *87*, 365–379.

12. Othman, M. Y., Ibrahim, A., Jin, G. L., Ruslan, M. H., & Sopian, K. (2013). Photovoltaic-thermal (PV/T) technology—The future energy technology. *Renewable Energy*, *49*, 171–174.
13. Ibrahim, A., Othman, M. Y., Ruslan, M. H., Alghoul, M. A., Yahya, M., & Zaharim, A. (2009). Performance of photovoltaic thermal collector (PV/T) with different absorbers design. *WSEAS Transactions on Environment and Development*, *5*, 321–330.
14. Sultan, S. M., & Ervina Efzan, M. N. (2018). Review on recent photovoltaic/thermal (PV/T) technology advances and applications. *Solar Energy*, *173*, 939–954.
15. YouGen. Energy made easy: Solar PV-T systems—what are the pros and cons? http://www.yougen.co.uk/blog-entry/2833/What+are+the+advantages+of+Hybrid+Solar+Panels%273F/ (Accessed on September 2019).
16. Abdul Hamid, S., Yusof Othman, M., Sopian, K., & Zaidi, S. H. (2014). An overview of photovoltaic thermal combination (PV/T combi) technology. *Renewable and Sustainable Energy Reviews*, *38*, 212–222.
17. Cunio, L. N., & Sproul, A. B. (2012). Performance characterisation and energy savings of uncovered swimming pool solar collectors under reduced flow rate conditions. *Solar Energy*, *86*, 1511–1517.
18. Klein, S. A. (1975). Calculation of flat-plate collector loss coefficients. *Solar Energy*, *17*, 79–80.
19. Kreider, J., Rabl, A., & Curtiss, P. (2017). *Heating and Cooling of Buildings* (3rd ed.). Boca Raton, Florida: CRC Press.
20. Klein, S. A. (1979). Calculation of monthly-average transmittance-absorptance product. *Solar Energy*, *23*, 547–551.
21. Axaopoulos, P., Ed. (2011). *Solar Thermal Conversion: Active Solar Systems. European Network of Education and Training in Renewable Energy Sources (EuroCENTRES).* Athens: Symmetria Editions.
22. Souka, A. F., & Safwat, H. H. (1966). Determination of the optimum orientations for the double-exposure, flat-plate collector and its reflectors. *Solar Energy*, *10*, 170–174.
23. Tesfamichael, T., & Wäckelgård, E. (2000). Angular solar absorptance and incident angle modifier of selective absorbers for solar thermal collectors. *Solar Energy*, *68*, 335–341.
24. Hill, J. E., Jenkins, J. E., & Jones, D. E. (1979). Experiment verification of a standard test procedure for solar collectors. *NBS Building Science Series 117. U.S. Department of Commerce.* https://www.govinfo.gov/content/pkg/GOVPUB-C13-23781ef67c8a78bca2bf3e0132ec1cf6/pdf/GOVPUB-C13-23781ef67c8a78bca2bf3e0132ec1cf6.pdf (Accessed on September 2019).
25. Henning, H.-M. (2007). Thermal systems and components. *Solar Assisted Air-Conditioning in Buildings: A Handbook for Planners (2nd ed.).* Berlin, Germany: Springer.
26. Axaopoulos, P., Pitsilis, G., & Panagakis, P. (2002). Multimedia education program for an active solar hot water system. *International Journal of Solar Energy*, *22*, 83–92.
27. Axaopoulos, P., & Pitsilis, G. (2007). Energy software programs for educational use. *Renewable Energy*, *32*, 1045–1058.
28. Centre of Renewable Energy Source (CRES), Department of Solar Thermal Systems. http://www.cres.gr/kape/present/labs/solar_uk.htm (Accessed on July 2019).
29. Chow, T. T. (2003). Performance analysis of photovoltaic-thermal collector by explicit dynamic model. *Solar Energy*, *75*, 143–152.
30. Kteniadakis M. (2010). *Heat Transfer Applications* (1st ed.). Athens, Greece: Cities Edition.
31. McAdams, W. H. (1954). *Heat Transmission* (3rd ed.). New York: McGraw-Hill.
32. Brandemuehl, M. J., & Beckman, W. A. (1980). Transmission of diffuse radiation through CPC and flat-plate collector glazings. *Solar Energy*, *24*(5), 511–513.
33. Rabl, A. (1976). Optical and thermal properties of compound parabolic concentrators. *Solar Energy*, *18*, 497–511.
34. Katsaprakakis, D. A., & Zidianakis, G. (2019). Optimized dimensioning and operation automation for a solar-combi system for indoor space heating. A case study for a school building in Crete. *Energies*, *12*, en12010177.
35. Mariyappan, J., & Anderson, D. (2001, October). Thematic Review of GEF-Financed Solar Thermal Projects. Monitoring and Evaluation Working Paper 7. London, UK: Imperial College of Science, Technology, and Medicine. http://citeseerx.ist.psu.edu/viewdoc/download;jsessionid=4B498CA0F6DC3F0F738264831F904107?doi=10.1.1.475.5608&rep=rep1&type=pdf (Accessed on November 2019).
36. BrightSource Energy: Ivanpah Solar Thermal Power Plant. Image gallery: http://www.brightsourceenergy.com/image-gallery#.Xcexjlcza00. (Accessed on November 2019).
37. Ong, S., Campbell, C., Denholm, P., Margolis, R., & Heath, G. (2013, June). Land-use requirements for solar power plants in the United States. *Technical Report NREL/TP-6A20-56290.* https://www.nrel.gov/docs/fy13osti/56290.pdf (Accessed on November 2019).

38. Zarza Moya, E. (2012). Chapter 7: Parabolic-trough concentrating solar power (CSP) systems. In: *Concentrating Power Technology (1st ed.).* Cambridge, England: Woodhead Publishing, 197–239.
39. Breeze, P. (2016). Chapter 3: Solar thermal power generation. In: *Solar Power Generation (1st ed.).* Academic Press, 17–24.
40. Lv, T., & Li, N. (2009). Study on the Continuous and Stable Running Mode of Solar Thermal Power Plant. Power and Energy Engineering Conference, APPEEC, Asia-Pacific.
41. Drosou, V., Kosmopoulos, P., & Papadopoulos, A. (2016). Solar cooling system using concentrating collectors for office buildings: A case study for Greece. *Renewable Energy, 97,* 697–708.
42. Zohuri, B. (2017, September). Chapter 8: Compact heat exchangers application in new generation of CSP. In: *Compact Heat Exchangers.* doi:10.1007/978-3-319-29835-1_8
43. Bachelier, C., Selig, M., Mertins, M., Stieglitz, R., Zipf, V., Neuhäuser, A., & Steinmann, W. D. (2015). Systematic analysis of Fresnel CSP plants with energy storage. *Energy Procedia, 69,* 1201–1210.
44. Themeinfo II/2013. Solar thermal power plants. Utilizing concentrated sunlight for generating energy. FIZ Karlsuhe. ISSN: 1610 – 8302. http://www.bine.info/fileadmin/content/Publikationen/Englische_Infos/themen_0213_engl_Internetx.pdf (Accessed on November 2019).
45. Torresol Energy: The Gemasolar Project. https://torresolenergy.com/gemasolar/ (Accessed on November 2019).
46. Agami Reddy, T., Kreider, J. F., Curtiss, P. S., & Rabl, A. (2015, July 26). *Heating and Cooling of Buildings: Principles and Practice of Energy Efficient Design* (3rd ed.). CRC Press. (ISBN: 9781439899892).
47. Erneuebare Energien und Klimaschutz.de: Solar thermal power plants. Technology Fundamentals. https://www.volker-quaschning.de/articles/fundamentals2/index.php (Accessed on November 2019).
48. Poullikkas, A. (2009). Economic analysis of power generation from parabolic trough solar thermal plants for the Mediterranean region. A case study for the island of Cyprus. *Renewable and Sustainable Energy Reviews, 13,* 2474–2484.
49. Sargent & Lundy LLC Consulting Group. (2003, October). Assessment of Parabolic Trough and Power Tower Solar Technology Cost and Performance Forecasts. NREL https://www.nrel.gov/docs/fy04osti/34440.pdf (Accessed on November 2019).
50. Hafez, A. Z., Soliman, A., El-Metwally, K. A., & Ismail, I. M. (2016). Solar parabolic dish Stirling engine system design, simulation, and thermal analysis. *Energy Conversion and Management, 126,* 60–75.
51. Alva, G., Lin, Y., & Fang, G. (2018). An overview of thermal energy storage systems. *Energy, 144,* 341–378.
52. Lu, T., & Li, N. (2009, March 27-31). Study on the Continuous and Stable Running Mode of Solar Thermal Power Plant. Power and Energy Engineering Conference, APPEEC, Wuhan, China.
53. Alva, G., Liu, L., Huang, X., & Fang, G. (2017). Thermal energy storage materials and systems for solar energy applications. *Renewable and Sustainable Energy Reviews, 68,* 693–706.
54. Roubaud, E. G., Osorio, D. P., & Prieto, C. (2017). Review of commercial thermal energy storage in concentrated solar power plants: steam vs. molten salts. *Renewable and Sustainable Energy Reviews, 80,* 133–148.
55. Kenda, E. S., N'Tsoukpoe, K. E., Ouédraogo, I. W. K., Coulibaly, Y., Py, X., & Ouédraogo, F. M. A. W. (2017). Jatropha curcas crude oil as heat transfer fluid or thermal energy storage material for concentrating solar power plants. *Energy Sustainable Development, 40,* 59–67.
56. Benoit, H., Spreafico, L., Gauthier, D., & Flamant, G. (2016). Review of heat transfer fluids in tube-receivers used in concentrating solar thermal systems: properties and heat transfer coefficients. *Renewable and Sustainable Energy Reviews, 55,* 298–315.
57. Jacob, R., Belusko, M., Fernandez, A. I., Cabeza, L. F., Saman, W., & Bruno, F. (2016). Embodied energy and cost of high temperature thermal energy storage systems for use with concentrated solar power plants. *Applied Energy, 180,* 586–597.
58. Ercole, D., Manca, O., & Vafai, K. (2017). An investigation of thermal characteristics of eutectic molten salt-based nano fluids. *International Communication in Heat and Mass Transfer, 87,* 98–104.
59. Felderhoff, M., & Bogdanović, B. (2009). High temperature metal hydrides as heat storage materials for solar and related applications. *International Journal of Molecular Sciences, 10,* 325–344.
60. Alonso, M. C., Verae Agullo, J., Guerreiro, L., Flore Laguna, V., Sanchez, M., & Collarese Pereira, M. (2016). Calcium aluminate based cement for concrete to be used as thermal energy storage in solar thermal electricity plants. *Cement and Concrete Research, 82,* 74–85.

61. Grirate, H., Zari, N., Elamrani, Iz., Couturier, R., Elmchaouri, A., Belcadi, S., & Tochon, P. (2014). Characterization of several Moroccan rocks used as filler material for thermal energy storage in CSP power plants. *Energy Procedia*, *49*, 810–819.
62. Calvet, N., Gomez, J. C., Faik, A., Roddatis, V. V., Meffre, A., Glatzmaier G. C., et al. (2013). Compatibility of a post-industrial ceramic with nitrate molten salts for use as filler material in a thermocline storage system. *Applied Energy*, *109*, 387–393.
63. Cárdenas, B., & León, N. (2013). High temperature latent heat thermal energy storage: phase change materials, design considerations and performance enhancement techniques. *Renewable and Sustainable Energy Review*, *27*, 724–737.
64. Stamatiou, A., Obermeyer, M., Fischer, L. J., Schuetz, P., & Worlitschek, J. (2017). Investigation of unbranched, saturated, carboxylic esters as phase change materials. *Renewable Energy*, *108*, 401–409.
65. Yuan, Y., Zhang, N., Tao, W., Cao, X., & He, Y. (2014), Fatty acids as phase change materials: a review. *Renewable and Sustainable Energy Reviews*, *29*, 482–498.
66. Alva, G., Huang, X., Liu, L., & Fang, G. (2017). Synthesis and characterization of microencapsulated myristic acid palmitic acid eutectic mixture as phase change material for thermal energy storage. *Applied Energy*, *203*, 677–685.
67. Solé, A., Neumann, H., Niedermaier, S., Martorell, I., Schossig, P., & Cabeza, L. F. (2014). Stability of sugar alcohols as PCM for thermal energy storage. *Solar Energy Materials and Solar Cells*, *126*, 125–134.
68. Myers, P. D. (2015). Additives for heat transfer enhancement in high temperature thermal energy storage media: selection and characterization. University of South Florida Graduate Theses and Dissertations. http://scholarcommons.usf.edu/etd/5749 (Accessed on November 2019).
69. Pardo, P., Deydier, A., Anxionnaz-Minvielle, Z., Rougé, S., Cabassud, M., & Cognet, P. (2014). A review on high temperature thermochemical heat energy storage. *Renewable and Sustainable Energy Reviews*, *32*, 591–610.
70. Pelay, U., Luo, L., Fan, Y., Stitou, D., & Rood, M. (2017). Thermal energy storage systems for concentrated solar power plants. *Renewable and Sustainable Energy Reviews*, *79*, 82–100.
71. Gil, A., Medrano, M., Martorell, I., Lázaro, A., Dolado, P., Zalba, B., et al. (2010). State of the art on high temperature thermal energy storage for power generation. Part 1: Concepts, materials and modellization. *Renewable and Sustainable Energy Reviews*, *14*, 31–55.
72. Wu, M., Li, M., Xu, C., He, Y., & Tao, W. (2014). The impact of concrete structure on the thermal performance of the dual-media thermocline thermal storage tank using concrete as the solid medium. *Applied Energy*, *113*, 1363–1371.
73. Katsaprakakis, D. A. (2019). Introducing a solar-combi system for hot water production and swimming pools heating in the Pancretan Stadium, Crete, Greece. *Energy Procedia*, *159*, 174–179.
74. Katsaprakakis, D. A. (2015). Comparison of swimming pools alternative passive and active heating systems based on renewable energy sources in Southern Europe. *Energy*, *81*, 738–753.
75. *2009 ASHRAE Handbook—Fundamentals (SI Edition)*. Atlanta, Georgia: American Society of Heating, Refrigerating and Air-Conditioning Engineers, Inc.

5 Cogeneration Power Plants

5.1 INTRODUCTION

The term *cogeneration* refers to the concurrent production of different final energy forms with the same energy transformation process and with the consumption of the common amount of primary energy source.

Cogeneration power plants have been widely introduced for the production of electricity and heat, and, in more rare applications, for cooling production. This combined production essentially differs from the conventional production processes, e.g., from electricity produced by centralized conventional or large-size hybrid power plants (presented in Chapters 2 and 3), or from heating and cooling produced by local, decentralized systems, installed and operated directly by the final users (e.g., central heaters, heat pumps). The concept of heat and electricity cogeneration is known as combined heat and power (CHP). In cases of the produced thermal energy utilization for heating and cooling, the cogeneration system is also referred to as a trigeneration system or as combined cooling, heating, and power (CCHP) system. The utilization of centralized heat production from a cogeneration power plant for the coverage of the heating or cooling needs of a nearby city, industrial zone, agricultural area, etc., is also known as district heating or cooling. District heating or cooling requires the availability of a considerable technical infrastructure for the transportation and the distribution of the produced heat from the source (cogeneration power plant) to the final users [1]. For this purpose, usually hydraulic pipelines networks are used. The alternative cogeneration applications are graphically analyzed in Figure 5.1.

The idea of cogeneration systems is not a new one. In fact, CHP first appeared in Europe and in the US at the end of the nineteenth century [2]. At the beginning of the twentieth century, most industries in the US and in Europe owned their proprietary electricity production units, mainly steam turbines operating with coal, exploited also for heat production, namely operated as cogeneration plants. In Europe, significant development of CHP is referred for Denmark, Finland, and The Netherlands [2]. In these countries, apart from the introduction of CHP units in big industries and power plants, exploited essentially for cities' district heating, there are also significant applications mentioned in tertiary (e.g., hospitals, hotels, and sport centers) and residential sectors.

The evolution of CHP systems and processes was the result of efforts and research focusing on the increase of the relatively low overall efficiency of conventional thermal generators. As presented in Chapter 2, conventional thermal generators exhibit overall efficiencies which, under most favorable conditions (i.e., for combined cycles), can reach values up to 50%, while in most cases of single generators, they can be close to 35% for steam and gas turbines and 45% for diesel generators. If we also account the transportation and final usage losses, the overall efficiency can be lower than 30% [3] (Figure 5.2). The low overall efficiency of a thermal generator can be increased with the concurrent exploitation of the produced thermal power, which, under conventional operation, is disposed in the ambient as waste.

CHP systems are precisely based on this very specific concept, namely the recuperation of the largest amount of the produced thermal energy in conventional power plants, in order to supply coverage of thermal energy needs. In other words, CHP systems aim to convert heat, which constitutes a waste of the production process in a conventional power plant, into useful final product. In this way, simultaneous electricity and heat production is achieved, without the consumption

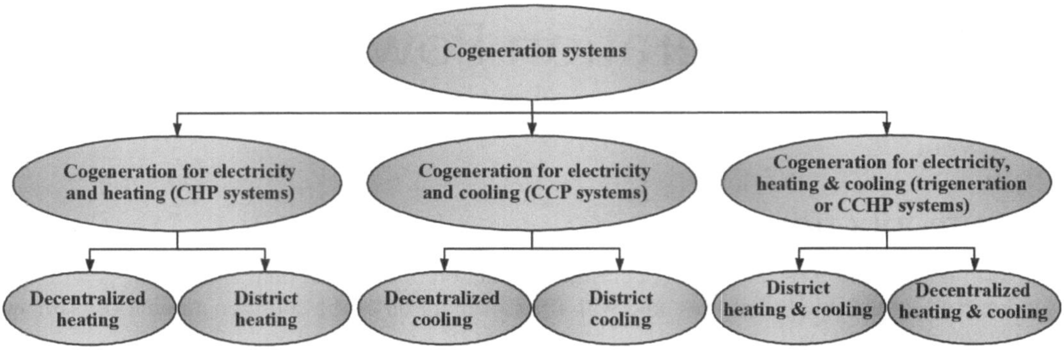

FIGURE 5.1 Cogeneration systems breakdown.

of additional primary energy, increasing, subsequently, the overall efficiency of the power plant (Figure 5.3).

From the above, we conclude that CHP does not constitute a new technology, but it is rather formulated as the combined application of electricity production and thermal energy recovery technologies, integrated with thermal energy transportation and distribution networks. In most cases, cogeneration systems constitute extensions of existing electricity production thermal or nuclear power plants, realized with the introduction of one or more heating recovery systems in the conventional electricity production process. These recovery systems can be introduced either in the exhausted gases disposal route or in the cooling system of the thermal generator, or in both of them. Nevertheless, exactly due to the remarkable contribution of cogeneration systems to the rational use of energy, the design and the manufacturing of new thermal generators (turbines and reciprocating engines), through intensive research, development, and pilot operation, supported by computational design tools and simulation methods, have led to the maximization of the efficiency of each generator technology.

The overall efficiency of a CHP system is formulated by the combined production of electricity and thermal energy. Specifically, assuming that the overall system can be distinguished into two sub-systems, namely the electricity production system (e.g., a conventional thermal generator) and the thermal energy recovery system (usually a recuperator or a heat exchanger), the overall efficiency will be derived by the efficiencies of the two discrete energy transformation processes. If P_{ch} is the chemical energy consumption rate of the consumed fossil fuel at the combustion chamber of

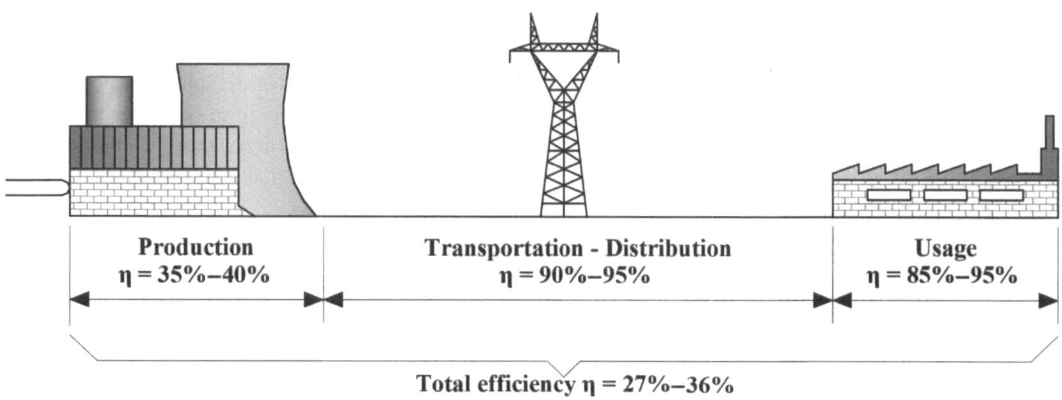

FIGURE 5.2 Energy flow in a conventional power production and transportation system.

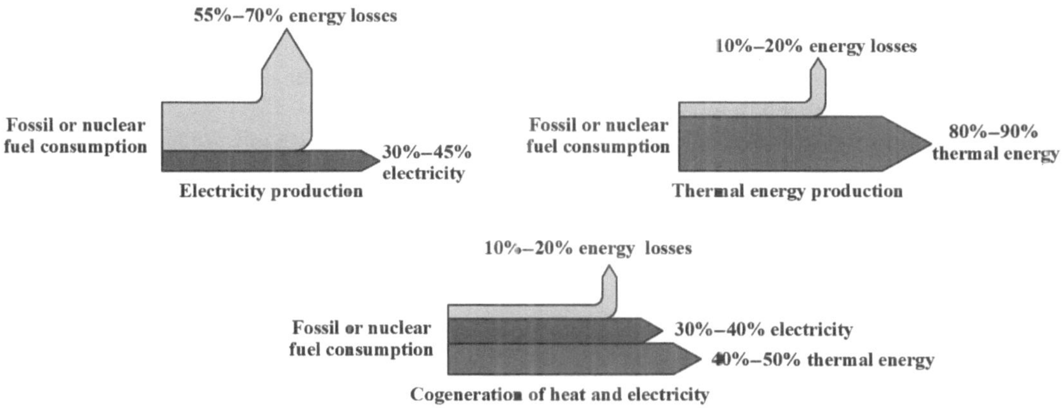

FIGURE 5.3 Energy flow in conventional electricity and thermal power systems and in cogeneration power plants.

the thermal generator (chemical energy), P_{el} is the electrical power production, and P_{th} the recovered thermal power, then:

- The instant efficiency η_{el} for the electricity production process will be:

$$\eta_{el} = \frac{P_{el}}{P_{ch}} \tag{5.1}$$

- The instant efficiency η_{th} for the thermal power production process will be:

$$\eta_{th} = \frac{P_{th}}{P_{ch}} \tag{5.2}$$

- The total efficiency η_{co} for the combined electricity and thermal energy production (cogeneration) will be:

$$\eta_{co} = \frac{P_{el-co} + P_{th}}{P_{ch}} \tag{5.3}$$

where P_{el-co} is the electrical power production when the thermal energy is also recovered, namely during the cogeneration process. As it will be explained later in this chapter, usually this power is slightly lower than the corresponding electrical power production P_{el} when all the available heat is disposed in the ambient. Namely, the recovery of the disposed heat, especially when it is required to be performed above a minimum acceptable temperature, may often cause subsequent reduction of the produced electricity. The efficiency η_{co}, which characterizes the cogeneration process, is also known as the energy utilization factor (EUF) [4, 5].

From the above relationships, the increase of the overall efficiency of the cogeneration process is obvious, compared to the efficiencies of single, independent electricity or thermal energy production processes. As shown in Figure 5.3, the overall efficiency of CHP systems may be as high as 85%. As it will be thoroughly analyzed in the next sections, cogeneration efficiencies above 80% can be approached as thermal power production increases versus electrical power production. Additionally, as stated previously, as the temperature of the recovered heat increases, the ratio of the produced thermal energy versus the produced electricity increases too, as well as the exergy ratio of the CHP process. Additionally, the higher temperature of the recovered heat contributes to the improved quality of the produced thermal energy, expressed thermodynamically by its exergy content. The increasing exergy imposes more effective transportation of the produced thermal energy and versatility to be utilized in more applications.

Given its improved overall efficiency, CHP exhibits a series of attractive features, such as:

- Increased efficiency of the overall cogeneration process.
- Reduction of the specific primary energy consumption per unit of final produced total electricity and thermal energy.
- Reduction of the specific gas emissions per unit of final produced total electricity and thermal energy.
- Minimization of the specific production cost per unit of final produced total electricity and thermal energy.
- Maximization of the viability of decentralized CHP systems, due to the concurrent electricity and thermal power production, increasing further the already high efficiency of the CHP process, because of the minimization of transportation losses, with regard to electricity transportation losses over long distances from centralized power plants. Moreover, the operation of CHP systems with biomass fuels leads to guaranteed power production from renewable energy sources (RES).

CHP systems can be essentially utilized in the following applications:

1. *Thermal power plants of large size*, which can be converted to cogeneration plants and contribute to the heating needs coverage of nearby cities, settlements, industries, agricultural activities, etc.
2. *Industrial sector.* A considerable potential for decentralized CHP systems introduction is met in the food and drink industries, in clothing and fabric manufacturing, in paper industries, in chemical factories, in refineries, in cement factories, and in fundamental metallurgic processes, for drying, water boiling, materials melting, and other uses.
3. *Building sector.* Three subcategories are classified: hotels and hospitals, large groups of residences, and blocks of offices. Each of these building categories is characterized by a specific daily profile of electrical and thermal power demand. Other types of buildings (e.g., schools, universities, malls) may exhibit daily electrical and thermal power demand profiles that can arise as a combination of the above three fundamental building categories.
4. *Agricultural sector.* The recovered heat from nearby thermal power plants can be exploited in a series of agricultural activities, such as products drying and heating of greenhouses.

5.2 BASIC CATEGORIES OF COGENERATION SYSTEMS

Before proceeding to the analytical presentation of the possible alternative implementations of cogeneration systems, we should first of all classify them in two main categories:

- *Centralized cogeneration systems.* Usually these are medium- or large-size systems, combining the production of electricity and thermal energy. They most commonly result from the upgrade of an electricity thermal or nuclear power plant, by introducing a thermal energy recovering system. They can also be designed and developed in parallel during the construction of electricity power plants. Such systems aim to the heating needs coverage of residential or commercial consumers in a nearby city, in industries, in agricultural activities, etc., via a district heating system. Obviously, the design and the dimensioning of an appropriate and adequate thermal energy transportation and distribution system is required.

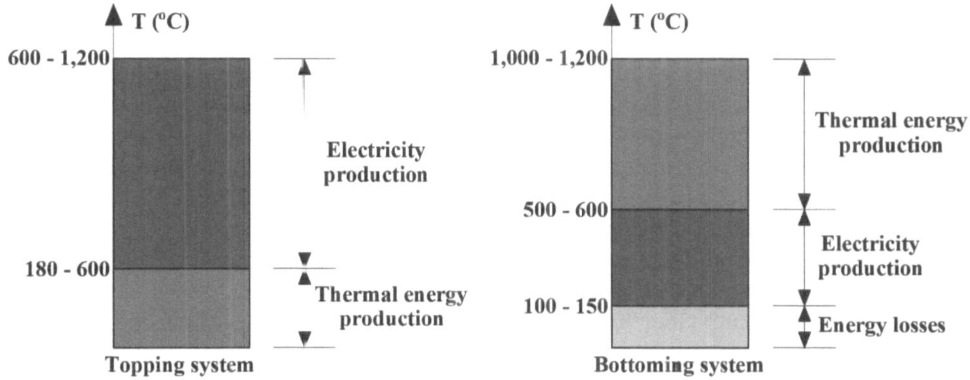

FIGURE 5.4 Temperature operation ranges for topping and bottoming cogeneration systems.

- *Decentralized cogeneration systems.* They are introduced in industries, sports centers, small towns and settlements, hospitals and hotels, universities, etc. These systems are obviously of smaller size compared to centralized cogeneration systems. In most cases the final products are electricity and thermal energy for heating needs of indoor space, warm water production, or industrial applications (drying, sterilizing, etc.). However, there is also the potential for exploiting heating for cooling needs. In such cases, so-called trigeneration systems are defined.

Moreover, with regard to the prior energy form production, cogeneration systems can be also classified as:

- *Topping systems.* In these systems the working medium is initially employed for electricity production, while the disposed thermal energy is recovered and used for heating or cooling needs coverage. These systems are the most regularly met ones.
- *Bottoming systems.* In these systems, thermal energy is initially produced and then the hot gases or steam is channeled to a steam turbine for the production of electricity. Such applications are mainly met in industries, such as steel mills, cement factories, and glass industries.

Characteristic temperature operation ranges are given in Figure 5.4 for these two categories of cogeneration systems.

5.2.1 Centralized Cogeneration Systems

As in conventional interconnected electrical systems, the operation algorithm and the dimensioning process of centralized, interconnected cogeneration systems are simpler, with regard to decentralized cogeneration systems, because the connection of the cogeneration system with a larger energy production and transportation grid/network maximizes its flexibility. Usually the centralized, interconnected cogeneration systems are realized by introducing the required technologies for the recovery of the disposed heat in existing thermal or nuclear power plants.

The operation of the power production units, namely the thermal generators, is in most cases determined by the electrical power demand, since the cogeneration systems are developed on the basis of an existing conventional electricity production plant. Theoretically, the thermal power production can be alternatively approached:

- As a by-product of the electricity production process, with the final thermal power production being determined by the electrical power production and not by the current thermal power demand. In this case, any possible thermal power production shortage with regard to the thermal power demand must be supplemented by the final consumers with decentralized thermal power production technologies (e.g., oil heaters, heat pumps). If the required electricity production imposes thermal power production higher than the current thermal power demand, then the thermal power production surplus can be disposed in the ambient, as in conventional electricity power plants.
- As the basic energy product, regardless of the electricity demand. In this case, the operation of the cogeneration system will be defined by the minimum from the electrical or thermal power demand. Specifically:
  - If the electricity demand imposes thermal power production lower than the thermal power demand, the thermal power production shortage can be covered either by centralized additional, auxiliary thermal power production units, or by the main cogeneration system itself, by increasing the thermal generator power output. Consequently, the power output of the cogeneration system is determined by the thermal load. The potential electricity production surplus can be, in best case, exported in neighboring electrical grids, or stored in storage units or, in the worst case, rejected in dump loads (electrical resistances).
  - If the thermal power demand imposes electricity production lower than the required, then the power production shortage can be covered by dispatching supplementary generators, not integrated in the cogeneration system or by increasing the power output of the cogeneration system itself. In this case, the additional thermal power production is most commonly rejected in the ambient.

Given the above, in centralized, interconnected cogeneration systems the thermal power production is practically determined by the electrical power demand. The maximum potential thermal power production is determined by the chemical power input contained in the consumed fossil or nuclear fuel, after the subtraction of the electrical power production and the involved energy losses (thermal, mechanical). More will be presented in the following sections.

5.2.2 Decentralized Cogeneration Systems

Decentralized cogeneration systems aim to cover the electricity, heating, and cooling needs of usually small, decentralized consumptions. They are usually introduced in residential or commercial buildings, industries and factories, sports premises, agricultural activities, etc. Their decentralized layout introduces particular peculiarities with regard to the technical requirements they should satisfy, their operation algorithm, and their dimensioning, just like in cases of hybrid power plants.

The main differences between a decentralized cogeneration system and a hybrid power plant are:

- In decentralized cogeneration systems, two final forms of energy are produced, namely electricity and thermal energy for heating or cooling, while in hybrid power plants for electricity (Chapter 3) or thermal energy (Chapter 4) production, there is always only one final form of energy produced.
- A cogeneration system can be based on a primary source of energy with guaranteed availability, such as the chemical energy of fossil or biomass fuels. Consequently, a decentralized cogeneration system does not focus on the guaranteed power production from nonguaranteed power plants, like hybrid power plants, but on the maximization of the efficiency of the overall system and the corresponding minimization of the consumed primary energy source, by combining the production of two different final energy forms.

The configuration of cogeneration hybrid power plants of course is not possible when wind parks or photovoltaic (PV) stations are employed as RES units, because there is not any thermal energy production stage within the whole energy transformation process in either of these two RES technologies. Certainly there is the chance to form a cogeneration power plant based on RES when the primary energy source is provided by biomass fuels or high enthalpy geothermal fields. However, in this case, because both geothermal potential and biomass fuels are considered guaranteed energy sources, there is no sense for speaking about hybrid power plants. The only case that a common cogeneration system and hybrid power plant can be considered is solar thermal power plants, which will be examined in the next chapter.

A decentralized cogeneration system is usually introduced in existing electrical systems; hence, there is always the option of power injection from the grid to the local consumption, in case of the cogeneration system's inadequacy to satisfy the electricity consumption. Additionally, there is also apparently the possibility of supplementary thermal power production from a guaranteed unit (e.g., an additional heater), in case the cogeneration system is unable to satisfy the thermal power demand.

The operation algorithm of a decentralized cogeneration system depends on the fundamental scope which the system aims to serve. The alternative options are:

- *Thermal power demand priority*

  In this case the thermal power demand coverage constitutes the main priority of the cogeneration system. This means that the operation point of the cogeneration system is defined by the thermal power demand, namely the thermal power production cannot be lower than the thermal power demand, obviously within the operation range of the cogeneration system. This, in turn, implies that in the general case of nonconstant thermal load, the cogeneration system should be able to follow the thermal power demand fluctuations.

  If the concurrently produced electricity is higher than the power demand, the excess electricity production is provided for the grid, according to the relevant contract between the producer and the operator. In the opposite case, complementary electrical power is injected from the grid to the local consumption.
- *Coverage of base thermal load*

  In this case, the system's dimensioning aims to provide the minimum required thermal power demand for a particular consumption, characterized as base thermal load. For the thermal power peak demand and the fluctuations of the thermal power demand above the base thermal load, additional heaters or boilers are usually used. The cogeneration system normally operates constantly under nominal or, at least, constant load. It is conceivable that the requirements of this specific implementation are not as demanding as in the previous case.

  Similarly with the previous case, if the electricity demand is higher than the constant electricity production from the cogeneration system, supplementary electricity is injected from the grid. On the contrary, any potential electricity production surplus can be provided for the grid.
- *Electricity demand priority*

  In this case, the electricity demand coverage is the first priority of the system. Consequently, the electricity power production is equal to the power demand, obviously without exceeding the operation range of the cogeneration system.

  If the coproduced thermal power is lower than the thermal load, auxiliary heaters can be utilized to cover the thermal power shortage. In the opposite case, the excess thermal power production can be disposed in the ambient. Alternatively, it can be distributed to neighboring consumers, within the frame of a smart grid, expanded also to thermal energy exchange (more on smart grids is presented in Chapter 7).

- *Coverage of base electricity demand*
  In this case the cogeneration system sizing aims to cover a minimum electricity demand. Normally, any electricity demand higher than this minimum consumption is covered with electricity injection from the grid. The thermal power demand of the decentralized consumption can be covered entirely by the cogeneration system, in case the demand is lower than the production, or with additional heaters, in the opposite case. If the thermal energy produced by the cogeneration system concurrently with the base electricity output exceeds the thermal power demand, the excess thermal power can be sold to neighboring consumers or disposed in the ambient.
- *Mixed coverage*
  In this case, for particular time periods during the year, the cogeneration system follows the thermal power demand priority or the base thermal load operation algorithm, while, in other time periods, the system's priority is the electricity power demand coverage or the base electricity production. The shift between the alternative operation modes is based on the estimation of specific operational parameters, such as the scale of the thermal or electrical load, the consumed fuels and the purchased electricity price, and any possible crucial thermal or electricity needs (e.g., in case of a swimming pool center or a hospital respectively). This flexible production can be adopted in case of smart grid integration, with the shifting between the alternative operation modes potentially determined also by the conditions and the prerequisites of the electricity wholesale market, in combination with the previously mentioned possibly involved parameters.
- *Autonomous operation*
  In case of autonomous operation, the system must be able to support the full coverage of the existing electricity and thermal energy demand continuously, without the support of electrical grids or external thermal power production units. This operation exhibits requirements similar to the ones of hybrid power plants for electricity production. However, there is always the essential difference, already mentioned previously, that the cogeneration system is based on guaranteed power production units. Yet, this feature does not relieve the cogeneration system from the requirement to maintain additional backup units for both electricity and thermal energy production, in order to be able to respond to the existing thermal or electrical power demand, in case of possible malfunctions of the main production units. Practically, the secure operation of a cogeneration system imposes the installation of a small thermal power plant, which in turn leads to the considerable increase of the system's setup and operation cost.

Generally, the operation of a cogeneration system under thermal power production priority implies more effective operation and, in other words, higher final energy production per unit of consumed primary energy source and more cost-effective operation, both for residential and industrial uses. However, the introduction of general rules for cogeneration systems is not the wiser approach. Every application has its particular features, because there is a plethora of cogeneration systems implementations, depending on the involved technologies, the size of the system, potential special requirements of the power demand, while, as the system's layout can be adapted on the final user's particular needs, the final operation algorithm can be also affected. Besides, several technical and economic parameters can vary during the life period of the cogeneration system, modifying the optimum operation mode. All these aspects impose the necessity for the continuous update of the cogeneration system operation mode, not on the basis of general rules only, but, most importantly, according to the use of regular optimization processes, with regard to both the design and the operation of the cogeneration system.

Depending on the type and the size of the particular consumption that a cogeneration system is designed to cover, the selection of the optimum technologies changes. Hence, for small size demand (e.g., detached residential buildings), for which, apart from the small power demand size, a crucial

TABLE 5.1
Proposed Decentralized Cogeneration Technologies for Different Categories of Applications (with due diligence of Technical Chamber of Greece) [6]

No.	Application	Electrical Power Consumption Range (kW_e)	Proposed Cogeneration System
1	Detached houses	5–50	• Otto engine • Stirling engine • Fuel cell • Compact cogeneration unit
2	Blocks of apartments	50–250	• Otto or diesel engine • Fuel cell • Compact cogeneration unit
3	Hospitals	500–2,000	• Otto or diesel engine • Gas turbine • Steam turbine
4	Hotels	200–2,000	• Otto or diesel engine • Gas turbine • Steam turbine
5	Blocks of offices	200–500	• Otto or diesel engine
6	Sports and swimming pool centers	100–300	• Otto or diesel engine • Gas turbine
7	Commercial centers, malls	200–1,000	• Otto or diesel engine • Gas turbine
8	Educational buildings	200–500	• Otto or diesel engine • Gas turbine

fact is most commonly the requirement for the minimization of the noise and gas emissions, among the optimum choices there are the installation of an Otto internal combustion engine, a Stirling engine, a compact cogeneration unit of small size, or even a fuel cell. As the electricity and thermal energy demands increase, the options of the Stirling engine and the fuel cell become economically nonfeasible, while thermal generator technologies of higher size emerge as more reasonable solutions, such as diesel engines, gas turbines, and steam turbines.

Table 5.1 gives a synopsis of the proposed types of decentralized cogeneration systems for different types of decentralized applications [6].

5.3 TECHNOLOGIES OF COGENERATION SYSTEMS

A cogeneration system consists of the following discrete components:

- *Power production part.* The fundamental power production is executed in this part. It normally consists of conventional thermal generators, nuclear reactors, or biomass burners. It can also be a geothermal high enthalpy field.
- *Heating recovery system.* This part of the cogeneration system aims at the recovery of the disposed heat from the thermal generators. It usually consists of thermal energy recuperators (heat exchangers). Depending on the final thermal energy requirements, a supplementary source of thermal power can be also involved, like a heater, heat pump, etc.
- *Cooling system.* The cooling system aims to exploit the recovered heat for the production of cooling. The components and the cooling production process executed in this particular part will be thoroughly presented in a later section.

- *Heating or cooling transportation and distribution system.* In case of district heating or cooling, a considerable infrastructure is required for the transportation and the distribution of the produced heating or cooling, in most cases in the form of hot/cold water or steam. This infrastructure consists of insulated pipelines, circulators, heat exchangers, measuring instruments, protection devices (expansion vessels or valves), etc. Cooling is not easy to transfer through long distances. For short distances, namely in the range of some hundreds or thousands of meters, a very well-insulated pipeline network is required for the transportation of the working medium.

The alternative cogeneration technologies are configured on the basis of the employed power production technologies:

- *Steam turbine cogeneration systems.* The main power production units are steam turbines. The primary energy source can be a solid fossil fuel (coal or lignite), heavy fuel, solid biomass, nuclear fuel, urban wastes, or a geothermal high enthalpy field. In most typical topping systems, heat is recovered from the exhausted steam. It can also be recovered from the disposed hot gases. Steam turbines can be introduced as compact units for small decentralized production and as large units, for centralized cogeneration or even trigeneration systems.
- *Reciprocating engine cogeneration systems.* If diesel generators are used for the basic power production, thermal energy can be recovered both from the engine's cooling loop and from the exhausted hot gases from the combustion chamber. The consumed fossil fuel can be heavy fuel, diesel oil, or natural gas. Diesel engines can be introduced both as compact units for small decentralized production and as large units, for centralized cogeneration systems.
- *Gas turbine cogeneration systems.* Diesel oil and natural gas are the only possible fossil fuels that can be consumed by a gas turbine. Similarly with steam turbines, the heat is recovered from the exhausted gases. Gas turbines are met only in large-scale, centralized cogeneration systems.
- *Combined cycle cogeneration systems.* The operation concept of combined cycles constitutes itself a type of cogeneration system, in the sense that the hot exhausted gases from the gas turbines are further exploited for the production of additional power in a serially connected steam turbine. Despite the low exergy content of the finally disposed gases, namely after the heat recovery in the steam turbine, further thermal energy may be recovered for use in applications with low temperature requirements.

In Table 5.2 there is a comprehensive presentation of the available cogeneration technologies and their characteristic features [6] for applications in buildings. In the following sub-sections, the possible technical implementations of centralized and decentralized cogeneration systems will be presented more thoroughly.

5.3.1 Steam Turbine Centralized Cogeneration Systems

Centralized cogeneration systems based on the usage of steam turbines are the most popular and common systems, appropriate for power production from 500 kW_e to 1,000 MW_e. Their most critical advantage, compared to the other available technologies, is based on their ability to operate with a large variety of primary energy sources, from solid fuels, such as coal, lignite, and solid biomass, to high enthalpy geothermal fields and nuclear energy. Additionally, given that steam turbines are employed as base units, which implies their uninterrupted operation, there is also the positive feature of continuous availability of thermal energy, for the coverage of heating loads in winter and cooling loads in summer. The overall efficiency of a steam turbine

TABLE 5.2
Categories of Cogeneration Systems and Characteristic Features for Buildings (with due diligence of Technical Chamber of Greece) [6]

Main Cogeneration System	Minimum Electrical Output (kW)	Maximum Electrical Output (kW)	Electrical Efficiency (%)	Thermal Efficiency (%)	Total Efficiency (%)	Electrical versus Thermal Power Output Ratio	Disposed Gases Temperature (°C)	Output Thermal Energy Quality
Otto engine	15	1,300	32–35	50–60	80–85	0.5–0.8	400–450	H.W.–L.P.S.
Diesel engine	100	20,000	35–45	40–45	70–80	0.7–0.9	320–450	H.W.–L.P.S.
Gas turbine with heat recovery boiler	100	30,000	25–35	40–50	70–80	0.25–0.8	400–600	L.P.S.
Micro turbine	25	200	25–35	40–50	70–80	0.5–0.8	200–300	H.W.–L.P.S.
Stirling engine	3	100	35–45	50–60	80–85	0.5–0.8	400–500	H.W.
Fuel cell	3	30–120	20–30	25–35	45–80	0.7–1	140–200	H.W.
Steam turbine	500	100,000	25–30	40–60	60–80	0.1–0.3	180–200	L.P.S.–M.P.S.

H.W.: Hot water; L.P.S.: Low-pressure steam; M.P.S.: Medium-pressure steam

cogeneration system ranges from 60% to 80%, depending on the operation of the system in partial or nominal load.

The three basic implementations of steam turbine cogeneration systems are presented below.

5.3.1.1 Cogeneration System with a Back-Pressure Steam Turbine

In these configurations, steam of high pressure (220–1,000 bar) and temperature (480°C–540°C), produced in the boiler, once expanded in the turbine for power production, is released in pressure and temperature appropriate for thermal processes. The term *back-pressure* refers to the feature that the steam's pressure led for heat recovery is higher than the atmospheric one (3–20 bar). The power and the steam flow diagram is presented in Figure 5.5.

Back-pressure steam turbine cogeneration systems exhibit the following advantages, compared to the other two alternative steam turbine cogeneration systems, presented in the next sections:

- simple layout and construction
- lower construction and installation cost

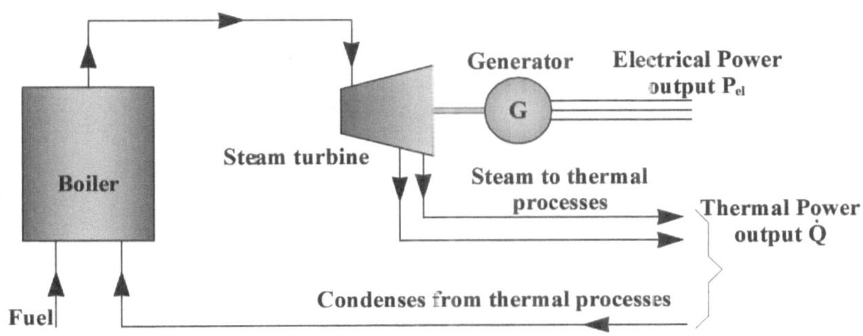

FIGURE 5.5 Power and steam flow diagram of a back-pressure steam turbine cogeneration system.

- reduced or null requirements for cooling water
- higher overall efficiency (around 80%–85%), mainly because no heat is disposed in the ambient through the cooling system.

On the other hand, back-pressure steam turbine cogeneration systems have the significant drawback that the produced electricity depends indissolubly on the required thermal energy. Consequently, thermal energy production is impossible without electricity production. Given this fact, the introduction of back-pressure steam turbine cogeneration systems is mainly favored in cases of large-size, centralized thermal power plants, where the electricity production is constant, rather than in small-size, decentralized energy systems.

The electrical over thermal power output ratio in a back-pressure steam turbine cogeneration systems ranges between 1:10 to 1:4.

5.3.1.2 Cogeneration System with an Extraction Steam Turbine

Also in this case, electricity production is executed in a similar way as in back-pressure systems. The difference between these two systems is that for extraction steam turbines, thermal energy is recovered with the steam extraction from one or more intermediate stages, both for the preheating of the supplied water to the boiler and for the thermal energy transition at the heat exchanger for the required thermal processes. The power and the steam flow diagram is presented in Figure 5.6.

The extraction steam turbine cogeneration systems are more expensive and exhibit lower efficiency than back-pressure steam turbines (around 80%), due to the heat disposal at the steam condenser. On the other hand, they have the advantage of independent (up to a maximum ratio) regulation of electrical and thermal power production. This is achieved with the regulation of the overall steam flow.

The extraction steam turbine cogeneration systems:

- are constructed with a nominal electrical power from 0.5 to 100 MW_e
- exhibit an electricity production efficiency at 25%–30%, thermal production efficiency at 40%–60%, and overall cogeneration efficiency at 65%–80%
- have electrical over thermal power output ratio ranges between 1:3 and 1:4
- exhibit an average life period of 30 years.

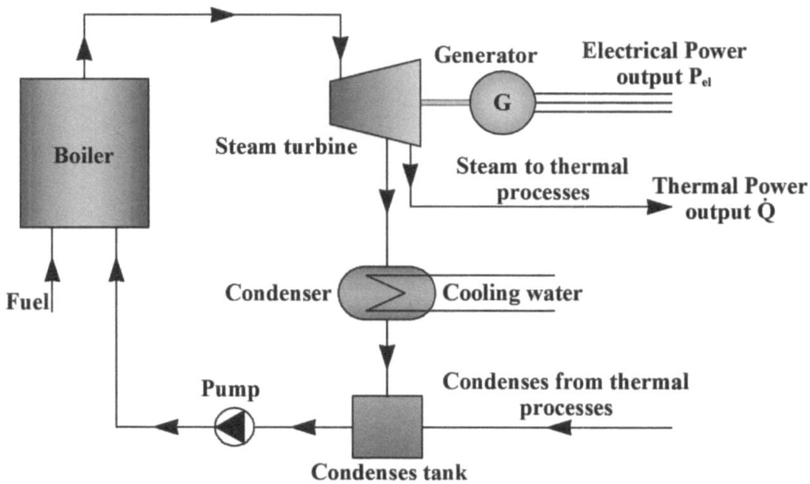

FIGURE 5.6 Power and steam flow diagram of an extraction steam turbine cogeneration system.

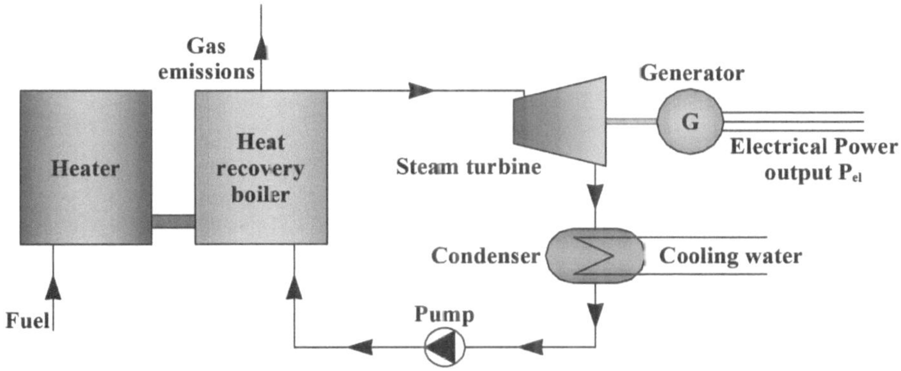

FIGURE 5.7 Power and steam flow diagram of a bottoming cycle steam turbine cogeneration system.

5.3.1.3 Cogeneration System with a Bottoming Cycle Steam Turbine

In this case, the initial product is thermal power, produced in a boiler, a heater, etc., within an industrial process, such as steel mills, glass industry, ceramics factories, cement factories, aluminum industry, or oil refineries. In these industrial activities, thermal and chemical processes are involved, executed within the production procedure, resulting in the release of hot gases. In cogeneration applications, these gases are most commonly led to a heat recovery boiler for the production of steam, which, in turn, is expanded in a turbine for power production. In this way, the thermal power production unit is upgraded to a CHP plant based on a bottoming cycle steam turbine system. A relevant steam and power diagram flow is provided in Figure 5.7.

In such types of applications, the electricity production efficiency ranges between 5%–15%. It is underlined, yet, that electricity production constitutes an additional product, based on the achieved heat recovery, which otherwise would have been disposed in the ambient. Consequently, even with this considerably low efficiency, electricity production can be feasible in such types of cogeneration systems.

5.3.2 Gas Turbine Cogeneration Systems

There are two basic types of CHP configurations with gas turbines, the open and closed type CHP systems. The consumed fuel is usually natural gas, liquefied petroleum gas (LPG), or diesel oil.

5.3.2.1 Cogeneration Systems with Open-Cycle Gas Turbines

Open-cycle gas turbines are the conventional gas turbines presented in Chapter 2. Air is suctioned from the ambient, compressed, and led to the combustion chamber. The produced hot gases after the ignition of fuel and the combustion of the fuel-air mixture are expanded, releasing power which is captured in the form of mechanical work by the turbine. The remaining hot gases are eventually disposed in the atmosphere with a temperature that can range from 300°C to 600°C.

The disposed heat exploitation from an open-cycle gas turbine in a cogeneration system can be approached with two alternative ways:

- With its direct uses in thermal processes (district heating, industrial thermal processes, etc.).
- By supplying it to heat recovery units, named *heat recovery boilers* or simply *gas boilers*. High enthalpy steam is then produced in these gas boilers, appropriate for various production processes, like thermal processes, or also to drive a steam turbine for additional electricity production. This last implementation is the classic case of a combined cycle for electricity production, presented thoroughly in Chapter 2.

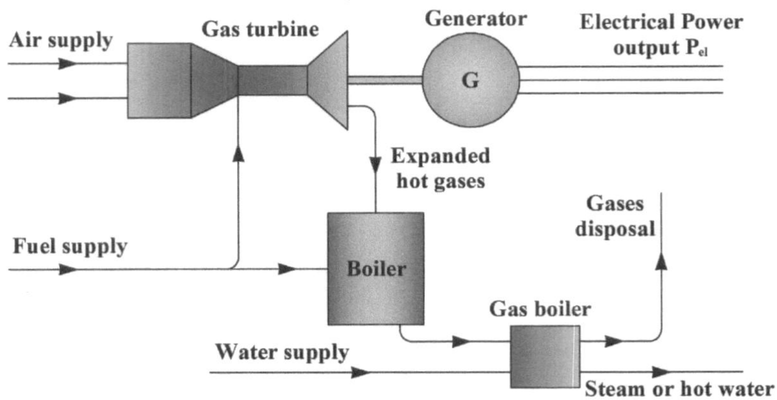

FIGURE 5.8 Power flow and layout of an open-cycle gas turbine cogeneration system.

In the above mentioned systems, there is the possibility to increase the hot gases' specific enthalpy, and, subsequently, the delivered thermal energy, due to their high oxygen content. This is achieved by inserting a combustion chamber between the gas turbine and the heat recovery boiler, where, with the consumption of additional fuel, the capturing process of the contained oxygen is integrated, creating improved combustion conditions and increasing the system's overall efficiency (Figure 5.8).

CHP systems with an open-cycle gas turbine and a heat recovery boiler:

- are constructed with nominal electrical power from 100 to 30,000 kW_e
- exhibit electricity production efficiency at 25%–35%, thermal production efficiency at 40%–50%, and overall efficiency at 70%–80%
- exhibit an electrical over thermal output power ratio range around 1:4–1:1.25
- exhibit an average life period of 15–20 years.

The open-cycle gas turbine cogeneration systems exhibit the drawback of low efficiency (25%–35%) for the electricity production process, due to the high power consumption in the compressor and the hot gases' high outlet temperature. Nevertheless, precisely due to the hot gases' high outlet temperature, open-cycle gas turbines are ideal for cogeneration applications, through which the overall efficiency for both electricity and thermal energy production processes may reach 70%–80%. The electricity production efficiency, compared to the corresponding efficiency of steam turbine CHP systems, is higher both in full- and in partial-load operation, yet it is more quickly reduced in partial loads. Additionally, the open-cycle gas turbine CHP systems exhibit higher electrical versus thermal energy production ratio. Gas turbines with inlet air preheating from the expanded gases exhibit higher electricity production efficiency but lower overall cogeneration efficiency.

5.3.2.2 Cogeneration Systems with Closed-Cycle Gas Turbines

In closed-cycle gas turbine cogeneration systems, the working medium (usually hellion or air) circulates in a closed loop. It is heated in the appropriate temperature in a heat exchanger, before it is led to the gas turbine, while it is cooled when it leaves the turbine. The basic power flow diagram of this system is presented in Figure 5.9.

This implementation has the advantage that the working medium, which comes in contact with the turbine's blades, remains clean and unaffected from the fuel and the ambient air in the combustion chamber, since it is not involved in the combustion process, and, subsequently, any mechanical or chemical corrosion of the gas turbine from the combustion's products is avoided. Additionally,

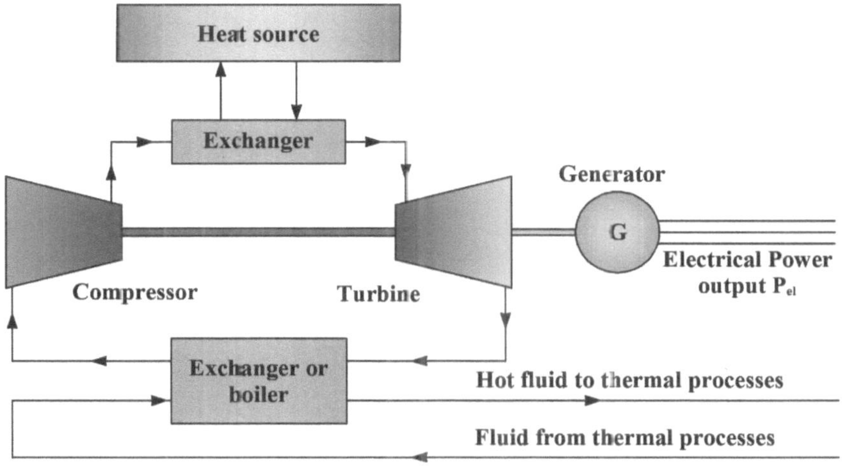

FIGURE 5.9 Power flow of a closed-cycle gas turbine cogeneration system.

the external combustion enables the use of any type of fuel, like coal, urban waste, or biomass, a fact that leads to the considerable reduction of electricity production-specific cost.

Closed-cycle gas turbines exhibit significant advantages compared to open-cycle gas turbines, such as:

- higher availability because of reduced maintenance requirements, due to the cleaner working medium
- the electricity production can be increased with the preheating of the inlet air by recovering the heat from the cooled working medium.

5.3.3 Cogeneration Systems with Reciprocating Engines

Reciprocating engines are the most efficient technology for the integration of CHP systems. This is due to the fact that, generally, internal combustion piston engines exhibit the highest electricity production efficiency of all the other thermal generators, as presented in Chapter 2. Although several types of reciprocating engines are commercially available, in practice two essential types have dominated the market and have been introduced in most applications and uses. These are the four-stroke spark-ignited Otto engine and the four-cycle or two-cycle, in case of electricity sector, compression-ignited diesel engine. For electricity production, especially in large-size, centralized power plants, diesel engines are used.

Reciprocating engines are usually used for small-size cogeneration systems, with nominal power up to 20–1,000 kW_e (Figure 5.10) while for larger systems the use of gas turbines is more common. More specifically, diesel engine cogeneration systems are classified in the following four categories:

1. small-size systems, with a gas engine or a diesel engine with nominal power from 75 to 250 kW_e
2. small-size systems with a gas engine or a diesel engine with nominal power up to 1,000 kW_e
3. medium-size systems with a gas engine or a diesel engine with nominal power up to 6,000 kW_e
4. large-size systems with a gas engine or a diesel engine of nominal power up to 6,000 kW_e.

Gas engines are internal combustion reciprocating engines that operate with a gas fuel, e.g., natural gas, LPG, biogas. In the next sections, the commercially available reciprocating engines are presented.

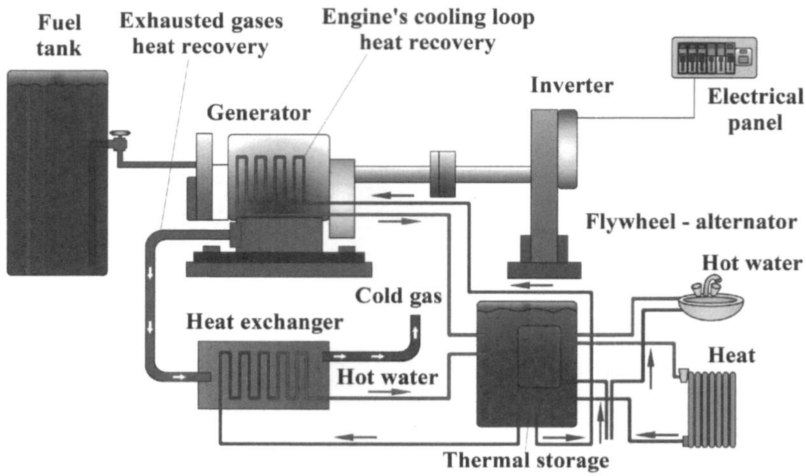

FIGURE 5.10 Typical structure of a decentralized cogeneration system for residential use with a reciprocating engine [7].

5.3.3.1 Cogeneration Systems with Otto Gas Engines

Cogeneration systems with Otto gas engines are usually small-size engines, with high power concentration. In most cases, they are former car engines converted to gas engines. This conversion has a slight effect on the engine's nominal mechanical power and a 15%–20% reduction on the mechanical power production efficiency. They have a relatively low procurement cost but also a relatively restricted life period (10,000–30,000 operation hours). The consumed fuel is usually natural gas or LPG. An inductive generator is used for electricity production while thermal energy is also recovered from the heat exchanger of the engine's cylindrical liners and from the expanded hot gases, through a heat recovery boiler.

Cogeneration systems with Otto gas engines:

- are constructed with nominal power from 15 to 1,300 kW_e
- exhibit electricity production efficiency at 32%–35%, thermal energy production efficiency at 50%–60% and overall cogeneration efficiency at 80%–85%
- exhibit electrical over thermal power output ratio between 1:2–1:1.25
- have an average life period of around 10 years.

5.3.3.2 Cogeneration Systems with Diesel Gas Engines

These reciprocating engines usually come from the conversion of diesel engines previously used in cars to diesel gas engines. This conversion usually does not affect the nominal mechanical power of the diesel engine and is achieved by modifying the pistons and the valves head, required because the ignition in the modified engine will not be executed with the compression of the ignited fuel–air mixture, but with a spark.

Cogeneration systems with a diesel gas engine:

- are constructed with nominal power from 100 to 20,000 kW_e
- exhibit electricity production efficiency at 35%–45%, thermal energy production efficiency at 40%–45% and overall cogeneration efficiency at 75%–80%
- exhibit electrical over thermal power output ratio between 1:1.4–1:1.1
- have an average life period of around 15–20 years.

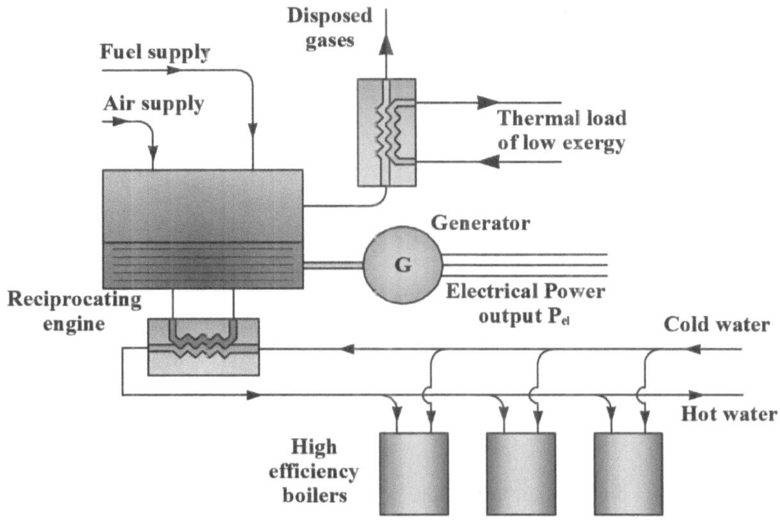

FIGURE 5.11 Cogeneration power plant layout and power flow diagram with diesel engines of large size [3].

5.3.3.3 Cogeneration Systems with Diesel Engines for Electricity Production

These systems are based on the conventional diesel engines presented in Chapter 2. They are heavy machines, with nominal power up to 50,000 kW_e. They have limited maintenance requirements and increased procurement cost. They are considered ideal for constant operation as base units. A typical power flow diagram and layout for a CHP system with a diesel engine is presented in Figure 5.11.

The disposed heat comes mainly from the high enthalpy hot gases, the engine's cylindrical liner, and the lubricants cooling system. Additionally, heat can also be recovered from the cooling systems of other fluids involved in the engine's operation, namely the motor's closed cooling loop and the turbocharged air, with the use of appropriate heat exchangers.

The electricity production efficiency for small- and medium-size diesel engines ranges from 35% to 45%, while for modern diesel engines of large size it may reach 50%. The overall efficiency of the cogeneration process may reach 80%. These systems have two major advantages: they exhibit generally high efficiency for the electricity production process, not seriously affected by the operation in partial load, and they respond quickly to power demand fluctuations. They have a life period of 15–20 years, depending on the size of the system, the consumed fuel quality, and the applied maintenance schedule. Reciprocating engines exhibit increased maintenance requirements compared to rotational engines (turbines), resulting in reduced availability over the year (80%–90%).

5.3.4 Cogeneration Systems with Combined Cycles

The combined cycle concept was analytically presented in Chapter 2. Combined cycles are introduced in conventional electricity production thermal power plants aiming at the increase of the efficiency of conventional gas turbines. On a theoretical basis, power production from combined cycles is analyzed in two thermodynamic cycles, connected to each other with a working medium and operating in different temperatures. Heat is released from the high-temperature cycle (top cycle) to be recovered in the low-temperature one (bottom cycle) for the production of additional electricity, increasing, thus, the total efficiency of the electricity production process.

In cases of cogeneration technologies introduced in combined cycles, thermal energy can be recovered from the remaining heat in the form of low enthalpy steam after the power production in the steam turbine or by the disposed hot gases from the gas turbines, after the gas boiler. A power

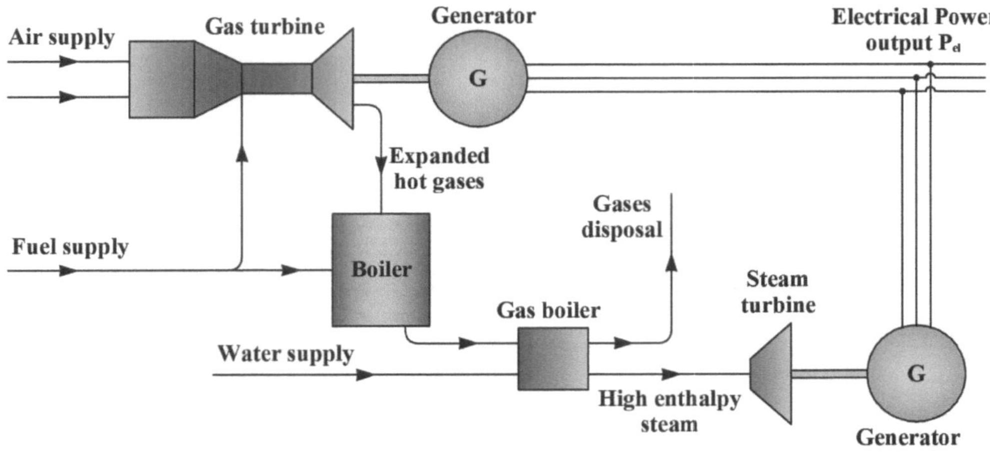

FIGURE 5.12 Cogeneration power plant layout and power flow diagram with combined cycle.

flow diagram and the fundamental layout of a cogeneration system with a combined cycle is presented in Figure 5.12.

The total efficiency of the cogeneration process with a combined cycle is by far much higher than the efficiency of cogeneration systems with conventional thermal generators, reaching values from 70% to 85%. Yet, the overall efficiency is significantly affected in partial-load operation. They also exhibit higher power concentration (mechanical and thermal power output per unit of volume) compared to the other cogeneration systems with steam or gas turbines. Cogeneration systems with combined cycles are developed with nominal power around 20–400 MW_e, although smaller units can also be constructed (from 4 to 11 MW_e). Usually the high oxygen content in the expanded hot gases from the gas turbines enables the consumption of supplementary fuel in a gas boiler, in case this is required, to increase the total power output of the system.

The consumed fuels in a cogeneration system with combined cycle coincide with the available options in case of gas turbines, namely they can be in the form of gas fuels (natural gas or LPG) or liquid diesel oil.

The installation of a cogeneration system with a combined cycle may require 2–3 years, yet the integration of the overall setup can be accomplished in two stages. Initially, the gas turbines are installed, a task that may be completed in 12–18 months, and the steam turbine may be installed then, while the gas turbines are in operation availability. Usually cogeneration combined cycle systems exhibit annual availability of 77%–85% and a life period of 15–25 years.

5.3.5 Compact Cogeneration Systems of Small Size

Compact cogeneration systems of small scale, ideal for decentralized applications, have been introduced to the market during the last decades. These units are easy to install and quickly ready to be used. They are usually constructed with nominal power of 10–1,000 kW_e. These small-scale cogeneration units have the following advantages:

- low procurement cost
- small volume
- quick and easy installation (normally the only required action is their connection to the local electrical grid and to the heating distribution hydraulic network)
- automatic operation, without the necessity for constant inspection from specialized staff.

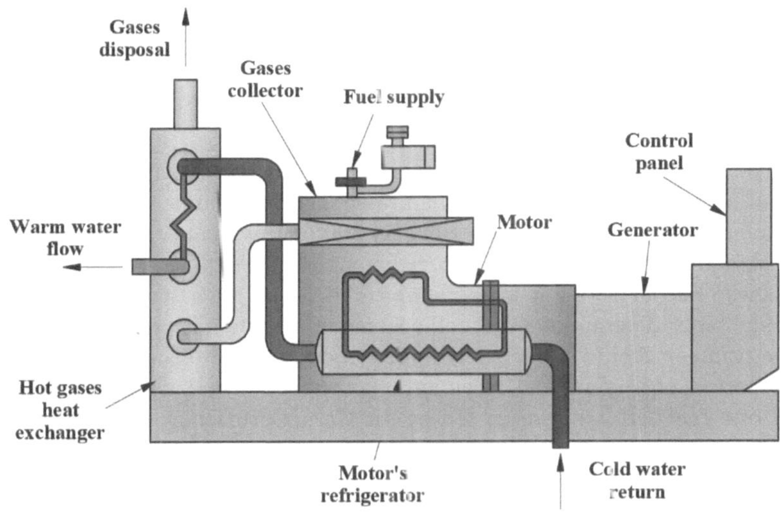

FIGURE 5.13 Graphical representation of a compact cogeneration unit with a reciprocating internal combustion engine.

These units are usually driven by a diesel engine, or, for nominal power up to 100 kW$_e$, Otto engines can also be used. For power output higher than 600 kW$_e$, gas turbines are employed. They usually operate with gas or liquid fuels (natural gas, LPG, or diesel oil). A drawing of such a system with a reciprocating internal combustion engine is presented in Figure 5.13.

Compact cogeneration systems with a diesel engine exhibit exceptional interest for commercial and residential building applications. Around 27%–35% of the supplied primary energy is transformed to electricity, while 50%–55% is transformed to thermal energy. Hence, the ratio of the electrical versus thermal power output ranges from 1:2–1:4, while the cogeneration process overall efficiency can reach 80%. The annual availability of these compact cogeneration systems is estimated at 90%. A considerable contribution to their commercial success should be accredited to their extensive automatic operation and control.

5.3.6 Other Types of Cogeneration Systems

5.3.6.1 Bottoming Cycles with Organic Fluids

Electrical or mechanical power production can be feasible with heat recovery from low enthalpy (or exergy) fluids, namely with temperatures from 80°C to 300°C, if instead of water, specific organic fluids are used, such as toluene, with boiling temperatures much lower than that of water. Consequently, with the use of the appropriate organic fluids, low exergy heat sources can be exploited in cogeneration systems, such as solar radiation, industrial or urban waste, medium enthalpy geothermal fields, the disposed heat with the exhausted gases, or the cooling system of thermal generators.

The electrical power output of these alternative systems can be from 2 kW$_e$ to 10 MW$_e$. The efficiency of the electricity production process is low and depends on the working medium's temperature, in other words on the exergy content of the organic fluid. For temperatures from 75°C to 425°C, the electricity production efficiency ranges at 5%–30%, most usually at 10%–20%. From a manufacturing point of view, considerable attention should be paid to the selection of the appropriate materials, which must exhibit strong anticorrosion protection, in order to withstand the corrosive action of the organic fluid, as well as at the adequate sealing of the casing of the whole construction, to avoid any potential leakage of the organic fluid to the ambient.

The availability and the reliability of this technology has not been evaluated so far, because these systems are relatively new. However, it is estimated that their annual availability should be expected between 80% and 90% and their life period may reach 20 years.

5.3.6.2 Fuel Cells

From the several available types of fuel cells presented in Chapter 3, only the phosphoric acid fuel cell (PAFC) can be considered technically mature enough for commercial electricity production. The operation temperature of PAFCs (approximately 200°C) sets an upper limit for the potential heat recovery. It is obvious that, the higher the operation temperature, the higher the potentially recovered thermal power. Fuel cells of low temperatures (<80°C) are not suitable for CHP applications.

A fundamental block diagram of a fuel cell cogeneration system is presented in Figure 5.14.

Because the primary chemical energy is directly converted to electricity, namely without the appearance of thermal energy and the execution of a thermodynamic cycle, the electricity production efficiency of a fuel cell is not restricted by the Carnot efficiency. Nevertheless, even though theoretically the electricity production process efficiency for a fuel cell can be expected to be significantly higher, the actual efficiency of PAFC usually reaches values in the area of 37%–45%. For partial-load operation above 50% of the fuel cell's nominal power, the electricity production process efficiency is equal to, or even slightly higher than, the efficiency under nominal operation. The cogeneration process overall efficiency can reach 85%–90%, while the ratio of the electrical versus the thermal power production ranges at 1:1.25–1:1 [8].

Molten carbonate fuel cells (MCFCs) and the solid oxide fuel cells (SOFCs) are theoretically more appropriate for CHP applications, due to their high operation temperature (up to 600°C). Their electricity production efficiency is estimated at 50%.

Fuel cells constitute theoretically an excellent option for decentralized cogeneration applications in industrial, commercial, and residential sectors, due to their very specific advantages, such as modular construction, which facilitates the development of units with precisely the desirable size, their operation with relatively high electricity production efficiency, even under partial-load operation, and the gas and noise low emissions.

So far, they have two major drawbacks, which affects their potential uses for commercial applications: high setup costs and short service lives.

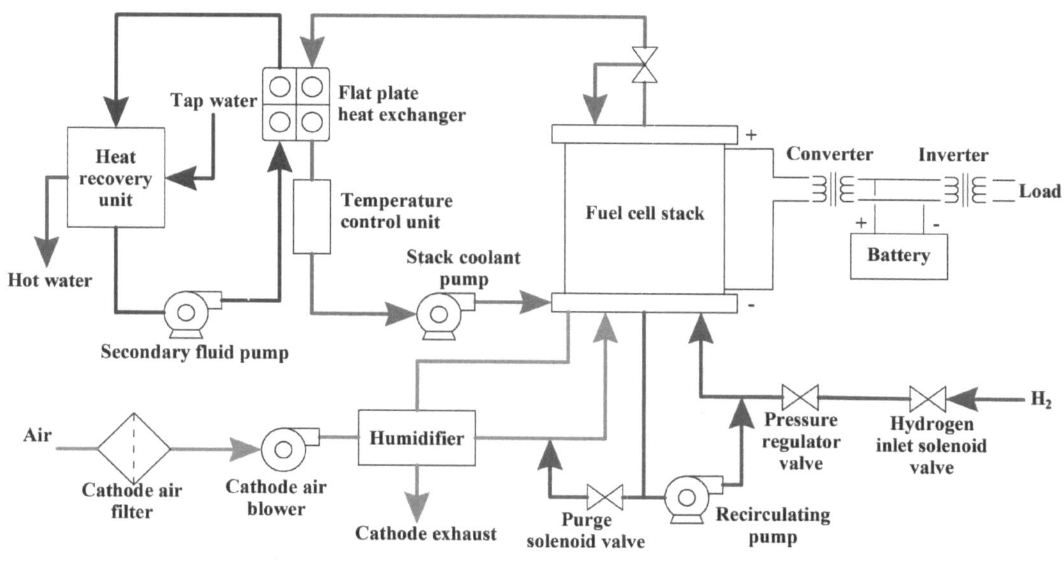

FIGURE 5.14 Simplified block diagram of a fuel cell cogeneration system [8].

Fuel cells are constructed with nominal power higher than 3 kW_e and have an average service life of 5 years.

On the condition that this technology becomes more technically mature and, subsequently, more cost-effective, it may be a highly promising option for the integration of CHP systems.

5.3.6.3 Cogeneration Systems with Stirling Engines

Cogeneration systems with Stirling engines, although not widely introduced until recently, have started to gain attention mainly because the Stirling engine's thermodynamic cycle better approaches the ideal Carnot cycle compared to diesel and Otto internal combustion engines (diesel and Otto cycle respectively), to gas turbines (Joule cycle), and to steam turbines (Rankine cycle).

The Stirling engine is considered an external combustion engine. The operation of a Stirling engine is based on the compression and the expansion of a gas (e.g., hydrogen, hellion). The working gas is heated in a heat exchanger, without being involved in the combustion. The external combustion in Stirling engines enables the use of different types of fuels, e.g., liquid or gas fuels, coal, biomass fuels, even urban wastes.

A simplified construction layout of a Stirling engine is presented in Figure 5.15a. Practically, the engine consists of a hot cylinder (hot heat exchanger) that is heated externally, typically by the exhausts of a combustion process, and a cold cylinder (cold heat exchanger), which is cooled externally, mostly by tap water. The working medium is compressed in the cold cylinder and expanded in the hot cylinder. Both processes result in the production of mechanical power. The basic operation concept of the Stirling engine can be analyzed as follows:

- First, heat is produced by an external source and transferred to the working medium at the expansion space.
- The heated gas is expanded, causing the displacer piston to move to the left, with regard to Figure 5.15a.
- Between the expansion and the compression space, there can be a regenerator, although this is not obligatory. The role of this regenerator is double. First, during the expansion process it stores temporary heat from the hot expansion space. Otherwise this heat would be disposed as waste in the ambient.
- The contained working medium in the compression space is compressed from the displacer piston and causes the power piston to move to the left too, again referring to Figure 5.15a.

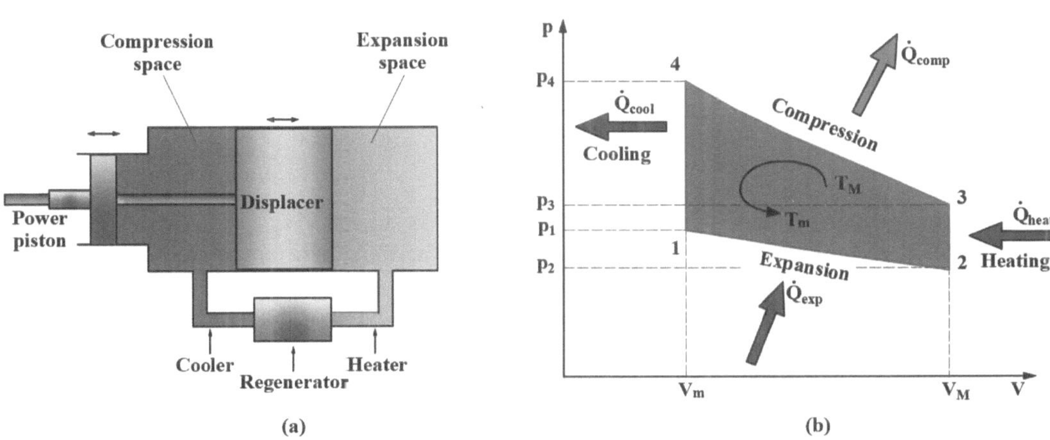

FIGURE 5.15 Simplified operation concept of a Stirling engine (a) and the corresponding thermodynamic cycle (b).

- The reciprocating motion of the two pistons is eventually converted to rotational motion through a typical connecting rod–crankshaft layout.
- After the compression process, the regenerator returns back the stored heat from the expansion and transfers it to the expanded gas in the expansion space. Namely the heat flow is reverted.
- The contained gas in the expansion space is heated again and the cycle is repeated.

The operation of the Stirling engine is approached with the thermodynamic cycle presented in Figure 5.15b. It consists of the following thermodynamic processes:

- Process 1–2: Isotherm compression at low temperature, during which heat is removed from the compressed gas towards the environment.
- Process 2–3: Isochoric compression, during which the heat stored in the regenerator is returned back to the working medium.
- Process 3–4: Isotherm expansion at high temperature, during which heat is added to the working medium from the external heat source.
- Process 4–1: Isochoric expansion, during which heat is removed from the expanded medium and stored in the regenerator.

It must be noted that heat recovery in the regenerator is incomplete because of heat losses and that the isochoric processes in practice are not free of work consumption, because of mechanical frictions between the moving components of the machine.

Among the major advantages of the Stirling engine we can mention its high efficiency, the high flexibility with regard to the fuel's selection, the improved operation in partial loads, the reduced gas emissions, and the low vibrations and noise emission level. Because of the external combustion and the closed loop of the working medium, the moving and vulnerable components of the engine are not exposed in the combustion's gases, restricting the potential wear of the machine. On the other hand, the proper operation of the engine requires adequate (and usually hard to achieve) sealing techniques to avoid leakages of either the high-pressure working gas to the ambient, or the engine's lubricant to its inner space. The development of adequate sealing solutions with satisfying life periods constitutes one of the main issues of the Stirling engine's manufacturing process.

The electricity production process efficiency of the Stirling engine can be as high as 50%, remaining almost constant even in partial-load operation. The cogeneration process overall efficiency usually ranges between 60% and 80%, while the ratio of the electrical versus the thermal power output can be from 1:0.8–1:0.6. Because the cogeneration systems with Stirling engines still remain in the development process, it is estimated that their availability can be comparable with that of the diesel engines.

The Stirling engines:

- are usually constructed with nominal power from 3 kW_e to 100 kW_e
- exhibit electricity production efficiency 35%–45%, thermal energy production efficiency 50%–60%, and cogeneration efficiency 60%–85%
- exhibit considerably reduced gas and noise emissions, though they are more expensive than internal combustion engines
- have limited maintenance requirements, due to the closed loop of the working medium.

Typically Stirling engines can be integrated with conventional boilers to form cogeneration systems, based on the recovery of the rejected heat from the Stirling engine during the compression stage of the working medium. A typical layout is provided in Figure 5.16.

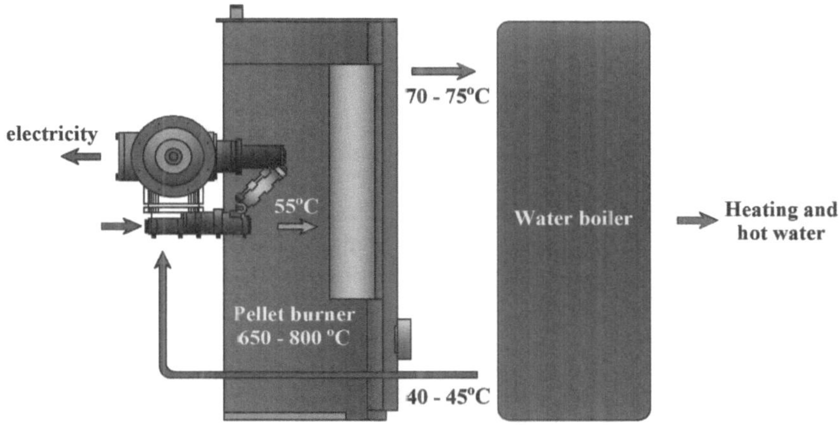

FIGURE 5.16 Integration of a Stirling engine with a biomass boiler to form a decentralized cogeneration system [9].

5.4 EFFICIENCY FACTORS OF COGENERATION SYSTEMS

Different factors have been developed for the quantitative evaluation of cogeneration power plants' effectiveness. The most popular of them are presented in this section.

First of all, the efficiency of the power generator of the cogeneration system (e.g., a gas turbine or a diesel engine) is defined as:

$$\eta_m = \frac{\dot{W}_s}{P_{ch}} = \frac{\dot{W}_s}{\dot{m}_f \cdot H_u} \tag{5.4}$$

where:

$\dot{W}_s$: the mechanical power at the shaft of the power generator
P_{ch}: the inlet chemical energy of the consumed fossil fuel
$\dot{m}_f$: the mass flow rate of the consumed fuel
H_u: the lowest heat capacity of the consumed fuel.

Based on Equation 5.4, Equation 5.1 for the electricity production process of the cogeneration system becomes:

$$\eta_{el} = \frac{P_{el}}{P_{ch}} = \frac{P_{el}}{\dot{m}_f \cdot H_u} \tag{5.5}$$

where P_{el} is the net electrical power output, arisen from the electrical power production of the involved generators minus the electrical power consumption required for the operation of the power plant.

Additionally, the thermal power production process of the cogeneration system is expressed by the following relationship:

$$\eta_{th} = \frac{\dot{Q}_{th}}{P_{ch}} = \frac{\dot{Q}_{th}}{\dot{m}_f \cdot H_u} \tag{5.6}$$

where $\dot{Q}_{th}$ is the final thermal power output of the cogeneration system.

Finally, given Equations 5.5 and 5.6, Equation 5.3 expressing the total efficiency of the overall cogeneration process becomes:

$$\eta_{co} = \eta_{el} + \eta_{th} = \frac{P_{el} + \dot{Q}_{th}}{P_{ch}} = \frac{P_{el} + \dot{Q}_{th}}{\dot{m}_f \cdot H_u} \tag{5.7}$$

A new basic index introduced for the performance evaluation of cogeneration systems is the power to heat ratio (PHR), also mentioned in the previous sections. This index expresses the ratio of the electrical power output versus the thermal power output from a cogeneration system, defined by the following relationship [10]:

$$\text{PHR} = \frac{P_{el}}{\dot{Q}_{th}} \tag{5.8}$$

Another index is the fuel savings rate (FSR), defined by the relationship [10]:

$$\text{FSR} = \frac{P_{ch\text{-}S} - P_{ch\text{-}co}}{P_{ch\text{-}S}} \tag{5.9}$$

where:

$P_{ch\text{-}S}$: the consumed chemical power of the fossil fuel for electrical and thermal power production through separate, independent processes

$P_{ch\text{-}co}$: the consumed chemical power of the fossil fuel for the same electrical and thermal power production through the cogeneration process.

A cogeneration system can constitute a sensible and feasible choice, with regard to the achieved energy saving, if FSR > 0.

From the above relationships we are led to the following ones:

$$\eta = \eta_{el} \cdot \left(1 + \frac{1}{\text{PHR}}\right) \tag{5.10}$$

$$\text{PHR} = \frac{\eta_{el}}{\eta_{th}} = \frac{\eta_{el}}{\eta - \eta_{el}} \tag{5.11}$$

which may be used for the definition of an acceptable range of the PHR, when the efficiency of the electricity production process is known. It must be noted that for every particular cogeneration application, the PHR constitutes one of the basic parameters for the selection and the configuration of the cogeneration system.

If we assume that a cogeneration system substitutes two distinguished electricity and thermal energy production processes with efficiencies η_{el} and η_{th} respectively, it can be proved that:

$$\text{FSR} = 1 - \frac{\text{PHR} + 1}{\eta \cdot \left(\frac{\text{PHR}}{\eta_{el}} + \frac{1}{\eta_{th}}\right)} \tag{5.12}$$

where the subscripts $_{el}$ and $_{th}$ designate the distinguished electricity and thermal energy production respectively. For example, if a cogeneration system with total efficiency $\eta = 0.80$ and PHR = 0.60

substitutes an electricity generator with $\eta_{el} = 0.35$ and a boiler with $\eta_{th} = 0.85$, then from Equation 5.12 it is calculated that FSR = 0.325. This result implies that the overall primary energy consumption is reduced at 32.5% with the cogeneration process, compared to the production of the same electricity and thermal energy amounts through separate, independent processes.

The efficiency of an energy system depends on several fluctuating parameters, such as the power demand, the ambient conditions, the technical specifications of the involved generators, etc. Moreover, the utilization factor of the produced final energy forms is affected by the initial design and layout of the system, the operation algorithm of the cogeneration system, the corresponding strategic control, and the chronicle coincidence of the power production and demand. For these reasons, the average indices for a time period, expressed in terms of energy, e.g., the annual efficiencies or indices, are often more important and exhibit more useful information than instant indices, expressed in terms of power, given that they provide a more representative image on the energy system's performance.

All the above definitions can also be expressed in terms of energy, instead of power. In this case, we take the average values of these indices for the specific time period during which the involved energy quantities were produced or consumed. For instance, Equation 5.7 can also be written as presented below, in terms of the electricity E_{el} and thermal energy Q_{th} produced and the required chemical energy consumption E_{ch}, over a particular time period:

$$\eta_{co} = \eta_{el} + \eta_{th} = \frac{E_{el} + Q_{th}}{E_{ch}} = \frac{E_{el} + Q_{th}}{m_f \cdot H_u} \tag{5.13}$$

where m_f is the fossil fuel's consumed mass during the examined time period. Equation 5.13 will then provide the overall, average efficiency of the cogeneration system for the investigated time period.

The energy in the form of heat is considered of lower quality than in the form of electricity. The quality of thermal energy decreases with the temperature of the working medium. This is expressed theoretically with the concept of exergy, which will be presented and explained in the next section. For example, the quality (or the exergy) of the thermal energy in the form of hot water is lower than in the form of steam. Hence, the evaluation of a cogeneration system exclusively in terms of the total efficiency of the energy transformation processes, by introducing in the numerator of Equation 5.7 two forms of energy that are not qualitatively equivalent, is not the most objective approach. Even if in the academic literature the energetic efficiencies are more often used, a more accurate and fair evaluation of the effectiveness of a cogeneration process, from a thermodynamic point of view, can be achieved through the concept of exergy and the corresponding exergy ratios of the evaluated processes. All these issues will be thoroughly presented in the next section.

5.5 FUNDAMENTAL THERMODYNAMIC CONCEPTS

5.5.1 Energetic Analysis

The most familiar analysis of thermal processes is based on the first thermodynamic law, which defines the conservation of energy. In general, the application of the first thermodynamic law for the analysis of thermal processes takes into account the balance between all the incoming and outgoing energy flows in a thermodynamic system. The general principle is that the total energy of a thermodynamic system is always maintained while being transformed from the one form to the other. The performance of the executed thermal processes is usually expressed through the introduced efficiencies or indices, such as the ones presented in the previous chapter, describing the ratio of the desirable finally delivered energy form (or forms of energy, in case of cogeneration) versus the consumed primary energy source.

However the analysis of thermal processes based exclusively on the concept of energy conservation exhibits some disadvantages. Specifically:

- The energetic analysis is only a quantitative approach, namely we can only retrieve information on the transformed amount of energy from the one form to the other. Yet, no information is given on the ideality of the executed process, or in thermodynamic terms, on the reversibility of the process. For example, when an amount of fossil fuel is burned, although energy is conserved and the efficiency of the combustion can be higher than 95%, the produced thermal energy cannot be fully transformed back to the primary chemical energy, contained in the consumed fossil fuel. As indicated in Chapter 1, thermal energy cannot be directly transformed to chemical energy. The only way to retrieve chemical energy from thermal energy is by converting thermal energy first to electricity, with an efficiency which in best cases can be as high as 55% (for combined cycles) and then using the produced electricity in an electrolysis device to produce hydrogen, with an efficiency that cannot exceed 70%. This means that the reverse procedure can be executed with an overall efficiency, in best cases at the range of $0.55 \times 0.70 = 0.385$. This irreversibility of the initial process cannot be expressed by the energetic evaluation.
- The energetic analysis does not provide any information on the quality of the final delivered form of energy. For example, when electricity is consumed in a hot plate or in an electrical resistance for cooking or heating respectively, the transformation efficiency can be close to 100%. However, the produced thermal energy is of considerable low quality in the sense that, because it is provided at a relatively low temperature, it cannot be utilized as an initial form of energy for its transformation to another energy form, e.g., for the production of mechanical work, as in an internal combustion engine. Practically, in the cases of a hot plate or an electrical resistance, the only way to exploit the produced thermal energy is to use it for heating, namely as a final form of energy. In other applications, under favorable conditions, thermal energy can be recovered, e.g., by the condensers of freezers in large systems for heating applications, or in cogeneration plants. The energetic analysis of the executed thermal processes cannot express the degradation of the quality of energy.

The above inadequacies of the energetic approach for the evaluation of thermal processes are treated with the introduction of the concept of exergy and the execution of the exergetic analysis.

5.5.2 The Concept of Exergy

Exergy is a new magnitude introduced in thermodynamics to compensate for all the above mentioned inadequacies related with the energetic analysis of thermal processes. While the energetic analysis is based on the first thermodynamic law, exergy is defined on the basis of the second thermodynamic law. What the second thermodynamic law says is that, although energy is conserved in a closed thermodynamic system, exergy is always degraded through the executed energy processes by being transformed in forms of energy less capable to be utilized for energy transformations. Actually, thermodynamic systems exhibit the trend to approach a state of equilibrium with the surrounding environment. When this equilibrium is reached, no further energy transformation is possible. In this sense, the terms of the "global energy problem" and the "energy crisis" actually should not be expressed as the "necessity for the conservation of energy," which, in fact, is certain, as a fundamental physical law. It would be more accurate to state it as the "necessity for the conservation of natural energy resources," or, in more specific phrasing, the "necessity to conserve energy forms appropriate for useful energy transformation processes."

The concept of exergy expresses exactly this feature of thermodynamic systems, namely their ability to be involved in energy transformation processes. More accurately, the exergy of a thermodynamic system is defined as the maximum possible obtainable work during a thermal process that brings the system into equilibrium with the surrounding environment. In other words, exergy expresses the ability of a thermodynamic system to cause a change until it comes into equilibrium with its surrounding environment, or, even more simply, exergy is the energy of a thermodynamic system that is available to be used. The term *exergy* also gives a clear aspect of what this magnitudes expresses. It is a composite word coming from the Greek words "ex" and "ergon," which mean "from" and "work" respectively; namely the term *exergy* actually means "from work," hence it gives a verbal expression of the definition of exergy.

Energy is always conserved. However, exergy is not conserved. The exergy of a thermodynamic system is consumed as the system is involved in irreversible thermal processes. In other words, exergy is destroyed and this destruction is proportional to the entropy increase of the system. This means that exergy and entropy are inverse magnitudes, both of them expressing the potential of a thermodynamic system to produce work, yet in an inverted mode (high potential of the system imposes high exergy and low entropy, and vice versa). The destroyed exergy through a thermal process is called anergy, a term also retrieved from the Greek words "an," which means "no,", and "ergon," hence the term anergy actually means "no work."

Exergy is defined versus a surrounding environment. This means that the same thermodynamic system can have different exergy in different environments. Hence, for the exergy calculation of a thermodynamic system, the surrounding environment must be defined. This is normally performed by assessing the temperature, the pressure, and the chemical composition of the environment. In any case, the exergy of a thermodynamic system, under equilibrium with its environment is zero.

The reference environment is in absolute stable equilibrium. This means that the reference environment is simultaneously in thermal equilibrium (determined by its temperature), in mechanical equilibrium (determined by its pressure), and in chemical equilibrium (determined by the chemical potential of the contained compounds). The reference environment behaves as an infinite system, namely it constitutes both a sink and a source for heat and matter. Only internal reversible processes can occur in the reference environment, while its intensive properties (pressure, temperature, chemical potential) remain constant, regardless of the executed thermal processes.

It is conceivable that real, natural environments defer in several attributes from the reference environment. Reversible processes are not met in natural environments, while their intensive properties do change. Despite these deviations, natural environments are usually considered as reference environments in simulation problems. This, in general, may not be such a dramatic approach, if the calculations are executed over a simulation step with relatively short duration (maximum 1 hour), during which the involved magnitudes can be considered stable and the behavior of the natural environment does not deviate significantly from the reference environment. Several simulation models of the reference environment have been developed. A typical natural environment model for the simulation of the reference environment is presented in Table 5.3 [11, 12]. The reference temperature and pressure are considered 298.15 K and 1 atm respectively. The chemical composition consists of saturated air with the condensed phases of water (H_2O), gypsum ($CASO_4{\cdot}2H_2O$), and limestone ($CaCO_3$).

5.5.3 The Energetic and Exergetic Analysis of a Thermal Process

The energetic and exergetic analysis of thermal processes is based on the formulation of the balance equations of the involved thermodynamic magnitudes, which are the working medium's mass, energy, and exergy flows. The general form of a balance equation is the following one:

$$\text{Input} + \text{Generation} - \text{Output} - \text{Consumption} = \text{Accumulation} \tag{5.14}$$

TABLE 5.3
A Typical Natural Environment Model for the Simulation of the Reference Environment

Composition	Air Constituents	Mole Fraction
Atmospheric air saturated at $T_o = 298.15$ K and $P_o = 1$ atm	N_2	0.7567
	O_2	0.2035
	H_2O	0.0303
	Ar	0.0091
	CO_2	0.0003
	H_2	0.0001
Condensed phases	**Components**	
	water (H_2O)	
	gypsum ($CASO_4 \cdot 2H_2O$)	
	limestone ($CaCO_3$)	

The meanings of the quantities in the above equation are:

- Input: any flows of the involved magnitudes (mass, energy, exergy) entering the thermodynamic system
- Output: any flows of the involved magnitudes exiting the boundaries of the thermodynamic system
- Generation: any quantities of the involved magnitudes generated inside the thermodynamic system
- Consumption: any quantities of the involved magnitudes consumed within the thermodynamic system
- Accumulation: any concentration of the involved magnitudes (positive or negative) that remains (or is removed) from the thermodynamic system.

The above general form of the balance equation can be written particularly for the mass, energy, and exergy of the thermodynamic system. Because mass and energy are neither produced nor consumed, given the corresponding mass and energy conservation laws, the above equation is written as:

$$\text{Mass Input} - \text{Mass Output} = \text{Mass Accumulation}$$

$$\sum_i \dot{m}_i - \sum_j \dot{m}_j = 0 \tag{5.15}$$

$$\text{Energy Input} - \text{Energy Output} = \text{Energy Accumulation}$$

$$\sum_i (U + P \cdot v)_i \cdot \dot{m}_i + \sum_b \dot{Q}_b - \sum_j (U + P \cdot v)_j \cdot \dot{m}_j - \dot{W} = 0 \tag{5.16}$$

where:

$\dot{m}_i$, $\dot{m}_j$:	the entering and exiting mass flow rates in and from the system respectively
U, P, v:	the specific internal energy, the absolute pressure, and the specific volume of the thermodynamic system
$\dot{Q}_b$:	the thermal energy rate transferred into the system from the environment through the boundary b

$\dot{W}$: the mechanical work transferred out of the system

$\sum_i (U + P \cdot v)_i \cdot \dot{m}_i$: the contained energy in the entering mass flow (energy input)

$\sum_j (U + P \cdot v)_j \cdot \dot{m}_j$: the contained energy in the exiting mass flow (energy output).

The term $\sum_i (U + P \cdot v)_i \cdot \dot{m}_i - \sum_j (U + P \cdot v)_j \cdot \dot{m}_j$ expresses the net energy rate associated with matter.

Similarly, regarding exergy, in a closed thermodynamic system exergy cannot be generated; hence, the corresponding exergy balance relationship is written as:

$$\text{Exergy Input} - \text{Exergy Output} - \text{Exergy Consumption} = \text{Exergy Accumulation}$$

$$\sum_i e_i \cdot \dot{m}_i + \sum_b \dot{E}_b^q - \sum_j e_j \cdot \dot{m}_j - \dot{W} - \dot{E}_d = 0 \tag{5.17}$$

where:

$\dot{m}_i$, $\dot{m}_j$: the entering and exiting mass flow rates in and from the system respectively
e_i, e_j: the specific exergy for the entering and exiting mass flows
$\dot{E}_b^q$: the exergy rate transferred into the system from the environment through the boundary b associated with the entering thermal energy rate $\dot{Q}_b$
$\dot{W}$: the mechanical work transferred out of the system
$\dot{E}_d$: the exergy consumed within the thermodynamic system

$\sum_i e_i \cdot \dot{m}_i$: the contained exergy in the entering mass flow (exergy input)

$\sum_j e_j \cdot \dot{m}_j$: the contained exergy in the exiting mass flow (exergy output).

The term $\sum_i e_i \cdot \dot{m}_i - \sum_j e_j \cdot \dot{m}_j$ expresses the net exergy rate associated with matter and the term $\sum_b \dot{E}_b^q$ expresses the exergy input associated with heat. The terms $\dot{W}$ and $\dot{E}_d$ represent exergy consumption for work production and exergy destruction rates respectively.

As known, the energetic efficiency of an energy transformation process is defined as:

$$\eta = \frac{\text{Energy in product outputs}}{\text{Energy in primary inputs}} \tag{5.18}$$

Similarly with the above general definition, the exergy efficiency η_{ex} of a thermal process is defined as:

$$\eta_{ex} = \frac{\text{Exergy in product outputs}}{\text{Exergy in inputs}} = \frac{E_{out}}{E_{in}} = 1 - \frac{E_d + E_{cons}}{E_{in}} \tag{5.19}$$

where E_d and E_{cons} are the destroyed and the consumed exergy respectively during the process.

5.5.4 Analytical Expressions of Exergy Quantities

As presented in the previous section, the exergy rate $\dot{E}_m$ related to a mass flow rate $\dot{m}$ is given by the following general relationship:

$$\dot{E}_m = e \cdot \dot{m} \tag{5.20}$$

where e is the specific exergy of the mass flow. This overall exergy quantity is analyzed in a number of exergy forms, with which the overall exergy is contained in the mass flow, as presented in the following relationship:

$$\dot{E}_m = e \cdot \dot{m} = \dot{E}_{ph} + \dot{E}_{ch} + \dot{E}_{kin} + \dot{E}_{pot} \tag{5.21}$$

where:

$\dot{E}_{ph}$, $\dot{E}_{ch}$, $\dot{E}_{kin}$, $\dot{E}_{pot}$ are the physical, the chemical, the kinetic, and the potential form of exergy respectively. These forms are given from the relationships below:

$$\dot{E}_{ph} = \dot{m} \cdot \left[h - h_o - T_o \cdot (s - s_o)\right] \tag{5.22}$$

$$\dot{E}_{ch} = \sum_i (\mu_i - \mu_{io}) \cdot \dot{N}_i \tag{5.23}$$

$$\dot{E}_{kin} = \dot{m} \cdot E_k \tag{5.24}$$

$$\dot{E}_{pot} = \dot{m} \cdot E_p \tag{5.25}$$

where:

- h, s: the specific enthalpy and entropy of the mass flow
- h_o, s_o: the specific enthalpy and entropy of the reference environment
- T_o: the absolute temperature of the reference environment
- μ_i: the chemical potential of the chemical compound i contained in the flow
- μ_{jo}: the chemical potential of the chemical compound i contained in the reference environment
- $\dot{N}_i$: the molar flow rate of the chemical compound i contained in the flow
- E_k: the kinetic energy of the mass flow
- E_p: the potential energy of the mass flow due to conservative force fields.

Beyond the exergy associated with mass flow, there are also the exergy related to heat transfer and the exergy related to work. The exergy rate $\dot{E}_b^q$ related to heat transfer rate $\dot{Q}$ is given by the relationship:

$$\dot{E}_b^q = \tau \cdot \dot{Q} \tag{5.26}$$

where τ is the exergetic temperature factor. This factor expresses the fraction of a heat rate that is entirely converted to work at temperature T and is given by the following relationship (T_o is again the absolute temperature of the reference environment):

$$\tau = 1 - \frac{T_o}{T} \tag{5.27}$$

The application of the above relationships will be extensively described in a relevant example on the calculation of exergy of an air stream, given in a later section.

5.5.5 Base Enthalpy and Chemical Exergy of Species

From the previous section, we see that for the execution of the presented energy and exergy analysis the specific enthalpy and the chemical exergy of the involved chemical compounds should be known. Specific enthalpies should be defined with regard to the reference environment. This means

TABLE 5.4
Base Enthalpy and Chemical Exergy Values of Selected Chemical Compounds

Component	Base Enthalpy (kJ/(gr·mol))	Chemical Exergy (kJ/(gr·mol))
Ammonia (NH_3)	382.585	2.478907·ln(x) + 337.861
Argon (Ar)	0.000	2.478907·ln(x) + 11.650
Benzene (C_6H_6)	3,301.511	2.478907·ln(x) + 3,253.338
Carbon (C)	393.505	410.535
Carbon dioxide (CO_2)	0.000	2.478907·ln(x) + 20.108
Carbon monoxide (CO)	282.964	2.478907·ln(x) + 275.224
Carbon oxysulfide (COS)	891.150	2.478907·ln(x) + 848.013
Ethane (C_2H_6)	1,564.080	2.478907·ln(x) + 1,484.952
Hydrogen (H_2)	285.851	2.478907·ln(x) + 235.153
Hydrogen sulfide (H_2S)	901.757	2.478907·ln(x) + 803.374
Methane (CH_4)	890.359	2.478907·ln(x) + 830.212
Methanol (CH_3OH)	764.018	2.478907·ln(x) + 721.500
Nitrogen (N_2)	0.000	2.478907·ln(x) + 0.693
Oxygen (O_2)	0.000	2.478907·ln(x) + 3.948
Phenol (C_6H_5OH)	3,122.226	2.478907·ln(x) + 3,090.784
Sulfur (S)	636.052	608.967
Sulfur dioxide (SO_2)	339.155	2.478907·ln(x) + 295.736
Water (H_2O)	44.001	2.478907·ln(x) + 8.595

that a chemical compound in the reference environment with temperature and pressure T_o and P_o respectively is defined to have specific enthalpy equal to 0. The specific enthalpies calculated with regard to the above assumption are defined as base enthalpies. For many fossil fuels, base enthalpies are usually equal to their highest heat capacity. Typical values for base enthalpies of several compounds, calculated with regard to the reference environment model presented in Table 5.3, are given in Table 5.4 [12].

With regard to chemical exergy, several methods have been proposed for its estimation for different chemical compounds and species. Among them, there are alternative methods for evaluating chemical exergy for solids, liquids, gases, and complex materials. The chemical exergy of gas flows can be calculated by assuming the atmospheric air and the involved gases as ideal gas mixtures and by applying the empirical relationships presented in Table 5.4 [12].

5.6 ENERGETIC AND EXERGETIC ANALYSIS OF COGENERATION PROCESSES

The entering and exiting power flows in a general cogeneration model are presented in Figure 5.17. In this figure it is seen that power is introduced in the cogeneration system with the form of fuel and air mass flows $\dot{m}_f$ and $\dot{m}_a$, respectively. Through the executed thermal processes, the entering power flows are converted to useful mechanical $\dot{W}_p$ and thermal $\dot{Q}_p$ power, which constitute the exploitable energy products of the system. Additionally, an amount of the inlet power is lost in the form of material $\dot{m}_w$ and heat $\dot{Q}_w$ waste respectively, disposed in the ambient.

According to the power inlet and outlet flows presented in Figure 5.17 and the general energy and exergy balances analyzed in the previous section, and neglecting the kinetic and the potential energy and exergy, the energy balance equation for the cogeneration plant is written as:

$$\dot{m}_f \cdot h_f + \dot{m}_a \cdot h_a = \dot{W}_p + \dot{Q}_p + \sum_w \dot{Q}_w + \sum_w \dot{m}_w \cdot h_w \tag{5.28}$$

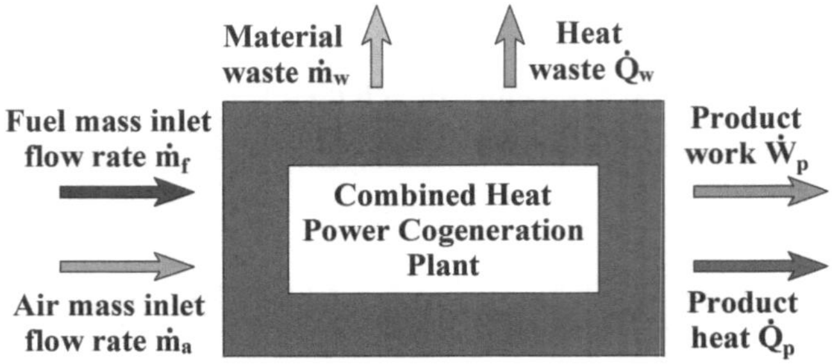

FIGURE 5.17 The inlet and outlet power flows in a cogeneration plant general model.

where:

h_f, h_a: the specific enthalpies of the inlet fuel and air respectively
h_w: the specific enthalpy of the exiting gases

Similarly, based on the same data and assumptions as above and following the general exergy balance presented in the previous section, the exergy balance for the examined CHP system is written as:

$$\dot{m}_f \cdot e_f + \dot{m}_a \cdot e_a = \dot{W}_p + \dot{Q}_p \cdot \tau_p + \sum_w \dot{Q}_w \cdot \tau_w + \sum_w \dot{m}_w \cdot e_w + \dot{E}_d \tag{5.29}$$

where:

e_f, e_a: the specific exergies of the inlet fuel and air respectively
τ_p, τ_w: the exergetic temperature factors for the heat flows $\dot{Q}_p$ and $\dot{Q}_w$ respectively, which, according to Equation 5.27, are functions of the corresponding temperatures T_p, T_w, respectively, and T_o
T_p, T_w, T_o: the temperatures associated with the heat flows $\dot{Q}_p$ and $\dot{Q}_w$ respectively and the temperature of the environment.

The terms in the left side of the above two equations represent the energy and exergy inputs in the cogeneration system. The first two terms in the right side of the equations represent the useful energy and exergy outputs, while the two summations represent the energy and exergy wastes associated with heat and mass. Finally, the term $\dot{E}_d$ represents the exergy destruction rate in the cogeneration system.

The energetic and exergetic analysis of a general cogeneration system is integrated with the presentation of a number of indices and efficiencies, which provide indications on the performance of the cogeneration system.

Specifically, the ratio:

$$q = \frac{\dot{Q}_p}{\dot{Q}_p + \sum_w \dot{Q}_w} \tag{5.30}$$

represents the fraction of the useful (exploitable) heat output versus the total heat output of the cogeneration system. This index can theoretically range from 0, for pure electricity production systems, to 1, for cogeneration plants with null heat waste.

Additionally, the energy efficiency given according to Equation 5.7, is presented below:

$$\eta_{co} = \frac{\dot{W}_p + \dot{Q}_p}{\dot{m}_f \cdot h_f} \tag{5.31}$$

and the exergy efficiency can be derived from the analysis of Equation 5.19:

$$\eta_{ex} = \frac{\dot{W}_p + \dot{Q}_p \cdot \tau_p}{\dot{m}_f \cdot e_f} \tag{5.32}$$

From the last two relationships, it is seen that although both energy and exergy efficiencies of the cogeneration process depend on the output heat rate $\dot{Q}_p$, only the exergy efficiency depends on the temperature T_p with which the output heat product rate is delivered. This is due to the involved exergetic temperature factor τ_p in the corresponding relationship. The same factor also introduces the dependence of the exergetic efficiency on the temperature T_o of the reference environment.

In many cases the specific enthalpy and exergy of the reference environment are both considered equal to zero: $e_a = h_a = 0$. In this case, the fuel's specific enthalpy is taken equal to its highest heat capacity at the reference environment conditions T_o and P_o, namely it is $h_f = H_u$. Also, in this case the fuel's specific enthalpy and exergy are approximately equal to each other, with a 15% maximum error.

From the last two relationships, it is seen that as the heat rate $\dot{Q}_p$ increases, both the energetic and exergetic efficiencies increase too. Yet, since $\dot{Q}_p$ is multiplied with τ_p ($0 \le \tau_p \le 1$ for $T_p \ge T_o$), the exergetic efficiency does not increase as fast as the energetic one. Generally, $\eta_{ex} \le \eta_{co}$ for $T_p \ge T_o$ and $e_f = h_f$. The difference between the two efficiencies is given by the relationship:

$$\eta_{co} - \eta_{ex} = \frac{\dot{Q}_p \cdot T_o}{\dot{m}_f \cdot h_f \cdot T_p} \tag{5.33}$$

From the above relationship, it is seen that for standard reference environment conditions, the difference between the two efficiencies decreases, as the produced heat rate $\dot{Q}_p$ decreases and the delivery temperature T_p increases. Higher temperature T_p implies higher potential for the delivered heat rate $\dot{Q}_p$ to produce work, namely it corresponds to higher quality of energy, hence higher exergy. From the last relationship it is also seen that the difference between the two efficiencies is independent of the produced work rate $\dot{W}_p$.

In some cases of conventional electrical generators, there is a little electrical output loss when they are integrated in cogeneration systems. In other words, the generator's electricity production output is slightly decreased when introduced in the cogeneration system, compared to its operation purely for electricity production. In order to express quantitatively this effect, the following index is introduced:

$$COP_{CHP} = \frac{\text{Produced heat output rate}}{\text{Electricity production rate reduction}} = \frac{\dot{Q}_p}{\dot{W}_r} \tag{5.34}$$

The denominator in the above ratio expresses the difference between the electricity production from the thermal generators when they operate exclusively for electricity production $\dot{W}_{p\text{-}el}$ and when they are employed as the base units of a cogeneration system $\dot{W}_p$:

$$\dot{W}_r = \dot{W}_{p\text{-}el} - \dot{W}_p \tag{5.35}$$

In most cases the electricity production reduction rate $\dot{W}_r$ is rather low, so the corresponding COP_{CHP} takes high values, approaching even infinity for $\dot{W}_r$ close to zero.

Example 5.1 Energy and Exergy Analysis of a Back-Pressure Steam Turbine

Perform the energy and exergy analysis of a back-pressure steam turbine under the following operation conditions:

Inlet steam flow properties	Flow $\dot{m}_{s\text{-}in}$ (kg/s)	40
	Pressure $p_{s\text{-}in}$ (bar)	100
	Temperature $T_{s\text{-}in}$ (°C)	500
High-pressure outlet flow	Flow $\dot{m}_{HP\text{-}out}$ (kg/s)	23
	Pressure $p_{HP\text{-}out}$ (bar)	35
	Temperature $T_{HP\text{-}out}$ (°C)	350
Low-pressure outlet flow	Flow $\dot{m}_{LP\text{-}out}$ (kg/s)	17
	Pressure $p_{LP\text{-}out}$ (bar)	1.5
	Temperature $T_{LP\text{-}out}$ (°C)	130

The output electrical power P_{el} of the steam turbine is 14 MW and the mechanical and electrical efficiencies η_m and η_g respectively of the turbine and the inductive generator are both equal to 98%. The temperature and pressure of the surrounding environment should be taken equal to T_o = 25°C and p_o = 1 atm.

SOLUTION

First of all, given the steam's temperature and pressure of the involved steam flows, the specific enthalpy and entropy for each one of them are found. An easy, quick, and accurate online calculator is available.[13] Specifically:

Inlet steam flow	Specific enthalpy $h_{s\text{-}in}$ (kJ/kg)	3,375.06
$p_{s\text{-}in}$ = 100 bar, $T_{s\text{-}in}$ = 500°C	Specific entropy $s_{s\text{-}in}$ (kJ/kgK)	6.60
High-pressure outlet flow	Specific enthalpy $h_{HP\text{-}out}$ (kJ/kg)	3,104.84
$p_{HP\text{-}out}$ = 35 bar, $T_{HP\text{-}out}$ = 350°C	Specific entropy $s_{HP\text{-}out}$ (kJ/kgK)	6.66
Low-pressure outlet flow	Specific enthalpy $h_{LP\text{-}out}$ (kJ/kg)	2,732.07
$p_{LP\text{-}out}$ = 1.5 bar, $T_{LP\text{-}out}$ = 130°C	Specific entropy $s_{LP\text{-}out}$ (kJ/kgK)	7.32

- *Energy analysis*

  As clearly seen above, there is one inlet energy flow (inlet steam flow) and two outlet energy flows (steam outlet flows from the high- and the low-pressure stages).

  The power input in the steam turbine with the inlet steam flow is hence given by the equation:

$$\dot{Q}_{in} = \dot{m}_{s\text{-}in} \cdot h_{s\text{-}in} \Rightarrow \dot{Q}_{in} = 40\frac{kg}{s} \cdot 3{,}375.06\frac{kJ}{kg} \Leftrightarrow \dot{Q}_{in} = 135{,}002.40\frac{kJ}{s}$$

The thermal power output from the steam turbine is the sum of the thermal power contained in the outlet steam flows from the high- and low-pressure stages of the steam turbine:

$$\dot{Q}_{out} = \dot{Q}_{HP\text{-}out} + \dot{Q}_{LP\text{-}out} \Leftrightarrow \dot{Q}_{out} = \dot{m}_{HP\text{-}out} \cdot h_{HP\text{-}out} + \dot{m}_{LP\text{-}out} \cdot h_{LP\text{-}out} \Rightarrow$$

$$\dot{Q}_{out} = 23\frac{kg}{s} \cdot 3{,}104.84\frac{kJ}{kg} + 17\frac{kg}{s} \cdot 2{,}732.27\frac{kJ}{kg} \Leftrightarrow \dot{Q}_{out} = 71{,}411.32\frac{kJ}{s} + 46{,}448.59\frac{kJ}{s} \Leftrightarrow$$

$$\dot{Q}_{out} = 117{,}859.91\frac{kJ}{s}$$

The output work rate is calculated from the difference of the input and output heat rates, given the fundamental energy balance in the thermodynamic system:

$$\dot{W}_{out} = \dot{Q}_{in} - \dot{Q}_{out} \Rightarrow \dot{W}_{out} = 135{,}002.40\frac{kJ}{s} - 117{,}859.91\frac{kJ}{s} \Leftrightarrow \dot{W}_{out} = 17{,}142.49\frac{kJ}{s}$$

The mechanical power $\dot{W}_{shaft}$ developed at the steam turbine's shaft can be calculated from the following relationship:

$$\eta_m \cdot \eta_g = \frac{P_{el}}{\dot{W}_{shaft}} \Leftrightarrow \dot{W}_{shaft} = \frac{P_{el}}{\eta_m \cdot \eta_g} \Rightarrow \dot{W}_{shaft} = \frac{14{,}000}{0.98 \cdot 0.98}\frac{kJ}{s} \Leftrightarrow \dot{W}_{shaft} = 14{,}577.26\frac{kJ}{s}$$

The efficiency of the steam turbine can now be calculated as shown below:

$$\eta_{st} = \frac{\dot{W}_{shaft}}{\dot{W}_{out}} \Rightarrow \eta_{st} = \frac{14{,}577.26}{17{,}142.49} \Leftrightarrow \eta_{st} = 85.04\%$$

The efficiency η_{el} of the electricity production process is:

$$\eta_{el} = \frac{P_{el}}{\dot{Q}_{in}} \Rightarrow \eta_{el} = \frac{14{,}000.00}{135{,}002.40} \Leftrightarrow \eta_{el} = 10.37\%$$

- *Exergy analysis*
  The corresponding exergy inlet and outlet flows coincide with the above clarified energy inlet and outlet flows. From Equations 5.21 and 5.22, neglecting the chemical, kinetic, and potential exergy of the steam flows and assuming the specific exergy and entropy of the reference environment equal to zero ($h_o = s_o = 0$), we can calculate the inlet and outlet steam flows exergy, as presented below:

$$\dot{E} = \dot{E}_{ph} = \dot{m} \cdot \left[h - h_o - T_o \cdot (s - s_o)\right] \Rightarrow \dot{E} = \dot{m} \cdot (h - T_o \cdot s)$$

The exergy contained in the inlet steam flow is:

$$\dot{E}_{s\text{-}in} = \dot{m}_{s\text{-}in} \cdot (h_{s\text{-}in} - T_o \cdot s_{s\text{-}in}) \Rightarrow \dot{E}_{s\text{-}in} = 40\frac{kg}{s} \cdot \left[3{,}375.06\frac{kJ}{kg} - 298K \cdot 6.60\frac{kJ}{kg \cdot K}\right] \Leftrightarrow \dot{E}_{s\text{-}in} = 56{,}330.40\frac{kJ}{s}$$

The exergy contained in the outlet steam flows is:

$$\dot{E}_{out} = \dot{m}_{HP\text{-}out} \cdot (h_{HP\text{-}out} - T_o \cdot s_{HP\text{-}out}) + \dot{m}_{LP\text{-}out} \cdot (h_{LP\text{-}out} - T_o \cdot s_{LP\text{-}out}) \Rightarrow$$

$$\dot{E}_{out} = 23\frac{kg}{s} \cdot \left[3{,}104.84\frac{kJ}{kg} - 298K \cdot 6.66\frac{kJ}{kg \cdot K}\right] + 17\frac{kg}{s} \cdot \left[2{,}732.27\frac{kJ}{kg} - 298K \cdot 7.32\frac{kJ}{kg \cdot K}\right] \Leftrightarrow$$

$$\dot{E}_{out} = 25{,}763.68\frac{kJ}{s} + 9{,}365.47\frac{kJ}{s} \Leftrightarrow \dot{E}_{out} = 35{,}129.15\frac{kJ}{s}$$

Following the essential exergy balance expressed with Equation 5.17, the exergy destruction rate $\dot{E}_d$ can be calculated:

$$\sum_i e_i \cdot \dot{m}_i + \sum_b \dot{E}_b^q - \sum_j e_j \cdot \dot{m}_j - \dot{W} - \dot{E}_d = 0 \Rightarrow$$

$$\dot{E}_{s\text{-}in} - \dot{E}_{s\text{-}out} - \dot{W} - \dot{E}_d = 0 \Leftrightarrow \dot{E}_d = \dot{E}_{s\text{-}in} - \dot{E}_{s\text{-}out} - \dot{W} \Rightarrow$$

$$\dot{E}_d = 56{,}330.40\frac{kJ}{s} - 35{,}129.15\frac{kJ}{s} - 14{,}577.26\frac{kJ}{s} \Leftrightarrow \dot{E}_d = 6{,}623.99\frac{kJ}{s}$$

Finally, the exergy efficiency of the executed thermal process is calculated according to:

$$\eta_{ex} = \frac{\text{Exergy in product outputs}}{\text{Energy in inputs}} = \frac{\dot{E}_{out}}{\dot{E}_{in}} \Rightarrow \eta_{ex} = \frac{35{,}129.15\frac{kJ}{s}}{56{,}330.40\frac{kJ}{s}} \Leftrightarrow \eta_{ex} = 62.36\%$$

5.7 DISTRICT HEATING AND COOLING

With the *term district heating or cooling,* we refer to the whole heating or cooling production and distribution system developed to produce heating or cooling centrally and distribute it to a number of buildings or houses, or even different types of consumptions, such as agricultural or industrial uses. Heating or cooling is distributed through a hydraulic network, by means of hot water or chilled water, consisting of pipelines, circulators, heat exchangers, pressure vessels, and several other safety and automation devices. District heating or cooling can be based on the exploitation of a variety of alternative primary energy sources, starting from conventional fossil fuel, in cogeneration CHP plants, expanding to biomass fuels and urban wastes and ending with advanced solar heating–cooling trigeneration systems. This flexibility makes district heating and cooling often a highly attractive solution for thermal energy applications, from an economic point of view, contributing also at the same time both to the increase of the overall efficiency of the energy transformation process (e.g., in case of cogeneration CHP plants) and to the reduction of fossil fuel consumption and greenhouse gases emissions.

District heating and cooling is best suited to highly populated areas, namely where the physical distances between the final consumers are not long, which, in turn, imposes reduction of the required infrastructure setup cost, as well as of the operation cost, due to the reduced thermal and flow losses of the working medium. In such cases, the economic benefit can be considerably attractive for both the operators of the system and the final consumers, who experience a relatively low final heating and cooling purchase price. Apart from the obvious economic benefits, district heating and cooling exhibit also a number of attractive features, such as:

- Reduction of air pollution and improvement of the atmosphere quality in urban environments. This is a direct effect from the fact that all the energy transformation processes are normally executed centrally, in a power plant located outside the urban zones. Practically, with district heating, hundreds or thousands of central heating burners, fire places, etc. are replaced by the central heating or the cogeneration power plant.
- Increase of supply security for final heating and cooling. All the required procedures for the production and the distribution of heating and cooling, including the procurement and the storage of the primary energy sources, the maintenance of the required systems and infrastructures, the distribution of the final heating and cooling, are centrally controlled and executed, normally by specialized staff and under the relevant quality standards. Any insecurities from the operation of small, decentralized heating and cooling systems separately from nonspecialized individuals and owners are eliminated.
- Increase of space in buildings, because no internal heating or cooling devices or systems are required.

- Increase of spare time for the buildings' owners, because all the procedures related to the operation and the maintenance of decentralized heating or cooling systems are no longer required.
- A multilevel economic benefit is achieved, configured by the elimination of setup costa for the required final heating and cooling disposal systems, and the minimization of the operation and maintenance cost.
- For district heating systems based on biomass fuels or solar cooling, namely in cases of district heating or cooling systems of relatively small size based on the exploitation of RES, the perspectives for the foundation of local energy cooperatives schemes are developed, creating the prerequisites for a sustainable economic and social development of the local communities.

5.7.1 Fundamental Layout of a District Heating System

There can be several different implementations of district heating systems, arisen mainly from the primary heating production plant. Specifically, as stated also previously, heating can be produced as a by-product of a CHP process, or by decentralized biomass power stations, or by urban waste utilization units or by solar collectors. All these alternative systems may lead to different realizations of the district heating system. In Figure 5.18 the general layout of a district heating system is presented. It consists of the following fundamental components:

- The combined heat and power plant, equipped with one or more thermal generators and a heat recovery system, which collects heats from the disposed hot gases or the engines' cooling system.
- A thermal storage tank, playing the role of a thermal buffer between the heating production and the heating load. The thermal storage tank should be dimensioned to undertake the heating load for at least 12 hours or more, depending on the availability of the thermal power production from the main CHP plant and the heating demand fluctuation. The thermal storage tank is actually constructed with an adequately insulated steel vessel, with an expansion space filled with nitrogen gas, capable to undertake the water volume changes and the subsequent pressure fluctuations.
- The heating distribution system, consisting of a number of hydraulic components.

We now discuss the most important of the involved hydraulic components of a district heating system.

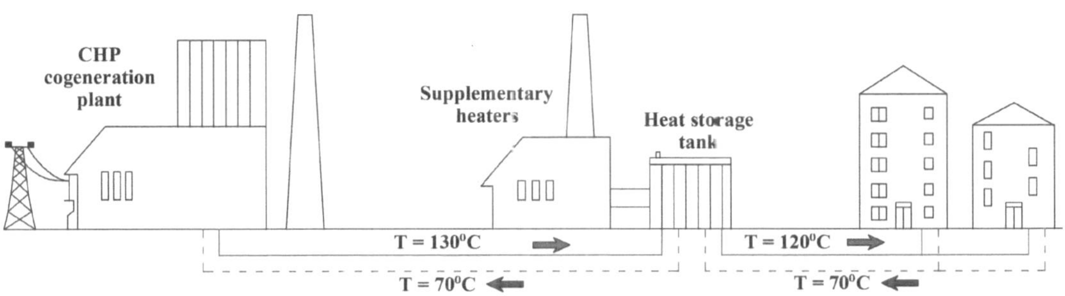

FIGURE 5.18 General layout of a district heating system (with due diligence of Danfoss) [14].

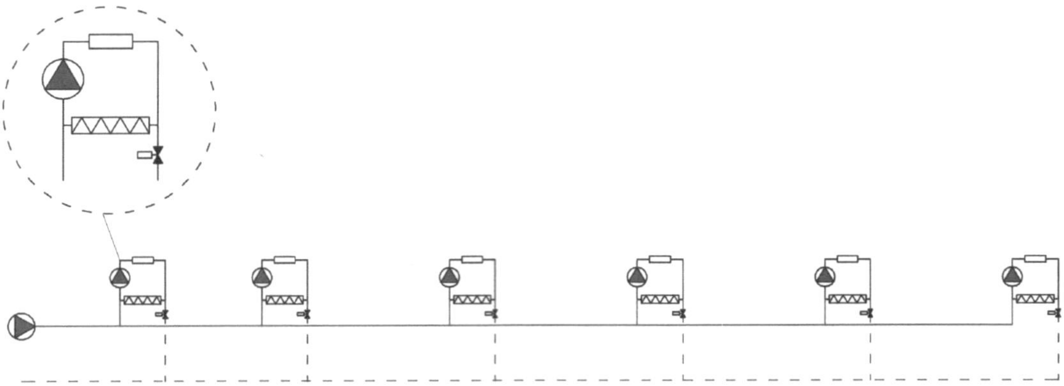

FIGURE 5.19 Heat exchangers' fundamental connectivity concept in a heating district system (with due diligence of Danfoss) [14].

5.7.1.1 Heat Exchangers

Heat exchangers are introduced in the distribution network to transfer heat from the main distribution loop to secondary final loops for the disposal of heat to the final users (see Figure 5.19). Heat exchangers are also used before and after the thermal storage tank, for the heat transfer from the primary heating production loop to the thermal tank and from the latter to the heating distribution network. For district heating systems, typical plate heat exchangers are used. The sizing of the heat exchangers is accomplished with the procedure presented in Chapter 4, based on the working temperature differences, the flows in the primary and secondary loop, and the transferred thermal power.

5.7.1.2 Expansion Systems

Usually expansion systems are closed expansion insulated vessels, appropriately dimensioned to undertake the working medium temperature fluctuations and the subsequent pressure increases in every closed hydraulic loop. This means that at least one expansion vessel should be installed in every closed loop, namely at the primary loop from the CHP plant to the thermal storage, at the heating distribution main loop, and at the heating disposal terminal loops for the final consumers.

5.7.1.3 Circulators

Circulators are installed, similarly to the heat exchangers, in every closed hydraulic loop, to supply the required power for the realization of the working medium required flow inside the loop. The required flow $\dot{m}_w$ of the working medium in the pipelines is determined by the current heating load $\dot{Q}$, calculated by the following essential relationship, given the temperature difference $T_i - T_o$ during the heat transfer process (c_w the thermal capacity of the working medium):

$$\dot{Q} = \dot{m}_w \cdot c_w \cdot (T_i - T_o) \tag{5.36}$$

Typical centrifugal circulators are used. Their sizing follows the fundamental procedure for the dimensioning of pumps in hydraulic networks, aiming to support the required flow and head, with the latter being practically configured by the flow losses in the hydraulic network. A relevant presentation of this procedure is provided in Chapter 3, in the presentation of the dimensioning of the hydraulic network in the pumped hydro storage system. The overall basic formulation of a district heating distribution network is presented in Figure 5.18.

5.7.1.4 Pipelines

For the transportation of the working medium, preinsulated steel pipelines are usually used, with high-quality insulation and a safe waterproof protective cover. The diameter of the pipelines is selected in order to ensure a maximum flow velocity lower than 2 m/s, a prerequisite which, in turn, will lead to the conservation of the flow losses in sensible level (see the presented procedure in Chapter 3). Typically, a polyethylene foam insulation is used, while for the waterproof external layer, a polyethylene pipe is also used.

In Figure 5.18, along with the essential construction layout of district heating, typical operating conditions are also presented. The steam is supplied for the thermal storage tank with a typical temperature of 130°C. The maximum temperature at the main distribution loop is normally not higher than 120°C. The return temperature from the final users and to the cogeneration heat and power plant ranges around 70°C. With regard to static pressure, steam is typically provided in the thermal storage tank with a pressure of 200 kPa. This pressure is also maintained in the storage tank with the expansion vessel. The static pressure in the most far away consumption should not be lower than 150 kPa. This is achieved with the installation of circulators within the route of the working medium, aiming to compensate for any pressure drops. The circulators are installed typically after every pressure drop of 15–20 kPa during the working medium flow.

5.7.2 Fundamental Layout of a District Cooling System

Usually, cooling is produced centrally by a number of chillers. In a more sophisticated approach, cooling can be alternatively produced by geothermal plants or solar collectors, with so-called solar cooling systems, leading to trigeneration applications. A typical diagram of a cooling district system is presented in Figure 5.20. Practically it follows the same layout as with district heating systems. Cooling is produced centrally in cooling chillers and distributed for the final consumers through a distribution network. A cooling storage tank may also be introduced to balance the fluctuations between the cooling production and the cooling demand. As with district heating systems, water is most commonly used as the working medium in the cooling distribution network. Consequently, the construction layout, the required equipment, and the dimensioning of a district cooling distribution network follow the typical principles of a common hydraulic network, roughly presented in the previous section.

In modern applications chillers are driven by electricity, while in larger applications primarily developed for cooling production, steam and natural gas can also be used. Three main chiller technologies have been developed: centrifugal chillers, steam turbine-driven chillers, and absorption chillers.

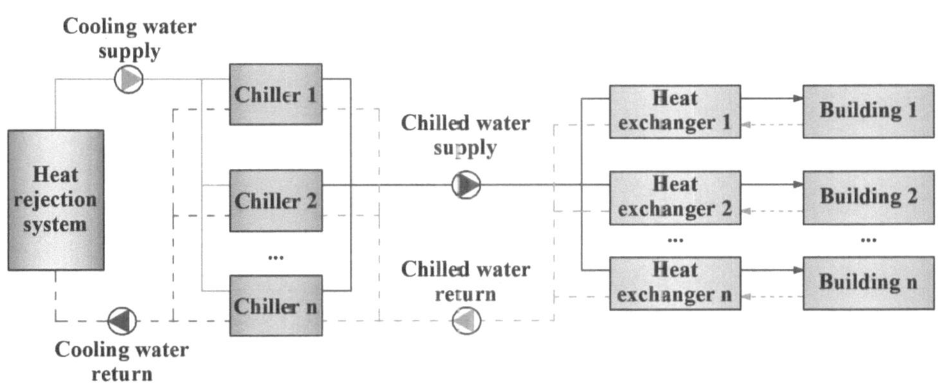

FIGURE 5.20 Fundamental layout concept in a cooling district system [15].

The term *centrifugal chillers* refers to the centrifugal compressors employed for the compression of the refrigerant in a common cooling cycle. The cooling production is based on the fundamental cooling cycle implemented through the passage of the refrigerant from the evaporator, the compressor, the condenser, and the expansion valve, which constitute the four basic components of the cooling device. In the evaporator, the refrigerant—a volatile fluid in liquid phase—is evaporated at ambient temperature by absorbing thermal energy from the ambient that is cooled. Hence, cooling is produced during this phase. After its exit from the evaporator, the refrigerant enters the compressor, where it is compressed. The compression stage is the first step towards the reliquefaction of the refrigerant. The gained heat from the compression process is disposed in the ambient in the condenser. At the exit of the condenser, the refrigerant has returned to its liquid phase; nevertheless it still remains in high pressure. Hence, before entering again the evaporator, an expansion valve is intervened, where the refrigerant's pressure is dropped back to the ambient one. After the expansion valve, the initial conditions of the refrigerant have been restored (low pressure, liquid phase) and it is so prepared to enter the evaporator for the absorption of further heat from the ambient and the repetition of the cooling cycle. The overall operation and layout of a cooling device is presented in Figure 5.21. In the whole cooling cycle, the compressor is the unique component where power is consumed. Centrifugal chillers are most frequently driven by electricity and, in special cases, by natural gas or steam.

The power flows in the basic cooling cycle, as presented in Figure 5.20, are:

- In the evaporator, thermal power $\dot{Q}_c$ is absorbed from the cooling space. This absorbed thermal power constitutes the useful cooling power of the cooling device.
- In the condenser, thermal power $\dot{Q}_r$ is rejected towards the surrounding space, which is heated.
- From the above it is seen that thermal power is absorbed from a cool indoor space and rejected to the ambient warm environment, namely conversely to the natural heat transfer flow. To achieve this, as described previously, the consumption of mechanical power $\dot{W}$ is required in the compressor of the cooling device.

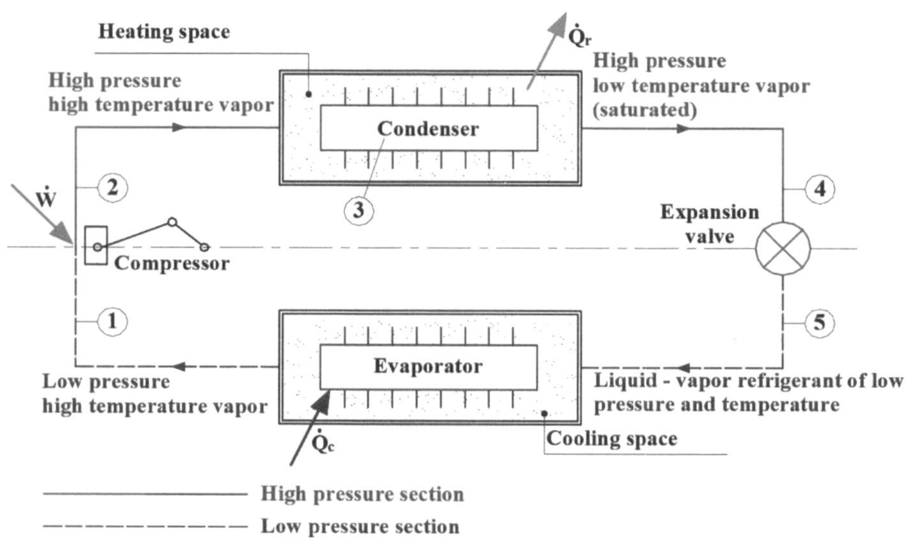

FIGURE 5.21 The cooling cycle.

During the implementation of the cooling cycle, the mechanical power $\dot{W}$ and the useful cooling power $\dot{Q}_c$ are offered from the external environment to the refrigerant. On the contrary, the rejected thermal power $\dot{Q}_r$ is disposed from the refrigerant to the environment. Consequently, the energy balance in the essential, theoretical cooling cycle is written as:

$$\dot{Q}_c + \dot{W} = \dot{Q}_r \tag{5.37}$$

From the executed energy flows in the cooling cycle, power consumption is executed only in the compressor, while the only useful power is the cooling power absorbed in the evaporator from the conditioned space. It is sensible to define the theoretical efficiency of a cooling device as:

$$\mathrm{COP} = \frac{\dot{Q}_c}{\dot{W}} \tag{5.38}$$

namely as the ratio of the produced cooling power versus the consumed mechanical power per cooling cycle iteration. With Equation 5.38 we introduce the theoretical coefficient of performance (COP_{th}) of the cooling device, which expresses the produced cooling power per unit of consumed mechanical power.

Natural gas engine-driven chillers use natural gas to fire an engine that runs the compressor. Electric centrifugal chillers use an electric motor to drive the compressor. The operation principle of steam turbine-driven chillers is the same with electric driven chillers, with the only difference that a steam turbine is employed to produce rotational power, rather than an electrical motor. Steam is supplied at high pressure (e.g., 9 bars) and expanded in a condensing turbine, producing thus mechanical energy that drives a compressor, practically the same as in the case of an electrical chiller.

Absorption chillers, on average, have much lower cooling capacities than centrifugal units. The two basic types of absorption chillers are single-stage and two-stage chillers. They will be described in detail in the section dedicated to solar cooling.

A very special case of district cooling is based on the disposal of the absorbed heat from the conditioned space in underground soil or water natural resources. This process is known as geothermal cooling. In case of the disposal of the absorbed heat in underground soil, a pipeline closed loop is buried in a horizontal or vertical layout below the ground's surface (Figure 5.22). This closed pipeline loop is known as a geothermal heat exchanger. Geothermal heating is certainly also possible with geothermal heat exchangers.

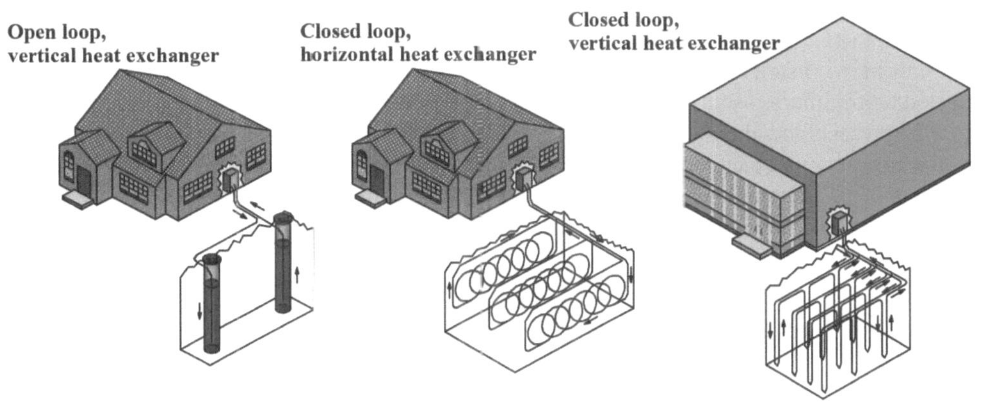

FIGURE 5.22 Alternative layouts for the exploitation of normal geothermy for indoor space conditioning.

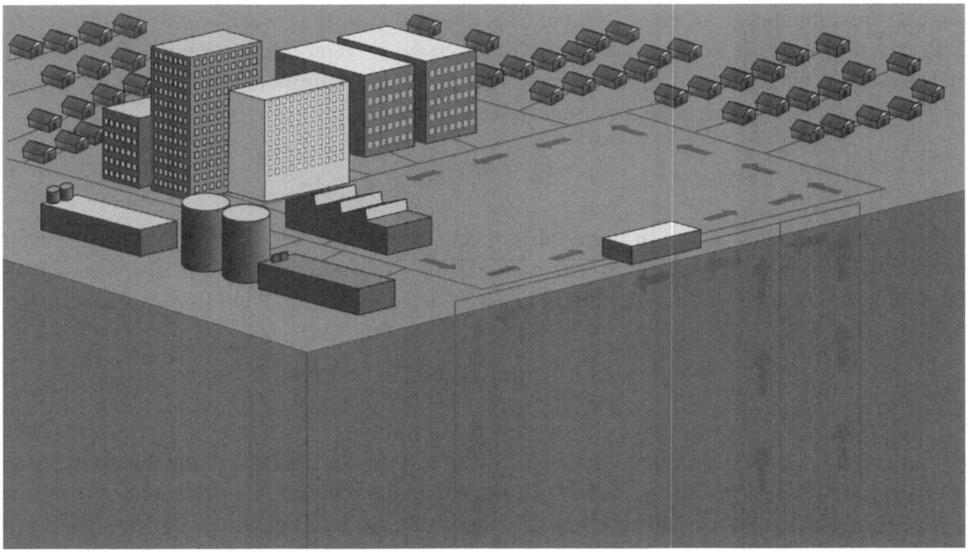

FIGURE 5.23 General layout of a district heating and cooling system based on normal geothermy [18].

The geothermal heat exchanger is properly dimensioned, with regard to its length and diameter, so as after a full passage of the flowing water inside the pipeline, its temperature will be equal to the temperature of the surrounding underground soil. In this way, at the end of the geothermal heat exchanger, on the condition of appropriate dimensioning, the working medium has disposed all the absorbed heat from the conditioned space underground; in other words, it has been cooled by the ground. It is so prepared to be led to the condenser of the cooling device, which in this case is known as a geothermal heat pump, for the absorption of further heat from the conditioned space and the repetition of the cooling cycle. Several methodologies for the dimensioning of a geothermal heat exchanger are presented in the relevant literature [16, 17]. The required data for the dimensioning of geothermal heat exchangers are the essential properties of soil regarding the heat transfer from and to it and the heating or cooling loads, while the basic results are the required pipeline length and diameter, the number of boreholes (in case of vertical heat exchangers), etc.

This procedure can be alternatively executed by disposing the absorbed heat in natural water resources (lake, sea, aquifer), instead of underground. In such cases, the geothermal heat exchanger is substituted by the involved natural water resources. Water is pumped from one borehole and led to the condenser to absorb the disposed heat of the refrigerant. The warm water, after its passage from the condenser, is returned back to its source, however with a second borehole, located at a distance from the location of the first one capable to ensure that the inlet temperature of the pumped water from the first borehole will not be affected (a minimum distance of 50 m is usually referred to as adequate in the relevant handbooks). Figure 5.22 presents the alternative layouts for the exploitation of so-called normal geothermy, namely the geothermal potential met in depths lower than 100 m from the earth's surface with a normal geothermal gradient of 1°C increase per every 40 m of depth increase from the ground surface.

A general layout of a district cooling and heating system based on normal geothermy is presented in Figure 5.23 for heating operation mode.

5.8 DISTRICT HEATING EXAMPLES

District heating has been widely recognized as one of the most viable alternatives towards the elimination of the use of conventional fossil fuels for heating applications. This can be achieved if the thermal power is produced with RES technologies, such as solar collectors, geothermal heat exchangers, biomass stations, or even with the exploitation of urban wastes. This approach has been so far introduced by several communities, located mainly in central Europe, Scandinavia, the US, and Australia. In this section, two indicative cases will be presented for the Austrian city of Güssing and the city of Milan in Italy.

5.8.1 Biomass District Heating in Güssing, Austria

The story of district heating for the city of Güssing goes back to the early 1990s, when the municipal authorities started an effort for community development, based on the exploitation of locally available energy sources, mainly forest products and residues. The plan was realized with the construction of a cogeneration power plant, with four thermal power generators and three cogeneration engines, covering in total more than 70% of the city's electricity and heating needs. Specifically, the local power plant produces on an annual basis 56 GWh_{th} of final thermal energy and 22 GWh_e of final electricity demand, out of 60 GWh_{th} and 50 GWh_e of final thermal energy and electricity annual consumption respectively [19]. Also, a reduction of 14,500 tn on the CO_2 annual emissions in the city was achieved.

All these heat and power generators are fed with biomass fuels, mainly wood wastes from the parquet factories in Güssing and wood chips delivered from the nearby forest of Burgenland, located approximately 30–40 km away. However, what is maybe more important, is that the cogeneration units are fed with a synthetic natural gas (BioSNG) or synthetic liquid fuel (gasoline or diesel) through an innovative wood gasification technology developed by the Technical University of Vienna [20–22]. Following this innovation, the European Center for Renewable Energy (EEE) was founded in Güssing, acknowledged as the top research center in Europe in the field of wood gasification and the production of second-generation biofuels.

Several benefits for the city of Güssing and its wider area were gained from the transition to biomass district heating. First of all, the citizens of Güssing enjoy a stable, secure, and cost-effective heating system. In Figure 5.24 the total net heating cost fluctuation for the whole city is presented since the late 1980s, alternatively with the use of oil and with the operating district heating system [23]. It is seen that the introduction of district heating has enabled the conservation of the heating total net cost at the level of the mid 1990s, completely independent from the fluctuations of the oil prices and remaining around 20%–25% of the corresponding total net heating cost with the use of oil, with regard to 2010s oil prices.

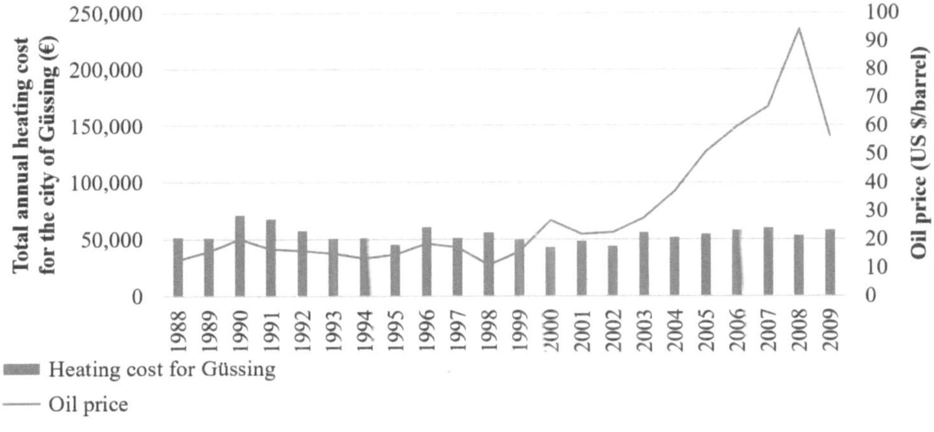

FIGURE 5.24 Comparison between the total net cost in the city of Güssing, Austria, with the operating of a biomass district heating system and with the use of oil [23].

More importantly, more than 1,100 occupation positions were created and 50 new enterprises were founded for the exploitation of forest by-products, the development and the application of new biomass heaters, and cogeneration engine technologies. New enterprises and industries were also moved to the area due to the availability of cheap heating. The total invested capital in the new founded firms exceeds €35.5 million.

5.8.2 Geothermal District Heating in Milan, Italy

Milan has a long history with the exploitation of geothermal energy. It was the first city in Europe where geothermal heat pumps were initially installed, in 1937–1938, for the conditioning of two large buildings at the center of the city, the press palace in Cavour Square and a bank headquarters. During the last years, new geothermal heat pump systems, operating with geothermal open loops, have been installed for the production of heating and cooling for several large and popular buildings of Milan, such as Regional Palace, Bocconi University, the historical Castello Sforzesco, the Scala Theater, the Museum of Natural Sciences, and the Archbishopric archives. All these projects have been developed on the basis of the rich underground water resources available in the aquifer, below the city of Milan.

Following this long tradition, AEM S.p.a., the local public energy company, so far responsible for covering a large part of the city's energy needs, including gas, electricity, and district heating, has undertaken a large and ambitious district heating/cooling project, based on the exploitation of the available geothermal potential in the city's aquifer. The project aims to cover the heating/cooling needs of 250,000 residents of Milan, a population that corresponds to more than 20% of the city's overall population. In total, around 650,000 MWh_{th}/year of final thermal energy will be produced for district heating for approximately 20,000,000 m^3 of indoor conditioned space [24]. It will be the second largest district heating/cooling project based on normal geothermal potential in the world, after Paris, in terms of connected clients, and the largest district heating/cooling system exploiting underground water resources.

The whole project consists of five stations, allocated in strategic locations around the city's center. The operation concept of the installed systems, the same for simplicity reasons at each location, is presented in Figure 5.25 [24]. It consists of the geothermal heat pump system, a CHP cogeneration engine, heat tanks, and conventional heaters. The geothermal heat pumps produce thermal power by exploiting the available geothermal potential in the aquifer water. The produced thermal power from the heat pumps is supplemented with the recovered thermal power from the CHP engine.

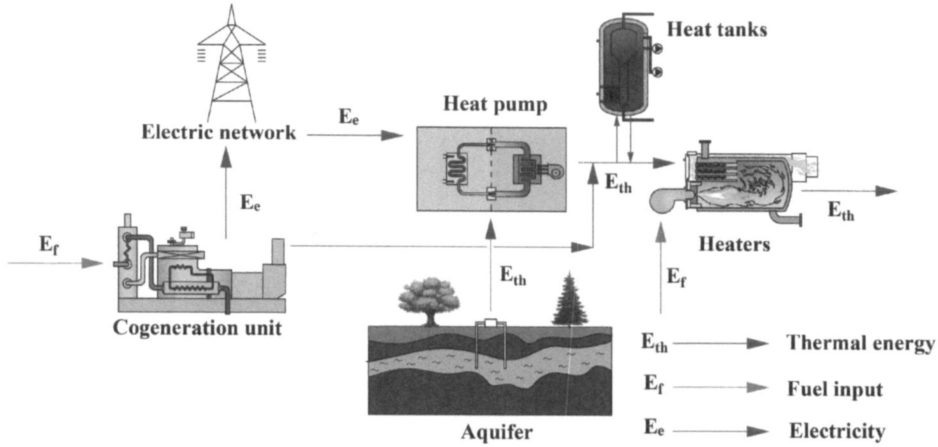

FIGURE 5.25 The operation concept of the district heating system in Milan [24].

The electricity production from the CHP engine is injected in the local electricity grid. The available thermal power is led directly to the heating distribution network. In case the thermal loads are lower than the concurrently produced thermal power, the latter is stored in the available heat tanks. Finally, if the produced thermal power from the geothermal heat pumps and the cogeneration plant is not adequate to cover the current heating demand, the conventional heaters are involved. It is conceivable that this system approaches the operation philosophy of the thermal hybrid power plants, yet with the essential difference that the availability of the primary production units, namely the geothermal heat pumps and the cogeneration plant, cannot be considered as nonguaranteed. In this case, the backup units, namely the conventional heaters, operate whenever the heating demand exceeds the total available thermal power production from the cogeneration system and the heat pumps. This has to do with the installed nominal capacity of the corresponding equipment and not with the unpredictable fluctuations of the primary energy sources, as in the case of hybrid power plants.

The simulation of the system's operation has led to the following dimensioning, for each one of the five installed stations:

- nominal thermal power output of the cogeneration engine: 15 MW_{th}
- nominal thermal power output of the heat pumps: 30 MW_{th}
- nominal thermal power output of the natural gas heaters: 30 MW_{th}
- thermal storage capacity of the heat tanks: 80 MWh_{th}–100 MWh_{th}.

It must be noted that the system was optimized essentially on economic criteria. Namely, the principal target of the dimensioning optimization procedure was to keep the total setup cost under an upper acceptable limit, which, in combination with the electricity and heating demand, will enable the payback of the invested capital within a reasonable time period with a cost-effective policy pricing for the produced electricity and heating, to benefit the citizens of Milan.

In Figure 5.26, the annual energy flows and the efficiencies of the involved energy transformation processes are given in the format of a flow-diagram chart.

The total fossil fuel annual saving upon the implementation of all five cogeneration–geothermal power plants is estimated at 40,900 toe, a figure which corresponds to 35% of the current fossil fuel consumption for electricity and heating needs for the city of Milan. Additionally, the expectations

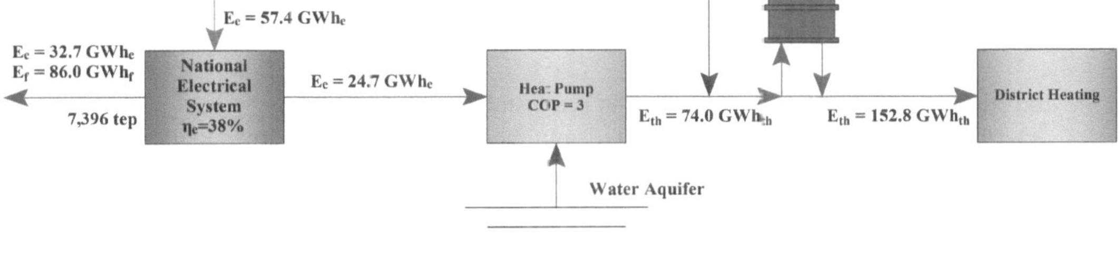

FIGURE 5.26 Annual energy flows and the efficiencies of the involved energy transformation processes in Milan [24].

for environmental benefits are also significant. Specifically, reduction percentages of 45%, 50%, and 99% for the annual CO_2, NO_x, and SO_2 emissions are estimated [24].

5.9 A CHP PLANT CASE STUDY

This section presents a case study of a small CHP plant, interconnected to the local electrical grid for the injection of the electricity produced, while the disposed heat will be exploited for the heating of water ponds utilized for the cultivation of algae. The CHP power plant will be based on a natural gas reciprocating Otto engine, with a nominal electrical power output of 475 kW_e. Heat is disposed by the CHP unit with the exhausted hot gases and with the cooling closed loop.

Given the technical specifications of a specific commercial model, the disposed heat from the above sources under nominal operation of the cogeneration unit is disposed with the following conditions:

- Heat disposed with the exhausted gases:
  - exhausted gases mass flow rate: $\dot{m}_g = 0.785$ kg/s
  - exhausted gases specific enthalpy: $h_g = 485$ kJ/kg

  Given the above data, the disposed heat with the exhausted hot gases from the cogeneration unit can be calculated as:

$$\dot{Q}_{dis\text{-}g} = \dot{m}_g \cdot h_g \Rightarrow \dot{Q}_{dis\text{-}g} = 0.785\text{kg/s} \cdot 485\text{kJ/kg} = 380.7 \text{ kW}$$

- Heat disposed with the cooling closed loop of the cogeneration unit:
  - water mass flow rate in the cooling loop: $\dot{m}_w = 2.390$ kg/s
  - water inlet temperature in the cooling loop: $T_{w\text{-}out} = 95°C$
  - water outlet temperature from the cooling loop: $T_{w\text{-}in} = 70°C$
  - water specific heat capacity: $c_p = 4.184$ kJ/kg·K.

  Given the above data, the disposed heat from the cogeneration unit with the cooling closed loop is calculated as:

$$\dot{Q}_{dis\text{-}w} = \dot{m}_w \cdot c_p \cdot (T_{w\text{-}out} - T_{w\text{-}in}) \Rightarrow \dot{Q}_{dis\text{-}w} = 2.390\text{kg/s} \cdot 4.184\text{kJ/kg} \cdot \text{K} \cdot (95 - 70) \cdot °\text{C} = 250.0 \text{ kW}.$$

Consequently, the total heat disposed by the cogeneration unit both with the hot gases and the cooling loop is calculated as:

$$\dot{Q}_{dis} = \dot{Q}_{dis\text{-}g} + \dot{Q}_{dis\text{-}w} = 380.7\text{kW} + 250.0\text{kW} = 630.7 \text{ kW}$$

The above operation conditions and the total calculated disposed heat are valid for nominal operation of the cogeneration unit, namely for electrical power output at 475 kW_e.

The scope of this case study is the proper dimensioning of the overall system, in order to maximize the electricity production from the cogeneration plant and the exploitation of the disposed thermal energy for the heating of the algae ponds. The steps towards the integration of this study are:

- introduction of the climate conditions in the installation area
- determination of the required algae ponds number (or, alternatively, covered area)
- calculation of the algae ponds heating loads
- configuration of the overall CHP plant layout
- computational simulation of the CHP plant annual operation
- annual energies calculation
- calculation of characteristic evaluation indices.

FIGURE 5.27 Location of the area of Pachino on the island of Sicily.

5.9.1 Location and Climate Conditions

This project is studied for the area of Pachino, in southern Sicily, Italy, an area with extensive agricultural activities and greenhouse crops. More than 8,000 ha of greenhouse covered area have been developed in the specific region. The area exhibits an average geographical latitude of 35° 45′, and an average longitude of 15° 5′. The location of the area of Pachino on the island of Sicily and the extensive greenhouse crops are clearly depicted in Figure 5.27.

In Figures 5.28 and 5.29 the annual fluctuation of the Global Horizontal Irradiance (GHI) and the ambient temperature are presented respectively for the under consideration area. The yearly GHI is calculated at 1,780 kWh/m^2. It is obvious that the selected area exhibits highly attractive climate conditions for the development of agricultural activities.

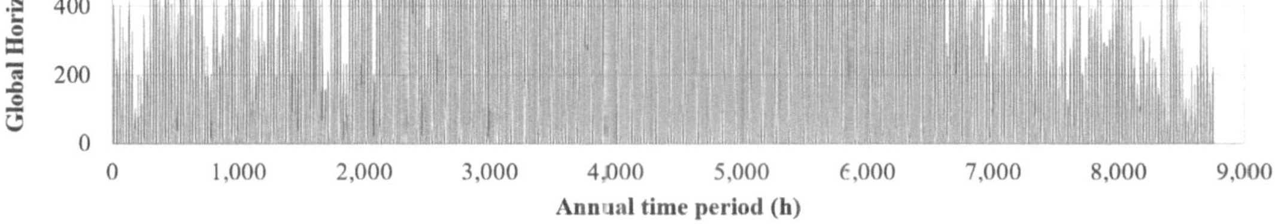

FIGURE 5.28 Annual time series of the available global horizontal irradiance for the under consideration area.

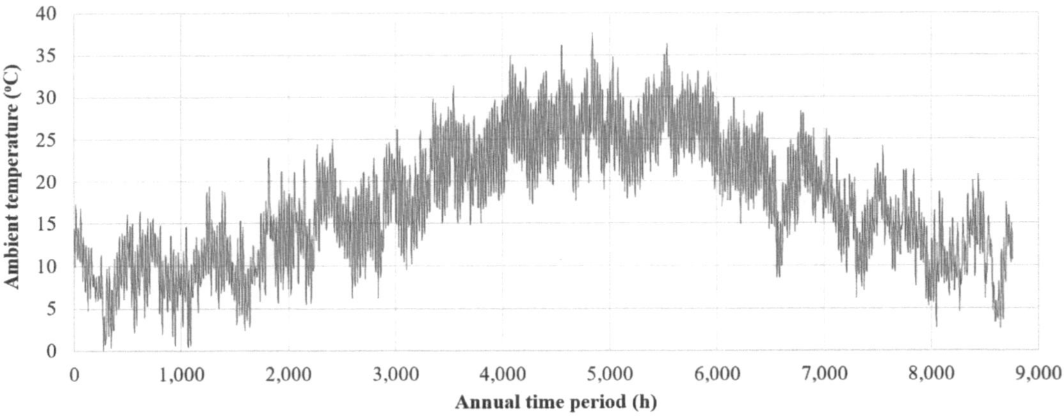

FIGURE 5.29 Annual fluctuation of the ambient temperature for the under consideration area.

5.9.2 Calculation of the Algae Ponds' Heating Loads

A greenhouse will be constructed for the housing of the algae ponds. A general aspect of the greenhouse construction is presented in Figure 5.30. The main constructive features will be a metallic bearing structure and double polyethylene sheets, with an average U-factor at the range of 4.0 W/m^{2}K [25]. For five algae ponds of 1,000 m^2 each, the total covered area will be 6,600 m^2, formulated by an orthogonal shape of 100 × 66 m, as shown in Figure 5.30. The interior volume of the greenhouse will be 24,663 m^3.

The main thermophysical properties of the greenhouse constructive elements, determining the heat transfer between the indoor space in the greenhouse and the ambient environment, are presented in Table 5.5.

The first step for the calculation of the algae ponds' heating loads is the calculation of the greenhouse indoor space temperature annual fluctuation. This is because the algae ponds' heating loads practically coincide with the heat losses from their upper, free surfaces to the indoor space environment. The indoor space temperature annual fluctuation is calculated through a typical heating loads calculation process of the greenhouse indoor space, accounting for all the involved heat transfer processes between the indoor space and the environment, the solar heat gains through the semitransparent greenhouse surfaces and the internal heat gains, mainly from the algae cultivation ponds themselves. The main prerequisite for the algae cultivation ponds is the maintenance of the water temperature constantly at the value of 35°C. This implies a constant heat flow from the ponds' upper surfaces to the indoor space, constituting the main internal heat gain, which,

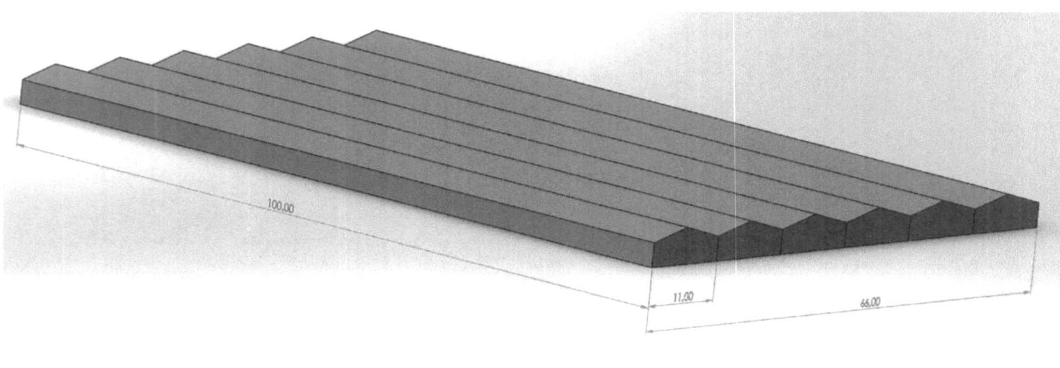

FIGURE 5.30 Three-dimensional aspect of the proposed greenhouse construction.

TABLE 5.5
Constructive Features and Thermophysical Properties of the Algae Ponds' Greenhouse

Constructive Feature	Material	Percentage Over the Overall Greenhouse Surface (%)	U-factor ($W/m^2{\cdot}K$)	Solar Gain Factor
Bearing structure	Steel beams	10	5.7	0.0
Main semitransparent surface	Double polyethylene sheet	90	4.0	0.8

obviously, cannot be neglected in the indoor space heating loads' calculation process, executed on the basis of the essential relevant theory and the data introduced by ASHRAE [26, 27]. By applying the process, the annual fluctuation of the indoor space temperature is calculated. The corresponding time series is presented in Figure 5.31. An important regulation is that in case the indoor space temperature exceeds 38°C, the side sheets of the greenhouse are opened, so the indoor space temperature becomes equal to the ambient one. This is to avoid overheating of the water in the algae ponds and to enable secure human presence inside the greenhouse.

Once the greenhouse indoor space temperature annual fluctuation has been calculated, the calculation of the algae ponds' annual heating loads is a simple and straightforward process. Specifically, the heat losses transfer rate $\dot{Q}_{sp}$ from the algae ponds' upper free surface to the greenhouse indoor space is given by the relationship:

$$\dot{Q}_{sp} = A_{sp} \cdot U \cdot (T_w - T_{in}) \tag{5.39}$$

where:

- A_{sp}: the total algae ponds' free surface (in m^2)
- U: the U-factor determining the heat transfer from the algae ponds to the greenhouse indoor space (in $W/m^2{\cdot}K$)
- T_w: the water temperature in the algae ponds (in °C)
- T_{in}: the greenhouse indoor space temperature, as calculated previously (in °C).

The thermal transfer factor U refers to the direct heat transfer from the algae ponds' free surface to the indoor space and is given by the relationship:

$$U = h_{rw} + h_c \tag{5.40}$$

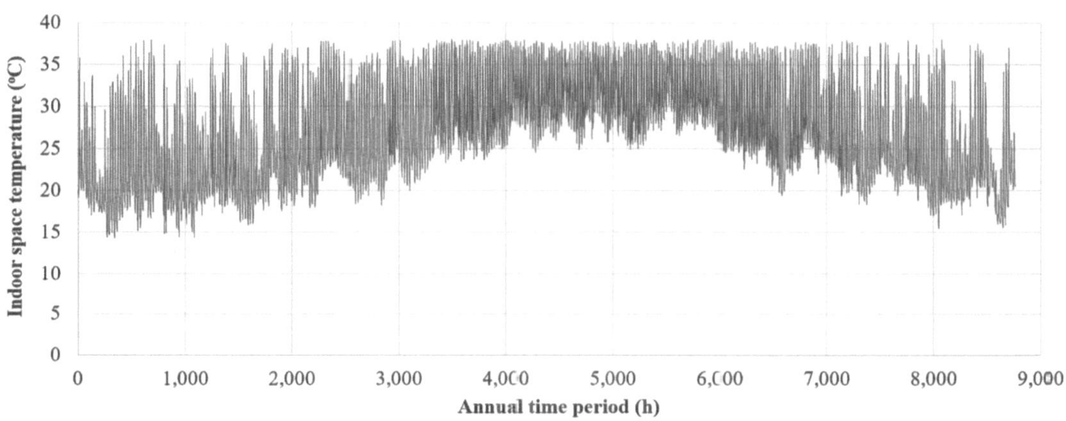

FIGURE 5.31 Annual fluctuation of the greenhouse indoor space temperature.

TABLE 5.6
Values of the Involved Magnitudes in the Algae Ponds' Heating Loads Calculation

Emissivity of water ε_w	0.957
Convective heat transfer factor h_c for indoor air flow above horizontal surfaces (W/m^2·K)	10
Total surface of algae ponds A_{sp} (m^2)	5,000
Desirable water temperature in the algae ponds T_w (°C)	35
Stefan-Boltzmann constant W/m^2·K^4	$5.67 \cdot 10^{-8}$

where h_c is the convective heat transfer factor for indoor air flow above horizontal surfaces and h_{rw} the heat transfer factor with radiation, which is given by the following relationship:

$$h_{rw} = 4 \cdot \varepsilon_w \cdot \sigma \cdot \left(\frac{T_w + T_{in}}{2} \right)^3 \tag{5.41}$$

where ε_w is the emissivity of water and σ the Stefan-Boltzmann constant. The ratio in the parentheses stands for the average temperature between the water in the algae ponds and the indoor space. The introduced values for the above magnitudes are summarized in Table 5.6. The total algae ponds' surface is formulated by a number of five ponds with 1,000 m^2 upper surface each. The total number of the algae ponds practically constitutes a result of the iterative simulation process, described below.

Following the above described process and introducing the data presented in Table 5.6, the annual fluctuation of the algae ponds heating loads is calculated and presented in Figure 5.32.

5.9.3 The CHP Plant Layout

The overall CHP plant layout is presented in Figure 5.33. It consists of the following fundamental components:

- the CHP cogeneration unit, an Otto gas engine
- a thermal storage tank, for heat storage whenever the disposed thermal power from the CHP unit exceeds the algae ponds' heating loads
- a backup, external heat source, which could be a conventional heater (preferably a biomass heater) or a heat pump, for the algae ponds heating in case of inadequate thermal power supply both from the CHP engine and the thermal storage tank

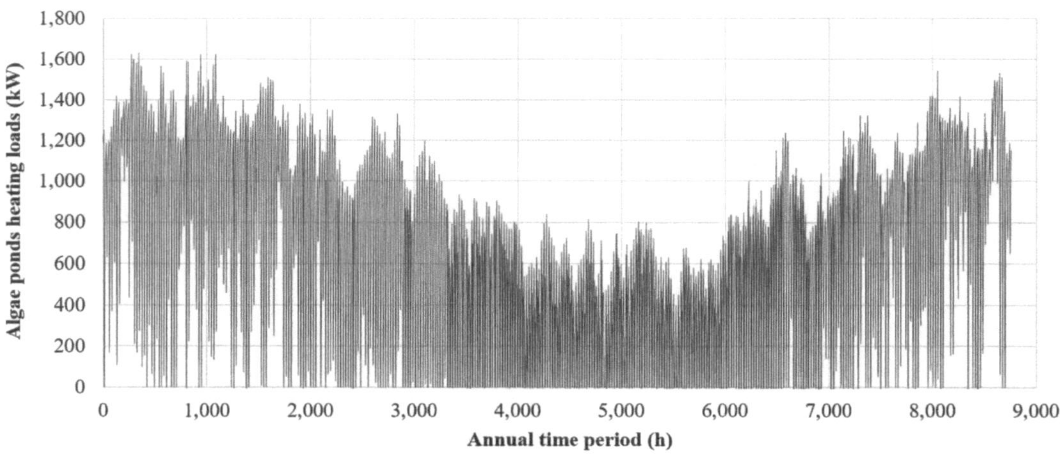

FIGURE 5.32 Annual fluctuation of the algae ponds' heating loads.

- a cell-tube heat exchanger for the exploitation of the disposed heat with the exhausted hot gases
- a first tube-type diathermic oil–water heat exchanger for the heat transfer from the secondary loop of the cell-tube heat exchanger to the algae ponds
- a second tube-type water–water heat exchanger for the heat transfer from the engine's cooling system secondary loop to the algae ponds
- a third tube-type water–water heat exchanger for the heat transfer from the heat storage tank to the algae ponds
- the hydraulic network required for the heat transfer from and to the involved system's components, consisting of pipelines, mechanic valves, pressure vessels, circulators, measuring instruments, etc.
- a central control unit, which will supervise and determine the operation of the system, based on the predefined operation algorithm, analyzed in the next section.

The dimensioning of the CHP plant will be executed on the basis of the following objectives:

- the constant nominal electricity power production from the CHP unit during the whole annual period, aiming at the maximization of the annual income from selling the produced electricity
- the maintenance of the water temperature in the algae ponds constantly at 35°C
- the achievement of the following values for the corresponding characteristic indices:

$$C = \frac{E_C}{H_{CHP}} \geq 75\%$$

where E_C the annual electricity production and H_{CHP} the annual thermal energy production absorbed for the algae ponds' heating needs. The ratio C is the power-to-heat ratio.

$$\eta = \eta_e + \eta_h = \frac{E_C}{F_C} + \frac{H_{CHP}}{F_C} \geq 75\% \quad (5.42)$$

where η the total cogeneration efficiency, η_e and η_h the electrical and the thermal efficiency of the cogeneration process and F_C the primary energy consumption during the cogeneration process.

$$PESR = \frac{F_E + F_H - F_C}{F_E + F_H} > 0 \quad (5.43)$$

where PESR is the primary energy saving ratio, F_E the primary energy consumption for the production of the same amount E_C of electricity, through a pure electricity production process (without concurrent thermal energy exploitation through a cogeneration process), and F_H the primary energy consumption for the production of the same amount H_{CHP} of thermal energy with a conventional heating production method (e.g., a conventional heater), namely not within a cogeneration process.

5.9.4 The CHP Plant Operation Algorithm

The operation algorithm of the examined CHP plant is described below. First, the following magnitudes are introduced:

- the total thermal power $\dot{Q}_{dis}$ disposed by the cogeneration unit, calculated equal to 630.7 kW for nominal operation
- the thermal power demand $\dot{Q}_D$ for the heating of the algae ponds, presented in Figure 5.32
- the involved heat exchanges efficiencies for the cell-tube and the tube-type exchangers, given equal to $\eta_{ct} = 92.0\%$ and $\eta_{tub} = 92.2\%$ respectively by the manufacturers.

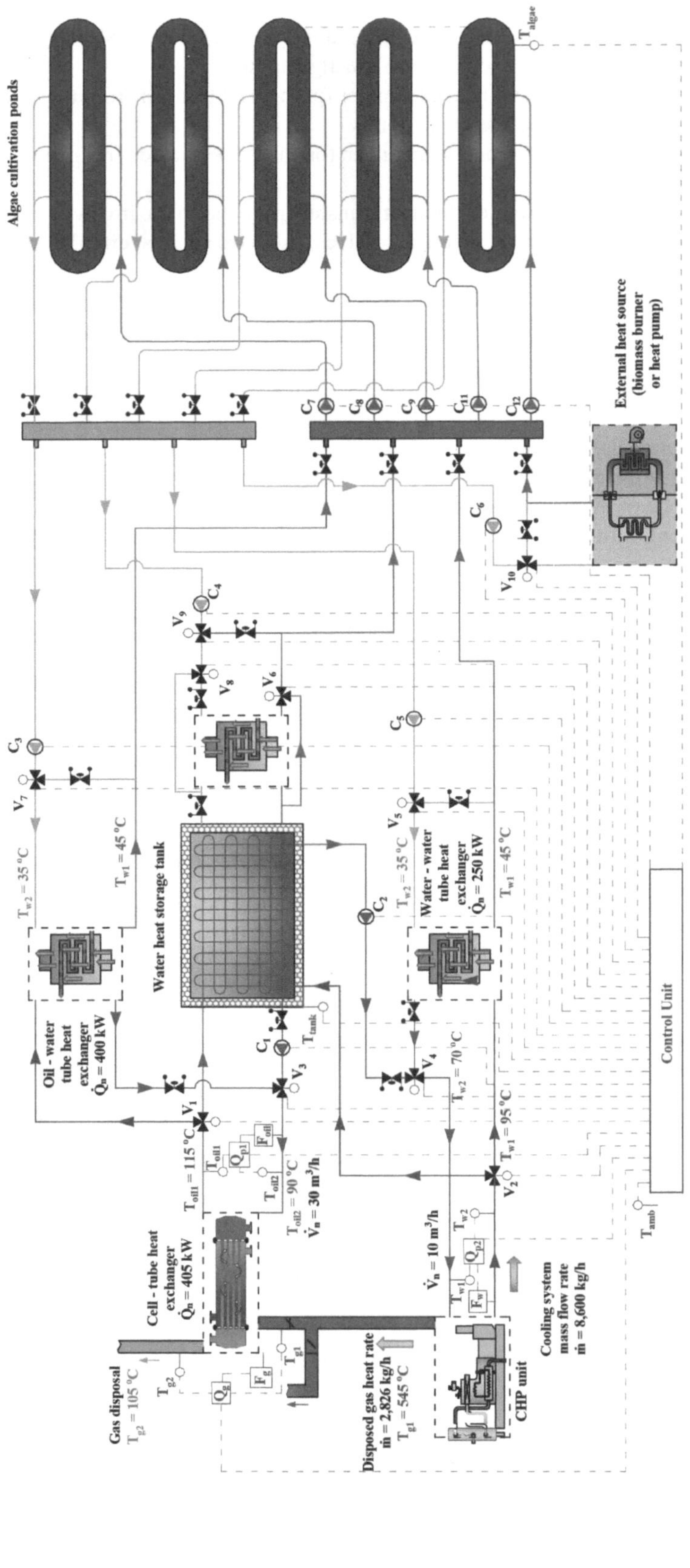

FIGURE 5.33 Layout of the proposed CHP plant.

Once the above involved magnitudes have been introduced, the following steps are executed for every calculation time step i:

1. Available disposed thermal power higher than the thermal power demand, accounting also for the heat transfer efficiencies

$$\dot{Q}_{dis} \geq \dot{Q}_D/\eta_p :$$

where:

$$\eta_p = \eta_{ct} \cdot \eta_{tub}$$

In this case, the total thermal power demand is totally covered by the disposed thermal power from the cogeneration unit. The absorbed thermal power for the algae ponds' heating will be:

$$\dot{Q}_{abs} = \dot{Q}_D/\eta_p$$

The thermal power production from the backup heat source will be null:

$$\dot{Q}_{bup} = 0$$

A thermal power surplus will be available, equal to:

$$\dot{Q}_{sur} = \dot{Q}_{dis} - \dot{Q}_{abs}$$

This thermal power surplus can be stored in the thermal storage tank, if it does not impose an increase of the water temperature in the tank higher than a maximum introduced value T_{max}, set equal to 90°C. In general, the water temperature in the thermal storage tank at the end of the current step can be calculated with the following relationship:

$$\left(\dot{Q}_{sur} \cdot \eta_{ct} - \dot{Q}_L\right) \cdot t = m \cdot c_p \cdot \left[T(i) - T(i-1)\right] \Leftrightarrow T(i) = T(i-1) + \frac{\left(\dot{Q}_{sur} \cdot \eta_{ct} - \dot{Q}_L\right) \cdot t}{m \cdot c_p} \quad (5.44)$$

where:

$\dot{Q}_L$: the heat losses from the thermal tank to the indoor ambient space
$T(i-1)$: the water temperature in the thermal tank from the previous calculation step $i-1$
$T(i)$: the water temperature in the thermal storage tank at the end of the current calculation step i, after the thermal power storage $\dot{Q}_{sur}$ and the heat losses $\dot{Q}_L$
m: the total water mass in the thermal storage tank
c_p: the specific heat capacity of water, equal to 4.184 kJ/kg·K
t: the calculation time step duration, equal to 3,600 s in case of hourly step.

Having calculated the temperature T(i), the following sub-cases are checked:

i. If $T(i) \leq T_{max}$, then the thermal power surplus can be stored in the thermal storage tank, hence the thermal power storage will be:

$$\dot{Q}_{st} = \dot{Q}_{sur} \cdot \eta_{ct}$$

and the water temperature T(i) in the thermal storage tank at the end of the current calculation time step will be given by Equation 5.44.

ii. If $T(i) > T_{max}$, then the thermal power surplus cannot be stored in the thermal storage tank, hence the thermal storage power will be null:

$$\dot{Q}_{st} = 0$$

and the water temperature T(i) in the thermal storage tank at the end of the current calculation time step will be given by the following relationship:

$$T(i) = T(i-1) - \frac{\dot{Q}_L \cdot t}{m \cdot c_p} \tag{5.45}$$

namely, it is practically affected solely by the heat losses from the tank to the ambient, given the absence of any other thermal flow from or to the tank.

2. The thermal power disposed by the cogeneration unit is lower than the thermal power demand for the algae ponds' heating, accounting also for the involved heat exchangers efficiencies:

$$\dot{Q}_{dis} < \dot{Q}_D / \eta_p :$$

In this case, obviously all the disposed heat by the cogeneration unit is exploited for the heating of the algae ponds. The finally absorbed thermal power for the algae ponds' heating will be:

$$\dot{Q}_{abs} = \dot{Q}_{dis} \cdot \eta_p$$

Obviously, there is no chance for thermal power storage:

$$\dot{Q}_{st} = 0$$

On the contrary, a thermal power production $\dot{Q}_{tank}$ will be required by the thermal storage tank, equal to:

$$\dot{Q}_{tank} = \frac{\dot{Q}_D - \dot{Q}_{abs}}{\eta_{tub}}$$

This thermal power production from the thermal storage tank will be possible if the water temperature in the thermal storage tank does not fall below a minimum acceptable value T_{min}, set equal to 45°C, given the operation parameters of the involved heat exchangers for the thermal power final transfer to the algae ponds (see Figure 5.33). In any case, the new temperature in the thermal storage tank is given by the following relationship:

$$\left(\dot{Q}_{tank} + \dot{Q}_L\right) \cdot t = m \cdot c_p \cdot \left[T(i-1) - T(i)\right] \Leftrightarrow T(i) = T(i-1) - \frac{\left(\dot{Q}_{tank} + \dot{Q}_L\right) \cdot t}{m \cdot c_p} \tag{5.46}$$

Once the temperature T(i) has been calculated, the following sub-cases are investigated:

i. If $T(i) \geq T_{min}$, the required thermal power can be supplied by the thermal storage tank. The thermal power supply from the thermal storage tank will be:

$$\dot{Q}_{tank} = \frac{\dot{Q}_D - \dot{Q}_{abs}}{\eta_{tub}}$$

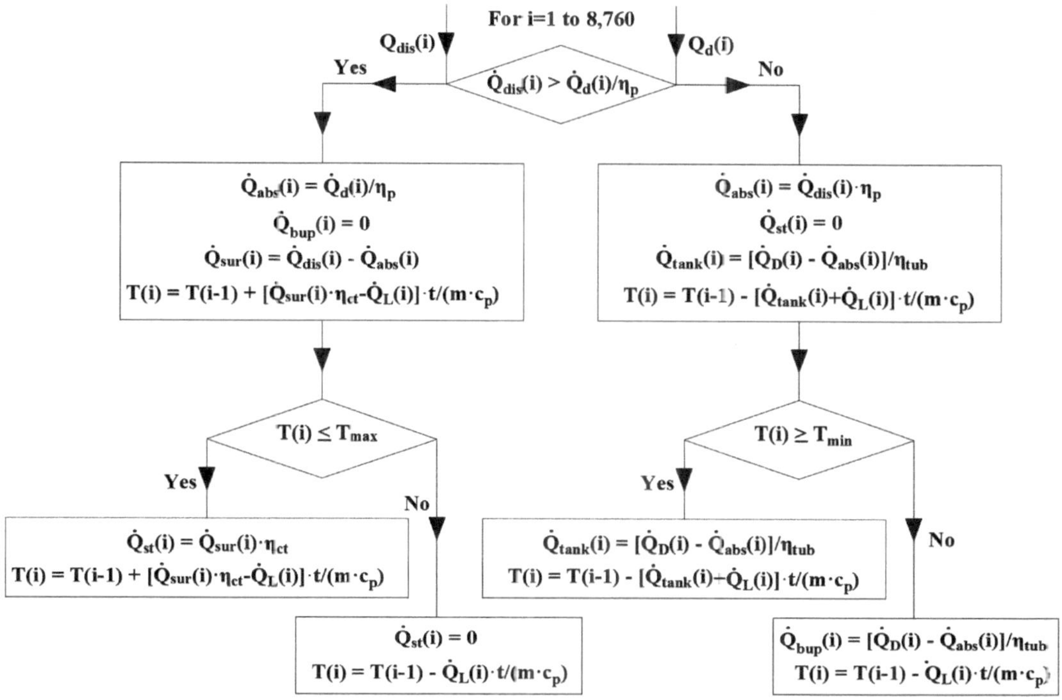

FIGURE 5.34 Operation algorithm of the under consideration CHP plant.

and the water temperature T(i) in the thermal storage tank at the end of the current calculation time step will be given by Equation 5.46.

The thermal power production from the backup heat source will be null:

$$\dot{Q}_{bup} = 0$$

ii. If $T(i) < T_{min}$, then the required thermal power cannot be covered by the thermal storage tank. Hence, the only available option is to supply this remaining power with the external heat source:

$$\dot{Q}_{bup} = \frac{\dot{Q}_D - \dot{Q}_{abs}}{\eta_{tub}}$$

The water temperature in the thermal storage tank is again calculated with Equation 5.45; namely, it is practically affected solely by the heat losses from the tank to the ambient, given the absence of any other thermal flow from or to the tank.

The above described operation algorithm is graphically presented in Figure 5.34.

5.9.5 Simulation Results

The above operation algorithm is computationally simulated, through the development of a relevant application, and executed iteratively for different sizes of the thermal storage tank capacity and the quantity of algae ponds. Eventually, it is calculated that at least five (5) algae ponds are necessary to absorb the required disposed thermal energy in order to fulfill the criteria related to the evaluation

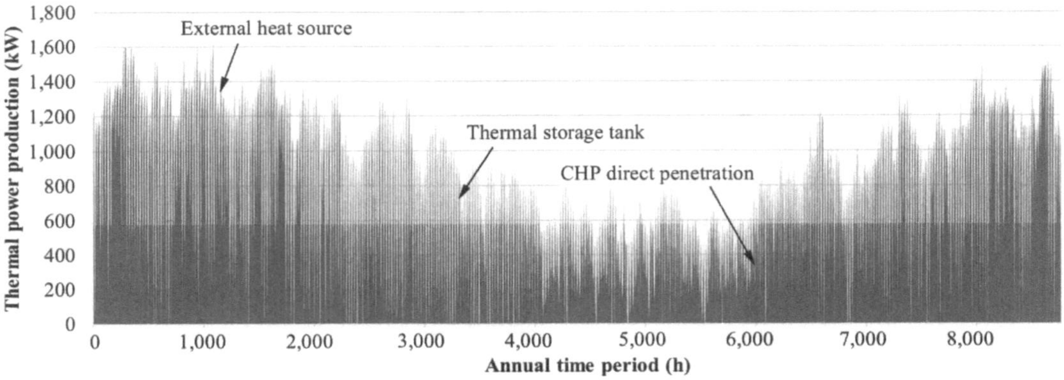

FIGURE 5.35 Annual thermal power production synthesis of the CHP plant.

indices C, η, and PESR presented previously. Additionally, a thermal storage tank with 100 m^3 storage capacity will be adequate to support the CHP plant.

With the concluded dimensioning, the annual thermal power production synthesis graph is presented in Figure 5.35, while in Figures 5.36 and 5.37, thermal power production synthesis graphs are presented for the periods from January 20 to January 30 and from March 1 to March 10 respectively. The alternative thermal power production sources, appearing in these graphs, are:

- direct heating load coverage with the disposed heat by the cogeneration unit
- heating load coverage from the thermal storage tank
- heating load coverage from the external, backup heat source.

In Figure 5.36 it is seen that even in winter, there are periods during the daytime with considerably low algae pond heating loads. During these periods, the excess thermal power disposal from the cogeneration unit is stored in the thermal storage tank, enabling the contribution of the latter to the thermal power demand coverage when the heating loads become again higher than the disposed heat. From March 1 to March 10 (Figure 5.37), the thermal storage tank is adequate to fully undertake the coverage of the algae ponds' heating loads, after full direct penetration of the disposed heat from the cogeneration unit. The corresponding thermal power storage graph from the same time

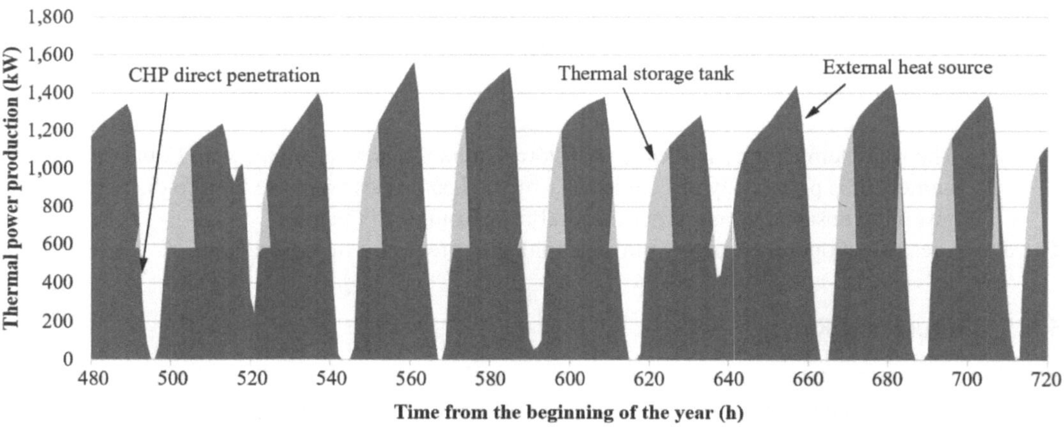

FIGURE 5.36 Thermal power production synthesis from January 20 to January 30.

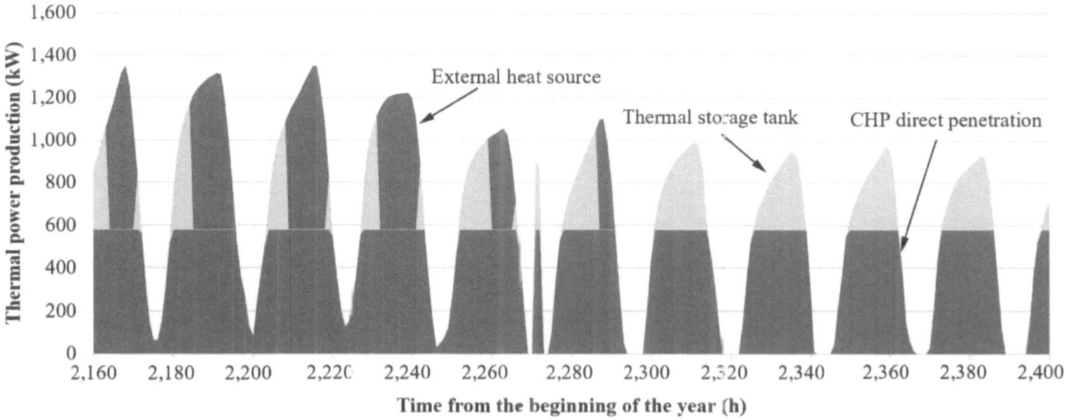

FIGURE 5.37 Thermal power production synthesis from March 1 to March 10.

period (March 1–10) is presented in Figure 5.38. These remarks highlight the necessity and the feasibility of the thermal storage tank in this specific CHP project.

Finally, in Figure 5.39, the annual fluctuation of the water temperature in the thermal storage tank is presented.

The case study is integrated with the presentation of the annual electricity and thermal energy production and storage, as well as the calculated evaluation indices. All of these values are summarized in Table 5.7. All the presented magnitudes with No. 1–7 have been calculated through the simulation process. The annual thermal energy demand has been calculated by integrating the heating load time series, presented in Figure 5.32. The annual electricity production is calculated by assuming constant electrical power output at 475 kW_e for the whole annual period. The annual disposed thermal energy from the cogeneration unit is calculated by assuming constant thermal power disposed at 630.7 kW_{th} for the whole annual period. The electrical efficiency η_e is assumed equal to 35% at nominal operation, based on relevant data provided by the manufacturers. For the primary energy consumption calculation F_H for pure thermal energy production H_{CHP} (not within a cogeneration process) a total thermal efficiency of 85% for a conventional heater was assumed.

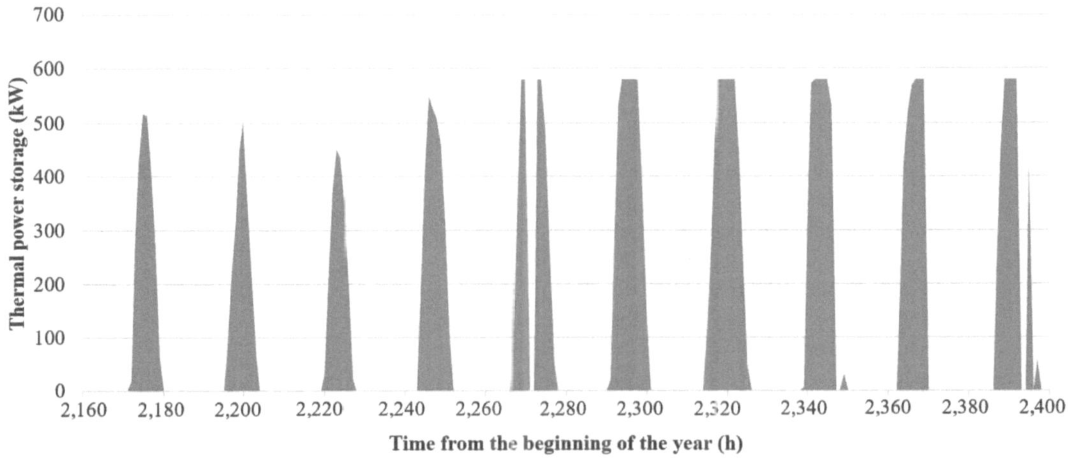

FIGURE 5.38 Thermal power storage fluctuation from March 1 to March 10.

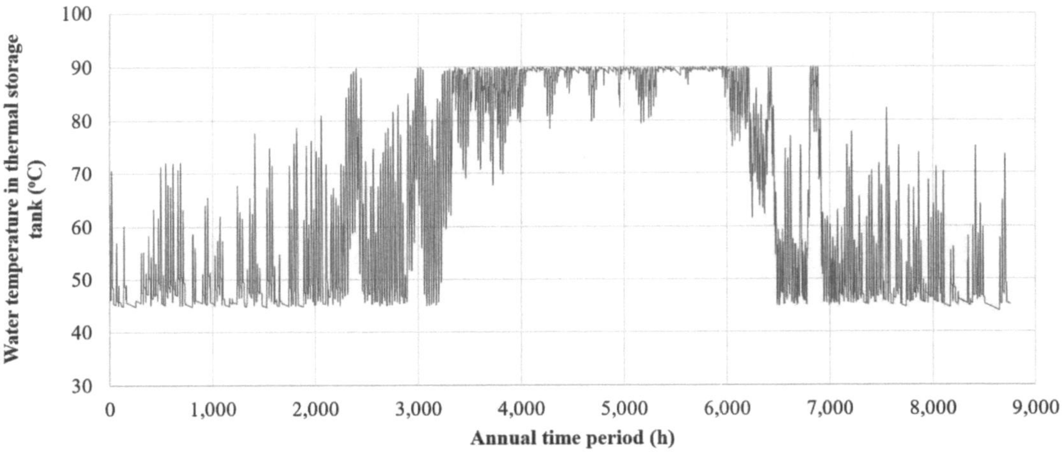

FIGURE 5.39 Annual fluctuation of the water temperature in the thermal storage tank.

TABLE 5.7
Annual Electricity and Thermal Energy Production and Storage—Evaluation Indices

No.	Description	Value
	Annual Thermal Energy Production and Storage	
1	Heating load coverage directly from the cogeneration unit $E_{CHP\text{-}th}$ (kWh)	3,921,658
2	Heating load coverage from the thermal storage tank $E_{tank\text{-}th}$ (kWh)	512,018
3	Heating load coverage from the external heat source $E_{ext\text{-}th}$ (kWh)	1,534,367
4	Thermal energy storage E_{stor} (kWh)	568,914
5	Thermal energy supply from the thermal storage tank $E_{stor\text{-}dis}$ (kWh)	556,542
6	Thermal energy surplus $E_{rej\text{-}th}$ (kWh)	592,608
7	Heat losses from the thermal storage tank $E_{loss\text{-}th}$ (kWh)	14,158
	Results Summary	
8	Annual thermal energy demand E_D (kWh)	5,968,043
9	Annual thermal energy totally produced from the CHP plant E_{prod} (kWh)	5,968,042
10	Annual electricity production from the CHP unit E_C (kWh)	4,161,000
11	Annual thermal energy disposed from the cogeneration unit $E_{CHP\text{-}dis}$ (kWh)	5,525,195
12	Annual thermal energy exploited from the cogeneration unit $E_{CHP\text{-}abs}$ (kWh)	4,881,056
	Annual Primary Energy Consumption	
13	Primary energy consumption from the cogeneration unit F_C (kWh)	11,888,571
14	Primary energy consumption F_E for the pure electricity production E_C (kWh)	11,096,000
15	Primary energy consumption F_H for pure thermal energy production H_{CHP} (kWh)	5,742,419
	Evaluation Indices	
16	Annual cogeneration power-to-heat ratio C (%)	85.2
17	Annual heating load coverage from the disposed thermal energy r_d (%)	74.3
18	Annual exploitation of the disposed thermal energy r_{abs} (%)	88.3
19	Annual average electricity production efficiency η_e (%)	35.0
20	Annual average thermal energy production efficiency η_h (%)	41.1
21	Total annual average cogeneration efficiency η (%)	76.1
22	Primary energy saving ratio (PESR)	0.3

Additionally, the other magnitudes in the table are calculated according to the following relationships:

1. *Annual thermal energy totally produced from the CHP plant E_{prod}:*

$$E_{prod} = E_{CHP\text{-}th} + E_{tank\text{-}th} + E_{ext\text{-}th}$$

2. *Annual thermal energy exploited from the cogeneration unit $E_{CHP\text{-}abs}$:*

$$E_{CHP\text{-}abs} = \frac{E_{CHP\text{-}th}}{\eta_{ct} \cdot \eta_{tub}} + \frac{E_{stor}}{\eta_{tub}} \tag{5.47}$$

3. *Primary energy consumption from the cogeneration unit F_C:*

$$F_C = \frac{E_C}{\eta_e}$$

4. *Primary energy consumption F_E for the pure electricity production E_C:*

$$F_E = \frac{E_C}{\eta_{el}}$$

where η_{el} is the electrical efficiency for pure electricity production, assumed equal to 37.5%.

5. *Primary energy consumption F_H for pure thermal energy production H_{CHP}:*

$$F_H = \frac{E_{CHP\text{-}abs}}{\eta_{th}} \tag{5.48}$$

where $\eta_{th} = 85\%$ for a conventional heater, as explained above.

6. *Power-to-heat ratio C:*

$$C = \frac{E_C}{H_{CHP}}$$

7. *Annual heating load coverage from the disposed thermal energy r_d:*

$$r_d = \frac{E_{CHP\text{-}th} + E_{tank\text{-}th}}{E_d} \tag{5.49}$$

8. *Annual exploitation of the disposed thermal energy r_{abs}:*

$$r_{abs} = \frac{E_{CHP\text{-}abs}}{E_{CHP\text{-}dis}} \tag{5.50}$$

9. *Primary energy saving ratio (PESR):*

$$PESR = \frac{F_E + F_H - F_C}{F_E + F_H}$$

It is seen that with the concluded layout, operation algorithm, and dimensioning, all the initially introduced parameters are fulfilled. The introduced CHP plant, with the proposed layout and operation algorithm, is properly dimensioned in order to fully undertake the algae ponds' heating loads and maintain the water temperature in the ponds constantly at the desirable value. On the other hand, the number of the algae ponds is adequate to absorb the required amount of the disposed heat from the cogeneration unit and ensure the achievement of acceptable values for the introduced evaluation indices.

5.10 SOLAR COOLING SYSTEMS

5.10.1 Trigeneration and Solar Cooling

High efficiencies are obtained in a CHP system practically only during the seasons of simultaneous electricity and thermal energy demand. For example, if the final thermal energy is exploited for the heating of buildings, the overall efficiency of the cogeneration process is maximized only during the winter period. Substantially, when there is no thermal energy demand, the system stops operating as a CHP system, because its operation is restricted purely for electricity production. In case of mild climate conditions during the winter period or relatively short winter duration (e.g., in warm climates), the economic feasibility of CHP systems can be seriously affected, due to reduced final thermal energy needs. In such unfavorable cases, focusing on the maximization of the operation period and the efficiency of the cogeneration system, the combined heating–power production in the winter period is switched to combined cooling–power production in the summer period, through which the produced thermal power during summer is exploited for the production of cooling with so-called absorption chillers or the open cycle/desiccant cooling systems, formulating, in this way, a trigeneration system.

Trigeneration is the production of electricity, heating, and cooling from the same primary energy source. Trigeneration has been widely introduced in the building sector, in hospitals, hotels, blocks of offices, commercial centers, and in heating–cooling district systems. Generally, as a rule of thumb, we can say that trigeneration is introduced when the electricity and heating and/or cooling period exceeds 4,500–5,000 hours annually. The trigeneration units are mainly based on reciprocating internal combustion engines (diesel generators) or micro-turbines, combined with an absorption chiller or an open cycle system for cooling production. A typical layout of a trigeneration system is presented in Figure 5.40.

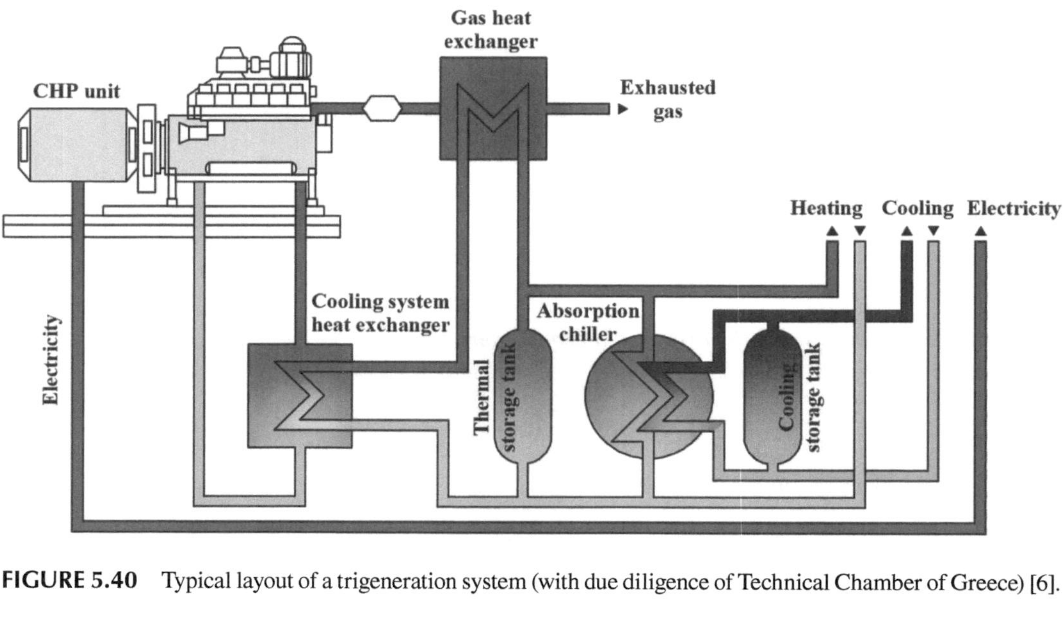

FIGURE 5.40 Typical layout of a trigeneration system (with due diligence of Technical Chamber of Greece) [6].

Particularly popular during the last decades is the production of the required thermal energy for the operation of an absorption chiller or an open cycle system from solar collectors. In this case, the term *solar cooling* has been introduced for the description of cooling production with the exploitation of solar radiation.

As shown in the next section, absorption chillers are thermally driven chillers that produce cooling in the form of chilled water. On the other hand, open cycle or desiccant cooling systems exploit the provided heat to produce directly conditioned air, which can then be supplied in a ventilation system, such as an air-duct network. Desiccant cooling systems are appropriate for applications in temperate climates. In hot climates, desiccant cooling systems are not adequate to fully undertake the cooling loads of buildings and infrastructures. For this reason, the dominant thermal-driven chillers' technology is the absorption chillers. In the following sections, a detailed presentation of the absorption cooling concept and the combined operation of absorption chillers with solar collectors is provided. The presentation of desiccant cooling systems is beyond the scope of this chapter.

5.10.2 Fundamental Principles of Absorption Cooling

Absorption cooling follows the essential evaporation–condensation processes of the basic cooling cycle. Similarly to a conventional cooling device, absorption chillers are also equipped with an evaporator and a condenser, where the corresponding processes are executed. Yet, instead of mechanical power provided by a compressor, absorption chillers exploit thermal energy, produced either directly from the combustion of a conventional fuel or solar collectors, or indirectly, with the use of steam, warm water, or the thermal energy surplus/recovery from another thermal process. Practically, absorption chillers are thermally driven chillers that produce chilled water that can be utilized in any type of cooling equipment.

Commercially available absorption chillers are supplied with steam, warm water, or the thermal energy contained in the exhausted gases from an internal combustion engine, which may by all means be produced by cogeneration systems. An absorption chiller consists of an evaporator, a condenser, an absorber, a generator, and a pump for the working solution. In an absorption chiller, the compression of the vaporized refrigerant is implemented with the combined operation of the absorber, the pump, and the generator.

The essential operation principle of an absorption chiller is presented in Figure 5.41. According to the well-known procedure from the basic cooling cycle, thermal energy from the conditioned space is absorbed in the evaporator, producing the required cooling and forcing the refrigerant to be vaporized. The refrigerant's vapor leaves the evaporator and enters the absorber, where it is sprayed with a strong sorbent solution that absorbs the refrigerant. The sorbent-refrigerant solution, called a weak solution, is pumped through a preheater (a heat exchanger) on to the generator, where it is provided with heat produced by an external source, causing the solution to boil and leading to the separation of the refrigerant from the absorbent in the form of high-pressure vapor. It is so conceivable that the combined operation of the absorber, the weak solution pump, and the generator leads to the production of high-pressure refrigerant vapor, similarly to the compression of the refrigerant vapor in the compressor of a centrifugal chiller. The refrigerant vapor is then led to the separator where it is completely separated from any possible sorbent residues. The remaining condensed solution of the sorbent in the separator is channeled through the preheater, where it is cooled by the weak solution. In this way, it is ready for the absorption of more refrigerant and the repetition of the process. At the same time, the regenerated refrigerant, called a strong solution, is led to the condenser, for the reduction of its temperature, causing it to condense and collect in the bottom of the condenser. After the condenser, there is an expansion device, aiming to the pressure drop of the refrigerant. After these two stages, the refrigerant has been restored back to liquid phase with

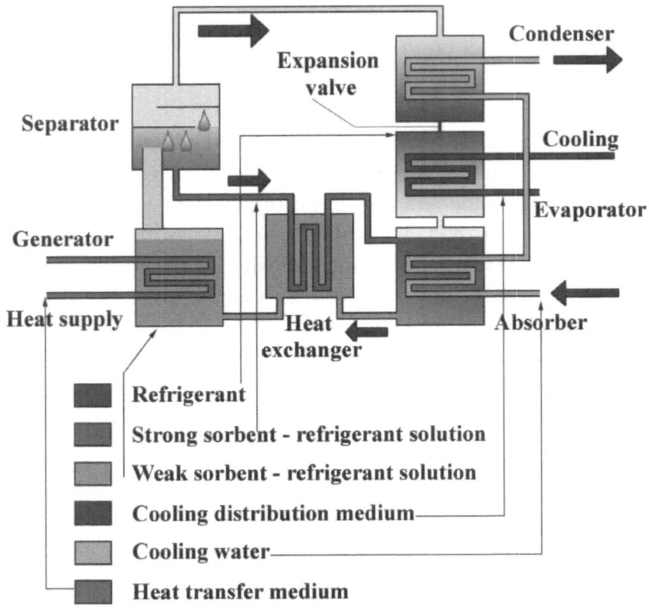

FIGURE 5.41 Operation principle of an absorption chiller (with due diligence of Technical Chamber of Greece) [28].

low pressure and low temperature, ready to be supplied again to the evaporator for its evaporation again and the repetition of the cooling cycle.

The thermal energy flows involved in the basic absorption cycle presented in Figure 5.41 are [29]:

a. thermal energy absorption in the evaporator from the conditioned space and cooling production, in low temperature
b. heat disposal in the absorber from the strong sorbent, implemented in medium temperature
c. heat supply in the generator for the weak solution, executed in high temperature level
d. heat disposal in the condenser from the refrigerant towards the environment, executed in medium temperature
e. mechanical power in the pump for the circulation of the sorbent-refrigerant solution.

These five distinct power flows are graphically depicted in the ideal absorption cycle presented in Figure 5.42 where the components that substitute the mechanical compressor of a conventional cooling cycle are also highlighted, called a *thermal compressor*. It should be underlined that the mechanical power consumed in the pump practically constitutes an auxiliary power supply, required for the circulation of the weak solution, which, yet, is not involved in the thermodynamic process of the absorption cycle.

In an absorption cycle, the sorbent and the refrigerant formulate the working fluid pairs. Several different working fluid pairs have been tested. The working fluid pairs that have prevailed are:

- solution of lithium bromide (Li-Br) as the sorbent, and water as the refrigerant
- solution of ammonia (NH_3) as the sorbent, and water as the refrigerant.

For single-stage absorption chillers operating with lithium bromide–water, the thermal power must be supplied at temperatures 70°C–90°C. For single-stage absorption chillers operating with

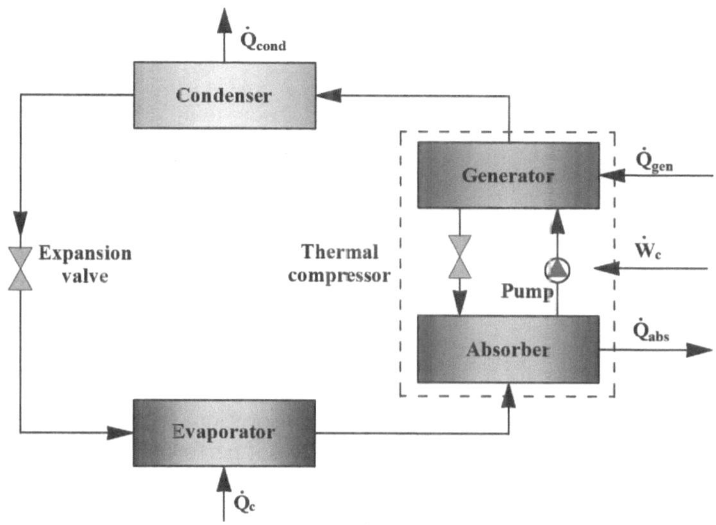

FIGURE 5.42 Basic absorption cycle layout and energy flows (with due diligence of CIBSE Journal) [29].

ammonia–water, the thermal power must be supplied at temperatures 100°C – 120°C. The lithium bromide–water solution is employed for indoor space cooling applications, with the temperatures required above 0°C. The ammonia–water solution is mainly used for freezers, with temperatures required below 0°C. The pressure of an absorption chiller with ammonia–water solution is usually higher than the atmospheric pressure, while the operation of lithium bromide–water chillers is executed under near vacuum conditions.

The two-stage absorption chillers employ two groups of generator–absorber integrated in series, aiming to utilize the provided thermal power twice. Thermal power is usually provided in temperatures around 150°C in the first generator and the disposed heat from the corresponding condenser is employed for the required thermal power supply for the second generator, at temperatures close to 100°C, as in single-stage absorption chillers. Practically, by dividing the generator into high- and low-temperature generators, higher heat efficiency is achieved in the condenser. Two-stage absorption units also use higher pressure steam. The efficiency of a two-stage absorption chiller is about 40%–50% higher than a single-stage unit and comparable to that of a centrifugal chiller.

The advantages of absorption chillers versus the conventional centrifugal chillers are:

- remarkably reduced power consumption
- very few moving parts, leading to fewer damages, reduced maintenance requirements, higher lifetime period, and higher reliability
- low noise and vibration emission levels
- environmentally friendly refrigerants, with low pollutants and harmless for the ozone layer compounds emissions.

The disadvantages of absorption chillers versus the conventional centrifugal chillers are:

- massive constructions with high installed thermal power
- relatively high setup cost
- water consumption in cooling towers
- low coefficient of performance (COP).

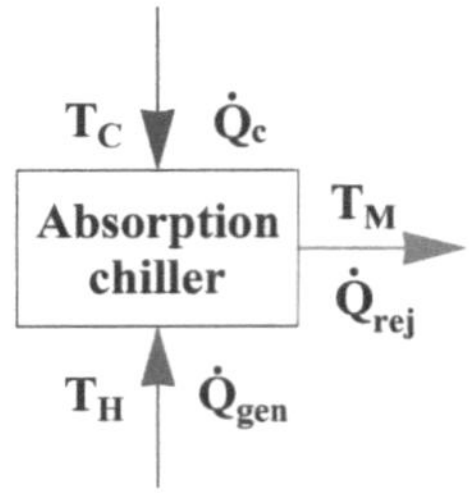

FIGURE 5.43 Thermal energy flows in an absorption chiller.

The thermal power flows balance, directly retrieved from Figure 5.43 can be written as:

$$\dot{Q}_{rej} = \dot{Q}_{abs} + \dot{Q}_{cond} = \dot{Q}_c + \dot{Q}_{gen} \tag{5.51}$$

where $\dot{Q}_{rej}$ is the total rejected thermal power from the absorption chiller (Figure 5.43). Similarly to the conventional cooling cycle, the theoretical coefficient of performance of an absorption cooling cycle is defined as the ratio of the thermal power absorbed in the evaporator, namely the final, useful cooling power $\dot{Q}_c$, versus the supplied thermal power in the generator $\dot{Q}_{th}$:

$$COP = \frac{\dot{Q}_c}{\dot{Q}_{gen}} \tag{5.52}$$

By applying the first and the second thermodynamic law in the thermal processes involved in an absorption chiller, we may express the maximum COP, named as the ideal COP, with the following relationship:

$$COP_{ideal} = \frac{T_C}{T_H} \cdot \frac{T_H - T_M}{T_M - T_C} \tag{5.53}$$

where:

T_C: the cooling medium temperature
T_H: the heat source temperature (generator or heat storage tank)
T_M: the intermediate temperature of the heat sink where the rejected heat from the chiller is rejected (this heat sink can most commonly be the ambient environment).

In Figure 5.44, typical fluctuation curves are presented for the ideal COP and actual COP for commercially available absorption chillers [30].

5.10.3 Solar Cooling

Solar cooling is approached with the supply of the thermal energy required for the absorption chiller operation from solar collectors. Solar cooling gained considerable attention during the beginning of the twenty-first century, due to its obvious favorable features, such as the reduction on the dependence of electricity consumption and the contribution to the transition to renewable and clean energy sources. However, with the improvement of the efficiency of VAV and VRV cooling systems, and given the considerably high setup cost of solar cooling systems, the installations of such projects globally still remains limited.

Solar cooling obviously is based on the availability of solar radiation. This feature offers both a major advantage and a critical drawback. Specifically:

- The maximization of solar radiation availability during the midday in summer normally coincides with the daily peak cooling loads. This feature provides a favorable and highly effective option to cover the peak cooling loads, minimizing the primary energy consumption and contributing to the approach of rational energy use.

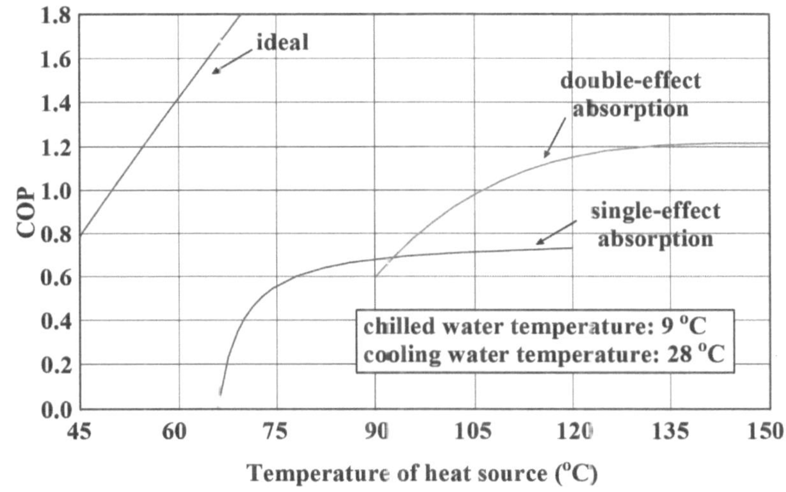

FIGURE 5.44 Ideal and actual COP fluctuation curves [30].

- On the other hand, it is obvious that solar radiation is not possible during night periods, due to the lack of sunlight. This practically means that a conventional cooling production technology should be available for the coverage of any cooling loads during night periods. In other words, solar cooling is not possible to guarantee on its own fully cooling loads coverage, in cases where they continue to exist during night periods (e.g., residential buildings, hotels). This inadequacy is eliminated in buildings with usage only during day periods, such as municipal, public, or commercial buildings.

Apart from the above essential characteristics, the fluctuation of solar radiation introduces also another peculiarity in the operation of solar cooling systems, operated with absorption chillers. Any mismatches between the available solar radiation and the cooling loads should be compensated, most commonly, with the introduction of a thermal storage tank. In this way, a sort of cooling hybrid power plant is formulated. Yet, thermal storage tanks, apart from the above obvious objective (matching of cooling loads and solar radiation availability) can also be utilized for another significant purpose, which is the maximization of the overall efficiency of the solar collectors–absorption chiller system. This can be easily conceived by observing Figure 5.45, in which the solar collectors' efficiency, the absorption chiller's COP, and the product of these two magnitudes' fluctuation curves are plotted versus the absorption chiller heat source temperature. In this figure, it is seen that the overall system's efficiency, namely the product of the collectors' efficiency and the chiller's COP, is maximized for a supply temperature close to 70°C. This means that a major target of the thermal storage tank could be the maintenance of the medium in the tank constantly at this temperature. The storage capacity of the thermal storage tank could then be formulated according to the size of the cooling loads.

A typical solar cooling system with an absorption chiller is presented in Figure 5.46.

As seen in this figure, the thermal power produced by the solar collectors is stored in the thermal storage tank. An external heat production device (e.g., a conventional heater), is also connected with the thermal storage tank, aiming at the maintenance of the temperature in the tank within the predefined range, which ensures maximization of the solar collectors–absorption chiller overall efficiency (Figure 5.45). A secondary closed hydraulic network is utilized for the transfer of the thermal power from the thermal storage tank to the absorption chiller. With this specific configuration, the thermal storage tank undertakes the role of the heat generator for the absorption chiller. The overall layout is integrated with two additional closed hydraulic loops. The first one connects the absorber and the condenser of the chiller with a cooling tower, aiming at the thermal

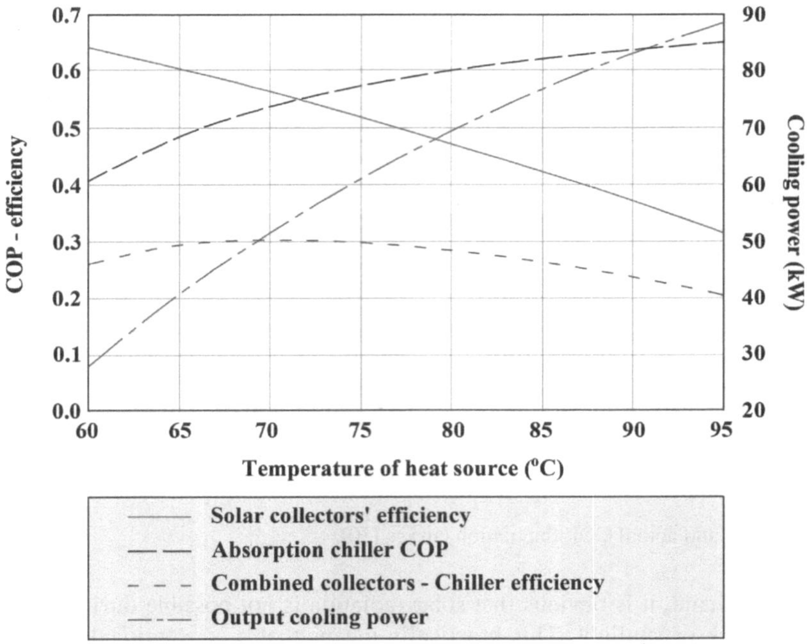

FIGURE 5.45 Ideal and actual COP fluctuation curves versus the absorption chiller heat source temperature [30].

power $\dot{Q}_{rej}$ disposal. The second one connects the evaporator of the absorption chiller with the cooling distribution network in the conditioned space, for the supply of the effective cooling power $\dot{Q}_c$. A chilled water tank can be also introduced in the return branch of the cooling distribution network, aiming at the storage of any cooling power surplus and the retrofitting of the evaporator.

Another crucial point is the operation concept of a solar cooling system with the specific configuration, strongly imposed by the availability of solar radiation only during daytime. A sensible

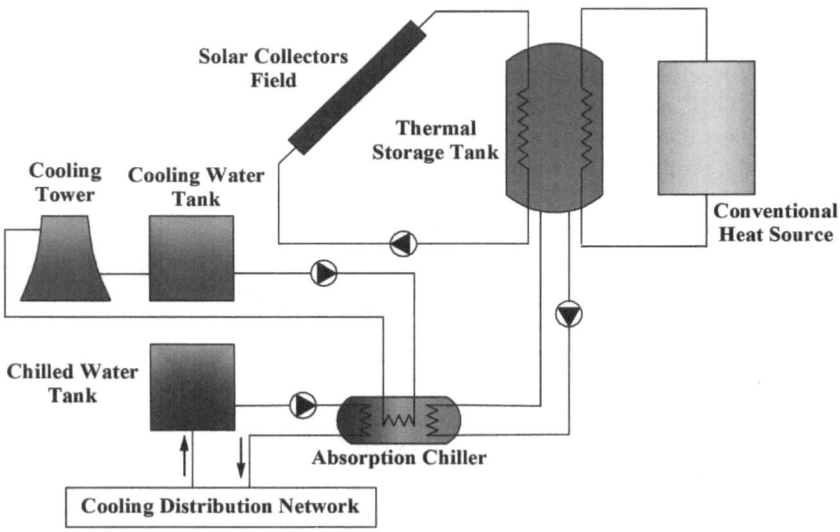

FIGURE 5.46 Typical layout of a solar cooling system, with a thermal storage tank and an external heater.

approach could be the cooling loads coverage from the solar cooling–absorption chiller system only during daytime. A secondary cooling system should be available for any potentially existing night-time cooling loads. In this way, any over-dimensioning of the solar collectors' field and the thermal storage tank is avoided, reducing the corresponding setup cost respectively.

Given the above presented parameters, a possible operation algorithm can be the following one, realized for every calculation time step. Let's adopt the system's layout presented in Figure 5.46, with the support of the thermal storage tank. The total potential thermal power flows transferred from and to the thermal storage tank are:

- any potential thermal power stored from the solar collectors $\dot{Q}_{sol}$
- any potential additional thermal power stored from the external heat source $\dot{Q}_{ext}$
- the thermal power offered from the storage tank to the absorption chiller $\dot{Q}_{gen}$
- any thermal power losses from the tank to the ambient $\dot{Q}_L$.

Let T_{opt} be the optimum temperature in the thermal storage tank, which ensures maximum overall efficiency of the solar collectors–absorption chiller system. If $\dot{Q}_c$ is the current cooling load, then it should be:

$$COP = \frac{\dot{Q}_c}{\dot{Q}_{gen}} \Leftrightarrow \dot{Q}_{gen} = \frac{\dot{Q}_c}{COP}$$

where COP is the coefficient of performance of the absorption chiller for the heat source temperature T_{opt}.

Let also m_{st} be the total mass capacity of the thermal storage tank, c_p the specific heat capacity of the thermal storage medium in the tank (most typically water), and T(i) the temperature in the thermal storage tank at the end of the current calculation time step i. The simulation of the system's daily operation follows the steps below:

1. $T_{out} > T(i-1)$

   First, the solar collectors' field outlet temperature T_{out} is checked to see if it is higher than the temperature $T(i-1)$ in the thermal storage tank from the previous calculation time step. In this case, the available thermal power $\dot{Q}_{sol}$ from the solar collectors can be stored in the thermal storage tank. The system's operation is initially checked without thermal power production from the conventional heater. In this case, the thermal power flows balance from and to the thermal storage tank gives:

$$m_{st} \cdot c_p \cdot \left[T(i) - T(i-1)\right] = \left[\dot{Q}_{sol} - \dot{Q}_{gen} - \dot{Q}_L\right] \cdot t = \left[\dot{Q}_{sol} - \frac{\dot{Q}_c}{COP} - \dot{Q}_L\right] \cdot t \Rightarrow$$

$$T(i) = \left[\dot{Q}_{sol} - \frac{\dot{Q}_c}{COP} - \dot{Q}_L\right] \cdot \frac{t}{m_{st} \cdot c_p} + T(i-1)$$

   where t is the duration of the calculation time step (t = 3,600 s in case of hourly calculation step).

   Two sub-cases are distinguished:

   i. $T(i) < T_{opt}$:

   In this case, the thermal power produced by the solar collectors is not adequate to provide the required thermal power $\dot{Q}_{gen}$ for the absorption chiller and maintain the temperature in the thermal storage tank at the desirable value T_{opt}. For this reason, a supplementary thermal power production from the external heat source is required, equal to:

$$\dot{Q}_{ext} = \frac{m_{st} \cdot c_p \cdot \left[T_{opt} - T(i)\right]}{t}$$

The solar collectors' thermal power absorbed (solar penetration) will be:

$$\dot{Q}_{sp} = \dot{Q}_{sol}$$

and the rejected (not absorbed) thermal power produced by the solar collectors will be null:

$$\dot{Q}_{rej} = 0$$

ii. $T(i) \geq T_{opt}$
In this case, the thermal power produced by the solar collectors is too high to be entirely absorbed by the thermal storage tank. In order to maintain the temperature in the thermal storage tank at T_{opt}, the solar collectors' absorbed power will be equal to:

$$\dot{Q}_{sp} = \dot{Q}_{sol} - \frac{m_{st} \cdot c_p \cdot \left[T(i) - T_{opt}\right]}{t}$$

Obviously, the required thermal power production from the conventional heater will be null:

$$\dot{Q}_{ext} = 0$$

and the rejected thermal power from the solar collectors will be:

$$\dot{Q}_{rej} = \frac{m_{st} \cdot c_p \cdot \left[T(i) - T_{opt}\right]}{t}$$

In both the above sub-cases the current cooling load is totally covered by the solar cooling system. The water temperature in the thermal storage tank at the end of the current calculation time step will be given in any case by the relationship:

$$T(i) = \left[\dot{Q}_{sp} + \dot{Q}_{ext} - \frac{\dot{Q}_c}{COP} - \dot{Q}_L\right] \cdot \frac{t}{m_{st} \cdot c_p} + T(i-1)$$

and will normally be equal to the desirable value T_{opt}.

2. $T_{out} \leq T(i-1)$
If the solar collectors' field outlet temperature T_{out} is lower than or equal to the temperature in the thermal storage tank from the previous calculation time step, no thermal power from the solar collectors can be stored in the thermal storage tank. In such a case, it is better not to employ the solar cooling system for the current cooling load coverage. An alternative active cooling system should be used. In this case it will obviously be:

$$\dot{Q}_{sp} = \dot{Q}_{ext} = 0$$

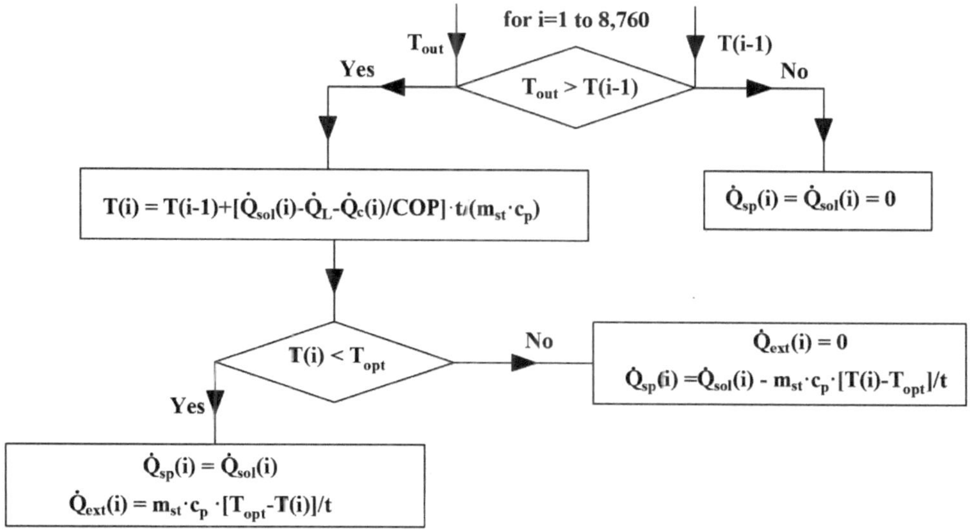

FIGURE 5.47 Operation algorithm of a solar cooling system.

The new temperature in the thermal storage tank at the end of the current calculation time step will be:

$$T(i) = T(i-1) - \dot{Q}_L \cdot \frac{t}{m_{st} \cdot c_p}$$

The above simulation algorithm is graphically presented in Figure 5.47.

The dimensioning of solar cooling systems refers to the sizing of the fundamental components of the system (solar collectors' field, absorption chiller, thermal storage tank, cooling tower) and can be accomplished with the implementation of the above operation algorithm. The goal could be the restriction of the thermal power production from the external heat source below an upper acceptable limit or the optimization of the economic indices of the overall project. Particularly regarding the sizing of the solar collectors' field, the experience gained by the installation of solar cooling systems gives a required specific collector's area around 3 m^2 per kW of cooling capacity, for absorption chillers [30]. For desiccant cooling systems, a typical value for this feature is close to 1.5 m^2/kW. However, it should be noted that these figures, retrieved by existing and operating solar cooling plants, are quite approximate, because several uncertainty parameters are involved, such as:

- depending on the introduced collector's model manufacturer, the collector's area can be alternatively defined as the collector's net absorber area, the aperture area, or the collector's gross area
- the dimensioning of the solar collectors' field may in some specific installations have been accomplished not only exclusively for the solar cooling application, but also for other heat production uses, according to which the size of the solar collectors' field may have been defined, appearing, thus, to be too high for the solar cooling needs
- in other implemented cases, the required heating for the absorption chiller operation may be mainly provided by a conventional biomass heater, with solar cooling being introduced only as a supplementary heat source, with the total installed solar collectors' number being relatively lower than the actual required.

A dimensioning example for a solar cooling system is provided with the case study presented below.

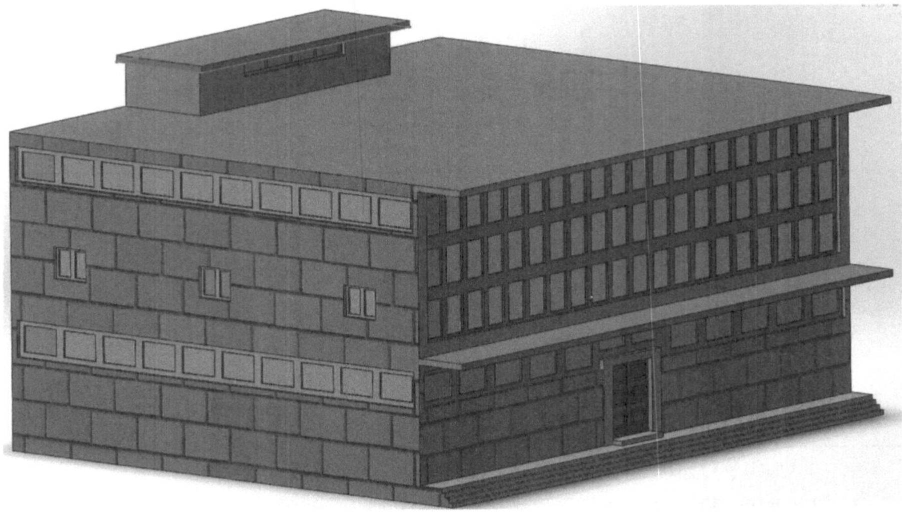

FIGURE 5.48 3D view of the solar cooling case study building.

5.10.4 A Solar Cooling Case Study

In this section a solar cooling system will be dimensioned for a new public building located in Las Palmas, in Canary Islands, Spain. A 3D aspect of the building is presented in Figure 5.48. It consists of four levels, including the basement, the ground floor, the first floor, and the second floor. The total covered area of the building is calculated at 1,270 m^2, with 970 m^2 of conditioned indoor space. The building is considered well-insulated, with average U-factor for the opaque surfaces at the range of 0.45 W/m^2·K. Similarly, the openings are considered of double glazing, with thermal break, and a total, average U-factor at the range of 1.3–1.9 W/m^2·K.

Introducing the annual GHI and the ambient temperature for the under consideration location, presented in Figures 5.49 and 5.50 respectively, the annual cooling load time series for the indoor

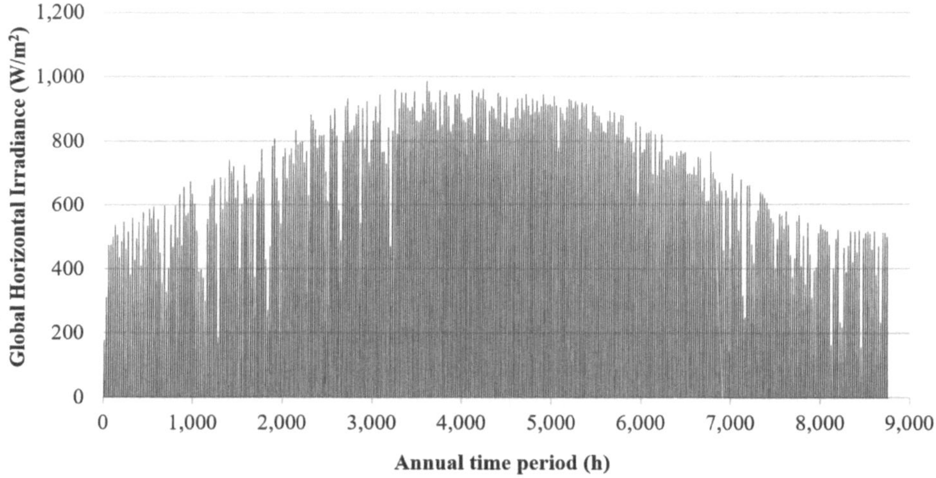

FIGURE 5.49 Annual global horizontal irradiance for the city of Las Palmas.

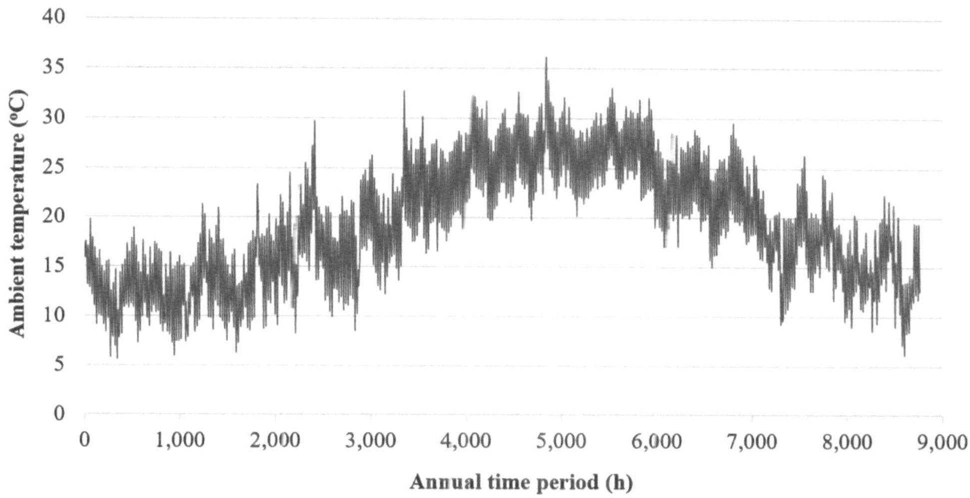

FIGURE 5.50 Annual ambient temperature fluctuation for the city of Las Palmas.

conditioned space is calculated, following the essential methodology presented in the relevant literature [26, 27]. It is presented in Figure 5.51. The yearly GHI for the specific geographical location is calculated at 1,750 kWh/m^2.

Given the above calculated annual cooling load fluctuation, a solar cooling system is studied. The system's annual operation is computationally simulated, following the operation algorithm presented in the previous section. The dimensioning of the solar cooling system will be executed on the basis of the cooling load. The exploitation of the system for heating production during winter is not examined in this case study. It has been covered within the case studies presented in Chapter 4. The annual peak cooling load is calculated at 62.9 kW, as seen in Figure 5.51. Given the above empirical conclusion with regard to the required 3 m^2 of solar collectors' area per kW of cooling power, approximately 190 m^2 of solar collectors should be installed at least.

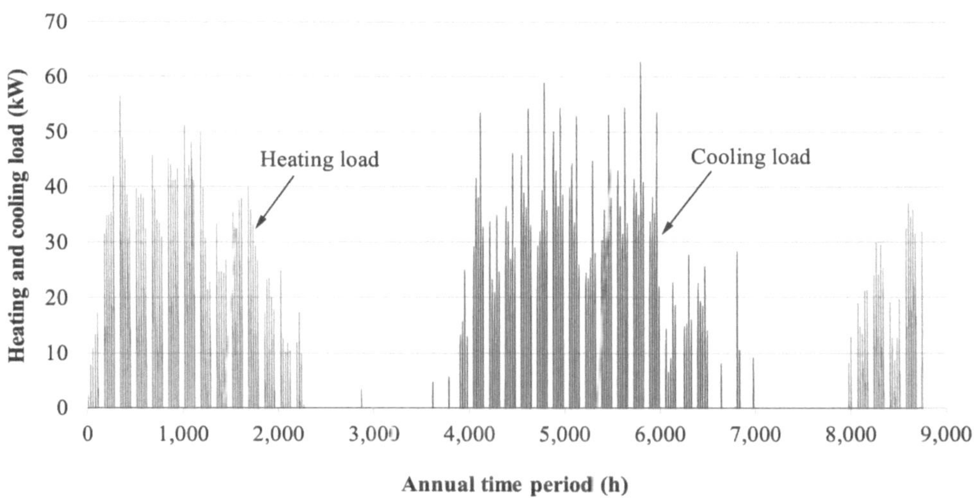

FIGURE 5.51 Heating and cooling load annual fluctuation for the solar cooling case study building.

TABLE 5.8
Results of the Executed Simulations for the Case Study Solar Cooling System

Investigated Scenario	4 × 20 solar collectors – 200 m²			5 × 20 solar collectors – 250 m²		
Temperature in the tank (°C)	60	70	80	60	70	80
Thermal energy stored from solar collectors (kWh)	11,219	7,104	3,515	10,888	8,209	4,654
Thermal energy stored from external heat source (kWh)	28,473	16,668	10,610	29,172	16,490	11,055
Total thermal energy stored to thermal tank (kWh)	39,692	23,772	14,125	40,061	24,699	15,709
Solar collectors/external heat percentage thermal storage (%)	28/72	30/70	25/75	27/73	33/67	30/70
Absorption chiller average, COP	0.40	0.56	0.62	0.40	0.56	0.62
Cooling load annual coverage percentage from solar cooling (%)	84	69	45	84	72	50

For this specific case study, two alternative scenarios will be investigated, with regard to the solar collectors' field and the total solar collectors' number. In the first one, in total 80 solar collectors are introduced, arranged in 20 parallel groups of 4 solar collectors each connected in series. In the second scenario, 100 solar collectors are introduced, arranged again in 20 parallel groups of 5 solar collectors each, connected in series. By introducing a solar collector model of 2.5 m^2 effective area, the total solar collectors' field effective area for the two alternative investigated layouts is calculated at 200 m^2 and 250 m^2 respectively. For each one of the two solar collectors' fields, three alternative scenarios are executed, with regard to the water temperature kept in the thermal storage tank, set at 60°C, 70°C, and 80°C, respectively. Consequently, in total six alternative scenarios are investigated. The results regarding the thermal energy stored in the thermal storage tank are presented in Table 5.8.

The annual fluctuation of the thermal power storage from the solar collectors and the external heat source in the thermal storage tank is presented in Figure 5.52, for the investigated scenario of 200 m^2 of solar collectors' installed surface and 60°C temperature in the thermal storage tank.

For the same investigated scenario, the annual fluctuation of the water temperature in the thermal storage tank and the outlet temperature from the solar collectors field are presented in Figures 5.53 and 5.54 respectively.

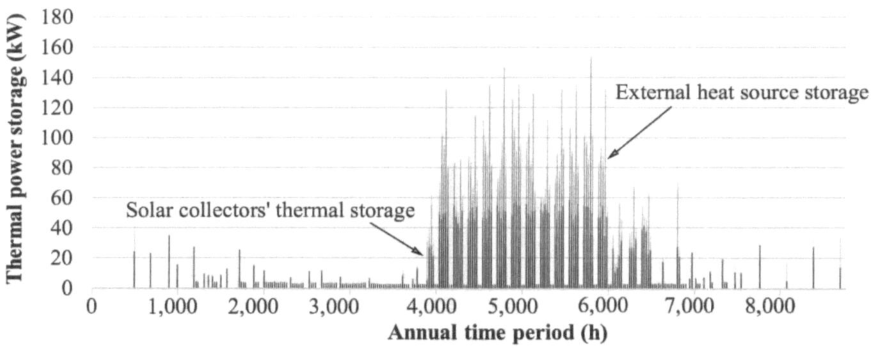

FIGURE 5.52 Annual fluctuation of the thermal power storage from the solar collectors and the external heat source in the thermal storage tank, for 200 m² of solar collectors' surface and 60°C temperature in the thermal storage tank.

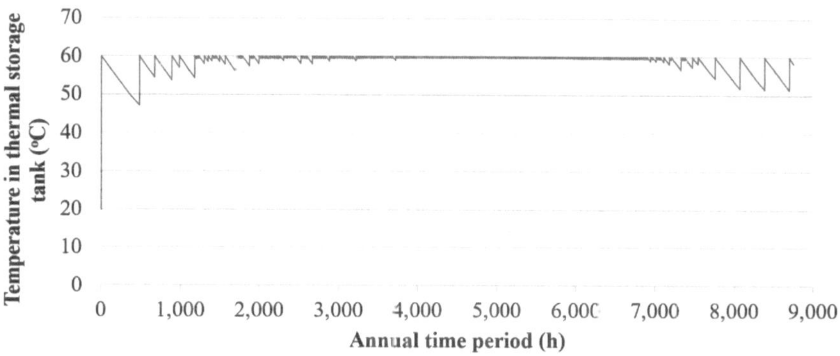

FIGURE 5.53 Annual fluctuation of the water temperature in the thermal storage tank.

By observing the annual energy results presented in Table 5.8 we can make the following conclusions:

- The lower the required temperature in the thermal storage tank, the higher the annual contribution of the solar cooling system to the annual cooling load coverage, maximized at 84% for water temperature 60°C in the thermal storage tank. However, this is achieved with remarkably high annual thermal energy stored from the external heat source.
- The highest contribution percentage of the solar collectors to the thermal energy stored is achieved for 250 m^2 of solar collectors' field and for 70°C temperature in the thermal storage tank, confirming the literature conclusions presented above. For the specific case study, the required thermal energy storage from the external heat source is also considerably restricted. Normally, this scenario should sensibly be the optimum one from an economic point of view, although the economic feasibility of the examined project is highly affected by the technology of the employed external heat source (e.g., a conventional oil or biomass heater or a heat pump) and the consumed primary energy source (e.g., oil, biomass, electricity).

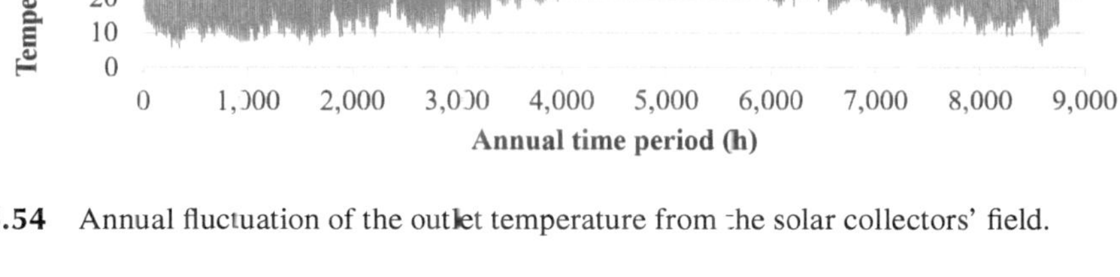

FIGURE 5.54 Annual fluctuation of the outlet temperature from the solar collectors' field.

REFERENCES

1. Rosen, M. A., & Koohi-Fayegh, S. (2016). *Cogeneration and District Energy Systems: Modelling, analysis and optimization*. Stevenage, UK: The Institution of Engineering and Technology (IET).
2. The European Association for the Promotion of Cogeneration—COGEN EUROPE. www.cogeneurope.eu (Accessed on May 2019).
3. Aristotle University Thessaloniki, Department of Mechanical Engineering, Laboratory of Heat Transfer and Environmental Engineering. Combined Heat and Power Cogeneration. http://aix.meng.auth.gr/lhtee/education/IAxBE7.pdf (Accessed on May 2019).
4. Horlock, J. H. (1996). *Cogeneration-Combined Heat and Power (Chp): Thermodynamics and Economics* (1st ed.). Malabar: Krieger Pub Co.
5. Breeze, P. (2017). *Combined Heat and Power (Power Generation)* (1st ed.). Cambridge, Massachusetts: Academic Press.
6. Technical Chamber of Greece: Technical Directive on Combined Heat and Power systems. http://portal.tee.gr/portal/page/portal/tptee/totee/TOTEE-20701-5-Final-%D4%C5%C5.pdf (Accessed on April 2019).
7. Wikidot.com: Thermal Systems, Small Scale Cogeneration Including Automotive Applications. http://me1065.wikidot.com/printer–friendly//small-scale-cogeneration-including-automotive-applications (Accessed on May 2019).
8. Jiang Hwang, J., & Lin Zou, M. (2010). Development of a proton exchange membrane fuel cell cogeneration system. *Journal of Power Sources, 195*, 2579–2585.
9. Crema, L., Alberti, F., Bertaso, A., & Bozzoli, A. (2011). Development of a pellet boiler with Stirling engine for m-CHP domestic application. *Energy, Sustainability and Society, 1*, 5.
10. Beith, R. (2011). *Small and Micro Combined Heat and Power (CHP) Systems: Advanced Design, Performance, Materials and Applications (1st ed.)*. Sawston, UK: Woodhead Publishing.
11. Gaggioli, R. A., & Petit, P. J. (1977). Use the second law first. *Chemtech, 7*, 496–506.
12. Rodriquez, L. S. J. (1980). Calculation of available energy quantities. In: R. A. Gaggioli (ed.), *Thermodynamics: Second Law Analysis. ACS Symposium Series, 122*, 39–60. Washington, DC: American Chemical Society.
13. Steam Tables Calculator—Steam Tables Online. https://www.steamtablesonline.com/steam97web.aspx (Accessed on May 2019).
14. Danfoss. Chapter 5: Instructions for designing district heating system. http://heating.danfoss.com/pcmfiles/1/master/other_files/library/heating_book/chapter5.pdf (Accessed on October 2019).
15. Gang, W., Wang, S., Xiao, F., & Gao, D. (2016). District cooling systems: Technology integration, system optimization, challenges and opportunities for applications. *Renewable and Sustainable Energy Reviews, 53*, 253–264.
16. Kavanaugh, S., & Rafferty, K. (1997). Ground Source Heat Pumps. Design of Geothermal Systems for Commercial and Institutional Buildings. Atlanta, Georgia.
17. Kavanaugh, S., & Rafferty, K. (2015). *Geothermal Heating and Cooling: Design of Ground-Source Heat Pump Systems* (1st ed.). Atlanta: ASHRAE.
18. Sabet Behrooz, B., Demollin, E., & Van Bergermeer, J.-J. (2008, February 6-8). Geothermal use of deep flooded mines. *Post-Mining*. Nancy, France.
19. European Post-Carbon Cities of Tomorrow. "Model Güssing:" a vision of energy self-sufficiency. https://pocacito.eu/sites/default/files/ModelG%C3%BCssing_G%C3%BCssing.pdf (Accessed on May 2019).
20. Carbo, M. C., Smit, R., van der Drift, B., & Jansen, D. (2011). Bio energy with CCS (BECCS): Large potential for BioSNG at low CO_2 avoidance cost. *Energy Procedia, 4*, 2950–2954.
21. Hellsmark, H., & Jacobsson, S. (2009). Opportunities for and limits to Academics as System builders—The case of realizing the potential of gasified biomass in Austria. *Energy Policy, 37*, 5597–5611.
22. Bermudez, J. M., & Fidalgo, B. (2016). Chapter 15: Production of bio-syngas and bio-hydrogen via gasification. *Handbook of Biofuels Production (2nd ed.)*, 431–494.
23. Keglovits, C. (2016). Güssing: An example for a sustainable energy supply. https://slideplayer.com/slide/7103314/ (Accessed on October 2019).
24. Sparacino, M., Camussi, M., Colombo, M., Carella, R., & Sommaruga, C. The world's largest geothermal district heating using ground water under construction in Milan (Italy): AEM unified heat pump project. International Geothermal Association. https://www.geothermal-energy.org/pdf/IGAstandard/EGC/2007/143.pdf (Accessed on May 2019).

25. Maslak, K. (2015). Thermal Energy Use in Greenhouses. The Influence of Climatic Conditions and Dehumidification. https://pub.epsilon.slu.se/12037/1/maslak_k_150407.pdf (Accessed on May 2019). Faculty Landscape Architecture, Horticulture and Crop Production Science, Department of Biosystems and Technology. Licentiate Thesis, Swedish University of Agricultural Sciences, Alnarp.
26. Kreider, J., Rabl, A., & Curtiss, P. (2017). *Heating and Cooling of Buildings* (3rd ed.). Boca Raton, Florida: CRC Press.
27. *2009 ASHRAE Handbook—Fundamentals (SI Edition).* Atlanta, Georgia: American Society of Heating, Refrigerating and Air-Conditioning Engineers, Inc.
28. SAVE Programme Action No 4.1031/Z/01-130/2001: TriGeMed – Trigeneration in the Mediterranean Countries – Technologies and Prospects in the Tertiary Sector.
29. CIBSE Journal. (2009, November). Module 10: Absorption refrigeration. https://www.cibsejournal.com/cpd/modules/2009-11/ (Accessed on October 2019).
30. Henning, H.-M. (2007). Solar assisted air conditioning of buildings—an overview. *Applied Thermal Engineering*, *27*, 1734–1749.

6 Smart Grids

6.1 BACKGROUND

So far, in the earlier chapters, the examined topics mainly focused on centralized power systems, especially as far as electricity production is concerned. Centralized power production has been the traditional and the most common method of electricity production globally. Central power stations installed in strategically selected geographical locations undertake the major part of electricity production, which is eventually provided for the final consumers through the transportation and the distribution grids. Centralized power production during the twentieth century actually was not a choice from the side of the utilities or the consumers. In fact, it was a mandatory path, given the lack of power production and storage technologies, as well as the communication facilities offered today by the available high-tech apparatus. Even today, it still remains the dominant electricity production approach, but now it is not due to the lack of the appropriate technology to support an alternative method for electricity production, but mainly, due to the lack of adequate awareness and cultivation of electricity final consumers, or in best cases, because the alternative approach, namely decentralized power production, is still attempting to make its first steps.

We may claim that centralized power production dominated during the twentieth century because of some very specific reasons:

- The lack of appropriate technologies, regarding the production and storage of energy and the communication facilities between the involved parties in a potential decentralized energy system.
- Centralized production offers comfort and security for the consumers that everything is done centrally, relieving them of any obligations or tasks, apart from paying the electricity bills.
- The idea of consumers' involvement, in one way or another, in the power production process, didn't even exist until recently.

Nevertheless, centralized power production is characterized by some considerable disadvantages:

- It requires the existence of electricity transportation and distribution networks, which constitute a major component of the electricity systems, accounting for a corresponding percentage of the invested capital and of the annual required maintenance process of the electrical systems
- Electricity transfer through transportation or distribution networks imposes losses, due to the ohmic or inductive resistance of the electrical wirings.
- In cases of intensive seasonal power demand variations (e.g., in tourist insular regions), the power production capacity required to undertake the annual peak power demand should be centrally available. This capacity determines the total installed nominal power of the involved generators, which, although practically employed only for a very short time period over the year, imposes additional maintenance cost for the utility, contributing, thus to higher electricity production fixed cost.
- Centralized power production systems are more vulnerable to grid or generator contingencies. A fault in the grid may result in a power loss for a whole portion of the electrical grid, and a fault in the generation (e.g., in a thermal power plant) may affect the whole electrical system. This vulnerability has become more crucial with the introduction of nonguaranteed power production units from renewable energy sources (RES), as thoroughly explained in Chapter 2.

In Figure 6.1, typical daily power demand profiles are presented for the state of California, in the US [1]. They refer to real recorded data from the utility from 2013 until 2016 and to projections until 2020.

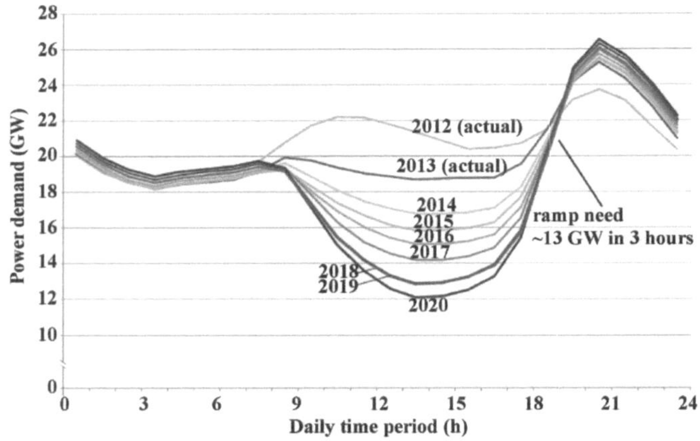

FIGURE 6.1 The effect of increasing photovoltaic power production in the state of California on the overall power demand daily profile [1].

It is seen that the photovoltaic (PV) stations' contribution to the power demand coverage until 2013 was almost negligible. Yet, since 2013, from year to year, with the increasing installations of PV stations in the state, it is seen that during the daytime period the utility will have to face an intensive and remarkable initial drop and eventual increase of the power demand, arisen from the PV stations' production during this period. As seen in this figure, within a period of 6.5 hours in the morning, the power production from the guaranteed units should be decreased from 18 to 11 GW, namely 7 GW, by shutting down units or reducing their power production close to their technical minimums. Even worse, while approaching sunset, within a period of 3 hours (from 3 p.m. to 6 p.m.), the PV power production is eliminated and this power production loss should be compensated with a corresponding increase of the thermal generators. This impact of the PV stations on California's power demand daily profile is known as the California duck curve, due to its shape [1].

Similar impacts are also observed in Figure 6.2, which refers to the autonomous insular system of the island of Crete, Greece [2]. In this figure, the power production synthesis is presented for 1 week.

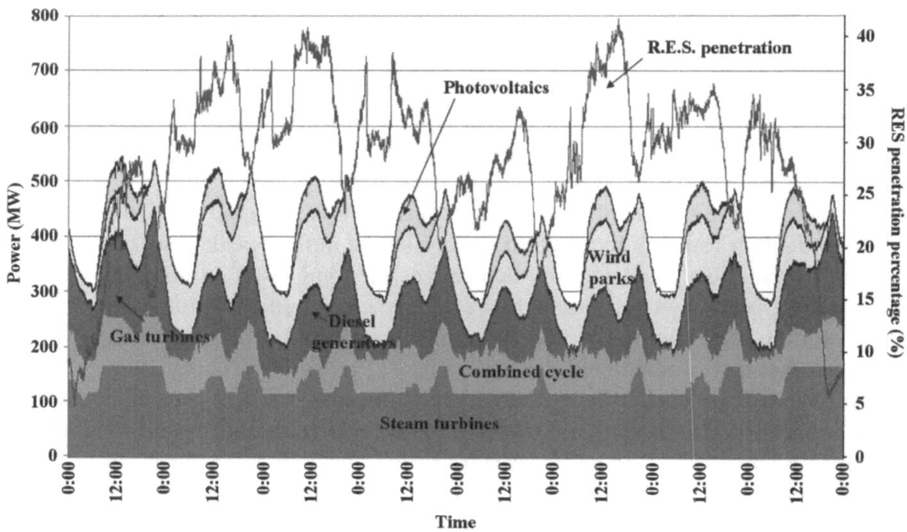

FIGURE 6.2 Real power production synthesis graph and instant RES penetration percentage fluctuation in the autonomous insular system of Crete, Greece from June 27 to July 4, 2012 [source: HEDNO].

Together with the power production time series of the dispatched units, the variation of the RES penetration percentage is also depicted in the right vertical axis of the diagram (wind parks and PV stations). This percentage varies during the under consideration period from 5% to 40% roughly. Again, especially during the sunset periods, when the PV power production rapidly drops, the RES total power penetration percentage is respectively affected. These intensive fluctuations in the power production from the RES units are handled with corresponding regulation of the combined cycle power output.

A more crucial event is presented in Figure 6.3, again for the system of Crete. In Figure 6.3a the total power production synthesis from all the wind parks installed on the island is presented for December 26, 2006. Suddenly, around 2:40 p.m. there is an abrupt and remarkable power production loss, most probably due to a fault in one of the dispatched wind parks, caused, perhaps, by adverse weather conditions. The initial incidence caused a sequel of gradual wind parks tripping,

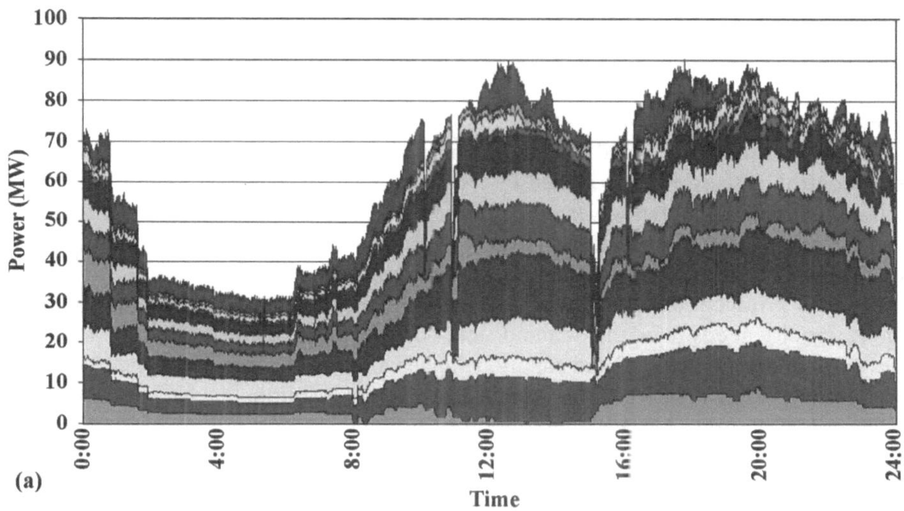

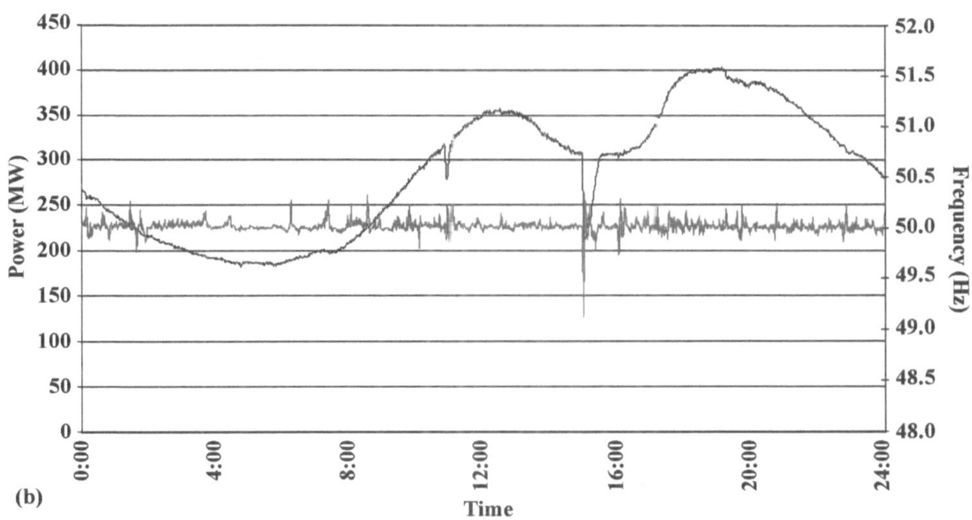

FIGURE 6.3 (a) Wind parks power production synthesis in the autonomous insular system of Crete, Greece during a sudden wind power production loss from June 27 to July 4, 2012, and (b) grid's fundamental frequency response [source: HEDNO].

given their sensitivity to the grid's voltage or frequency fluctuations (see Chapter 2). In total, around 100 MW of wind park power production were lost, as shown in Figure 6.3b, corresponding to 31% of the total power demand (320 MW), resulting in a grid's frequency drop at 49.3 Hz (1.4% versus the grid's nominal frequency). The operator was eventually able to restore the dynamic balance between the power demand and production within a period of 40 minutes, obviously because of the adequate maintenance of fast spinning reserve and/or, perhaps, the cut-off of power supply for a particular part of the grid.

The above crucial events or unfavorable operation conditions constitute characteristic examples of inadequacies met in centralized power production systems. These inadequacies could be handled or relieved with several techniques and strategies introduced in the overall frame of smart grids, such as dispersed power production or power demand shifting from peak to low demand periods. The concept of smart grids is presented in the next section.

6.2 THE CONCEPT OF SMART GRIDS

The ultimate target of either conventional electrical systems or smart grids is the continuous maintenance of the power production and demand dynamic balance. In the first case, this is approached through the control of the power produced by a relatively small number of centralized power production plants, according to the power demand introduced in the system through the distribution grids. This balance is maintained through highly sophisticated automations and facilities, installed both in the transmission and distribution networks. Transmission grids represent the supply side, because all power production plants are connected to them. Distribution grids communicate to the system the electricity needs from the demand side, because all final consumers are served by them. The maintenance between demand and supply balance is, therefore, translated into the conservation of a corresponding balance between the transmission and distribution grids. The successful fulfillment of this task often requires the intervention of highly experienced and skilled staff from the operator's side.

On the other hand, the dynamic balance between power demand and supply in smart grids is not restricted any more strictly to the conservation of the demand imposed by the distribution grids and the supply offered by the transmission grids. New power production, storage, and management technologies, introduced in the consumers' side and integrated within the frame of wholesale electricity markets, offer new options for the achievement of the above essential target directly inside the distribution grids. This potential constitutes the main distinguishing feature between smart grids and conventional electrical systems.

With the term *smart grid*, a dynamic electrical system is conceived, aiming to deliver electricity to the final consumers through an interactive process between them and the grid's operator. The ultimate objective of smart grids is the secure electricity supply for the final consumers through a flexible, efficient, and cost-effective procedure.

Smart grids do not have a unique definition, universally accepted. They have been described in alternative ways, either with simple, descriptive terms, or with less thorough and more complex approaches. Formally, smart grids have so far been given various definitions by different organizations, institutes, etc. Some characteristic definitions are discussed below:

- International Energy Agency (IEA) [3]

> "A smart grid is an electricity network that uses digital and other advanced technologies to monitor and manage the transport of electricity from all generation sources to meet the varying electricity demands of end-users. Smart grids co-ordinate the needs and capabilities of all generators, grid operators, end-users and electricity market stakeholders to operate all parts of the system as efficiently as possible, minimizing costs and environmental impacts while maximizing system reliability, resilience and stability."

- European Commission (EC) [4]

"Smart grids are energy networks that can automatically monitor energy flows and adjust to changes in energy supply and demand accordingly. When coupled with smart metering systems, smart grids reach consumers and suppliers by providing information on real-time consumption. With smart meters, consumers can adapt—in time and volume—their energy usage to different energy prices throughout the day, saving money on their energy bills by consuming more energy in lower price periods.

Smart grids can also help to better integrate renewable energy. While the sun doesn't shine all the time and the wind doesn't always blow, combining information on energy demand with weather forecasts can allow grid operators to better plan the integration of renewable energy into the grid and balance their networks. Smart grids also open up the possibility for consumers who produce their own energy to respond to prices and sell excess to the grid."

- United States Office of Electricity Delivery and Energy Reliability [5]

"'Smart grid' generally refers to a class of technology that people are using to bring utility electricity delivery systems into the 21st century, using computer-based remote control and automation. These systems are made possible by two-way communication technology and computer processing that has been used for decades in other industries. They are beginning to be used on electricity networks, from the power plants and wind farms all the way to the consumers of electricity in homes and businesses. They offer many benefits to utilities and consumers—mostly seen in big improvements in energy efficiency on the electricity grid and in the energy users' homes and offices."

- International Electrotechnical Commission [6]

"The general understanding is that the Smart Grid is the concept of modernizing the electric grid. The Smart Grid comprises everything related to the electric system in between any point of generation and any point of consumption. Through the addition of Smart Grid technologies the grid becomes more flexible, interactive and is able to provide real time feedback.

A smart grid is an electricity network that can intelligently integrate the actions of all users connected to it—generators, consumers and those that do both—in order to efficiently deliver sustainable, economic and secure electricity supplies. A smart grid employs innovative products and services together with intelligent monitoring, control, communication, and self-healing technologies to: facilitate the connection and operation of generators of all sizes and technologies; allow consumers to play a part in optimizing the operation of the system; provide consumers with greater information and choice of supply; significantly reduce the environmental impact of the whole electricity supply system; deliver enhanced levels of reliability and security of supply."

- Japan Smart Community Alliance [7]

"In the context of Smart Communities, smart grids promote the greater use of renewable and unused energy and local generation of heat energy for local consumption and contribute to the improvement of energy self-sufficiency rates and reduction of CO_2 emissions. Smart grids provide stable power supply and optimize overall grid operations from power generation to the end user."

- National Institute of Standards and Technology [8]

"The Smart Grid is a grid system that integrates many varieties of digital computing and communication technologies and services into the power system infrastructure. It goes beyond smart meters for homes and businesses as the bidirectional flows of energy and the two-way communication and control capabilities can bring in new functionalities."

If we attempt to merge the above definitions into one, we could say that smart grids are simply intelligent grids. Unlike traditional electrical systems, which can only transmit and distribute electricity in only one direction (from production to consumption), in smart grids electricity can be transferred between the final consumers and it can be stored, while the involved actors in the grid can communicate between them and decisions for alternative actions can be made versus the grid's current operational conditions and parameters. A better understanding of smart grids can be made by comparing conventional and smart grids. A synopsis of the main features of the two alternative electrical systems is provided in Table 6.1 [9, 10]. More elements on the capabilities offered by smart grids, presented in this table, are given in the next sections of this chapter.

TABLE 6.1
Comparison Between the Main Characteristic Features of Conventional and Smart Grids

Conventional Electrical Grids	Smart Grids
Mechanization	Digitization
One-way communication	Two-way, real-time communication
Centralized power production	Distributed power production
Radial network layout	Dispersed network
Limited data involved	Large volume of data involved
Small number of sensors	Large number of sensors
Limited or no automatic monitoring	Extensive automatic monitoring
Manual control and recovery	Automatic control and recovery
Limited dynamic security and stability, increased vulnerability especially in weak grids	Improved dynamic security and stability, even in small, weak systems
Concurrent production and consumption	Increased use of direct or indirect energy storage
Limited control	Extensive control systems
Fewer final consumers' choices	Vast final consumers' choices

6.2.1 Functionalities of Smart Grids

Smart grids offer a vast variety of functionalities, which can be implemented at the scale of a city, at the national or regional level, or even at the scale of small settlements, introducing, thus, the concepts of mini or micro-grids. They can be categorized as presented below:

- Demand side management (DSM)

  A major innovation of smart grids compared with conventional systems, a common point clearly revealed by the above definitions, is the electricity final consumers' involvement in the power production and management process. This can be done in several ways, stimulated by predefined provided incentives (in most cases economic) and supported with advanced bidirectional communication facilities, which enable information transfer concurrently from the operator to the consumers, and vice versa.

  The involvement of final consumers in the power production and management process is implemented with four basic strategies: dispersed power production, peak load shaving (or shifting), load curtailments, and improvement of the electricity usage efficiency. The above strategies constitute, in total, so-called demand side management (DSM), and they incorporate all the possible available alternatives for the final consumers' involvement in smart grids' functionalities. DSM provides final consumers with the chance to be involved in the electrical grid's operation and optimize the use of their own electrical apparatus, from both a technical and economic point of view. DSM strategies will be extensively presented in a later section.
- Exploitation of distributed energy resources

  Dispersed production, using either RES or conventional power production plants can be one of the DSM strategies, if it is provided by the final consumers. Dispersed power production units can be PV stations, small wind turbines, biomass stations, combined heat and power units, and even thermal generators of small size, such as compact steam turbines, diesel gensets, and gas turbines, most commonly owned by large- or medium-size industrial or commercial consumers, employed as backup units in case of power supply interruptions. Dispersed power production can contribute towards the increasing and secure RES penetration, due to the smoothness of the power production fluctuation curve from the dispersion of small individual power production units in wide geographical areas with different climate conditions.

- Introduction of advanced metering devices and systems

  Advanced metering devices include smart meters and automation, aiming at the measuring of the transferred electricity between the consumers or between the consumers and the operator. Smart meters enable two-way communication between the involved parties. They ensure accurate measurements and bills and they offer control for the consumers over their consumptions.

  Smart meters consist of sensors, consumption or production metering, and power quality assessment mechanisms. Metering devices provide the required decision criterion for the placement of specific orders or requests, from the operator's side, for dispersed power injection in the grid, power curtailment, etc., or for the final consumers, regarding their potential involvement in power curtailments or in power storage aiming to peak load shifting, avoiding high electricity pricing during high power demand periods. Of course, the metering infrastructure can also be utilized to determine the final payments between the consumers or from the consumers to the operator.
- Advanced electricity and thermal energy storage and peak shaving technologies

  The introduction and integration of electricity and thermal storage technologies and flexible, manageable loads can be potentially utilized for power peak shaving. Storage technologies can also constitute a part of DSM strategies, for dispersed storage devices of small size (e.g., residential electrochemical batteries) employed within the frame of peak load shifting. Yet, storage technologies of larger size, owned by producers, can also be exploited as components of smart grids.

Additionally, flexible and manageable loads, according to the power production available capacity, can also be introduced, such as electric vehicles or desalination plants, offering a useful tool for effective power production management and improvement of the grid's dynamic security and stability. At first glance, the introduction of electric vehicles to electrical grids requires new approaches on control systems, because, if the charge process of the vehicles is not managed properly, it can lead to significant peak power demand rise, which, in turn, will require additional infrastructure on generation capacity and grid assets, to ensure the system's reliability.

However, on the other hand, electric vehicles (EVs), plug-in hybrid electric vehicles (PHEVs), or fuel cell electric vehicles can remain connected to the grid when parked, offering the stored energy in their batteries to the system in emergency cases or during peak power demand periods. This approach is known as vehicle-to-grid (V2G) [11]. With the appropriate scheduling of V2G power injection in the grid, the daily power demand profile can be effectively smoothed. At the same time, by offering an additional power production source in the grid, particularly available during peak demand periods, EVs can significantly contribute to the enforcement of the electrical system's reliability and efficiency. EVs, as controllable loads, can also support the secure integration of stochastic distributed generation from RES technologies and contribute to the system's stability, with the provision of ancillary services. From the owner's point of view, the EVs procurement and operation cost can be significantly compensated with the V2G programs. The realization of V2G programs is supported with advanced, two-way communication networks and smart metering.

- Integration of smart appliances and consumer devices

  Smart appliances and devices constitute high-tech pieces of equipment, capable of communicating directly with the grid and executing several tasks and functions, according to the requirements of the grid and for the benefit of the final consumers. Such tasks can be, for example, the switching off and on of specific, flexible-to-use devices during peak and off-peak demand periods, respectively. Smart appliances, through bidirectional communication technologies, enable direct access from the operator's side to the final consumers' load control. In this way, they can act as operating reserves for the operator, e.g., within the frame of a smart building, enabling load shifting and curtailments.

- The utilization of Internet and cyber technologies
  Internet and cyber technologies can be utilized for the supervision and the management of the smart grid by both sides (consumers and operators). Through advanced, online applications, final consumers can place their bids regarding their available power production from their power production units (e.g., PV or biomass stations, storage devices), or declare their capacity for power demand curtailments, energy storage, etc. Additionally, they can also declare the selected provider from which they decide to purchase the electricity required for the coverage of their needs, determine the electricity procurement price, pay for their consumption, etc.
- Interoperability of involved actors and timely communication flow
  Within a dynamically varied electrical system, in a wholesale electricity market environment, information should be transferred at the appropriate time between the involved actors and the executed functions should be executed effectively. To this end, interoperability between the involved actors constitutes a crucial factor due to the increasing complexity of smart grids, the number of them, and the many different types of them. On the other hand, timely data and information transfer is also essential for the proper and adequate operation of smart grids, because, actually, both the electricity flows and the corresponding selling and procurement prices are based on this information and the delivery time to the involved recipients.

All the above facilities are based on the development of adequate electrical transportation and communication networks. Two-way, high-tech communication networks ensure the interoperability between the involved actors in smart grids, based on reliable and on-time communication flows between them. This infrastructure includes power and communication lines, wireless and wired networks, and distribution servers (concentrators). In the most effective layout, the communication network is developed following a treelike structure, with all the information from the final consumers gathered gradually to a central control and management server. More examples provided in Section 6.4.

Figure 6.4 presents an indicative smart grid structure. All involved technologies mentioned previously, including dispersed and centralized production and electricity or thermal energy storage

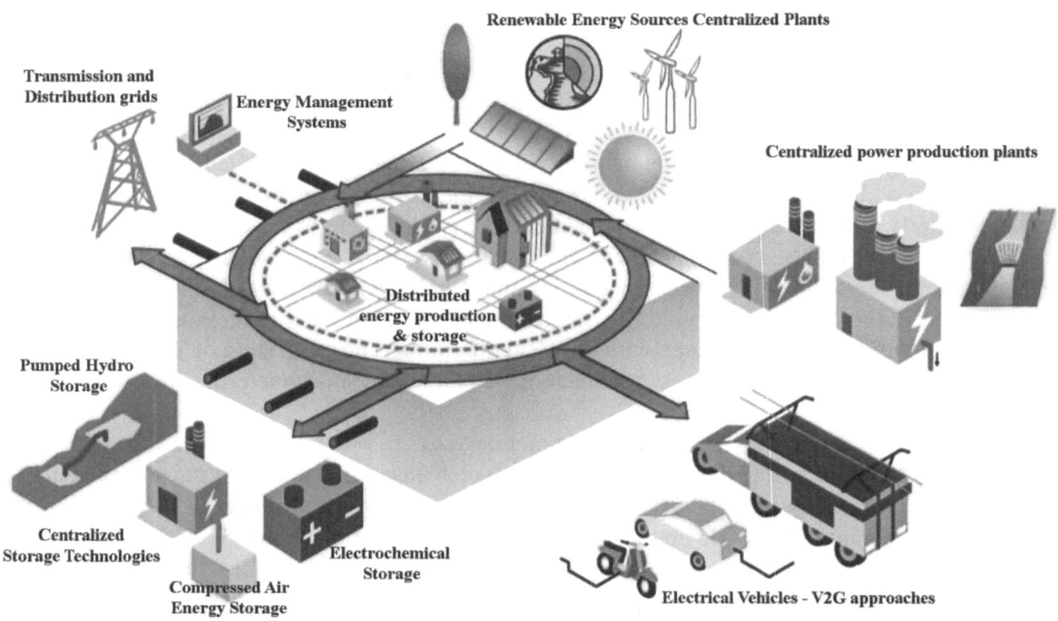

FIGURE 6.4 Smart grid fundamental structural and operating philosophy (with due diligence of International Energy Agency) [12].

technologies, are integrated in an interactive energy system, equipped with the above described advanced technical infrastructure, enabling the application of all those facilities that determine the smart grid functionality. All the above functions, infrastructure, and processes aim at the following objectives:

- the development of electrical systems with improved stability and dynamic security, capable to recover with internal processes from power disturbances and contingencies
- the active participation of final consumers in power production and management and in the configuration of the electricity wholesale market
- resilient operation against physical and cyber attacks
- exploitation of all power production and storage alternatively available technologies, at the most appropriate scale versus the conditions met at each particular geographical region
- design and development of new products, services, and markets
- optimization of the offered assets, system efficiency, and cost-effective operation.

6.2.2 Evolution of Smart Grids

The origination of the smart grids' concept is rather unclear. Practically, the need for control and management of the electricity flows begins with the development of the first distribution grids. The essential target has always been the reliability and the cost-effective operation of the electrical systems. During the last decades, with the increasing share of RES in the electricity production mixture, the conservation of the electrical grids' reliability becomes more and more challenging. At the same time, the emerging technologies on electricity production, storage, and control create the prerequisites for advanced participation of final consumers in the electrical systems' processes, regarding production, transfer, storage, and management of electricity.

A first approach towards the investigation of the smart grid history is their association with smart meters. In this sense, their origination coincides with the invention of two-way communication smart meters, which is placed back in the 1970s [13]. Smart meters were widely used in the 1980s. The first scientific article on "Grids get smart protection and control" is found in 1997, while there is another one in 2003, two in 2005, and one in 2008 [10].

Another approach is the connection of smart grids with the development of sensors and control technologies, already introduced in the 1930s. Nevertheless, the first wireless network with a layout and operation relatively close to a modern wireless sensor network (WSN) was the sound surveillance system developed by the US military in the 1950s. The second milestone was the transition of WSN awareness and experience into the academia and nonmilitary research within the distributed sensor network (DSN) program of the US Defense Advanced Research Projects Agency in 1980 [10]. According to this initiative, "A DSN is a set of spatially scattered intelligent sensors that can obtain measurements from the environment, to abstract relevant information from data gathered, and to derive appropriate inferences from the data gathered" [14]. The DSN program paved the way towards the introduction of smart control and management applications based on signal transfers and processing in industries, in wastewater treatment, and in electrical systems.

The evolution of smart grids is mainly configured by the provided capabilities by the applied metering system on the demand side. The evolution of the relevant technology is graphically depicted in Figure 6.5. The initial approach was the application of one-way automated meter reading (AMR) systems, which enable the data reading from the consumers' side towards the grids' operators [15].

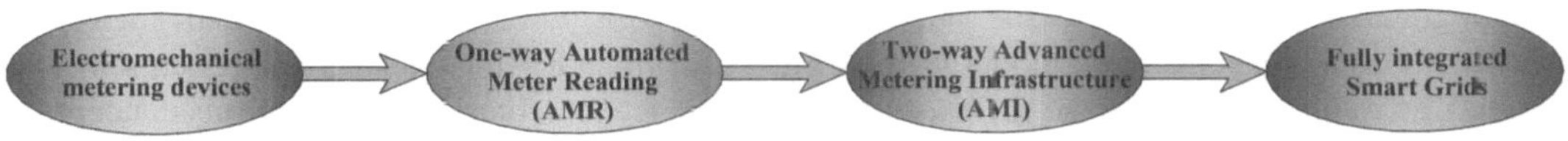

FIGURE 6.5 Main stages of the smart grid evolution process.

However, precisely due to its one-way only data transfer, the development of smart grids cannot be based on AMR systems, because those systems did not allow any control options from the operators' side towards the consumers' side.

A second approach is the development of so-called advanced metering infrastructure (AMI). AMI has been developed in order to enable two-way, bidirectional data transfer between the utilities and the consumers. With AMI, the measured information can be transferred from the consumers' to the operators' side, while, at the same time, requests and orders can be transmitted from the operators to the consumers. Actually, this is the first integrated approach towards the development of smart grids.

A third stage is the further development of AMI to a technology infrastructure level that will enable direct control and management actions from the operators. With this approach, operators will be able to directly perform distributed commands on the demand side, such as load curtailments, in a pervasive way, across the overall geographical topology of the grid and for all the involved components and procedures.

Three different smart grid generations can be identified during the evolution of the relevant technology. Smart Grid 1.0 generation (existing smart grid stage) enables the implementation of specific fundamental services, such as prepaid metering, intelligent disconnection, in-home displays of several domestic and grids' data, accurate load control by both the consumers' and the operators' side, bidirectional metering and info transmission, demand response strategies, and advanced outage management. Smart Grid 2.0 generation (grid resident intelligence) will enable the introduction of additional services with a more integrated and intelligent approach, such as intelligent street lighting versus the usage of streets, integration of EVs in electrical grids and their exploitation for the implementation of specific DSM strategies (see Section 6.3.2), advanced control and secure penetration of distributed generation, including RES penetration combined with energy storage, demand and production capacity prediction, fault prediction and outages prevention, and automatic demand response.

Smart Grid 3.0 (grid leveraged applications) will lead to a fully integrated grid management and automated operation. It will enable electricity transfer between the consumers themselves without the intervention of the operators through fully automated functions, real-time inspection of available power production capacity and demand, the currently configured electricity prices in wholesale electricity markets, and the implementation of relevant economic transactions, etc. Smart Grid 3.0 will extend beyond an advanced technical infrastructure for the effective delivery of electricity to an innovative transactional platform that will foster social and economic development with the introduction of new business opportunities, new professional activities, new services, and economic models.

The first smart grid commercial application is considered the ENEL's Telegestore Project in Italy. ENEL is the largest utility company in Italy and the second largest in Europe, with regard to installed power. Currently, the Telegestore Project is the pioneer smart metering application internationally. It consists of 32 million electronic smart meters, more than 350,000 data concentrators, and thousands of smart meters in secondary substations [16].

6.2.3 Smart Grid Conceptual Model

From the previous section, we see that for successful transition from conventional electricity generation and distribution to smart grids, new technologies are required and considerable infrastructure upgrades of existing facilities should be implemented. A first approach towards this direction is the definition of a conceptual model for smart grids in which all the involved components will be defined, as well as their roles and the executed functions between them. The role of this conceptual model is to standardize the involved stakeholders and their roles in smart grids and to provide a high-level view of the system, understandable by all of them. It is anticipated to constitute a useful tool for regulators, grid operators, aggregators, and final consumers at all levels to determine best practices towards the maximization of electricity production and distribution efficiency and the minimization of the corresponding arisen costs, so as to stimulate investments in upgrading and modernizing the national electrical systems. Aggregators are new entities, practically service providers, with the aim to integrate smaller participants (producers and consumers) and represent them in the processes executed within a smart grid.

So far, several smart grid conceptual models have been developed by different standardization bodies. The most widely accepted and used in relevant references and scientific works is the conceptual model introduced by the National Institute of Standards and Technology (NIST) of the US [17]. According to its updated version, this conceptual model identifies seven discrete domains in smart grids: a. customers, b. markets, c. service providers, d. operations, e. generation including distributed energy resources, f. transmission, and g. distribution. A general description of each domain along with its role and related services, according to its updated version, is presented in Table 6.2 [17].

Each domain with its sub-domains consists of actors and processes. Actors may include devices, programs, software applications, and systems, all of them cooperating and exchanging the information required for the realization of processes and the integration of smart grids. Processes, on the other hand, include tasks and applications that actors execute. For example, metering infrastructure, RES or conventional power production units, storage devices, management systems, etc., constitute the actors in a smart grid, while electricity production and storage, load curtailments, home automation, and load shifting constitute the executed processes. Additionally, the executed tasks and

TABLE 6.2
Domains Defined in the NIST Updated Conceptual Model, Along with Their Assigned Roles and Services

No.	Domain	Roles/Services in the Domain
1	Customer	The final consumers of electricity. They may also generate, store, and manage electricity. Three customer types are discussed, each with its own sub-domain: residential, commercial, and industrial.
2	Markets	An economic mechanism that provides services for effecting electricity procurement and selling, using supply and demand balance to define prices. This is where grid assets are exchanged. Actors are the operator and the participants in the electricity markets.
3	Service providers	Entities that offer support services to electrical customers, producers, and distributors. Actors are organizations that provide services to electrical customers and to utilities.
4	Operations	Any tasks regarding the proper operation of the electrical system, including the management of electricity generation, storage, transmission, distribution, and final consumption. Actors are the managers of the electricity transactions.
5	Generation including distributed energy resources	The production of electricity from either conventional or RES power plants, bulk or distributed generation, including storage plants.
6	Transmission	Transmission of high-voltage electricity over long distances through transportation grids. It may also include production and storage. Actors are the carriers of electricity over long distances.
7	Distribution	Distribution of medium- or low-voltage electricity from and to the final consumers, including consumption metering, distributed generation, and storage. Actors are the distributors of electricity to and from customers.

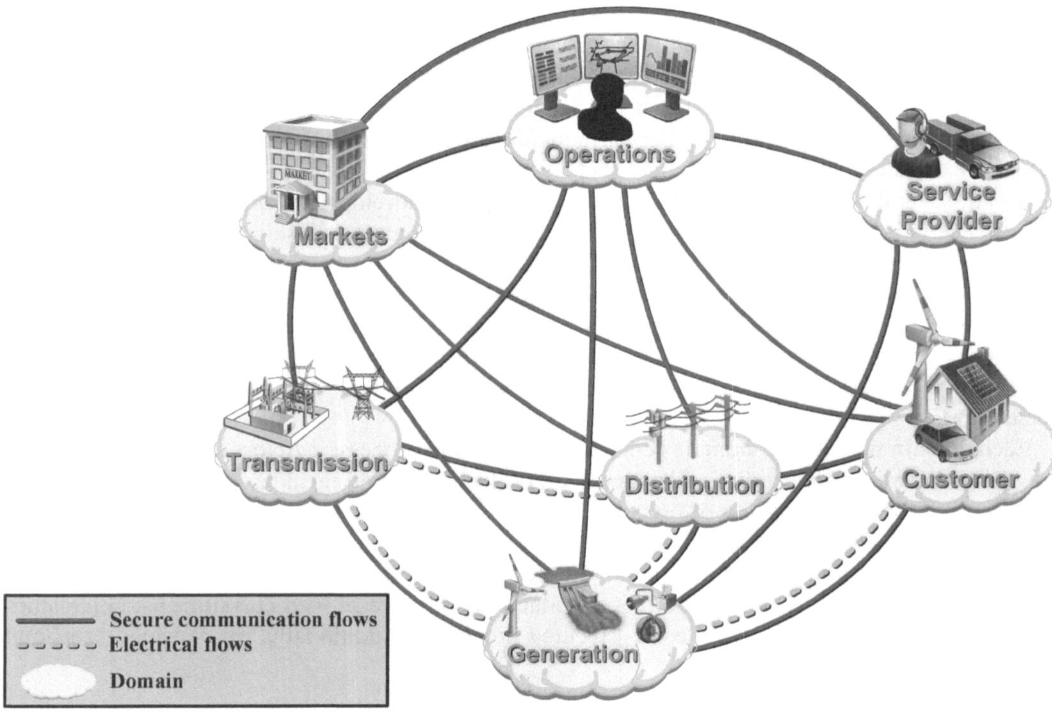

FIGURE 6.6 The NIST conceptual model [17].

processes within a smart grid may require the interoperability of roles and functions between the domains. The NIST conceptual model is presented in Figure 6.6.

An updated version of the NIST conceptual model was provided by the CEN/CENELEC/ETSI strategic partnership [18]. The fundamental update was the addition of a new domain, related to the distributed energy resources. As seen in Table 6.2, this domain has been introduced in the "generation including distributed energy resources" domain of the updated NIST model.

The introduction of the conceptual model by NIST is based on two essential concepts:

- The roles and the responsibilities assigned to the actors involved in a domain depend on this domain. This means that the role and the operating functions of the same actor involved in different domains may be different versus the domain. The same stands for the expected benefits. For example, a PV station will operate under different operating status and parameters when installed on the roof of a residential building, or on the roof of an industrial building, or as a centralized power production plant for bulk electricity generation.
- While the energy transfer processes remain relatively simple, as they depend on a few simple physical electrical connections, realized on the existing transportation or distribution grids, the communication network and infrastructure required for the development and the expansion of smart grids, is characterized with a growing complexity. Indeed, electricity transportation and distribution is always based on the same fundamental design and construction principles, depicted in the well-known layout presented in Chapter 2. Conversely, as final consumers are more and more involved in the electricity production and management processes, enabled and supported by the upgrade of the relevant technology (power electronics, sensors, software applications, data transfer grids), the corresponding communication networks and infrastructure are continuously expanding.

A short description of the defined domains is provided below.

6.2.3.1 Customers Domain

The customers domain constitutes the domain that the whole smart grid is developed to service and support, because this is the domain where electricity is consumed. The design and the development of smart grids aim to provide advanced generation and management opportunities for the customer domain, through the involved actors and processes, such as dispersed power production and energy storage (see Section 6.3.2) and flow information transfer between the customers domain and other domains (e.g., service providers, markets). Typically, the boundaries of the customers domain are defined by utility meters or the local energy management systems (EMS).

The EMS constitutes a secure interface between the utility or the service provider and the final consumers. It acts as the main communication node for the consumers with the smart grid, integrating in the grid any probably involved building automation system from the consumers' side, such as local smart meters, sensors, and the home area network, which is presented in another section. The EMSs can communicate with other domains or the grid's operator with the advanced metering infrastructure or the Internet.

A general layout of the customer's domain is presented in Figure 6.7. The customers domain is divided into three sub-domains: residential domain, commercial domain, and industrial domain, with typical power demand below 20 kW for residential consumers, between 20 and 200 kW for commercial consumers, and over 200 kW for industrial consumers. Several actors and applications are involved in each sub-domain, which may be also present in other sub-domains. The main entrance–connection point of each sub-domain with the smart grid is a meter actor with an EMS.

The customers domain has direct communication with the markets, the service providers, the operations, the generation including distributed energy resources (DER), and the distribution domains. More than one communication entry can be available for each final consumer. These entries can

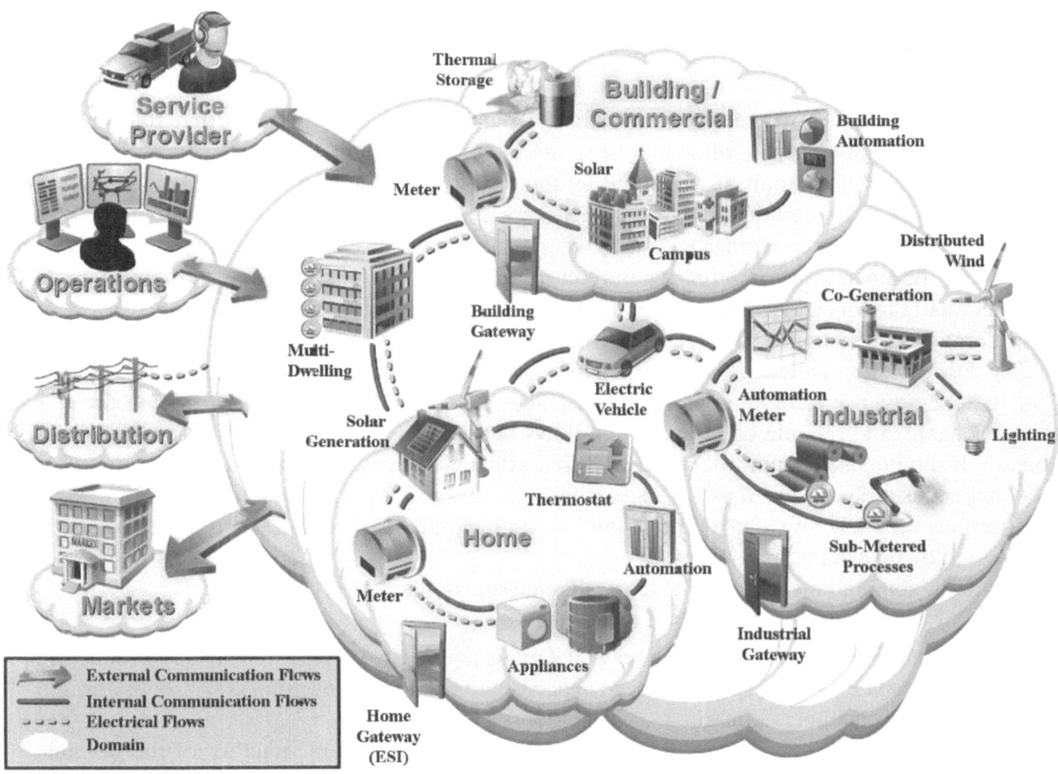

FIGURE 6.7 The customers domain, according to the NIST conceptual model [17].

provide access for the utility, the aggregator, etc. to the remote load control, the monitoring and the management of the dispersed power production, the monitoring of the current power consumption, and control over each different involved type of electrical load (air conditioning, lighting, etc.) and, finally, integration in the grid of any potentially existing EMS in the consumer's premises. They may also provide auditing/logging for cyber security purposes.

The main tasks and processes executed in the customers domain can be:

- *Residential buildings automation:* this process can be implemented by a building energy management system (BEMS), which is capable of controlling all the involved different type of loads and the corresponding systems and devices, versus specific, predefined operational parameters or goals to be achieved (e.g., desirable thermal comfort conditions, maximum power consumption at specific time periods, available luminous flux).
- *Industrial automation:* a process similar to the automation executed by a BEMS, but adapted to the needs, the specifications, and the relevant involved applications and processes of industrial units (e.g., thermal processes, motorized machinery, stocking).
- *Distributed production:* includes any type of decentralized, small-scale power generation (e.g., PV, small wind turbines, biomass units, diesel gen-sets). This dispersed power generation can be controlled and monitored via the available communication infrastructure.
- *Storage processes:* include any type of electricity storage in small dispersed storage units (most typically electrochemical storage devices), aiming to power peak shaving, load curtailment, etc.

More on the above alternative processes executed in customers domain is presented in Section 6.3.2, dedicated to demand side management strategies.

6.2.3.2 Markets Domain

The markets domain is where electricity and the related services are bought and sold, in a fully liberalized electricity regime, following the terms and conditions defined by the supply and demand balance. The markets domain is delimited by the edge of the domains that supply assets and services, which are the generation including DER domain; transmission domain; distribution domain; customers domain, which defines the place where the provided assets and services are offered and consumed; the operations domain, where control and management of the overall smart grid are executed; and the service providers domain. In other words, the markets domain interfaces with all other domains of the smart grid (see Figure 6.8).

The successful and adequate integration and operation of the markets domain within the smart grid require advanced communication infrastructure, which will ensure on-time, secure, auditable, traceable, and reliable communication flows between the involved domains and the markets domain. It should also support e-commerce standards for integrity and nonrepudiation. Finally, further challenges are introduced by the increasing share of DER in the electricity production balance. This implies very short response time of the info on the markets domain related to any offers, bids, etc., posted by distributed producers.

The main applications and challenges in the markets domain are [19]:

- Communicating of electricity prices and signals coming from DER (electricity production offers, load curtailments, and storage capacity), to each of the customers sub-domains and among the customers domain.
- Simplifying rules regarding the electricity pricing and the functionality of the markets domain.

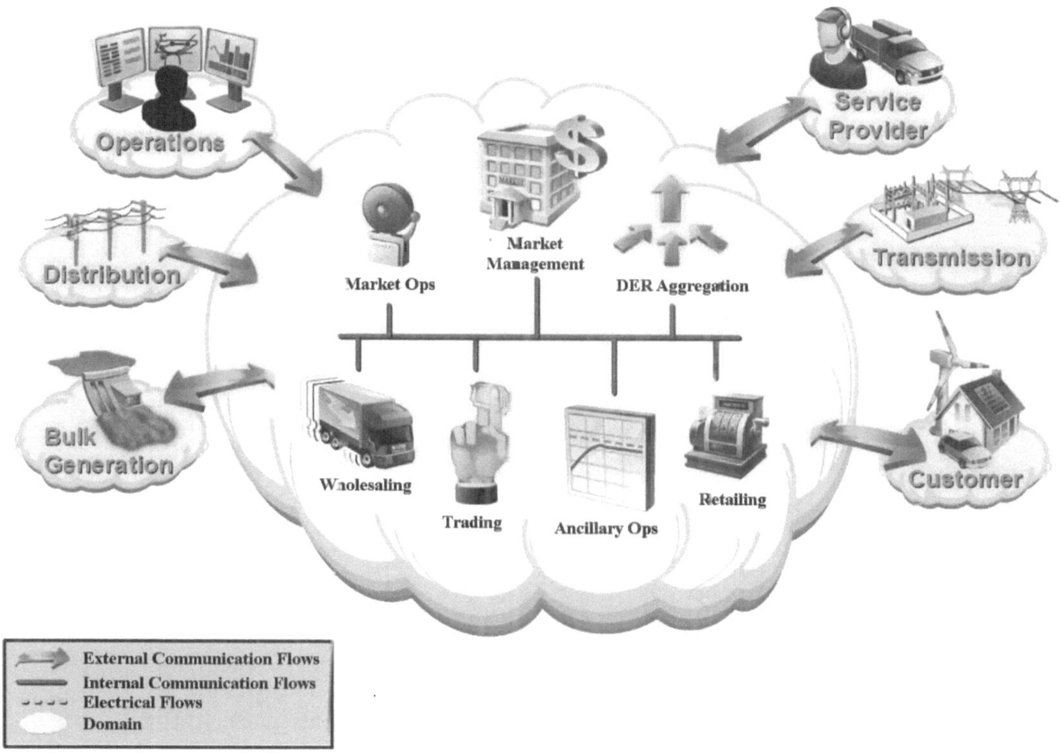

FIGURE 6.8 The markets domain, according to the NIST conceptual model [17].

- Facilitating the services provided by the aggregators to enable DER in the wholesale electricity market.
- Supporting interoperability across all involved domains.
- Managing and inspecting the growth (and regulation) of the retail and wholesale electricity market.

6.2.3.3 Service Providers Domain

The role of the involved actors in the service providers domain is to support the executed processes in the wholesale electricity market between the power producers, the distributors, and the consumers. The provided support ranges from conventional offered utility services, like electricity billing and consumers' accounts management, to the management of newly introduced services in smart grids, such as distributed generation and DSM. In a general approach, service providers develop and offer new and innovative services and assets to support the emerging opportunities and challenges arisen from the continuously evolving smart grids. These services can be provided by the grid's operator, by the aggregators, or by third parties arisen from any new business models. Service providers constitute the contact point of the customers domain, the operation domain, and the markets domain (Figure 6.9).

In practice, the above presented objectives of the service provider domain are realized with the development of key interfaces and standards that will support the operation of a dynamic market-driven electrical system, providing, at the same time, the required security and protection for both

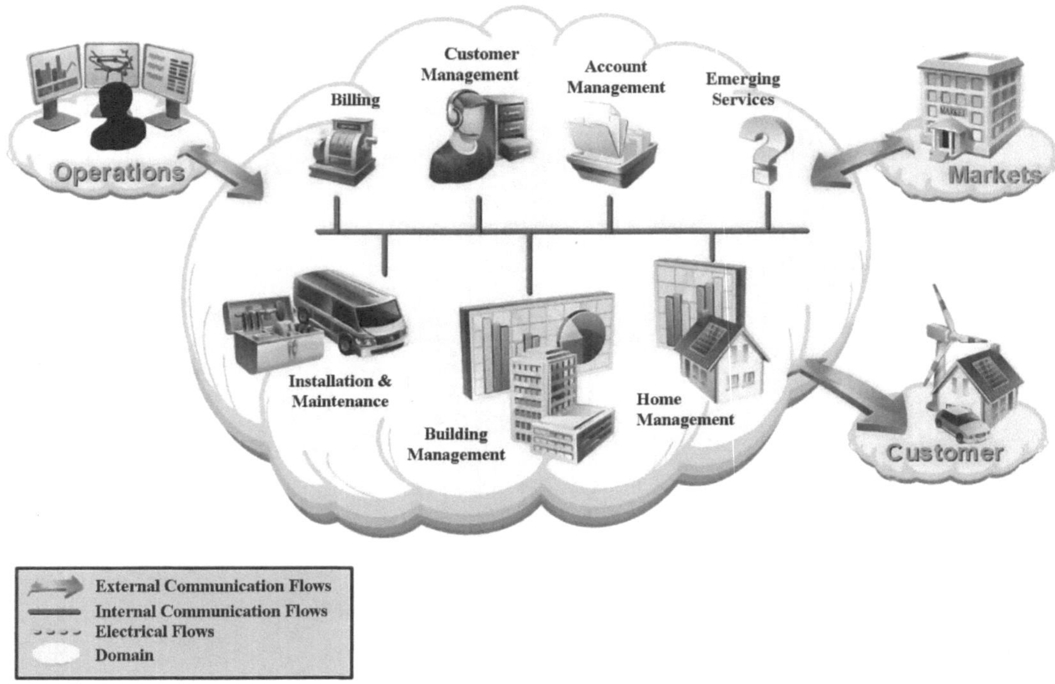

FIGURE 6.9 The service providers domain, according to the NIST conceptual model [17].

the power production and distribution infrastructure and the executed tasks and processes in the frame of the liberalized electricity market. These requirements imply that the reliability, the integrity, and the cyber security of the employed online systems, as well as the safety and the stability of the electrical power network, constitute critical issues, not negotiable within a smart grid system.

The main applications performed in the service providers domain are:

- *Customers' management:* management of the customers' relationships between them and with the operator, aggregator, etc., by providing a point-of-contact and resolutions for any potential customer issues, queries, and problems.
- *Installation and maintenance of the required infrastructure:* installation, operation, and maintenance of the required technical infrastructure within the operation of the smart grids.
- *Buildings' management:* monitoring and control of building energy consumption and response to any signals sent by the operator, minimizing, at the same time, any impacts on the building's users.
- *Energy management:* monitoring and management of offered products and assets (energy production, storage capacity, ancillary services, etc.), perhaps by different providers and at different locations, and optimization of the smart grid's operation for the involved different customers, operators, etc.
- *Billing and accounts management:* management of customers' billing information, including posting of billing statements, payment processing, etc.

6.2.3.4 Operations Domain

The main objective of the operations domain is the smooth operation of the smart grid. The tasks of the operations domain are normally executed in conventional electrical systems by the utility. Within the transition to smart grids, these tasks are transferred to service providers. Due to the evolution of the markets and the service providers domains, the following applications will need to be provided in the frame of the operations domain:

- *Monitoring–control:* the monitoring of the smart grid's operation, including the supervision of the network topology, the connectivity and loading conditions (breakers and switch states), the control of the equipment status, the field crew location and status, the substations, the local automatic and manual control of the distribution systems, etc.
- *Faults management:* fast detection of the location of a fault, fault's identification and classification, coordination of the workforce dispatch and actions towards the fast restoration of the system. Additionally, information posting on the fault for the customers, compilation of the possible causes, and statistic records.
- *Analysis:* analysis of the electrical system's operation, statistical records based on real-time operation, analysis and compilation of network incidents, connectivity status, seasonal analysis of the power demand fluctuation, optimization of the system's scheduled maintenance.
- *Reporting and statistics:* annual statistics on production, fuels' consumption, and RES penetration in the electrical systems, feedback analysis on system's annual average efficiencies and reliability, online posting of annual technical bulletins on the system's operation.
- *Network calculations:* real-time measurements and calculations of characteristic grid's magnitudes (nominal frequency and RMS voltage), and through them assessment of the system's smooth operation, reliability, security, and stability.
- *Operation planning:* computational simulations of the smart grid's operation and planning of specific tasks and functions based on the results of the simulations, such as short- or long-term planning of the system's operation, scheduling of switching actions, scheduling the power import from neighboring grids to avoid high production cost, for instance during peak power demand periods.
- *Maintenance and construction:* inspection, cleaning, adjustment, and maintenance of the electrical system's equipment, design, and organization of the construction of new assets; allocation and schedule of the maintenance and construction work, capture of records gathered by field technicians on the performed tasks.
- *Extension planning:* development of long-term plans for the electrical grid extension and the enforcement of the system's reliability, design, and definition of the projects particularly required for the grid's extension (new transportation or distribution lines, feeders, switches, transformers, etc.), monitoring of the construction cost and the system's performance, management and scheduling of the construction.
- *Customer support:* support customers to purchase, install, and operate power production, storage or management equipment, handle and troubleshoot any operation problems or malfunctions, relay customers' problems and keep relevant records.
- *State estimation:* estimation of the current operation state of the electrical system based on algorithms applied to real-time measurements from specific locations of the electrical grid and accomplishment of corresponding actions towards the optimization of the system's operation.

An overview of the operations domain is given in Figure 6.10.

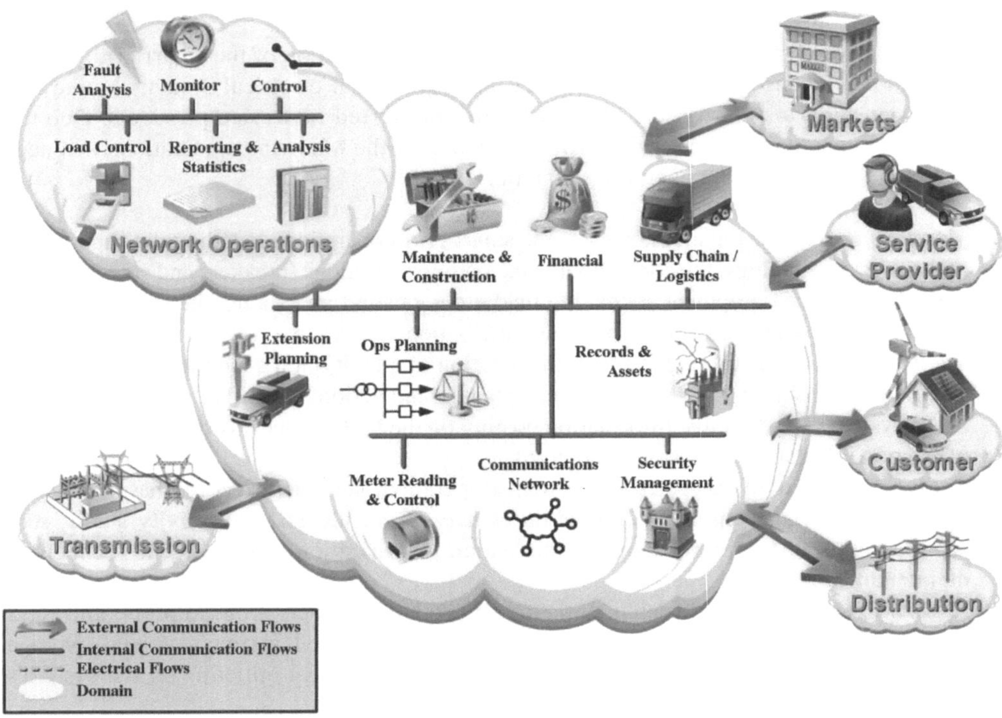

FIGURE 6.10 The operations domain, according to the NIST conceptual model [17].

6.2.3.5 Generation Including Distributed Energy Resources Domain

This domain includes everything that has to do with the electricity production from primary energy resources, starting from centralized power plants (e.g., thermal, nuclear, hydro power plants), expanding to large-size RES plants (e.g., wind parks, PV stations, geothermal power plants) and ending at distributed power production from small-size stations (e.g., small wind turbines, PVs, biomass stations). The generation including DER domain constitutes the primary electricity supply for the smart grid. As such, it can be connected to the transmission, the distribution, or the customer domains, while it can share communication interfaces with the operations, markets, transmission, and distribution domains (Figure 6.11).

Traditionally, electricity was supplied by large power plants, connected only with the high-voltage transmission grids. Yet, the scalability and modularity of modern power production technologies (e.g., PVs) change the physical relationship between production and demand, as well as the coupling nodes between generation assets and the grid. Hence, the generation including DER domain has been respectively adapted and updated to reflect and enable direct electrical interconnection with the distribution system of smaller and distributed power production stations, maximizing the flexibility of the grid.

An essential parameter for the adequate integration of the generation including DER domain in the smart grid is the communication with the involved domains, namely the transmission, distribution, customers, and the markets domains. Especially in the case of DER, the adequate status of the offered communication infrastructure is significantly crucial.

Emerging requirements for the generation including DER domain may include specific priorities, depending on each different geographical region and the applied directives, such as reduction of greenhouse gas emissions from the energy sector, achievement of RES annual penetration shares on the electricity production balance, involvement of storage plants to manage the fluctuation of the electricity production from RES (mainly wind parks and PV), etc. The achievement of these

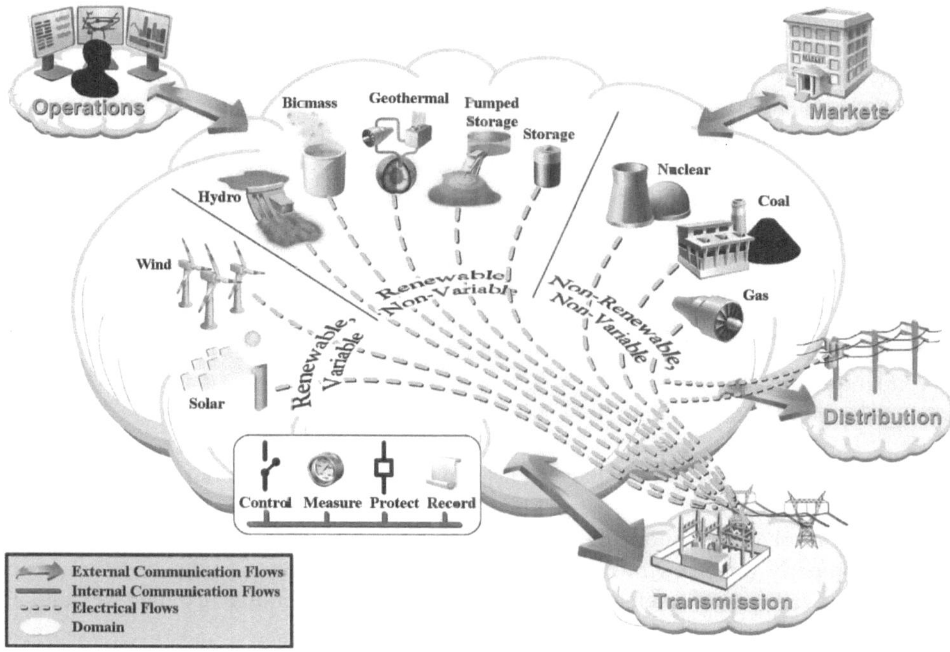

FIGURE 6.11 The generation including DER domain, according to the NIST conceptual model [17].

specific objectives may require coordination across multiple domains. This complexity and associated interoperability requirements should be handled through the conceptual model communication flows. Some typical functions within the generation including DER domain that depend on communication flows and require interoperability between domains are:

- *Control:* tasks performed within the operations domain aiming at the management of the power flow and the reliability of the system. A typical example is the use of phase-angle regulators in a substation to control power flows between two neighboring power production plants.
- *Measure:* measurements performed in strategic points of the grid to provide inspection on the power flow and the overall operation conditions of the electrical systems. A typical example is the measurement gathering of remote supervisory control and data acquisition (SCADA) systems in a thermal power plant, a wind park, etc., and the collection of all these measurements in a control center of the operations domain. In the future, increasing needs for measurements can be implemented by installing a network of field measuring individual devices in the grid.
- *Protect:* fast reactions to system faults that may result in power supply interruptions, voltage or frequency sags, and other contingencies. Protection can be provided centrally or locally in terms of the so-called ancillary services, aiming at the maintenance of the system's dynamic security and stability.

6.2.3.6 Transmission Domain

The transmission domain includes the required technical infrastructure and the performed tasks with regard to the electricity transportation (or transmission) from the power plants to the substations that constitute the coupling points between the transportation and the distribution networks. Typically, responsible for the construction, the maintenance, and the operation of transmission networks are the transmission-owning utilities, known either as a regional transmission operator (RTO) or independent system operator (ISO) or transmission system operator (TSO), whose essential responsibility is to maintain the dynamic security on the electrical system by balancing power production and demand across the transmission network.

A transmission network typically consists of towers, power lines, and field telemetry and is usually monitored and controlled through a SCADA system that uses a communication network, field monitoring devices, and control devices. The transmission grids are integrated in the transformers installed in the transmission network substations, where the high voltage in the transmission grid is transformed to the medium voltage of the distribution grid. Examples of physical actors involved in the transmission domain can be remote terminal units, substation meters, protection relays, power quality monitors, phasor measurement units, sag monitors, fault recorders, and substation user interfaces. The transmission domain may also contain DER, such as electricity storage devices or peak power production units (Figure 6.12).

Actors in the transmission domain typically perform the following applications:

- *Substation:* high- to medium-voltage transformation, control and monitoring of electricity flows, and the performed processes within the substation.
- *Storage:* control and monitoring the charge and discharge processes of electricity storage devices properly allocated in the transmission grid, in order to bridge temporary mismatches between power production and demand, given the available installed capacity and the grid's topology.
- *Measurement and control:* any measurement and control process and equipment employed to measure, record, and control the operation of the grid, aiming at the secure electricity supply, the protection of the equipment, and the optimization of the grid's operation.

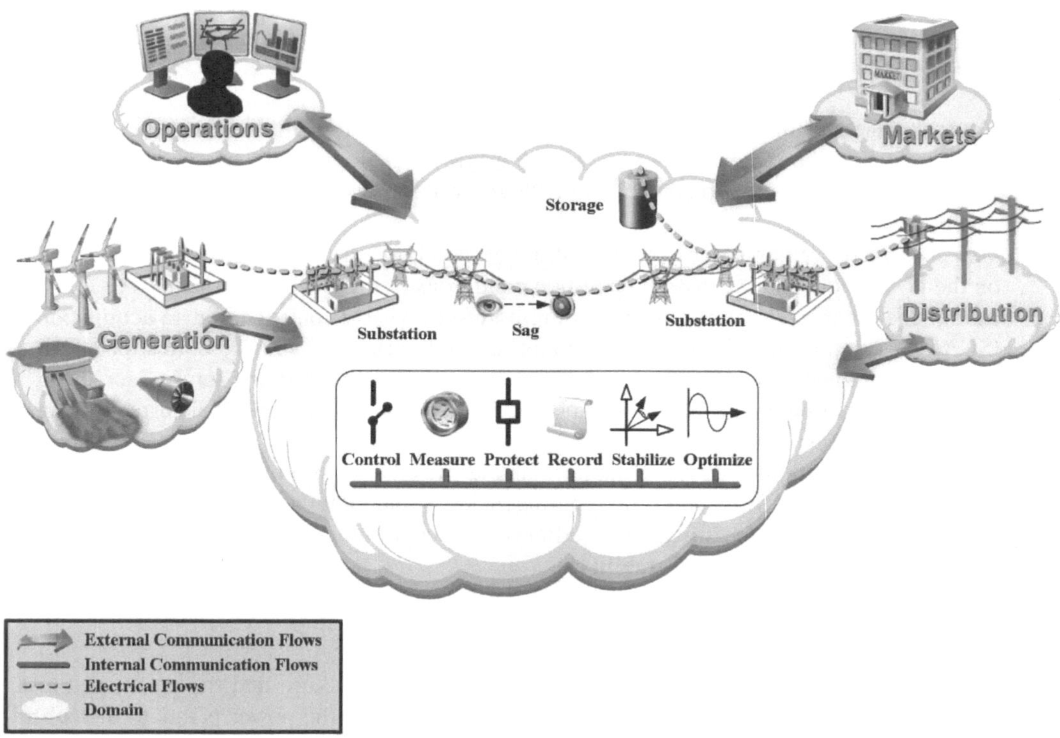

FIGURE 6.12 The transmission domain, according to the NIST conceptual model [17].

6.2.3.7 Distribution Domain

The distribution domain constitutes the electrical interconnection between the transmission domain and the customers domain. It contains the metering points for the electricity consumption, the distribution grid lines, the medium- to low-voltage substations, and any possibly existing distributed storage and generation units.

The electrical distribution systems may be arranged in a variety of structures, including radial, looped, or meshed, with the radial layout being the most common. The reliability of the distribution grid depends on its structure, the control devices and their configuration, and the applied communication flows between these devices and with actors involved in the other domains. The integration of effective and adequate smart grids requires bidirectional communication flows, while the use of high-speed communications is also essential to manage and optimize power flow and electricity generation and consumption in real-time, especially in grids with high DER penetration.

Within smart grids, increased sensing, control, and communication facilities are required with the operations domain in real-time, to manage the complex power and communication flows associated with the new employed technologies, the intensively dynamic markets domain in the wholesale electricity market, as well as other environmental and security-based factors. In general, the effective operation of smart grids under this dynamic environment implies the necessity for improving observability and inspection of the distribution grid operation conditions, which can be approached with additional sensing devices (e.g., fault circuit indicators) as well as domain operational functions (e.g., stabilize). The necessity for all the above is obviously enforced by the fact that the distribution domain is the main connection domain with the customers domain, namely this is the port of the customers domain with the other domains of the smart grid (Figure 6.13).

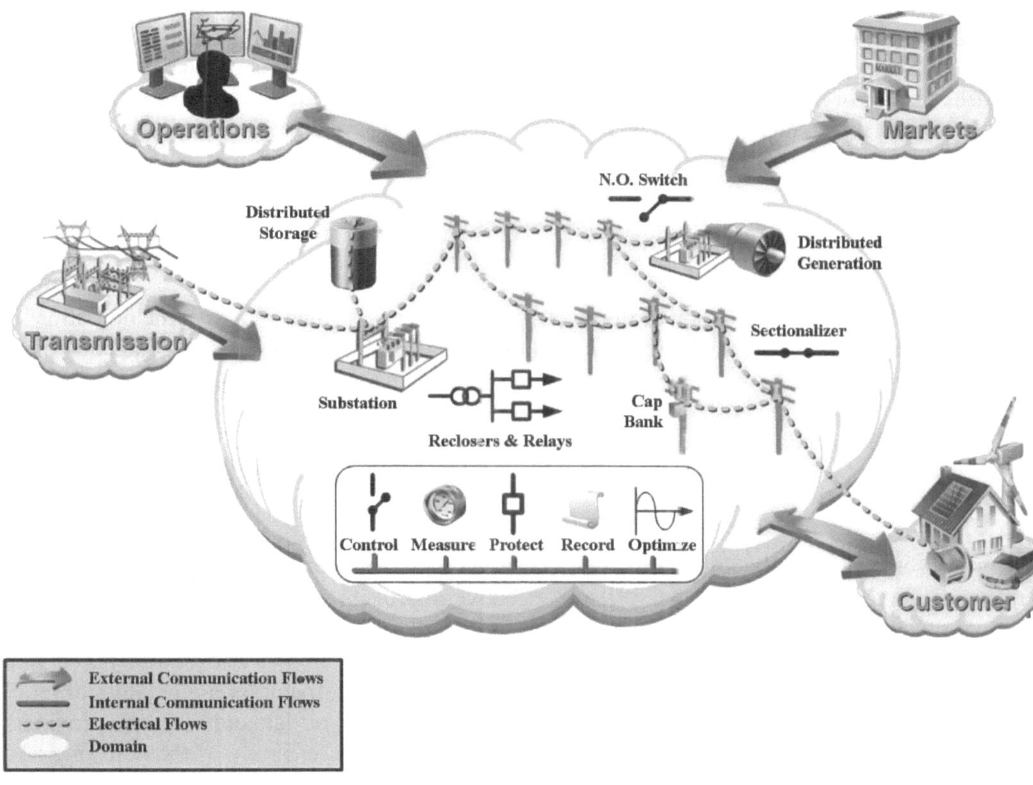

FIGURE 6.13 The distribution domain, according to the NIST conceptual model [17].

Typical actors and functions executed in the distribution domain are:

- *Substation:* control and inspection systems within a substation from the medium-voltage side.
- *Storage:* control and monitoring the charge and discharge processes of electricity storage devices properly allocated in the distribution grid, in order to bridge temporary mismatches between power production and demand, given the available installed capacity and the grid's topology.
- *Distributed generation:* power production plants located at the distribution grid.
- *DER:* distributed power production units of small size, owned either by the customers or the distribution grid operator.
- *Measurement and control:* any measurement and control system introduced in the distribution grid to measure, record, and control power flows, aiming, ultimately, at the protection and the optimization of the grid's operation.

6.3 DEMAND SIDE MANAGEMENT

Demand side management (DSM) includes everything performed on the demand side of an electrical system and it constitutes an integral part of smart grids. DSM aims to improve the flexibility, the reliability, and the efficiency of smart grids, leading also to subsequent reduction of the production cost, by applying actions on the demand side, directly by the consumers themselves, or by the grid's operator, of course with the approval and the consent of the consumers. Both consumers and utilities have specific potential benefits from the implementation of DSM actions, which will be analyzed below in this section.

The Federal Energy Regulatory Commission of the US defines demand response as "changes in electric usage by demand-side resources from their normal consumption patterns in response to changes in the price of electricity over time, or to incentive payments designed to induce lower electricity use at times of high wholesale market prices or when system reliability is jeopardized" [20].

DSM refers to any changes in the power demand profiles of the final consumers, in order to claim lower electricity prices, according to the pricing policy in effect, or given the grid's security and stability requirements. Practically, final consumers are requested or motivated by the operator to modify and adapt their power consumption profile according to the existing operation conditions of the power system, in order to:

- facilitate the power production of the electrical system, given some very specific operation parameters, such as the available total generation capacity, the type and the efficiency of the dispatched generators, the conservation of the system's dynamic security, the electricity production specific cost, etc.
- approach lower electricity procurement cost for themselves, given predefined economic incentives between them and the provider or to claim low electricity prices under a fully liberalized electricity wholesale market.

DSM processes can be realized either directly between the consumers and the utility, or through an intermediary party. In a more flexible and effective approach, especially in fully liberalized wholesale electricity markets, final consumers are organized in groups by intermediaries, which play the role of the consumers' representatives to the utility. They are known as curtailments service providers (CSP), because they contribute to the power demand curtailment, or aggregators of retail customers (ARC), because they aggregate the customers in groups, or demand response providers (DRP), because they cumulatively provide the utility with the total capacity of the consumers to respond to the utility's current requests, e.g., for dispersed power production, or the consumers' capacity to store energy or to reduce their power consumption.

6.3.1 Consumers' Classification

Final electricity consumers (or customers) are classified in the following discrete groups, versus their size, as it is depicted in their overall electricity consumption [21]:

- large commercial and industrial (C&I) consumers
- small commercial and industrial (C&I) consumers
- residential consumers
- individual or fleet of plug-in electric vehicles (PEVs).

The size and the type of the above consumers' classes offer different options and opportunities for the application of DSM strategies. For example, large-size commercial and industrial consumers are usually already equipped with advanced loads' control and management and metering infrastructure. They are also highly aware of the importance of energy consumption and its contribution to the total cost of their running production lines. Moreover, the more energy-consuming their production processes are, the higher the anticipated economic benefits will be from the application of DSM strategies. This potential can also justify the economic feasibility for further investments serving specific DSM strategies, such as the procurement of storage devices, which, in case of residential usage, as it will be shown in the next section, does not seem to be feasible, at least with the currently existing battery procurement prices. Consequently, large-size commercial and industrial consumers are both convinced and prepared, from the technical point of view, to introduce DSM strategies in their facilities.

The most common loads met in commercial consumers of small size refer to their heating, cooling, and ventilation active systems and the lighting equipment. Small industries can be involved in several DSM strategies. Depending on the type and the scheduling of their production lines, they can possibly implement dispersed production, load shifting, energy efficiency, and even load curtailment. In case of merchant consumers (e.g., shops, malls) the usage of their facilities from their customers restricts their options for the application of DSM strategies. Nevertheless, an important feature of these consumers is that most of them have on-site small power production units installed, typically diesel gen-sets, for emergency backup generation or for auxiliary power production. These backup units can be used, under specific contracts with the utility, as dispersed power production units, in case of inadequate centrally installed power production capacity, offering cold stand-up reserves and avoiding, thus, the installation and the maintenance of additional production generators by the utility, which most probably will be dispatched only for very limited time periods during the peak power demand season.

Residential users do not seem yet mature and motivated to be involved in DSM strategies, maybe with the exception of some countries. This is due to the lack of awareness, on the one hand, or adequate technical infrastructure, on the other, especially with regard to the required communication and automation technology. Usually residential consumers are accustomed to the currently existing electricity supply regime, from a unique utility or through a number of involved providers, at a fixed, predefined price on a contract signed between them and their providers. They are not familiar with the process of being involved in retail electricity markets and claiming lower prices for the electricity they consume or higher prices for the electricity they could produce. Additionally, residential consumers are usually characterized by relatively small and very specific type of loads, typically indoor space conditioning, lighting, and motor power. Hence, they are rather reluctant to invest in the application of DSM strategies that will result in a more rational use of energy in their facilities. Consequently, as a result, the so far implemented applications of DSM strategies in residential consumers are rather limited.

However, all the above status regarding residential consumers will sensibly change in the forthcoming years. This is due to the opportunities and the obligations expected from the full liberalization of the electricity markets. Consumers will be obliged to adapt to a new electricity supply and demand regime, being, themselves, an active part of this process. Additionally, the new advances in the relevant technology, especially with regard to the required communications, automation, and metering systems, will enable a more secure, user-friendly, and reliable environment for all the involved parties, with advanced supported actions and offered facilities, such as direct inspection of demand and production from all producers, real-time configured average electricity price, fast transfer of orders, offers, and requests by the utilities, the aggregators, the producers or any other involved parties with online alerts, etc. More on these emerging technologies is presented in a later section.

Finally, plug-in electric vehicles (PEVs) constitute a new type of electrical load, expected to increase considerably in the next years, mainly because of the anticipated cost reductions in lithium-ion battery manufacturing and procurement [22]. The massive introduction of PEVs will lead to an analogous increase of the power consumption for the electrical grids. However, given their flexibility with regard to the charge schedule over the 24-hour period, they can be exploited to support load shifting and/or load curtailments, aiming at a more balanced power demand between nights and days. In a more organized and controlled approach, PEVs can be charged following a general proposed schedule by the utilities.

6.3.2 Demand Side Management Strategies

DSM can be realized with four alternative strategies: peak load shaving (load shifting), dispersed power production, load curtailments, and energy usage efficiency. These DSM strategies are analyzed in the following paragraphs.

6.3.2.1 Load Shifting (Peak Load Shaving)

Power shifting can be implemented with two alternative ways. The first one is based on the direct shifting of the use of specific, flexible loads from the peak demand periods to the "valleys" of the daily consumption profile (off-peak demand periods). Such loads or activities can be washing machines, ironing, even cooking and indoor space conditioning, by precooling or preheating the conditioned space during periods of low power demand, or by introducing heating or cooling active systems with low operating temperatures of the working medium and continuous operation, even when the building is not used for short time periods (e.g., during the morning when the occupants may be away from home). A typical example of such systems is the combined operation of heat pumps with underfloor heating distribution pipelines.

The second approach is by introducing distributed energy storage devices. In this case the concept is also simple and similar to the process of peak power demand shaving, analyzed in Chapter 2. However, instead of constructing a central storage plant of large size, DSM peak shaving is based on the introduction of small-size, dispersed energy storage devices (e.g., electrochemical batteries) from the consumers. Then, the process is quite the same: electricity is absorbed from the grid during low power demand periods and electricity procurement prices, in order to be injected back to the grid during peak power demand periods.

Dispersed energy storage can significantly contribute to the reduction of the peak power demand of an electrical system. Let's return to the case of Sifnos Island, Greece, presented in Chapter 3, Section 3.8.2, and assume that 200 out of the 800 permanent families on the island install a battery unit of 5 kW charge/discharge power and 14 kWh of storage capacity. This will give a total discharge power of 1 MW and storage capacity of 2.8 MWh on the island. The total storage capacity cannot be considered as an important contribution to the average daily electricity consumption, which, given the annual electricity consumption of 18,500 MWh (see Chapter 3), is calculated at 50.7 MWh. Moreover, comparing the total storage capacity of 860 MWh of the pumped hydro storage (PHS) system examined in Chapter 3 for the same island, we could say that the additional storage capacity of the dispersed storage units is almost negligible. Nevertheless, the total discharge

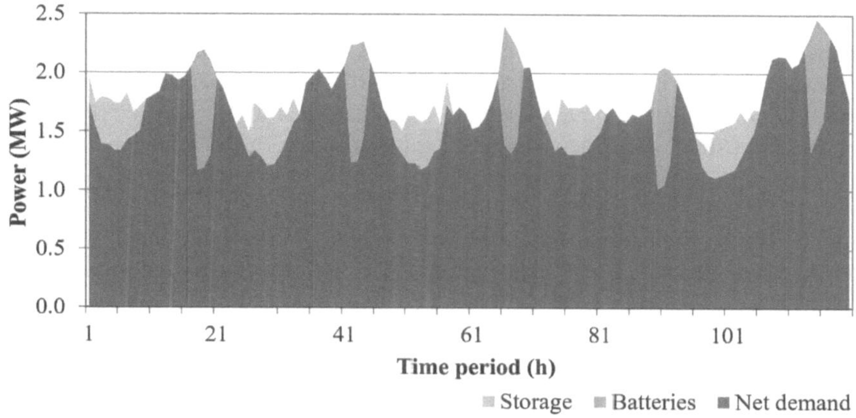

FIGURE 6.14 Load shifting realized with distributed energy storage in the power production synthesis for the island of Sifnos, Greece.

capacity of 1 MW corresponds to 40% and 15% of the daily peak power demand on the island during the low and the peak power demand season, respectively. Hence, by storing electricity during night periods and returning it back to the grid during peak demand periods, the system can benefit from a significant reduction of the peak power demand, as depicted in Figure 6.14. This, in turn, will increase the security of the insular, weak grid and will facilitate the target of the local energy cooperative towards energy independence.

However, the economic feasibility for the introduction of dispersed energy storage units does not seem be quite obvious. Continuing with the example for the island of Sifnos, a typical current procurement cost of lithium-ion batteries is 400 €/kWh of storage capacity, so, each one of the above mentioned batteries units, with a total storage capacity of 14 kWh, costs €5,600. If we assume 80% battery discharge depth and 90% total charge–discharge cycle efficiency, then the annual electricity production from a battery unit will be:

$$(0.8 \times 0.9 \times 14 \text{ kWh/day}) \times 365 \frac{\text{days}}{\text{year}} = 3{,}679 \text{ kWh/year.}$$

Furthermore, if we adopt a total economic benefit for the final consumers of €0.085 per each kWh shifted from peak to low power demand periods (this value corresponds to 50% of the final, gross electricity price for residential consumers in Greece), then the annual economic benefit will be:

$$3{,}679 \text{ kWh/year} \times 0.085 \text{ €/kWh} = 312.7 \text{ €/year}$$

which implies a payback period of the total procurement cost of:

$$5{,}600 \text{ €}/312.7 \text{ €/year} = 17.9 \text{ years}$$

With a foreseeable lifetime period of the lithium-ion battery at 10 years, the above calculated time period essentially means that the investment will not ever be paid back.

However, in case of a procurement price drop from 400 €/kWh to 100 €/kWh, as currently predicted for the next 10–15 years, the total procurement price for each battery unit will be €1,400. With the annual economic benefit at 312.7 €/year, the payback period of the investment is now calculated at 4.5 years, which seems to be quite attractive.

6.3.2.2 Dispersed Power Production

Dispersed power production refers to the electricity production from dispersed units of small size, owned by consumers or by individual producers. These dispersed units most commonly involve PV stations, small wind turbines, biomass stations, combined heat and power (CHP) units, and even thermal generators of small size. Such small-size stations can be implemented under two alternative incentives:

- Refunding of the producers for the electricity injected in the grid, under alternative pricing policies: predefined, fixed, and guaranteed electricity prices (feed-in-tariff regime), or varied prices, determined on the basis of bids posted online by the producers within liberalized, wholesale electricity markets.
- Operation under net-metering mode: the power production station is related to a specific consumer. The total electricity production on a monthly or, most usually, an annual basis is compensated with the total electricity consumption of the related consumer. No reward for any excess electricity production is provided for the producer at the end of the under consideration time period, while the consumer will have to pay only for the excess electricity consumed, after being compensated with the total electricity production.

Dispersed power production, even for PVs or small wind turbines, namely of nonguaranteed power production units, is characterized with relevant reliability, precisely due to the reduced power production fluctuations obtained as a result of the power production units dispersion. Additionally, because these dispersed stations are connected to the low-voltage distribution grids and, normally, very close to the final consumers, electricity transportation and transformation losses are avoided or minimized.

The most common technology involved in dispersed power production is PVs. The significant reduction of PV panel construction and procurement cost has led to a subsequent drop of the total setup cost of PV stations at values lower than 1,000 €/kWp. This can guarantee the economic feasibility of such investments even in cases of installations under net-metering mode. For example, a typical household in Europe exhibits average daily electricity consumption at 12 kWh, which gives a total annual electricity consumption of 4,380 kWh. Assuming an annual, final capacity factor of a PV station at 16.5% (fixed installation), the required nominal power of a PV station for annual electricity production equal to the above estimated consumption per household will be 3 kW, with a total setup cost of €3,000 (adopting a setup specific cost of 1,000 €/kWp). Accounting a net electricity procurement price (without taxes, special fees, etc.) for the final consumers at 0.12 €/kWh, the compensation of annual electricity production of 4,380 kWh will result in an annual economic benefit of €525.6 for each household. This implies a payback period of the total setup cost of 5.7 years.

Additionally, dispersed power production can have a significant contribution towards the elimination of the peak power demand. As an example, we will go back to the case of the insular system in Sifnos. Following the assumptions from the previous example, if 200 families on the island install a 3 kW PV station on their houses' roofs (in total 600 kW), the power production from these stations practically will result in the day peak power demand shaving, as depicted in Figure 6.15. In Figure 6.15, the cumulative impact on the daily power demand profile is depicted, including also the load shifting impact from distributed energy storage. What is observed from this figure is that with the combined introduction of dispersed power production and storage units, both day and night peaks can be eliminated.

6.3.2.3 Load Curtailment

Energy curtailment refers to the reduction of the end-users' power consumption, regardless of their energy needs, at specific time periods and following a relevant request of the operator or an aggregator. Obviously it aims at the reduction of the total power demand in the electrical grid, in

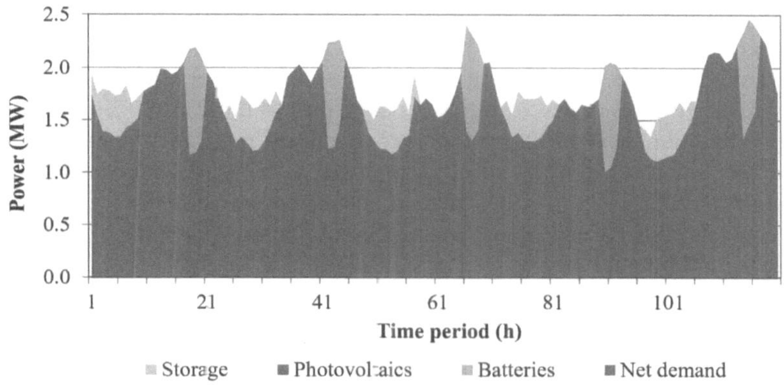

FIGURE 6.15 The effect of distributed power production and load shifting with distributed energy storage in the power production synthesis for the island of Sifnos, Greece.

order to avoid dispatching thermal generators with higher electricity production-specific cost (e.g., gas turbines) or due to limited available power production capacity. This strategy can be realized with several alternative approaches and programs. The simplest one is through a direct load control (DLC) program. DLC programs are based on agreements between the utility and the consumers that define the terms and the conditions under which the utility, or an aggregator, can remotely control and manage the use of specific electrical appliances. DLC is mainly applied to control the use of lighting appliances and thermal comfort equipment operation and, secondarily, other types of consumption, such as refrigerators, freezers, and pumps. In practice, control over lighting equipment is realized with the reduction of the luminous flux emitted by lighting bulbs, performed with the installation of dimmable apparatus, while control over heating or cooling equipment is simply realized with the decrease or the increase, respectively, of the desirable indoor conditioned space temperature. All these are defined in the above agreements, together with other parameters required for the full realization of a DLC program, such as the time periods during a 24-hour interval or the seasons of the DLC application (e.g., for a summer tourist destination, during the peak power demand periods particularly in summer), the maximum allowed power curtailment for each different type of load involved, the economic refunding of the consumers or any other type of alternative incentives for being involved in the DLC program, etc.

Alternatively, load curtailments can also be implemented on a voluntary and individual approach by the consumers themselves, under a smart pricing program. Smart pricing simply refers to the reduction of the electricity consumption in periods of high electricity pricing. The most popular available options are critical peak pricing (CPP), the time of use rates (ToU), and real-time pricing (RTP). With the CPP approach, a predefined electricity price is triggered by the utility if the power demand exceeds an also predefined upper level and is kept for as long as this condition exists. According to ToU, different electricity pricing is predefined for specific time periods during the day. Finally, RTP refers to continuously varied electricity prices, typically redefined on an hourly basis, in response to the wholesale electricity liberalized market. Consumers are free to adapt their consumption given the applied pricing policy by the utility, in order to eventually achieve the lowest possible electricity pricing, according to their needs. More on the available alternative electricity pricing options is presented in a later section dedicated to DSM programs.

6.3.2.4 Energy Efficiency

The energy efficiency strategy imposes the consumption of less electricity—or, in a more general approach, primary energy resources—for the coverage of the same amount of final energy needs (e.g., lighting, indoor space conditioning, motor drive power, etc.). Practically, this strategy implies that less electricity should be consumed without modifying the power consumption patterns and habits, which, in turn, determines the same final energy needs. The only way to achieve this is with the replacement of old, ineffective electrical devices (e.g., refrigerators, TV sets, washing machines, lighting apparatus), with new ones of higher energy performance rank. To encourage consumers towards this direction, several alternative incentives and programs, mainly of economic and financial objectives, are offered by the utilities, the regulators, or the responsible state authorities. Perhaps the most common incentive for the final consumers towards the improvement of energy efficiency of their facilities is funding up to a maximum eligible percentage, which can be varied based on the applicant's annual income, of the required capital for the energy performance upgrade of their buildings, including the replacement of old electrical devices with new ones, with improved energy performance.

State-of-the-art domestic appliances not only offer higher efficiency, but they also exhibit a series of innovative features to approach higher energy savings and to facilitate the grid's effective operation. These are the so-called smart appliances, with monitoring and control functionalities, enabling communication with other devices, smart meters, and energy management systems. The Association of Home Appliance Manufacturers (AHAM) defines smart appliances as "a modernization of the electricity usage system of a home appliance so that it monitors, protects and automatically adjusts its operation to the needs of its owner" [23].

Several typical electric devices can be equipped with technical features that enable their operation as smart appliances. Indicatively, smart appliances can comprise typical white goods, such as refrigerators and freezers, washing machines, tumble dryers and dishwashers, ovens and stoves, as well as air conditioners, circulation pumps for central heating systems, electric storage heating, and water heaters. The main operation feature of these devices that characterizes them as smart appliances is the employment of a smart management algorithm that aims to optimize their induced load on the electrical system, according to received signals by the utility. Such algorithms can introduce rescheduling of the operation of some smart appliances (e.g., washing machines, dryers, or dishwashers) to avoid operation during peak demand periods. They can also include intermittent operation of specific appliances or the use of other appliances as a sort of energy storage devices, in order to avoid their power consumption during peak power demand periods (e.g., precooling or preheating of indoor space or refrigerators and freezers operation in advance). In other operation approaches, smart appliances can also detect power quality disturbances or security contingencies in the grid and adapt respectively their operation, to support the stabilization of the grid's normal operation. For example, they can be automatically switched off for a few minutes, in case of under-frequency detection, to support the recovery of the grid's fundamental frequency to its nominal value.

6.3.3 Demand Side Management Programs

DSM programs are clusters of measures or incentives set by the utilities or the grid's operators, aiming at the motivation and the stimulation of the final consumers to participate in one or more DSM strategies. To this end, customers should clearly understand the benefits anticipated from their involvement in DSM strategies both for them and the electrical systems. The main benefits for the final consumers can be monetary savings and increasing control and awareness of their own electricity production and their household facilities. Additionally, they could also be motivated by the cultivation of a sense of responsibility to contribute to the improved reliability of the electrical system with more effective electricity production and management. On the other hand, several parameters can be discouraging for them to participate in DSM programs. These can be:

- their habit to act solely as passive consumers, with regard to any electricity production and management processes, combined maybe with their overloaded daily programs and lack of time

- their uncertainty to participate in price response programs and the corresponding economic viability, in the frames of the electricity wholesale market, given that they are accustomed to the security of fixed, and often quite low, electricity procurement prices, regardless of the scale and the daily or seasonal fluctuation of their power consumption profile
- the lack of familiarization, especially for older consumers, with modern technologies and their reluctance to get used to the hardware equipment and software applications required for the integration of DSM strategies
- their willingness to maintain their comfort level and the subsequent reluctance to undertake a reduction of their consumption during a contingency in the electrical system, e.g., by participating in load curtailment strategies.

The above deterring factors can be handled with the design of smart and flexible DSM programs, which allow the consumers to adapt their participation according to their willingness and concerns and with the provision of extensive support for them, consisting of a set of centrally coordinated services, configured versus the DSM strategies they are involved in and their familiarization level with modern technologies. For example, because individual residential or commercial customers of small or medium size may not feel secure in establishing commercial relationships with electricity providers, aggregators, etc., they may decide to participate in some DSM programs only if utilities work with them on designing and implementing these programs, at least until they are used to the executed processes and tasks.

Several DSM programs have been proposed in the literature, classified according to various criteria [24, 25]. In general, these can be classified in three groups, according to the party that initiates the executed task:

- *Rate-based or price DSM programs:* the electricity procurement prices vary versus time and customers are motivated to transfer their consumption during periods of low electricity prices. Electricity prices may be fixed in advance at preset time periods or may vary dynamically versus the time period of the 24-hour interval, the specific day of the week, or the season of the year, according to the configured balance between the available power production capacity, the power demand, and the existing reserve margin. In cases of dynamically varied pricing, prices can be set on a daily or an hourly basis, namely 1 day or 1 hour in advance, or even in real-time, and customers are requested to react respectively to the electricity prices fluctuation. Regardless of the pricing policy, customers will obviously pay higher electricity prices during peak power demand periods and lower prices during off-peak periods. The three alternative pricing options presented in the previous section, namely the ToU, the CPP, and the RTP options, fall into this category of DSM programs. ToU rates refer to the definition of fixed electricity price blocks for specific time periods over the day (higher prices during peak power demand periods and vice versa). CPP includes a prespecified increased price triggered by the utility if the power demand becomes higher than a preset threshold. This price is maintained at this increased level either for a limited number of hours or until the power demand falls again below the predefined threshold. Finally, in a RTP regime, electricity prices vary continuously, typically they are reset on hourly basis, in response to the wholesale electricity market.
- *Incentive of event-based DSM programs:* according to these DSM programs, customers are rewarded for reducing their power consumption upon request or for giving up control access to a maximum predefined level to the program's administrator over the usage of specific electric equipment. In the first case, voluntary requests or mandatory orders are sent by the utility or the aggregator to the participating consumers in the form of demand reduction signals. These signals can be invoked after a variety of triggering conditions, such as local or regional grid congestion, increasing production cost, grid's security and stability issues, etc. In case of incentive of event-based programs, there are upper limits set for the maximum duration of each individual event and the maximum number of hours per

year that a particular customer can be involved, which, indicatively, can be between 40 and 100 [26]. These limitations are imposed by both the willingness of the final consumers to suffer from load curtailments and the capabilities offered by the current DSM technologies. The most typical programs in this category are:

- *Direct load control:* customers are rewarded for allowing utilities to obtain control up to a maximum level over the usage of certain electrical equipment.
- *Emergency demand response programs:* customers receive incentive payments to reduce their power consumption in emergency cases with regard to the electrical grid's stability and dynamic security.
- *Capacity market programs:* customers receive incentive payments for load curtailments as substitutes for the electrical system's overall available production capacity.
- *Interruptible/curtailable:* customers receive a reduced electricity price (discount rate) for agreeing to reduce load upon request.
- *Ancillary services market programs:* customers receive payments from the grid's operator in order to reduce their power consumption upon request to support the secure operation of the electrical system. In a sense, this substitutes the provision of ancillary services.

• *Demand reduction bids:* customers that participate in this category of programs send power demand reduction bids to the utility or the aggregator, according to their capacity to reduce their power consumption. In these, the offered power demand reduction and the requested prices are placed. This program mainly stimulates customers of large size to participate in load curtailment strategies. On the other hand, through their participation in this process, they are enabled to recognize the load amount they would be willing to curtail at the announced price. Obviously, the most economically favorable periods for placing bids are the peak power demand period, when the electricity production cost is maximized and the margins for higher rewards for the curtailed load increase.

An alternative classification of the DSM programs can be based on the type of the source motivation, particularly whether it is driven by economic parameters of smart grids (i.e., real-time pricing, price signals, and incentives) or by the physical operation of smart grids and the issues regarding their reliability and security (i.e., grid management and emergency signals). In the first case, DSM programs are activated by the economics of smart grids, namely the reduction of the electricity production and distribution cost, which, in turn, leads to the reduction of the electricity procurement price for the final consumers. DSM aiming at reducing electricity production costs and prices are activated by price-based, market-led, and price-response programs. In the second case, physical DSM programs depend on the smart grid's reliability requirements. These tasks are activated by emergency-based, system-led, load-response, incentive-based, and direct-load control programs.

6.3.4 Demand Side Management Benefits

The potential benefits from the implementation of DSM programs in smart grids can vary, depending on the layout of the smart grid and the introduced technologies, the applied DSM strategies, and the participating type of consumers. A typical classification of the DSM benefits is presented below, in terms of whether they refer directly to participants or to some or all groups of electricity consumers [27].

6.3.4.1 Bill Savings for Customers Involved in DSM Programs

These can come from monetary savings on electricity bills due to increase of energy efficiency, load curtailment, dispersed production, or load shifting, or direct incentive payments from the utilities to the customers for the implementation of some or all of the above DSM strategies. For example, consumers can decrease their electricity procurement cost by shifting loads from peak to low power demand periods. They can also benefit from a direct drop in their electricity bills by reducing their consumption, either through load curtailments, improved energy efficiency, or through compensation of their consumption with their production from remote power plants, operating under net-metering mode.

6.3.4.2 Bill Savings for Customers Not Involved in DSM Programs

Due to the implementation of DSM programs in smart grids, reduced electricity production cost can be achieved for the utilities. This can be the result of peak power demand shaving (load shifting) or load curtailments, especially during peak demand periods. Both the above strategies can result in a flattening of the daily power demand profile, which, in turn, will lead to a series of positive impacts, such as the elimination of the peak generators operation, reduced requirements for spinning reserve maintenance, with a final result being the lower electricity power production cost. If the achieved drop of the production cost is transferred as a discount in the electricity selling price from the providers to all the final consumers, even those consumers not involved in DSM programs may benefit from reduced electricity bills.

6.3.4.3 Reliability Benefits for All Customers

This category refers to any kind of arisen benefit related to the achieved improved reliability of the electrical system. Improved reliability means reduced probability for power supply interruptions and other contingencies, higher power quality, and better system performance. All these can potentially imply reduced possibilities for income loss, for industrial and commercial consumers, and a series of societal benefits for all civilians. Improved reliability through DSM strategies is approached through a sequel of positive impacts, analyzed in the point below related to improved system security and performance.

6.3.4.4 Market Performance

DSM increases the competitiveness of the wholesale electricity market and improves the technical and economic flexibility and effectiveness of the system by offering more choices to maintain the power demand and supply balance and to handle crucial and emergency events. In this way, DSM prevents the monopolies of the electricity market from a small number of power producers and providers. Practically, the increased options for direct or indirect (through load reduction) power production in smart grids reduce the concentration of power production, make collusions more difficult, and help to avoid the potential for energy monopolies to exercise the market.

Market-driven DSM programs are normally implemented in the form of time-varying tariffs, allowing an active participation of the demand side in the market. DSM can contribute to the reduction of wholesale market prices. In fact, DSM programs can act as a damper during periods of high electricity prices, contributing to the amortization of the effects of extreme system's events. This is simply approached through the adjustment of the power demand by the final consumers in response to price signals. In this way, electricity is consumed in reverse proportion to its cost, leading, automatically, to the maintenance of an optimum balance between consumption and offered prices. Customers and utilities are protected from the risk to be exposed to price volatility and system emergencies.

The flattening of the power demand daily profile, also mentioned previously, leads to electricity production cost reduction too, translated to a direct economic benefit for both utilities and consumers. The same results are also approached by the improved flexibility of the power demand, through load curtailments and shifting, limiting the extent and number of price spikes.

6.3.4.5 Improved System Security and Performance

More options and choices are provided for the system's operator to handle system stability and security issues. For example, in case of power supply interruptions, the capacity for direct (e.g., dispersed production) or indirect (e.g., load curtailments) power supply through DSM strategies can be employed to contribute towards the recovery of the electrical system to precontingency levels. For the same reason, DSM strategies can also be considered as alternative means of delivering ancillary services for system operators, such as voltage support, active and reactive power balance, frequency regulation, and power factor correction [28]. This is easily conceivable if we consider load as a virtual (or negative) spinning reserve. If load is reduced (e.g., through energy efficiency, load curtailments, or load shifting strategies) the available spinning reserve in the system increases, and vice versa. This can be achieved if the power demand is associated in a smart way with the grid state (i.e., a droop control).

In a similar approach, loads centrally controlled by the utility or an aggregator through a SCADA system, can also behave as virtual storage, through load shifting. The aggregation and the management of a considerable amount of loads available to be involved in load shifting processes, can in total offer alternative storage options for the electricity wholesale markets, competing, in this way, with traditional storage technologies, with the major advantages of no required setup or procurement cost and no imposed maintenance or replacement cost, as with real storage devices [29].

Additionally, DSM strategies, especially dispersed production, load shifting, and load curtailment, by contributing to the maintenance of the power demand and supply balance, can also be considered as a significant means to mitigate any problems caused by the increased, uncertain power production of nonguaranteed power plants (e.g., RES units). The same DSM strategies, through their support on the system capability to react effectively on the occasion of a system's fault or contingency, may also relieve the necessity for spinning reserve maintenance. This, in turn, will lead to higher operation efficiency of the overall system and cost-effective performance.

Finally, all four main DSM strategies, through either direct dispersed production, the reduction of power demand, or load shifting can contribute to reduced power transmission losses, through the transmission or the distribution grids, namely to another means of increasing the efficiency and the cost-effective operation of the electrical system. For the same reason, DSM strategies can also help to relieve the grid's constraints or avoid outages in case of contingencies or congestion in the transmission or distribution grid.

6.3.4.6 System Expansion

As already stated in the above analysis, DSM can contribute in several ways to the flattening of the daily power demand profile through power demand peak shaving and, in general, to the reduction of the power demand peaks. This is considerably important, because, since electrical systems are built and dimensioned for the expected peak power demand, this can result in the postponement or even the cancellation of new investments on the construction and the development of additional infrastructure on the power production, transmission, and distribution facilities of the electrical system, without affecting the reliability of the system and its adequacy to support the power demand [30].

6.4 ENABLING TECHNOLOGIES FOR SMART GRIDS

It is obvious that the adequate and integrated development of smart grids can only be performed with the support of innovative enabling technologies on control systems, communication and information facilities, and electronic circuits. Additionally, crucial factors towards the expansion and the realization of smart grids constitute the parallel deployment of the buildings' energy management systems (BEMS), as well as the final consumers' gradual cultivation and awareness of the exploitation of the possibilities offered by these new technologies. In short, enabling technologies for smart grid implementation consist of, but are not limited, to:

- *Two-way monitoring interval electricity consumption meters:* these devices offer bidirectional information flow to the final consumer and the utility, revealing, in both directions, the actual final energy usage pattern. They can also be used to adjust the energy consumption according to electricity price fluctuations, e.g., the reduction of a required indoor space temperature through a smart thermostat.
- *On-site generation and storage equipment:* small wind turbines, PVs, electrochemical batteries, etc., with increasing reliability and reduced procurement cost and maintenance

requirements offer the choice for distributed production and energy management through storage processes, with all the imposed benefits listed in the previous section.

- *Load control systems—BEMS:* these integrated systems offer complete control on the final consumers' load, both for the consumers and the system's operator. They constitute the contact point with the utility or the aggregator. Through these systems, orders or requests from the grid's side pass to the demand side for potential load curtailment or power injection in the grid. They provide real-time information on the current electricity consumption for both sides too. Control systems are also introduced for the management of DER.
- *Information and communication transfer networks:* This essential infrastructure can consist of a combination of wired (e.g., optical fibers) or wireless networks. They can also include communication substations, where the information is gathered to be forwarded centrally to the main servers. The major advance in this sector is the two-way communication networks, which make possible the concurrent bidirectional information exchange between the customers and the utilities, enabling the realization of all the tasks and processes described previously within the frame of smart grids.

The above main technology categories and the way they cooperate in the frame of a smart grid are presented in Figure 6.16. Given that the innovative dispersed production and storage technologies

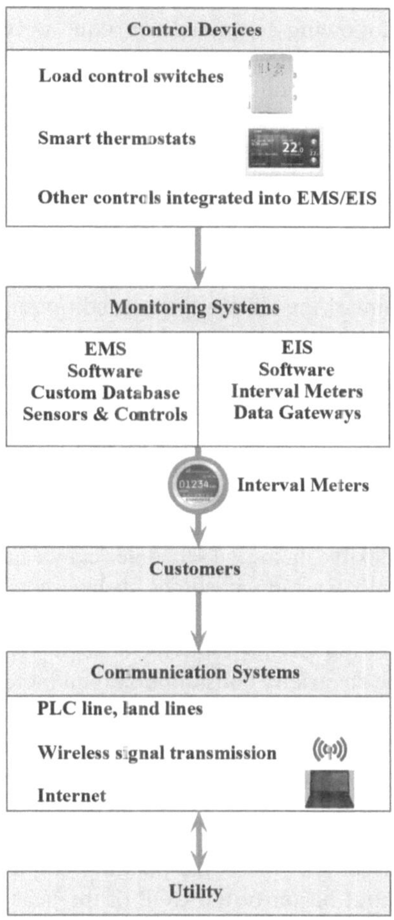

FIGURE 6.16 Typical input and output information flows involved in an energy management system.

have been, more or less, examined in previous chapters, the remaining three categories of smart grid enabling technologies, namely the monitoring devices, the control systems, and the communication systems are presented in the following sections.

6.4.1 Control Devices and DSM

Control devices for DSM strategies can be stand-alone electronic devices, or integrated into energy management systems (EMS), in more advanced implementations. In most cases they consist of load switches, for the remote control of specific loads, such as motors, compressors, and lighting, and thermostats, for the automatic control of the active systems employed for indoor space conditioning. Control devices can be either remotely controlled by the utility, by means of communication systems, or locally controlled by the consumers. Smart thermostats and switches, instead of being on–off switching devices, allow flexible programming of the operation of the controlled devices or systems based on the fluctuation of essential operation parameters, such as the indoor space temperature or the available luminous flux inside a building from natural lighting sources. Additionally, some new-generation thermostats can operate as a version of a small-scale EMS with limited functions, because they can also act as repeaters, forwarding prices and event signals from the utility to other appliances too.

Control devices can also be integrated in EMSs, enabling bidirectional functionality for both the consumer and the utility. EMSs can offer an automatic way to manage all types of loads in a remote consumption, such as lighting, motors, air conditioning, as well as the dispersed electricity generation and storage units, following commands or requests sent and received by the utility or the aggregator. Yet, it should be noted that the most common targets of the executed energy saving functions by an EMS are air conditioning and the lighting consumptions. This is because these two load categories are usually coincident with peak power demand periods, both in summer and in winter, while they also represent the two main loads for commercial and residential buildings. Additionally, as small modifications to the daily usage pattern of the air conditioning and lighting systems do not affect significantly the offered sense of comfort for the final consumers, especially when these modifications are introduced in a smart way and during specific time periods, these particular consumptions offer potential margins for load curtailments. For example, a slight rise or drop of the temperature set-point of the indoor space conditioning active system, under cooling or heating operation mode respectively, can cause considerable electricity saving, without affecting the achieved thermal comfort level. A different approach is to preheat or precool the conditioned indoor space during off-peak demand periods, in order to reduce the subsequently required power consumption during peak demand periods (load shifting).

Apart from air conditioning active systems and lighting equipment, EMSs can also control the operation of additional electric or electronic devices, versus the essential decision parameters of smart grids, such as the current electricity production cost, the security status of the system, etc. For example, several devices with more flexible operation schedules can be arranged to operate during off-peak periods. Such devices are typically washing machines, dishwashers, dryers, water heaters, etc.

The automatic control of the current electricity load implies that the consumption is adjusted with respect to the received signals or requests, without the necessity for human action. For example, the EMS will proceed to reduce the electricity consumption if an increased electricity price or an emergency security signal is received. Respectively, it will initiate electricity storage processes if the electricity price falls below a maximum acceptable and preset threshold. Another common function is the selection of the most cost-effective active system for indoor space conditioning, based on essential parameters that determine the final thermal energy production-specific cost. Such parameters are the procurement prices of the alternatively available energy resources (e.g., oil, gas, or electricity), the operation efficiency of the corresponding employed active systems (e.g., overall efficiency of the oil or gas central heater or the COP of the heat pump), the outdoor temperature, and the indoor desirable thermal comfort conditions. A typical input and output information flows scheme involved in an EMS is presented in Figure 6.17.

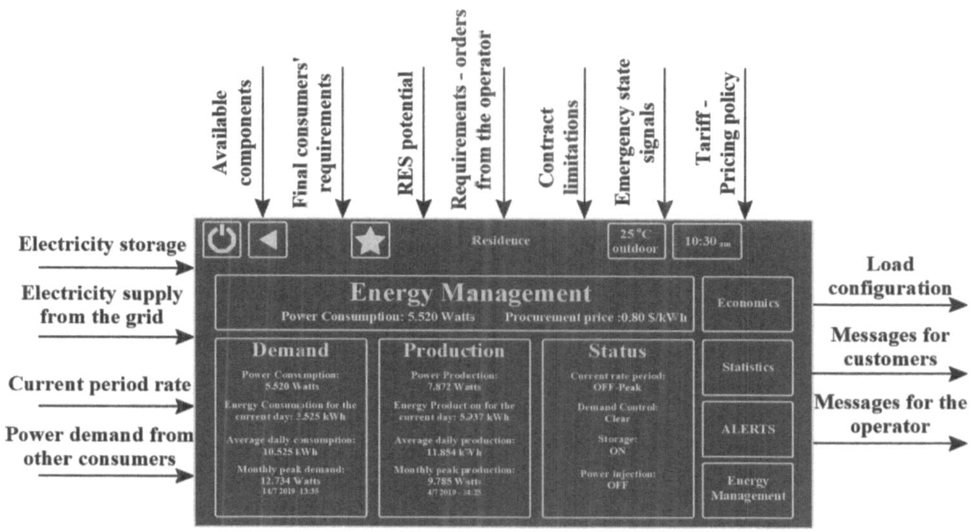

FIGURE 6.17 Typical input and output information flows involved in an EMS.

6.4.2 Control Devices and DER

The extensive introduction of distributed power production units creates the necessity for their inspection, control, and management. To this end, the main employed control systems are power electronics, multi-agent systems, advanced fault management, and virtual power plant (VPP) control technologies.

Power electronics are included in the inverters that accompany and support the secure penetration of small-size generation units, such as PV stations or small wind turbines, offering the options for autonomous or grid-connected operation. These devices constitute the contact point between the distributed generation unit and the grid, providing direct control on the operation and the management of the unit.

Multi-agent systems (MASs) practically constitute decentralized control systems, substituting the centralized control systems at local scale. This is because it is not possible to handle large numbers of distributed generation units centrally. On the contrary, MASs are designed and adapted respectively to the needs of a local, small-scale grid with increased power production from distributed units. They also offer the option of a seamless transition from grid connected operation to an island-grid, in case upstream outages or faults are detected [30].

Advanced fault management (AFM) systems are realized with the coordination of local automation and monitoring, locally controlled switchgears, and relay protection. The inspection of the normal operation of smart grids and the application of the system's diagnostics are the first steps of AFM systems. The on-time detection of any faults or malfunctions in the electrical systems constitutes perhaps the most important step for the adequate reaction of AFMs. A common reaction following the appearance of a system's fault is the islanding operation of a specifically selected part of the grid, which may serve sensitive and important consumers (e.g., hospitals) [31].

Virtual power plants (VPPs) are nothing more than cloud-based power plants that aggregate the available power production from distributed generation units to provide a secure and reliable overall power injection in the electrical grid. In a sense, VPPs represent an "Internet of energy." Distributed generator units are connected with means of communication systems with the central VPP control system, transferring information regarding their production capacity. VPP control systems, by following specific operation algorithms, aiming either to the maximization of DER penetration,

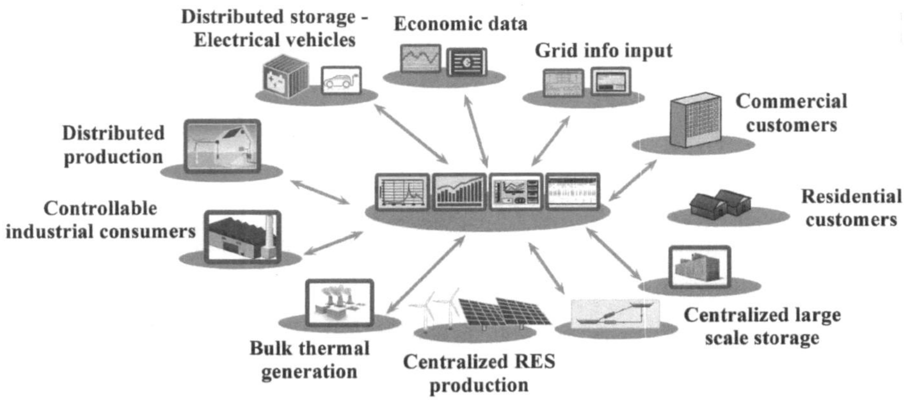

FIGURE 6.18 Typical layout of a virtual power plant.

or to the improved cost-effective grid's operation, etc., determine the absorbed power production synthesis from the DER units. Within electricity wholesale markets, VPPs also act as intermediaries between the DER units and the market, trading the produced electricity by the DER units to the market, on behalf of their owners. From the point of view of other market participants, a VPP behaves as an integrated power plant, although, in practice, it constitutes a cluster of many, diverse distributed generation units [32]. A typical layout of VPPs is given in Figure 6.18.

Finally, crucial components in a smart grid control system are the sensors, which constitute small detection nodes that enable remote systems, equipment, and generation unit monitoring and control. The advanced technology in this category are the phasor measurement units (PMUs) or synchrophasors, which are high-speed sensors employed for the measurement of real-time phasors of voltages and currents. They are usually employed in electrical systems for advanced monitoring, control, and protection processes. PMUs are 100 times faster than typical SCADA systems and are able to perform measurements on grid magnitudes with high accuracy.

6.4.3 Monitoring Systems

Monitoring systems can be smart meters, advanced metering infrastructure (AMI), and energy management systems (EMSs).

6.4.3.1 Smart Metering

According to the directive 2009/72/EC, European member states are required to install smart meters for at least 80% of the electricity consumers located in their territories by 2020. This percentage corresponds to around 250 million smart meters. Usually, the installation of smart meters is ordinated by local distributed systems operators (DSOs) and constitutes a part of a wider program regarding the development of smart grids, incorporating also additional applications.

Smart meters are composed of an electronic device equipped with a communication link. They measure for specific, predefined time intervals, typically hourly time periods, the consumers' electricity consumption and, possibly, other consumption parameters and related magnitudes, and they transfer these data, through the communication network, to the utility and any other involved actor for billing, load factor control, peak load requirements, and the development of pricing strategies based on consumption information. The same information can also be shared with end-user devices, so the final consumers are kept informed about their own consumption patterns and the associated costs. Smart meters can also provide other information, such as historical consumption data, greenhouse gas emissions, tariff options, demand response rates, tax credits, remote connect/disconnect of users, appliance control and monitoring, power quality monitoring, switching, and prepaid metering [15].

There are several types of smart meters, distinguished according to their technical specifications, such as the data-storage capability, one- or two-way communication capabilities, etc. [33]. The requirements of the smart meter's available specifications, such as the duration of the measuring intervals and the time resolution (typically from 15 to 60 min) according to which any inspected magnitude (such as electricity consumption and production) should be measured, are set by the provider's electricity pricing policy. Measurement accuracy constitutes another crucial issue, handled according to the defined specifications and processes in IEC61036. Moreover, the issue of the final consumers' reimbursement should also be considered, in case of low power quality and system deficiencies. To this end, future smart meters should also incorporate power quality measurements [34].

6.4.3.1.1 *Advanced Metering Infrastructure*

Advanced metering infrastructure (AMI) is composed of a number of different technologies and subsystems, such as smart meters, meter data management systems (MDMS), systems for data integration into software applications, operational gateways, and a cluster of alternative communication networks, like wide area networks (WAN), home (or local) area networks (HANs or LANs), and neighborhood area networks (NANs). The essential layout of an AMI is presented in Figure 6.19 [35]. AMI represents a metering and data storage system that, either on request or following a predefined measuring schedule, measures several magnitudes involved in the electrical system, like electricity production, consumption and storage, grid's voltage and frequency, or even final customers' data. These data, measured by the appropriately installed grid meters, are transferred to central servers using various communication networks, to be saved for further processing or for the system's logistics.

Starting from final consumers' level, smart meters are connected with several controllable devices, such as smart thermostats, light sensors, etc., implementing energy management tasks. On the other hand, they communicate the data collected by the control devices and additional measuring devices, such as distributed power production or storage units, EMSs, etc., to both the final consumer and the operator through the HAN. HANs may also offer a smart interface to the market and support security monitoring.

NANs are networks used for collecting the measured and transferred data through HANs. NANs are composed by the communication networks, which can be either wired or wireless, and central

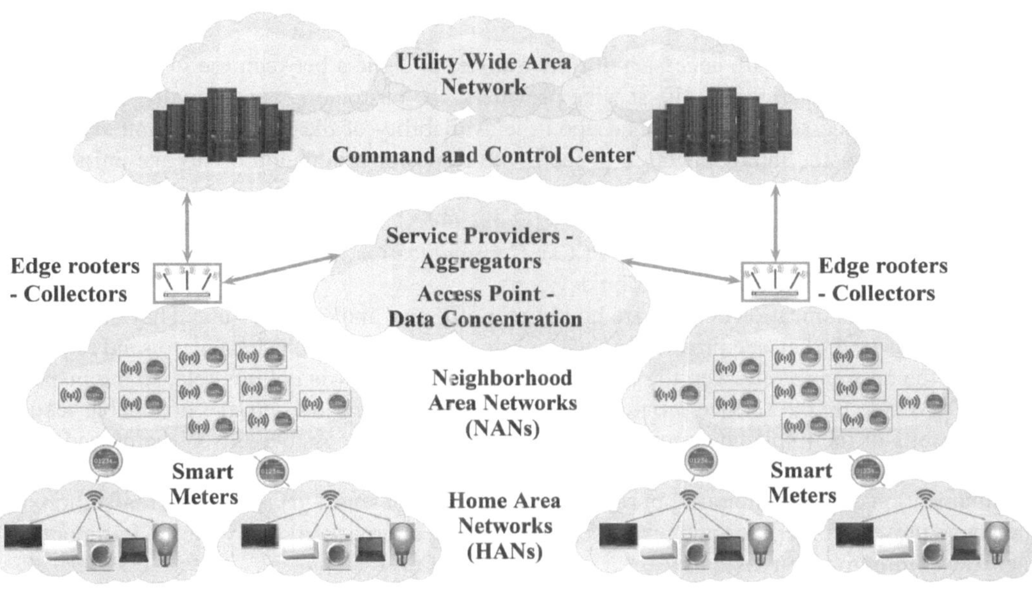

FIGURE 6.19 Typical input and output information flows involved in an AMI [35].

servers employed for the development of corresponding databases. The collected data are used by the utility to analyze and optimize the electrical system's usage, to trace any power quality instances or system's malfunctions, to calculate the arisen costs, and to optimize the provided services for the final consumers.

The smart grid's tree-like layout is integrated with the WAN and the MDMS. The MDMS constitutes a database for the validation and the editing of the data captured and collected with the AMI, aiming to guarantee that the provided data are accurate and complete. MDMS is also equipped with advanced systems and applications, which enable the operation of the AMI within a WAN, with further management processes executed, such as advanced operations on the distribution and transmission grids, control and management of the DER, outage management, distribution geographical information systems, etc.

6.4.3.1.2 Energy Management Systems

Energy management systems (EMSs) are integrated, smart electronic systems that enable the control and the management, the monitoring and the operation analysis of systems, equipment, devices, and processes included in a building, by means of sensors, meters, switches, controls, and operation algorithms introduced with relevant software applications. EMSs are the communication gate between the HAN and the smart grid. An EMS transfers all the gathered information and data from the final consumers' side to the NANs. At the same time, EMSs perform control on the consumers' devices and processes by passing the orders or requests from the grid side to the remote consumptions. Hence, all the potential DSM strategies are implemented through EMSs. For the implementation of DSM strategies, a time resolution of 1 min is most commonly applied for all the executed measurement and monitoring tasks. Longer time intervals (e.g., 10 min) can be used for simulation and verification purposes.

EMSs can also constitute a remote data center for the final consumers, which can collect and store in local disks data and relevant statistics after essential editing, regarding the total electricity consumption and production locally, daily or seasonal consumption and production patterns, statistics on the involvement of the customer in DSM strategies, consumptions particularly from specific appliances or for the coverage of specific needs, and hints for further energy saving.

6.4.4 Communication Systems

Communication systems are necessary for the transfer of signals between the involved actors in a smart grid, namely the market, the service providers, the customers, etc. Actually, all smart grid strategies and processes are established upon the availability of the required communication network facilities, for the interconnection of the involved power production or storage units, the network sensors and the smart meters. Several data transfer technologies can be employed, from simple telephone lines and radio media to wireless networks (GSM, Internet) and wired grids, like optical fibers and power line communications (PLCs). Communication networks can be developed on the basis of one- or two-way communication devices.

One-way communication devices are highly cost-effective and simple to use. They are employed by the utilities or the service providers to alert customers, e.g., on a DSM event, to send requests or orders to the customers, etc. Yet, the diverse communication flow is not possible, so the inspection of the DSM side by the service providers or the aggregators and the retrieve of reliable feedback on the results of the implementation of DSM strategies are not possible. Given the precedent presentation of the potential operations and capabilities of a smart grid, it is conceivable that one-way communication devices cannot support the full integration of smart grids with the realization of all the above presented facilities and processes. This can be approached only with the introduction of two-way communication systems.

Two-way communication systems are more expensive; however, they enable the bidirectional communication flow between the utilities or the service providers and the customers. In this way, apart from placing orders and requests from the grid's side towards the customers' domain, it is also possible for the utilities to get reply confirmations from the customers with a cluster of information regarding the realization of tasks in the frame of DSM strategies. For instance, utilities can proceed to real-time monitoring of the facilities/devices involved at each final customer and get an accurate picture of the level of the load's diversity. In case of the utilization of smart meters, after the application of a DSM strategy, measured load response is automatically recorded and sent back to the utility. This, in turn, enables accurate and real-time estimation of the effect of each alternatively applied DSM strategy and the achieved power consumption drop. The secure penetration of distributed production and storage resources is also possible through two-way communication systems. This will enable further increase of DER in the grid, with subsequent benefits on the electrical system's economics, the support of ancillary services, and, eventually, the dynamic security improvement. All the above, of course, will also contribute to increasing possibilities for the reduction of electricity procurement prices for the final consumers.

The overall layout of a two-way communication system with a central smart grid controller (server) follows the tree-like structure presented in Figure 6.19. It starts with the HANs, which consist of devices and appliances in final consumers' facilities (e.g., a home, a commercial building, an industry). The introduced smart meters with the HANs are initially connected with the communication gateway. The communication system is then expanded to NANs, which collect data from the multiple HANs connected with each particular NAN, and deliver them to local distributed servers. WAN comes next, which gathers the data collected at the local NAN servers to the central control server. Finally, the system is completed with a gateway, which collects information from the HANs and communicates them to the interested parties [36].

For the realization of the above communication network, several alternative technologies can be used. However, the most usually applied systems are the combination of PLCs for the communication between the smart meter and the distributed servers in the NANs (concentrators) and the use of optical fibers or GMS/GPRS wireless networks for the communication between the NANs concentrators and the MDMS server [37].

The integration into a smart grid of intelligent and complex electronic devices (IED), software applications, and data transfer mechanisms between the multiple actors and entities has determined, in fact, the requirement of advanced network architectures. These networks should be able to analyze, store, integrate, and process the increasing quantity of information available in the various devices. Additionally, a common format that covers all the areas of data exchange in the electrical power domain should be established. In any case, the design and the construction of the communication network should guarantee a high speed of data transfer and the expandability of a peer-to-peer network, capable to support the needs of DERs and electricity distribution grids. To meet these requirements, both wireless and wired communication grids are, currently, designed and constructed according to the specifications and norms set in the IEC 61850 standard [38].

Wireless communication systems can be optionally used for HANs, NANs, and WANs, while they constitute a unique option in case of vehicle-to-grid (V2G) communications. For HANs and NANs, the networks IEEE 802.15.4 (ZigBee) and IEEE 802.11 (Wi-Fi) are the most appropriate technologies, because they can offer a coverage range from some tens to some hundreds of meters [39]. On the other hand, the coverage requirements for WANs, which can be in the range of tens of kilometers, imply the use of cellular wireless networks, like GPRS, GSM, UMTS, LTE, or broadband wireless access networks like IEEE 802.16m (WiMax). Two new options for wireless communications, especially utilized for metering tasks in smart grids, are provided by the wireless sensor networks (WSN), a technology offered in the frame of the Internet of Things (IoT), and the machine-to-machine communications [40, 41].

With regard to the available wired technologies, PLCs and fiber optic lines are the most favorable options, the first one for distances up to some hundreds of meters, in local HANs and NANs networks, and the second one for WANs. PLCs are normally used in cases of HANs and NANs, namely at the heart of the smart grid, where most of the complexity and the intelligence of the smart

TABLE 6.3
Smart Grid Communication Applications

Technology	Spectrum	Data Rate	Coverage Range	Applications	Limitations
GSM	900–1,800 MHz	≤ 14.4 kbps	1–10 km	AMI, DSM, HAN	Low data rates
GPRS	900–1,800 MHz	≤ 170 kbps	1–10 km	AMI, DSM, HAN	Low data rates
WiMAX	2.5, 3.5, 5.8 GHz	≤ 75 Mbps	10–50 km (LOS) 1–5 km (NLOS)	AMI, DSM	Not widespread
PLC	1 – 30 MHz	2–3 Mbps	1–3 km	AMI, fraud detection	Harsh, noisy channel environment
ZigBee	2.4 GHz, 868–915 MHz	250 kbps	30–50 m	AMI, HAN	Low data rates, short range

grid is included. In these micro-scale smart grids, where requirements for speed, reliability, and accuracy are maximized, PLCs offer a guaranteed and robust technology, without any requirement for further infrastructure. PLCs are distinguished in two main categories, narrowband and broadband PLCs.

Narrowband PLCs exhibit limited bandwidth at low frequencies, limited bit rate, and attenuation of very few dBs per km, constituting an appropriate choice for signal transmission for longer distances at relatively low data flow speeds. Therefore, they are used for data gathering, monitoring, management of remote consumptions, and control of distributed power production units. On the other hand, broadband PLCs operate at higher frequencies, offering higher bit rates (higher data flow speeds) but lower coverage ranges (up to some hundreds of meters), due to the intensive signal attenuation at this higher frequency range. Hence, this category of PLCs is mainly suitable for in-home applications. However, by keeping the bit rate at values lower than 4 Mbps and regulating some modulation parameters to robust mode choices, broadband PLCs can also be used for outdoor communication in NANs.

A summary of the available communication technologies with their main technical specifications is presented in Table 6.3 [41].

6.5 SMART GRID BENEFITS

In Section 6.3.4, a detailed presentation of the potential benefits of DSM strategies was given. In a more generic approach, this section aims to give a synopsis of smart grid benefits, not limited only to those arisen from the application DSM measures. They can be distinguished in the following main categories:

- *Incentives for final consumers:* final consumers receive a series of incentives relevant to either economic or energy terms, aiming to modify their power consumption profile, in order to contribute to the maintenance of the dynamic balance between power production and demand. This, in turn, leads to increasing reliability of electrical systems, due to the expanded options for demand response services.
- *Advanced distributed power generation and storage:* the options offered by the two-way communication AMI enable effective management and control functions of distributed generation and storage units. This results in increasing and more secure RES penetration, effective load shifting/peak shaving through power storage, and enforcement of the system's flexibility and reliability.

- *New products and services:* smart grids create the prerequisites for the development of new products and services for the final consumers, while they also enable new market opportunities for third parties. For example, final consumers are given the option to choose various services offered by different providers, while private investors are invited to take over the management of independent grid variables (e.g., energy, capacity, location, time, rate of change, and quality).
- *Improved power quality and secure operation:* all the involved functions in a smart grid lead to a common resultant: the improvement of the system's dynamic security, stability, and power quality. Advanced management and control, DSM strategies, distributed power production, V2G functionality, etc., contribute to the improvement of the system's stability and dynamic security, because any potential system fault can be handled with a cluster of alternative reactions. Additionally, the grid's diagnostics applied with AFM systems enable on-time detection of events, faults, and malfunctions, such as lightning, switching surges, line faults, and harmonic sources that can impact power quality.
- *Optimized assets utilization:* it is obvious that the use of the available assets in smart grids is optimized in terms of efficient and cost-effective operation. For example, with load curtailments and shifting, the dispatched thermal generators can be adjusted to operate close to their maximum efficiency. Advanced management on distributed generation and storage, on V2G functionality, etc., contributes to eliminate grid congestions and to minimize the electricity transfer losses in transmission and, probably, in distribution grids.
- *Resilience:* smart grids, by their nature and their basic structure, exhibit increased resilience against any physical disasters (e.g., lightning on transmission grid) and intentional attacks on both the physical facilities (e.g., power plants, substations, pylons, electricity lines, transformers) and the cyber infrastructure (e.g., markets, communication systems, software applications). This is because the overall structure and operation concept of smart grids enable the isolation of the problematic elements, while the rest of the system can be restored to its normal operation.
- *Lower production and maintenance costs for the utilities:* sensibly, the reduction of peak power demand, the increasing options for ancillary services from the consumer side, imposing reduced necessity for spinning reserve maintenance at centralized power plants, the higher overall efficiency of the system, the reduction of the electricity transmission losses due to the increased penetration of DER, etc., contribute, in common, to the reduction of the electrical system's operation and maintenance cost and, eventually, to the reduction of the total electricity production-specific cost.
- *Lower costs and electricity prices for the final consumers:* the reduced electricity production-specific costs for the utilities and the involvement of the final consumers in the electricity wholesale market set the basis and create the prerequisites for them to claim and approach lower electricity procurement prices.
- *Primary energy resource savings:* primary energy savings are approached in smart grids with load curtailments, increasing energy efficiency, dispersed production, peak power demand shaving, and advanced monitoring and control capabilities, which enable detection of higher consumptions and abnormal operations, as well as optimization of the consumption profile according to the dynamically varied electricity prices and the corresponding technical parameters of electrical systems. According to the so far implemented programs, annual energy saving percentages from 10% to 25% have been achieved [42–44]. An estimation by the US Department of Energy on the anticipated CO_2 emission reduction and energy saving due to smart grids until 2030 is given in Table 6.4 [45].

TABLE 6.4
Energy and Emissions Savings in 2030 due to Smart Grid Technologies (with due diligence of US Department of Energy)[45]

Mechanism	Electricity Consumption and CO_2 Emission Saving Percentages (%) Direct	Indirect
Conservation effect of consumer information and feedback systems	3	—
Joint marketing of energy efficiency and demand response programs	—	0
Deployment of diagnostics in residential and small/medium commercial buildings	3	—
Measurement and verification (M&V) for energy efficiency programs	1	0.5
Shifting load to more efficient generation	< 0.1	—
Support additional electric vehicles and plug-in hybrid electric vehicles	3	—
Conservation voltage reduction and advanced voltage control	2	—
Support penetration of renewable wind and solar generation (25% renewable portfolio standard)	< 0.1	5
Total reduction	**12**	**6**

6.6 SMART GRID BARRIERS

As in any new, emerging technology, the deployment of smart grids faces some very specific barriers, which can be distinguished in five main categories: inadequate awareness of potential involved actors (especially customers); additional required costs for new infrastructures, equipment, etc.; the lack of complete and adequate legislation framework; the continuously updated standards and directives; and the required safety and security facilities that will ensure the secure operation of smart grids, regarding both the physical assets and the electronic, online electricity and economic transactions.

Lack of awareness always constitutes an important barrier towards the realization of new systems and processes on a specific territory, especially when these require vast involvement of the local population in the under consideration geographical area. This is also the case of smart grids, with final consumers being a major actor involved in the customers domain. In most cases, final consumers do not have a clear picture, or maybe not a picture at all, of the structure and the implemented processes in an electrical generation, transmission, and distribution system. This simply implies that they sensibly are not in the position to conceive the opportunities and the potential benefits, described in the previous section, expected by the realization of smart grids. However, not only final consumers, but in several cases even potential investors, utilities, and state regulators, are not convinced of the potential positive impacts from smart grids realization, due to the lack of performance data on cost reduction and energy savings. These barriers are expected to be overcome with the implementation and the operation of the first smart grids and the dissemination of the arisen results, not only in academia, but also, and most importantly, in human communities.

The second major barrier comes from the required capital for the construction of the technical infrastructure and the electronic applications and systems, necessary for the realization of all the processes and actions involved in smart grids. Although the anticipated economic benefits are far from negligible, the scale of the required capital always remains an important issue, especially in periods of economic recession. This barrier is currently handled with the financial support offered by several funding actions (e.g., in EU the Horizon 2020 projects). In the future, with the gradually expanding smart grid infrastructure, this barrier will sensibly become less and less crucial.

Another essential issue is the existence of complete regulation frameworks, on which all the involved physical processes are clearly defined, regarding the electricity flows between the involved actors, as well as the corresponding electronic transactions on the wholesale electricity market. Additional issues that should be defined are the determination of the involved actors' operations (e.g., smart meters), the ownership of the measured data and the market entities, the provided incentives for the final consumers, etc. Finally, further regulatory issues are the inconsistent policy between different countries in the EU or among the states in the US.

The offered state-of-the-art technologies enable the realization of almost all the above described smart grid systems and processes. Yet, smart grids cannot be developed with full exploitation of the anticipated benefits, due to the lack of relevant standards for the integration, the interoperability, and the homogenization of the involved processes, applications, and systems. Both the regulatory framework and the relevant standards will be sensibly developed with the deployment of smart grids, through a dynamic, bidirectional process.

Finally, security and privacy also constitute two major, key issues for successful deployment of smart grids. From the utility point of view, security issues may range from single individual frauds to full collapse of the overall physical or electronic infrastructure. From the customers' point of view, security issues can include disclosure of sensitive personal data, such as personal habits that can be revealed with the daily consumption profile. These issues are handled with relevant norms and directives issued by the European Network and Information Security Agency (ENISA) [46] for Europe and the NIST [47], for the US, which address specific strategies on cyber security, policies on privacy issues, security processes on data transmission protection, and potential vulnerabilities on concrete security problems. Security and privacy issues can include user identity theft, user tracking in order to predict user activities or behavior patterns, malware distribution, real-time surveillance, consumption data manipulation, selfish behaviors from owners of RES power plants, remote smart meter manipulation by hackers, and flooding of communication and computation resources.

6.7 SMART GRID IMPLEMENTATION EXAMPLES

6.7.1 The Smart Grid of Azienda Elettrica di Massagno

Azienda Elettrica di Massagno (AEM), is a small Swiss utility that explores the possibilities for the implementation of a smart grid that will monitor sensitive parameters from the end users with the installation of smart meters, realize new business models in the frame of the wholesale electricity market, and promote vast dispersed electricity production and storage from the final consumers. The core of this initiative is based on a new algorithm, which aims to maintain a predetermined load profile by connecting data gathered from different sources, i.e., the consumption, the power flow at the transmission grid substations, the climate conditions, the currently available bulk generation capacity, and the storage capacity in the future. This load profile will lead to the reduction of the electricity production cost, enabling a subsequent drop in the offered electricity prices for the final consumers and an increased competitiveness of the AEM. Ancillary services by the final consumers for the grid are also envisaged, such as voltage support by injecting dispersed power production or by introducing load curtailments, when required. The whole infrastructure will be supported with a broadband bidirectional communication network.

The smart grid integration is based on an operation algorithm designed and realized by the Centre for Studying Artificial Intelligence (www.idsia.ch). The first prototype was available for validation tests in winter 2018. Based on high-frequency data flow provided by the installed smart meters, the broadband communication system is able to collect data, such as bulk generation, dispersed production capacity, and power demand, at 1-minute intervals, integrating them with voltage data collected at the distribution grid. This high-resolution demand–production collected data will eventually enable the utility to split its main grid into many mini-grid entities, which will be supported by dispersed self-production and electricity storage and managed by blockchain procedure.

To realize the data collection process, a high-performance communication network is essential. The high-resolution data flow is generated on the consumption site by Landis+Gyr E450 smart meters. The data are then transferred to a central unit using a hybrid system that utilizes G3-PLC broadband radio frequencies and optic fibers. The data grid is able to download and upload data orders to any consumption and production location. The parameters of the installed smart meter will be programmed by Optimatik, a Swiss-based service provider.

Upon completion, AEM is aiming to achieve three goals:

- Aligning the distribution grid's load profile consistently with TSO requirements, to optimize costs and technical efficiency at the national high-voltage point of connection. At a first approach, this means reducing peak loads.
- Managing power flow on the low-voltage grid to utilize available dispersed production capacity, limiting redundant investments, controlling voltage levels and avoiding over/under voltage situations, particularly on low-voltage grids where a higher density of PV plants is gradually expected. The projection–target of AEM is the installation of a 3 kW PV plant at each one of the 10,000 customers' houses of AEM, namely a total installed capacity of 30 MW.
- Developing new business models, and particularly new tariff schemes, based on a grid split into small units extensively supported by dispersed self-production and a basket of services to be provided for the customers.

6.7.2 Duke Energy Carolinas Grid Modernization Projects

Duke Energy received $4 million from the US Department of Energy in 2009 for the implementation of smart grid pilot projects in North and South Carolina [48]. These projects aim to the deployment of an interoperable, two-way communication grid, comprising smart meters, distributed generation, distributed automation systems, dynamic pricing programs, and introduction of EV plug-in technologies. The overall cost of these pilot projects is estimated up to $7.5 million.

The objectives of this smart grid are:

- to implement distribution automation to help prevent and shorten outages
- to enable AMR and reduce the need for estimated bills
- to enable remote service connections and disconnections for faster customer service
- to capture and post daily energy usage data online so customers can make wiser energy decisions
- to incorporate more renewable, distributed generation into the grid.

According to the program's guidelines, the installation of smart meters is mandatory for all customers. The adequate and accurate operation of the installed smart meters is assured by Duke Energy, through enhanced testing procedures. The overall effort started with a pilot program conducted in Charlotte, North Carolina, with which the abilities of home energy management with 100 volunteer residential customers were tested. The realization of the program was facilitated with a call center established by Duke Energy to provide support for the residential customers. It received roughly five to eight calls each month from customers with questions about using the system. It was revealed that customers needed more robust training to use the system effectively. These initial 100 participants will be expanded to 8,300, including residential, commercial, and industrial customers, with the installation of approximately 17,000 digital smart meters and other automated equipment. In short, the main equipment components are:

- smart meters capable of two-way communication, with PLC technology
- new distribution automation equipment including electronic breakers, digital sensors, 45 new phasor measurement units, and automated switching devices
- new distribution system communication nodes
- home energy management systems, on an IP-based, open system network platform
- development of dynamic pricing programs
- technical infrastructure for the support of the deployment of PEVs.

The EMS consists of a countertop, touchscreen device, supported on an IP-based, open system network. A cluster of household appliances, like electrical outlets, heating and cooling equipment, washing machines, dryers and dishwashers, water heaters, and PEVs are connected to and controlled by the EMS. Altogether, more than 150 residential customers tested Duke Energy's first-generation EMS in 2009 and 2010.

6.7.3 The Smart Micro-Grid on the Island of Tilos

The smart micro-grid in the Aegean Sea island of Tilos, Greece, was implemented in the frame of the European Union Horizon 2020 project "TILOS," engaging 13 participating enterprises and institutes from 7 European countries (Germany, France, Greece, UK, Sweden, Italy, and Spain). The project's main goal is to demonstrate the potential of local/small-scale battery storage to serve a multipurpose role within an island micro-grid that also interacts with a main electricity network. Among others, the project aims to achieve large-scale RES penetration and asset value maximization through the optimum integration of a hybrid RES (wind and PV) power station together with advanced battery storage; distributed, domestic heat storage; smart metering; and DSM.

Project objectives will be accomplished through the development and operation of an integrated, smart micro-grid on the island of Tilos, a small island of the Dodecanese complex in the Eastern Aegean Sea, Greece. Tilos is connected to the electrical grid of the neighboring islands of Kos and Kalymnos, yet, as a whole, the overall insular system constitutes an autonomous, non-interconnected electrical system.

The main objective of TILOS is the development and operation of a prototype battery system based on $NaNiCl_2$ batteries, supported by an optimum, real-environment smart grid control system and coping with the challenge of supporting multiple tasks, including:

- micro-grid energy management
- maximization of RES penetration
- grid stability
- export of guaranteed energy
- ancillary services to the main grid of Kos.

The battery system will support both stand-alone and grid-connected operation, while proving its interoperability with the rest of the micro-grid components, such as DSM aspects and distributed, residential heat storage in the form of domestic hot water. In addition, different operation strategies will be tested in order to define the optimum system integration.

The main realization steps of the TILOS project are:

- Conduction of a complete system design study in order to develop the optimum system layout by considering aspects of the RES power station, the local grid structure, the demand side features, the battery storage, and the proposed micro-grid.
- Definition, design, and installation of the SCADA system together with the monitoring of the RES side.
- Installation and testing of the main demand side/end use components in the island, along with an assessment of the local demand side characteristics. At the same time, DSM strategies will be developed and tested, engaging also the local population.
- Development and manufacturing of a prototype $NaNiCl_2$ battery storage system (storage capacity in the order of 2 MWh and charge/discharge rate in the order of 800 kW) together with the appropriate inverters will be accomplished.
- Development and validation of forecasting models will be undertaken, regarding RES power generation and the local electricity demand for the proposed micro-grid.
- Development and commissioning of the EMS employed on Tilos island and the development of a micro-grid simulator for simulation-based development of the EMS for Tilos as well as for simulations of other micro-grid scenarios.
- Emphasis is given to the training, study, and engagement of the local population. It concerns the long-term observation of the islanders in order to both filter their needs in relation to the system operation and also measure their gradual engagement and willingness to actively support the operation of the micro-grid.
- The micro-grid is integrated with the developed battery storage system delivered to the island and connected to the rest of the system and the SCADA operation center, followed also by the commissioning of the EMS system so the micro-grid can be fully operational.
- The developed extended micro-grid simulator will be used in order to examine both stand-alone and interconnected micro-grid case studies under the application of different RES, battery energy storage, and interconnector configurations.
- Dissemination of the project results to the wider public through leaflets, web channels, and through an actual info-kiosk for local residents.
- The gradual installation and grid connection of the main RES components (i.e., wind power in the order of 300 kW and PV power in the order of 700 kW), attracting a private investment of ~1 M€.

More specifically, with regard to smart grid actions, first of all, inspection, assessment, and certification of existing electrical boards and infrastructures were carried out by the local grid's operator, ensuring that all safety measures are implemented properly. Secondly, a smart meter-DSM device prototype has been developed, with the respective end-to-end testing being under completion. Next, almost 150 smart metering and DSM panels were shipped to Tilos in December 2016, and their rollout was started immediately in the residential sector of the island, in public buildings, and in central municipality loads, such as pumping stations. Each panel has the ability to remotely monitor and control up to three electrical loads per residence, aiming to test the functionality of the prototype in representative households of Tilos.

To validate the energy management algorithms of the TILOS system operation, a dedicated micro-grid simulator has been developed. In a parallel work stream, the TILOS team also looks at the development of the extended micro-grid simulator, suitable for the conduction of feasibility studies and able to simulate different kinds of RES storage configurations, for both stand-alone and market-based applications. The main goal of the extended micro-grid simulator is to provide a user-friendly tool that can enable local communities to evaluate the potential for the establishment of energy schemes similar to the one of TILOS.

REFERENCES

1. California ISO time-of-use periods analysis. (2016, January 22). http://docs.cpuc.ca.gov/PublishedDocs/Efile/G000/M157/K905/157905349.pdf (Accessed on August 2019).
2. Gigantidou, A. (2013, August). Renewable energy sources in Crete. Bulk Power System Dynamics and Control—IX Optimization, Security and Control of the Emerging Power Grid (IREP). 2013 IREP Symposium. DOI:10.1109/IREP.2013.6629344
3. International Energy Agency Smart Grid Roadmap. https://www.iea.org/publications/freepublications/publication/smartgrids_roadmap.pdf (Accessed on August 2019).
4. European Commission. http://ec.europa.eu/energy/en/topics/markets-and-consumers/smart-grids-and-meters (Accessed on August 2019).
5. United States Office of Electricity Delivery & Energy Reliability. http://energy.gov/oe/services/technology-development/smart-grid (Accessed on August 2019).
6. International Electrotechnical Commission. www.iec.ch/smartgrid/background/explained.htm (Accessed on August 2019).
7. Japan Smart Community Alliance. www.smart-japan.org/english/index.html (Accessed on August 2019).
8. Office of the National Coordinator for Smart Grid Interoperability. (2012). NIST framework and roadmap for smart grid interoperability standards. https://www.nist.gov/sites/default/files/documents/smartgrid/NIST_Framework_Release_2-0_corr.pdf (Accessed on August 2019).
9. Yu, Y., Yang, J., & Chen, B. (2012). Smart grids in China—a review. *Energies*, *5*, 1321–1338.
10. Lorena Tuballa, M., & Lochinvar Abundo, M. (2016). A review of the development of Smart Grid technologies. *Renewable and Sustainable Energy Reviews*, *59*, 710–725.
11. University of Delaware. (2007, December 9). Car prototype generates electricity and cash. *Science Daily*. https://www.sciencedaily.com/releases/2007/12/071203133532.htm (Accessed on August 2019).
12. United Nations Economic Commission for Europe (UNECE). Electricity system development: a focus on smart grids. Overview of activities and players in smart grids. https://www.unece.org/fileadmin/DAM/energy/se/pdfs/eneff/eneff_h.news/Smart.Grids.Overview.pdf (Accessed on August 2019).
13. Energy and Technology History Wiki. The history of making the grid smart. https://ethw.org/The_History_of_Making_the_Grid_Smart (Accessed on August 2019).
14. Iyengar, S. S., Boroojeni, K. G., & Balakrishnan, N. (2014). *Mathematical Theories of Distributed Sensor Networks*. Berlin, Germany: Springer.
15. Moura, P. S., López, G. L., Moreno, J. I., & De Almeida, A. T. (2013). The role of Smart Grids to foster energy efficiency. *Energy Efficiency*, *6*, 621–639.
16. International Confederation of Energy Regulators. (2012, April). *Report on Experiences on the Regulatory Approaches to the Implementation of Smart Meters*. Annex 4–Smart meters in Italy.
17. NIST Framework and Roadmap for Smart Grid Interoperability Standards, Release 3.0. https://www.nist.gov/system/files/documents/smartgrid/NIST-SP-1108r3.pdf (Accessed on August 2019).
18. CEN/CENELEC/ETSI. (2012, November). Smart Grids Coordination Group Technical Report Reference Architecture for the Smart Grid, Version 2.0. https://ec.europa.eu/energy/sites/ener/files/documents/xpert_group1_reference_architecture.pdf (Accessed on August 2019).
19. Chiu, A., Ipakchi, A., Chuang, A., Qiu, B., Hodges, B., Brooks, D., et al. (2009). Framework for integrated demand response (DR) and distributed energy resources (DER) models. http://www.naesb.org/pdf4/smart_grid_ssd111709reqcom_pap9_a1.doc (Accessed on August 2019).
20. Federal Energy Regulatory Commission. (2018, November). Reports on demand response and advanced metering. https://www.ferc.gov/legal/staff-reports/2018/DR-AM-Report2018.pdf (Accessed on August 2019).
21. Siano, P. (2014). Demand response and smart grids—a survey. *Renewable and Sustainable Energy Reviews*, *30*, 461–478.
22. Zubi, G., Dufo-López, R., Carvalho, M., & Pasaoglu, G. (2018). The lithium-ion battery: State of the art and future perspectives. *Renewable and Sustainable Energy Reviews*, *89*, 292–308.
23. Smith, C., et al. (2009, December). Smart Grid White Paper: The Home Appliance Industry's Principles & Requirements for Achieving a Widely Accepted Smart Grid. Washington D.C.: Association of Home Appliance Manufacturers. https://edecade.files.wordpress.com/2011/08/summary-smart-grid-white-paper-110520.pdf (Accessed on August 2019).
24. Conchado, A., & Linares, P. (2012). The economic impact of demand-response programs on power systems. A survey of the state of the art. *Handbook of Networks in Power Systems I Energy Systems*, 281–301.

25. Palensky, P., & Dietrich, D. (2011). Demand side management: demand response, intelligent energy systems, and smart loads. *IEEE Trans Ind Inf, 7*, 381–388.
26. Reid, M., Levy, R., & Silverstein, A. (2010, January). Coordination of energy efficiency and demand response. Ernest Orlando Lawrence Berkeley National Laboratory, Charles Goldman. https://emp.lbl.gov/sites/all/files/report-lbnl-3044e.pdf (Accessed on August 2019).
27. Crossley, D. (2008). Assessment and development of network-driven demand-side management measures. Task XV, Research Report No. 2. Energy Futures Australia Pty Ltd. NSW, Australia: IEA Demand Side Management Programme. http://www.ieadsm.org/wp/files/Tasks/Task%2015%20-%20Network%20Driven%20DSM/Publications/IEADSMTaskXVResearchReport2_Secondedition.pdf (Accessed on August 2019).
28. Kirschen, D. S. (2003). Demand-side view of electricity markets. *IEEE Trans Power Syst, 18*, 520–527.
29. Federal Energy Regulation Commission. (2011, February). Assessment of Demand Response and Advanced Metering Staff Report Docket. https://www.ferc.gov/legal/staff-reports/2010-dr-report.pdf (Accessed on August 2019).
30. Pipattanasomporn, M., Feroze, H., & Rahman, S. (2009, March). Multi-Agent Systems in a Distributed Smart Grid: Design and Implementation. Seattle, Washington: Proc. IEEE PES 2009 Power Systems Conference and Exposition (PSCE'09).
31. Popovic, D. S., & Boskov E. E. (2008). Advanced fault management as a part of smart grid solution. Smart Grids Distribution. 2008 IET – CIRED, 1-4. Frankfurt: CIRED Seminar.
32. Othman, M. M., Hegazy, Y. G., & Abdelaziz, A. Y. (2017). Electrical energy management in unbalanced distribution networks using virtual power plant concept. *Electric Power Systems Research, 145*, 157–165.
33. Kärkkäinen, S., & Oy, E. (2012). Task XVII: Integration of demand side management, distributed generation, renewable energy sources and energy storages — Subtask 5, Report No. 5: Smart metering. http://www.ieadsm.org/wp/files/Exco%20File%20Library/Key%20Publications/SmartMetering_final.pdf (Accessed on August 2019).
34. Katsaprakakis, D. A., Christakis, D. G., Zervos, A., & Voutsinas, S. (2008). A power quality measure. *IEEE Transactions on Power Delivery, 23*, 553–561.
35. Parvez, I., Sarwat, A. I., Wei, L., & Sundararajan, A. (2016). Securing metering infrastructure of smart grid: a machine learning and localization based key management approach. *Energies, 9*, 691.
36. Kärkkäinen, S. (2008). Task XVII: Integration of demand side management, distributed generation, renewable energy sources and energy storages. *Main Report, 1*. http://www.ieadsm.org/wp/files/Exco%20File%20Library/Key%20Publications/SynthesisFinalvol1.pdf (Accessed on August 2019).
37. Di Fazio, A. R., Erseghe, T., Ghiani, E., Murroni, M., Siano, P., & Silvestro, F. (2013). Integration of renewable energy sources, energy storage systems and electrical vehicles with smart power distribution networks. *Journal of Ambient Intelligence and Humanized Computing, 4*, 663–672.
38. Adamiak, M., & Baigent, D. (2003). IEC 61850-1: Communication networks and systems in substations—Part 1: Introduction and overview.
39. Pilloni, V., & Atzori, L. (2011). Deployment of distributed applications in wireless sensor networks. *Sensors, 11*, 7395–7419.
40. Wu, G., Talwar, S., Johnsson, K., Himayat, N., & Johnsson, K. (2011). M2M: from mobile to embedded Internet. *IEEE Communications Magazine, 49*, 36–43.
41. Bayindir, R., Colak, I., Fulli, G., & Demirtas, K. (2016). Smart grid technologies and applications. *Renewable and Sustainable Energy Reviews, 66*, 499–516.
42. Darby, S. (2006, April). The effectiveness of feedback on energy consumption. A review for DEFRA of the literature on metering, billing and direct displays. Environmental Change Institute, University of Oxford. https://www.eci.ox.ac.uk/research/energy/downloads/smart-metering-report.pdf (Accessed on August 2019).
43. Ehrhardt-Martinez, K., Donnelly, K. A., & Laitner, J. A. (2010, June). Advanced Metering Initiatives and Residential Feedback Programs: A Meta-Review for Household Electricity-Saving Opportunities. American Council for an Energy-Efficient Economy. Report No. E105. https://www.smartgrid.gov/files/ami_initiatives_aceee.pdf (Accessed on August 2019).
44. Foster, B., & Mazur-Stommen, S. (2012, February). Results from recent real-time feedback studies. American Council for an Energy-Efficient Economy. Research Report B122. https://aceee.org/sites/default/files/publications/researchreports/b122.pdf (Accessed on August 2019).
45. United States Department of Energy. (2009, December). The Smart Grid: An Estimation of the Energy and CO2 Benefits. https://www.smartgrid.gov/files/The_Smart_Grid_Estimation_Energy_CO2_Benefits_201011.pdf (Accessed on August 2019).

46. ENISA. (2012, April). Smart Grid Security. Annex II: Security aspects of the smart grid. https://www.enisa.europa.eu/topics/critical-information-infrastructures-and-services/smart-grids/smart-grids-and-smart-metering/ENISA_Annex%20II%20-%20Security%20Aspects%20of%20Smart%20Grid.pdf (Accessed on August 2019).
47. NIST Report. (2010, September). Introduction to NISTIR 7628 Guidelines for Smart Grid Cyber Security. https://www.nist.gov/sites/default/files/documents/smartgrid/nistir-7628_total.pdf (Accessed on August 2019).
48. United States Energy Information Administration. (2011, December). Smart Grid Legislative and Regulatory Policies and Case Studies. https://www.eia.gov/analysis/studies/electricity/pdf/smartggrid.pdf (Accessed on August 2019).

7 Energy as a Consumptive Product

7.1 INTRODUCTION

A book dedicated to energy and its transformation technologies, like this one, must approach the concept of energy not only on a technical and natural magnitude, but as a consumptive product as well, which, most importantly, in the modern developed and developing world, constitutes a primary good without any substitutes. The last feature, namely that energy has no substitutes, determines precisely its attribute as an irreplaceable medium required for the implementation of several tasks and works, which cannot be substituted by any other alternative; consequently, without its availability, the implementation of these tasks becomes impossible. For example, to read a book at night the energy availability in the form of light energy is required. Similarly, for the transportation or the elevation of a body, mechanical energy should be available. None of the above tasks can be implemented without the availability of the corresponding required form of energy.

The fact that energy has no substitutes imposes specific peculiarities. First of all, energy becomes essential for the approach of a modern, comfortable way of life, often for our own survival. More accurately, the availability of energy in specific forms constituted a fundamental prerequisite not only for the survival, but also for the evolution, of humankind, during the last century. For example, the discovery of fire by the Homo heidelbergensis was a milestone in the history of the humanity. On the one hand, it offered the required heat that facilitated the survival of our ancestors under adverse weather conditions, enabling the gradual increase of the average human life period. On the other hand, it contributed to the cultivation of the human imagination, an essential prerequisite for the interpretation of nature and the development of knowledge and sciences.

Furthermore, the importance of energy as a consumptive product is vividly depicted through the international struggles of states and nations over the control of the available primary energy resources, practically over the control of the wealth arisen from their exploitation. Indeed, the uniqueness and the nonsubstitution of energy as a consumptive product imply huge amounts of material wealth for the owner of the natural primary energy reserves. Sequels of international and civil wars have broken out in the twentieth century over the control of the dominating primary energy form during the same period, namely of oil reserves. Enormous economic interests are currently and continuously evolved in the field of the exploitation of renewable energy sources (RES) for electricity production during the last two decades, particularly for the occupation and the control of land with the highest available RES potential. The above facts are further intensified by the energy consumption rate in developed countries. In Figure 7.1 the evolution of the total annual energy consumption on earth is presented, exhibiting until 2016 a constant increase rate at the range of 1% [1], despite the global campaign towards the rational use of energy.

Another fact that very few of us have considered is the size of energy consumption in the modern world. The annual primary energy consumption in 2016 in the US was 28,545 TWh [2], while in the EU-28 in 2014 it was 18,677 TWh [3]. Given the populations of the US and the EU-28 (approximately 325,000,000 and 510,000,000 respectively), these figures imply an annual average power consumption per American or European inhabitant of 10.0 kW and 4.2 kW respectively. This means that an electricity, thermal, and mechanical energy multigenerator with the above nominal power has to operate continuously to cover the energy needs of each American or European inhabitant. A direct image of the above energy consumption rates can be obtained if we consider that the average power of a middle-aged

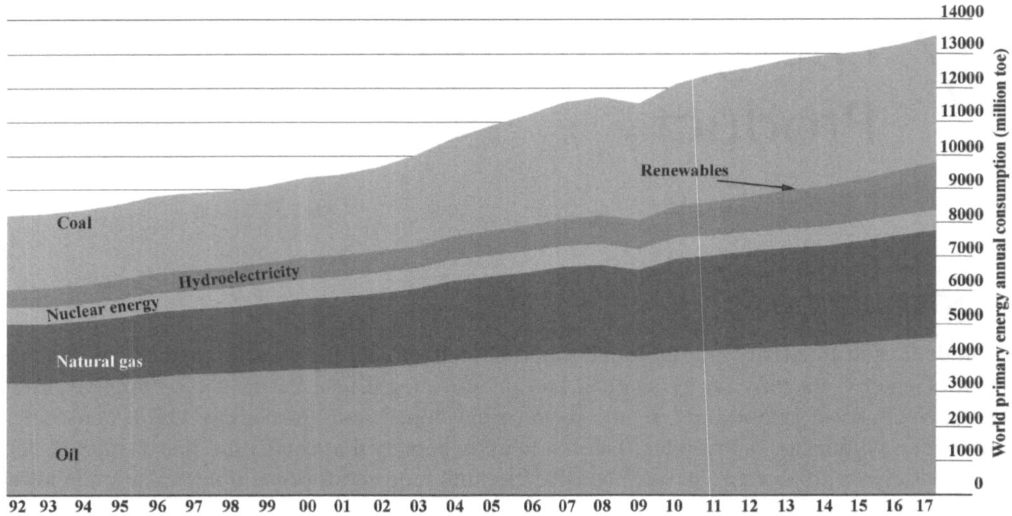

FIGURE 7.1 World primary energy annual consumption [1].

male adult is around 100 W. Given this fact, the above constant annual power consumptions are equivalent to the availability of 100 and 42 "energy human slaves" for each American and European inhabitant, who offer their services under an uninterrupted basis for each one of them during the whole year.

The modern generations consider the availability of energy, with the simple turning-on of a switch, as something obvious and absolutely expectable. However, most of us cannot realize, for example, that in order to heat up 50 lt of water required for a shower in London, 10 kg of coal must be extracted and transferred from North Whales or Yorkshire, namely more than 250 km away, or 1 Nm^3 of natural gas must cross the whole European continent from the faraway Siberia, covering a distance of some thousands of km. What's more, much fewer people perceive that behind the turning-on of the electrical lights there is the fight of the strongest and most developed states of the planet over the control of the primary energy reserves and the raging antagonism of the large multinational enterprises. This antagonism, on the one hand boosts the technological progress, while, on the other hand, constitutes the main cause of serious international crises and destructive conflicts around the world. The conservation of the current system of values in the developed world and its expansion to the other developing countries leads to endless increases of the annual energy consumption and to rapid exhaustion of fossil fuel reserves.

Energy affects and configures global trends and determines our way of living and our culture more than any other product or good does. To conceive these claims, just imagine how different our lives would be if, for example, electricity hadn't been invented. A number of electrical devices of the typical modern household would not exist. The lack of telecommunication devices would cause direct consequences on the internalization and the globalism of the way of living and information diffusion would be considerably restricted. Can we today imagine ourselves without direct access to information ensured by the Internet? Can we imagine that without the support of the modern electronic technology, even the sending of a simple message would require a period of some days to be delivered? Or can we perhaps imagine that we would not be able to conserve our foods without refrigerators or freezers? The impossible realization of the above tasks without energy, which today we take for granted, gives us the chance to realize how different our lives would be, our way of thinking and our habits would have been configured, if energy would not have been available in the various final energy forms as it is today.

From the above arguments, we see that energy affects and configures the way of thinking and acting of ordinary people, as well as the national policies of states and nations. At the same time, the final disposed energy forms, as a common good produced by energy producers and provided through commercial transactions by energy providers for the final consumers, may constitute a mean for economic and, by extension, social development of states and local communities. This may be a direct result of the potential, even partial, exploitation of the net profits arisen from the commercial supply of energy. As an example, we may mention the impressive development achieved during the last decades in the Middle East, based entirely on exports of the available domestic oil reserves.

On the other hand, every production activity, more or less, always constitutes a potential source of impacts for both the environment and human activities. Energy conversion processes do not constitute an exemption from this rule. On the contrary, some of them are responsible for some of the most serious environmental impacts compared to any other activities executed by humans.

In this last chapter a critical evaluation is attempted of the available energy transformation and disposal technologies, based on the following criteria:

- their contribution towards the development of the national states and, particularly, the local communities, on the area in which they are installed
- their environmental footprint from their installation and operation.

Through this critical presentation, the role of energy emerges as a potential source of wealth and development, as a parameter of international policy and economy configuration, and as the origination of potential serious environmental impacts. The ultimate target of this chapter is the formulation of holistic awareness, required for the configuration of integrated critical judgment from the professional engineer, the researcher, or the scientist, as well as every single and unskilled interested individual, on the energy conversion technologies and projects and the developmental, social, and environmental roles they can undertake.

7.2 OIL AND DEVELOPMENT

7.2.1 Brief Historical Background

Since the second half of the twentieth century, oil has constituted the main primary energy source on the planet. Its modern history goes back to 1848, when the Scottish chemist James Young discovered a natural petroleum seepage in Derbyshire, from which he managed to distill a light thin oil that was used as lamp oil and as a machinery lubricant. During the next 15 years, the first commercial oil wells were drilled and exploited in Pennsylvania (US), Germany, Poland, Canada, and Romania. The Russian Empire was the lead producer of crude oil by the end of the nineteenth century [4]. By the middle 1960s, however, the countries of the Middle East, having 80% of the oil's accessible reserves globally, had already emerged as the top region on earth for oil production, with the Soviet Union following after them [4].

Great Britain was the first country in the world that expressed practical interest in the exploitation of the Middle East's oil reserves, after the discovery in 1908 of the first oil fields in Kuwait from the Anglo-Iranian Oil Company (now known as BP). However, the low oil demand did not justify the development of the required infrastructure and activities for the commercial exploitation of the discovered reserves. Through a series of international political and commercial pressures, the US was the second state that developed oil exploitation activities mainly in Iraq and Kuwait in the second decade of the twentieth century [5].

Until 1939 the contribution of oil to the global economy was rather restricted, with the Middle East's reserves holding a 5% share over the global oil demand. Oil exports from the Middle East were mainly focused on the neighboring countries and, through the Suez Canal, on the countries of

Western Europe [5]. The major contribution of all this progress achieved before 1939 on oil reserve exploitation was that it prepared the ground for the wide expansion of oil use after 1945.

After World War I, the demand for oil was considerably increased, as a consequence of the increased energy needs resulting from the high development rates of the European states. Nevertheless, due to the general insecure, war climate during this era, the discovery of new oil reserves in the Middle East was met with hesitation for new investments on the research and the mining of oil in the specific geographical area. On the contrary, huge amounts of money were invested on the creation of the required infrastructure for the secure and cost-effective oil transportation from the Middle East to Europe, including the construction of new oil transportation pipelines and modification works in the Suez Canal to enable the passage of larger oil tankers. All these investments were mainly accomplished by seven multinational oil companies, known as the Seven Sisters, which dominated the global petroleum industry from the mid-1940s to the mid-1970s. These are the Anglo-Iranian Oil Company (now BP), Gulf Oil (later part of Chevron), Royal Dutch Shell, Standard Oil Company of California (SoCal, now Chevron), Standard Oil Company of New Jersey (Esso, later Exxon), Standard Oil Company of New York (Socony, later Mobil, now part of ExxonMobil), and Texaco (later merged into Chevron) [6].

Oil demand and production recorded significant increases after the end of World War II, with the Seven Sisters being the only companies active on the exploitation of the discovered oil fields until 1953. From 1953 to 1973, more than 300 new companies entered the oil industry, playing a substantial role in international oil markets and technology.

In the modern history of oil, the first oil crisis in 1973 is considered a major landmark. In this year, Israel was attacked at the same time from Egypt, Syria, and another 10 Arabic countries to retrieve their lost land from the 1967 war. The US re-equipped Israel with military material and, as a retaliation, the OAPEC countries (the OPEC countries together with Egypt, Syria, and Tunisia) proclaimed an oil embargo against Canada, Japan, Holland, United Kingdom, and the US, causing a recession in the global economy. Crude oil prices increased globally from \$2.5 to \$12 per barrel, with direct negative effects on European and Japanese economies, which were 75%–80% dependent on oil imports from the Middle East.

The revolution in Iran, the second largest oil-producing country in the world, and the Soviet Union invasion in Afghanistan led to the second oil crisis in 1979. The following war between Iraq and Iran in 1980 caused the postponement of almost any oil production from these two countries. The result was that oil prices increased to \$35 per barrel. After 1980, there were 6 years of economic recession. Eventually, in 1986, forced by the low oil demand resulting from the global economic recession, OPEC announced reduced oil prices at the range of \$10 per barrel, in order to provoke an increase in oil consumption.

The tension between Iraq and Kuwait in the area of the Persian Gulf in 1990 was another crucial time period in the modern history of oil. The oil prices raised from \$17 to \$36 per barrel in October 1990. The US reacted immediately by setting at the disposal of the oil market their oil military reserves and achieving, in this way, a vertical drop of oil price at \$20 per barrel. The effects on oil prices were much more modest; however, it took several years for both Iraq and Kuwait to recover from the consequences of the war and regain a total oil production at its former level.

Oil prices ranged below \$25 per barrel until 2003. Since then, a constant rise in oil prices has occurred, starting from \$30 per barrel in 2003, \$60 per barrel in 2005, and reaching the maximum value of \$147 per barrel in 2008 [7]. The parameters that caused these continuous price increases are the long-lasting tensions in the Middle East, the developing Chinese economy and the corresponding rise in oil demand, worries and uncertainties regarding the remaining oil reserves and the possible oil peak production, and, of course, financial speculation.

The economic crisis that broke out in 2008 caused a considerable reduction on the primary energy sources demand, with a subsequent drop of oil prices from \$147 to \$32 per barrel at the end of 2008. From 2009 until the end of 2014, oil prices stabilized roughly between \$70 and \$120 per barrel, before settling at levels below \$50 per barrel since 2015 and after.

Oil price fluctuations since 1946 are depicted in Figure 7.2 [8].

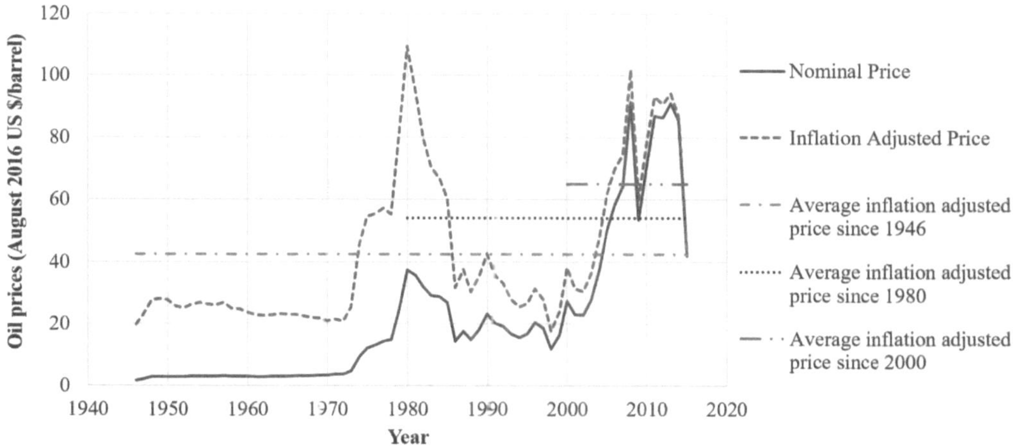

FIGURE 7.2 Annual average nominal and inflation-adjusted oil prices since 1946 [8].

7.2.2 Effects of Oil Prices on International and National Macro Economies

Oil prices constitute an essential parameter on the global economy. High oil prices can cause extensive negative effects on the international economy and, especially, on national economies with strong dependence on oil imports. For example, in the European Union (EU), 53% of the annual primary energy consumption in 2014 was covered by imported energy sources. Specifically, the EU imports [9]:

- 90% of its crude oil
- 66% of its natural gas
- 42% of its coal and other solid fuels
- 40% of its uranium and other nuclear fuels.

The above figures reveal that the EU economy is potentially highly affected by any fluctuations of the imported energy sources prices, including crude oil. A probable oil price increase leads to corresponding income transfer from the importing to the exporting countries, with subsequent economic effects. The higher the oil prices and the longer time period they remain, the more serious the economic long-term effects will be for national economies. The national trade balances are deranged. The oil-importing countries face deterioration on the balance of payments, affecting negatively the exchange rates. As a result, the imports become more expensive, implying a collapse of the gross national income (GNI).

On the other hand, for the exporting countries, a significant oil price increase leads to a direct increase of the real GNI. Higher GNI leads to increasing inflation, increasing production cost, further increase of oil prices, and, eventually, reduction of oil demand and economic recession.

A number of studies have shown that oil price fluctuations seem to have a considerable effect on the real economy, as well as on specific economic indicators and parameters. Indicatively, James Hamilton proved, by using data from the period 1950–1980, covering the first two serious oil crises, that there is a strong connection between high oil prices and economic recession [10]. According to Hamilton, all the economic recessions of the American economy were caused by abrupt increases in oil prices. He concluded that high oil prices have strong negative effects on oil-importing countries, which become more significant after 9–12 months.

Another interesting study comes from Carlos de Miguel et al. [11] In their work investigating the effects of oil price shocks in the European economies, they conclude that the national economies of the European South countries (Spain, Portugal, Italy, and Greece) are more vulnerable to oil price shocks. By examining the oil crises of the 1970s and 1980s, they found the highest losses in prosperity levels were 7.9% and 8.1% in Greece and Portugal respectively.

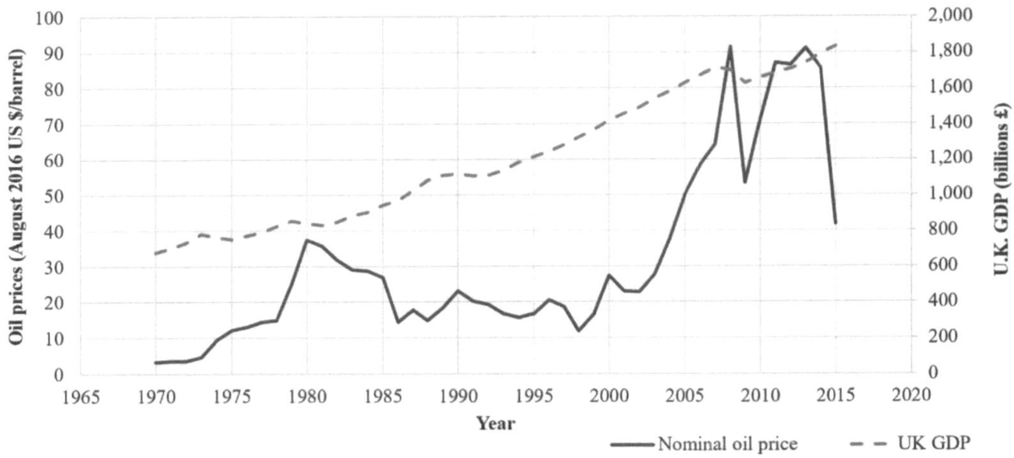

FIGURE 7.3 Effects on GDP in the UK versus oil price fluctuations.

Hooker investigated the relationship between oil price fluctuations and inflation [12]. He used data from the period 1962–2000 and he concluded that oil price fluctuations had a stronger effect on inflation before 1980. After 1980, it seems that the national economies had found a way to keep inflation stable even amid oil price shocks. This conclusion is also verified by LeBlanc and Chin [13], who studied the effects of oil price changes on the national inflation of the US, UK, France, Germany, and Japan. They concluded that a 10% increase in oil prices can cause direct inflationary increases in the range of 0.1%–0.8%.

The effects of oil price fluctuations on real economic activity are investigated for the main industrialized Organization for Economic Co-operation and Development (OECD) countries in a study conducted and funded by the European Central Bank [14]. The study is based on the estimation of the gross domestic product (GDP) for the under consideration countries versus oil price fluctuations, using different linear and nonlinear arithmetic models. What is clear from this study, is that the real GDP growth of oil-importing economies is negatively affected by oil price increases. For example, in the US, a GDP loss from 3.5% to 5% is estimated as a result from a 100% rise of oil price, depending on the employed simulation model. Similarly, the GDP loss from the same oil price shock ranges at 2%–5% for the Euro-zone countries and less than 1% in Canada. Conclusively, oil price shocks have considerable negative effects on national real GDPs. The effect of oil price shocks on the national GDP of UK is presented in Figure 7.3. In this figure, the GDP drops during the three main oil price shocks and the economic crisis in 2008 are characteristically depicted [1, 15].

7.2.3 Oil and Development of Local and National Economies

Oil doubtlessly constitutes a potential source of wealth. If we only consider that all this impressive technical and technological evolution and progress achieved during the twentieth century was mainly based on oil, either as the primary energy source for the industries and the transportation, or as the raw material for the production of a large number of goods and products, we would expect that the oil-exporting countries should be the richest countries in the world with the highest GDP and standard of living. However, our practical experience and awareness from what has happened in reality is rather far from this assumption. In the dawn of the twenty-first century, on the one hand we can find countries with truly remarkable economic development and astonishing technical achievements based exclusively on oil exports, such as the Middle East countries, but, on the other hand, we can also find underdeveloped countries in Africa and in Latin America, still remaining below poverty status, despite their considerable oil exports. So, the natural question is: What has happened in these countries, leading to such intensive disparities regarding the achieved development in the one versus the other? The answer to the above question is not so simple and it goes back to the beginning of the modern history of oil.

Before the discovery of oil and its introduction as the most popular primary energy source, the Middle East countries, such as Iraq, Iran, Saudi Arabia, and Kuwait, used to be poor and underdeveloped countries. In practice, they were small kingdoms in the desert with very few and poor natural resources, inadequate for the conservation of the population. When oil was discovered at the beginning of the twentieth century, these countries did not have the technical infrastructure and the knowledge required for its mining and exploitation [16]. The large multinational oil companies took this inability as a chance to increase their own economic profits and began negotiations with the involved states towards the exploitation of the discovered reserves. They eventually established agreements with the Middle East states for the exclusive research and exploitation of oil reserves in their territory. These agreements referred to specific geographical areas for determined time periods. The companies would undertake all the expenses and the investment risk. At the same time, the states maintained the right to claim their rights on the exploitation of a part of the produced oil [16]. The results of these negotiations were that the above mentioned seven companies known as the Seven Sisters, took control of the largest percentage of the so-far discovered oil reserves globally.

The agreements between the states and the Seven Sisters placed the latter in an advantageous position, often allowing them even to intervene in national affairs and legislation. Among others, the companies claimed and succeeded to achieve the execution of any required arbitration between them and the states on any probable disagreements, by a third person, rather than the official states' justice. Hence, although these agreements were initially welcome by the oil-exporting states, the common opinion gradually swayed, at the beginning by groups of nationalists, who spread the idea that these private companies hold and control the national wealth of their countries. They claimed that the relevant agreements were unfair for their states, because they could not be modified, they gave the companies exclusive rights on enormous amounts of land, and their validity was prolonged over long time periods. These aspects cultivated the common opinion that the state was exploited by the private companies and led to the creation of a unified front in favor of the oil reserves nationalization. The oil reserves nationalization process practically imposed the confiscation of the oil mining companies from the local governments, aiming at the maximization of the state income from the exploitation and the exports of the locally available oil reserves.

However, the reality turned out not to be so easy. Among the pioneering countries in the oil reserves nationalization were Mexico in 1938 and Iran in 1951. These movements, manifested before the major changes in the oil industry in the 1950s, had a strong effect on the other oil-exporting countries. The nationalization in Mexico, although it proved that such a target is feasible, also led to international isolation from the oil market, which, at this early time, was still under the exclusive control of the seven multinational companies. Also, the nationalization in Iran did not succeed because the Iranian company was not able to establish collaboration with the international oil market. These two unsuccessful efforts showed the other oil-exporting countries that oil reserves nationalizations are accompanied by high investment risk and considerable failure probability, as long as the international oil market is controlled by a small number of private companies.

Nevertheless, things began to change in the early 1950s from a political, economic, and technical point of view, for the oil-exporting countries, most of them belonging to the so-called Third Word. First of all, the gradually increasing nationalism and the prominence of a common conscience for Third World countries led to the end of colonialism in the 1950s and 1960s. This common conscience also led to the foundation of the Organization of the Petroleum Exporting Countries (OPEC), aiming at the establishment of a common tactic, at the cultivation of a fertile communication and at synchronized efforts for common actions at the end of the 1960s. The foundation of OPEC offered the oil-exporting countries a medium for quick, direct, and valid information exchange.

Additionally, the technological progress and the developing science on business and companies management after World War II, contributed to the enforcement of negotiating skills of the oil-exporting countries.

The oil-exporting countries gradually achieved an essential progress, based even on the agreements that were initially unfavorable for them. Following this progress, their requirements for higher public income increased, making them demand higher shares of the economic benefits from their oil reserve exploitation, control on the oil pricing policy, and on the mining rates. On the other hand, this new trend impinged the confidence of foreign investors, which became less willing to invest money in the oil industry in these countries.

Finally, the last crucial event that triggered the nationalization movements was the companies' policy on the oil prices. The official oil posted price was initially the fundamental parameter for the calculation of the taxes that the companies should pay to the states, as a compensation for their rights to oil reserve exploitation. This term was highly favorable for these companies, because the oil posted price was configured by them and no one else. Practically, the companies could raise the real oil market price, without modifying the official posted one, avoiding, in this way, the payment of corresponding increasing taxes to the states. The latter did not realize this double oil pricing from the side of the companies, until the reduction on oil production cost recorded at the end of 1950s, a fact that gave the excuse to the companies to decrease the official oil posted prices. From their side, the oil-exporting countries reacted negatively, as they were likely to do, against the oil price reduction exclusively from the companies, without any former coordination. These reactions peaked in 1960, when a 10%–15% reduction in oil prices led to a subsequent 5%–7.5% drop for the states' annual income.

The first country that successfully achieved the nationalization of oil reserves was Algeria, which in 1971 nationalized 51% of the French oil companies. Algeria, later, was able to increase this percentage to 100%. The nationalization in Algeria boosted Libya to nationalize British Petroleum also in 1971, as well as the other private companies in 1974. A domino of nationalizations followed in the next 2 years, starting with countries with more aggressive policies, such as Iraq, to be expanded to more conservative countries, such as Saudi Arabia. Until 1976, almost every major oil-exporting state in the Middle East, Asia, Africa, and Latin America had succeeded, more or less, with a form of nationalization, in order to either glean a percentage of the economic profits from the mining and the exporting of oil, or to fully take control over the domestic oil reserves.

After the total or the partial nationalization of oil reserves, oil was often used from oil-exporting countries as a medium to ask international pressure and a major negotiation tool to perform or impose their foreign policies. Another direct consequence was that the initially existing monopoly on the oil market from the seven multinational companies was eliminated. Specifically, the OPEC countries gained full control of the mining and promotion of crude oil, while the companies maintained the oil transportation and refinery, as well as the distribution and the selling of the oil final products to the consumers. The involvement of both states and companies in the overall oil mining and final product distribution line led to long-term signed agreements between them.

Another impact of the oil reserves nationalization and the foundation of OPEC was that some countries, such as Great Britain, the US, and Canada decided to invest in the research and exploitation of possible domestic available oil reserves, establishing an alternative oil market. Gradually, this alternative market expanded to cover, apart from crude oil mining, the refinery process and the disposal in the market of the relevant products. The size of this parallel market was constantly increasing, especially during the two oil crises in the 1970s, with a direct consequence the liquidity of the international oil prices, which, in turn, raised the investment risk of oil reserves research and exploitation projects.

All the above contributed to the rise of the competition in oil markets and the limitation of the oligopoly. Additionally, the OPEC countries:

- often broke the signed agreements between them and the private sector, due to the overall uncertainty in the oil market, in cases where these were considered unfavorable
- established stronger relationships between them, especially whenever there was a global trend towards the reduction of oil prices, in order to claim in common the most possible favorable terms for the oil distribution in the international markets, through the formulation of a common and holistic negotiation policy.

Despite the above achievements, the anticipated development was not approached for all the oil-exporting countries. As mentioned at the beginning of this section, in the modern world we have examples of impressive development progress in oil-exporting countries, but we can also still see underdeveloped communities, despite their considerable oil exports. Below we will present some of these cases.

7.2.3.1 Iran

The oil industry in Iran passed through alternative periods of progress and recession. The rapid development period during World War I was followed by a period of decadence right after the breakout of World War II, to give its position back to progress in 1943, with the restoration of the oil supply process from Iran to Great Britain. The control and the exploitation of oil reserves was managed until the end of the 1940s by the Anglo-Iranian Oil Company (AIOC).

Iran claimed its independence from the British policy and its exploitation from AIOC. The negotiations between the Iranian government and the AIOC in 1951 failed, leading the Iranian government to the nationalization of the oil reserves. The British reacted with a boycott against oil exports from Iran and with the military isolation of the area of Abadan from 1951 to 1954, a rich area in oil reserves and in oil mining infrastructure, a fact that remained known in modern history as the Abadan crisis. These two measures from the British side led to the near elimination of oil production from Iran. Additionally, following the British initiative, the CIA overthrew the Prime Minister Mossadegh of Iran in the mission Ajax in 1953, with a coup. Although the nationalization of the oil reserves still remained, in practice a new participation of both the Iranian State and the AIOC on the oil reserves exploitation was founded at a 50%–50% share.

The above facts left Great Britain a considerable share over its total offshore property assets, and, at the same time, they hampered the smooth transition towards the Iranian democracy, bequeathing their consequences to the next decades and, according to some estimations, to the Iranian revolution in 1979 and the second serious oil crisis. After the Iranian revolution, the oil reserves in the country were nationalized again.

Today the largest percentage of the country's economy is controlled by the state itself, with oil being the essential economic source. Iran is the fourth largest oil-producer country globally, with a total annual oil production that corresponds to 11% of the total oil production from the OPEC countries. The oil exports correspond to 85% of the total exports in the country. Yet, the nominal GDP per capita is estimated at $5,400 in 2017, ranking Iran in the 96th position globally. The economy of Iran is characterized as a transition economy, and the country is considered as semi-developed. However, it is estimated by economic forums that Iran has the potential to become one of the largest ten economies in the twenty-first century, of course based on oil exports, on the condition of political policy and currency stability.

7.2.3.2 Saudi Arabia

Since 1950, Saudi Arabia has been designated as an exceptionally successful oil-exporting geographical area, also maintaining an even larger unexploited potential of oil reserves. Due to favorable geological conditions and to the neighboring oil-pumping drillings with the coastline, the oil mining cost in the area was configured at relatively low levels. The oil selling price was initially agreed, between the private American company and the state, to be configured on the basis of the desirable profit level, rather than the oil mining cost. Initially the profits from the oil exploitation were divided in equal shares between the private firm and the state. Eventually, Saudi Arabia gained full control of the available oil reserves in 1980.

7.2.3.3 Kuwait

The oil history of Kuwait constitutes a success story example. It began in 1934 with the signing of the first concession agreement between the state and the Kuwait Oil Company, formed by the Gulf Oil Corporation and the Anglo-Iranian Company. After a period of investigation, oil was first discovered in Kuwait 4 years later, in 1938, while in 1946 the first oil export was implemented. The oil reserves and mining infrastructures were nationalized by the state of Kuwait in 1975, following the general trend of the Arabic countries, after a period of negotiations that lasted 5 years (since 1970). The transition towards the complete nationalization was performed peacefully. Kuwait's economy today is fully based on petroleum. The state is considered the fourth richest country in the world, with an average annual GDP of $70,000 per capita.

7.2.3.4 Mexico

The modern history of oil in Mexico began in 1901, with the opening of the Ebano oil field along the Mexican Central Railway. By 1935, any oil exploitation project in Mexico was performed by foreign companies, particularly by the Mexican Eagle Petroleum Company (a subsidiary of Royal Dutch Shell) and the American firms Jersey Standard and Standard Oil of California. Mexico nationalized the domestic oil reserves in 1938, without permitting the involvement of any private company in the country's oil industry since then. The state company Petróleos Mexicanos was founded that same year, having the exclusive rights over the investigation, the mining, the refining, and the distribution of oil products in Mexico.

The reaction of the American, Dutch, and English companies against the nationalization of the Mexican oil reserves was direct. All of them, with the support of the official states, claimed compensation for their nationalized investments. Soon after, Jersey Standard and Royal Dutch Shell began a boycott against Mexico, supported by several American companies, by trying to prevent Mexico from obtaining the required chemicals and machinery for the refinery of oil. The reaction of Mexico was timely. Students at the Instituto Politécnico Nacional and the National Autonomous University of Mexico synthesized tetraethyl lead, a gasoline additive used in this particular era to raise the fuel's octane ratings. Furthermore, American suppliers of the required equipment for the mining and the refining of oil gradually changed their attitude toward Mexico, given the option available for Petróleos Mexicanos to proceed with the procurement of this equipment alternatively from Europe. The situation was totally reversed with the start of World War II. Given the priority of the US to establish an alliance against fascism, a Good Neighbor Policy was signed between them and Mexico in 1971, defining, on the one hand, the compensation level of the Mexican government for the American companies and restoring, on the other hand, the oil trading terms and conditions between the two countries.

New discoveries of significant oil reserves in Mexico during the 1970s led to subsequent increases in oil production and exports, accompanied incidentally with oil price shock periods. Mexico today is the country with the highest oil exports in Latin America and the eleventh largest producer of oil globally.

7.2.3.5 Russia

The exploitation of oil in the territory of the former Russian Empire began in the second half of the nineteenth century. The first fields were discovered in this era at the region of Baku, in Krasnodar Krai, at the banks of the river Ukhta and in Cheleken peninsula (Turkmenistan). These first discovered oil fields were exploited by Royal Dutch Shell. By the turn of the century, the Russian oil production corresponded to more than 30% of global oil production.

The oil production in Russia was temporarily affected negatively by the October Revolution in 1917 and the nationalization of the oil fields. Yet, the Communists were not totally opposed to the foreign capital inflow for the Russian oil industry. Private companies, such as Vacuum and Standard Oil of New York, invested in the oil industry of Russia, helping Russian oil production to come back. By 1923, the oil exports of the country had been restored to their levels before the revolution of 1917. Oil, and particularly the oil reserves in the Caspian basin and the North Caucasus, became the driving force for Russian industrialization until the late 1940s. By 1950, the new discovered oil fields in Volga and the Urals accounted for 45% of total oil production in the USSR.

The discovery of the new oil fields and the opening of the new refinery in Omsk led to a considerable rise of Soviet oil production, enabling also the development of export activities. In the early 1960s,

USSR was the second largest oil producer globally. The disposal of cheap Russian oil in the global market led to a decrease in oil prices, and, eventually, to the foundation of OPEC. This situation was prolonged with the discovery of new oil reserves in the middle 1960s in Siberia. USSR ceased being the top oil-producing country in the world in 1988. By that time, a continuous decrease of the annual oil production started and lasted for the next 10 years, due to poor oil reservoir management and the economic crisis that followed the collapse of the Soviet Union.

Since 2000, there has been a restoration of the oil reserves nationalizations in Russia. The control over the major exploitation project of the oil fields on the island of Sakhalin has been gained by the state company Gazprom. Additionally, the private company Yukos has been absorbed by the state company Rosneft. These changes have shocked the confidence of the multinational oil companies for Russia. The oil industry of Russia today has been restored again as one of the largest in the world; however, the proven oil reserves in the country are not as high as in other geographical regions. Additionally, with the collapse of the Soviet Union, Russia has been detached from former accessible regions, rich in oil reserves (Caspian Sea, Central Asia States, and Azerbaijan).

7.2.3.6 Nigeria

Oil was first discovered in Nigeria in 1956 by Royal Dutch Shell and British Petroleum. This discovery, unlike the paradigms from the previously presented countries, triggered the beginning of a period of conflicts and civil wars between the indigenous population for the control over the discovered oil reserves, rather than a period of progress and prosperity. The oil production began by the above two companies in 1958. Oil reserves in Nigeria were exploited exclusively by the private sector, maximizing their economic benefits, without submitting any fair compensation to the official state. Massive reactions and criticism on the lost benefits for the country from the oil industry in Nigeria were raised during the 1960s and the 1970s, while most of the population used to survive with a living cost lower than $1 per day.

At the same time, inappropriate practices often led to considerable environmental pollution in the area, such as the oil spill in 1970 in Ogoniland, in the southeast of Nigeria, causing thousands of tons of oil spilling in the neighboring agricultural areas and rivers. A fine of 26,000,000 pounds was imposed to Shell by the Nigerian government 30 years later. More than 7,000 oil spills between 1970 and 2000 have occurred in Nigeria, according to the official state.

Corruption has also played an important role in the oil exploitation processes in Nigeria. A considerable amount of oil is illegally distributed on the black market, controlled by organized dealers in collaboration with smuggling syndicates [17]. In 1990, peaceful reactions were raised by the local population in Ogoniland against the government's announcement for the concession of a new set of licenses for oil field exploitation, due to the pale compensation and the high risk of environmental contamination. A few years later, in 1995, the demonstrators' leader and author Ken Saro-Wiwa was charged with incitement to murder. In the end he was executed by Nigeria's military government. Among others, Shell was eventually accused of collaborating in the execution of Ken Saro-Wiwa and eight more campaign leaders and agreed to pay a fine of 9,600,000 pounds out of court.

During recent years, there has been an escalation of reactions against the environmental degradation, the low standard of living of the local population, and the low development status of the country, with pipelines sabotaged by armed groups of activists, kidnapping of oil companies' staff, etc. Today Nigeria remains a poor country, with a large amount of the population still without electricity, lack of fundamental infrastructure (water supply networks, roads etc.), inaccessibility to education systems, etc. (Figure 7.4).

7.2.3.7 Venezuela

The first attempts towards the investigation of oil reserves in Venezuela were accomplished in 1908 by the Caribbean Petroleum Company (later acquired by Royal Dutch Shell). World War I came to interrupt the successful efforts for the discovery of oil fields in the country and the foundation of the first refineries. Oil was first included in Venezuela's exporting products in 1918, 10 years after

FIGURE 7.4 Life and environmental conditions in the oil mining area at the Delta of Niger, in Nigeria (with due diligence of Climate Reporters) [18].

the first research projects. In 1922, the blowout of the Barroso 2 well boosted the beginning of the new era for the country. By the end of 1928, Venezuela had already become the largest oil producer in the world, after the massive invasion in the country of tens of companies aiming to gain a portion on the exploitation process of the country's oil reserves.

The abrupt development of oil industry in Venezuela had a direct, negative impact on other pre-existing economic activities, such as farming and agriculture. This is due to the phenomenon that there was a massive turn of the local population to the easy and much promising benefits of oil mining. In 1920 agriculture accounted for the one-third of the GDP in Venezuela, while by 1950, it had fallen to one-tenth. The country was then unable to create new professional activities or to maintain the traditional ones. At the same time, essential social provisions were ignored, such as education, health, and technical infrastructures, leading to a gradual degradation of the standard of living. This whole situation is known as Dutch Disease.

In 1943, with the introduction of the new Hydrocarbon Law, the state gained a share of 50% over the total annual profits from the oil industry. This was the first step towards the full nationalization of the oil reserves, which was integrated in 1976. Prior to this, Venezuela was among the founding member states of OPEC in 1960. The oil price drop during the 1950s and 1960s contributed to the restoration of a balance in the economic activities in the country. However, a short period of prosperity was recorded after the two oil crises to lead to a second degradation of the basic economy in Venezuela in the mid 1980s.

Increasing oil prices during the first decade of the twenty-first century led to the third Dutch Disease. The state income from oil exports increased from 51% in 2000 to 56% in 2006. The national economy was 96% dependent on the oil industry, becoming extremely vulnerable to any potential international oil price fluctuations. At the same time, an overspending policy of the revenues from the oil exports on social programs and provisions did not allow the state to recover from the heavy national debt. Despite the rich oil resources and the huge oil production from the country during the second half of the twentieth century, Venezuela did not approach a constantly improved standard of living, mainly due to inadequate policies, bad management of the national economic affairs, and corruption.

7.2.3.8 Canada

The history of oil in Canada goes back to 1850, when the first oil reserves were discovered in the area of Ontario. The first enterprises were established for the mining of oil and the production of asphalt material and kerosene. From 1880 and since the middle twentieth century, several private companies and oil pioneers have conducted a sequel of oil research projects in Western Canada, in the area of Alberta. The modern era of oil industry in Canada begins in 1947 with the discovery of oil reserves in Leduc. Reaching peak oil production in 1973, more than 90% of oil and gas production companies were under foreign control, mostly American.

After the oil embargo from the OPEC countries in 1973, Canada took the initiative to gain control over the domestic oil reserves, resulting in the foundation of the state company Petro-Canada. The company Petro-Canada suggested specific targets, such as the state's increased involvement over the oil reserves control, the exploitation of the reserves available in geographical territories mainly in the north regions and offshore, improvement on safety production policy, independence from the large multinational oil companies, and increases in the economic benefits for the state from the oil industry's revenues.

As sensibly expected, several of the active private companies in the Canadian oil industry were opposed to the foundation of Petro-Canada. Today the vast majority of Canada's oil industry remains under the ownership of foreign private companies.

During the passage of the twentieth century, Canada consistently remains among the third or fourth largest oil and natural gas production countries in the world, together with Saudi Arabia, Russia, and Venezuela.

7.2.3.9 United Arab Emirates

The first surveys on oil reserves were executed in 1936 in the area of Abu Dhabi. Due to the physical difficulties met in the area (lack of accessible infrastructure, lack of geological data) and the premature available technology, the initial surveys were not successful. After World War II interrupted any research efforts, the investigation for oil reserves began again in 1946 with several unsuccessful drillings onshore. It was not until 1958 when the Abu Dhabi Marine Areas Ltd. (ADMA), jointly owned by British Petroleum and Compagnie Française des Pétroles (later Total), drilled the first successful oil well offshore, with the support of Jacques Cousteau and his research ship *Calypso,* which undertook the development of the sea bed geological map. In the next year, the second oil field in the Emirate was discovered onshore by Petroleum Development Trucial Coast (PDTC), a firm already founded since 1936. During the next two decades, new oil fields were found, bringing the daily oil production close to 60,300 barrels per day in 1979.

In 1965, the gradual nationalization of oil reserves in the Emirate began. First, Sheikh Shakhbut and the ADPC (the former PDTC) signed an agreement in which each of them obtained an equal share of oil exports profits. A similar agreement was also signed with ADMA in 1966. In 1971 the Abu Dhabi National Oil Company was established in the newly created state of UAE. In December 1974, the state company gained a 60% share in ADPC and ADMA.

UAE possesses the sixth largest proven oil reserves globally. The oil history in UAE constitutes another example of a success story. The UAE economy rapidly grew, especially after the oil price shocks in 1973 and 1979. During the last 25 years, the standard of living in UAE has become one of the highest in the world. The oil industry profits are effectively reinvested in creating exciting technical infrastructure in transportation (roads, harbors, and airports), leisure and activities, accommodation and other tourism facilities (luxury hotels, restaurants, shopping centers), desalination, agriculture, etc. (Figures 7.5a, b).

From the above presentation of indicative national histories of oil industries and the exploitation of oil reserves, we may conclude that the development of local and national economies based on oil is a multi-parameter approach, affected by economic, political, social, and environmental factors. From the success stories in UAE and Kuwait, to the maintained underdevelopment and poverty in Nigeria and Venezuela, we can see how different policies and circumstances have contributed

(a)

(b)

FIGURE 7.5 The results from the impressive development in UAE from the exploitation of the domestic oil reserves.

positively or negatively to the effective exploitation of the available domestic oil reserves for the common benefit and prosperity. For example:

- Nationalization of oil reserves and industries had a definite positive impact in Kuwait and UAE, while similar contributions are expected within the first decades of the twenty-first century for Iran. However, wrong and inadequate policies have deprived Venezuela of the prospect to approach a better living standard, even though oil reserves in the country have been nationalized since 1976.
- On the other hand, despite the vast majority of oil companies in Canada remaining under the control of foreign companies, Canada has the tenth top national economy in the world, remains one of the world's richest nations, and is a member of the Organization for Economic Co-operation and Development and of Group of Seven (G7).
- In Iran, the regular wars in the area until the 1980s have also delayed the development and the growth of the local economy, despite the large proven oil reserves and their nationalization.

- In Nigeria, the inability of the state to negotiate effectively with the private companies, coupled with corruption, has sentenced the country to an extended period of underdevelopment with its people in poverty and indigence.

Keeping in mind the above facts, we can now explore the main prerequisites for the effective approach of national development based on the exploitation of oil reserves. Specifically:

- The collaboration of the state and the private sector seems to be, if not necessary, at least a good potential. The success stories from the Middle East and Canada may be used as a template for the establishment of relevant agreements.
- The secure and stable economic and geographical environment and the unimpeachable political scene certainly constitute essential prerequisites too. Obviously, corruption and venality phenomena do not facilitate any developmental effort.
- A crucial parameter is also the presence of gifted politicians, with a long-term, effective plan and a realistic vision for their country. The profits from oil exports should be reinvested in creating infrastructure and perspectives for the development of further economic activities in the country. Again, UAE can be used as a template, where people in charge are well aware that in the near future oil supplies will be depleted. To prevent their country from dropping suddenly from the paradise of wealth and prosperity to the hell of recession, they have already tried to develop parallel economic activities, to strengthen their economy and to provide it with alternative sources of national income. Consequently, the target from the oil exploitation should not be the economic profits themselves, but their effective investment to foster the creation of new economic activities.
- Finally, it is certain that every oil exploitation project should be performed with caution and respect for the conservation of the natural environment and existing human activities.

From 100 to 150 years of oil history on earth, humanity has already gained valuable knowledge on the perspectives but also on the consequences arisen from the oil industry. Let's hope that this knowledge will be utilized for the optimum exploitation of the remaining oil reserves for the common benefit and for a better world for all.

7.3 NUCLEAR ENERGY AND DEVELOPMENT

Nuclear energy is often presented as a clear and economically competitive form of energy, compared to conventional fossil fuels. A detailed evaluation on the environmental impacts from the use and the exploitation of nuclear energy is given in a later section. In this section, the contribution of nuclear energy to the development of countries will be investigated. We refer exclusively to the contribution of nuclear energy to development at the scale of countries, and not local communities, because, unlike electricity production plants from RES, the size of nuclear power plants (NPP) is too big, specifically in the range of some hundreds of MWs or even GWs, to be considered as an option for the development of small, decentralized human communities, through, for example, the implementation of NPPs of small size. Such an option is not feasible with NPPs.

Every electricity production technology should fulfill the following major prerequisites, so it can substantially contribute to the economic and social development of a state or a community:

- the construction technology of the required power plants should be simple and safe, with the minimum possible risks for the human communities and the natural environment
- the added value for the state and, more specifically, for the wider geographical area of installation, during both the installation and operation stage, should be the maximum possible
- both the setup and operation cost of the power plant should be low and competitive, in order to ensure the maximization of the project's economic profits, through its cost-effective operation.

The required technology of a NPP is far from simple. The high technological requirements for the construction and the operation a NPP are mainly imposed by the non-negotiable necessity to ensure the maximum possible safe operation level of the power plant. This implies the introduction of advanced electronic systems for the automatic operation, control, and supervision of the NPP, as well as the construction of the NPP under strict safety standards, starting from the nuclear reactor's containment building and ending at the outer shell of the NPP, normally consisting of reinforced concrete with a thickness that may be higher than 2 m, capable of surviving the impact of a falling aircraft. Of course, beyond the safety requirements, advanced technological systems are also employed for the construction and operation of the NPP core, namely the nuclear reactor, where the main power production process is performed, as well as for the removal and the secure disposal of the remaining nuclear fuel after the nuclear fission, the so-called nuclear wastes.

From the above, we see that the construction and operation of a NPP are highly demanding tasks, requiring highly specialized staff for their development, starting with the design, the construction, and the procurement of the electromechanical equipment, and ending with the operation of the power plant. All these steps are normally taken over by specialized large firms, while the collaboration of local engineers, professionals, enterprises, and workers is rather limited. It follows, then, that the added value from the construction and operation of a NPP for the local and national economies is restricted in most states in the world, apart from those which have developed their own nuclear energy experience and technological infrastructure, such as France, the US, Canada, Japan, and Russia.

A direct consequence of the above facts is the significant increase of the NPP setup and operation cost. In order to approach an analysis for the setup cost of an NPP, the following can be mentioned:

- setup cost of NPP (overnight cost—no interest included during construction): from 2,021 $/kW in Korea to 6,215 $/kW in Hungary [19]
- total set-p cost (including setup capital interest charging during the construction period): 8,500–10,500 $/kW
- construction period: 7 years
- infrastructure cost for waste disposal: 950 $/kW
- decommissioning cost after 40 years: 700–1,800 $/kW (15.5–40 $/kW in today's money values)
- NPP life period: 40 years
- operation cost: 0.05 $/kWh
- average capacity factor of NPP: 85%.

Given the above data, the electricity production levelized cost from a NPP is estimated at 0.18 – 0.208 $/kWh. This estimation agrees with the results from the real data presented shown in Figures 7.6 and 7.7 [20].

From the above results it is revealed that the electricity production-specific cost from an NPP is rather high, compared with the same features of conventional thermal power plants (see Chapter 2).

To justify the above arguments, the setup-specific costs for different electricity production power plants are presented in Table 7.1. The setup-specific cost of an NPP is higher than the corresponding feature of technologies of much smaller size that are still considered not particularly popular or mature, such as hybrid power plants. The figures presented in this table come from simulations and studies of several projects.

Finally, in Figure 7.8, a comparative diagram of the electricity production-specific cost for different power plants is provided for 2018 by Lazard [21]. According to this figure, the production-specific cost from an NPP in 2018 can be higher than the corresponding figure of a wind park or a conventional coal thermal power plant.

It should also be noticed that the electricity production-specific cost from an NPP is obviously affected by the nuclear fuel, practically the procurement price of uranium oxide (UO_2). The uranium oxide price, like any other primary source of energy, is not stable; on the contrary,

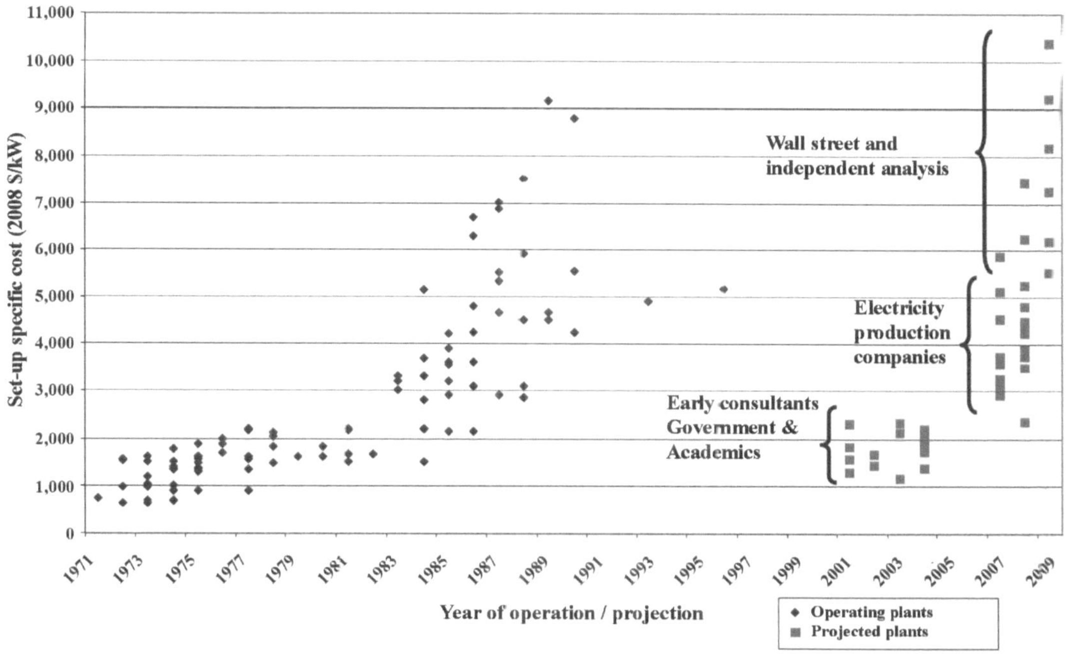

FIGURE 7.6 Evolution of NPP setup-specific cost versus time.

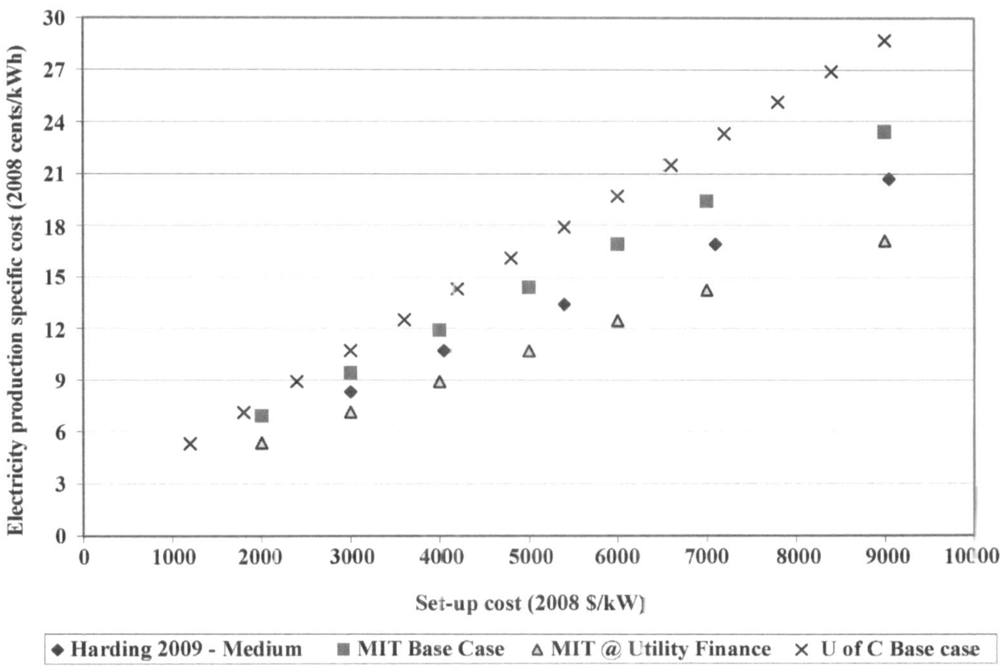

FIGURE 7.7 Evolution of NPP operation-specific cost versus time.

TABLE 7.1
Comparison of Setup-Specific Costs for Different Electricity Production Technologies

Electricity Production Technology	Setup Average Specific Cost ($/kW)	Average Annual Electricity Production (kWh/kW)	Operation Average Specific Cost ($/kWh)	Life Period (years)
Wind parks	1,500	2,500–3,500	0.025–0.040	25
Conventional thermal power plants	2,000	8,000	0.500–1.200	30
Biomass plants	3,000	7,500	0.450–0,900	25
Photovoltaic stations	1,300	1,200–1,400	0.030–0.050	25
Residential photovoltaic stations	2,000	1,300	0.040–0.060	25
Hybrid power plants of large size	2,000	3,000–3,500	0,120–0,180	25
Hybrid power plants of small size	4,000	5,000	0.150–0.020	25

it exhibits fluctuations, often intensive, versus time. Figure 7.9 presents the evolution of the uranium oxide price from 1995 to 2014 [22].

It is so designated, following the precedent analysis and documentation, that nuclear power exploitation for electricity production, on the one hand requires the utilization of advanced technology to approach the maximum possible secure operation of the NPP, while, on the other hand, the electricity production-specific cost is estimated higher even than rather premature alternative production technologies. Consequently, nuclear power cannot be considered as a cost-effective option that will contribute to the sustainable development of a country.

What's more, the large size of NPPs and the subsequent considerably high setup cost, do not permit the involvement of small investors in nuclear power projects. These projects are normally funded by large funds and implemented by large companies, in best case in collaboration with the states. Any type of involvement of the local economy in such nuclear power projects is rather impossible.

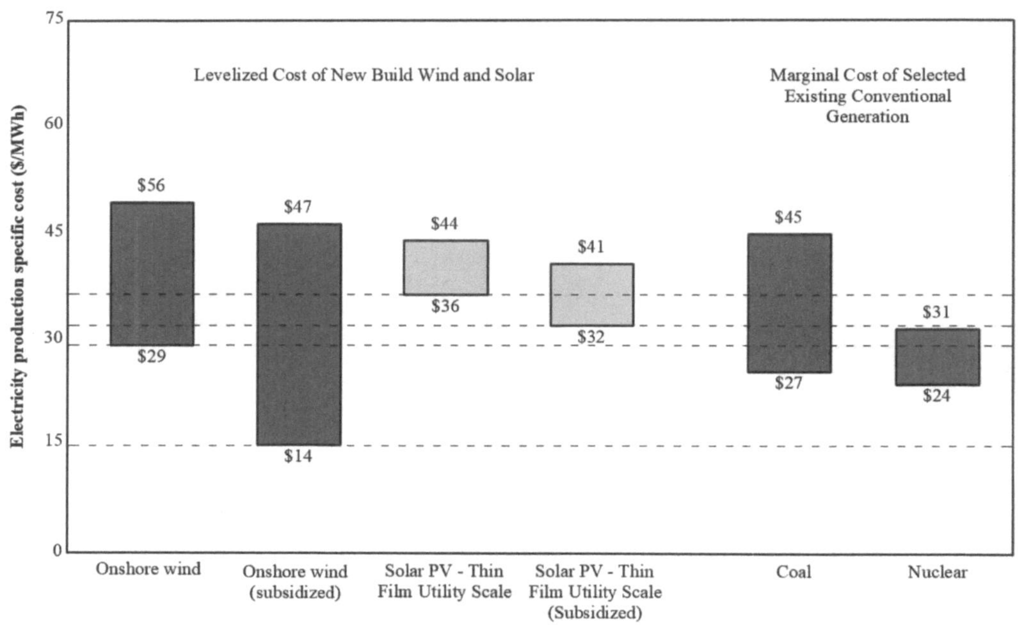

FIGURE 7.8 Estimated configuration of electricity production-specific cost in 2020 from different technologies [21].

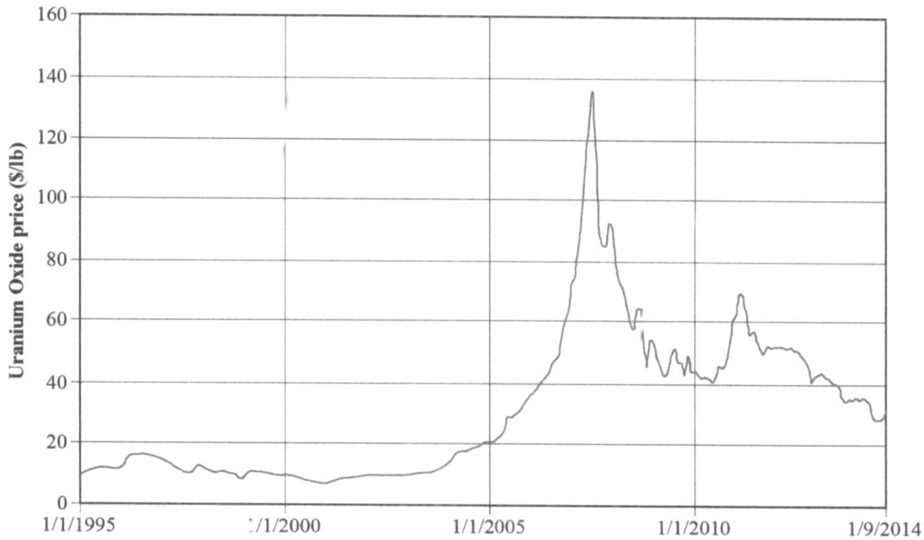

FIGURE 7.9 Evolution of uranium oxide price versus time [22].

As also explained previously, the added value from the construction and the operation of NPPs for the local economy cannot be considered as important, compared to the size and the budget of the corresponding investment, due to the fact that such projects are usually implemented as a whole by specialized engineering and construction firms.

Today only a few countries worldwide base their electricity production on nuclear power. The US features first among them, while France holds the second position, with 75% of the total electricity production coming from NPP. China and South Korea feature as emerging forces in the nuclear power sector. However, the French government already approved in July of 2015 a plan for the gradual reduction of its dependence on nuclear power from 75% to 50%.

7.4 RENEWABLE ENERGY SOURCES AND DEVELOPMENT

Electricity production technologies from RES are promoted as an ideal energy and developmental prospect, strongly connected to environmental conservation. Unlike NPPs, the possibility to adapt the size of an electricity production power plant from RES according to the funding capacity of potential investors or any other type of possible limitations (environmental, licensing, etc.), enables the development of RES projects of various sizes. Hence, an electricity production plant from a RES technology can exhibit nominal power of some tens or hundreds of MWs, so it theoretically contributes to the development of a state or an extensive geographical area, or it may also have a nominal power of some kWs or a few MWs, so it may certainly be developed by a small investing group for the benefit of a limited geographical territory.

Installations of RES electricity projects usually require an extensive amount of land with potential corresponding impacts on the natural environment or the existing human activities, examined in a later section. For this reason, during the design and the implementation of such projects, some very specific prerequisites should be fulfilled to ensure that the project will have a positive contribution to the development of a geographical region rather than to its degradation, such as:

- The siting of the project should be accomplished based on a holistic study, focused not only on the maximization of the expected annual electricity production from the RES station, by trying to exploit the maximum available RES potential, but also on the elimination of any possible impacts from the construction or the normal operation of the station on the neighboring natural and human environment.

- A major target of a RES project should be the maximization of the economic, and not only, direct and indirect benefits for the local communities, an approach similar to the efforts of states towards the nationalizations of available oil reserves, presented in a previous section.

The above two essential prerequisites are analyzed in a later section in detail.

As with the development of the oil industry, the exploitation of RES potential for the implementation of electricity production plants was not always approached from the optimum direction regarding the achievement of the above two essential prerequisites. On the contrary, the more unprepared a state was regarding the development of RES projects, mainly from a legislation point of view, the more the RES electricity projects were designed without following the most favorable and beneficial procedures for the nations and the local communities. Greece constitutes such an indicative case, which will be presented as an example to avoid in the following section. Through the analysis of what has happened in Greece on the development of RES electricity projects, the looming dangers are designated and the best practices for rational and effective development of RES projects are proposed. The conclusions from this analysis are obviously extended to any geographical area on earth.

7.4.1 Development of RES Electricity Production Projects in Greece

Greece, as most regions in the world with extensive insular territories (e.g., Caribbean and Mediterranean Sea, Pacific Ocean countries) is blessed with high RES potential. Certified measurements since the 1980s from hundreds of meteorological stations indicate the existence of plenty of sites with average annual wind velocities higher than 9 m/s and annual global solar irradiation higher than 1,700 kWh/m^2. The annually available primary RES in Greece on the one hand more than cover the domestic energy needs (electricity, heating, and transportation), while, on the other hand, can constitute an essential export product, capable to be the driving lever for the recovery of the national economy.

The rich available RES potential in Greece naturally attracted the interest of large- and small-size investors, indigenous and foreign, mainly from the private sector, for the implementation of electricity production projects, particularly wind parks and photovoltaics (PV). However, based on the incomplete and inadequate legislation framework, a massive amount of applications for the licensing of RES projects, without any central siting plan or RES development policy, was submitted, leading to a rather anarchic and problematic approach. These applications are often of very large size in small insular territories, with several violations with regard to the existing legislation and the relevant licensing procedures, submitted without any former promotion or communication campaign for the local communities and neglecting the land ownerships.

Table 7.2 characteristically presents the licensed RES projects in Greece, categorized per technology [23]. The table shows that the total licensed power for RES electricity projects exceeds 30 GW, while the annual peak power demand for the whole country remains at the range of 11 GW. This fact vividly depicts the intensive competition between the investors for the installation of RES projects in Greece.

Most of the submitted applications or issued licenses refer to RES projects of very large size proposed for disproportionally small insular territories, with regard to both their geographical and

TABLE 7.2
Total Licensed Power per Electricity Production Technology from RES Projects in Greece in 2015

Wind Parks	Biomass Stations	Geothermal Plants	Solar Thermal Plants
23,250.66	479.23	8.00	481.70
Small Hydro	**Photovoltaics Stations**	**Hybrid Power Plants**	**Total**
968.07	4,422.04	421.85	**30,031.54**

TABLE 7.3
Indicative Submitted Applications and Issued Licenses for Wind Parks on Non-Interconnected Insular System and Typical Power Demand Features

	Population	Area (km²)	Maximum Annual Power Demand (MW)	Annual Electricity Consumption (MWh)
	License Case 1: 317.4 MW			
Anafi	271	39.0	0.553	1,179
Astypalaia	1,334	96.9	2.250	6,670
Amorgos	1,859	121.5	2.900	9,072
Ios (interconnected with Paros, Naxos, etc.)	2,030	108.7	62.400	194,740
Total	5,494	365.7	68.103	211,661
	License Case 2: 348 MW			
Kalymnos	16,179	110.6	90.500	352,984
Kos	33,388	290.3		
Leros	7,917	54.1		
Total	57,484	455.0	90.500	352,984
	License Case 3: 330 MW			
Ikaria	8,423	255	7.380	27,613
	Application Case 1: 1,047 MW			
Milos	5,129	150.6	11.500	45,402
Kimolos	838	37.4		
Sikinos	260	41.7	90.500	352,984
Folegandros	780	32.4		
Astypalaia	1,334	96.9	2.250	6,670
Total	8,341	359.0	92.750	359,654

electricity demand size. All these applications are accompanied by simultaneous plans for the interconnection of the currently autonomous insular grids with the mainland system, for the distribution of the produced electricity. Some characteristic cases are presented in Table 7.3 [23, 24].

The above situation was practically enabled due to specific points in the relevant legislation framework, such as:

- Since 2011, after a relevant revision of the legislation, the applicant is not obliged any more to justify the ownership of the land required for a RES project applied for the issuance of the very first license, the so-called power production license. However, at the same time, this very first application imposes full priority of the applicant for the occupied installation site. This means that any other upcoming RES projects applied for licensing at the same region are not examined, unless the first project is rejected.
- The land ownership will have to be justified by the applicant in a next licensing stage. However, even if the land owner is not willing to proceed to a contract with the applicant for the concession of the land, the applicant has the right to expropriate it, since RES projects are considered as projects of public benefit, contributing to the reduction of fossil fuel consumption and gas emissions.
- Although the applicant is obliged to justify the available wind potential for wind park projects with certified measurements, the distance of the mast's installation point from the wind park's site is not determined. Based on this inaccuracy, there are licenses issued based on

wind potential measurements captured 25 km away from the wind park's installation site. Given the intensive land morphology of the Greek insular territories, implying a corresponding variance of the wind potential, this approach lacks in any scientific background.

Practically, given the above legislative simplifications, every potential investor could submit an application for a RES project everywhere, without any trace of licensing maturity, gaining full priority over the captured land on the one and only condition that he or she applies first for the specific site. This fact turned the case of RES project development in Greece into a safari for the capturing of as many land areas as possible mainly from large investing companies. The consequences of this configured RES projects development framework were tragic.

The siting of very large RES projects in limited insular territories, apart from the violations of several restrictions defined in the licensing procedure, regarding the conservation of the environment and the cultural and historical heritage, certainly affects the existing human activities and, much more, converts the traditional insular aspect to an electricity production industrial zone. Indicatively, in Figure 7.10 the geographical depiction of wind parks applications and licenses is presented for the small islands of Astypalaia and Amorgos [23, 24], in the central Aegean Sea. (Amorgos was included among the islands where some scenes were filmed for the popular movie "The Big Blue.") Similar siting plans are met for most of the Greek islands.

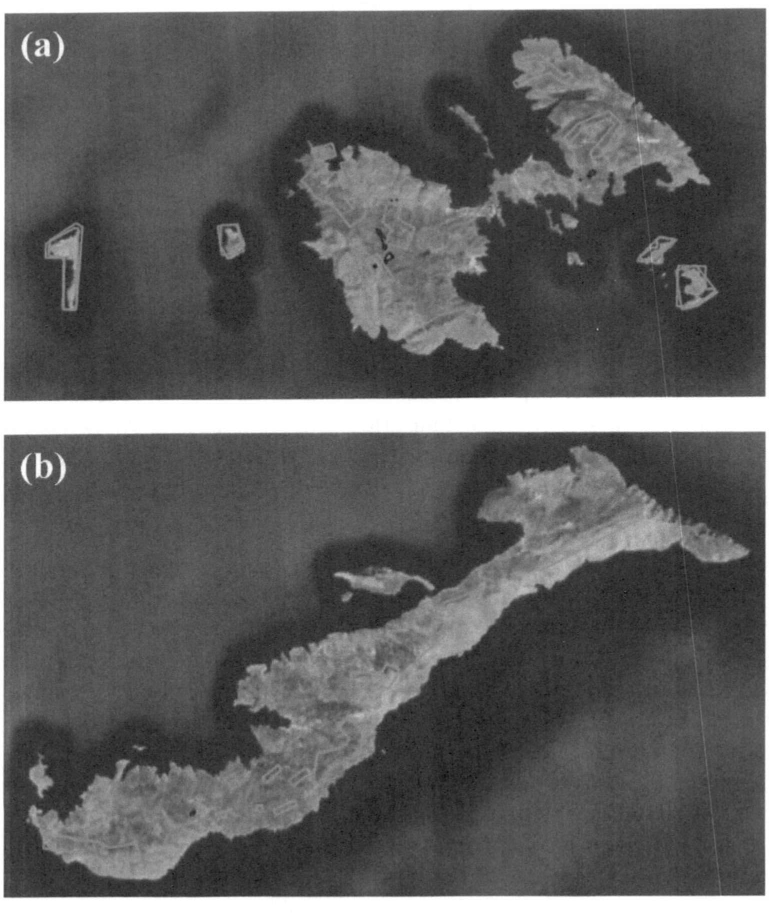

FIGURE 7.10 Geographical depiction of wind park applications and licenses for the islands of (a) Astypalaia and (b) Amorgos.

A first direct result of these large-size applications was the reversion of the highly positive common opinion on the exploitation of RES, recorded in statistical surveys before 2008. Since 2010, for first time in Greece we see massive demonstrations and strong reactions from whole communities against the development of RES electricity projects in their territory. Given the lack of any communication and approach campaign from the applicants for the local communities, these applications were viewed mainly as invasions rather than development projects.

Additionally, due to the huge land areas required for the installations of the applied large-size RES projects, it was practically impossible for the applicants to approach and inform the corresponding land owners and claim for a land concession, on the basis of a land rent or divestment contract. As a sensible consequence from this lack of communication, the land owners, being indirectly informed from several other sources about the development of RES projects on their properties, feel threatened by forces that they neither know, nor are able to detect, while their right to the properties is revoked, as well as their sense of pride and independence [24].

The unhampered implementation of any project requires a positive common opinion. The opposed common attribute, combined with the size and the budget of the submitted applications, actually makes the successful implementation of these large projects considerably doubtful.

Another indirect negative consequence arises from the above mentioned term in the legislation that the application of a RES project which is submitted first for a specific site gains full priority over any other applications that come later for the same site, even if the latter exhibit more advanced licensing level (e.g., justification of the land ownership, certified wind potential measurements, approvals from the involved authorities in the licensing procedure). Many cases of small, realistic, and mature RES projects cannot be evaluated by the authorities in charge, because there is a land overlap with a previously applied large-size, immature, and practically unfeasible RES project, for the reasons mentioned above (e.g., violations with the existing legislation, environmental and human activities impacts, negative common attribute, doubtful funding). Because they have been applied first, these large projects continue to capture extensive land areas, preventing other realistic, feasible, and rational RES projects from being applied, licensed, and implemented.

Summarizing the above, the consequences from the application of large-size RES projects in Greece had an overall negative contribution regarding the introduction of RES technologies in the country and the reduction of imported fossil fuels use, which may be distinguished in the following impacts:

- A significant negative opinion on the application of RES projects in Greece has emerged since 2010, mainly in the insular and countryside communities, despite the fact that before 2010 the vast majority of the population in Greece was strongly in favor of RES [25].
- The introduction of RES technologies in the Greek energy systems has been considerably delayed, due to the capture of extensive land areas from large-size applied unfeasible projects with relatively low implementation possibilities. The remaining geographical and energy (for autonomous grids) space for the development of small-size, realistic, feasible, and mature RES projects is rather restricted.

Keeping the configured situation in Greece as an example to avoid, in the next sub-section we will try to formulate some essential prerequisites for the rational implementation of RES projects and the maximization of their contribution on the national and local communities' economic and social development.

7.4.2 Rational Development of RES Projects and Maximization of Common Benefits

From the case of Greece, we have seen that the rational introduction of RES projects in a country is not obvious. On the contrary, the usually met economic interests of private investors and multinational companies can destroy the developmental prospects of a country by exploiting the available RES, while also negatively affecting other existing activities. To eliminate this probability and claim a more rational approach, a cluster of measures is proposed in this section.

7.4.2.1 A Clear, Objective, and Effective Legislation Framework

The existence of a clear and effective legislation framework is essential for the rational and fast development of RES projects. Obviously, the processes defined in the legislation framework should:

- Aim at the acceleration of the licensing procedures, however by treating objectively the environmental issues, with respect to the culture and the history of a place, to the human properties and the existing economic activities. For example, the land ownership should be justified from the very beginning of the licensing procedure. Furthermore, national or local RES siting plans must be accomplished in order to define areas of priority and exclusion for the installation of RES projects.
- Be clearly defined, without ambiguities and inaccuracies. For example, the installation position of a wind potential meteorological station should be within a predefined radius from the wind park's installation site. Additionally, any licensing stage of a RES project should be completed within a predefined, known-beforehand time period. Immature projects, or projects unable to comply with the requirements of the legislation, should be rejected after the end of this period, without any extension.
- Be applied effectively and objectively. Any deadlines, obligations, and terms should be kept and applied equally for all.

7.4.2.2 Public Rates for Local Municipalities

Usually electricity production projects from RES are charged with a small percentage over the annual investment's revenues for the benefit of local municipalities. This public rate has the sense that a percentage from the exploitation of a common good, like the wind or the sun, should be returned back to the official representative of the local community, namely the municipality, for the creation of works and infrastructures of common benefit. The question is how much this compensation percentage should be.

Coming back to Greece, for example, this percentage is set at 3%. Given this value and the general conditions under which most wind parks currently under operation in Greece have been developed, namely:

- annual final capacity factor of 40% (plenty of sites in most insular systems in the world exhibit even higher capacity factors)
- setup cost at 1,100 €/kW
- produced electricity selling price at 0.090 €/kWh
- annual wind power curtailments due to penetration restrictions at 10%
- funding scheme 40% private capital and 60% bank loan with a payback period of 10 years and a loan rate of 7%

A wind park investment exhibits a payback period of 3 years and an internal rate of return (IRR) of 27%, calculated over the investment's private capital. If we maintain all the above design and economic terms, but the public rate percentage increases from 3% to 15%, then the payback period increases to 5.5 years and the IRR drops to 18%. Consequently, after a 500% increase of the project's public rates, the investment still exhibits high economic efficiency. The public rate percentage can be even higher if the investment has been implemented with any sort of subsidy (setup cost contribution, tax exemption for specific period, etc.).

The above example is presented indicatively. In general a public rate percentage should be provided for the local municipality. The value of this percentage should be case-specific, depending on various involved parameters, such as the project's capacity factor, the employed technology, etc.

7.4.2.3 Support of Local Entrepreneurship for the Development of RES Projects

The social and economic benefits for the local communities and, subsequently, for the states and the national economies, are expected to be maximized with the active involvement of local investors, municipalities, and energy cooperatives in the implementation of RES electricity production projects. Tens of such examples regarding the involvement of energy cooperatives or municipal companies in RES projects are met in Europe, the US, and Australia [26–30]. More details are given in the following sub-section for some characteristic cases. The positive economic and social impacts of such local community-based projects are maximized, because the net profits, and certainly a considerable portion of the gross revenues, return back to individuals and legal entities that remain and perform their activities in the community.

Obviously the active participation of local investors, cooperatives, and municipalities in RES investments should be centrally supported with a cluster of measures, aiming mainly at the:

- So-called capacity building, namely the cultivation and the awareness of local communities on the opportunities and the potential for the development of RES projects, through promotion and educational actions.
- Acceleration of the licensing procedures through the introduction of specific priorities and modules, such as fast-track licensing regimes.
- Availability of attractive funding tools. Actually the funding of the relatively expensive RES projects, given the economic abilities of local investing groups, constitutes maybe the major problem. The availability of state subsidies or bank loans with favorable terms, such as low rates and long payback periods, can constitute essential supporting measures.
- Finally, the local entrepreneurship, private or public, should be protected against investments on RES projects of very large size, which can cover high percentages of the RES power that can be potentially installed in a geographical territory, defined by either the available physical or energy space, especially when these large-size projects are proposed for restricted geographical territories, such as islands, interconnected or not with mainland grids. Particularly for islands, there should be, for example, an upper limit on the maximum permissible power of a single RES project, which can be defined as a percentage versus the insular grid's annual peak power demand. Of course, this restriction cannot be applied for systems of small size, namely with peak power demand lower than 10 MW, because projects of very small size may not be economically feasible.

7.4.2.4 Protection of the Environment, Respect to Existing Domestic and Commercial Activities

The construction of RES projects should not affect negatively any existing infrastructure and already in place domestic and commercial activities, so any expected progress should not interfere with achieved development. For example, there is no use of trying to claim the development of a local economy with the installation of a new wind park in the vicinity of a popular tourist destination, which may affect negatively the aesthetics of the area. Similarly, any potential negative impacts on the natural environment from the installation of RES projects may have cumulative adverse consequences which, in total, can eventually negate any anticipated benefit from the operation of the project.

The above undesirable influences can be easily predicted and avoided with the careful and thoughtful design of siting plans for central or decentralized RES projects. In these plans, areas of priority and exclusion can be defined versus the RES technology under consideration, creating a secure investment environment for the potential investors and, at the same time, introducing an effective means with which to ensure the protection of the physical and the structured environment.

7.4.2.5 Cultivation of a Positive Common Attitude

The conservation of a positive common opinion on RES projects constitutes an essential prerequisite for their fast and unobstructed development. The necessary restoration of a positive common attribute on RES projects depends on the growth of a trustful environment of the overall RES implementation frame, including the official state and the relevant legislation, the involved authorities in the licensing procedure, the funding organizations, and the private investors. From the case of Greece, it is obvious that the majority of people do not oppose the RES generally, but the development of RES projects of very large size, without any sort of planning and by neglecting the participation of local communities. Consequently, the fundamental conditions for the cultivation of a positive common opinion can be found in all the measures presented and justified above for the rational development of RES projects and the maximization of the local communities' benefits, namely:

- the definition of a clear, objective, and fair legislative framework
- the design of a central siting plan for the RES projects and the definition of areas of priority and exclusion
- the maximization of the economic compensation for the local municipalities
- the protection of the natural environment and the existing human activities, domestic or commercial
- the support of the local entrepreneurship.

Once the above measures are applied in practice, it should be expected that the common opinion on RES projects will gradually become strongly favorable, fostering the exploitation of the locally available RES.

7.4.3 Examples from RES and Development

In this section, we will present some characteristic examples of success stories regarding achieved development of local communities based on RES projects, or anticipated plans under implementation. The investigated projects have been designed or already accomplished by national utilities companies, small private investors, unifications of individuals or small companies in the form of energy cooperatives, or collaborations of the above physical and legal entities. The common feature met in all these efforts is that the involved individuals or companies come from the local communities to play an active role in the project, starting from their right to have an opinion and vote for or against a particular project or any crucial decision and ending with their expected direct economic benefit given their share in the cooperative's equity capital. Another important issue from these community-based initiatives is the perspective that opens by reinvesting a part of the RES project's profits to create infrastructure to support new economic activities in the area, an approach that can boost the local social and economic development. This general concept, namely the situation under which local communities have the responsibility for the production of the energy they consume and the management of the required power plants is described in the modern literature as energy democracy.

The various projects implemented so far by energy cooperatives aim either to the coverage of the local energy consumption and the reduction of the energy procurement price for the final consumers, or to the large-scale electricity production and its distribution to the local grids. Below are some characteristic successful or highly promising cases.

7.4.3.1 Hydroelectricity in Norway

We have several success stories on the exploitation of RES projects from Scandinavian countries. First of all in Norway in 2011, 122 TWh out of 128 TWh of final produced electricity, namely a percentage of 95%, was produced from hydro power plants (Figure 7.11), while the remaining 6 TWh were covered by thermal generators (4.8 TWh) and wind parks (1.2 TWh). The annual domestic electricity consumption in the country was 114 TWh, which means that Norway produces 107% more than the

(a)

(b)

FIGURE 7.11 Impressive architecture harmonized with the exquisite landscape in the hydro power plant of Øvre Forsland in Norway (a) during daytime and (b) during sunset (with due diligence of Bjørn Leirvik) [31].

consumed electricity from RES plants. The extreme significance of the availability of cheap electricity from domestic primary energy sources towards the development of the country is easily conceivable. Large amounts of national income remain in the country to be invested in the creation of infrastructure, in research and education, in support of young entrepreneurship and in social provisions, instead of being spent for the import of expensive, polluting, and nonrenewable energy sources.

Another major fact is that the electricity consumption per annum and capita in Norway is approximately four times higher than the average consumption in the European Union (EU). Specifically, in 2014 the annual electricity consumption in Norway was at 23 MWh per capita [32], while in EU the same figure for the same year was 5.9 MWh per capita. The above observation reveals that the use of electricity in Norway is preferred over the use of other energy sources, such as oil, especially for heating production, namely one of the major final forms of energy consumed in the country, while the use of electrical vehicles has already begun. This is precisely due to the fact that electricity production is cheap, clean, and renewable, rather than using conventional fossil fuels.

7.4.3.2 Wind Power in Denmark

Denmark constitutes a pioneer on the wind power penetration in the national electrical grid. The annual contribution of wind parks in the electricity production share for the whole country in 2018 exceeded 41% [33], a figure which constitutes a global record. With Vestas, the top wind turbine manufacturer, with regard to the installed wind turbines globally, founded and operating in Denmark, the wind industry has emerged as one of the main pylons of the Danish economy.

In Denmark there is a particular policy focused on the active participation of individuals and community-based firms in the development of RES projects. For example, there are specific tax exemptions measures for families and households that are actively involved in the electricity production for their community from RES technologies. Since 2001 more than 100,000 inhabitants participate in community-based legal entities for the development of wind parks. As a result, 86% of the currently operating wind turbines belong to energy cooperatives, small investors, and individuals. A subsequent result of this policy is the achievement of a wide acceptance of wind parks and wind power in the country.

A short reference should be also made for the island of Samsø, where a project towards 100% RES penetration in the island started in 1997. First of all, 11 onshore wind turbines of 1 MW nominal power each were installed in 1999–2000. In 2002, 10 more offshore wind turbines of 2.3 MW nominal power each were installed, leading to a total installed wind power on the island of 34 MW. At the same time, a cluster of measures was designed aiming at the installation of biomass stations and solar collectors for the heating needs of the island. All these projects were totally funded and implemented by local farmers. Today 100% of the electricity needs on the island is covered by the wind parks and 75% of the heating loads is covered by biomass stations and solar collectors (Figure 7.12).

7.4.3.3 RES Penetration in Iceland

The electricity production in Iceland follows the pattern of Norway. It is exclusively based on renewables, particularly on hydroelectricity (approximately 70% contribution) and geothermal fields of high enthalpy (approximately 30%). Both involved technologies are considered as guaranteed power production units, a feature of significant importance regarding their secure penetration in the autonomous national electrical system. Thermal generators are employed for less than 0.2% of the annual electricity production balance. Iceland features in the first position in the EU-28 with regard to the annual electricity consumption per capita, which is configured close to 54 MWh per annum and capita, namely almost ten times higher than the average value in the EU [32]. Of course this feature is due to the adverse, polar weather conditions during the whole year and the low production cost for electricity. Iceland, like Norway, sensibly depends every domestic and commercial energy need on electricity, eliminating the use of oil and other types of conventional fossil fuels.

(a)

(b)

FIGURE 7.12 The offshore wind park (a) and a biomass burner operating with straw (b) in Samsø (with due diligence of Samsoe Energy Academy and Huffington Post) [34, 35].

7.4.3.4 Wind Power in Germany

In Germany hundreds of thousands of citizens have invested in the installation of community-based wind parks, while thousands of small-medium enterprises are active in the wind industry (engineering, procurement, and construction), which in 2008 employed around 90,000 people. Wind power in Germany supplied more than 21% of the annual electricity consumption on a national level in 2018 [33], with a total installed wind power of 59 GW. As in Denmark, through the massive participation of individuals and energy cooperatives in the wind industry in Germany, a highly positive common opinion on wind power and wind parks has been cultivated. Characteristically, in the German county of North Frisia more than 60 wind parks have been installed with a total nominal power of 700 MW and with more than 90% participation of local small-size investors.

7.4.3.5 Wind Power in United Kingdom

Until 2012 there were 43 community-based companies active in the sector of electricity production in UK. All these companies were founded by and belong to individuals, most of them living and active in the surrounding areas of the RES project installation sites. Wind parks, PV stations, and hydro power plants have been so far developed for electricity production for their local communities and to foster local development.

The first community-based wind park in the UK, with a nominal power of 2.5 MW, was constructed in 1997 by the company Baywind Energy Co-operative, the first energy cooperative in UK, founded in 1996. In Scotland the wind park at Isle of Gigha from the energy cooperative Gigha Renewable Energy began its commercial operation in 2005. It initially consisted of three wind turbines with a combined capacity of 675 kW, known as The Dancing Ladies. A fourth wind turbine of 330 kW has been also added in a second stage. The wind park is totally owned by the residents of Gigha, who reinvest the profits of its operation in developmental works for their area (Figure 7.13).

(a)

(b)

FIGURE 7.13 Wind parks at Westmill (a) and at Isle of Gigha (b), in United Kingdom (with due diligence of Scottish Islands Federation and Falck Renewables) [36, 37].

7.4.3.6 The Energy Cooperative of Sifnos Island, Greece

Sifnos is a small Greek island with a permanent population of 2,500 inhabitants, located in the Western Aegean Sea. Unlike the previously described situation in Greece, Sifnos Island Cooperative (SIC) was established in December 2013, aiming to claim the island's energy independence and to create the prerequisites for a sustainable economic and social development for the local community. These targets are approached with a cluster of actions, containing energy-saving measures, especially in the tourism sector, introduction of electrical vehicles, and demand side management policies. All these measures are plugged in a wind park–pumped hydro storage (WP-PHS) power plant, which is envisaged to constitute the main support pylon for the approach of the SIC's main targets (Figure 7.14). The proposed WP-PHS aims at 100% annual electricity production in the autonomous insular system of Sifnos. If this target is achieved, Sifnos will be the first non-interconnected, 100% energy-independent island in the world [38].

The construction and the operation of the hybrid power plant will simultaneously lead to a reduction of the annual expenses for the electricity production, currently afforded by the local network operator and to the implementation of a feasible and profitable investment, with an annual net income for the local energy cooperation of €3,000,000.

Additionally, the potential power surplus from the wind park, estimated at 10 GWh, could be exploited in newly introduced loads, such as desalination or electrical vehicles. Adopting an average specific electricity consumption for potable water production through a reverse osmosis process at kWh/m^3, it is estimated that these 10 GWh of electricity surplus, once exploited in desalination units, will permit an annual potable water production of 2,500,000 m^3. This amount of available fresh water for a community of 2,500 permanent residents will create a huge potential for the development of new activities on the island, based on the exploitation of the locally available sources and the possibilities offered by the excellent local climate conditions, such as the biological growth of agricultural products and the implementation of pilot farms of biological husbandry, beekeeping, etc.

The introduction of these new activities to the island's local economy will require the disposal of initial capital for the construction of the required infrastructure (e.g., potable water storage reservoirs, hydraulic networks, housing of the new enterprises). These funds can be provided by the hybrid power plant's net profits. Consequently, by reinvesting a percentage of the annual net profits of the hybrid power plant for the creation of the required infrastructure, new trades and professions can be introduced in currently unexploited sectors, creating thus serially new occupation positions and strengthening the local economy, reducing also its almost exclusive dependence on tourism.

Two more major targets of SIC are the transition from conventional transportation means to electrical vehicles within the next 10 years and the procurement of a ship to establish a regular transportation line from Sifnos to Athens, Crete, and the largest neighboring islands (Milos and Santorini), powered either by electrochemical batteries or fuel cells charged by the central WP-PHS. All these actions create the prerequisites for a secure and sustainable social and economic development of the insular community, with infrastructure and activities based on the locally available energy sources and works with maximized added value. All this concept is depicted graphically in Figure 7.15 [38].

7.4.3.7 Faroe Islands, 100% Energy Autonomy by 2030

SEV, the local national utility of the Faroe Islands, has initiated an attempt for the full energy independence of the country by 2030. Generally, the Faroe Islands is a geographical region blessed with abundant wind potential and rainfalls, available on constant basis for 10 months per year. However, there is a 2-month period during summer, when the availability of wind and hydro is remarkably restricted. This is the main obstacle towards 100% RES penetration for the local energy needs coverage. On the other hand, luckily the energy demand during this summer period reduces as well, mainly due to the milder weather conditions.

Currently there are 18 MW of wind parks and 39 MW of hydro power plants already installed, contributing close to 60% of the annual electricity production (Figures 7.16a,b). Heating and transportation

(a)

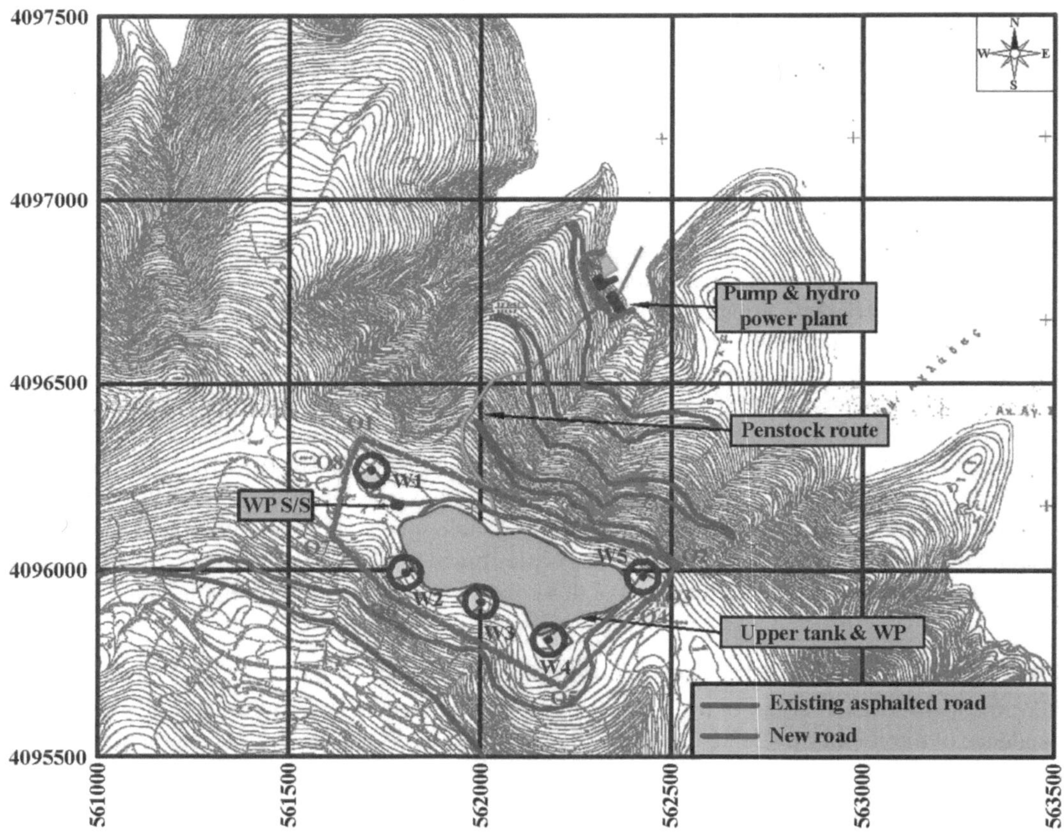

(b)

FIGURE 7.14 (a) Three-dimensional depiction of the wind turbine locations on a wind potential background and (b) siting of the WP-PHS.

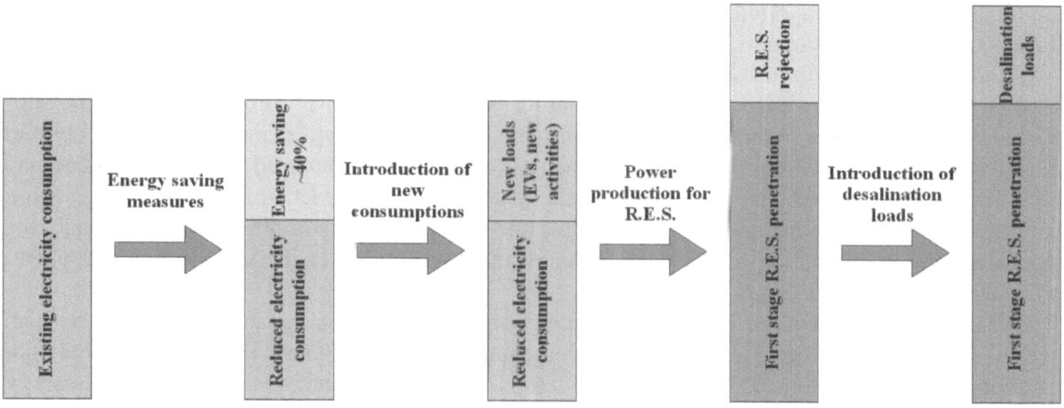

FIGURE 7.15 Energy balance between energy saving, energy production from RES, and new loads.

FIGURE 7.16 Wind park (a) and hydro power plant (b) in the Faroe Islands (left photography: Bjarti Thomsen).

are totally covered by imported oil. All these onshore energy needs are planned to be transferred into the electricity sector, similar to Norway and Iceland.

Wind and hydro seem to be the fundamental RES towards the achievement of 100% RES penetration for electricity, transportation, and heating needs. The RES power plants should be supported with one or more pumped storage power plants. Due to the long period of low RES potential during summer, the total required storage capacity is considerably increased. This requirement may be relieved with the introduction of small-scale decentralized electrochemical storage (e.g., li-ion batteries) in residential and commercial buildings or in wind parks, smoothing simultaneously the stochastic power production of the wind turbines and contributing to the stability and the power quality of the local grid.

In case of full upper reservoir periods of the pumped storage systems, met during winter time, energy storage can be approached in a complementary mode, with thermal storage in either special concrete constructions or in pressurized water. These implementations enable thermal energy storage in temperatures higher than 300°C, which, in turn, provides the possibility of combined electricity and heat production, an option strongly significant in cold climates, like the one in Faroe.

From the above, we see that the whole procedure constitutes a complex problem to be solved, with a number of different parameters that should be taken into account.

7.5 ENVIRONMENTAL IMPACTS FROM THERMAL POWER PLANTS

The most important environmental impacts of thermal power plants come from gas emissions. The basic gases disposed by a thermal generator are carbon oxides (monoxide CO and dioxide CO_2), sulfur oxides (monoxide SO and dioxide SO_2), nitrogen oxides (NO_x), incombustible hydrocarbons (HC), and solid particles (ashes).

In this section, the gas emissions from thermal power plants will be presented only as a source of atmosphere pollution. Any potential impact of them on the greenhouse effect and climate change will not be investigated, because all those contributions still remain an unproved theory.

The specific gas emissions from the combustion of the most commonly used fossil fuels are presented in Table 7.4 [39]. The table shows that the main emitted gas from the combustion of fossil fuels is CO_2, which actually constitutes the basic product, together with water, from the perfect combustion of hydrocarbons.

The extensive use of fossil fuels in industrial activities and particularly in the energy production sector has led to considerable gas emissions, mainly CO_2. In Chapter 2, for example, with the annual operation simulation of the electrical systems of Crete and the Faroe Islands, the annual CO_2 emissions were calculated at 1,725,113 tn and 222,438 tn respectively, which correspond to specific emissions were 2.97 tn and 4.61 tn per capita respectively. It is clarified that these emissions come

TABLE 7.4
Specific Gas Emissions from the Combustion of Hydrocarbons (with due diligence of European Environment Agency)[39]

	Gas Emissions (gr/GJ of produced electricity)					
Fuel	**CO_2**	**SO_2**	**NO_x**	**CO**	**Volatile Organics**	**Particles**
Asphalted coal	94,600	765	292	89.1	4.92	1,203
Lignite	101,000	1,361	183	89.1	7.78	3,254
Heavy fuel oil	77,400	1,350	195	15.7	3.70	16
Diesel oil	74,100	228	129	15.7	3.24	1.91
Natural gas	56,100	0.68	93.3	14.5	1.58	0.1

TABLE 7.5
Annual CO_2 Specific Emissions per Geographical Area in tn Per Capita [40]

Continent / Region / Country	2009	2010	2011	2012	2013
North, Central, and South America					
United States	17.19	17.48	17.02	16.29	16.39
Canada	14.84	14.49	14.48	13.86	13.53
Mexico	3.86	3.75	3.88	3.94	3.95
Brazil	1.87	2.11	2.19	2.32	2.47
Argentina	4.41	4.56	4.60	4.57	4.46
Europe					
Russian Federation	11.02	11.73	12.37	12.82	12.47
Germany	8.81	9.28	8.95	9.19	9.22
United Kingdom	7.57	7.86	7.08	7.33	7.13
Italy	6.80	6.84	5.70	6.21	5.72
France	5.43	5.42	5.07	5.07	5.05
Spain	6.22	5.82	5.79	5.66	5.08
Croatia	4.81	4.63	4.69	4.26	4.16
Greece	8.28	7.51	7.27	7.24	6.31
Middle East					
Iran, Islamic Rep.	8.06	8.15	8.24	8.45	8.00
Saudi Arabia	17.10	18.53	17.39	19.19	17.93
United Arab Emirates	21.80	19.31	18.27	19.25	18.71
Iraq	3.49	3.63	4.20	4.64	4.92
Africa					
South Africa	10.06	9.34	9.22	9.02	8.86
Egypt, Arab Rep.	2.57	2.47	2.59	2.53	2.43
Nigeria	0.49	0.58	0.59	0.59	0.55
Asia and Oceania					
China	5.72	6.55	7.23	7.42	7.55
India	1.43	1.40	1.48	1.60	1.59
Japan	8.62	9.15	9.32	9.64	9.76
Australia	18.20	16.92	16.86	16.52	16.35

only from the electricity production sector; namely, the contributions of the transportation and the heating sector are not included.

In Table 7.5 the annual specific CO_2 emissions are presented for selected countries per continent and geographical region [39]. By observing Table 7.5, we should underline the definite decreasing trend on CO_2 annual emissions, recorded between 2009 and 2013 in the developed parts of the world, such as North America, Europe, and Australia. On the contrary, there is an increase trend in the CO_2 annual emissions for underdeveloped regions. More specifically:

- The annual CO_2 emissions in Europe are limited below 10 tn per capita, apart from Russia. The average value of CO_2 annual emissions in EU-28 has decreased from 7.11 in 2008 to 6.82 tn per capita in 2013.
- The countries with the highest CO_2 per capita are the UAE, Saudi Arabia, and the US, with Australia, Canada, and Russia coming next.
- The CO_2 emissions remain considerably low in India, while in China, they range at the level of the EU.

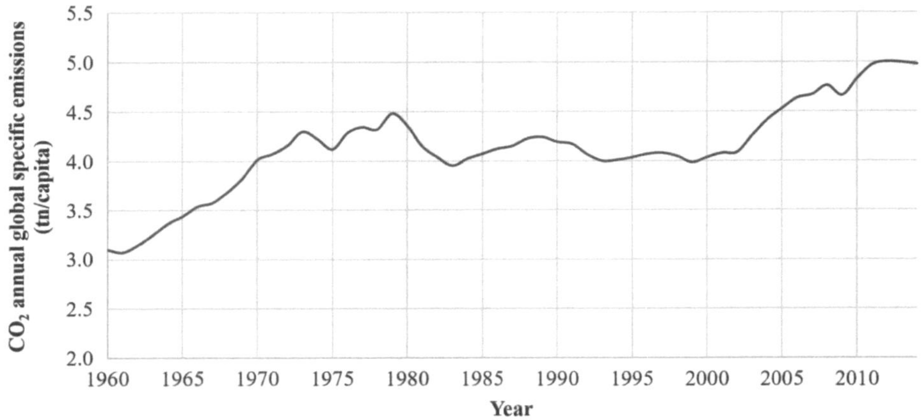

FIGURE 7.17 Evolution of the global CO_2 annual specific emissions from 1960 to 2013 [40].

- The countries of Africa (apart from the Republic of South Africa), Central and Latin America, and most of the Asian countries exhibit considerably low annual CO_2 emissions.
- The annual CO_2 emissions on a global level exhibit a 17.6% increase during a period of 10 years, namely from 4.25 in 2003 to 5.00 tn per capita in 2013. The evolution of the global CO_2 annual specific emissions from 1960 to 2013 is depicted in Figure 7.17.

From the presented data in Table 7.5 we can say that the CO_2 emissions reveal the living standard of each country, particularly in cases like the US, Canada, Saudi Arabia, and UAE. The reverse conclusion can be also derived for the underdeveloped countries in Africa, Latin America, and Asia.

The overall accomplished efforts towards the reduction of the dependence on the use of fossil fuels, through the CO_2 emissions trade, the new rational energy use technologies, and the extensive installations of RES projects for electricity and thermal energy production, have already led to measured results in the developed countries. The European countries, despite the high achieved living standard, exhibit relatively low CO_2 emissions, with a constant reduction trend. In the US, Canada, and Australia there is also a certain drop in the annual CO_2 specific emissions. However, the availability of domestic, cheap oil and coal in these countries and the high living standard, still maintain the CO_2 specific emissions at high levels. Obviously, for these regions, as well as for the Middle East countries, further efforts are required to approach more rational standards of energy consumption.

Air pollution can cause significant diseases for human and animals, leading even to death. Indicatively we can mention several diseases of the respiratory system, such as emphysema, bronchitis, pneumonia, and asthma, caused mainly by sulfur dioxide and nitrogen oxides. High-level exposure to sulfur dioxide and nitrogen oxides can cause pulmonary edema, while in cases of people with lung or heart diseases, it can lead to death. Finally, carbon monoxide is a highly poisonous gas. From 1999 to 2012 there were 6,136 CO poisoning fatalities recorded in the US alone and more than 450 fatal events per year [41]. It should be made clear that these fatalities did not originate from CO emitted by the electricity production sector. This feature is only given to indicate the lethal action of this particular gas.

Apart from air pollution from gas emission in the atmosphere, the use of fossil fuel has additional important impacts. A brief presentation is provided below.

7.5.1 Landscape Degradation

This impact refers to the surroundings of coal or oil mines and refineries. Large areas are occupied and affected by these infrastructures and the executed activities. These areas are considerably degraded, especially in the case of coal mines. They are totally converted to large industrial, mining areas. Any other activity cannot be executed in the proximity of these areas.

Today 60% of the coal consumed globally is strip-mined. The whole procedure implies the employment of large-scale excavators, which usually dig more than 60 m deep to remove the covering soil and reach the coal. This large-scale earth removal results in a harshly scarred land. Also, the natural vegetation is totally eliminated.

Underground coal mines exhibit a 40% contribution to the total coal production globally. However, this percentage is expected to increase, because the available locations for on ground coal mines gradually run out. A common impact of the underground coal mines on the landscape is the land subsidence, due to the underground cavities configured with the removal of coal. So far, approximately 25% of the underground mines have caused some type of land subsidence. Such events can cause a series of different consequences, depending on the location of the underground mine and, in the end, what actually exists above ground. About 7% of land subsidence has occurred close to cities, causing damages to the structural environment (buildings, roads, etc.), which usually are of large scale. On the other hand, land subsidence in the countryside affects the rainwater drainage routes, the uniformity of the land terrain, and the appropriateness of land for agriculture.

Finally, the atmosphere close to mines is heavily polluted with coal dust and airborne particles, seriously affecting public health conditions. The inhabitance of such areas is practically impossible (Figures 7.18a, b). Even in towns or settlements located more than 10 km from the coal mines, the quality of the atmosphere is degraded. Black lung disease, most commonly met in miners, named after the picture of their lungs, is the worst impact on human health for people living or working close to mines. According to statistics, the anticipated average life period for coal miners is 3 years less than the rest of the population of the same socioeconomic status.

7.5.2 Leaks Through the Drilling and Transportation Processes

Another source of serious impacts from the use of fossil fuels is the unavoidable risk of leaks during the mining/drilling and transportation processes. Leaks obviously refer to liquid and gas fuels, namely oil and natural gas.

The major natural gas content is methane (CH_4). Hence, by talking about natural gas leaks during drilling, extraction, transportation, and storage, we are actually referring to methane leaks, namely an 86-times stronger gas at trapping heat than CO_2 over a period of 20 years. This means that the leakage of 1 kg of methane in the atmosphere corresponds to the emission of 86 kg of CO_2.

However, the impacts from oil leaks are considered more important, especially at the sea. During the second half of the twentieth century, we witnessed a sequel of oil tanker accidents offshore, leading to tremendous disasters of the marine ecosystem. The water-soluble elements of oil consist of a variety of chemical compounds, which are toxic for a wide group of marine species with fish eggs, larva, and young, more sensitive fishes. These toxic compounds introduce disturbances in the fishes' physiology, behavior, and normal growth, leading eventually to early death.

As the oil spill is transferred by the winds and the sea streams to the shore, the severity of the imposed impacts increases. The liquid oil is stuck on the coastal land and will remain there for long time periods, as it is difficult to be unstuck with physical processes from either sand or rocky formations. If it still remains in liquid phase, it is absorbed underground, where, given the lack of oxygen, it cannot be decomposed, maintaining, thus, its toxic properties for extensive time periods.

With regard to the sea flora, oil sticks to the leaves of the sea plants and the algae and it is hardly washed out with the sea waves. In cases of mild oil leaks, the plants are able to recover after a period of 3 weeks; however, after extensive oil leaks and subsequent contamination, the plants die. Ecosystems with plants of annual life cycle seem to require 2 to 3 years to fully recover, while perennial plants exhibit a variety of reactions, from strong resistance against the toxic action of oil to death, depending on marine conditions and the severity of the contamination.

(a)

(b)

FIGURE 7.18 Pictures of strip mining areas in West Macedonia, Greece (a) at the vicinity of the thermal power plant and (b) currently exploited in long distance from the thermal power plant (with due diligence of Vice.com) [42].

7.5.3 Impacts on Water Resources

The use of fossil fuels can also cause serious impacts on water resources, both during the extraction process and during thermoelectricity generation. Actually, what may be unknown to most people is that thermoelectricity is the largest water user in the world. The water used for all the required processes for thermoelectricity production corresponds to 49% of the annual consumed water.

First of all, water is used for all the fossil fuel extraction processes. It is used during coal mining, transport and storage, refining process, dust suppression and post-mining activities, such as land reclamation and revegetation. Additionally, large water volumes are withdrawn during natural gas and oil extraction processes. This water is re-injected into the aquifer, but, obviously, with its quality certainly affected. Water is also used in the refineries and in thermal power plants, for the thermal generators cooling, for the production of steam, for cleaning, etc.

Apart from the use of water as a means for fossil fuel extraction, transportation, and refinery, all the processes involved in the thermoelectricity production sequel imply serious risks for the contamination of surface and underground water resources. A major source of water contamination comes from abandoned coal mines. Rain water seeps into them and reacts with sulfur compounds to produce sulfuric acid, which, in turn, is carried by the water streams out of the mine and eventually lands in rivers and streams, making them acidic. The effects of the contaminated water on fishes are fatal, while, obviously, it is no longer suitable for drinking, swimming, or many industrial applications.

As explained previously, strip mining operations result in a significant modification of the land topography and elimination of natural vegetation. This affects the natural water drainage and may lead to floods and landslides.

The aquifer and surface water can be contaminated from all fossil fuels leaks or inadequate extraction and transportation processes (defective installations, accidents, or even sabotages).

The effects of water contamination can be critical on the adequacy of water resources, on the conservation of natural aquatic biotopes, and on public health.

7.5.4 Acid Deposition

Acid rain is a general term used to describe the acid content of atmospheric water, configured mainly by the emission of sulfur oxides, nitrogen oxides, and ammonia. A more accurate term is *acid deposition,* which consists of two parts, the liquid and the gas deposition.

Liquid deposition refers to acid rain, mist, and snow. As acid rain falls to the ground, it affects the plants, the soil, the animals, and the humans. The impacts from acid rain depend on several parameters, such as the pH of the acid rain, snow, or mist, the ability of ground and surface water to absorb and dissolve acid precipitations, and the fauna and flora species.

Gas deposition refers to acid particles and gases, transferred by the air and landing on any physical or technical surface. During the first stage of rain, water sweeps this gas deposition, increasing its acidification. Gas deposition can be carried with air over distances up to 1,000 km from their emission location, affecting, in this way, geographical territories that are not responsible for gas emissions.

Sulfur oxides are emitted mainly by the combustion of coal. More than 60% of the emitted sulfur oxides in the atmosphere originates from the electricity production sector. Nitrogen oxides are mainly emitted by the transportation sector.

The most important impacts of acid deposition are:

- water pH drop in lakes and rivers
- plants and trees slower growth, sparser foliage, even death events in soils with pH lower than 5.1
- negative effects on animal and human health, either directly, through the respiratory system, or indirectly, through the food chain

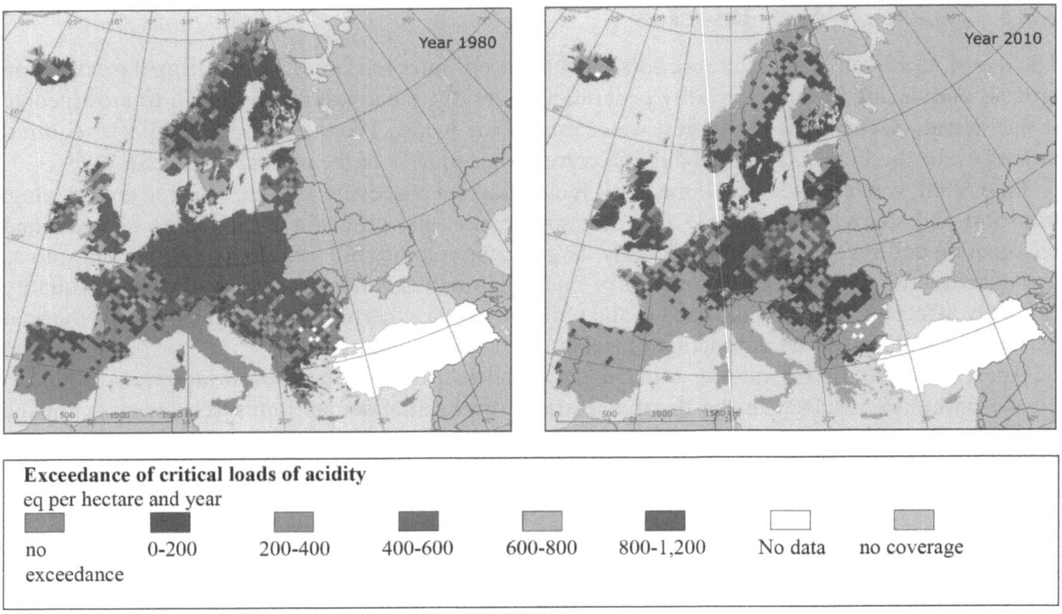

FIGURE 7.19 Exceedance of the critical loads of acidity in Europe in 1980 and in 2010 [43].

- for pH higher than 5.5 there are effects on aquatic species, especially small ones
- effects on the food chain of some specific species, due to decomposition of nutrients contained in their diet or dissolution of particular toxic mineral elements
- corrosion of metallic materials and wear of some structural materials, like stones, marbles, etc., an effect of considerable importance with regard to the conservation of cultural monuments
- visibility restriction: under sulfur atmospheric concentration above 0.1 ppmv (parts per million by volume), visibility is reduced to 8 km.

However, it should be stated that acid deposition has been considerably reduced since 1980, due to the extensive efforts on the rational use of energy and the improvement of thermal power plants and vehicle technology. Figure 7.19 shows two maps with exactly the achieved progress in the reduction of acid deposition [43]. As seen in these maps, in 1980 there were large parts in the European continent with acidification higher than the critical loads, namely higher than 1,200 CO_2 equivalent per hectare and year. This situation was considerably improved by 2010.

7.6 IMPACTS FROM THE USE OF NUCLEAR POWER

Nuclear energy is released from the fission of heavy nuclei. During a nuclear fission, an instable atomic nucleus, created by the bombardment of the initial available nucleus of the nuclear fuel by a neutron and the absorption of the neutron from the nucleus, splits into two or more smaller fragments and by-products (e.g., additional neutrons). In controlled nuclear reactions, such as the ones performed inside a nuclear reactor, this nucleus fission is accompanied by the release of energy in the form of thermal energy, because the arisen nuclei fragments have a total mass slightly lower than the mass of the initial nucleus. The released thermal energy can be exploited for the production of electricity. However, the above procedure is characterized by significant impacts, which will be presented in the following sub-sections.

7.6.1 Nuclear Wastes

Unlike coal, oil, or gas, nuclear fuels cannot be fully burned inside the nuclear reactor. This is because during the process of a nuclear reaction a number of the nuclear fuel's initial available nuclei do not shift to instable condition after the absorption of the bombarding neutron, which, in turn, will eventually lead to the nuclear fission. With the nuclear reaction evolution with time, the number of the unburned nuclei increases, preventing the conservation of the chained reaction. When the nuclear reaction is completed, the remaining unburned nuclear fuel do not function as a fuel any more, but as a highly radioactive nuclear waste. The unburned heavy nuclei will still continue to split in lighter nuclear fragments with very slow rates, emitting high amounts of radioactivity. For some nuclear wastes, the time required for the reduction in half of the initial amount of the unburned nuclei, known as half-life, ranges at the scale of some tens or even hundred thousands of years. This implies the requirement for a special treatment regarding the storage of these wastes for considerable time periods. Practically, the necessity for the disposal of the produced nuclear wastes constitutes perhaps the most crucial issue with regard to the use of nuclear power, leading often to serious social, political, and regulatory problems.

Nuclear wastes are distinguished in wastes of low, medium, and high radioactivity. Depending precisely on this categorization, nuclear wastes should be stored under full inspection for some tens or some thousands of years. However, under the intensively fragile and flexible international political and economic balances, how could it be possible to ensure that the nuclear wastes disposed today will be adequately stored for the next 20,000 years? Actually, the use of nuclear energy today creates a truly heavy mortgage for the next generations.

A typical nuclear power plant with a nominal power of 1,000 MW produces an annual quantity of 30 tn of high radioactivity nuclear wastes, 300 tn and 450 of medium and low radioactivity, respectively. Every produced tn of nuclear wastes contains10 kg of plutonium, a quantity capable for the construction of an atomic bomb. The depleted uranium contained in the nuclear wastes can also be used for the construction of conventional and radioactive, highly effective nuclear weapons. Additionally, the nuclear factories themselves constitute another particularly significant source of nuclear wastes. The average service life of a nuclear power plant is set at 30 years. After the end of this period, the power plant should be decommissioned and treated as nuclear waste.

From the operation of nuclear power plants so far, approximately 150,000 tn of nuclear wastes have been produced, from which only one-third have been specially treated and stored. Eventually, all the nuclear wastes end, or should end, to be stored in underground rocky formations far deep from the earth surface. The controlled storage process, obligatory for high and medium radioactivity wastes, is practically very expensive and difficult, often leading to intentional accidents, to justify the disposal of the nuclear wastes at the bottom of the sea or in far away, remote areas on the planet, causing almost always strong reactions worldwide from environmental organizations and activists. Sweden and Finland feature as the pioneers in nuclear waste disposal technologies, while research projects are executed in France, Germany, Switzerland, and UK. Countries like Russia, Ukraine, and Estonia, although highly experienced in nuclear energy, seem incapable of securely managing the problem of nuclear waste disposal due to lack of the required funds.

7.6.2 Risk of a Nuclear Accident

Nuclear power plants are presented as totally safe from the International Atomic Energy Agency (IAEA), exhibiting the lowest index of accidents compared to any other type of industrial production and any other type of human technological activities. Even if we accept that this statistic is valid, the probability for the event of an accident does not constitute by itself an objective scientific index of safety. The actual parameter that can depict from a technocratic point of view the issue of safety of a technical procedure is the so-called mathematical danger, which, in short terms, is the product of the accident's probability by the accident's expected effects. Defining the real danger with the above approach, it comes that nuclear power plants are the most insecure industrial installations in the modern technical world, of course due to their tremendous and extensive effects on the natural and structural environment.

No international organization or academic institute can guarantee the inexistence of any probability for the occurrence of a nuclear accident. However, even the reduction of this probability requires the investment of huge amounts, leading to a subsequent conflict between the achievement of the required safety standards and the economic feasibility of the nuclear project. The recent history of the three most serious nuclear accidents in Three Mile Island in March 1979 (US), in Chernobyl in April 1986 (Ukraine, former USSR), and in Fukushima in March 2011 (Japan) has proven that actually nuclear accidents can happen anywhere, namely either in the former communist world, or in the core of the technological progress, caused by technical faults, human oversight, or natural forces. Regardless of the origination of the nuclear accident, the consequences always are severe and long-term. For the integrity of the present section and as an indicative case, below some brief details are provided on the nuclear accident in Chernobyl.

7.6.2.1 The Nuclear Accident in Chernobyl

The nuclear accident in Chernobyl occurred on April 26, 1986, at nuclear reactor No. 4 of the nuclear power plant of Chernobyl of the former USSR, located in the territory of Ukraine. The accident was classified as a category 7 "major accident" of the International Nuclear and Radiological Event Scale (INES), namely the category with the most severe consequences. It affected significantly the economic and social conditions of the neighboring geographical areas and caused serious impacts on the natural environment and public health. Two of the workers in the factory were immediately killed, while in the next 4 months 28 firefighters also died due to their burns and exposure to the disposed radioactivity. Nineteen more deaths were recorded until 2004. Additionally, it is estimated that the health of hundreds of thousands of people was affected due to the environmental burden from the emitted radioactivity. A percentage increase of cancer diseases of 15% was estimated for the nearby population, a figure that translates to thousands of deaths within the next 20 years.

The accident in Chernobyl had serious impacts for most countries in Europe, with the western, eastern, and northern parts of the continent receiving the largest portion of the emitted radioactivity (it is estimated that more than half of the emitted radioactivity was transferred in geographical regions outside the USSR). Former Yugoslavia, Finland, Sweden, Germany, Bulgaria, Norway, Romania, Austria, and Poland received more than 1 PBq (1 PBq = 10^{15} Bq) of Cesium-137. The total European land was polluted with more than 4,000 Bq/m^2, while a land percentage of 2.3% received more than 40,000 Bq/m^2 (Figure 7.20). It is estimated that 36% of the total emitted radioactivity from the nuclear accident was disposed in the land of Russia, Ukraine, and Belarus and 53% in the other European countries.

It is estimated that half of the Iodine-131 released from the nuclear plant of Chernobyl was disposed outside USSR, an element with a direct connection to thyroid cancer. A considerable increase of such incidences was recorded in UK and Czech Republic. According to other statistics, a relevant increase of child leukemia events was observed in Western Germany, Greece, and Belarus. Keeping in mind that most cancer types take 20 to 60 years from the victim's exposure to the disease's source until the appearance of the disease, it is still too early to attempt to estimate the real dimensions of the Chernobyl accident's effects on public health.

Finally, the accident of Chernobyl had specific effects on the psychology of the neighboring populations. From research conducted by the Hellenic Psychiatric Association, more than 2,500 abortions were executed in 1986 in Greece from couples being afraid of a possible effect of the radioactivity on the embryo.

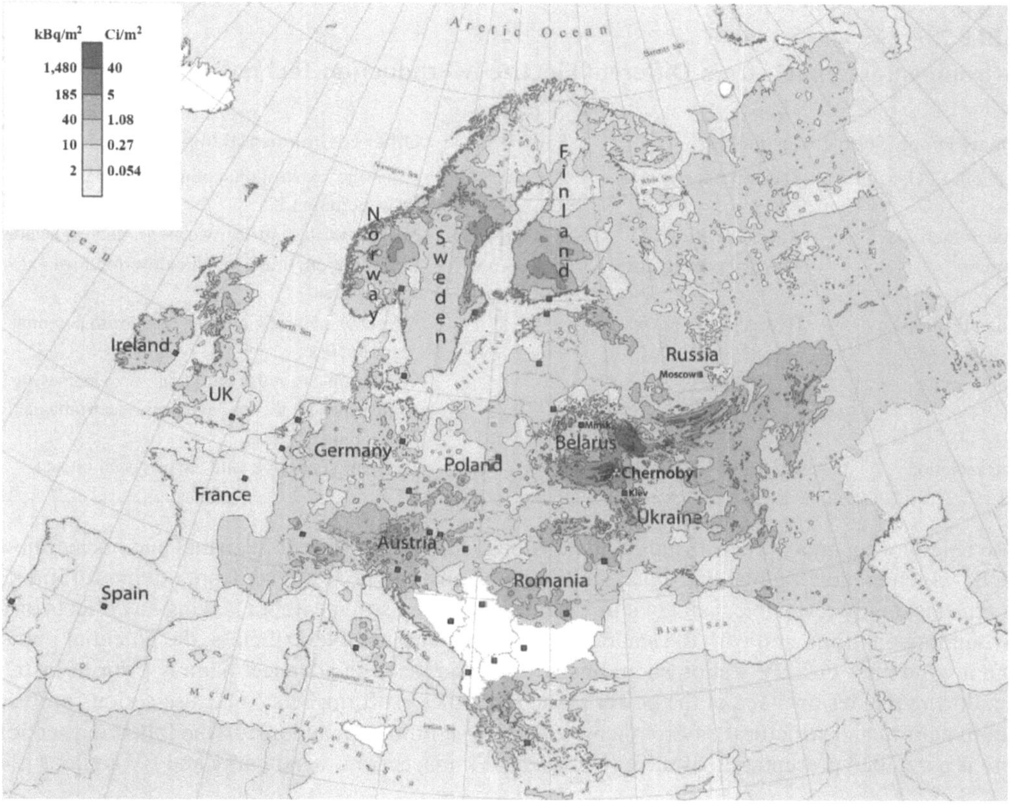

FIGURE 7.20 Surface contamination with 137-Cs in Europe after the Chernobyl nuclear accident [44].

7.7 IMPACTS FROM WIND PARKS AND PHOTOVOLTAIC STATIONS

Wind turbines and PV panels exploit the renewable kinetic energy of wind and solar radiation respectively. During their operation, there are no gas emissions or any other type of emissions in the atmosphere, the earth's ground, the water resources, etc., and no fuel transportation is required, during which there is always a probability for environmental pollution. The energy spent for the construction of a wind turbine is produced by the wind turbine within a time period of 3 to 9 months, depending on the available wind potential in the installation site. Similarly, this time period for PV panels ranges between 1 to 4 years, based on the employed PV technology and the available solar radiation. Additionally, a wind turbine or a PV panel is easily decommissioned at the end of its life period, without anything remaining in the natural environment, with most of their constructive materials being completely recycled. Today, electricity production from wind turbines or PV panels should be considered the most environmentally friendly electricity production technology. Any other production technology is accompanied with potential more serious impacts on the natural environment and human communities.

In Table 7.6 the most important environmental impacts from the most common electricity production technologies are summarized for comparison [25].

The contribution of wind turbines and PV panels to environmental conservation is approached with the imposed decrease of conventional thermal generators operation and the corresponding reduction of gas emissions. In short, wind energy and solar radiation can be considered as clean and soft energy sources. All their potential impacts are local, namely their effects are restricted to the

TABLE 7.6
Environmental Impacts from Different Electricity Production Technologies

Primary Energy Source	Production Technology	Gas Emissions	Other Environmental Impacts
Coal, oil, LPG, natural gas	Thermal power plants	CO_2, NO_x, SO_x, VOC, ash	Fossil fuel exhaustion, pollution during transportation
Nuclear fuel	Nuclear reactors	—	Nuclear wastes, nuclear weapons, nuclear accidents
Biomass	Thermal power plants	CO_2, SO_x, VOC, cinder	Deforestation, limited land cultivation for food production
Hydraulic drop	Hydroelectric power plants	—	Effects on natural wetlands, risks from potential earthquakes
Wind energy	Wind parks	—	Visual impacts, noise emissions, accidents with birds, shadow flicker, land uses, electromagnetic interference
Solar radiation	Photovoltaic stations	—	Occupation of large land areas, visual impact

close neighboring area of their installation positions, unlike the effects of thermal and nuclear power plants, which exhibit a global attribute. Additionally, any potential impact from a wind turbine or a PV panel is totally reversible, namely it immediately stops existing, without leaving any effects on the surrounding environment, with the shutdown of the power plant. Nevertheless, the potential impacts from a wind park or a PV station should be studied and faced in advance, namely prior to installation, during the design stage of the power plant. Even these soft impacts can cause serious problems to both natural and structural environments and raise significant reactions. In the following sections, there is a detailed presentation of the environmental impacts from wind parks and PV stations [24].

7.7.1 Visual Impact

The visual impact from a wind park or a PV station comes from the spoiling of the natural landscape through their installation on areas without any other human interferences. The visual impact is integrated with the installation of the power plant's interconnection network with the local electrical grid.

The large size of the installed wind turbines, combined probably with the installation site (e.g., on the top of hills or on mountain ridges), makes them visible from far-away locations, over long distances. Consequently, the visual impact from a wind park or a PV station on the landscape's aesthetics and attributes is doubtless. However, it is impossible to absolutely define whether this effect should be considered of minor or major importance, because such an attempt seems to be based on rather subjective criteria. So, wind turbines or PV panels can be considered as huge and ugly constructions, converting a natural landscape into an industrial area, but they also can be seen as elegant, clever, and, consequently, acceptable machines, exploiting the free and renewable offered power of nature for the production of useful electricity, or as a unique feasible option for the substitution of thermal generators and nuclear power plants, a fact that makes them absolutely necessary and irreplaceable.

The effects from the installation of a wind park or a PV station on the landscape's aesthetics are very difficult to evaluate. Beyond the subjective attribute of the subject, different installations have, objectively, different visual impacts. The general objective factors for the evaluation of a wind park's or a PV station's visual impacts are:

- The installation position. The installation, for example, of a wind park in a flat area of central Europe or in the vast desert areas in Texas, implies more restricted zone of visual impact compared to the installation of a wind park on a mountain ridge on an island. Similarly, the installation of large PV stations in closed areas (e.g., valleys) affects more intensively the landscape aesthetics, than the same installation in an open area in the Middle East.

- The attribute of the installation area. The installation of a wind park or a PV station in a forest area, or in an area of exceptional natural beauty and environmental importance, combined with the construction of the required infrastructure works (access roads, interconnection grid, etc.), will certainly cause strong reactions. On the contrary, the installation of the same power plants in a rocky or infertile area with negligible flora, will most probably not have any negative ramifications.
- The existence of human activities in the surrounding area. Generally people are anxious about any possible effects of wind turbines or PV panels on their existing professional or recreation activities. For example, shepherds and farmers are afraid that the RES projects will capture their land, tourism entrepreneurs are afraid that the attractiveness of the area on tourists will be affected due to the degradation of the natural landscape, the noise emissions, etc., and inhabitants believe that the quality of their life will be degraded. In general, RES projects should be installed in locations with the minimum possible effects on any existing activities in the structured, human environment.
- The normal operation of the wind turbines. Normally operating wind turbines are more likely to be accepted, because they are perceived by humans as useful machines, serving a particular scope. On the contrary, if most of the wind turbines in a wind park remain still, due to faults, while the wind is blowing, the observer's appreciation for their usefulness is diminished. In this case, the wind turbines are considered as huge, ugly machines that spoil the landscape and the sense of the negative visual impact becomes stronger.
- The type of the installed wind turbine and its color. It is generally accepted that the installation of tubular pylons, colored according to the surrounding environment seems to exhibit an improved optical performance compared to the use of lattice towers, similar to the ones installed in high-voltage transportation grids. Additionally, the optical uniformity also contributes to the harmonic integration of the wind turbine in the environment. For example, wind turbines with three blades have a more harmonic optical picture, while the coloring of the turbine's tower has also a particular contribution towards the harmonic integration of the turbine in the natural landscape. White and grey colors seem to be more easily accepted. Another approach is the gradual transition from a green color, at the base of the tower, to white (Figures 7.21a, b).
- The distance from the RES plant installation site. The optical impact of a wind turbine degrades with the distance. According to an empirical rule, the significant optical impact of a wind turbine extends to a distance equal to ten times its height, Hence, a wind turbine with a 50 m height will optically dominate an area with a radius of 500 m, although under conditions of atmosphere clarity and lack of technical or physical obstacles, it can be visible from distances equal to 400 times its height, namely from locations 20 km away. However, at distances longer than 5 km, the wind turbine is gradually absorbed by the landscape without affecting the aesthetics (Figure 7.22).

PV stations have the advantage that normally they are not visible from long distances, due to their low installation heights from the ground, unless they are installed in trackers. Yet, from short distances, the effect on the natural landscape is more intensive compared to a wind park.

7.7.2 Noise Emission

Noise emission obviously refers only to wind turbines. Noise emission from wind turbines is usually not important for areas located farther than a certain distance from the wind park's installation site and, usually, it is covered by other noise produced by the wind and the surroundings. Besides, wind turbines are usually installed in windy areas, where strong winds often blow for long periods. As a matter of common sense, normally there are no settlements or human activities close to such areas, which could suffer from the noise emission from a wind park.

FIGURE 7.21 Wind turbines colored according to the environment [45].

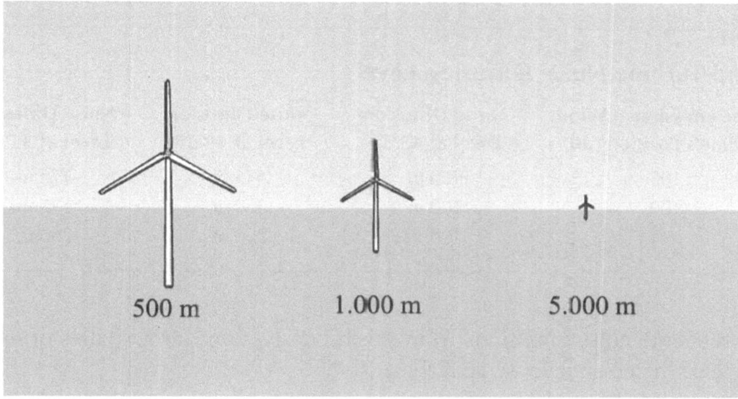

FIGURE 7.22 The effect on the landscape aesthetics versus the distance of the wind turbine [25].

The noise produced by a wind turbine is distinguished as aerodynamic noise and mechanical noise.

The aerodynamic noise is related to the wind velocity and the aerodynamic design of the wind blade. The aerodynamic noise should be faced during the aerodynamic design of the wind turbine's blades. It consists of the rotation noise and the turbulence noise. The rotation noise includes all the noises with frequencies harmonics of the rotor's rotational frequency. The noise emission level increase with the rotor's diameter, the blade tip's linear speed, and the aerodynamic load of the blade (increase of the absorbed wind power).

The turbulence noise is related to the existence of turbulence at the airfoil's trailing edge and the overall turbulence field behind the rotor. To reduce the turbulence noise, the blade tip's linear speed should be reduced, leading also to a subsequent reduction of the absorbed wind power.

Since 2000, there has been significant attention and efforts towards the reduction of the aerodynamic noise emission from wind turbines, leading to considerable results. For example, the aerodynamic noise emission from the last-generation wind turbines has been reduced more than 90% compared to the emitted noise from the wind turbines of the 1980s.

Mechanical noise is caused by the moving electromechanical parts of the wind turbine, with the most important sources the switch gear, the inductive generator, and the bearings. The mechanical noise can be treated either directly at its source, or during its propagation stage. The mechanical noise can be reduced at its source by special measures on the noisiest components, e.g., by using helical gears instead of spur gears, or by installing a noise insulation coating in the inner side of the turbine's nacelle. The mechanical noise can also be reduced during its propagation, by using noise insulation materials or antivibration mountings.

Noise emission from modern wind turbines ranges from 95 to 105 dB and mainly consists of aerodynamic noise. The mechanical noise has been remarkably restricted, either through the installation of noise insulation materials and antivibration mountings or by the removal of the switch gears from wind turbines. The mechanical noise from a modern wind turbine can be perceivable only in case of a malfunction on a specific component. The aerodynamic noise from wind turbines is constantly reduced by their manufacturers, based on the improved aerodynamic design of the wind turbine blades.

For the calculation of the noise emission from wind turbines, there are several methodologies developed. For an accurate calculation, according to most methodologies, the proposed experimental process should be executed during night time, when any other natural or technical noises are reduced. In Table 7.7 the noise diffusion level is presented for new-generation wind turbines versus the distance from the source [25].

TABLE 7.7
Wind Turbine Noise Diffusion Level

Noise Emission at Wind Turbine's Position (dB)	Noise Diffusion Level at 45 dB	Noise Diffusion Level at 40 dB	Noise Diffusion Level at 35 dB
105	350 m	575 m	775 m
100	200 m	350 m	575 m
95	120 m	200 m	350 m

Table 7.8 presents the noise emissions from several daily, familiar activities or devices, in order to provide a sense of the noise level scale [25].

The upper permissible level of wind turbine noise emission level in inhabited areas varies from country to country. In Denmark it is set at 45 dB, while in Sweden and in Greece at 40 dB. In UK, the noise diffusion from a wind turbine in an inhabited area is not allowed to be higher than 5 dB from the average noise in the same area.

Wind turbine noise can be perceivable only under certain circumstances. For very low wind velocities, the wind turbines remain inoperative, so there is no noise produced. On the other hand, when the wind velocity is higher than 8 m/s (wind intensity higher than 5 Beaufort), the wind turbine noise is covered by the wind itself and all the noises produced precisely due to the blowing wind (tree leaves, electrical wires, etc.). Hence, wind turbine noise can be heard at distances longer than 500 m only when the wind velocity ranges from 3 to 8 m/s. Another important fact is that the wind noise diffusion is detectable only towards the wind blowing direction. The noise diffusion in different directions with regard to the wind blowing one is significantly restricted.

Finally, another crucial parameter that affects the sense of the diffused noise on nearby people, is whether there is a direct visual contact with the wind park or not. Relevant research articles have shown that a direct optical contact between a settlement and a wind park increases the sense of the public disturbance from the noise emitted by the wind turbines.

7.7.3 Impacts on Birds

As with noise emission, impacts on birds also exclusively refers to wind turbines. Environmental impacts of wind turbines related to fatalities of birds and bats first appeared in the US in the late 1980s. Specifically, it was observed that endemic bird species, particularly endangered golden eagles and hawks, were killed by the wind turbines or the high-voltage transportation lines at the wind parks of Altamont, in California (Figure 7.23), built in the path of many migratory birds and rarer birds of prey. Significant birds' fatalities were also recorded in the area of the town of Tarifa, Spain, at the Strait of Gibraltar, another important migration passage for emigrant birds across the Mediterranean.

TABLE 7.8
Noise Emission Level from Various Activities or Devices

Activity	Noise Emission Level (dB)
Human speech	65
Electrical refrigerator	35–40
Center of a city	75
Night club	100
Bedroom during quiet times	30

FIGURE 7.23 A view of some of the wind parks in Altamont, California, US (with due diligence of CK Vango) [46].

The installation and operation of wind parks can potentially affect birds in the following ways:

- fatalities due to electrical shocks or collisions of birds with the rotating blades
- installations on areas important for birds, such as prey or procreation areas
- installations on areas that constitute passages for migratory birds
- degradation or even loss of birds' habitats.

There are strong bonds between birds and their natural habitats. Many bird species strongly depend on their habitat and are considerably vulnerable to any possible modifications of it. On the other hand, there is also a strong dependence of a specific geographical area and the perspective for wind turbines installations in this, which is mainly based on the availability of remarkable wind potential. The possibility of a conflict between these two priorities is obviously the main reason that may affect the bird wildlife in the area.

A considerable number of studies have been accomplished on the potential impacts of wind turbines on birds. A common conclusion is that wind turbines, in general lines, can constitute a threat for birds. Areas with strong wind streams are usually used as corridors by migratory species. The installation of wind turbines in such places can imply a serious risk for the safety of the passing-by birds. Similar risks may also appear from the installation of wind turbines in biotopes inhabited by endangered birds' species.

On the other hand, another conclusion supported by a large number of studies is that wind turbines do not constitute a serious risk for birds, because birds have the ability to learn from the signals they receive from their environment and adapt their routes with regard to the existence of wind turbines, avoiding the collision with them. Another usually raised argument from the supporters of wind turbines is that they constitute a potential much lower danger for birds than other human constructions or activities, given the statistics on fatalities of birds from their collision with glass buildings and moving vehicles, from the electricity transportation and distribution grids, etc.

In any occasion, both categories of studies conclude that, in order to eliminate or, at least, minimize the probability of fatal accidents on birds from wind turbines, wind parks should be studied and designed appropriately, taking into account any potential risk on the existing bird fauna beforehand. Locations with high importance for the conservation of endangered species should be avoided for wind turbine installations. If necessary, special measures for the protection of birds should be introduced, such as the installation of special radars for the detection of birds or bats in the area close to the wind park, or the placement of bird feeders in the proximity of the wind parks for predators or carrion-eating birds, a measure that aims to provide secure food for these species, reducing their food-seeking activities and the subsequent collision risk with the turbines.

7.7.4 Shadow Flicker

During specific seasons over the year, and for particular intervals during a 24-hour period, wind turbines can impose a very annoying impact on the neighboring settlements, the so-called shadow flicker. Shadow flicker is caused by the shadow of the turbine's spinning blades, as it falls periodically on a specific geographical point where this phenomenon can be disturbing, like the window of a nearby house.

The risk of shadow flicker is significant in cases of wind park installations very close to settlements. According to most national regulations and directives, the minimum distance of a wind park from a settlement cannot be less than 500 m. Yet, under specific conditions, a wind park installed in the wrong position and too close to a neighboring settlement, can cause shadow flicker, an impact truly disturbing.

The shadow of a turbine's blade fades with distance in the atmosphere. Theoretically the shadow of a 22 m long turbine blade can be visible from distances up to 4.8 km. This can happen right after the sunrise and shortly before the sunset. In practice, the shadow of the spinning blade of a wind turbine with a nominal power of 2 or 3 MW, with 45 m length and 2 m width, can have an effective shadow at distances up to 1.4 km, although some weak shadow effects can be sensed from points 2 km away.

Another important parameter, apart from the distance from the wind turbine, is the shadow flicker frequency. Flicker frequency can be critical for serious human diseases and should be lower than three blade passes per second, hence 60 rotations per minute. Additionally, the turbine's blades should not be reflective.

From the sunrise to the sunset, the shadow of a wind turbine follows the route of a sundial, starting from west and ending at east. As the sunrise and sunset time change through the year, the route of the wind turbine's shadow will alter for the same geographical location. Hence, it is obvious that a particular geographical point will suffer from a blade's shadow flicker only for a very specific time period during a year.

The sunrise and sunset location for a specific geographical point are well-known for every day of the year, and the route of the shadow of the physical or technical body can be easily calculated, based on the fundamentals of solar geometry. This means that the total time period that a specific point will suffer from a wind turbine's shadow flicker can be calculated in advance, with the support of simple computational applications. If required, the necessary measures can be taken, such as the re-siting of the wind turbines, in order to totally avoid any shadow effect on sensitive positions.

Shadow flicker cannot be avoided after the installation of a wind park, unless the disturbing wind turbines are kept still. Shadow flicker can be predicted and easily avoided with the proper design and study of the wind park, tasks that should be accomplished before any installation. The minimum distances for close settlements should be kept.

Finally it is obvious that shadow flicker can be caused only if the position of the wind park is between the sun and the settlement. That means there is a potential risk for shadow flicker only when there is a settlement north of the wind park, for the north hemisphere, or south of the wind park, for the south hemisphere.

7.7.5 Land Occupation

It is often mentioned that electricity production technologies from RES, such as wind parks or PV stations, occupy large amounts of land compared to the conventional thermal or nuclear plants. Nevertheless, a more detailed investigation on this issue will be enough to persuade that things are not exactly so. Indeed, both thermal and nuclear power plants occupy large areas during the whole production procedure, from the fuel's mining, refinery, preparation, transportation, storage, the power production, and, eventually, the disposal of the wastes. Through this overall sequel, mines, pump stations, oil refineries, harbors, storage stations, and production plants are employed. Apart from all this land occupied by these facilities, large areas around the main power production plants are affected by their operation, making impossible the development of any other activities on them.

The captured land by a 2 or 3 MW wind turbine is a rectangle of 40 m × 50 m = 2,000 m^2, around the foundation of the wind turbine. This area is cleaned and flattened for the assembly and the erection of the wind turbine. This is the only area that no other activity will be able to take place. All the other areas, although they may officially belong to the wind park's installation site, will totally remain available for the implementation of any activity (Figures 7.24a, b and 7.25).

Unlike wind parks, PV stations exhibit considerable impact on the occupation of land, precisely due to their low power density (approximately 25,000 m^2 per installed MW). This feature can cause serious social and economic consequences, especially if the PV station is chosen to be installed in areas with a different, formerly productive activity, such as agriculture.

The required land occupation from the different technologies involved in the electricity production is summarized, for comparison reasons, in Table 7.9 [50]. As seen in this table, wind parks and PVs seem to exhibit the lowest impact on land occupation, while, on the other hand, biomass and nuclear power plants require the most extensive land amounts, starting from the primary energy source collection or mining process, extending to its transportation and the installation of the main power plant, and ending in the disposal of the production unit's by-products (nuclear wastes). The figures presented in Table 7.9 include also the time period required for the occupied land restoration. This time period, for example, can be from 30 to 250 years, depending on the type of the occupied ecosystem for conventional coal thermal power plants, while for nuclear power plants it is considered equal to 10,000 years, namely equal to the half-life of the nuclear wastes' radiation from fission processes.

7.7.6 Electromagnetic Interference

The potential impact of wind turbines related to electromagnetic interference areas is distinguished in two main categories:

- effects on the signal transmission from radio or television broadcast stations, due to the proximity of the wind turbines with them
- electromagnetic emission from the wind turbines.

TABLE 7.9
Land Occupation for Electricity Production from Different Technologies

	Biomass	Nuclear	Coal	Natural Gas	Hydro	PV 30-Year Lifetime	Wind 30-Year Lifetime
Land occupation (m^2·years/ GWh of produced electricity)	380,000	300,000	1,290 –25,200	4,200	2,350 (50-year lifetime) – 25,000 (30-year lifetime)	9,900	1,030–3,230

FIGURE 7.24 Agricultural activities inside the installation site of wind parks (a) in North Dakota and (b) in Europe (with due diligence of North Dakota Tourism and National Geographic) [47, 48].

FIGURE 7.25 Cows and sheep grazing close to wind parks in Australia (photography: David Simmonds) [49].

The problem with the interference of the wind turbines on the transmitted radio or television signal was more intensive with first-generation wind turbines, which used to be constructed with metallic blades with high reflective factors, given the fact that, in general, the radio or TV signal transition, especially through FM broadcast frequencies, can be affected by physical or technical obstacles. This problem, however, is rather negligible for modern wind turbines, since their blades are constructed with synthetic materials with minimum impacts to the transmitted signals. In any case, any potential problems can be easily predicted and treated with the appropriate siting of the wind turbines. In worst case, an additional transmitter mast can be installed, with negligible cost for the wind park's investor.

With regard to the electromagnetic emission from wind turbines, the only components that can possibly contribute to such an impact are the inductive generator and the low/medium voltage transformer. In any case, the electromagnetic field of these components is very weak, restricted to a short area around the wind turbine, within a maximum radius of 50 m. Given the fact that the electric generator is installed at 50 m height above ground, the electromagnetic field is negligible even at the base of the wind turbine's tower.

7.8 IMPACTS FROM HYDROELECTRIC POWER PLANTS

In this section our reference on the impacts from the electricity production technologies on the natural environment and human activities is integrated with the presentation of hydroelectric power plants. The impacts from hydro power plants can be significant, although remaining restricted to the vicinity of the installation site. A thorough description is provided below [51].

7.8.1 Impacts on Ground

The first impacts of hydro power plants refer to the ground. When the hydro power plant is accompanied by the construction of a dam or a reservoir, the most important potential impacts can be:

- with the construction of the water reservoir and the accompanied works, usually large agricultural or forest areas are occupied, imposing direct effects on the existing land usage
- a considerable qualitative distortion on the natural environment is also observed, based on the transition of ground land and river biotopes into lake ecosystems, while the previous natural status is also transformed to a structured environment

- modifications on the land morphology and the natural terrain from the excavation works for the configuration of the water reservoir and the construction of the dam itself
- during the water reservoir filling, there may be slight seismic activities or landslides, as well as possible effects on the statics of the ground, caused by the increasing water surface level in the aquifer.

In hydro power plants with no dams or water reservoirs, the following impacts may occur:

- modifications of the ground from the possible construction of a spoil pit for the disposal of any excavation soil for the penstock trench construction and any other possible excavation works
- modification of the land morphology and the natural environment from all the required technical works.

The measures that can be taken for the restriction of the above potential impacts are:

- limitation of the occupied land
- use the reservoir basin itself as a borrow pit for any probably required soil or rock materials (e.g., for the construction of a gravity dam)
- use of a possibly already existing quarry
- implementation of detailed geotechnical studies and execution of all the required experimental tests during the design stage, in order to investigate any probability for possible seismic vibrations and landslides during the construction or the operation of the power plant and figure out the necessity for the construction of accompanying supporting works
- disposal of the debris probably kept by the dam to the neighboring crops as a fertilizer or to the downstream side of the river
- trees and vegetation planting.

7.8.2 Water

Important impacts can be also caused by a hydro power plant on the water resources at the area of installation, with regard to both the available water quantity and quality.

The potential impacts on the quantities of the available water resources by a hydro power plant combined with the construction of a water reservoir can be:

- changes on the water flow downstream from the reservoir's location
- deceleration of the river's flow before the reservoir and acceleration after it
- changes on the aquifer of the area.

The first two impacts can be also met even in cases of hydro plants without the support of a water reservoir, although of minor scale.

With regard to the impacts on the water resource quality, the water overflown from the reservoir is poor in debris, because it is kept by the dam, causing a possible corrosion risk for the existing river bed downstream. Additionally:

- In case the reservoir's basin is not deforested before filling with water, a reduction of the water's oxygen natural content is observed because of the biodegradability of the organic organizations and the emission of methane due to the imposed anaerobic conditions at the reservoir's bottom.
- The developed anaerobic bacteria can transform the harmless inorganic mercury contained in the ground to methylmercury, a toxic and bio-accumulative compound, transmitted to superior organizations through the food chain.

- Additionally, different water temperatures can be observed versus the depth from the reservoir's free surface, with higher temperatures prevailing in upper layers and vice versa. This fact, combined with the reduction of the water oxygen content, which of course will be more intensive close to the reservoir's bottom too, implies that if the water released to the physical stream comes from the reservoir's lower layers, either directly or after the hydro power plant, it will be both colder and with lower contained oxygen, most possibly inadequate for the conservation of the existing flora and fauna at the river's biotope, especially of fishes, which require very specific conditions with regard to the water's temperature and oxygen content.

  Of course this problem has a locally restricted impact for only the downstream area after the reservoir.
- Finally, there may be a drop of the water pH due to the decay of the existing biomass inside the reservoir, blurriness, and floating solid particles from the bottom erosion, as well as probable water salinization.

The first impact can also be observed for hydro power plants without the support of a water reservoir. The suggested measures for the treatment of the above possible qualitative and quantitative impacts on the water resources are:

- the conservation of a minimum required water flow in the river bed downstream
- create a hierarchy for the reservoir's water use
- reduction of the water's maximum time in the reservoir
- for the improvement of the water quality, potential solutions can be technical heating and conditioning systems, water and air mixers placed inside the reservoir contributing to the increase of the contained oxygen and the construction of spillways in different heights of the dam, so an average water temperature can be maintained downstream by releasing water from different positions following a cycling schedule
- deforestation of the existing vegetation in the reservoir's basin before its filling with water.

7.8.3 Fish Fauna

The impacts on fish fauna from the construction of a hydro power plant can be the most important from any other category of potential impacts. Specifically:

- New fish species are favored, particularly lake fishes, while there is a considerable disturbance on any other former existing fish fauna in the river, leading to its reduction or even elimination. This is due to the water temperature and the oxygen content changes downstream with regard to the reservoir's position, the stream's flow drop, and the seasonal fluctuations of the river's free surface level.
- There are considerable problems for the migratory fish species, the anadromous (e.g., salmon), the catadromous (e.g., eels), and the amphidromous (e.g., some mullet species) ones. The migratory fishes require different natural environments during the discrete stages of their life. Their life cycle takes place partially in marine and in river environment. The dam and the reservoir constitute technical obstacles regarding the migration of these species, which can possibly:
  a. force them to pass through the hydro turbines, putting them under static pressure's abrupt changes, collision risk with the turbine's mechanical components, and possibly causing a fatal event, an injury, or, at least, a serious disturbance in their migratory trip
  b. delay them, leading to possible fish concentration before the dam, making them to remain perhaps in unfavorable thermal zones in the deeper water layers for long time periods and increasing the risk for becoming easy preys of illegal fishing or other fish predators.

The above potential consequences on the fish fauna from hydro power plants can be faced with the following measures:

- Installation of effective, attractive, and safe fish passages through the dam, with appropriate siting, aiming to create technical conditions for the fishes' passage similar to the natural ones, with regard to the water flow, the water quality, and the bottom's features. These passages can be constructed in the form of consecutive ramps, stairs, or inclined surfaces. Additionally, they should be protected with grating and locks, to avoid the chance of illegal fishing.
- For the safe fish passage through the dam and in order to eliminate the risk of entering the hydro turbines, several techniques are often applied, such as passages close to the water surface or technical lights or sounds that may attract fishes and force them to follow a specific safe route, avoiding, thus, the hydro turbines.
- Collaboration of engineers and biologists for the interpretation and the simulation of the fish reaction against the dam construction, aiming to approach alternative methods and ways towards the secure fish passage across the dam.
- Creation of technical fish crops.
- Installation of ventilation and heating units for the regulation of the water temperature and oxygen content in the former favorable conditions for the conservation of the existing fish fauna.
- Careful cleaning of the riverside zone.
- The conservation of a minimum stream flow in the existing river bed downstream.

7.8.4 Other Fauna

With regard to the other fauna species, apart from fishes, the potential impacts of hydro power plants can be:

- the existence of the reservoir is favorable for some birds, mammals, and reptiles, mainly predators
- the reservoir or the dam constitutes a technical obstacle for some species, mainly mammals, making their transportation difficult
- due to increase of the air's humidity and the, generally, milder environmental conditions in the area, some insect species are favored.

The impacts from the construction of a hydro power plant in the existing fauna apart from fishes can be faced with the following measures:

- careful cleaning of the riverside zone
- tree planting
- establishment of protection zones for sensitive fauna species.

7.8.5 Biotope: Flora and Vegetation

The impacts on flora and vegetation of the local ecosystem can be:

- With the construction of the dam and the filling of the reservoir's basin with water, all the existing ecosystem is lost. Subsequent changes are so caused to the river and riverside ecosystems, before and after the reservoir. At the same time, the existing ecosystem is substituted with a technical lake, with intensive and unnatural fluctuations of the water free surface, leading to probable loss of specific species of the riparian flora and fauna. Important degradation of sea coasts is also possible, due to changes on the river's estuary, with subsequent impacts on sea fishes and bird populations.

- All the initially carried sediment by the river's flow, before the construction of the reservoir, is now kept by the dam. Since all this amount of solid material is not transferred downstream, the existing river environment is affected, mainly in the mouth of the river or even over a long distance upwards. It is so verified the argument that dams do not only keep water, but also their content.
- The switching between periods of floods and droughts can cause the elimination of the riverside vegetation.

The last of the above impacts can also be observed for hydro power plants without the support of a water reservoir.

The above impacts can be faced with the following measures:

- the conservation of a minimum water flow for the existing river
- tree planting
- conservation of a constant surface level of the reservoir with the construction of multiple reservoirs with daily or monthly charge–discharge cycles, so the main reservoir approaches the attribute of a natural lake
- to face the problem with the sediment holding from the dam and the subsequent impacts, the following measures can be suggested:
    a. the construction of canals before the reservoir for sediment disposal downstream, as well as the construction of small reservoirs before the main one, for the deposit of the debris and its disposal through specially constructed canals
    b. regular cleaning of the floating debris and disposal downstream.

7.8.6 Landscape

The construction of a hydro power plant certainly brings changes to the natural landscape. Particularly:

- There is a local change of the landscape, because it is partially converted to a structured environment. The long, rambling, and savage river environment is converted to a lake, mild landscape. The forests and the existing vegetation give their places to the lake, the canals, the new access roads, the power plant, and the new electrical grid.
- Due to the reduced downstream water flow and the usually eroded river bed, the riverside landscape is often considerably changed through the whole river route.

The potential impacts on the landscape from the construction and the operation of a hydro power plant can be faced with the following measures:

- Tree and vegetation planting.
- Construction of the required civil works with the use of materials harmonized with the surrounding environment (e.g., use of locally available rocky material for the construction of the required buildings).
- If the dam is very big, an outer coating with locally available stones or rocks can contribute towards an acceptable optical aspect.
- Underground construction of the power plant's building and the water penstock, to avoid any optical interference on the existing aesthetic.
- A minimum required water flow in the physical river bed must be maintained after the construction of the hydro plant, in order to eliminate any impacts on the existing biotope and the landscape attribute.

7.8.7 Microclimate

Impacts on the microclimate of the hydro plant's area can be:

- The construction of the reservoir implies changes on the hydrological cycle locally, leading to a subsequent rise of the relative humidity. The phenomenon of the appearance of mist during the early morning hours is usual.
- The climate becomes milder. Small temperature rise is observed, while the wind is affected as well, since the construction of the lake imposes a modification of the boundary conditions of the wind flow, which now blows over a flat, smooth liquid surface, instead of above inclined ground slopes or through vegetation and trees.

Impacts on microclimate are treated with an increase of tree planting.

7.9 IMPACTS FROM GEOTHERMAL POWER PLANTS

Geothermal energy is the heat contained underground in relatively shallow depths. Normally, the earth's temperature increases at a rate of 1°C for every 30 m of depth from the ground. Yet, there are cases that this rate is higher than normal. This creates the so-called geothermal potential, namely the potential for the exploitation of this heat for heating or electricity production. Depending on the earth's—or the geothermal fluid's—temperature, the geothermal fields are distinguished in the following categories:

- High-enthalpy geothermal fields, with temperatures above 150°C. These fields are exploited for electricity production and co-generation.
- Medium-enthalpy geothermal fields, with temperatures between 90°C and 150°C. These fields are exploited for electricity and heating production.
- Low-enthalpy geothermal fields, with temperatures between 25°C and 90°C. These fields are suitable for heating production for large-scale applications (district heating, agricultural, and industrial usage).
- Normal geothermy fields, with temperatures lower than 25°C. This geothermy is met in all regions on earth and it comes as the normal outcome of the earth's heating by the sun. It is restricted in low depths (at the scale of some hundred meters). In the first 10 meters, the earth's temperatures are affected by the ambient climate conditions. It is exploited with geothermal heat exchangers and water-to-water geothermal heat pumps for indoor air conditioning.
- Permafrost geothermy, with temperatures constantly close to or lower than 0°C. These fields are met in the coldest regions on earth (arctic zones) and are not suitable for energy purposes.

Although geothermal applications are environmentally friendly, there are some potential impacts both on human activities and the natural environment, which, if not handled properly, can be of considerable importance. The most significant of them are met, as sensibly expected, in the exploitation of high- or medium-enthalpy geothermal fields for electricity or heating production. During these processes, geothermal fluid (gas or liquid) is pumped on the earth's surface and then disposed back to it, through drills of some kilometers depth. Significant impacts from geothermal power plants can be potentially mainly caused during their development phase, which contain both the geothermal field research and the construction of the power plant, and during the normal operation. The most typical of them can be noise emission, disposal in the atmosphere of geothermal wastes, geologic hazards, pollution of natural water resources, degradation of air quality, land use issues, impacts on biological resources, and degradation of the natural environment's aesthetics. The above impacts can cause serious problems and consequences to the natural ecosystems, the local economies, and the existing structured environment if appropriate measures are not taken beforehand for their mitigation. All of them are analyzed in the following paragraphs.

7.9.1 Impacts on Air Quality

The impacts on air quality from geothermal power plants come from the potential gas emissions. Compared to conventional thermal power plants, gas emissions from geothermal power plants are relatively lower. This is because in geothermal power plants there is no hydrocarbon combustion, which results in significant CO_2 emissions. Additionally, in conventional thermal power plants with water-based cooling towers, the content of the emitted steam in particulates is high if cooling water rich in total dissolved solids (TDS) is used. On the contrary, cooling water of low TDS content is used in the cooling tower of geothermal power plants, resulting in considerably reduced particulate emissions. Typical specific gas emissions and particulates per produced MWh of electricity for conventional and geothermal power plants are summarized in Table 7.10 [52].

The potentially released gas of highest concern from geothermal power plants is hydrogen sulfide (H_2S). H_2S is a highly toxic gas, dangerous for human life and health at concentrations higher than 142 mg/kg of ambient air. H_2S can be detected by a normal human nose at concentrations as low as 0.0047 ppm H_2S. The concentration of the disposed H_2S with steam from geothermal power plants can vary, depending on the technology and the specific geologic conditions, reaching up to 2% by weight in the separated-steam phase.

H_2S can be released in the atmosphere both during the construction phase or the normal operation of geothermal power plants. During the construction phase, H_2S can escape with geothermal fluids during drilling and testing processes. Under the normal operation phase, H_2S can be released in the atmosphere during startups and shutdowns with venting steam, from well leakages, in the cooling tower drift, and in the condenser (Figure 7.26). The H_2S can be controlled and captured with several abatement systems, which can result in 99% reduction of the emitted H_2S. Among these, thermal oxidation converts H_2S to SO_2, via incineration at 1,100°C. Alternatively, with chemical oxidation liquefied H_2S is converted into solid sulfur, which can be used as a fertilizer for agricultural processes. Another approach is the removal of the noncondensable gases from the working steam and their disposal back in the geothermal field.

Regular air-quality monitoring should be executed during the geothermal plant's normal operation, as well as before its development, in order to establish a baseline for the evaluation of any impact on the H_2S content in the ambient air by the geothermal power plant.

7.9.2 Impacts on Water Resources

As normal in thermal power plants, water resources are also involved in geothermal power plants, imposing potential impacts with regard both to the available resources and their quality. Yet, although the potential impacts on water resources are always of major concern, it should be stated that, in

TABLE 7.10
Gas Emissions from Different Thermal Power Plant Technologies (kg/MWh)

Emission	Geothermal—Dry Steam	Geothermal—Flashed Steam	Geothermal—Binary Cycle	Oil	Natural Gas	Coal
CO_2	27.1339	179.7587		873.6189	390.5884	997.9032
CH_4				0.0301	0.0076	0.1144
$PM_{2.5}$ (1)					0.0499	0.2676
PM_{10} (2)					0.0544	0.3266
SO_2	0.0002	0.1588		5.4204	0.0020	8.5049
N_2O				0.0060	0.0008	0.0166

(1) $PM_{2.5}$: Particulate matter that is 2.5 microns or less in width
(2) PM_{10}: Particulate matter that is about 10 microns or less in width

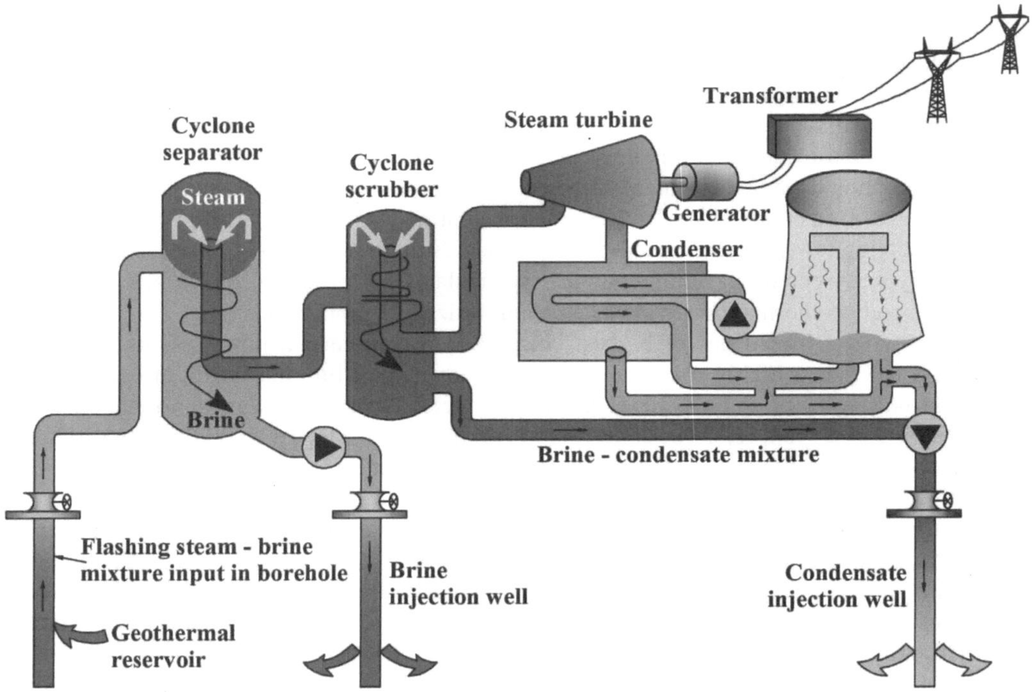

FIGURE 7.26 Typical layout of a flashed steam geothermal power plant with brine disposal back to the geothermal field.

general, the development of geothermal power plants has not been hindered by such type of issues. This is because it is normally expected that the quality of nonthermal water resources in the vicinity of high-enthalpy geothermal fields has already been affected in a natural way due to the physical recirculation of the geothermal fluid in the aquifer, and not because of the human interference. Additionally, the usage of qualitative construction techniques and materials for the casing of the geothermal wells, such as nonhazardous cements, can ensure the isolation of geothermal fluids from nonthermal aquifers.

The most dangerous source for water resource contamination comes from the disposal of the geothermal fluid, if it is utilized in the form of liquid brine. In this case, it can contain a series of hazardous elements, such as arsenic, chlorine, boron, silicon dioxide, iodine, sodium, cesium, potassium, antimony, and rubidium. Although the most effective and sensible way to treat this liquid after its exploitation for power production is its disposal back in the geothermal repository (Figure 7.26), there are cases in which, instead, it was disposed in nearby streams or in the ocean (Philippines, Tiwi, El Salvador, Ahuachapan). In other cases, brine pumped from the underground during tests of new geothermal wells was simply disposed on ground, affecting shallow water aquifers and contaminating the nearby crops (the case of Milos Island, Greece). All these incidences did happen, despite the fact that dilution of these brines with cool water, available from the water supply network, can drop the concentration of the hazardous elements below acceptable levels. In any case, the return of the utilized geothermal fluid back to the underground reservoir constitutes the most effective method, both from an environmental and power production point of view. The optimum environmental treatment is obvious, because the pumped brine is returned back to its physical location, without any impact on the surface ecosystem. Additionally, in this way the pressure of the geothermal field is maintained, contributing to the prolongation of the life period of the geothermal resource.

Contamination of water resources can be detected with regular monitoring. Water resource monitoring is highly important if natural water resources or hot springs exist close to the geothermal power plant. A baseline should be established regarding the regular water conditions before the development of the power plant, so contamination can be detected if a divergence is detected on the captured measurements. In case of contamination detection, additional measurements should be executed in water resources located farther from the contaminated area, so the flow and the dispersion of the contamination versus distance can be understood. The measurements should be repeated regularly, to evaluate the evolution of the contamination versus time.

7.9.3 Geologic Hazard

Significant impacts of geologic nature can also be caused by geothermal power plants. These can include volcanic activities, subsidence, hydrothermal explosions, landslides, soil erosion, and earthquakes of local scale.

Volcanic activity constitutes a direct consequence of the fact that geothermal power plants can be located in the vicinity of active volcanos. Geothermal reservoirs can have direct underground access to the volcanic zones, offering a route, through the geothermal wells, for the volcanic magma towards the earth's ground. In this way, small volcanic eruptions or hydrothermal explosions can occur.

Hydrothermal explosions are the result of the violent and abrupt decompression of small, high-pressure vapor enclaves, when, somehow, their pressure exceeds the lithostatic pressure. The pumping of geothermal fluid within a geothermal power plant's operation can cause a drop of the lithostatic pressure in the specific area, making it easier for the existing fumaroles or acid springs to explode. Yet, it should be noted that hydrothermal explosion craters are created where fumaroles already exist, namely these come as the result of physical processes, rather than as impacts of geothermal power plants. On the contrary, hydrothermal explosions often pockmark the locations of considerable geothermal potential. At the same time, these areas are not favorable for any other type of exploitation (e.g., residential development or agricultural crops), precisely due to the unstable ground conditions or the possible acid environment. Hence, they are normally offered for geothermal exploitation. However, there are incidences in poor countries where inhabitants have chosen to establish settlements on or close to such areas, due to the natural and free of charge availability of heat, which is utilized for all the common heating needs of a household (warm water production, indoor space heating, cooking). This has led, although rarely, to important natural disasters, with fatal accidents, buildings' destructions, etc., when violent hydro thermal explosions caused showers of mud, steam, and rocks that destroyed and buried whole settlements (e.g., in Ahuachapan, El Salvador [53]).

Landslides can also occur in geothermal power plants, in case of intensive land morphology, with steep slopes and high relief, as a result of the reaction between the geothermal fluids and the hosting rocks, which leads to unstable ground (clays, mud, and silica residues). On the one hand, landslides can constitute flat areas suitable for building erections, or, due to their chemical content, fertile and flat ground, appropriate for agricultural activities. Yet, on the other hand, landslides related to geothermal power plants, have destroyed several technical infrastructures (roads, bridges, etc.), or the geothermal power plant itself, and have buried inhabitants due to abrupt collapse (Zunil, Guatemala, January 5, 1991 [54]).

Long-term operation of geothermal power plants can cause earth subsidence due to the removal of large amounts of geothermal fluid from the repository and its replacement with colder water from neighboring aquifers. Earth subsidence can affect roads and, rarely, building facilities. Under extreme earth removal, seismic activities of local, limited scale can also be caused.

Potential geologic hazards should be investigated in advance, during the development phase of geothermal power plants This investigation is based on essential geological research of the under consideration area, examining any pre-existing volcanic activity, the earth's geological features, the land morphology, and the potential of landslides due to ground erosion, existing fumaroles, or hydrothermal explosions, etc. All these involved parameters should be taken into account for the final siting of the geothermal power plant.

7.9.4 Wastes

Geothermal power plants can potentially impose a series of solid and liquid wastes during both the development and the normal operation phase. Some of them have been already mentioned in the earlier paragraphs.

Indicatively, regarding the creation of solid wastes, the following can be mentioned:

- drilling mud residues and rock cuttings during the well's construction
- chemical wastes from the power plant's operation and the cooling tower
- wastes from the H_2S treatment processes
- other solid wastes scales from the wells and filters cleaning process, filters' materials, drilling additives, etc.

The largest amount of solid wastes comes from the H_2S treatment. Yet, through a specific H_2S treatment process, a sulfur cake is produced. The geothermal promotion campaign has successfully introduced this product as a nonhazardous earth fertilizer.

As far as liquid wastes are concerned, the most important waste is the utilized geothermal fluid, which, in the most effective approach, can be reinjected in the geothermal reservoir.

7.9.5 Noise

Noise always constitutes a major impact of every industrial activity. Noise in geothermal power plants is emitted by the heavy tracks traffic, unabated steam venting, and drilling works, during the construction period. Noise can be reduced with sound barriers around the noisiest machinery (e.g., drills).

7.9.6 Biological Resources

Impacts on biological resources are related to loss of vegetation, loss of wildlife habitats, and direct loss of wildlife. Loss of vegetation is a direct impact of land's cleaning for the erection of industrial facilities. Additional vegetation loss can be a result of the disposal of dust, or solid or liquid wastes in the neighboring environment of the geothermal power plant, by solid erosion, by land subsidence, and landslides.

Extensive loss of vegetation can result in loss of wildlife habitat, with a direct impact on the extinction of wildlife species from the installation area. Indirect loss of wildlife can also be caused by increased noise or extensive human activities. Direct loss of wildlife can be caused by the contamination of water resources with geothermal brine, by landslides, and subsidence.

Impacts on wildlife can be mitigated with several measures. Revegetation is one of them, offering protection against wildlife habitat loss and against landslides and soil erosion. Additionally, restricting sites of human activities to the minimum required and reclaiming areas that are no longer necessary can contribute to the minimization of wildlife habitat loss.

REFERENCES

1. BP. (2018). Statistical Review of World Energy, Workbook (xlsx), London. https://www.bp.com/content/dam/bp/business-sites/en/global/corporate/pdfs/energy-economics/statistical-review/bp-stats-review-2018-full-report.pdf (Accessed on August 2019).
2. Independent Statistics and Analysis. U.S. Energy Information Administration. https://www.eia.gov/energyexplained/?page=us_energy_home (Accessed on August 2019).
3. Eurostat. Statistics Explained. Archive: Consumption of Energy. http://ec.europa.eu/eurostat/statistics-explained/index.php/Consumption_of_energy#Consumption (Accessed on August 2019).
4. Wikipedia: Petroleum. https://en.wikipedia.org/wiki/Petroleum (Accessed on August 2019).

5. Odell, P. R. (1968). The significance of oil. *Journal of Contemporary History, 3*, 93–110.
6. Wikipedia: Seven Sisters (oil companies). https://en.wikipedia.org/wiki/Seven_Sisters_(oil_companies) (Accessed on August 2019).
7. Futures. http://tfc-charts.com/chart/QM/W (Accessed on August 2019)
8. Historical Crude Oil Prices. https://inflationdata.com/Inflation/Inflation_Rate/Historical_Oil_Prices_Table.asp (Accessed on August 2019).
9. European Commission: Energy Security. https://ec.europa.eu/energy/en/topics/imports-and-secure-supplies (Accessed on August 2019).
10. Hamilton, J. D. (1983). Oil and the macroeconomy since World War II. *The Journal of Political Economy, 91*, 228–248.
11. de Miguel, C., Manzano, B., & Martin-Moreno, J. M. (2003). Oil price shocks and aggregate fluctuations. *The Energy Journal, 24*, 47–61.
12. Hooker, M. (2020). Are oil shocks inflationary? Asymmetric and nonlinear specifications versus changes in regime. *Journal of Money, Credit, and Banking, 34*, 540–561.
13. Cuñado, J., & Pérez de Gracia, F. (2003). Do oil price shocks matter? Evidence for some European countries. *Energy Economics, 25*, 137–154.
14. Jiménez-Rodríguez, R., & Sánchez, M. (2004, May). Oil price shocks and real G.D.P. growth. Empirical evidence for some OECD countries. European Central Bank. Working paper series. No. 362. https://www.ecb.europa.eu/pub/pdf/scpwps/ecbwp362.pdf?b35d2a5fd0bae52378b274ce13a956c4 (Accessed on August 2019).
15. U.K. Office of National Statistics. https://www.ons.gov.uk/ (Accessed on August 2019).
16. Bird, A., & Brown, M. (2005, June 2). E297c Ethics of Development in a Global Environment. *The History and Social Consequences of a Nationalized Oil Industry.*
17. Ikelegbe, A. (2005). The economy of conflict in the oil rich Niger delta region of Nigeria. *Nordic Journal of African Studies, 14*, 208–234.
18. Climate Reporters. The Ogoni cleanup: Going forward. https://climatereporters.com/2016/06/the-ogoni-cleanup-going-forward/ (Accessed on September 2019).
19. World Nuclear Association. (2015, September). The economics of nuclear power. http://www.world-nuclear.org/info/Economic-Aspects/Economics-of-Nuclear-Power/ (Accessed on August 2019).
20. Cooper, M. (2009, August). *The economics of nuclear reactors: renaissance or relapse?* No. 692–693. World Information Service on Energy – Nuclear Information and Resource Service. https://www.nirs.org/wp-content/uploads/mononline/nm692_3.pdf (Accessed on August 2019).
21. Lazard. (2018, November 8). Levelized Cost of Energy and Levelized Cost of Storage Analysis. https://www.lazard.com/perspective/levelized-cost-of-energy-and-levelized-cost-of-storage-2018/ (Accessed on October 2019).
22. http://www.infomine.com (Accessed on August 2019).
23. Regulatory Authority of Energy. www.rae.gr (Accessed on August 2019).
24. Katsaprakakis, D. A., & Christakis, D. G. (2016). The exploitation of electricity production projects from Renewable Energy Sources for the social and economic development of remote communities. The case of Greece: An example to avoid. *Renewable and Sustainable Energy Reviews, 54*, 341–349.
25. Katsaprakakis, D. A. (2012). A review of the environmental and human impacts from wind parks. A case study for the Prefecture of Lasithi, Crete. *Renewable and Sustainable Energy Reviews, 16*, 2850–2863.
26. Community wind power. http://www.communitywindpower.co.uk/ (Accessed on August 2019).
27. Westmill Wind Farm Co-operative. https://www.westmill.coop/ (Accessed on August 2019).
28. Hepburn wind. https://www.hepburnwind.com.au/ (Accessed on August 2019).
29. Sifnos Island Cooperative. www.sifnosislandcoop.gr/ (Accessed on August 2019).
30. European Federation of Renewable Energy Cooperatives. https://rescoop.eu/ (Accessed on August 2019).
31. Øvre Forsland og Bjørnstokk Hydraulic Power Stations. https://architizer.com/projects/ovre-forsland-og-bjornstokk-hydraulic-power-stations/ (Accessed on August 2019).
32. World Bank—Indicator. http://data.worldbank.org/indicator (Accessed on August 2019).
33. Wind Energy in Europe in 2018. Trends and statistics. Wind Europe. https://windeurope.org/wp-content/uploads/files/about-wind/statistics/WindEurope-Annual-Statistics-2018.pdf (Accessed on August 2019).
34. Energi Akademiet: Samsoe's green transition—Version 2. https://energiakademiet.dk/en/transition/island-2-0/objective-2-that-the-decentralized-and-flexible-energy-system-for-renewable-energy-production-are-maintained-and-further-developed/ (Accessed on October 2019).

35. Samso: World's First 100 Percent Renewable Energy-Powered Island Is a Beacon for Sustainable Communities. http://www.huffingtonpost.com/stefanie-penn-spear/samso-worlds-first-100-re_b_5303237.html (Accessed on August 2019).
36. Energy4All: wind power. https://energy4all.co.uk/wind-power/featured-image-falck-assel-valley-landscape/ (Accessed on August 2019).
37. The Scottish Islands Federation. http://www.scottish-islands-federation.co.uk/good-luck-to-gigha/ (Accessed on September 2019).
38. Katsaprakakis, D. A., & Voumvoulakis, M. (2018). A hybrid power plant towards 100% energy autonomy for the island of Sifnos, Greece. Perspectives created from energy cooperatives. *Energy*, *161*, 680–698.
39. EEA technical report No 4/2008. Air pollution from electricity-generating large combustion plants. Copenhagen: EEA, 2008. https://www.eea.europa.eu/publications/technical_report_2008_4 (Accessed on August 2019).
40. The World Bank Data. CO_2 emissions (metric tons per capita). http://data.worldbank.org/indicator/EN.ATM.CO2E.PC (Accessed on August 2019).
41. Sircar, K., Clower, J., Shin, M. K., Bailey, C., King, M., & Yip, F. (2015). Carbon monoxide poisoning deaths in the United States, 1999 to 2012. *The American Journal of Emergency Medicine*, *33*, 1140–1145.
42. Vice.com: PPC's mines swallow villages in Ptolemaida. https://www.vice.com/gr/article/53z378/ta-oryxeia-tis-dei-katapinoun-xoria-stin-ptolemaida (Accessed on September 2019).
43. European Environment Agency: Exceedance of critical loads of acidity. https://www.eea.europa.eu/data-and-maps/figures/exceedance-of-critital-loads-of-acidity (Accessed on September 2019).
44. Steinhauser, G., Brandl, A., & Johnson, T. E. (2014). Comparison of the Chernobyl and Fukushima nuclear accidents: A review of the environmental impacts. *Science of the Total Environment*, *470–471*, 800–817.
45. Wind-turbines-models.com. https://en.wind-turbine-models.com/turbines/113-enercon-e-112-45.114 (Accessed on September 2019).
46. CK Vango: Altamont Wind Farm to Power Google Offices. http://www.ckvango.com/2016/03/altamont-wind-farm-to-power-google-offices/ (Accessed on December 2019).
47. travelandleisure.com: Where to Find the Most Vibrant Sunflower Bloom in the Country. https://www.travelandleisure.com/trip-ideas/nature-travel/north-dakota-sunflower-bloom (Accessed on September 2019).
48. National Geographic: Planting Wind Energy on Farms May Help Crops, Say Researchers. https://www.nationalgeographic.com/news/energy/2011/12/111219-wind-turbines-help-crops-on-farms/ (Accessed on September 2019).
49. Acciona, Australia: Mt Gellibrand Wind Farm. https://www.acciona.com.au/projects/energy/wind-power/mt-gellibrand-wind-farm/ (Accessed on September 2019).
50. Fthenakis, V., & Kim, H. C. (2009). Land use and electricity generation: A life-cycle analysis. *Renewable and Sustainable Energy Reviews*, *13*, 1465–1474.
51. Lampropoulou, V., Karageorgopoulos, A., Kornaros, M., & Tsoutsos, Th. (2004). Environmental impacts of small hydro power plants—The experience in Greece. *Technical Moments*, *1–2*, 9–24. Scientific Edition of Technical Chamber of Greece III.
52. Matek, B. (2013). Promoting geothermal energy: air emissions comparison and externality analysis. *Geothermal Energy Association*. Washington, DC, USA. https://www.geothermal.org/Policy_Committee/Documents/Air_Emissions_Comparison_and_Externality_Analysis_Publication_May_2013.pdf (Accessed on December 2019).
53. Handar, S., & Barrios, L. A. (2004). Hydrothermal eruptions in El Salvador: a review. *Geological Society of America*, *375*, 245–255.
54. Flynn, T., Goff, F., Van Eeckhout, E., Goff, S., Ballinger, J., & Suyama, J. (1991, January 5). Catastrophic landslide at Zunil I Geothermal Field, Guatemala. *Geothermal Resources Council Transactions*, *15*, 425–433.

Index

Note: Locators in *italics* represent figures and **bold** indicate tables in the text.

A

B

I

J

K

L

M

N

U

V

W

Y